General data and fundamental constants

Quantity	Symbol	Value	Power of ten	Units
Speed of light	c	2.997 924 58	10^8	$m\,s^{-1}$
Elementary charge	e	1.602 177	10^{-19}	C
Faraday constant	$F = N_A e$	9.6485	10^4	$C\,mol^{-1}$
Boltzmann constant	k	1.380 66	10^{-23}	$J\,K^{-1}$
Gas constant	$R = N_A k$	8.314 51		$J\,K^{-1}\,mol^{-1}$
		8.314 51	10^{-2}	$L\,bar\,K^{-1}\,mol^{-1}$
		8.205 78	10^{-2}	$L\,atm\,K^{-1}\,mol^{-1}$
		6.2364	10	$L\,Torr\,K^{-1}\,mol^{-1}$
Planck constant	h	6.626 08	10^{-34}	$J\,s$
	$\hbar = h/2\pi$	1.054 59	10^{-34}	$J\,s$
Avogadro constant	N_A	6.022 14	10^{23}	mol^{-1}
Atomic mass unit	u	1.660 54	10^{-27}	kg
Mass electron	m_e	9.109 39	10^{-31}	kg
proton	m_p	1.672 62	10^{-27}	kg
neutron	m_n	1.674 93	10^{-27}	kg
Vacuum permittivity	ε_0	8.854 19	10^{-12}	$J^{-1}\,C^2\,m^{-1}$
	$4\pi\varepsilon_0$	1.112.65	10^{-10}	$J^{-1}\,C^2\,m^{-1}$
Magneton Bohr	$\mu_B = e\hbar/2m_e c$	9.274 02	10^{-24}	$J\,T^{-1}$
nuclear	$\mu_N = e\hbar/2m_p c$	5.050 79	10^{-27}	$J\,T^{-1}$
Bohr radius	$a_0 = 4\pi\varepsilon_0\hbar^2/m_e e^2$	5.291 77	10^{-11}	m
Rydberg constant	$R = m_e e^4/8h^3 c\varepsilon_0^2$	1.096 77	10^5	cm^{-1}
Standard acceleration of free fall	g	9.806 65		$m\,s^{-2}$

Th
of
Ch

Thi

Pet

Professo
and Fell

OXFORD
UNIVERSITY PRESS

OXFORD

UNIVERSITY PRESS

Great Clarendon Street, Oxford OX2 6DP
Oxford University Press is a department of the University of Oxford.
It furthers the University's objective of excellence in research, scholarship,
and education by publishing worldwide in

Oxford   New York

Athens Auckland Bangkok Bogotá Buenos Aires Chennai
Cape town Dar es Salaam Delhi Florence Hong Kong Istanbul Karachi
Kolkata Kuala Lumpur Madrid Melbourne Mexico City Mumbai Nairobi
Paris São Paulo Shanghai Taipei Tokyo Toronto Warsaw

with associated companies in Berlin Ibadan

Oxford is a registered trade mark of Oxford University Press
in the UK and in certain other countries

First edition 1992

Second edition 1996

Third edition 2001, reprinted 2001, 2002

A catalogue record for this book is available from the British Library
Library of Congress Cataloging in Publication Data

ISBN 019 879290 5

Typeset by J&L Composition, Filey North Yorkshire
Printed in Great Britain
on acid-free paper by
The Bath Press, Bath

PREFACE

This book began life as a shortened version of the fourth edition of my *Physical Chemistry* text. Since then, both books have come a long way. The new edition of this text has perhaps undergone the most extensive change, and has acquired a personality of its own. The most obvious changes are the greater number of chapters and the more comprehensive coverage; but there are many other changes beneath the surface.

The overall aim of the text is to provide an accessible description of what I judge to be the core material for a course in modern physical chemistry. The subject has a reputation for being disagreeably mathematical and therefore difficult. I have tried to alter that perception by showing that key ideas can be presented simply and that mathematics, judiciously but not overwhelmingly deployed, clarifies and provides power. I am conscious that the current flow of the river of physical chemistry is towards biological applications, and have added a great deal of material to cover this exciting development. However, there are other key users, including those who need a straightforward account of the essentials of this key component of chemistry and those with interests in the environment, chemical engineering, and other fields that draw on chemistry.

There are many more chapters (21 in place of 11). That does not imply an out-of-control inflation. Rather, I have broken the existing chapters down into smaller bite-sized pieces to make the text more flexible and the material more readily digestible. However, there is new material, as I considered that the text needed certain additions to make it fit the plan I have for it. For instance, magnetic resonance got short shrift in the second edition but has its own chapter here. I have also included a chapter on statistical thermodynamics: not to quell the last remnants of hope but to show how a single simple idea helps us to link the microscopic to the bulk. Almost every chapter is now equipped with a Box that summarizes an important, often but not always biological, application of the material. Each of those boxes has a pair of assignable Exercises based on material provided by Professor Dixie Goss of Hunter College, CUNY. To ensure that the material is covered evenly, I have nearly doubled the number of end-of-chapter Exercises. The answers to these have been provided by Professor Charles Trapp of the University of Louisville, KY. He is the author of the fully revised and expanded *Solutions Manual* for this text.

As implied by my remarks above, I have had to be cautious with my deployment of mathematics. I make no apology for including mathematics, for without it physical chemistry lose its voice. However, I have been cautious with its presentation. particularly calculus. Thus, I cut to a minimum what I judge to be essential, and take great care to justify each step. A technique I have found welcome in other contexts is to raise a question in the text, and then to show how mathematics can be marshalled to find an answer. These *Derivations* not only help to show the extraordinary power of even simple mathematics, but also help to hold detailed work away from the main thrust of the text. Almost all the calculus is confined to these *Derivations*, so in a pre-calculus course there should be no difficulty in stepping past them.

There is a lot of pedagogical support for users. Thus, there are *Worked Examples* in every chapter, each equipped with a *Strategy*, *Solution*, and immediately answered *Self-test*. Where I judge that the full panoply of a *Worked Example* is inappropriate, I include an *Illustration* to show how a concept or equation is used. There are also numerous *Self-tests* floating in appropriate locations throughout the chapters. The ten *Further Information* sections at the end of the book should provide support for some of the concepts drawn on in the text.

There is ample electronic support for the text, as may be discovered by visiting the web site at www.oup.com/echem and www.whfreeman.com/elements. I draw attention to two major components of the site. First, all the graphs accompanied by the icon are *Living Graphs* in the sense that you are invited to explore how they change with a change in parameter. These living graphs have been developed by Sumanas Inc on the basis of material supplied by me. They should be of great usefulness for exploring the consequences of changing variables. MathCad8 files for the same material are also available on the site and may be downloaded. The second major component of the site is one popular with users of the second edition; I have provided full colour versions of all the line art for adopters to use in their lectures (the normal conventions of copyright for other purposes are retained). Other items on the web site include pointers to the sites where other numerical data and the protein structures may be found.

I always depend extensively on colleagues, formally commissioned or informal correspondents, who have read the drafts or who have simply written in with comments. This time I would like to thank the following;

David Andrews, University of East Anglia
David Larsen, Univ of Missouri at St. Louis
Frederick Kundell, Saisbury State Univ
Gareth Price, Bath University
Gerald Manning, Rutgers University
John Atherton, Memorial University
John Albright, Texas Christian University
Kathleen Howard, Swarthmore College
Lynn James, University of Northern Colorado
Marjorie Langell, Univ of Nebraska
Maurice Rigby, King's College London
Paul Monk, Manchester Metropolitan University
Roseanne Sension, University of Michigan
Steven Roser, University of Bath
Sydney Bluestone, California State University at Fresno
Thorsten Dieckmann, University of California at Davis
Werner Horsthemke, Southern Methodist University

My special thanks go to Valerie Walters who read the whole book with an eye on keeping my science straight. Claire Eisenhandler once again acted as my careful copy editor.

Oxford 2000   P.W.A

CONTENTS

Chapter 0

Introduction

CHEMISTRY is the science of matter and the changes it can undergo. The branch of the subject called **physical chemistry** is concerned with the physical principles that underlie chemistry. Physical chemistry seeks to account for the properties of matter in terms of fundamental concepts such as atoms, electrons, and energy. It provides the basic framework for all other branches of chemistry—for inorganic chemistry, organic chemistry, biochemistry, geochemistry, and chemical engineering. It also provides the basis of modern methods of analysis, the determination of structure, and the elucidation of the manner in which chemical reactions occur. To do all this, it draws on two of the great foundations of modern physical science, thermodynamics and quantum mechanics. This text introduces the central concepts of these two subjects and shows how they are used in chemistry. This *Introduction* reviews material fundamental to the whole of physical chemistry, but which should be familiar from introductory courses.[1]

We begin by thinking about matter in bulk. The broadest classification of matter is into one of three **states of matter**, or forms of bulk matter, namely, gas, liquid, and solid. Later we shall see how this classification can be refined, but these three broad classes are a good starting point.

---

[1] The mathematical techniques required in physical chemistry are reviewed in *Further information 1* at the end of this book.

0.1 The states of matter

We distinguish the three states of matter by noting the behaviour of a substance enclosed in a container:

A **gas** is a fluid form of matter that fills the container it occupies.

A **liquid** is a fluid form of matter that possesses a well-defined surface and (in a gravitational field) fills the lower part of the container it occupies.

A **solid** retains its shape independent of the shape of the container it occupies.

One of the roles of physical chemistry is to establish the link between the properties of bulk matter and the behaviour of the particles—atoms, ions, or molecules—of which it is composed. A physical chemist formulates a **model**, a simplified description, of each physical state and then shows how the state's properties can be understood in terms of this model. The existence of different states of matter is a first illustration of this procedure, for the properties of the three states suggest that they are composed of particles with different freedom of movement. Indeed, as we work through this text, we shall gradually establish and elaborate the following models:

A gas is composed of widely separated particles in continuous rapid, disordered motion. A particle travels several (often many) diameters before colliding with another particle. For most of the time the particles are so far apart that they interact with each other only very weakly.

A liquid consists of particles that are in contact but are able to move past each other in a restricted manner. The particles are in a continuous state of motion, but travel only a fraction of a diameter before bumping into a neighbour. The overriding image is one of movement, but with molecules jostling one another.

A solid consists of particles that are in contact and unable to move past one another. Although the particles oscillate around an average location, they are essentially trapped in their initial positions, and typically lie in ordered arrays.

The essential difference between the three states of matter is the freedom of the particles to move past one another. If the average separation of the particles is large, there is hardly any restriction on their motion, and the substance is a gas. If the particles interact so strongly with one another that they are locked together rigidly, then the substance is a solid. If the particles have an intermediate mobility between these extremes, then the substance is a liquid. We can understand the melting of a solid and the vaporization of a liquid in terms of the progressive increase in the liberty of the particles as a sample is heated and the particles become able to move more freely.

0.2 Physical state

The term 'state' has many different meanings in chemistry, and it is important to keep them all in mind. We have already met one meaning in the expression 'the states of matter' and, specifically, 'the gaseous state'. Now we meet a second: by **physical state** (or just 'state') we shall mean a specific condition of a sample of matter that is described in terms of its physical form (gas, liquid, or solid) and the volume, pressure, temperature, and amount of substance present. (The precise meanings of these terms are described below.) So, 1 kg of hydrogen gas in a container of volume 10 L at a specified pressure and temperature is in a particular state. The same mass of gas in a container of volume 5 L is in a different state. Two samples of a given substance are in the same state if they are the same state of matter (that is, are both present as gas, liquid, or solid) *and* if they have the same mass, volume, pressure, and temperature.

To see more precisely what is involved in specifying the state of a substance, we need to define the terms we have used. The **mass**, m, of a sample is a measure of the quantity of matter it contains. Thus, 2 kg of lead contains twice as much matter as 1 kg of lead and indeed twice as much matter as 1 kg of anything. An average man contains more matter than an average woman. The SI unit of mass is the **kilogram** (kg), with 1 kg currently defined as the mass of a certain block of platinum–iridium alloy preserved at Sèvres, outside Paris. For typical laboratory-sized samples it is usually more convenient to use a smaller unit and to express mass in grams (g), where 1 kg = 10^3 g.

The **volume**, V, of a sample is the amount of space it occupies. Thus, we write $V = 100 \text{ cm}^3$ if the sample occupies 100 cm^3 of space. The units used to express volume (which include cubic metres, m^3; litres, L; millilitres, mL), and units and symbols in general, are reviewed in *Further information 2*.

The other properties we have mentioned (pressure, temperature, and amount of substance) need more introduction for, even though they may be familiar from everyday life, they need to be defined carefully for use in science.

0.3 **Pressure**

Pressure, p, is force, F, divided by the area, A, on which the force is exerted:

$$\text{Pressure} = \frac{\text{force}}{\text{area}} \qquad p = \frac{F}{A} \qquad (0.1)$$

When you stand on ice, you generate a pressure on the ice as a result of the gravitational force acting on your mass and pulling you towards the centre of the Earth. However, the pressure is low because the downward force of your body is spread over the area equal to that of the soles of your shoes. When you stand on skates, the area of the blades in contact with the ice is much smaller so, although your downward *force* is the same, the *pressure* you exert is much greater (Fig 0.1). The pressure may be so great, in fact, that it modifies the arrangement of water molecules at the surface of the ice, and hence allows you to slide smoothly over the surface.

Pressure can arise in ways other than from the gravitational pull of the Earth on an object. For example, the impact of gas molecules on a surface gives rise to a force and hence to a pressure. If an object is immersed in the gas, it experiences a pressure over its entire surface because molecules collide with it from all directions. In this way, the atmosphere exerts a pressure on all the objects in it. We are incessantly battered by molecules of gas in the atmosphere, and experience this battering as the atmospheric pressure. The pressure is greatest at sea level because the density of air, and hence the number of colliding molecules, is greatest there. The atmospheric pressure is very considerable: it is the same as would be exerted by

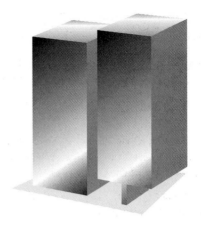

Fig 0.1 These two blocks of matter have the same mass. They exert the same force on the surface on which they are standing, but the block on the right exerts a higher pressure because it exerts the same force over a smaller area than the block on the left.

loading 1 kg of lead (or any other material) on to a surface of area 1 cm^2. We go through our lives under this heavy burden pressing on every square centimetre of our bodies. Some deep-sea creatures are built to withstand even greater pressures: at 1000 m below sea level the pressure is 100 times greater than at the surface. Creatures and submarines that operate at these depths must withstand the equivalent of 100 kg of lead loaded on to each square centimetre of their surfaces. The pressure of the air in our lungs helps us withstand the relatively low but still substantial pressures that we experience close to sea level.

When a gas is confined to a cylinder fitted with a movable piston, the position of the piston adjusts until the pressure of the gas inside the cylinder is equal to that exerted by the atmosphere. When the pressures on either side of the piston are the same, we say that the two regions on either side are in **mechanical equilibrium**. The pressure of the confined gas arises from the impact of the particles: they batter the inside surface of the piston and counter the battering of the molecules in the atmosphere that is pressing on the outside surface of the piston (Fig 0.2). Provided the piston is weightless (that is, provided we can neglect any gravitational pull on it), the gas is in mechanical equilibrium with the atmosphere

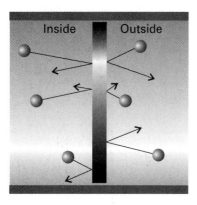

Inside   Outside

Fig 0.2 A system is in mechanical equilibrium with its sur-roundings if it is separated from them by a movable wall and the external pressure is equal to the pressure of the gas in the system.

whatever the orientation of the piston and cylinder, because the external battering is the same in all directions.

The SI unit of pressure is called the **pascal**, Pa:

$$1 \text{ Pa} = 1 \text{ kg m}^{-1}\text{s}^{-2}$$

The pressure of the atmosphere at sea level is about 10^5 Pa (100 kPa). This fact lets us imagine the magnitude of 1 Pa, for we have just seen that 1 kg of lead resting on 1 cm² on the surface of the Earth exerts about the same pressure as the atmosphere; so $1/10^5$ of that mass, or 0.01 g, will exert about 1 Pa. We see that the pascal is rather a small unit of pressure. Table 0.1 lists the other units commonly used to report pressure.[2] One of the most important in modern physical chemistry is the **bar**, where 1 bar = 10^5 Pa exactly. Normal atmospheric pressure is close to 1 bar.

Example 0.1 *Converting between units*

A scientist was exploring the effect of atmospheric pressure on the rate of growth of a lichen, and measured a pressure of 1.115 bar. What is the pressure in atmospheres?

Strategy Whenever making a conversion between units reporting the same physical property (in this case pressure), write the relation between them

Units given = units required

in the form of a conversion factor:

$$\text{Conversion factor} = \frac{\text{units required}}{\text{units given}}$$

Then write

Quantity (in units required) = conversion factor
× quantity
(in units given)

Treat the units like numbers that can be multiplied and cancelled.

Solution From Table 0.1 we have

1.013 25 bar = 1 atm

The conversion factor from units given (bar) to units required (atm) is therefore

$$\text{Conversion factor} = \frac{1 \text{ atm}}{1.013\ 25 \text{ bar}}$$

The calculation then takes the form

$$\text{Pressure (in atm)} = \frac{1 \text{ atm}}{1.013\ 25 \text{ bar}} \times 1.115 \text{ bar}$$
$$= 1.100 \text{ atm}$$

Notice how the units (bar in this case) cancel, just like numbers. Also notice that the number of significant figures in the answer (4) is the same as the number of significant figures in the data; the conversion factor itself is exact.

Self-test 0.1

The pressure in the eye of a hurricane was recorded as 723 Torr. What is the pressure in kilopascals?

[*Answer*: 96.4 kPa]

Table 0.1 *Pressure units and conversion factors*

SI unit: pascal (Pa)	1 Pa = 1 N m⁻²
bar	1 bar = 10^5 Pa
atmosphere	1 atm = 101.325 kPa = 1.013 25 bar
torr*	760 Torr = 1 atm
	1 Torr = 133.32 Pa

*The name of the unit is torr; its symbol is Torr.

2 See *Further information 2* for a fuller description of the units.

Atmospheric pressure (a property that varies with altitude and the weather) is measured with a **barometer**, which was invented by Torricelli, a student of Galileo. A mercury barometer consists of an inverted tube of mercury that is sealed at its upper end and stands with its lower end in a bath of mercury.[3] The mercury falls until the pressure it exerts at its base is equal to the atmospheric pressure (Fig 0.3). Therefore, provided we can find the relation between height and pressure, by measuring the height of the mercury column we can calculate the atmospheric pressure.

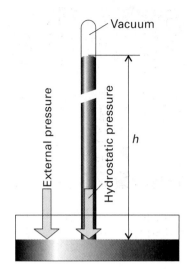

Fig 0.3 The operation of a mercury barometer. The space above the mercury in the vertical tube is a vacuum, so no pressure is exerted on the top of the mercury column; however, the atmosphere exerts a pressure on the mercury in the reservoir, and pushes the column up the tube until the pressure exerted by the mercury column is equal to that exerted by the atmosphere. The height, h, reached by the column is proportional to the external pressure, so the height can be used as a measure of this pressure.

Derivation 0.1 *Hydrostatic pressure*

Consider Fig 0.4. The volume of a cylinder of liquid of height h and cross-sectional area A is hA. The mass, m, of this cylinder of liquid is the volume multiplied by the mass density,[4] ρ (rho), of the liquid, or $m = \rho \times hA$. The downward force exerted by this mass is mg, where g is the acceleration of free fall, a measure of the Earth's gravitational pull on an object ($g = 9.81$ m s^{-2} at sea level). Therefore, the force exerted by the column is $\rho \times hA \times g$. This force acts over the area A at the foot of the column so, according to eqn 0.1, the pressure at the base is ρhAg divided by A, or

$$p = g\rho h \qquad (0.2)$$

Illustration 0.1

The pressure at the foot of a column of mercury of height 760 mm (0.760 m) and density 13.6 g cm^{-3} (1.36×10^4 kg m^{-3}) is

$$p = (9.81 \text{ m s}^{-2}) \times (1.36 \times 10^4 \text{ kg m}^{-3}) \times (0.760 \text{ m})$$
$$= 1.01 \times 10^5 \text{ kg m}^{-1}\text{ s}^{-2} = 1.01 \times 10^5 \text{ Pa}$$

This pressure corresponds to 101 kPa (1.00 atm).

We can measure the pressure of a gas *inside* a container by using a pressure gauge. The simplest type of gauge is a **manometer**, which is a

---

[3] In an *aneroid barometer*, the pressure is monitored by noting the change in size of an evacuated flexible metal container.

[4] Mass density (commonly just 'density') is the mass of a sample divided by the volume it occupies: $\rho = m/V$.

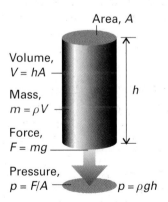

Area, A

Volume, $V = hA$

Mass, $m = \rho V$

Force, $F = mg$

Pressure, $p = F/A$

h

$p = \rho gh$

Fig 0.4 The calculation of the hydrostatic pressure exerted by a column of height h and cross-sectional area A.

U-tube containing liquid (water is sometimes used), with one limb connected to the container and the other either sealed (Fig 0.5a) or open to the atmosphere (Fig 0.5b). The difference in heights of the liquid in the two arms of the open-tube manometer is proportional to the difference in pressure between the gas in the container and the external atmosphere.

Illustration 0.2

According to eqn 0.2, the hydrostatic pressure of a column of water of height 10.0 cm (0.100 m) and density 1.00 g cm^{-3} (1.00 × 10^3 kg m^{-3}) is

$$p = (9.81 \text{ m s}^{-2}) \times (1.00 \times 10^3 \text{ kg m}^{-3}) \times (0.100 \text{ m})$$
$$= 9.81 \times 10^2 \text{ kg m}^{-1} \text{ s}^{-2}$$

or 0.981 kPa. Therefore, if the pressure of the atmosphere at the time of the experiment is 100.021 kPa, and the column of water is higher on the apparatus side of the open-tube manometer, which indicates that the pressure is lower in the apparatus than outside, then the pressure in the apparatus is 100.021 − 0.981 kPa = 99.040 kPa.

0.4 Temperature

In everyday terms, the temperature is an indication of how 'hot' or 'cold' a body is. In science, **temperature**, T, is the property of an object that determines in which direction energy will flow when the object is in contact with another object. Energy flows from higher temperature to lower temperature. When the two bodies have the same temperature, there is no net flow of energy between them. In that case we say that the bodies are in **thermal equilibrium** (Fig 0.6).

Temperature in science is measured on either the Celsius scale or the Kelvin scale. On the **Celsius scale**, in which the temperature is expressed in degrees Celsius (°C), the freezing point of water at 1 atm corresponds to 0°C and the boiling point at 1 atm corresponds to 100°C. This scale is in widespread everyday use. Temperatures on the Celsius scale are denoted by the Greek letter θ (theta) throughout this text. However, it turns out

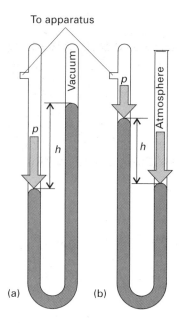

Fig 0.5 Two versions of a manometer used to measure the pressure of a sample of gas. (a) The height difference, h, of the two columns in the sealed-tube manometer is directly proportional to the pressure of the sample. (b) The difference in heights of the columns in the open-tube manometer is proportional to the difference in pressure between the sample and the atmosphere. In this case, the pressure of the sample is lower than that of the atmosphere.

to be much more convenient in many scientific applications to adopt the **Kelvin scale** and to express the temperature in kelvin (K; note that the degree sign is not used for this unit). *Whenever we use T to denote a temperature, we shall mean a temperature on the Kelvin scale.* The Celsius and Kelvin scales are related by

$$T \text{ (in kelvin)} = \theta \text{ (in degrees Celsius)} + 273.15$$

That is, to obtain the temperature in kelvins, add 273.15 to the temperature in degrees Celsius. Thus, water at 1 atm freezes at 273 K and boils at 373 K; a warm day (25°C) corresponds to 298 K.

A more sophisticated way of expressing the relation between T and θ, and one that we shall use in other contexts, is to regard the value of T as the product of a number (such as 298) and a unit (K), so that T/K (that is, the temperature divided by K) is a pure number. For example, if $T = 298$ K, then $T/\text{K} = 298$.

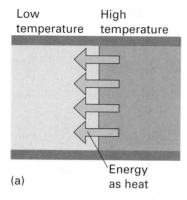

Low temperature   High temperature

(a)   Energy as heat

Equal temperatures

(b)

Fig 0.6 The temperatures of two objects act as a signpost showing the direction in which energy will flow as heat through a thermally conducting wall. (a) Heat always flows from high temperature to low temperature. (b) When the two objects have the same temperature, although there is still energy transfer in both directions, there is no net flow of energy.

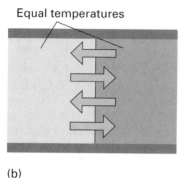

Likewise, $\theta/°C$ is also a pure number. For example, if $\theta = 25°C$, then $\theta/°C = 25$. With this convention, we can write the relation between the two scales as

$$T/K = \theta/°C + 273.15 \qquad (0.3)$$

This expression is a relation between pure numbers.[5]

Self-test 0.2

Use eqn 0.3 to express body temperature, 37°C, in kelvins.

[*Answer*: 310 K]

0.5 Amount of substance

Mass is a measure of the quantity of matter in a sample regardless of its chemical identity. In chemistry, where we focus on the behaviour of atoms, it is usually more useful to know the quantity of each specific kind of atom, molecule, or ion in a sample rather than the mass itself. However, because even 10 g of water consists of about 10^{23} H_2O molecules, it is clearly appropriate to define a new unit that can be used to express such large numbers simply. As will be familiar from introductory chemistry, chemists have introduced the **mole** (mol; the name is derived, ironically, from the Latin word meaning 'massive heap') which is defined as follows:

1 mol of specified particles is equal to the number of atoms in exactly 12 g of carbon-12.

We find that number by dividing that mass by the mass of one atom of carbon-12 determined by using a mass spectrometer. The result is 6.022×10^{23}, so this is the number of particles in 1 mol of any substance.[6] For example, a sample of hydrogen gas that contains 6.022×10^{23} hydrogen molecules consists of 1.000 mol H_2, and a sample of water that contains 1.2×10^{24} (= $2.0 \times 6.022 \times 10^{23}$) water molecules consists of 2.0 mol H_2O.

We always specify the identity of the particles when using the unit mole, for that avoids any ambiguity. If, improperly, we said that a sample consisted of 1 mol of hydrogen, it would not be clear whether it consisted of 6×10^{23} hydrogen atoms (1 mol H) or 6×10^{23} hydrogen molecules (1 mol H_2).

The mole is the unit used when reporting the value of the physical property called the **amount of substance**, n, in a sample. Thus, we can write $n = 1$ mol H_2 or $n_{H_2} = 1$ mol, and say that the amount of hydrogen molecules in a sample is 1 mol. The term 'amount of substance', however, has not yet found wide acceptance among chemists and in casual conversation they commonly refer to 'the number of moles' in a sample. The term **chemical amount**,

---

[5] Equation 0.3, in the form $\theta/°C = T/K - 273.15$, also defines the Celsius scale in terms of the more fundamental Kelvin scale.

[6] The currently accepted value is $6.022\ 136\ 7 \times 10^{23}$; see the end papers.

however, is becoming more widely used as a convenient synonym for amount of substance, and we shall often use it in this book.

There are various useful concepts that stem from the introduction of the chemical amount and its unit the mole. One is the **Avogadro constant**, N_A, the number of particles (of any kind) per mole of substance:

$$N_A = 6.022\ 136\ 7 \times 10^{23}\ mol^{-1}$$

The Avogadro constant makes it very simple to convert from the number of particles N (a pure number) in a sample to the chemical amount n (in moles) it contains:

$$\text{Number of particles} = \text{chemical amount (in moles)}$$
$$\times\ \text{number of particles}$$
$$\text{per mole}$$

$$N = n \times N_A \tag{0.4}$$

Illustration 0.3

A sample of copper containing 8.8×10^{22} Cu atoms corresponds to

$$n\,(\text{Cu}) = \frac{N(\text{Cu})}{N_A} = \frac{8.8 \times 10^{22}\,\text{Cu}}{6.022 \times 10^{23}\,\text{mol}^{-1}} = 0.15\ \text{mol Cu}$$

Notice how much easier it is to report the amount of Cu atoms present rather than their actual number.

The second very important concept that should be familiar from introductory courses is the **molar mass**, M, the mass per mole of substance: that is, the mass of a sample of the substance divided by the chemical amount of atoms, molecules, or formula units it contains. When we refer to the molar mass of an element we always mean the mass per mole of its *atoms*. When we refer to the molar mass of a compound, we always mean the molar mass of its molecules or, in the case of ionic compounds, the mass per mole of its formula units. The molar mass of a typical sample of carbon, the mass per mole of carbon atoms (with carbon-12 and carbon-13 atoms in their typical abundances), is 12.01 g mol^{-1}. The molar mass of water is the mass per mole of H_2O molecules, with the isotopic abundances of hydrogen and oxygen those of typical samples of the elements, and is 18.02 g mol^{-1}. The unit **dalton** (1 Da) is used as an abbreviation for 1 g mol^{-1}, especially in biophysical applications. The molar mass of a biological macromolecule measured as 1.2×10^4 g mol^{-1}, for instance, could be reported as 12 kDa (where 1 kDa = 1 kg mol^{-1}).

The terms *atomic weight* (AW) or *relative atomic mass* (RAM) and *molecular weight* (MW) or *relative molar mass* (RMM) are still commonly used to signify the numerical value of the molar mass of an element or compound, respectively. More precisely (but equivalently), the RAM of an element or the RMM of a compound is its average atomic or molecular mass relative to the mass of an atom of carbon-12 set equal to 12. The atomic weight (or RAM) of a natural sample of carbon is 12.01 and the molecular weight (or RMM) of water is 18.02.

The molar mass of an element is determined by mass spectrometric measurement of the mass of its atoms and then multiplication of the mass of one atom by the Avogadro constant (the number of atoms per mole). Care has to be taken to allow for the isotopic composition of an element, so we must use a suitably weighted mean of the masses of the atoms. The values obtained in this way are printed on the periodic table inside the back cover. The molar mass of a compound of known composition is calculated by taking a sum of the molar masses of its constituent atoms. The molar mass of a compound of unknown composition is determined experimentally by using mass spectrometry in a similar way to the determination of atomic masses, but allowing for the fragmentation of molecules in the course of the measurement.

Molar mass is used to convert from the mass, m, of a sample (which we can measure) to the amount of substance, n (which, in chemistry, we often need to know):

$$\text{Mass of sample (g)} = \text{chemical amount (mol)} \times$$
$$\text{mass per mole (g mol}^{-1})$$

$$m = n \times M \tag{0.5}$$

Illustration 0.4

To find the amount of C atoms present in 21.5 g of carbon, given that the molar mass of carbon is 12.01 g mol^{-1}, we write

$$n_C = \frac{m}{M_C} = \frac{21.5\ g}{12.01\ g\ mol^{-1}} = 1.79\ mol$$

That is, the sample contains 1.79 mol C.

Self-test 0.3

What amount of H_2O molecules is present in 10.0 g of water?

[*Answer*: 0.555 mol H_2O]

Exercises

0.1 Express (a) 110 kPa in torr,[7] (b) 0.997 bar in atmospheres, (c) 2.15×10^4 Pa in atmospheres, (d) 723 Torr in pascals.

0.2 Calculate the pressure in the Mindañao trench, near the Philippines, the deepest region of the oceans. Take the depth there as 11.5 km and for the density of sea water use 1.10 g cm^{-3}.

0.3 The atmospheric pressure on the surface of Mars, where $g = 3.7$ m s^{-2}, is only 0.0060 atm. To what extent is that low pressure due to the low gravitational attraction and not to the thinness of the atmosphere? What pressure would the same atmosphere exert on Earth, where $g = 9.81$ m s^{-2}?

0.4 What pressure difference must be generated across the length of a 15 cm vertical drinking straw in order to drink a water-like liquid of density 1.0 g cm^{-3}?

0.5 The water in the open tube of a manometer connected to an incubator was 3.55 cm lower than the water on the manometer side, and the atmospheric pressure was 758 Torr. What is the pressure inside the incubator?

0.6 Given that the Celsius and Fahrenheit temperature scales are related by $\theta_{Celsius}/°C = \frac{5}{9}(\theta_{Fahrenheit}/°F - 32)$, what is the temperature of absolute zero ($T = 0$) on the Fahrenheit scale?

0.7 The *Rankine scale* is used in some engineering applications. On it, the absolute zero of temperature is set at zero but the size of the Rankine degree (°R) is the same as that of the Fahrenheit degree (°F). What is the boiling point of water on the Rankine scale?

0.8 The molar mass of the oxygen-storage protein myoglobin is 16.1 kDa. How many myoglobin molecules are present in 1.0 g of the compound?

0.9 The mass of a red blood cell is about 33 pg, and it contains typically 3×10^8 haemoglobin molecules. Each haemoglobin molecule is a tetramer of a myoglobin-like molecule (see preceding exercise). What fraction of the mass of the cell is due to haemoglobin?

[7] Remember that the name of the unit is torr; its symbol is Torr.

The properties of gases

Contents

Aᴸᵀᴴᴼᵁᴳᴴ gases are simple, both to describe and in terms of their internal structure, they are of immense importance. We spend our whole lives surrounded by gas in the form of air and the local variation in its properties is what we term 'weather'. To understand the atmospheres of this and other planets we need to understand gases. As we breathe, we pump gas in and out of our lungs, where it changes composition and temperature. Many industrial processes involve gases, and both the outcome of the reaction and the design of the reaction vessels depend on a knowledge of their properties. The interiors of stars, although dense and at first sight quite unlike any gas we meet on Earth, can be described by the laws that summarize the behaviour of gases.

Equations of state

Wᴱ can specify the state of any sample of substance by giving the values of the following properties:

V, the volume the sample occupies

p, the pressure of the sample

T, the temperature of the sample

n, the amount of substance in the sample

However, an astonishing experimental fact of nature is that *these four quantities are not independent of one another*. For instance, we cannot arbitrarily choose to have a sample of 0.555 mol H_2O in a volume of 100 cm^3 at 100 kPa and 500 K: it is found *experimentally* that the state simply does not exist. If we select the amount, the volume, and the temperature, then we find that we have to accept a particular pressure (in this case, close to 230 kPa). The same is true of all substances, but the pressure in general will be different for each one. This experimental generalization is summarized by saying the substance obeys an **equation of state**, an equation of the form $p = f(n,V,T)$ that relates one of the four properties to the other three.

The equations of state of most substances are not known, so in general we cannot write down the mathematical relation between the four properties that define a state. However, certain equations of state are known. In particular, the equation of state of a low-pressure gas is known, and proves to be very simple and very useful. This equation is used to describe the behaviour of gases taking part in reactions, the behaviour of the atmosphere, as a starting point for problems in chemical engineering, and even in the description of the structures of stars.

1.1 The perfect gas equation of state

The equation of state of a low-pressure gas was among the first results to be established in physical chemistry. The original experiments were carried out by Robert Boyle in the seventeenth century and there was a resurgence in interest later in the century when people began to fly in balloons. This technological progress demanded more knowledge about the response of gases to changes of pressure and temperature.

Boyle's and his successors' experiments led to the formulation of the following **perfect gas equation of state**:

$$pV = nRT \qquad (1.1)$$

In this equation, which is probably the most important in the whole of physical chemistry, R is an ex-

Table 1.1 *The gas constant in various units*

$$R = 8.314\,51\ \text{J K}^{-1}\ \text{mol}^{-1}$$
$$8.314\,51\ \text{kPa L K}^{-1}\ \text{mol}^{-1}$$
$$8.205\,78 \times 10^{-2}\ \text{L atm K}^{-1}\ \text{mol}^{-1}$$
$$62.364\ \text{L Torr K}^{-1}\ \text{mol}^{-1}$$
$$1.987\,22\ \text{cal K}^{-1}\ \text{mol}^{-1}$$

perimentally determined constant with the same value for all gases. It is known as the **gas constant**, and depending on the units selected for the pressure and volume, has the value given in Table 1.1.[1]

The perfect gas equation of state—more briefly, the 'perfect gas law'—is so-called because it is an idealization of the equations of state that gases actually obey. Specifically, it is found that all gases obey the equation ever more closely as the pressure is reduced towards zero. That is, eqn 1.1 is an example of a **limiting law**, a law that becomes increasingly valid as the pressure is reduced and is obeyed exactly in the limit of zero pressure.

A hypothetical substance that obeys eqn 1.1 at all pressures is called a **perfect gas**.[2] From what has just been said, an actual gas, which is termed a **real gas**, behaves more and more like a perfect gas as its pressure is reduced. It behaves exactly like a perfect gas when the pressure has been reduced to zero. In practice, normal atmospheric pressure at sea level ($p \approx 100$ kPa) is already low enough for most real gases to behave almost perfectly, and unless stated otherwise we shall always assume in this text that the gases we encounter behave like a perfect gas. The reason why a real gas behaves differently from a perfect gas can be traced to the attractions and repulsions that exist between actual molecules and which are absent in a perfect gas (Chapter 16).

The perfect gas law summarizes three sets of experimental observations. One is **Boyle's law**:

At constant temperature, the pressure of a fixed amount of gas is inversely proportional to its volume.

---

[1] The gas constant may be determined by evaluating $R = pV/nT$ as the pressure is allowed to approach zero. Other techniques are available, such as measurement of the speed of sound (which depends on R) and its determination from Boltzmann's constant (Chapter 20).

[2] The term 'ideal gas' is also widely used.

Mathematically:

Boyle's law: at constant temperature, $p \propto \dfrac{1}{V}$

We can easily verify that eqn 1.1 is consistent with Boyle's law by treating n and T as constants, when it becomes pV = constant, and hence $p \propto 1/V$. Boyle's law implies that, if we compress a fixed amount of gas at constant temperature into half its original volume, then its pressure will double. Figure 1.1 shows the graph obtained by plotting experimental values of p against V for a fixed amount of gas at different temperatures and the curves predicted by Boyle's law.[3] Each curve is called an **isotherm** because it depicts the variation of a property (in this case, the pressure) at a single constant temperature. It is hard, from this graph, to tell whether Boyle's law is valid. However, when we plot p against $1/V$, we get straight lines, just as we would expect from Boyle's law (Fig 1.2). It is generally the case that a proposed relation is easier to verify if we plot the experimental data in a form that a relation predicts should give a straight line.

The second experimental observation summarized by eqn 1.1 is **Charles's law**:

At constant pressure, the volume of a fixed amount of gas varies linearly with the temperature.

Mathematically:

Charles's law: at constant pressure, $V = A + B\theta$

where θ is the temperature on the Celsius scale. Figure 1.3 shows typical plots of volume against temperature for a series of samples of gases at different pressures and confirms that (at low pressures, and for temperatures that are not too low) the volume varies linearly with the Celsius temperature. We also see that all the volumes extrapolate to zero as θ approaches the same very low temperature ($-273.15\,^{\circ}C$, in fact), regardless of the identity of the gas. Because a volume cannot be negative, this common temperature must represent the **absolute zero** of temperature, a temperature below which it is impossible to cool an object. Indeed, the Kelvin

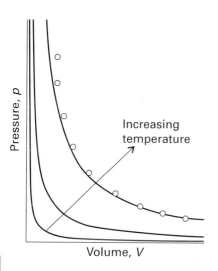

Fig 1.1 The volume of a gas decreases as the pressure on it is increased. For a sample that obeys Boyle's law and that is kept at constant temperature, the graph showing the dependence is a hyperbola, as shown here. Each curve corresponds to a single temperature, and hence is an isotherm.

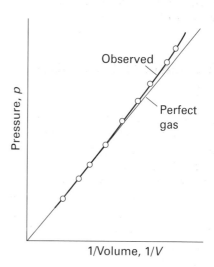

Fig 1.2 A good test of Boyle's law is to plot the pressure against $1/V$ (at constant temperature), when a straight line should be obtained. This diagram shows that the observed pressures (the heavy line) approach a straight line as the volume is increased and the pressure reduced. A perfect gas would follow the straight line at all pressures; real gases obey Boyle's law in the limit of low pressures.

---

[3] These curves are hyperbolas, graphs of xy = constant.

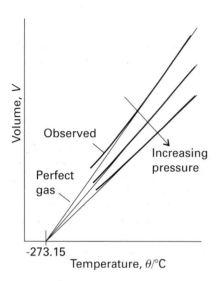

Fig 1.3 This diagram illustrates the content and implications of Charles's law, which asserts that the volume occupied by a gas (at constant pressure) varies linearly with the temperature. When plotted against Celsius temperatures (as here), all gases give straight lines that extrapolate to $V = 0$ at $-273.15°C$. This extrapolation suggests that $-273.15°C$ is the lowest attainable temperature.

scale ascribes the value $T = 0$ to this absolute zero of temperature. In terms of the Kelvin temperature, therefore, Charles's law takes the simpler form

Charles's law: at constant pressure, $V \propto T$

It follows that doubling the temperature (on the Kelvin scale, such as from 300 K to 600 K, corresponding to an increase from 27°C to 327°C) doubles the volume, provided the pressure remains the same. Now we can see that eqn 1.1 is consistent with Charles's law. First, we rearrange eqn 1.1 into $V = nRT/p$, and then note that, when the amount n and the pressure p are both constant, we can write $V \propto T$, as required.

The third feature of gases summarized by eqn 1.1 is **Avogadro's principle**:

At a given temperature and pressure, equal volumes of gas contain the same numbers of molecules.

That is, 1.00 L of oxygen at 100 kPa and 300 K contains the same number of molecules as 1.00 L of carbon dioxide at the same temperature and pressure. The principle implies that, if we double the number of molecules, but keep the temperature and pressure constant, then the volume of the sample will double. We can therefore write:

Avogadro's principle: at constant temperature and pressure, $V \propto n$

This result follows easily from eqn 1.1 if we treat p and T as constants. Avogadro's suggestion is a principle rather than a law (a direct summary of experience), because it is based on a model of a substance, in this case as a collection of molecules.

The **molar volume**, V_m, of any substance (not just a gas) is the volume the substance occupies per mole of molecules. It is calculated by dividing the volume of the sample by the amount of molecules it contains:

$$\text{Molar volume} = \frac{\text{volume of sample}}{\text{amount of substance}}$$

$$V_m = \frac{V}{n} \qquad (1.2)$$

Avogadro's principle implies that the molar volume of a gas should be the same for all gases at the same temperature and pressure. The data in Table 1.2 show that this conclusion is approximately true for most gases under normal conditions (normal atmospheric pressure of about 100 kPa and room temperature).

Table 1.2 *The molar volumes of gases at 25°C and 1 bar*

Gas	$V_m/(\text{L mol}^{-1})$
Perfect gas	24.7897*
Ammonia	24.8
Argon	24.4
Carbon dioxide	24.6
Nitrogen	24.8
Oxygen	24.8
Hydrogen	24.8
Helium	24.8

* At STP, $V_m = 22.4141$ L mol^{-1}.

Box 1.1 *The gas laws and the weather*

The biggest sample of gas readily accessible to us is the atmosphere, a mixture of gases with the composition summarized in the table. The composition is maintained moderately constant by diffusion and convection (winds, particularly the local turbulence called eddies) but the pressure and temperature vary with altitude and with the local conditions, particularly in the troposphere (the 'sphere of change'), the layer extending up to about 11 km.

One of the most variable constituents of air is water vapour, and the humidity it causes. The presence of water vapour results in a *lower* density of air at a given temperature and pressure, as we may conclude from Avogadro's principle. The numbers of molecules in 1 m³ of moist air and dry air are the same (at the same temperature and pressure), but the mass of an H_2O molecule is less than that of all the other major constituents of air (the molar mass of H_2O is 18 g mol⁻¹, the average molar mass of air molecules is 29 g mol⁻¹), so the density of the moist sample is less than that of the dry sample.

The pressure and temperature vary with altitude. In the troposphere the average temperature is 15°C at sea level, falling to –57°C at the bottom of the tropopause at 11 km. This variation is much less pronounced when expressed on the Kelvin scale, ranging from 288 K to 216 K, an average of 268 K. If we suppose that the temperature

has its average value all the way up to the tropopause, then the pressure varies with altitude, h, according to the *barometric formula*:

$$p = p_0 e^{-h/H}$$

where p_0 is the pressure at sea level and H is a constant approximately equal to 8 km. More specifically, $H = RT/Mg$, where M is the average molar mass of air and T is the temperature. The barometric formula fits the observed pressure distribution quite well even for regions well above the troposphere (see the first illustration). It implies that the pressure of the air and its density fall to half their sea-level value at $h = H \ln 2$, or 6 km.

Local variations of pressure, temperature, and composition in the troposphere are manifest as 'weather'. A small region of air is termed a *parcel*. First, we note that a parcel of warm air is less dense than the same parcel of cool air. As a parcel rises, it expands adiabatically (that is, without transfer of heat from its surroundings), so it cools. Cool air can absorb lower concentrations of water vapour than warm air, so the moisture forms clouds. Cloudy skies can therefore be associated with rising air and clear skies are often associated with descending air.

The composition of the Earth's atmosphere

Substance	Percentage	
	By volume	By mass
Nitrogen, N_2	78.08	75.53
Oxygen, O_2	20.95	23.14
Argon, Ar	0.93	1.28
Carbon dioxide, CO_2	0.031	0.047
Hydrogen, H_2	5.0×10^{-3}	2.0×10^{-4}
Neon, Ne	1.8×10^{-3}	1.3×10^{-3}
Helium, He	5.2×10^{-4}	7.2×10^{-5}
Methane, CH_4	2.0×10^{-4}	1.1×10^{-4}
Krypton, Kr	1.1×10^{-4}	3.2×10^{-4}
Nitric oxide, NO	5.0×10^{-5}	1.7×10^{-6}
Xenon, Xe	8.7×10^{-6}	3.9×10^{-5}
Ozone, O_3: summer:	7.0×10^{-6}	1.2×10^{-5}
winter:	2.0×10^{-6}	3.3×10^{-6}

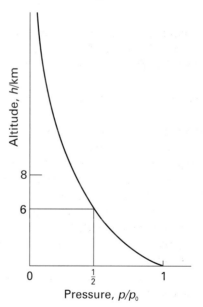

The variation of atmospheric pressure with altitude as predicted by the barometric formula.

The motion of air in the upper altitudes may lead to an accumulation in some regions and a loss of molecules from other regions. The former result in the formation of regions of high pressure ('highs' or anticyclones) and the latter result in regions of low pressure ('lows', depressions, or cyclones). These regions are shown as H and L on the weather map below. The lines of constant pressure—differing by 4 mbar (400 Pa, about 3 Torr)—marked on it are called *isobars*. The elongated regions of high and low pressure are known, respectively, as *ridges* and *troughs*.

In meteorology, large-scale vertical movement is called *convection*. Horizontal pressure differentials result in the flow of air that we call *wind*. Because the Earth is rotating from west to east, winds are deflected towards the right in the northern hemisphere and towards the left in the southern hemisphere. Winds travel nearly parallel to the isobars, with low pressure to their left in the northern hemisphere and to the right in the southern hemisphere (see illustration). At the surface, where wind speeds are lower, the winds tend to travel perpendicular to the isobars from high to low pressure. This differential motion results in a spiral outward flow of air clockwise in the northern hemisphere around a high and an inward counterclockwise flow around a low.

The air lost from regions of high pressure is restored as an influx of air converges into the region and descends. As we have seen, descending air is associated with clear skies. It also becomes warmer by compression as it descends, so regions of high pressure are associated with high surface temperatures. In winter, the cold surface air may prevent the complete fall of air, and result in a temperature *inversion*, with a layer of warm air over a layer of cold air. Geographical conditions may also

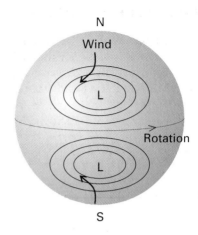

The horizontal flow of air relative to an area of low pressure in the northern and southern hemispheres.

trap cool air, as in Los Angeles, and the photochemical pollutants we know *as smog* may be trapped under the warm layer. A less dramatic manifestation of an inversion layer is the presence of hazy skies, particularly in industrial areas. Hazy skies also form over vegetation that generate aerosols of terpenes or other plant transpiration products. These hazes give rise to the various 'Blue Mountains' of the world, such as the Great Dividing Range in New South Wales, the range in Jamaica, and the range stretching from central Oregon into southeastern Washington, which are dense with eucalyptus, tree ferns, and pine and fir, respectively. The Blue Ridge section of the Appalachians is another example.

Exercise 1 Balloons were used to obtain much of the early information about the atmosphere and continue to be used today to obtain weather information. In 1782, Jacques Charles used a hydrogen-filled balloon to fly from Paris 25 km into the French countryside. What is the mass density of hydrogen relative to air at the same temperature and pressure? What mass of payload can be lifted by 10 kg of hydrogen, neglecting the mass of the balloon?

Exercise 2 Atmospheric pollution is a problem that has received much attention. Not all pollution, however, is from industrial sources. Volcanic eruptions can be a significant source of air pollution. The Kilauea volcano in Hawaii emits 200–300 t (a metric tonne, 1 t, is 1000 kg) of SO_2 per day. If this gas is emitted at a temperature of 800°C and 1.0 atm, what volume of gas is emitted?

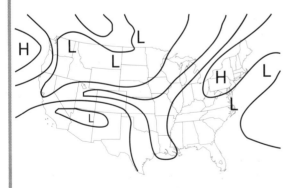

A typical weather map; this one for the continental United States on 14 July 1999. Regions of high pressure are denoted H and those of low pressure L.

1.2 Using the perfect gas law

Here we review three elementary applications of the perfect gas equation of state (eqn 1.1). The first, is the prediction of the pressure of a gas given its temperature, its chemical amount, and the volume it occupies. The second is the prediction of the change in pressure arising from changes in the conditions. The third is the calculation of the molar volume of a perfect gas under any conditions.

Example 1.1 *Predicting the pressure of a sample of gas*

A chemist is investigating the conversion of atmospheric nitrogen to usable form by the bacteria that inhabit the root systems of certain legumes, and needs to know the pressure in kilopascals exerted by 1.25 g of nitrogen gas in a flask of volume 250 mL at 20°C.

Strategy For this calculation we need to arrange eqn 1.1 into a form that gives the unknown (the pressure, p) in terms of the information supplied:

$$p = \frac{nRT}{V}$$

To use this expression, we need to know the amount of molecules (in moles) in the sample, which we can obtain from the mass and the molar mass (from $n = m/M$) and to convert the temperature to the Kelvin scale. To avoid rounding errors, it is best to leave all numerical calculations to the final stage. Select the value of R from Table 1.1 using the units that match the data and the information required (pressure in kilopascals and volume in litres).

Solution The amount of N_2 molecules (of molar mass 28.02 g mol^{-1}) present is

$$n_{N_2} = \frac{m}{M_{N_2}} = \frac{1.25\ g}{28.02\ g\ mol^{-1}} = \frac{1.25}{28.02}\ mol$$

The temperature of the sample is

$$T/K = 20 + 273.15$$

Therefore,

$$p = \frac{\overbrace{(1.25/28.02)\ mol}^{n} \times \overbrace{(8.31451\ kPa\ L\ K^{-1}mol^{-1})}^{R} \times \overbrace{(20+273.15\ K)}^{293\ K}}{\underbrace{0.250\ L}_{250\ mL}}$$

$$= 435\ kPa$$

Note how all units (except kPa in this instance) cancel like ordinary numbers.

Self-test 1.1

Calculate the pressure exerted by 1.22 g of carbon dioxide confined to a flask of volume 500 mL at 37°C.

[*Answer*: 143 kPa]

In some cases, we are given the pressure under one set of conditions and are asked to predict the pressure of the same sample under a different set of conditions. In this case we use the perfect gas law as follows. Suppose the initial pressure is p_1, the initial temperature is T_1, and the initial volume is V_1. Then from eqn 1.1 we can write

$$\frac{p_1 V_1}{T_1} = nR$$

Suppose now that the conditions are changed to T_2 and V_2, and the pressure changes to p_2 as a result. Then under the new conditions eqn 1.1 tells us that

$$\frac{p_2 V_2}{T_2} = nR$$

The nR on the right of these two equations is the same in each case, because R is a constant and the amount of gas molecules has not changed. It follows that we can combine the two equations into a single equation:

$$\frac{p_1 V_1}{T_1} = \frac{p_2 V_2}{T_2} \tag{1.3}$$

This expression is known as the **combined gas equation**. We can rearrange it to calculate one unknown (such as p_2, for instance) in terms of the other variables.

Self-test 1.2

What is the final volume of a sample of gas that has been heated from 25°C to 1000°C and its pressure increased from 10.0 kPa to 150.0 kPa, given that its initial volume was 15 mL?

[*Answer*: 4.3 mL]

Finally, we see how to use the perfect gas law to calculate the molar volume of a perfect gas at any temperature and pressure. Equation 1.2 expresses the molar volume in terms of the volume of a sample; eqn 1.1 in the form $V = nRT/p$ expresses the volume in terms of the pressure. When we combine the two, we get

$$V_m = \frac{V}{n} = \frac{nRT}{np} = \frac{RT}{p} \qquad (1.4)$$

This expression lets us calculate the molar volume of any gas (provided it is behaving perfectly) from its pressure and its temperature. It also shows that, for a given temperature and pressure, all gases have the same molar volume.

Chemists have found it convenient to report much of their data at a particular set of standard conditions. By **standard ambient temperature and pressure** (SATP) they mean a temperature of 25°C (more precisely, 298.15 K) and a pressure of 1 bar. The **standard pressure** of 1 bar is denoted $p^{\ominus}$, so $p^{\ominus}$ = 1 bar exactly. The molar volume of a perfect gas at SATP is 24.79 L mol^{-1}, as can be verified by substituting the values of the temperature and pressure into eqn 1.4. This value implies that, at SATP, 1 mol of perfect gas molecules occupies about 25 L (a cube about 1 ft, 30 cm, on a side). An earlier set of standard conditions, which is still encountered, is **standard temperature and pressure** (STP), namely 0°C and 1 atm. The molar volume of a perfect gas at STP is 22.41 L mol^{-1}.

1.3 Mixtures of gases: partial pressures

Scientists are often concerned with mixtures of gases, such as when they are considering the properties of the atmosphere in meteorology, the composition of exhaled air in medicine, or the mixtures of hydrogen and nitrogen used in the industrial synthesis of ammonia. They need to be able to assess the contribution that each component of a gaseous mixture makes to the total pressure.

In the early nineteenth century, John Dalton carried out a series of experiments that led him to formulate what has become known as **Dalton's law**:

The pressure exerted by a mixture of perfect gases is the sum of the pressures that each gas would exert if it were alone in the container at the same temperature:

$$p = p_A + p_B + \ldots \qquad (1.5)$$

where p_J is the pressure that a gas J would exert if it were alone in the container at the same temperature. Dalton's law is strictly valid only for mixtures of perfect gases (or for real gases at such low pressures that they are behaving perfectly), but it can be treated as valid under most conditions we encounter.

Illustration 1.1

Suppose we were interested in the composition of inhaled and exhaled air, and we knew that a certain mass of carbon dioxide exerts a pressure of 5 kPa when present alone in a container, and that a certain mass of oxygen exerts 20 kPa when present alone in the same container at the same temperature. Then, when both gases are present in the container, the carbon dioxide in the mixture contributes 5 kPa to the total pressure and oxygen contributes 20 kPa; according to Dalton's law, the total pressure of the mixture is the sum of these two pressures, or 25 kPa (Fig 1.4).

For any type of gas (real or perfect) in a mixture, the **partial pressure**, p_J is defined as

$$p_J = x_J \times p \qquad (1.6)$$

where x_J is the mole fraction of the gas J in the mixture. The **mole fraction** of J is the amount of J molecules expressed as a fraction of the total amount of molecules in the mixture. In a mixture that consists of n_A A molecules, n_B B molecules, and so on (where

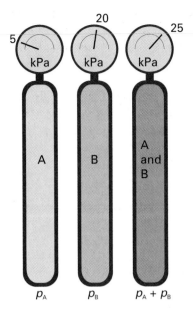

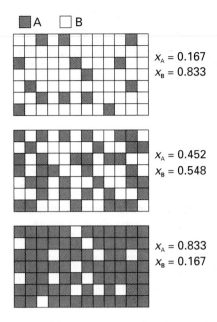

$x_A = 0.167$
$x_B = 0.833$

$x_A = 0.452$
$x_B = 0.548$

$x_A = 0.833$
$x_B = 0.167$

Fig 1.4 The partial pressure p_A of a perfect gas A is the pressure that the gas would exert if it occupied a container alone; similarly, the partial pressure p_B of a perfect gas B is the pressure that the gas would exert if it occupied the same container alone. The total pressure p when both gases simultaneously occupy the container is the sum of their partial pressures.

Fig 1.5 A representation of the meaning of mole fraction. In each case, a small square represents one molecule of A (dark squares) or B (white squares). There are 84 squares in each sample.

the n_J are amounts in moles), the mole fraction of J (where $J = A, B, \ldots$) is

$$\text{Mole fraction of J} = \frac{\text{amount of J molecules}}{\text{total amount of molecules}}$$

$$x_J = \frac{n_J}{n} \qquad (1.7)$$

where $n = n_A + n_B + \ldots$. For a **binary mixture**, one that consists of two species, this general expression becomes

$$x_A = \frac{n_A}{n_A + n_B} \qquad x_B = \frac{n_B}{n_A + n_B} \qquad x_A + x_B = 1 \quad (1.8)$$

When only A is present, $x_A = 1$ and $x_B = 0$. When only B is present, $x_B = 1$ and $x_A = 0$. When both are present in the same amounts, $x_A = \frac{1}{2}$ and $x_B = \frac{1}{2}$ (Fig 1.5).

Self-test 1.3

Calculate the mole fractions of N_2, O_2, and Ar in dry air at sea level, given that 100.0 g of air consists of 75.5 g of N_2, 23.2 g of O_2, and 1.3 g of Ar. (*Hint*. Begin by converting each mass to an amount in moles.)

[*Answer*: 0.780, 0.210, 0.009]

For a mixture of perfect gases, we can identify the partial pressure of J with the contribution that J makes to the total pressure. Thus, if we introduce $p = nRT/V$ into eqn 1.6, we get

$$p_J = x_J \times \frac{nRT}{V} = \frac{n_J RT}{V}$$

The value of $n_J RT/V$ is the pressure that an amount n_J of J would exert in the otherwise empty container.

Illustration 1.2

From Self-test 1.3, we have $x_{N_2} = 0.780$, $x_{O_2} = 0.210$, and $x_{Ar} = 0.009$. It then follows from eqn 1.6 that, when the total atmospheric pressure is 100 kPa, the partial pressure of nitrogen is

$$p_{N_2} = x_{N_2}\, p = 0.780 \times (100\ \text{kPa}) = 78.0\ \text{kPa}$$

Similarly, for the other two components we find $p_{O_2} = 21.0$ kPa and $p_{Ar} = 0.9$ kPa. Provided the gases are perfect, these partial pressures are the pressures that each gas would exert if it were separated from the mixture and put in the same container on its own.

The kinetic model of gases

W^E saw in the *Introduction* that a gas may be pictured as a collection of particles in continuous, random motion (Fig 1.6). Now we develop this model of the gaseous state of matter to see how it accounts for the perfect gas law. One of the most important functions of physical chemistry is to convert qualitative notions into quantitative statements that can be tested experimentally by making measurements and comparing the results with predictions. Indeed, an important component of science as a whole is its technique of proposing a qualitative model and then express-

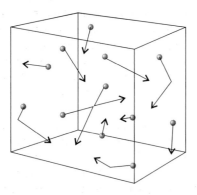

Fig 1.6 The model used for discussing the molecular basis of the physical properties of a perfect gas. The point-like molecules move randomly with a wide range of speeds and in random directions, both of which change when they collide with the walls or with other molecules.

ing that model mathematically. The 'kinetic model' of gases is an excellent example of this procedure: the model is very simple, and the quantitative prediction (the perfect gas law) is experimentally verifiable.

The **kinetic model of gases** is based on three assumptions:

1 A gas consists of molecules in ceaseless random motion.
2 The size of the molecules is negligible in the sense that their diameters are much smaller than the average distance travelled between collisions.
3 The molecules do not interact, except during collisions.

The assumption that the molecules do not interact unless they are in contact implies that the potential energy of the molecules (their energy due to their position) is independent of their separation and may be set equal to zero. The total energy of a sample of gas is therefore the sum of the kinetic energies (the energy due to motion) of all the molecules present in it.[4] It follows that, the faster the molecules travel (and hence the greater their kinetic energy), the greater the total energy of the gas.

1.4 The pressure of a gas according to the kinetic model

The kinetic theory accounts for the steady pressure exerted by a gas in terms of the collisions the molecules make with the walls of the container. Each collision gives rise to a brief force on the wall but, as billions of collisions take place every second, the walls experience a virtually constant force, and hence the gas exerts a steady pressure.

For the details of the calculation of the pressure exerted by the gas on the basis of this model see *Further information 4*. The outcome of the calculation for a gas of molar mass M in a volume V is

$$p = \frac{nMc^2}{3V} \tag{1.9}$$

---

[4] The various contributions to the energy are reviewed in *Further information 3*.

Here c is the **root-mean-square speed** (r.m.s. speed) of the molecules; for a sample consisting of N molecules with speeds $s_1, s_2, \ldots, s_N$

$$c = \left(\frac{s_1^2 + s_2^2 + \ldots + s_N^2}{N} \right)^{1/2} \tag{1.10}$$

The r.m.s. speed enters kinetic theory naturally as a measure of the average kinetic energy of the molecules. The kinetic energy of a molecule of mass m travelling at a speed v is $E_K = \frac{1}{2}mv^2$, so the mean kinetic energy is the average of this quantity, or $\frac{1}{2}mc^2$. Therefore:

$$c = \left(\frac{2E_K}{m} \right)^{1/2} \tag{1.11}$$

Therefore, wherever c appears, we can think of it as a measure of the mean kinetic energy of the molecules of the gas. The r.m.s. speed is quite close in value to another and more readily visualized measure of molecular speed, the **mean speed**, $\bar{c}$ of the molecules:

$$\bar{c} = \frac{s_1 + s_2 + \ldots + s_N}{N} \tag{1.12}$$

For samples consisting of large numbers of molecules, the mean speed is slightly smaller than the r.m.s. speed, the precise relation being

$$\bar{c} = \left(\frac{8}{3\pi} \right)^{1/2} c \approx 0.921\,c \tag{1.13}$$

For elementary purposes, and for qualitative arguments, we do not need to distinguish between the two measures of average speed, but for precise work the distinction is important.

Self-test 1.5

Cars pass a point travelling at 45.00 (5), 47.00 (7), 50.00 (9), 53.00 (4), 57.00 (1) km h^{-1}, where the number of cars is given in parentheses. Calculate (a) the r.m.s speed and (b) the mean speed of the cars. (*Hint.* Use the definitions directly; the relation in eqn 1.13 is unreliable for such small samples.)

[*Answer:* (a) 49.06 km h^{-1}, (b) 48.96 km h^{-1}]

Equation 1.9 already resembles the perfect gas equation of state; we can rearrange it into

$$pV = \tfrac{1}{3}nMc^2 \tag{1.14}$$

which resembles $pV = nRT$. This conclusion is a major success of the kinetic model, for the model implies an experimentally verified result.

1.5 The average speed of gas molecules

We now suppose that the expression for pV derived from kinetic theory is indeed the equation of state of a perfect gas. That being so, we can equate the expression on the right of eqn 1.14 to nRT, which gives

$$\tfrac{1}{3}nMc^2 = nRT$$

The ns now cancel. The great usefulness of this expression is that we can rearrange it into a formula for the r.m.s. speed of the gas molecules at any temperature:

$$c = \left(\frac{3RT}{M} \right)^{1/2} \tag{1.15}$$

Substitution of the molar mass of O_2 (32.0 g mol^{-1}) and a temperature corresponding to 25°C (that is, 298 K) gives an r.m.s. speed for these molecules of 482 m s^{-1}. The same calculation for nitrogen molecules gives 515 m s^{-1}. Both these values are not far off the speed of sound in air (346 m s^{-1} at 25°C). That similarity is reasonable, because sound is a wave of pressure variation transmitted by the movement of molecules, so the speed of propagation of a wave should be approximately the same as the speed at which molecules can adjust their locations.

The important conclusion to draw from eqn 1.15 is that *the r.m.s. speed of molecules in a gas is proportional to the square root of the temperature.* Because the mean speed is proportional to the r.m.s. speed, the same is true of the mean speed too. Therefore, doubling the temperature (on the Kelvin scale) increases the mean and the r.m.s. speed of molecules by a factor of $2^{1/2} = 1.414 \ldots$.

1.6 The Maxwell distribution of speeds

So far, we have dealt only with the *average* speed of molecules in a gas. Not all molecules, however, travel at the same speed: some move more slowly than the average (until they collide, and get accelerated to a high speed, like the impact of a bat on a ball), and others may briefly move at much higher speeds than the average, but be brought to a sudden stop. There is a ceaseless redistribution of speeds among molecules as they undergo collisions. Each molecule collides once every nanosecond (1 ns = 10^{-9} s) or so in a gas under normal conditions.

The mathematical expression that tells us the fraction of molecules that have a particular speed at any instant is called the **distribution of molecular speeds**. Thus, the distribution might tell us that at 20°C a fraction 19 in 1000 O_2 molecules have a speed in the range between 300 and 310 m s^{-1}, that 21 in 1000 have a speed in the range 400 to 410 m s^{-1}, and so on. The precise form of the distribution was worked out by James Clerk Maxwell towards the end of the nineteenth century, and his expression is known as the **Maxwell distribution of speeds**. According to Maxwell, the fraction f of molecules that have a speed in a narrow range between s and $s + \Delta s$ (e.g. between 300 m s^{-1} and 310 m s^{-1}, corresponding to $s = 300$ m s^{-1} and $\Delta s = 10$ m s^{-1}) is

$$f = F(s)\,\Delta s \quad \text{with} \quad F(s) = 4\pi\left(\frac{M}{2\pi RT}\right)^{3/2} s^2 e^{-Ms^2/2RT} \quad (1.16)$$

This is the formula used to calculate the figures quoted above.

Although eqn 1.16 looks complicated, its features can be picked out quite readily. One of the skills to develop in physical chemistry is the ability to interpret the message carried by equations. Equations convey information, and it is far more important to be able to read that information than simply to remember the equation. Let's read the information in eqn 1.16 piece by piece.

1 Because f is proportional to Δs, we see that the fraction in the range Δs increases in proportion to the width of the range. If at a given speed we increase the range of interest (but still ensure that it is narrow), then the fraction in that range increases in proportion.

2 Equation 1.16 includes a decaying exponential function (a function of the form e^{-x}, with x proportional to s^2 in this case). Its presence implies that the fraction of molecules with very high speeds will be very small because e^{-x} becomes very small when x is large.

3 The factor $M/2RT$ multiplying s^2 in the exponent is large when the molar mass, M, is large, so the exponential factor goes most rapidly towards zero when M is large. That tells us that heavy molecules are unlikely to be found with very high speeds.

4 The opposite is true when the temperature, T, is high: then the factor $M/2RT$ in the exponent is small, so the exponential factor falls towards zero relatively slowly as s increases. This tells us that, at high temperatures, a greater fraction of the molecules can be expected to have high speeds than at low temperatures.

5 A factor s^2 (the term before the e) multiplies the exponential. This factor goes to zero as s goes to zero, so the fraction of molecules with very low speeds will also be very small.

The remaining factors (the term in parentheses in eqn 1.16 and the 4π) simply ensure that, when we add together the fractions over the entire range of speeds from zero to infinity, we get 1.

Figure 1.7 is a graph of the Maxwell distribution, and shows these features pictorially for the same gas (the same value of M) but different temperatures. As we deduced from the equation, we see that only small fractions of molecules in the sample

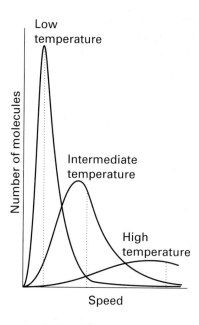

Fig 1.7 The Maxwell distribution of speeds and its variation with the temperature. Note the broadening of the distribution and the shift of the r.m.s. speed (denoted by the locations of the vertical lines) to higher values as the temperature is increased.

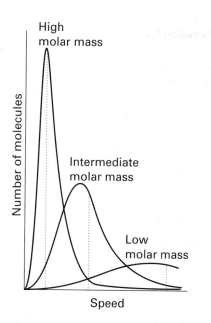

Fig 1.8 The Maxwell distribution of speeds also depends on the molar mass of the molecules. Molecules of low molar mass have a broad spread of speeds, and a significant fraction may be found travelling much faster than the r.m.s. speed. The distribution is much narrower for heavy molecules, and most of them travel with speeds close to the r.m.s. value (denoted by the locations of the vertical lines).

have very low or very high speeds. However, the fraction with very high speeds increases sharply as the temperature is raised, as the tail of the distribution reaches up to higher speeds. This feature plays an important role in the rates of gas-phase chemical reactions, for (as we shall see in Section 10.10) the rate of a reaction in the gas phase depends on the energy with which two molecules crash together, which in turn depends on their speeds.

Figure 1.8 is a plot of the Maxwell distribution for molecules with different molar masses at the same temperature. As can be seen, not only do heavy molecules have lower average speeds than light molecules at a given temperature, but they also have a significantly narrower spread of speeds. That narrow spread means that most molecules will be found with speeds close to the average. In contrast, light molecules (such as H_2) have high average speeds and a wide spread of speeds: many molecules will be found travelling either much more slowly or much more quickly than the average. This feature plays an important role in determining the composition of planetary atmospheres, because it

means that a significant fraction of light molecules travel at sufficiently high speeds to escape from the planet's gravitational attraction. The ability of light molecules to escape is one reason why hydrogen (molar mass 2.02 g mol^{-1}) and helium (4.00 g mol^{-1}) are very rare in the Earth's atmosphere.

1.7 Diffusion and effusion

Diffusion is the process by which the molecules of different substances mingle with each other. The atoms of two solids diffuse into each other when the two solids are in contact, but the process is very slow. The diffusion of a solid through a liquid solvent is much faster but mixing normally needs to be encouraged by stirring or shaking the solid in the liquid (the process is then no longer pure diffusion). Gaseous diffusion is much faster. It accounts for the largely uniform composition of the atmosphere for, if a gas is produced by a localized source (such as carbon dioxide from the respiration of animals, oxygen from photosynthesis by green plants, and pollutants from vehicles and industrial sources), then the molecules of gas will diffuse from the source and in due course be distributed throughout the atmosphere. In practice, the process of mixing is accelerated by winds: such bulk motion of matter is called **convection**. The process of **effusion** is the escape of a gas through a small hole, as in a puncture in an inflated balloon or tyre (Fig 1.9).

The rates of diffusion and effusion of gases increase with increasing temperature, for both processes depend on the motion of molecules, and molecular speeds increase with temperature. The rates also decrease with increasing molar mass, for molecular speeds decrease with increasing molar mass. The dependence on molar mass, however, is

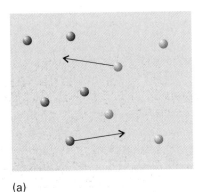

(a)

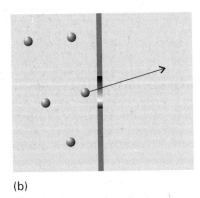

(b)

Fig 1.9 (a) Diffusion is the spreading of the molecules of one substance into the region initially occupied by another substance. Note that molecules of both substances move, and each substance diffuses into the other. (b) Effusion is the escape of molecules through a small hole in a confining wall.

simple only in the case of effusion. In effusion, only a single substance is in motion, not the two or more intermingling gases involved in diffusion.

The experimental observations on the dependence of the rate of effusion of a gas on its molar mass are summarized by **Graham's law of effusion**, proposed by Thomas Graham in 1833:

At a given pressure and temperature, the rate of effusion of a gas is inversely proportional to the square root of its molar mass:

$$\text{Rate of effusion} \propto \frac{1}{M^{1/2}} \qquad (1.17)$$

Rate in this context means the number (or number of moles) of molecules that escape per second.

Illustration 1.4

The rates at which hydrogen (molar mass 2.016 g mol⁻¹) and carbon dioxide (44.01 g mol⁻¹) effuse under the same conditions of pressure and temperature are in the ratio

$$\frac{\text{Rate of effusion of } H_2}{\text{Rate of effusion of } CO_2} = \left(\frac{M_{CO_2}}{M_{H_2}}\right)^{1/2}$$

$$= \left(\frac{44.01 \text{ g mol}^{-1}}{2.016 \text{ g mol}^{-1}}\right)^{1/2}$$

$$= \left(\frac{44.01}{2.016}\right)^{1/2} = 4.672$$

The *mass* of carbon dioxide that escapes in a given interval is greater than the mass of hydrogen because, although nearly 5 times as many hydrogen molecules escape, each carbon dioxide molecule has over 20 times the mass of a molecule of hydrogen.

The high rate of effusion of hydrogen and helium is one reason why these two gases leak from containers and through rubber diaphragms so readily. The different rates of effusion through a porous barrier are employed in the separation of uranium-235 from the more abundant and less useful uranium-238 in the processing of nuclear fuel. The process depends on the formation of uranium hexafluoride, a volatile solid. However, because the ratio of the molar masses of $^{238}UF_6$

and $^{235}UF_6$ is only 1.008, the ratio of the rates of effusion is only $(1.008)^{1/2} = 1.004$. Thousands of successive effusion stages are therefore required to achieve a significant separation. The rate of effusion of gases was once used to determine molar mass by comparison of the rate of effusion of a gas or vapour with that of a gas of known molar mass. However, there are now much more precise methods available, such as mass spectrometry.

Graham's law is explained by noting that the r.m.s. speed of molecules of a gas is inversely proportional to the square root of the molar mass (eqn 1.15). Because the rate of effusion through a hole in a container is proportional to the rate at which molecules pass through the hole, it follows that the rate should be inversely proportional to $M^{1/2}$, which is in accord with Graham's law.

1.8 **Molecular collisions**

The average distance that a molecule travels between collisions is called its **mean free path**, λ (lambda). The mean free path in a liquid is less than the diameter of the molecules, because a molecule in a liquid meets a neighbour even if it moves only a fraction of a diameter. However, in gases, the mean free paths of molecules can be several hundred molecular diameters.

The **collision frequency**, z, is the average rate of collisions made by one molecule. Specifically, z is the average number of collisions one molecule makes in a time interval divided by the length of the interval. It follows that the inverse of the collision frequency, $1/z$, is the **time of flight**, the average time that a molecule spends in flight between two collisions. As we shall see, this average time of flight is typically about 1 ns at 1 atm and room temperature.

Because speed is distance travelled divided by the time taken for the journey, the r.m.s. speed c, which we can loosely think of as the average speed, is the average length of the flight of a molecule between collisions (that is, the mean free path, λ) divided by the average time of flight ($1/z$). It follows that the mean free path and the collision frequency are related by

$$c = \frac{\text{mean free path}}{\text{time of flight}} = \frac{\lambda}{1/z} = \lambda z \qquad (1.18)$$

Therefore, if we can calculate either λ or z, then we can find the other from this equation.

To find expressions for λ and z we need a slightly more elaborate version of the kinetic model. The basic kinetic model supposes that the molecules are effectively point-like; however, to obtain collisions, we need to assume that two 'points' score a hit whenever they come within a certain range d of each other, where d is the diameter of the molecules (Fig 1.10). The **collision cross-section**, σ (sigma), the target area presented by one molecule to another, is therefore the area of a circle of radius d, so $\sigma = \pi d^2$. When this quantity is built into the kinetic model, we find that

$$\lambda = \frac{RT}{2^{1/2} N_A \sigma p} \qquad z = \frac{2^{1/2} N_A \sigma c p}{RT} \qquad (1.19)$$

Illustration 1.5

From the information in Table 1.3 we can calculate that the mean free path of O_2 molecules in a sample of oxygen at SATP (25°C, 1 bar) is

$$\lambda = \frac{\overbrace{(8.314\,51\ \text{Pa m}^3\ \text{K}^{-1}\text{mol}^{-1})}^{R} \times \overbrace{(298\ \text{K})}^{T}}{2^{1/2} \times \underbrace{(6.022 \times 10^{23}\ \text{mol}^{-1})}_{N_A} \times \underbrace{(0.40 \times 10^{-18}\ \text{m}^2)}_{\sigma}} \\ \times \underbrace{(1.00 \times 10^5\ \text{Pa})}_{p}$$

$$= 7.3 \times 10^{-8}\ \text{m}$$

or 73 nm. We have used R in one of its SI unit forms: this form is usually appropriate in calculations based on the kinetic model. Under the same conditions, the collision frequency is $6.6 \times 10^9\ \text{s}^{-1}$, so each molecule makes 6.6 billion collisions each second.

Table 1.3 lists the collision cross-sections of some common atoms and molecules.

Once again, we should *interpret the essence* of the two expressions in eqn 1.19 rather than trying to remember them.

1 Because $\lambda \propto 1/p$, we see that *the mean free path decreases as the pressure increases.* This decrease is

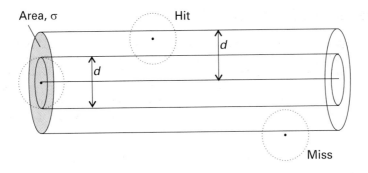

Fig 1.10 To calculate features of a perfect gas that are related to collisions, a point is regarded as being surrounded by a sphere of diameter d. A molecule will hit another molecule if the latter lies within a cylinder of radius d. The collision cross-section is the area of the tube, and is πd^2.

Table 1.3 *Collision cross-sections of atoms and molecules*

Species	σ/nm^2
Ar	0.36
C_2H_4	0.64
C_6H_6	0.88
CH_4	0.46
Cl_2	0.93
CO_2	0.52
H_2	0.27
He	0.21
N_2	0.43
O_2	0.40
SO_2	0.58

$1 \text{ nm}^2 = 10^{-18} \text{ m}^2$.

a result of the increase in the number of molecules present in a given volume as the pressure is increased, so each molecule travels a shorter distance before it collides with a neighbour. For example, the mean free path of an O_2 molecule decreases from 73 nm to 36 nm when the pressure is increased from 1.0 bar to 2.0 bar at 25°C.

2 Because $\lambda \propto 1/\sigma$, *the mean free path is shorter for molecules with large collision cross-sections.* For instance, the collision cross-section of a benzene molecule (0.88 nm^2) is about four times greater than that of a helium atom (0.21 nm^2), and at the same pressure and temperature its mean free path is four times shorter.

3 Because $z \propto p$, *the collision frequency increases with the pressure of the gas.* This dependence follows from the fact that, provided the temperature is the same, each molecule takes less time to travel to its neighbour in a denser, higher-pressure gas. For example, although the collision frequency for an O_2 molecule in oxygen gas at SATP is $6.6 \times 10^9 \text{ s}^{-1}$, at 2.0 bar and the same temperature the collision frequency is doubled, to $1.3 \times 10^{10} \text{ s}^{-1}$.

4 Because eqn 1.19 shows that $z \propto c$, and we know that $c \propto 1/M^{1/2}$, providing their collision cross-sections are the same, *heavy molecules have lower collision frequencies than light molecules.* Heavy molecules travel more slowly on average than light molecules do (at the same temperature), so they collide with other molecules less frequently.

Box 1.2 *The Sun as a ball of perfect gas*

The kinetic theory of gases is valid when the size of the particles is negligible compared with their mean free path. It may seem absurd, therefore, to expect the kinetic theory and, as a consequence, the perfect gas law, to be applicable to the dense matter of stellar interiors. In the Sun, for instance, the density is 1.50 times that of liquid water at its centre and comparable to that of water about halfway to its surface. However, we have to realize that the state of matter is that of a *plasma*, in which the electrons have been stripped from the atoms of hydrogen and helium that make up the bulk of the matter of stars. As a result, the particles making up the plasma have diameters comparable to those of nuclei, or about 10 fm. Therefore, a mean free path of only 0.1 pm satisfies the criterion for the validity of the kinetic theory and the perfect gas law. We can therefore use $pV = nRT$ as the equation of state for the stellar interior.

As for any perfect gas, the pressure in the interior of the Sun is related to the mass density, $\rho = m/V$, by

$$p = \frac{nRT}{V} = \frac{mRT}{MV} = \frac{\rho RT}{M}$$

The problem is to know the molar mass to use. Atoms are stripped of their electrons in the interior of stars so, if we suppose that the interior consists of ionized hydrogen atoms, the mean molar mass is one-half the molar mass of hydrogen, or 0.5 g mol^{-1} (the mean of the molar mass of H^+ and e^-, the latter being virtually 0). Halfway to the centre of the Sun, the temperature is 3.6 MK and the mass density is 1.20 g cm^{-3} (slightly denser than water); so the pressure there is

$p =$

$$\frac{(1.20 \times 10^3 \text{ kg m}^{-3}) \times (8.3145 \text{ J K}^{-1} \text{ mol}^{-1}) \times (3.6 \times 10^6 \text{ K})}{0.50 \times 10^{-3} \text{ kg mol}^{-1}}$$

$$= 7.2 \times 10^{13} \text{ Pa}$$

or 720 Mbar (about 720 million atmospheres).

We can combine this result with the expression for the pressure from kinetic theory ($p = \frac{1}{3} nMc^2/V$). Because the total kinetic energy of the particles is $E_K = \frac{1}{2} Nmc^2$, we can write $p = \frac{2}{3} E_K/V$. That is, the pressure of the plasma is related to the *kinetic energy density*, $\rho_K = E_K/V$, the kinetic energy of the molecules in a region divided by the volume of the region, by

$$p = \tfrac{2}{3} \rho_K$$

It follows that the kinetic energy density halfway to the centre of the Sun is

$$\rho_K = \tfrac{3}{2} p = \tfrac{3}{2} \times (7.2 \times 10^{13} \text{ Pa}) = 1.1 \times 10^{14} \text{ J m}^{-3}$$

or 0.11 GJ cm^{-3}. In contrast, on a warm day (25°C) on Earth, the (translational) kinetic energy density of our atmosphere is only 1.5×10^5 J m^{-3} (corresponding to 0.15 J cm^{-3}).

Exercise 1 Stars eventually deplete some of the hydrogen in their core, which contracts and results in higher temperatures. The increased temperature results in an increase in the rates of nuclear reactions, some of which result in the formation of heavier nuclei, such as carbon. The outer part of the star expands and cools to become a red giant. Assume that halfway to the centre a red giant has a temperature of 3500 K, is composed primarily of fully ionized carbon atoms and electrons, and has a mass density of 1.20 g cm^{-3}. What is the pressure at this point?

Exercise 2 If the above red giant consisted of neutral carbon atoms, instead of ionized carbon atoms and electrons, what would the pressure be at the same point under the same conditions?

Real gases

So far, everything we have said applies to perfect gases, in which the average separation of the molecules is so great that they move independently of one another. In terms of the quantities introduced in the previous section, a perfect gas is a gas for which the mean free path, λ, of the molecules in the sample is much greater than d, the separation at which they are regarded as being in contact (Box 1.2). This condition is written $\lambda \gg d$. As a result of this large average separation, a perfect gas is a gas in which the only contribution to the energy comes from the kinetic energy of the motion of the molecules and there is no contribution to the total energy from the potential energy arising from the interaction of the molecules with one another. However, in fact, all molecules do interact with one another provided they are close enough together, so the 'kinetic energy only' model is only an approximation.

1.9 Intermolecular interactions

There are two types of contribution to the interaction between molecules. At relatively long distances (a few molecular diameters), molecules attract each other. This attraction is responsible for the condensation of gases into liquids at low temperatures. At low enough temperatures the molecules of a gas have insufficient kinetic energy to escape from each other's attraction and they stick together. Second, although molecules attract each other when they are a few diameters apart, as soon as they come into contact they repel each other. This repulsion is responsible for the fact that liquids and solids have a definite bulk and do not collapse to an infinitesimal point.

Intermolecular interactions—the attractions and repulsions between molecules—give rise to a potential energy that contributes to the total energy of a gas. Because attractions correspond to a lowering of total energy as molecules get closer together, they make a *negative* contribution to the potential energy. On the other hand, repulsions make a positive contribution to the total energy as the molecules squash together. Figure 1.11 illustrates the general form of the variation of the intermolecular potential energy. At large separations, the energy-lowering interactions are dominant, but at short distances the energy-raising repulsions dominate.

The intermolecular interactions affect the bulk properties of a gas. For example, the isotherms of real gases have shapes that differ from those implied by Boyle's law, particularly at high pressures and low temperatures. Figure 1.12 shows a set of experimental isotherms for carbon dioxide. They should be compared with the perfect-gas isotherms shown in Fig 1.1. Although the experimental isotherms resemble the perfect-gas isotherms at high tempera-

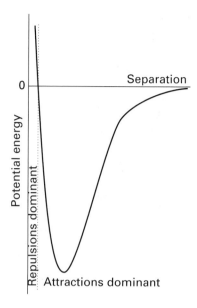

Fig 1.11 The variation of the potential energy of two molecules with their separation. High positive potential energy (at very small separations) indicates that the interactions between them are strongly repulsive at these distances. At intermediate separations, where the potential energy is negative, the attractive interactions dominate. At large separations (on the right) the potential energy is zero and there is no interaction between the molecules.

tures (and at low pressures, off the scale on the right of the graph), there are very striking differences between the two at temperatures below about 50°C and at pressures above about 1 bar.

1.10 The critical temperature

To understand the significance of the isotherms in Fig 1.12, let's begin with the isotherm at 20°C. At point A the sample is a gas. As the sample is compressed to B by pressing in a piston, the pressure increases broadly in agreement with Boyle's law, and the increase continues until the sample reaches point C. Beyond this point, we find that the piston can be pushed in without any further increase in pressure, through D to E. The reduction in volume from E to F requires a very large increase in pressure. This variation of pressure with volume is exactly what we expect if the gas at C condenses to a

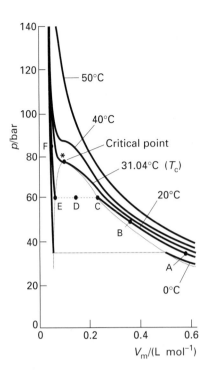

Fig 1.12 The experimental isotherms of carbon dioxide at several temperatures. The critical isotherm is at 31.04°C.

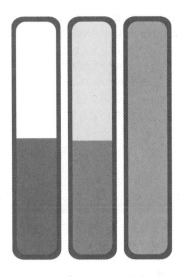

Fig 1.13 When a liquid is heated in a sealed container, the density of the vapour phase increases and that of the liquid phase decreases, as depicted here by the changing density of shading. There comes a stage at which the two densities are equal and the interface between the two fluids disappears. This disappearance occurs at the critical temperature. The container needs to be strong: the critical temperature of water is at 373°C and the vapour pressure is then 218 atm.

compact liquid at E. Indeed, if we could see the sample, we would see it begin to condense to a liquid at C, and the condensation would be complete when the piston was pushed in to E. At E, the piston is resting on the surface of the liquid. The subsequent reduction in volume, from E to F, corresponds to the very high pressure needed to compress a liquid into a smaller volume. In terms of intermolecular interactions, the step from C to E corresponds to the molecules being so close on average that they attract each other and cohere into a liquid. The step from E to F represents the effect of trying to force the molecules even closer together when they are already in contact, and hence trying to overcome the strong repulsive interactions between them.

If we could look inside the container at point D, we would see a liquid separated from the remaining gas by a sharp surface (Fig 1.13). At a slightly higher temperature (at 30°C, for instance), a liquid forms, but a higher pressure is needed to produce it. It might be difficult to make out the surface because the remaining gas is at such a high pressure that its

density is similar to that of the liquid. At the special temperature of 31.04°C (304.19 K) the gaseous state appears to transform continuously into the condensed state and at no stage is there a visible surface between the two states of matter. At this temperature, which is called the **critical temperature**, T_c, and at all higher temperatures, a single form of matter fills the container at all stages of the compression and there is no separation of a liquid from the gas. We have to conclude that *a gas cannot be condensed to a liquid by the application of pressure unless the temperature is below the critical temperature.*

Table 1.4 lists the critical temperatures of some common gases. The data there imply, for example, that liquid nitrogen cannot be formed by the application of pressure unless the temperature is below 126 K (−147°C).

The dense fluid obtained by compressing a gas when its temperature is higher than its critical temperature is not a true liquid, but it behaves like a liquid in many respects—it has a similar density, for instance, and can act as a solvent. However, despite its density, the fluid is not strictly a liquid because it

Table 1.4 *The critical temperatures of gases*

	Critical temperature/°C
Noble gases	
He	−268 (5.2 K)
Ne	−229
Ar	−123
Kr	−64
Xe	17
Halogens	
Cl_2	144
Br_2	311
Small inorganic molecules	
H_2	−240
O_2	−118
H_2O	374
N_2	−147
NH_3	132
CO_2	31
Organic compounds	
CH_4	−83
CCl_4	283
C_6H_6	289

never possesses a surface that separates it from a vapour phase. Nor is it much like a gas, because it is so dense. It is an example of a **supercritical fluid**. Supercritical fluids are currently being used as solvents; for example, supercritical carbon dioxide is used to extract caffeine in the manufacture of decaffeinated coffee where, unlike organic solvents, it does not result in the formation of an unpleasant and possibly toxic residue.

1.11 **The compression factor**

A useful quantity for discussing the properties of real gases is the **compression factor**, Z, which is the ratio of the actual molar volume of a gas to the molar volume of a perfect gas under the same conditions:

$$\text{Compression factor} = \frac{\text{molar volume of gas}}{\text{molar volume of perfect gas}}$$

$$Z = \frac{V_m}{V_m^{perfect}} \qquad (1.20a)$$

The molar volume of a perfect gas is RT/p (recall eqn 1.2), so we can rewrite this definition as

$$Z = \frac{V_m}{RT/p} = \frac{pV_m}{RT} \qquad (1.20b)$$

where V_m is the molar volume of the gas we are studying. For a perfect gas, $Z = 1$, so deviations of Z from 1 are a measure of how far a real gas departs from behaving perfectly.

When Z is measured for real gases, it is found to vary with pressure as shown in Fig 1.14. At low pressures, some gases (methane, ethane, and ammonia, for instance) have $Z < 1$. That is, their molar volumes are smaller than that of a perfect gas, suggesting that the molecules cluster together slightly. We can conclude that, for these molecules and these condi-

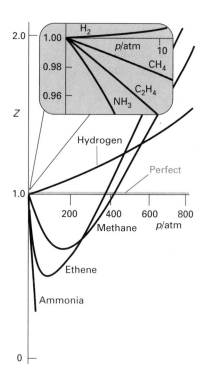

Fig 1.14 The variation of the compression factor, Z, with pressure for several gases at 0°C. A perfect gas has $Z = 1$ at all pressures. Of the gases shown, hydrogen shows positive deviations at all pressures (at this temperature); all the other gases show negative deviations initially but positive deviations at high pressures. The negative deviations are a result of the attractive interactions between molecules and the positive deviations are a result of the repulsive interactions.

tions, the attractive interactions are dominant. The compression factor rises above 1 at high pressures whatever the identity of the gas, and for some gases (hydrogen in Fig 1.14) $Z > 1$ at all pressures.[5] The observation that $Z > 1$ tells us that the molar volume of the gas is now greater than that expected for a perfect gas of the same temperature and pressure. This behaviour can be traced to the dominant repulsive forces. These forces tend to drive the molecules apart when they are forced to be close together at high pressures. For hydrogen, the attractive interactions are so weak that the repulsive interactions dominate even at low pressures.

1.12 **The virial equation of state**

We can use the deviation of Z from its 'perfect' value of 1 to construct an *empirical* (observation-based) equation of state. To do so, we suppose that, for real gases, the relation $Z = 1$ is only the first term of a lengthier expression, and write instead

$$Z = 1 + \frac{B}{V_m} + \frac{C}{V_m^2} + \dots \tag{1.21}$$

The coefficients $B, C, \dots$, are called **virial coefficients**; B is the second virial coefficient, C, the third, and so on; the unwritten $A = 1$ is the first.[6] They vary from gas to gas and depend on the temperature. This technique, of taking a limiting expression (in this case, $Z = 1$, which applies to gases at low pressures) and supposing that it is the first term of a more complicated expression, is quite common in physical chemistry. The limiting expression is the first approximation to the true expression, whatever that may be, and the additional terms take into account the secondary effects that the limiting expression ignores.

The most important additional term on the right in eqn 1.21 is B. From the graphs in Fig 1.14, it follows that, for the temperature to which the data apply, B must be positive for hydrogen (so that $Z > 1$) but negative for methane, ethane, and ammonia

(so that for them $Z < 1$). However, regardless of the sign of B, the positive term C/V_m^2 becomes large at high pressures (when V_m^2 is very small) and the right-hand side of eqn 1.21 becomes greater than 1, just as in the curves for the other gases in Fig 1.14. The values of the virial coefficients for many gases are known from measurements of Z over a range of pressures and fitting the data to eqn 1.21 by varying the coefficients until a good match is obtained.

To convert eqn 1.21 into an equation of state, we combine it with eqn 1.20b, which gives

$$\frac{pV_m}{RT} = 1 + \frac{B}{V_m} + \frac{C}{V_m^2} + \dots$$

We then multiply both sides by RT/V_m, to get an expression for p in terms of the other variables:

$$p = \frac{nRT}{V}\left(1 + \frac{nB}{V} + \frac{n^2C}{V^2} + \dots\right) \tag{1.22}$$

(We have also replaced V_m by V/n, throughout.) Equation 1.22 is the **virial equation of state**. At very low pressures, when the molar volume is very large, the terms B/V_m and C/V_m^2 are both very small, and only the 1 inside the parentheses survives. In this limit (of p approaching 0), the equation of state approaches that of a perfect gas.

1.13 **The van der Waals equation of state**

Although it is the most reliable equation of state, the virial equation does not give us much immediate insight into the behaviour of gases and their condensation to liquids. The **van der Waals equation**, which was proposed in 1873 by the Dutch physicist Johannes van der Waals, is only an approximate equation of state but it has the advantage of showing how the intermolecular interactions contribute to the deviations of a gas from the perfect gas law. We can view the van der Waals equation as another example of taking a soundly based qualitative idea and building up a mathematical expression that can be tested quantitatively.

---

[5] The type of behaviour exhibited depends on the temperature.
[6] The word 'virial' comes from the Latin word for force, and it reflects the fact that intermolecular forces are now significant. Virial coefficients are also denoted B_2, B_3, etc. in place of B, C, etc.

The repulsive interaction between two molecules implies that they cannot come closer than a certain distance. Therefore, instead of being free to travel anywhere in a volume V, the actual volume in which the molecules can travel is reduced to an extent proportional to the number of molecules present and the volume they each exclude (Fig 1.15). We can therefore model the effect of the repulsive, volume-excluding forces by changing V in the perfect gas equation to $V - nb$, where b is the proportionality constant between the reduction in volume and the amount of molecules present in the container. With this modification, the perfect gas equation of state changes from $p = nRT/V$ to

$$p = \frac{nRT}{V - nb}$$

This equation of state—it is not yet the full van der Waals equation—should describe a gas in which repulsions are important. Note that, when the pressure is low, the volume is large compared with the volume excluded by the molecules (which we write $V \gg nb$). The nb can then be ignored in the denominator and the equation reduces to the perfect gas equation of state. It is always a good plan to verify that an equation reduces to a known form when a plausible physical approximation is made.

The effect of the attractive interactions between molecules is to reduce the pressure that the gas exerts. We can model the effect by supposing that the attraction experienced by a given molecule is proportional to the concentration, n/V, of molecules in the container. Because the attractions slow

the molecules down, the molecules strike the walls less frequently *and* strike it with a weaker impact. We can therefore expect the reduction in pressure to be proportional to the *square* of the molar concentration, one factor of n/V reflecting the reduction in frequency of collisions and the other factor the reduction in the strength of their impulse. If the constant of proportionality is written a, we can write

$$\text{Reduction in pressure} = a \times \left(\frac{n}{V}\right)^2$$

It follows that the equation of state allowing for both repulsions and attractions is

$$p = \frac{nRT}{V - nb} - a\left(\frac{n}{V}\right)^2 \tag{1.23a}$$

This expression is the **van der Waals equation of state**. To show the resemblance of this equation to the perfect gas equation $pV = nRT$, eqn 1.23a is sometimes rearranged into

$$\left(p + \frac{an^2}{V^2}\right)(V - nb) = nRT \tag{1.23b}$$

We have built the van der Waals equation using physical arguments about the volumes of molecules and the effects of forces between them. It can be derived in other ways, but the present method has the advantage of showing how to derive the form of an equation out of general ideas. The derivation also has the advantage of keeping imprecise the significance of the **van der Waals parameters,** the constants a and b: they are much better regarded as empirical parameters than as precisely defined molecular properties. The van der Waals parameters depend on the gas, but are taken as independent of temperature (Table 1.5).

We can judge the reliability of the van der Waals equation by comparing the isotherms it predicts, which are shown in Fig 1.16, with the experimental isotherms already shown in Fig 1.12. Apart from the waves below the critical temperature, they do resemble experimental isotherms quite well. The waves, which are called **van der Waals' loops,** are unrealistic because they suggest that under some conditions an increase of pressure results in an in-

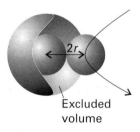

Fig 1.15 When two molecules, each of radius r and volume $v_{mol} = \frac{4}{3}\pi r^3$ approach each other, the centre of one of them cannot penetrate into a sphere of radius $2r$ and therefore volume $8v_{mol}$ surrounding the other molecule.

Excluded volume

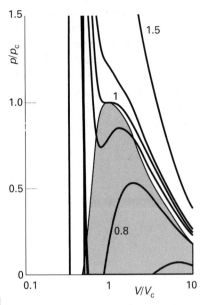

Fig 1.16 Isotherms calculated by using the van der Waals equation of state. The axes are labelled with the reduced pressure, p/p_c, and reduced volume, V/V_c, where $p_c = a/27b^2$ and $V_c = 3b$. The individual isotherms are labelled with the reduced temperature, T/T_c, where $T_c = 8a/27Rb$. The isotherm labelled 1 is the critical isotherm (the isotherm at the critical temperature).

crease of volume. The wave-like regions are therefore trimmed away and replaced by horizontal lines (Fig 1.17).[7] The van der Waals parameters in Table 1.5 were found by fitting the calculated curves to experimental isotherms.

Perfect-gas isotherms are obtained from the van der Waals equation at high temperatures and low pressures. To confirm this remark, we need to note that, when the temperature is high, RT may be so large that the first term on the right in eqn 1.23a greatly exceeds the second, so the latter may be ignored. Furthermore, at low pressures, the molar volume is so large that $V - nb$ can be replaced by V. Hence, under these conditions (of high temperature and low pressure), eqn 1.23a reduces to $p = nRT/V$, the perfect gas equation.

---

[7] Theoretical arguments show that the horizontal line should trim equal areas of loop above and below where it lies.

Table 1.5 Van der Waals parameters of gases

	$a/(L^2\ atm\ mol^{-2})$	$b/(L\ mol^{-1})$
Air	1.4	0.039
Ammonia	4.17	0.037
Argon	1.35	0.032
Carbon dioxide	3.59	0.043
Ethane	5.49	0.064
Ethene	4.47	0.057
Helium	0.034	0.024
Hydrogen	0.244	0.027
Nitrogen	1.39	0.039
Oxygen	1.36	0.032
Xenon	4.19	0.051

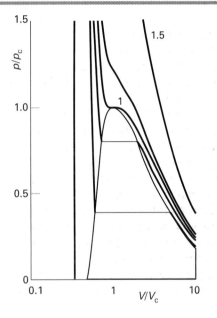

Fig 1.17 The unphysical van der Waals loops are eliminated by drawing straight lines that divide the loops into areas of equal size. With this procedure, the isotherms strongly resemble the observed isotherms.

1.14 The liquefaction of gases

A gas may be liquefied by cooling it below its boiling point at the pressure of the experiment. For example, chlorine at 1 atm can be liquefied by cooling it to below $-34°C$ in a bath cooled with dry ice (solid carbon dioxide). For gases with very low boiling points (such as oxygen and nitrogen, at $-183°C$ and $-196°C$, respectively), such a simple technique is not practicable unless an even colder bath is available.

One alternative and widely used commercial technique makes use of the forces that act between molecules. We saw earlier that the r.m.s. speed of molecules in a gas is proportional to the square root of the temperature (eqn 1.15). It follows that reducing the r.m.s. speed of the molecules is equivalent to cooling the gas. If the speed of the molecules can be reduced to the point that neighbours can capture each other by their intermolecular attractions, then the cooled gas will condense to a liquid.

To slow the gas molecules, we make use of an effect similar to that seen when a ball is thrown into the air: as it rises it slows in response to the gravitational attraction of the Earth and its kinetic energy is converted into potential energy. Molecules attract each other, as we have seen (the attraction is not gravitational, but the effect is the same), and, if we can cause them to move apart from each other, like a ball rising from a planet, then they should slow. It is very easy to move molecules apart from each other: we allow the gas to expand, which increases the average separation of the molecules. To cool a gas, therefore, we allow it to expand without allowing any heat to enter from outside. As it does so, the molecules move apart to fill the available volume, struggling against the attraction of their neighbours. Because some kinetic energy has been converted into potential energy, they travel more slowly. Therefore, because the average speed of the molecules has been reduced, the gas is now cooler than before the expansion. This process of cooling a real gas by expansion through a narrow opening called a 'throttle' is called the **Joule–Thomson effect**.[8]

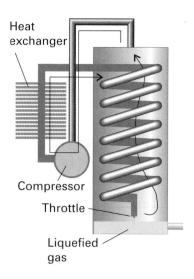

Fig 1.18 The principle of the Linde refrigerator. The gas is recirculated and cools the gas that is about to undergo expansion through the throttle. The expanding gas cools still further. Eventually, liquefied gas drips from the throttle.

The procedure works only for real gases in which the attractive interactions are dominant, because the molecules have to climb apart against the attractive force in order for them to travel more slowly. For molecules under conditions when repulsions are dominant (corresponding to $Z > 1$), the Joule–Thomson effect results in the gas becoming warmer.

In practice, the gas is allowed to expand several times by recirculating it through a device called a *Linde refrigerator* (Fig 1.18). On each successive expansion the gas becomes cooler and, as it flows past the incoming gas, the latter is cooled further. After several successive expansions, the gas becomes so cold that it condenses to a liquid.

---

[8] The effect was first observed and analysed by James Joule (whose name is commemorated in the unit of energy) and William Thomson (who later became Lord Kelvin).

Exercises

Treat all gases as perfect unless instructed otherwise.

1.1 What pressure is exerted by a sample of nitrogen gas of mass 2.045 g in a container of volume 2.00 L at 21°C?

1.2 A sample of neon of mass 255 mg occupies 3.00 L at 122 K. What pressure does it exert?

1.3 Much to everyone's surprise, nitrogen monoxide (NO) has been found to act as a neurotransmitter. To prepare to study its effect, a sample was collected in a container of volume 250.0 mL. At 19.5°C its pressure is found to be 24.5 kPa. What amount (in moles) of NO has been collected?

1.4 A domestic water-carbonating kit uses steel cylinders of carbon dioxide of volume 250 mL. They weigh 1.04 kg when full and 0.74 kg when empty. What is the pressure of gas in the cylinder at 20°C?

1.5 The effect of high pressure on organisms, including humans, is studied to gain information about deep-sea diving and anaesthesia. A sample of air occupies 1.00 L at 25°C and 1.00 atm. What pressure is needed to compress it to 100 cm^3 at this temperature?

1.6 You are warned not to dispose of pressurized cans by throwing them on to a fire. The gas in an aerosol container exerts a pressure of 125 kPa at 18°C. The container is thrown on a fire, and its temperature rises to 700°C. What is the pressure at this temperature?

1.7 Until we find an economical way of extracting oxygen from sea-water or lunar rocks, we have to carry it with us to inhospitable places, and do so in compressed form in tanks. A sample of oxygen at 101 kPa is compressed at constant temperature from 7.20 L to 4.21 L. Calculate the final pressure of the gas.

1.8 To what temperature must a sample of helium gas be cooled from 22.2°C to reduce its volume from 1.00 L to 100 cm^3?

1.9 Hot-air balloons gain their lift from the lowering of density of air that occurs when the air in the envelope is heated. To what temperature should you heat a sample of air, initially at 340 K, to increase its volume by 14 per cent?

1.10 At sea level, where the pressure was 104 kPa and the temperature 21.1°C, a certain mass of air occupied 2.0 m^3. To what volume will the region expand when it has risen to an altitude where the pressure and temperature are (a) 52 kPa, –5.0°C, (b) 880 Pa, –52.0°C?

1.11 A diving bell has an air space of 3.0 m^3 when on the deck of a boat. What is the volume of the air space when the bell has been lowered to a depth of 50 m? Take the mean density of sea water to be 1.025 g cm^{-3} and assume that the temperature is the same as on the surface.

1.12 A meteorological balloon had a radius of 1.0 m when released at sea level at 20°C and expanded to a radius of 3.0 m when it had risen to its maximum altitude where the temperature was –20°C. What is the pressure inside the balloon at that altitude?

1.13 A gas mixture being used to simulate the atmosphere of another planet consists of 320 mg of methane, 175 mg of argon, and 225 mg of nitrogen. The partial pressure of nitrogen at 300 K is 15.2 kPa. Calculate (a) the volume and (b) the total pressure of the mixture.

1.14 The vapour pressure of water at blood temperature is 47 Torr. What is the partial pressure of dry air in our lungs when the total pressure is 760 Torr?

1.15 A determination of the density of a gas or vapour can provide a quick estimate of its molar mass even though for practical work mass spectrometry is far more precise. The density of a gaseous compound was found to be 1.23 g L^{-1} at 330 K and 25.5 kPa. What is the molar mass of the compound?

1.16 In an experiment to measure the molar mass of a gas, 250 cm^3 of the gas was confined in a glass vessel. The pressure was 152 Torr at 298 K and the mass of the gas was 33.5 mg. What is the molar mass of the gas?

1.17 A vessel of volume 22.4 L contains 2.0 mol H_2 and 1.0 mol N_2 at 273.15 K. Calculate (a) their partial pressures and (b) the total pressure.

1.18 The composition of planetary atmospheres is determined in part by the speeds of the molecules of the constituent gases, because the faster moving molecules can reach escape velocity and leave the planet. Calculate the mean speed of (a) He atoms, (b) CH_4 molecules at (i) 77 K, (ii) 298 K, (iii) 1000 K.

1.19 At what pressure does the mean free path of argon at 25°C become comparable to the diameter of a spherical vessel of volume 1.0 L that contains it? Take $\sigma = 0.36$ nm^2.

1.20 At what pressure does the mean free path of argon at 25°C become comparable to 10 times the diameter of the atoms themselves? Take $\sigma = 0.36$ nm^2.

1.21 When we are studying the photochemical processes that can occur in the upper atmosphere, we need to know how often atoms and molecules collide. At an altitude of 20 km the temperature is 217 K and the pressure 0.050 atm. What is the mean free path of N_2 molecules? Take $\sigma = 0.43$ nm^2.

1.22 How many collisions does a single Ar atom make in 1.0 s when the temperature is 25°C and the pressure is (a) 10 bar, (b) 100 kPa, (c) 1.0 Pa?

1.23 Calculate the total number of collisions per second in 1.0 L of argon under the same conditions as in Exercise 1.22.

1.24 How many collisions per second does an N_2 molecule make at an altitude of 20 km? (See Exercise 1.21 for data.)

1.25 The spread of pollutants through the atmosphere is governed partly by the effects of winds but also by the natural tendency of molecules to diffuse. The latter depends on how far a molecule can travel before colliding with another molecule. Calculate the mean free path of diatomic molecules in air using $\sigma = 0.43$ nm^2 at 25°C and (a) 10 bar, (b) 103 kPa, (c) 1 Pa.

1.26 Use the Maxwell distribution of speeds to estimate the fraction of N_2 molecules at 500 K that have speeds in the range 290 to 300 m s^{-1}.

1.27 How does the mean free path in a sample of a gas vary with temperature in a constant-volume container?

1.28 Calculate the pressure exerted by 1.0 mol C_2H_6 behaving as (a) a perfect gas, (b) a van der Waals gas when it is confined under the following conditions: (i) at 273.15 K in 22.414 L, (ii) at 1000 K in 100 cm^3. Use the data in Table 1.5.

1.29 How reliable is the perfect gas law in comparison with the van der Waals equation? Calculate the difference in pressure of 10.00 g of carbon dioxide confined to a container of volume 100 cm^3 at 25.0°C between treating it as a perfect gas and a van der Waals gas.

1.30 Express the van der Waals equation of state as a virial expansion in powers of $1/V_m$ and obtain expressions for B and C in terms of the parameters a and b. The expansion you will need is

$$\frac{1}{1-x} = 1 + x^2 + x^3 \ldots$$

1.31 Measurements on argon gave $B = -21.7$ cm^3 mol^{-1} and $C = 1200$ cm^6 mol^{-2} for the virial coefficients at 273 K. What are the values of a and b in the corresponding van der Waals equation of state?

1.32 Show that there is a temperature at which the second virial coefficient, B, is zero for a van der Waals gas, and calculate its value for carbon dioxide. (*Hint*. Use the expression for B derived in Exercise 1.30.)

Thermodynamics: the First Law

Contents

THE branch of physical chemistry known as **thermodynamics** is concerned with the study of the transformations of energy, and in particular the transformation of energy from heat into work and vice versa. That concern might seem remote from chemistry; indeed, thermodynamics was originally formulated by physicists and engineers interested in the efficiency of steam engines. However, thermodynamics has proved to be of immense importance in chemistry. Not only does it deal with the energy output of chemical reactions, but it also helps to answer questions that lie right at the heart of everyday chemistry, such as why reactions reach equilibrium, what their composition is at equilibrium, and how reactions in electrochemical (and biological) cells can be used to generate electricity.

Classical thermodynamics, the thermodynamics developed during the nineteenth century, stands aloof from any models of the internal constitution of matter: we could develop and use thermodynamics without ever mentioning atoms and molecules. However, the subject is greatly enriched by acknowledging that atoms and molecules do exist and interpreting thermodynamic properties and relations in terms of them. Wherever it is appropriate, we shall cross back and forth between thermodynamics, which provides useful relations between observable properties of bulk matter, and the properties of atoms and molecules, which are ultimately responsible for these bulk properties. The theory of the connection between atomic and bulk thermodynamic properties is called **statistical thermodynamics** (see Chapter 20).

Thermodynamics has a number of branches. **Thermochemistry** is the branch that deals with the heat output of chemical reactions. As we elaborate the content of thermodynamics, we shall see that we can discuss the output of energy in the form of work. This leads us into the fields of **electrochemistry**, the interaction between electricity and chemistry, and **bioenergetics**, the deployment of energy in living organisms. The whole of equilibrium chemistry—the formulation of equilibrium constants, and the very special case of the equilibrium composition of solutions of acids and bases—is an aspect of thermodynamics.

The conservation of energy

ALMOST every argument and explanation in chemistry boils down to a consideration of some aspect of a single property: the *energy*. We shall find that energy determines what molecules may form, what reactions may occur, how fast they may occur, and (with a refinement in our conception of energy) in which direction a reaction has a tendency to occur.

Energy is central to chemistry, yet it is extraordinarily difficult to give a satisfactory account of what it is.[1] For the purposes of thermodynamics, **energy** *is the capacity to do work*. As we see in more detail below, **work** is motion against an opposing force. This definition implies that a raised weight has more energy than one on the ground because the former has a greater capacity to do work: it can do work as it falls to the level of the lower weight. The definition also implies that a gas at high temperature has more energy than the same gas at a low temperature: the hot gas has a higher pressure and can do more work in driving out a piston.

People struggled for centuries to create energy from nothing, for they believed that, if they could

create energy, then they could produce work (and wealth) endlessly. However, without exception, despite strenuous efforts, many of which degenerated into deceit, they failed. As a result of their failed efforts, we have come to recognize the **conservation of energy**—that energy can be neither created nor destroyed but merely converted from one form into another or moved from place to place. The conservation of energy is of great importance in chemistry. Most chemical reactions release energy or absorb it as they occur; so, according to the conservation of energy, we can be confident that all such changes must involve only the *conversion* of energy from one form to another or its transfer from place to place, not its creation or annihilation.

2.1 Systems and surroundings

We need to keep track of the location of energy. A **system** is the part of the world in which we have a special interest. The **surroundings** are where we make our observations (Fig 2.1). The surroundings, which can be pictured as a huge water bath, remain at constant temperature regardless of how much energy flows into or out of them. They are so huge that they have either constant volume or constant pressure regardless of any changes that take place to the system. Thus, even though the system might expand, the surroundings remain effectively the same size.

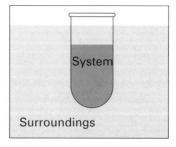

Fig 2.1 The sample is the system of interest; the rest of the world is its surroundings. The surroundings are where observations are made on the system. They can often be modelled, as here, by a large water bath.

---

[1] A physicist might say that energy is curved spacetime.

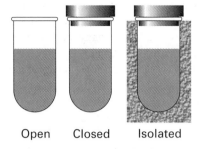

Open   Closed   Isolated

Fig 2.2 A system is *open* if it can exchange energy and matter with its surroundings, *closed* if it can exchange energy but not matter, and *isolated* if it can exchange neither energy nor matter.

We need to distinguish three types of system (Fig 2.2):

An **open system** is a system that can exchange both energy and matter with its surroundings.

A **closed system** is a system that can exchange energy but not matter with its surroundings.

An **isolated system** is a system that can exchange neither matter nor energy with its surroundings.

An example of an open system is a flask that is not stoppered and to which various substances can be added. A biochemical cell is an open system because nutrients and waste can pass through the cell wall. You and I are open systems: we ingest, respire, perspire, and excrete. An example of a closed system is a stoppered flask: energy can be exchanged with the contents of the flask because the walls may be able to conduct heat. An example of an isolated system is a sealed flask that is thermally, mechanically, and electrically insulated from its surroundings.

2.2 **Work and heat**

Energy can be exchanged between a closed system and its surroundings as work or as heat. **Work** is a transfer of energy that can cause motion against an opposing force. We can identify when a process produces work by noting whether the process can be used to change the height of a weight somewhere in the surroundings. **Heat** is a transfer of energy as a result of a temperature difference between the system and its surroundings. Walls that permit the pas-

(a) Diathermic   (b) Adiabatic

Fig 2.3 (a) A diathermic wall permits the passage of energy as heat; (b) an adiabatic wall does not, even if there is a temperature difference across the wall.

sage of energy as heat are called **diathermic** (Fig 2.3). A metal container is diathermic. Walls that do not permit heat to pass through even though there is a difference in temperature are called **adiabatic**.[2] The double walls of a vacuum flask are adiabatic to a good approximation.

As an example of these different ways of transferring energy, consider a chemical reaction that produce gases, such as the reaction of an acid with zinc:

$$Zn(s) + 2\ HCl(aq) \rightarrow ZnCl_2(aq) + H_2(g)$$

Suppose first that the reaction takes place inside a cylinder fitted with a piston; then the gas produced drives out the piston and raises a weight in the surroundings (Fig 2.4). In this case, energy has migrated to the surroundings as work because a weight has been raised in the surroundings. Some energy also migrates into the surroundings as heat. We can detect that energy by immersing the reaction vessel in an ice bath, and noting how much ice melts. Alternatively, we could let the same reaction take place in a vessel with a piston locked in position. No work is done, because no weight is raised. However, because it is found that more ice melts than in the first experiment, we can conclude that more energy has migrated to the surroundings as heat.

A process that releases heat into the surroundings is called **exothermic**. A process that absorbs heat from the surroundings is called **endothermic**. All combustions are exothermic. Endothermic reactions are much less common. The endothermic dissolution of ammonium nitrate in water is the basis of the instant cold-packs that are included in some

---

[2] The word is derived from the Greek words for 'not passing through'.

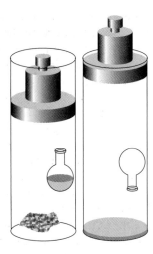

Fig 2.4 When hydrochloric acid reacts with zinc, the hydrogen gas produced must push back the surrounding atmosphere (represented by the weight resting on the piston), and hence must do work on its surroundings. This is an example of energy leaving a system as work.

first-aid kits. They consist of a plastic envelope containing water dyed blue (for psychological reasons) and a small tube of ammonium nitrate, which is broken when the pack is to be used.

The clue to the molecular nature of work comes from thinking about the motion of a weight in terms of its component atoms. When a weight is raised, all its atoms move in the same direction. This observation suggests that *work is the transfer of energy that achieves or utilizes uniform motion in the surroundings* (Fig 2.5). Whenever we think of work, we can always think of it in terms of uniform motion of some kind. Electrical work, for instance, corresponds to electrons being pushed in the same direction through a circuit. Mechanical work corresponds to atoms being pushed in the same direction against an opposing force.

Now consider the molecular nature of heat. When energy is transferred to an ice bath and some of the ice melts, the H_2O molecules in the ice oscillate more rapidly around their positions and those at the surface may escape into the surrounding liquid. The key point is that the motion stimulated by the arrival of energy from the system is disorderly, not uniform as in the case of work. This observation suggests that *heat is the transfer of energy that achieves or utilizes disorderly motion in the surroundings* (Fig 2.6).

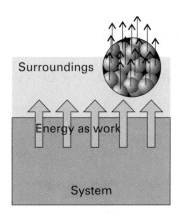

Fig 2.5 Work is transfer of energy that causes or utilizes uniform motion of atoms in the surroundings. For example, when a weight is raised, all the atoms of the weight (shown magnified) move in unison in the same direction.

A fuel burning, for example, generates disorderly molecular motion in its vicinity.

An interesting social point is that the molecular difference between work and heat correlates with the chronological order of their application. The liberation of heat by fire is a relatively unsophisticated procedure because the energy emerges in a disordered fashion from the burning fuel. It was developed—stumbled upon—early in the history of civilization. The generation of work by a burning fuel, in contrast, relies on a carefully controlled transfer of energy so that myriads of molecules move in unison. It was

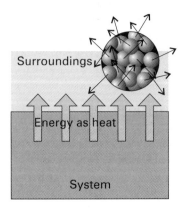

Fig 2.6 Heat is the transfer of energy that causes or utilizes chaotic motion in the surroundings. When energy leaves the system (the shaded region), it generates chaotic motion in the surroundings (shown magnified).

achieved thousands of years later, with the development of the steam engine.

2.3 **The measurement of work**

To calculate the work done by a given process we use the definition given in physics, that work is equal to the product of the distance moved and the force opposing the motion:

Work = distance × opposing force

It follows that moving through a long distance against a strong opposing force (like climbing a high mountain) requires a lot of work. If the force is the gravitational attraction of the Earth on a mass m, the force opposing raising the mass vertically is mg where g is the acceleration of free fall (9.81 m s^{-2}). Therefore, the work needed to raise the mass through a height h on the surface of the Earth is

$$\text{Work} = h \times mg = mgh \qquad (2.1)$$

For example, raising a book like this one (of mass about 1.0 kg) from the floor to the table 75 cm above requires

$$\text{Work} = (1.0 \text{ kg}) \times (9.81 \text{ m s}^{-2}) \times (0.75 \text{ m}) = 7.4 \text{ kg m}^2 \text{ s}^{-2}$$

The unit used to report energy (and therefore both work and heat) is the **joule** (J), which is named after James Joule, the Manchester brewer who made a detailed study of heat and work in the nineteenth century:

$$1 \text{ J} = 1 \text{ kg m}^2 \text{ s}^{-2}$$

The work we have just calculated would be reported as 7.4 J. Each beat of the human heart does work equal to about 1 J, so about 100 kJ of energy is expended each day driving the blood around your body.

Self-test 2.1

A useful relation between joules and pascals is $1 \text{ J} = 1 \text{ Pa m}^3$. Confirm this relation from the definition of pascals in *Further information 2*.

[*Answer*: $1 \text{ Pa} \times 1 \text{ m}^3 = 1 \text{ kg m}^{-1} \text{ s}^{-2} \times 1 \text{ m}^3 = 1 \text{ kg m}^2 \text{ s}^{-2} = 1 \text{ J}$]

When energy *leaves* a system as work, which happens when the system does work on the surroundings, such as by causing a weight to rise or forcing an electric current through a circuit, the work, w, is reported as a negative quantity. For instance, if a system raises a weight in the surroundings and in the process does 100 J of work (that is, 100 J of energy leaves the system as work), we would write $w = -100 \text{ J}$. When energy *enters* a system as work (for example, when we wind a spring inside a clockwork mechanism) w is reported as a positive quantity. We would write $w = +100 \text{ J}$ to signify that we had done 100 J of work on the system (that is, that 100 J of energy had been added to the system by doing work). The sign convention is easy to follow if we think of changes to the energy of the system: its energy decreases (w is negative) if energy leaves it as work, and its energy increases (w is positive) if energy enters it as work (Fig 2.7). We use the same convention for energy transferred as heat, q. We write $q = -100 \text{ J}$ if 100 J of energy leaves the system as heat, so reducing its energy, and $q = +100 \text{ J}$ if 100 J of energy enters the system as heat.

One very important type of work in chemistry is **expansion work**, the work done when a system expands against an opposing pressure. The action of acid on zinc illustrated in Fig 2.4 is an example of a reaction in which expansion work is done. Our first task is to discover how the work done depends on the change in volume and the opposing pressure.

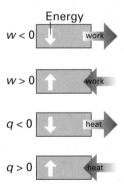

Fig 2.7 The sign convention in thermodynamics: w and q are positive if energy enters the system (as work and heat, respectively), but negative if energy leaves the system.

Derivation 2.1 *Expansion work*

To calculate the work done when a system expands from an initial volume V_i to a final volume V_f, a change $\Delta V = V_f - V_i$, we consider a piston of area A moving out through a distance h (Fig 2.8). The force opposing the expansion is the constant external pressure p_{ex} multiplied by the area of the piston (because force is pressure times area, Section 0.3). The work done is therefore

Work done by system = distance × opposing force
$$= h \times (p_{ex} \times A) = p_{ex} \times (h \times A)$$
$$= p_{ex} \times \Delta V$$

The last equality follows from the fact that hA is the volume of the cylinder swept out by the piston as the gas expands, so we can write $hA = \Delta V$. That is, for expansion work,

Work done by system $= p_{ex}\Delta V$

Now consider the sign. A system does work and thereby loses energy (that is, w is negative) when it expands (when ΔV is positive). We need a negative sign in the equation to ensure that w is negative when ΔV is positive, so we write

$$w = -p_{ex}\Delta V \tag{2.2}$$

According to eqn 2.2, the *external* pressure determines how much work a system does when it expands through a given volume: the greater the external pressure, the greater the opposing force and the greater the work that a system does. When the external pressure is zero, $w = 0$. In this case, the system does no work as it expands as it has nothing to push against. Expansion against zero external pressure is called **free expansion**.

Self-test 2.2

Calculate the work done by a system in which a reaction results in the formation of 1.0 mol $CO_2(g)$ at 25°C and 100 kPa. (*Hint*. The increase in volume will be 25 L under these conditions (if the gas is treated as perfect); note the relation between pascals and joules established in Self-test 2.1.)

[*Answer*: 2.5 kJ]

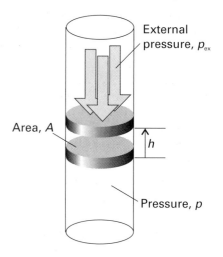

Fig 2.8 When a piston of area A moves out through a distance h, it sweeps out a volume $\Delta V = Ah$. The external pressure p_{ex} opposes the expansion with a force $p_{ex}A$.

Equation 2.2 shows us how to get the *least* expansion work from a system: we just reduce the external pressure to zero. But how can we achieve the *maximum* work for a given change in volume? According to eqn 2.2, the system does maximum work when the external pressure has its maximum value. The force opposing the expansion is then the greatest and the system must exert most effort to push the piston out. However, that external pressure cannot be greater than the pressure, p, of the gas inside the system, for otherwise the external pressure would compress the gas instead of allowing it to expand. Therefore, *maximum work is obtained when the external pressure is only infinitesimally less than the pressure of the gas in the system*. In effect, the two pressures are the same. In Chapter 1 we called this balance of pressures a state of mechanical equilibrium. Therefore, we can conclude that *a system in mechanical equilibrium does maximum expansion work*.

There is another way of expressing this condition. Because the external pressure is infinitesimally less than the pressure of the gas, the piston moves out. However, suppose we increase the external pressure so that it becomes infinitesimally greater than the pressure of the gas; now the piston moves in. That is, *when a system is in a state of mechanical equilibrium, an infinitesimal change in the pressure results in op-*

posite directions of change. A change that can be reversed by an *infinitesimal* change in a variable—in this case, the pressure—is said to be **reversible**. In everyday life 'reversible' means a process that can be reversed; in thermodynamics it has a stronger meaning—it means that a process can be reversed by an *infinitesimal* modification in some variable (such as the pressure).

We can summarize this discussion by the following remarks:

1 A system does maximum expansion work when the external pressure is equal to that of the system ($p_{ex} = p$).

2 A system does maximum expansion work when it is in mechanical equilibrium with its surroundings.

3 Maximum expansion work is achieved in a reversible change.

All three statements are equivalent, but they reflect different degrees of sophistication in the way the point is expressed.

We cannot write down the expression for maximum expansion work simply by replacing p_{ex} in eqn 2.2 by p (the pressure of the gas in the cylinder) because, as the piston moves out, the pressure inside the system falls. To make sure the entire process occurs reversibly, we have to adjust the external pressure to match the internal pressure at each stage. To calculate the work, we need to take into account the fact that the external pressure must change as the system expands.

Derivation 2.2 *Reversible, isothermal expansion work*

We think of the expansion as taking place in an infinite number of infinitesimally small steps. When the system expands through an infinitesimal volume dV, the infinitesimal work, dw, done is:[3]

For an infinitesimal expansion $dw = -p_{ex}\,dV$

This is eqn 2.2, rewritten for an infinitesimal expansion. However, at each stage, we set the external pressure equal to the current pressure, p, of the gas (Fig 2.9):

For an infinitesimal reversible expansion: $dw = -p\,dV$ (because $p_{ex} = p$)

The total work when the system expands from V_i to V_f is the sum (integral) of all these infinitesimal changes, which we write

For a measurable reversible expansion $w = -\int_{V_i}^{V_f} p\,dV$

To evaluate the integral, we need to know how p changes as the gas expands. For this step, we suppose that the gas in the system is perfect, in which case we can use the perfect gas law to write[4]

$$p = \frac{nRT}{V}$$

For isothermal expansion, T is constant. Isothermal expansion can be achieved in practice by immersing the cylinder of gas in a constant-temperature water bath. Then

$$w \overset{\text{Reversible}}{=} -\int_{V_i}^{V_f} p\,dV \overset{\text{Perfect gas}}{=} -\int_{V_i}^{V_f} \frac{nRT}{V}\,dV$$

$$\overset{\text{Isothermal}}{=} -nRT \int_{V_i}^{V_f} \frac{1}{V}\,dV$$

$$\overset{\text{Standard integral}}{=} -nRT \ln\frac{V_f}{V_i}$$

The standard integral we have used in the last step is

$$\int \frac{dx}{x} = \ln x + \text{constant}$$

That is, the maximum work of isothermal expansion of a perfect gas at a temperature T is

$$w = -nRT \ln\frac{V_f}{V_i} \tag{2.3}$$

Equation 2.3 is very important and will turn up in various disguises throughout this text. Once again, it is important to be able to interpret it rather than just remember it. First, we note that in an expansion $V_f > V_i$, so $V_f/V_i > 1$ and the logarithm is positive ($\ln x$ is positive if $x > 1$). Therefore, in an

---

[3] For a note on the use of calculus, see *Further information 1*. As indicated there, the replacement of Δ by d always indicates an infinitesimal change.

[4] If we were considering a real gas, we could use, for instance, the van der Waals equation of state introduced in Section 1.13.

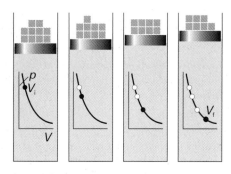

Fig 2.9 For a gas to expand reversibly, the external pressure must be adjusted to match the internal pressure at each stage of the expansion. This matching is represented in this illustration by gradually unloading weights from the piston as the piston is raised and the internal pressure falls. The procedure results in the extraction of the maximum possible work of isothermal expansion.

expansion, w is negative. That is what we should expect: energy *leaves* the system as the system does expansion work. Second, for a given change in volume, we get more work the higher the temperature of the confined gas (Fig 2.10). That is also what we should expect: at high temperatures, the pressure of the gas is high, so we have to use a high ex-

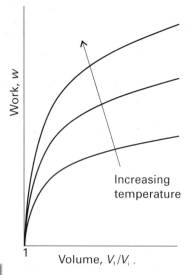

Fig 2.10 The work of reversible, isothermal expansion of a perfect gas. Note that, for a given change of volume and fixed amount of gas, the work is greater the higher the temperature.

ternal pressure, and therefore a stronger opposing force, to match the internal pressure at each stage.

Self-test 2.3

Calculate the work done when 1.0 mol Ar(g) confined in a cylinder of volume 1.0 L at 25°C expands isothermally and reversibly to 2.0 L.

[*Answer:* 1.7 kJ]

2.4 The measurement of heat

The supply of heat to a system typically results in a rise in temperature, so one way to measure the value of q is to use a **calorimeter** (Fig 2.11).[5] A calorimeter consists of a container, in which the reaction or physical process occurs, a thermometer, and a surrounding water bath. The entire assembly is insulated from the rest of the world. The first step is to calibrate the calorimeter (the entire assembly, the reaction vessel and the water bath) by comparing the observed change in temperature with a change in temperature brought about by a known quantity of heat. One procedure is to heat the calorimeter electrically by passing a known current for a measured time through a heater, and record the increase in temperature. The heat supplied is given by

$$q = I\mathcal{V}t \tag{2.4}$$

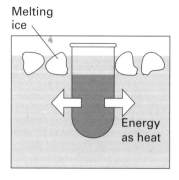

Fig 2.11 The loss of energy into the surroundings can be detected by noting whether ice melts as the process proceeds.

[5] The name comes from 'calor', the Latin word for heat.

where I is the current (in amperes, A), $\mathcal{V}$ is the potential of the supply (in volts, V), and t is the time (in seconds, s) for which the current flows. The observed rise in temperature lets us calculate the **heat capacity**, C, of the calorimeter from[6]

$$\text{Heat capacity} = \frac{\text{heat supplied}}{\text{temperature rise}} \quad C = \frac{q}{\Delta T} \quad (2.5)$$

Then we use the heat capacity to interpret a temperature rise in a combustion experiment in terms of the heat released by the reaction. An alternative procedure is to calibrate the calorimeter by using a reaction of known heat output, such as the combustion of benzoic acid (C_6H_5COOH), for which the heat output is 3227 kJ per mole of C_6H_5COOH consumed.

Example 2.1 *Calibrating a calorimeter and measuring a heat transfer*

In an experiment to measure the heat released by the combustion of a sample of nutrient, the compound was burned in an oxygen atmosphere inside a calorimeter and the temperature rose by 3.22°C. When a current of 1.23 A from a 12.0 V source flowed through a heater in the same calorimeter for 123 s, the temperature rose by 4.47°C. What is the heat released by the combustion reaction?

Strategy We calculate the heat supplied by using eqn 2.4. Use the relation 1 A V s = 1 J, which follows from the definition of the electrical units. Then use the observed rise in temperature to find the heat capacity of the calorimeter. We then use that heat capacity to convert the temperature rise observed for the combustion into a heat output by writing $q = C\Delta T$ (or $q = C\Delta\theta$ if the temperature is given on the Celsius scale).

Solution The heat supplied during the calibration step is[7]

$$q = (1.23\ \text{A}) \times (12.0\ \text{V}) \times (123\ \text{s}) = 1.23 \times 12.0 \times 123\ \text{J}$$

This product works out as 1.82 kJ, but we save the numerical work to the final stage. The heat capacity of the calorimeter is

$$C = \frac{q}{\Delta\theta} = \frac{1.23 \times 12.0 \times 123\ \text{J}}{4.47°C} = \frac{1.23 \times 12.0 \times 123}{4.47}\ \text{J °C}^{-1}$$

The numerical value of C is 406 J °C^{-1}. The heat output of the combustion is therefore

$$q = \left(\frac{1.23 \times 12.0 \times 123}{4.47}\ \text{J °C}^{-1} \right) \times (3.22°C) = 1.31\ \text{kJ}$$

Self-test 2.4

In an experiment to measure the heat released by the combustion of a sample of fuel, the compound was burned in an oxygen atmosphere inside a calorimeter and the temperature rose by 2.78°C. When a current of 1.12 A from a 11.5 V source flowed through a heater in the same calorimeter for 162 s, the temperature rose by 5.11°C. What is the heat released by the combustion reaction?

[Answer: 1.1 kJ]

We can also speak of the heat capacity of a single substance. Heat capacity is an **extensive property**, a property that depends on the size (the 'extent') of the sample. It is more convenient to report the heat capacity of a substance as an **intensive property**, a property that is independent of the size of the sample. We therefore use either the **specific heat capacity**, C_s,[8] the heat capacity divided by the mass of the sample ($C_s = C/m$, in joules per kelvin per gram, J K^{-1} g^{-1}) or the **molar heat capacity**, C_m, the heat capacity divided by the amount of substance ($C_m = C/n$, in joules per kelvin per mole, J K^{-1} mol^{-1}). For reasons that will be explained shortly, the heat capacity depends on whether the sample is maintained at constant volume (like a gas in a sealed vessel) as heat is supplied, or whether the sample is maintained at constant pressure (like water in an open container) and is free to change its volume. The latter is a more common arrangement, and the values given in Table 2.1 are for the **heat capacity at constant pressure**, C_p. The **heat capacity at constant volume** is denoted C_V.[9]

---

[6] In the context of calorimetry, the empirically determined heat capacity of the calorimeter is also called the *calorimeter constant*. The temperature change may be expressed in kelvins (ΔT) or degrees Celsius ($\Delta\theta$); the same numerical value is obtained.

[7] We avoid rounding errors by doing all numerical work in one step at the end. That is like using a calculator and not writing down intermediate results.

[8] In common usage, the specific heat capacity is often called the *specific heat*.

[9] You might also come across the name *isobaric heat capacity* for C_p and, much more rarely, *isochoric heat capacity* for C_V.

Table 2.1 *Heat capacities of some materials*

Substance	Specific heat capacity $C_{p,s}/(J\,K^{-1}\,g^{-1})$	Molar heat capacity* $C_{p,m}/$ $(J\,K^{-1}\,mol^{-1})$
Air	1.01	29
Benzene, C_6H_6	1.05	136.1
Brass	0.37	
Copper, Cu	0.38	
Ethanol, C_2H_5OH	2.42	111.5
Glass (Pyrex)	0.78	
Granite	0.80	
Marble	0.84	
Polyethylene	2.3	
Stainless steel	0.51	
Water, H_2O solid	2.03	37
liquid	4.18	75.29
vapour	2.01	33.58

* Molar heat capacities are given only for air and well defined, pure substances. See also Appendix 1.

Illustration 2.1

The molar heat capacity of water at constant pressure, $C_{p,m}$, is $75\,J\,K^{-1}\,mol^{-1}$. It follows that the increase in temperature of 100 g of water (5.55 mol H_2O) when 1.0 kJ of heat is supplied to a sample free to expand is approximately

$$\Delta T = \frac{q}{C_p} = \frac{q}{nC_{p,m}} = \frac{1.0 \times 10^3\,J}{(5.55\,mol) \times (75\,J\,K^{-1}\,mol^{-1})}$$
$$= +2.4\,K$$

The heat capacity lets us interpret an observed rise in temperature in terms of the heat supplied.

So, it provides an *experimental* route to the determination of q. The question arises, though, whether we can also calculate q like we calculated w for the expansion of a perfect gas. To answer this question, it is helpful to consider the changes at a molecular level that take place when a gas undergoes a change of state.

The simplest case is that of a perfect gas undergoing isothermal expansion. Because the expansion is isothermal, the temperature of the gas is the same at the end of the expansion as it was initially. Therefore, the mean speed of the molecules of the gas is the same before and after the expansion. That implies in turn that the total kinetic energy of the molecules is the same. But, for a perfect gas, the *only* contribution to the energy is the kinetic energy of the molecules (recall Section 1.4), so we have to conclude that the *total* energy of the gas is the same before and after the expansion. Energy has left the system as work; therefore, a compensating amount of energy must have entered the system as heat. We can therefore write:

For the isothermal expansion of a perfect gas: $q = -w$
$$(2.6)$$

For instance, if we find that $w = -100\,J$ for a particular expansion (so, 100 J has left as work), then we can conclude that $q = +100\,J$ (that is, 100 J must enter as heat). Likewise, because eqn 2.3 is an expression for the work done when a perfect gas expands isothermally *and reversibly*, we can immediately write

$$q = nRT \ln \frac{V_f}{V_i} \qquad (2.7)$$

for that type of expansion. When $V_f > V_i$, as in an expansion, the logarithm is positive and we conclude that $q > 0$, as expected: heat flows into the system to make up for the energy lost as work.

Internal energy and enthalpy

H EAT and work are *equivalent* ways of transferring energy into or out of a system. Once the energy is inside, it is stored as 'energy'. Regardless of how the energy was supplied, as work or as heat, it can be released in either form. The experimental evidence for this **equivalence of heat and work** goes all the way back to the experiments done by James Joule, who showed that the same rise in temperature of a sample of water is brought about by transferring a given quantity of energy either as heat or as work.

2.5 The internal energy

We need some way of keeping track of the energy changes in a system as energy is transferred in and out but at the same time acknowledge that heat and work are equivalent. This is the job of the property called the **internal energy**, U, of the system, a measure of the 'energy reserves' of the system. The internal energy is the sum of all the kinetic and potential contributions to the energy of all the atoms, ions, and molecules in the system: it is the grand total energy of the system. When energy is transferred into the system by heating it or doing work on it, the increased energy is stored in the increased kinetic and potential energies of the molecules.

The internal energy is an extensive property of a sample: 2 kg of iron has twice the internal energy of 1 kg of iron. The **molar internal energy**, $U_m = U/n$, the internal energy per mole of material, is an intensive property with a value that depends on the temperature and, in general, the pressure.

In practice, we do not know and cannot measure the total energy of a sample, because it includes the kinetic and potential energies of all the electrons and all the components of the atomic nuclei. Nevertheless, there is no problem with dealing with the *changes* in internal energy, ΔU, because we can determine those by monitoring the energy supplied or lost as heat or work. All practical applications of thermodynamics deal with ΔU, not with U itself. A change in internal energy is written

$$\Delta U = w + q \qquad (2.8)$$

where w is the energy transferred to the system as work and q the energy transferred to it as heat. The internal energy is an accounting device, like a country's gold reserves for monitoring transactions with the outside world (the surroundings) using either currency (heat or work).

Illustration 2.2

When a system releases 10 kJ of energy to the surroundings as work (that is, when $w = -10$ kJ), the internal energy of the system decreases by 10 kJ, and we write $\Delta U = -10$ kJ. The minus sign signifies the reduction in internal energy that has occurred. If the system releases 20 kJ of energy as heat (so $q = -20$ kJ), we write $\Delta U = -20$ kJ. If the system releases 10 kJ as work *and* 20 kJ as heat, as in an inefficient internal combustion engine, the internal energy falls by a total of 30 kJ, and we write $\Delta U = -30$ kJ. On the other hand, if we do 10 kJ of work on the system ($w = +10$ kJ), for instance, by winding a spring it contains or pushing in a piston to compress a gas (Fig 2.12), then the internal energy of the system increases by 10 kJ, and we write $\Delta U = +10$ kJ. Likewise, if we supply 20 kJ of energy as heat ($q = +20$ kJ), then the internal energy increases by 20 kJ, and we write $\Delta U = +20$ kJ.[10]

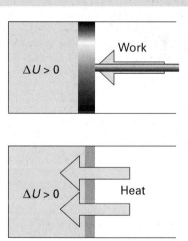

Fig 2.12 When work is done on a system, its internal energy rises. The internal energy also rises when energy is transferred into the system as heat.

---

[10] Notice that ΔU always carries a sign explicitly, even if it is positive: we never write $\Delta U = 20$ kJ but always +20 kJ.

We have seen that a feature of a perfect gas is that, for any *isothermal* expansion, the total energy of the sample remains the same, and that $q = -w$. That is, any energy lost as work is restored by an influx of energy as heat. We can express this property in terms of the internal energy, for it implies that the internal energy remains constant when a perfect gas expands isothermally: from eqn 2.8 we can write

Isothermal expansion of a perfect gas: $\Delta U = 0$ (2.9)

In other words, *the internal energy of a sample of perfect gas is independent of the volume it occupies.* We can understand this independence by realizing that when a perfect gas expands isothermally the only feature that changes is the average distance between the molecules; their average speed and therefore total kinetic energy remains the same. However, as there are no intermolecular interactions, the total energy is independent of the average separation, so the internal energy is unchanged by expansion.

Example 2.2 *Calculating the change in internal energy*

Nutritionists are interested in the use of energy by the human body and we can consider our own body as a thermodynamic 'system'. Calorimeters have been constructed that can accommodate a person to measure (non-destructively!) their net energy output. Suppose that in the course of an experiment someone does 622 kJ of work on an exercise bicycle and loses 82 kJ of energy as heat. What is the change in internal energy of the person? Disregard any matter loss by perspiration.

Strategy This example is an exercise in keeping track of signs correctly. When energy is lost from the system, w or q is negative. When energy is gained by the system, w or q is positive.

Solution To take note of the signs we write $w = -622$ kJ (622 kJ is lost as work) and $q = -82$ kJ (82 kJ is lost as heat). Then eqn 2.8 gives us

$$\Delta U = w + q = (-622 \text{ kJ}) + (-82 \text{ kJ}) = -704 \text{ kJ}$$

We see that the person's internal energy falls by 704 kJ. Later, that energy will be restored by eating.

Self-test 2.5

An electric battery is charged by supplying 250 kJ of energy to it as electrical work, but in the process it loses 25 kJ of energy as heat to the surroundings. What is the change in internal energy of the battery?

[*Answer*: +225 kJ]

An important characteristic of the internal energy is that it is a **state function**, a physical property that depends only on the current state of the system and is independent of the path by which that state was reached. If we were to change the temperature of the system, then change the pressure, then adjust the temperature and pressure back to their original values, the internal energy would return to its original value too. A state function is very much like altitude: each point on the surface of the Earth can be specified by quoting its latitude and longitude, and (on land areas, at least) there is a unique property, the altitude, that has a fixed value at that point.

The fact that U is a state function implies that *a change, ΔU, in the internal energy between two states of a system is independent of the path between them* (Fig 2.13). Once again, the altitude is a helpful analogy. If we climb a mountain between two fixed points, we

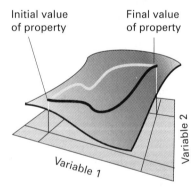

Fig 2.13 The curved sheet shows how a property (for example, the altitude) changes as two variables (for example, latitude and longitude) are changed. The altitude is a state function, because it depends only on the current state of the system. The change in the value of a state function is independent of the path between the two states. For example, the difference in altitude between the initial and final states shown in the diagram is the same whatever path (as depicted by the dark and light lines) is used to travel between them.

make the same change in altitude regardless of the path we take between the two points. Likewise, if we compress a sample of gas until it reaches a certain pressure and then cool it to a certain temperature, the change in internal energy has a particular value. If, on the other hand, we changed the temperature and then the pressure, but ensured that the two final values were the same as in the first experiment, then the overall change in internal energy would be exactly the same as before. This path independence of the value of ΔU is of the greatest importance in chemistry, as we shall soon see.

Suppose we now consider an isolated system. Because an isolated system can neither do work nor supply heat, it follows that its internal energy cannot change. That is,

The internal energy of an isolated system is constant.

This statement is the **First Law of thermodynamics**. It is closely related to the law of conservation of energy, but remember that internal energy is expressed in terms of work and heat. Unlike thermodynamics, mechanics does not deal with the concept of heat.

The experimental evidence for the First Law is the impossibility of making a 'perpetual motion machine', a device for producing work without consuming fuel. As we have already remarked, try as people might, they have never succeeded. No device has ever been made that creates internal energy to replace the energy drawn off as work. We cannot extract energy as work, leave the system isolated for some time, and hope that when we return the internal energy will have become restored to its original value.

The definition of ΔU in terms of w and q points to a very simple method for measuring the change in internal energy of a system when a reaction takes place. We have seen already that the work done by a system when it pushes against a fixed external pressure is proportional to the change in volume. Therefore, if we seal the reaction into a container of constant volume, the system can do no expansion work and we can set $w = 0$. Then eqn 2.8 simplifies to

At constant volume: $\Delta U = q$ (2.10)

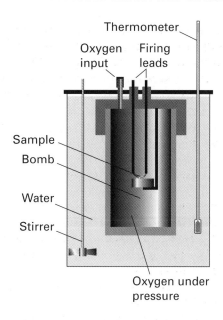

Fig 2.14 A constant-volume bomb calorimeter. The 'bomb' is the central, sturdy vessel, which is strong enough to withstand moderately high pressures. The calorimeter is the entire assembly shown here. To ensure that no heat escapes into the surroundings, the calorimeter may be immersed in a water bath with a temperature that is continuously adjusted to that of the calorimeter at each stage of the combustion.

It follows that, to measure a change in internal energy, we should use a calorimeter that has a fixed volume, and monitor the heat released ($q < 0$) or supplied ($q > 0$). A **bomb calorimeter** is an example of a constant-volume calorimeter: it consists of a sturdy, sealed, constant-volume vessel in which the reaction takes place, and a surrounding water bath (Fig 2.14). To ensure that no heat escapes unnoticed from the calorimeter, it is immersed in a water bath with a temperature adjusted to match the rising temperature of the calorimeter. The fact that the temperature of the bath is the same as that of the calorimeter ensures that no heat flows from one to the other. That is, the arrangement is adiabatic.

2.6 **The enthalpy**

Much of chemistry, and most of biology, takes place in vessels that are open to the atmosphere and subjected to constant pressure, not maintained at constant volume.[11] In general, when a change takes place in a system open to the atmosphere, the volume of the system changes. For example, the thermal decomposition of 1.0 mol $CaCO_3$ at 1 bar results in an increase in volume of 89 L at 800°C on account of the carbon dioxide gas produced. To create this large volume for the carbon dioxide to occupy, the surrounding atmosphere must be pushed back. That is, the system must perform expansion work. Therefore, although a certain quantity of heat may be supplied to bring about the endothermic decomposition, the increase in internal energy of the system is not equal to the energy supplied as heat because some energy has been used to do work of expansion (Fig 2.15). In other words, because the volume has increased, some of the heat supplied to the system has leaked back into the surroundings as work.

Another example is the oxidation of a fat, such as tristearin, to carbon dioxide in the body. The overall reaction is

$$2\ C_{57}H_{110}O_6\,(aq) + 163\ O_2(g) \rightarrow 114\ CO_2(g) + 110\ H_2O(l)$$

In this exothermic reaction there is a net *decrease* in volume equivalent to the elimination of $(163 - 114)\ mol = 49\ mol$ of gas molecules. The decrease in volume at 25°C is about 600 mL for the consumption of 1 g of fat. Because the volume of the system decreases, the atmosphere does work *on* the system as the reaction proceeds. That is, energy is transferred as work from the surroundings to the system as it contracts.[12] For this reaction, the decrease in the internal energy of the system is less than the energy released as heat because some energy has been restored as work.

We can avoid the complication of having to take into account the work of expansion by introducing the 'enthalpy'. This property will be at the centre of our attention throughout the rest of the chapter and will recur throughout the book. The **enthalpy**, H, of a system is defined as

------

[11] Care must be taken, though, with applying this remark to individual biological cells, which may act as constant-volume containers within organisms.

[12] In effect, a weight has been lowered in the surroundings, so the surroundings can do less work after the reaction has occurred. Some of their energy has been transferred into the system.

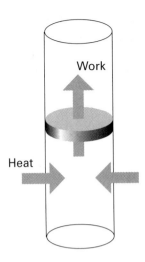

Fig 2.15 The change in internal energy of a system that is free to expand or contract is not equal to the energy supplied as heat because some energy may escape back into the surroundings as work. However, the change in enthalpy of the system under these conditions *is* equal to the energy supplied as heat.

$$H = U + pV \qquad (2.11)$$

That is, the enthalpy differs from the internal energy by the addition of the product of the pressure, p, and the volume, V, of the system; because $pV > 0$, the enthalpy of a system is always greater than its internal energy. Enthalpy is an extensive property. The **molar enthalpy**, $H_m = H/n$, of a substance, an intensive property, differs from the molar internal energy by an amount proportional to the molar volume, V_m, of the substance:

$$H_m = U_m + pV_m \qquad (2.12)$$

For a perfect gas we can write $pV_m = RT$, so

$$For\ a\ perfect\ gas: \ H_m = U_m + RT \qquad (2.13)$$

This relation tells us that the difference between the molar enthalpy and molar internal energy of a perfect gas increases with temperature. At 25°C, $RT = 2.5\ kJ\ mol^{-1}$, so the molar enthalpy of a perfect gas differs from its molar internal energy by $2.5\ kJ\ mol^{-1}$. Because the molar volume of a solid or liquid is typically about 1000 times less than that of a gas, we can also conclude that the molar enthalpy of a solid or liquid is only about 2.5 J mol^{-1} (note: joules, not kilojoules) more than its molar internal energy, so the numerical difference is negligible.

Although the enthalpy and internal energy of a sample may have similar values, the introduction of the enthalpy has very important consequences in thermodynamics. First, notice that, because H is defined in terms of state functions (U, p, and V), *the enthalpy is a state function*. The implication is that the change in enthalpy, ΔH, when a system changes from one state to another is independent of the path between the two states. Second, we shall now see that by focusing on the enthalpy of a system we automatically take into account the energy lost and gained as expansion work.

Derivation 2.3 *Heat transfers at constant pressure*

Consider a system open to the atmosphere, so that its pressure p is constant and equal to the external pressure p_{ex}. Initially, the enthalpy is

$$H_i = U_i + pV_i$$

After a reaction or other process has taken place at constant pressure, the internal energy and the volume of the system are different, and the enthalpy becomes

$$H_f = U_f + pV_f$$

The change in enthalpy is the difference between these two quantities, and is

$$\Delta H = H_f - H_i = U_f - U_i + p(V_f - V_i)$$

or

$$\Delta H = \Delta U + p\Delta V \qquad (2.14)$$

However, we know that the change in internal energy is given by eqn 2.8 ($\Delta U = w + q$) with $w = -p_{ex}\Delta V$. When we substitute that expression into this one we obtain

$$\Delta H = (-p_{ex} \Delta V + q) + p \Delta V$$

At this point we write $p_{ex} = p$ (because the system and the surroundings have the same pressure), which gives

$$\Delta H = (-p \Delta V + q) + p \Delta V = q$$

We can therefore conclude that, as with the enthalpy defined as in eqn 2.11, the change in enthalpy is equal to the heat absorbed at constant pressure:

$$\text{At constant pressure: } \Delta H = q \qquad (2.15)$$

The result expressed by eqn 2.15, that *at constant pressure we can identify the heat transferred with a change in enthalpy of the system*, is enormously powerful. It relates a quantity we can measure (the heat transfer at constant pressure) to the change in a state function (the enthalpy). Dealing with state functions rather than heat and work, which are not state functions, greatly extends the power of thermodynamic arguments, because we don't have to worry about how we get from one state to another: all that matters is the initial and final states.

Illustration 2.3

Equation 2.15 means that, if 10 kJ of heat is supplied to the system that is free to change its volume at constant pressure, then the enthalpy of the system increases by 10 kJ regardless of how much energy enters or leaves as work, and we write $\Delta H = +10$ kJ. On the other hand, if the reaction is exothermic and releases 10 kJ of heat when it occurs, then $\Delta H = -10$ kJ regardless of how much work is done. For the particular case of the combustion of tristearin mentioned at the beginning of the section, in which 90 kJ of energy is released as heat, we would write $\Delta H = -90$ kJ.

An endothermic reaction ($q > 0$) taking place at constant pressure results in an increase in enthalpy ($\Delta H > 0$) because energy enters the system as heat. On the other hand, an exothermic process ($q < 0$) taking place at constant pressure corresponds to a decrease in enthalpy ($\Delta H < 0$) because energy leaves the system as heat. All combustion reactions, including the controlled combustions that contribute to respiration, are exothermic and are accompanied by a decrease in enthalpy. These relations are consistent with the name 'enthalpy', which is derived from the Greek words meaning 'heat inside': the 'heat inside' the system is increased if the process is endothermic and absorbs heat from the surroundings; it is decreased if the process is exothermic and releases heat into the surroundings.[13]

2.7 The temperature variation of the enthalpy

We have seen that the internal energy of a system rises as the temperature is increased. The same is true of the enthalpy, which also increases when the temperature is raised (Fig 2.16). For example, the enthalpy of 100 g of water is greater at 80°C than at 20°C. We can measure the change by monitoring the energy that we must supply as heat to raise the temperature through 60°C when the sample is open to the atmosphere (or subjected to some other constant pressure); it is found that $\Delta H \approx +25$ kJ in this instance.

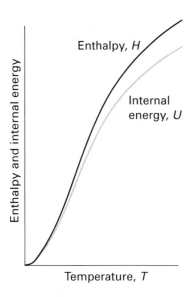

Fig 2.16 The enthalpy of a system increases as its temperature is raised. Note that the enthalpy is always greater than the internal energy of the system, and that the difference increases with temperature.

---

[13] But heat does not actually 'exist' inside: only energy exists in a system; heat is a means of recovering that energy or increasing it. Heat is energy in transit, not a form in which energy is stored.

To find the relation between the change in enthalpy and the change in temperature, we combine the statement that $\Delta H = q$ at constant pressure with the definition of the constant-pressure heat capacity C_p (the expression $q = C_p\Delta T$). Provided the heat capacity is constant over the range of temperatures of interest, it follows that

$$\Delta H = C_p \Delta T \qquad (2.16)$$

Illustration 2.4

When the temperature of 100 g of water (5.55 mol H_2O) is raised from 20°C to 80°C (so $\Delta T = +60$ K) at constant pressure, the enthalpy of the sample changes by

$$\Delta H = C_p \Delta T = nC_{p,m} \Delta T$$
$$= (5.55 \text{ mol}) \times (75.29 \text{ J K}^{-1}\text{mol}^{-1}) \times (60 \text{ K}) = +25 \text{ kJ}$$

The greater the temperature rise, the greater the change in enthalpy and therefore the more heat required to bring it about. Note that this calculation is only approximate, because the heat capacity depends on the temperature, and we have used an average value for the temperature range of interest.

The heat capacity at constant volume is the slope of a plot of internal energy against temperature. Likewise, the heat capacity at constant pressure is the slope of a plot of enthalpy against temperature (Fig 2.17, Table 2.1). However, because we know that the difference between the enthalpy and internal energy of a perfect gas depends very simply on the temperature (as in eqn 2.13), we can suspect that we ought to be able to find a relation between the two heat capacities.

Derivation 2.4 *The relation between heat capacities*

The molar internal energy and enthalpy of a perfect gas are related by eqn 2.13, which we can write in the form

$$H_m - U_m = RT$$

When the temperature increases by ΔT, the molar enthalpy increases by ΔH_m and the molar internal energy increases by ΔU_m, so

$$\Delta H_m - \Delta U_m = R\,\Delta T$$

Now divide both sides by ΔT, which gives

$$\frac{\Delta H_m}{\Delta T} - \frac{\Delta U_m}{\Delta T} = R$$

The first term on the left is the slope of molar enthalpy against temperature, or the molar constant-pressure heat capacity, $C_{p,m}$; the second term is the slope of the molar internal energy against temperature, or the molar constant-volume heat capacity, $C_{V,m}$. Therefore, this relation can be written

$$C_{p,m} - C_{V,m} = R \tag{2.17}$$

The relation in eqn 2.17 shows that the molar heat capacity of a perfect gas is greater at constant pressure than at constant volume. This difference is what we should expect. At constant volume, all the heat supplied to the system remains inside, and the temperature rises accordingly. At constant pressure, though, some of the energy supplied as heat escapes back into the surroundings as work when the system expands. As less energy remains in the system, the temperature does not rise as much, which corresponds to a greater heat capacity. The difference is significant for gases (for oxygen, $C_{V,m} = 20.8\ \mathrm{J\ K^{-1}\ mol^{-1}}$ and $C_{p,m} = 29.1\ \mathrm{J\ K^{-1}\ mol^{-1}}$), which undergo large changes of volume when heated, but is negligible for most solids and liquids.

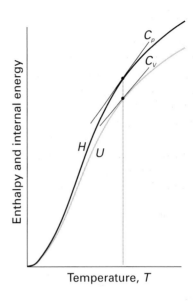

Fig 2.17 The heat capacity at constant pressure is the slope of the curve showing how the enthalpy varies with temperature; the heat capacity at constant volume is the corresponding slope of the internal energy curve. Note that the heat capacity varies with temperature (in general), and that C_p is greater than C_V.

Exercises

Assume all gases are perfect unless stated otherwise.

2.1 Calculate the work that a person must do to raise a mass of 1.0 kg through 10 m on the surface of (a) the Earth ($g = 9.81\ \mathrm{m\ s^{-2}}$) and (b) the Moon ($g = 1.60\ \mathrm{m\ s^{-2}}$).

2.2 When we are interested in biological energy resources and metabolism, we need to know, among other things, the work that an organism has to do to carry out the nor-

mal activities of being alive. How much metabolic energy must a bird of mass 200 g exert to fly to a height of 20 m? Neglect all losses due to friction, physiological imperfection, and the acquisition of kinetic energy.

2.3 Calculate the work needed for a person of mass 65 kg to climb through 4.0 m on the surface of the Earth.

2.4 The centre of mass of a cylindrical column of liquid lies halfway along its length. Calculate the work required to raise a column of mercury (density $13.6 \, g \, cm^{-3}$) of diameter 1.00 cm through 760 mm on the surface of the Earth ($g = 9.81 \, m \, s^{-2}$).

2.5 Calculate the work of expansion accompanying the complete combustion of 1.0 g of glucose to CO_2 and water vapour at 20°C when the external pressure is 1.0 atm.

2.6 We are all familiar with the general principles of operation of an internal combustion engine: the combustion of fuel drives out the piston. It is possible to imagine engines that use reactions other than combustions, and we need to assess the work they can do. A chemical reaction takes place in a container of cross-sectional area 100 cm^2; the container has a piston at one end. As a result of the reaction, the piston is pushed out through 10.0 cm against a constant external pressure of 100 kPa. Calculate the work done by the system.

2.7 The work done by an engine may depend on its orientation in a gravitational field, because the mass of the piston is relevant when the expansion is vertical. A chemical reaction takes place in a container of cross-sectional area 55.0 cm^2; the container has a piston of mass 250 g at one end. As a result of the reaction, the piston is pushed out (a) horizontally, (b) vertically through 155 cm against an external pressure of 105 kPa. Calculate the work done by the system in each case.

2.8 A sample of methane of mass 4.50 g occupies 12.7 L at 310 K. (a) Calculate the work done when the gas expands isothermally against a constant external pressure of 200 Torr until its volume has increased by 3.3 L. (b) Calculate the work that would be done if the same expansion occurred isothermally and reversibly.

2.9 In the isothermal reversible compression of 52.0 mmol of a perfect gas at 260 K, the volume of the gas is reduced from 300 mL to 100 mL. Calculate w for this process.

2.10 A sample of blood plasma occupies 0.550 L at 0°C and 1.03 bar, and is compressed isothermally by 0.57 per cent by being subjected to a constant external pressure of 95.2 bar. Calculate w.

2.11 A strip of magnesium metal of mass 12.5 g is dropped into a beaker of dilute hydrochloric acid. Given

that the magnesium is the limiting reactant, calculate the work done by the system as a result of the reaction. The atmospheric pressure is 1.00 atm and the temperature 20.2°C.

2.12 A current of 1.34 A from a 110 V source was passed through a heater for 5.0 min. The heater was immersed in a water bath. What quantity of heat was transferred to the water?

2.13 What is the heat capacity of a sample of liquid that rose in temperature by 5.23°C when supplied with 124 J of heat?

2.14 The high heat capacity of water is ecologically benign because it stabilizes the temperatures of lakes and oceans: a large quantity of energy must be lost or gained before there is a significant change in temperature. The molar heat capacity of water is $75.3 \, J \, K^{-1} \, mol^{-1}$. What energy is needed to heat 250 g of water (a cup of coffee, for instance) through 40°C?

2.15 When 229 J of energy is supplied as heat to 3.00 mol Ar(g) at a constant volume, the temperature of the sample increases by 2.55 K. Calculate the molar heat capacities at constant volume and constant pressure of the gas.

2.16 The heat capacity of air is much smaller than that of water, and relatively modest amounts of heat are needed to change its temperature. This is one of the reasons why desert regions, though very hot during the day, are bitterly cold at night. The heat capacity of air at room temperature and pressure is approximately $21 \, J \, K^{-1} \, mol^{-1}$. How much energy is required to raise the temperature of a room of dimensions $5.5 \, m \times 6.5 \, m \times 3.0 \, m$ by 10°C? If losses are neglected, how long will it take a heater rated at 1.5 kW to achieve that increase given that $1 \, W = 1 \, J \, s^{-1}$?

2.17 In an experiment to determine the calorific value of a food, a sample of the food was burned in an oxygen atmosphere and the temperature rose by 2.89°C. When a current of 1.27 A from a 12.5 V source flowed through the same calorimeter for 157 s, the temperature rose by 3.88°C. What is the heat released by the combustion?

2.18 The transfer of energy from one region of the atmosphere to another is of great importance in meteorology for it affects the weather. Calculate the heat needed to be supplied to a parcel of air containing 1.00 mol air molecules to maintain its temperature at

300 K when it expands reversibly and isothermally from 22.0 L to 30.0 L as it ascends.

2.19 A laboratory animal exercised on a treadmill which, through pulleys, raised a 200 g mass through 1.55 m. At the same time, the animal lost 5.0 J of energy as heat. Disregarding all other losses, and regarding the animal as a closed system, what is its change in internal energy?

2.20 In preparation for a study of the metabolism of an organism, a small, sealed calorimeter was prepared. In the initial phase of the experiment, a current of 15.22 mA from a 12.4 V source was passed for 155 s through a heater inside the calorimeter. What is the change in internal energy of the calorimeter?

2.21 Carbon dioxide, although only a minor component of the atmosphere, plays an important role in determining the weather and the composition and temperature of the atmosphere. (a) Calculate the difference between the molar enthalpy and the molar internal energy of carbon dioxide regarded as a perfect gas at 298.15 K. (b) Is the molar enthalpy increased or decreased when intermolecular forces are taken into account? For the latter calculation, treat carbon dioxide as a van der Waals gas and use the data in Table 1.5.

2.22 A sample of a serum of mass 25 g is cooled from 290 K to 275 K at constant pressure by the extraction of 1.2 kJ of energy as heat. Calculate q and ΔH and estimate the heat capacity of the sample.

2.23 When 3.0 mol $O_2(g)$ is heated at a constant pressure of 3.25 atm, its temperature increases from 260 K to 285 K. Given that the molar heat capacity of O_2 at constant pressure is 29.4 J K^{-1} mol^{-1}, calculate q, ΔH, and ΔU.

2.24 The molar heat capacity at constant pressure of carbon dioxide is 29.14 J K^{-1} mol^{-1}. What is the value of its molar heat capacity at constant volume?

2.25 Use the information in Exercise 2.24 to calculate the change in (a) molar enthalpy, (b) molar internal energy when carbon dioxide is heated from 15°C (the temperature when air is inhaled) to 37°C (blood temperature, the temperature in our lungs).

2.26 The heat capacity of a substance is often reported in the form

$$C_{p,m} = a + bT + \frac{c}{T^2}$$

Use this expression to make a more accurate estimate of the change in molar enthalpy of carbon dioxide when it is heated from 15°C to 37°C (as in the preceding exercise), given $a = 44.22$ J K^{-1} mol^{-1}, $b = 8.79 \times 10^{-3}$ J K^{-2} mol^{-1}, and $c = -8.62 \times 10^5$ J K mol^{-1}. (*Hint.* You will need to use calculus to integrate $dH = C_p dT$.)

Chapter 3

Thermochemistry

Contents

Tʜɪꜱ chapter is an extended illustration of the role of enthalpy in chemistry. There are three features to keep in mind. One is that a change in enthalpy can be identified with the heat supplied at constant pressure. Second, enthalpy is a state function, so we can calculate the change in its value between two specified initial and final states by selecting the most convenient path between them. Third, the slope of a plot of enthalpy against temperature is the constant-pressure heat capacity of the system. All the material in this chapter is based on these three features.

Physical change

Fɪʀꜱᴛ, we consider physical change, such as when one state of matter changes into another state of matter of the same substance. We shall also include changes of a particularly simple kind, such as the ionization of an atom or the breaking of a bond in a molecule.

3.1 The enthalpy of phase transition

A **phase** is a specific state of matter that is uniform throughout in composition and physical state. The liquid and vapour states of water are two of its phases. The term phase is more specific than 'state of matter' because a substance may exist in more than one solid form, each one of which is a solid phase. Thus, the element sulfur may exist as a solid. However, as a solid it may be found as rhombic sulfur or as monoclinic sulfur; these two solid phases differ in the manner in which the crown-like S_8 molecules stack together. No substance has more than one gaseous phase, so 'gas phase' and 'gaseous state' are effectively synonyms. The only substance that exists in more than one liquid phase is helium.[1] Most substances exist in a variety of solid phases. Carbon, for instance, exists as graphite, diamond, and a variety of forms based on fullerene structures; calcium carbonate exists as calcite and aragonite; there are at least eight forms of ice.

The conversion of one phase of a substance to another phase is called a **phase transition**. Thus, vaporization (liquid → gas) is a phase transition, as is a transition between solid phases (such as rhombic sulfur → monoclinic sulfur). Most phase transitions are accompanied by a change of enthalpy, for the rearrangement of atoms or molecules usually requires energy.[2]

The vaporization of a liquid, such as the conversion of liquid water to water vapour when a pool of water evaporates at 20°C or a kettle boils at 100°C, is an endothermic process, because heat must be supplied to bring about the change. At a molecular level, molecules are being driven apart from the grip they exert on one another, and this process requires energy. One of the body's strategies for maintaining its temperature at about 37°C is to use the endothermic character of the vaporization of water, because the evaporation[3] of perspiration requires heat and withdraws it from the skin.

The energy that must be supplied as heat at constant pressure per mole of molecules that are vaporized is called the **enthalpy of vaporization** of the liquid, and is denoted $\Delta_{vap}H$ (Table 3.1).[4] For example, 44 kJ of heat is required to vaporize 1 mol $H_2O(l)$ at 25°C, so $\Delta_{vap}H = 44$ kJ mol^{-1}. All enthalpies of vaporization are positive, so the sign is not normally given. Alternatively, we can report the same information by writing the **thermochemical equation**[5]

$$H_2O(l) \rightarrow H_2O(g) \qquad \Delta H = +44 \text{ kJ}$$

A thermochemical equation shows the enthalpy change (including the sign) that accompanies the conversion of an amount of reactant equal to its stoichiometric coefficient in the accompanying chemical equation (in this case, 1 mol H_2O). If the stoichiometric coefficients in the chemical equation are multiplied through by 2, then the thermochemical equation would be written

$$2 H_2O(l) \rightarrow 2 H_2O(g) \qquad \Delta H = +88 \text{ kJ}$$

This equation signifies that 88 kJ of heat is required to vaporize 2 mol $H_2O(l)$.

Example 3.1 *Determining the enthalpy of vaporization of a liquid*

Ethanol, C_2H_5OH, is brought to the boil at 1 atm. When an electric current of 0.682 A from a 12.0 V supply is passed for 500 s through a heating coil immersed in the boiling liquid, it is found that 4.33 g of ethanol is vaporized. What is the enthalpy of vaporization of ethanol at its boiling point?

Strategy Because the heat is supplied at constant pressure, we can identify the heat supplied, q, with the change in enthalpy of the ethanol when it vaporizes. We need to calculate the heat supplied and the amount of ethanol molecules vaporized. Then the enthalpy of vaporization is the heat supplied divided by

---

[1] Evidence is accumulating that water might also have two liquid phases.

[2] Note the 'most' and the 'usually'; there are exceptions.

[3] Evaporation is virtually synonymous with vaporization, but commonly denotes vaporization to dryness.

[4] The attachment of the subscript vap to the Δ is an international convention; however, the older convention in which the subscript is attached to the H, as in ΔH_{vap}, is still widely used.

[5] Unless otherwise stated, all data in this text are for 298.15 K.

Table 3.1 *Standard enthalpies of physical change*[*]

Substance	Formula	Freezing point T_f/K	$\Delta_{fus}H^{\ominus}$/(kJ mol^{-1})	Boiling point T_b/K	$\Delta_{vap}H^{\ominus}$/(kJ mol^{-1})
Acetone	CH_3COCH_3	177.8	5.72	329.4	29.1
Ammonia	NH_3	195.3	5.65	239.7	23.4
Argon	Ar	83.8	1.2	87.3	6.5
Benzene	C_6H_6	278.7	9.87	353.3	30.8
Ethanol	C_2H_5OH	158.7	4.60	351.5	43.5
Helium	He	3.5	0.02	4.22	0.08
Mercury	Hg	234.3	2.292	629.7	59.30
Methane	CH_4	90.7	0.94	111.7	8.2
Methanol	CH_3OH	175.5	3.16	337.2	35.3
Water	H_2O	273.2	6.01	373.2	40.7

[*] Values correspond to the transition temperature. For values at 25°C, use the data in Appendix 1.

There are some striking differences in enthalpies of vaporization: although the value for water is 41 kJ mol^{-1}, that for methane, CH_4, at its boiling point is only 8 kJ mol^{-1}. Even allowing for the fact that vaporization is taking place at different temperatures, the difference between the enthalpies of vaporization signifies that water molecules are held together in the bulk liquid much more tightly than methane molecules are in liquid methane.[6] The high enthalpy of vaporization of water has profound ecological consequences, for it is partly responsible for the survival of the oceans and the generally low humidity of the atmosphere. If only a small amount of heat had to be supplied to vaporize the oceans, the atmosphere would be much more heavily saturated with water vapour than is in fact the case.

Another common phase transition is **fusion**, or melting, as when ice melts to water or iron becomes molten. The enthalpy per mole of molecules that accompanies fusion is called the **enthalpy of fusion**, $\Delta_{fus}H$. Its value for water at 0°C is 6.01 kJ mol^{-1} (all enthalpies of fusion are positive, and the sign need not be given), which signifies that 6.01 kJ of energy is needed to melt 1 mol H_2O(s) at 0°C. Notice that the enthalpy of fusion of water is much smaller than its enthalpy of vaporization. In the latter transition,

the amount. The heat supplied is given by eqn 2.4 ($q = I\mathcal{V}t$; recall that 1 A V s = 1 J). The amount of ethanol molecules is determined by dividing the mass of ethanol vaporized by its molar mass ($n = m/M$).

Solution The energy supplied as heat is

$$q = I\mathcal{V}t = (0.682\ A) \times (12.0\ V) \times (500\ s)$$
$$= 0.682 \times 12.0 \times 500\ J$$

This value is the change in enthalpy of the sample. The amount of ethanol molecules (of molar mass 46.07 g mol^{-1}) vaporized is

$$n = \frac{m}{M} = \frac{4.33\ g}{46.07\ g\ mol^{-1}} = \frac{4.33}{46.07}\ mol$$

The molar enthalpy change is therefore

$$\Delta_{vap}H = \frac{0.682 \times 12.0 \times 500\ J}{(4.33/46.07)\ mol} = 4.35 \times 10^4\ J\ mol^{-1}$$

corresponding to 43.5 kJ mol^{-1}.

Self-test 3.1

In a similar experiment, it was found that 1.36 g of boiling benzene, C_6H_6, is vaporized when a current of 0.835 A from a 12.0 V source is passed for 53.5 s. What is the enthalpy of vaporization of benzene at its boiling point?

[*Answer:* 30.8 kJ mol^{-1}]

---

[6] We shall see in Chapter 16 that the interaction responsible for the low volatility of water is the hydrogen bond.

the molecules become completely separated from each other, whereas, when a solid melts, the molecules are merely loosened without separating completely (Fig 3.1).

The reverse of vaporization is **condensation** and the reverse of fusion is **freezing**. The enthalpy changes are, respectively, the negative of the enthalpies of vaporization and fusion, because the heat that is supplied to vaporize or melt the substance is released when it condenses or freezes.[7] It is always the case that *the enthalpy change of a reverse transition is the negative of the enthalpy change of the forward transition* (under the same conditions of temperature and pressure):

$$H_2O(s) \rightarrow H_2O(l) \qquad \Delta H = +6.01 \text{ kJ}$$
$$H_2O(l) \rightarrow H_2O(s) \qquad \Delta H = -6.01 \text{ kJ}$$

and in general

$$\Delta_{\text{forward}} H = -\Delta_{\text{reverse}} H \qquad (3.1)$$

This relation follows from the fact that H is a state property (which itself is a consequence, via the internal energy, of the First Law), so it must return to the same value if a forward change is followed by the reverse of that change (Fig 3.2). The high enthalpy of vaporization of water (44 kJ mol^{-1}), signifying a strongly endothermic process, implies that the condensation of water (−44 kJ mol^{-1}) is a strongly exothermic process. That exothermicity is the origin of the ability of steam to scald severely, because the energy is passed on to the skin.

The direct conversion of a solid to a vapour is called **sublimation**. The reverse process is called **vapour deposition**. Sublimation can be observed

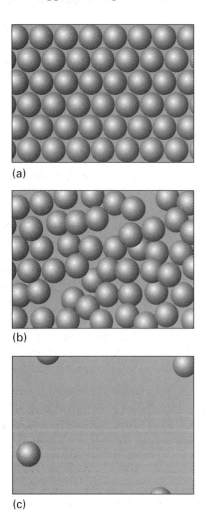

(a)

(b)

(c)

Fig 3.1 When a solid (a) melts to a liquid (b), the molecules separate from one another only slightly, the intermolecular interactions are reduced only slightly, and there is only a small change in enthalpy. When a liquid vaporizes (c), the molecules are separated by a considerable distance, the intermolecular forces are reduced almost to zero, and the change in enthalpy is much greater.

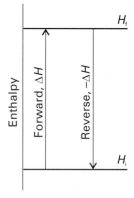

Fig 3.2 An implication of the First Law is that the enthalpy change accompanying a reverse process is the negative of the enthalpy change for the forward process.

[7] This relation is the origin of the obsolescent terms 'latent heat' of vaporization and fusion for what are now termed the enthalpy of vaporization and fusion.

on a cold, frosty morning, when frost vanishes as vapour without first melting. The frost itself forms by vapour deposition from cold, damp air. The vaporization of solid carbon dioxide ('dry ice') is another example of sublimation. The molar enthalpy change accompanying sublimation is called the **enthalpy of sublimation**, $\Delta_{sub}H$. Because enthalpy is a state property, the same change in enthalpy must be obtained both in the *direct* conversion of solid to vapour and in the *indirect* conversion, in which the solid first melts to the liquid and then that liquid vaporizes (Fig 3.3):

$$\Delta_{sub}H = \Delta_{fus}H + \Delta_{vap}H \qquad (3.2)$$

The two enthalpies that are added together must be for the same temperature, so to get the enthalpy of sublimation of water at 0°C we must add together the enthalpies of fusion and vaporization for this temperature. Adding together enthalpies of transition for different temperatures gives a meaningless result.

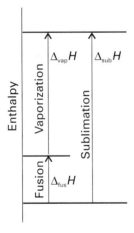

Fig 3.3 The enthalpy of sublimation at a given temperature is the sum of the enthalpies of fusion and vaporization at that temperature. Another implication of the First Law is that the enthalpy change of an overall process is the sum of the enthalpy changes for the possibly hypothetical steps into which it may be divided.

Self-test 3.2

Calculate the enthalpy of sublimation of ice at 0°C from its enthalpy of fusion at 0°C (6.01 kJ mol^{-1}) and the enthalpy of vaporization of water at 0°C (45.07 kJ mol^{-1}).

[*Answer:* 51.08 kJ mol^{-1}]

3.2 Atomic and molecular change

One group of enthalpy changes we employ quite often in the following pages are those accompanying changes to individual atoms and molecules. Among the most important is the **enthalpy of ionization**, $\Delta_{ion}H$, the molar enthalpy change accompanying the removal of an electron from a gas-phase atom (or ion). For example, because

$$H(g) \rightarrow H^+(g) + e^-(g) \qquad \Delta H = +1312 \text{ kJ}$$

the enthalpy of ionization of hydrogen atoms is reported as 1312 kJ mol^{-1}.[8] This value signifies that, at constant pressure (and 298 K), 1312 kJ of heat must be supplied to ionize 1 mol H(g). Table 3.2 gives values of the ionization enthalpies for a number of elements.

We often need to consider a succession of ionizations, such as the conversion of magnesium atoms to Mg$^+$ ions, the ionization of these Mg$^+$ ions to Mg^{2+} ions, and so on. The successive molar enthalpy changes are called, respectively, the **first ionization enthalpy**, the **second ionization enthalpy**, and so on. For magnesium, these enthalpies refer to the processes

$$Mg(g) \rightarrow Mg^+(g) + e^-(g) \qquad \Delta H = +738 \text{ kJ}$$
$$Mg^+(g) \rightarrow Mg^{2+}(g) + e^-(g) \qquad \Delta H = +1451 \text{ kJ}$$

Note that the second ionization enthalpy is larger than the first: more energy is needed to separate an electron from a positively charged ion than from the neutral atom. Note also that enthalpies of ionization refer to the ionization of the gas-phase atom or ion, not to the ionization of an atom or ion in a solid. To determine the latter, we need to combine two or more enthalpy changes.

---

[8] All enthalpies of ionization are positive. The enthalpy of ionization is closely related to the ionization energy; see Section 13.15.

Table 3.2 *First and second (and some higher) ionization enthalpies of the elements in kilojoules per mole (kJ mol^{-1})* *

H 1312							**He** 2370 5250
Li 519 7300	**Be** 900 1760	**B** 799 2420 14 800	**C** 1090 2350 3660 25 000	**N** 1400 2860	**O** 1310 3390	**F** 1680 3370	**Ne** 2080 3950
Na 494 4560	**Mg** 738 1451 7740	**Al** 577 1820 2740 11 600	**Si** 786	**P** 1060	**S** 1000	**Cl** 1260	**Ar** 1520
K 418 3070	**Ca** 590 1150 4940	**Ga** 577	**Ge** 762	**As** 966	**Se** 941	**Br** 1140	**Kr** 1350
Rb 402 2650	**Sr** 548 1060 4120	**In** 556	**Sn** 707	**Sb** 833	**Te** 870	**I** 1010	**Xe** 1170
Cs 376 2420 3300	**Ba** 502 966 3390	**Tl** 812	**Pb** 920	**Bi** 1040	**Po** 812	**At** 920	**Rn** 1040

* Strictly, these values are the values of ΔU at $T = 0$. For precise work, use $\Delta H(T) = \Delta U(0) + \frac{5}{2}RT$, with $\frac{5}{2}RT = 6.20$ kJ mol^{-1} at 298 K.

Example 3.2 *Combining enthalpy changes*

The enthalpy of sublimation of magnesium at 25°C is 148 kJ mol^{-1}. How much heat (at constant temperature and pressure) must be supplied to 1.00 g of solid magnesium metal to produce a gas composed of Mg^{2+} ions and electrons?

Strategy The enthalpy change for the overall process is a sum of the steps, sublimation followed by the two stages of ionization, into which it can be divided. Then the heat required for the specified process is the product of the overall molar enthalpy change and the amount of atoms; the latter is calculated from the given mass and the molar mass of the substance.

Solution The overall process is

$$Mg(s) \rightarrow Mg^{2+}(g) + 2\,e^-(g)$$

The thermochemical equation for this process is the sum of the following thermochemical equations:

		ΔH/kJ
Sublimation:	$Mg(s) \rightarrow Mg(g)$	+148
First ionization:	$Mg(g) \rightarrow Mg^+(g) + e^-(g)$	+738
Second ionization:	$Mg^+(g) \rightarrow Mg^{2+}(g) + e^-(g)$	+1451
Overall (sum):	$Mg(s) \rightarrow Mg^{2+}(g) + 2\,e^-(g)$	+2337

These processes are illustrated diagrammatically in Fig 3.4. It follows that the overall enthalpy change per mole of Mg is +2337 kJ mol^{-1}. Because the molar mass of magnesium is 24.31 g mol^{-1}, 1.0 g of magnesium corresponds to

$$n_{Mg} = \frac{m_{Mg}}{M_{Mg}} = \frac{1.00\ \text{g}}{24.31\ \text{g mol}^{-1}} = \frac{1.00}{24.31}\ \text{mol}$$

Therefore, the heat that must be supplied (at constant pressure) to ionize 1.00 g of magnesium metal is

$$q = \left(\frac{1.00}{24.31}\,\text{mol}\right) \times (2337\ \text{kJ mol}^{-1}) = +96.1\ \text{kJ}$$

This quantity of heat is approximately the same as that needed to vaporize about 43 g of boiling water.

Self-test 3.3

The enthalpy of sublimation of aluminium is 326 kJ mol^{-1}. Use this information and the ionization enthalpies in Table 3.2 to calculate the heat that must be supplied to convert 1.00 g of solid aluminium metal to a gas of Al^{3+} ions and electrons at 25°C.

[*Answer:* +203 kJ]

The reverse of ionization is **electron gain**, and the corresponding molar enthalpy change is called the **electron gain enthalpy**, $\Delta_{eg}H$.[9] For example, because experiments show that

$$Cl(g) + e^-(g) \rightarrow Cl^-(g) \qquad \Delta H = -349\ \text{kJ}$$

it follows that the electron gain enthalpy of Cl atoms is −349 kJ mol^{-1}. Notice that electron gain by

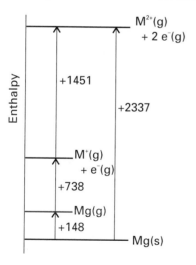

Fig 3.4 The contributions to the enthalpy change treated in the Example 3.2.

---

[9] This term is closely related to the electron affinity; see Section 13.15.

Cl is an exothermic process, so heat is released when a Cl atom captures an electron and forms an ion. It can be seen from Table 3.3, which lists a number of electron gain enthalpies, that some electron gains are exothermic and others are endothermic, so we need to include their sign. For example, electron gain by an O$^-$ ion is strongly endothermic because it takes energy to push an electron on to an already negatively charged species:

$$O^-(g) + e^-(g) \rightarrow O^{2-}(g) \qquad \Delta H = +844\ \text{kJ}$$

The final atomic and molecular process to consider at this stage is the **dissociation**, or breaking, of a chemical bond, as in the process

$$HCl(g) \rightarrow H(g) + Cl(g) \qquad \Delta H = +431\ \text{kJ}$$

The corresponding molar enthalpy change is called the **bond enthalpy**, so we would report the bond enthalpy of H—Cl as 431 kJ mol^{-1}.[10]

Some bond enthalpies are given in Table 3.4. Note that the nitrogen–nitrogen bond in molecular nitrogen, N$_2$, is very strong, at 945 kJ mol^{-1}, which helps to account for the chemical inertness of nitrogen and its ability to dilute the oxygen in the atmosphere without reacting with it. In contrast, the fluorine–fluorine bond in molecular fluorine, F$_2$, is relatively weak, at 155 kJ mol^{-1}; the weakness of this bond contributes to the high reactivity of elemental fluorine. Bond enthalpies are not the full reason for, although the bond in molecular iodine is even weaker, I$_2$ is less reactive than F$_2$. The strengths of the bonds that the elements can make to other elements in the products of a reaction is an additional factor.

A complication when dealing with bond enthalpies is that their value depends on the molecule in which the two linked atoms occur. For instance, the total enthalpy change for the atomization (the complete dissociation) of water

$$H_2O(g) \rightarrow 2\ H(g) + O(g) \qquad \Delta H = +927\ \text{kJ}$$

is not twice the O—H bond enthalpy in H$_2$O even though two O—H bonds are dissociated. There are,

---

[10] All bond enthalpies are positive.

Table 3.3 *Electron-gain enthalpies of the main-group elements, $\Delta_{eg}H/(kJ\ mol^{-1})$* *

H							He
−73							>0
Li	Be	B	C	N	O	F	Ne
−60	+18	−27	−122	+7	−141	−328	>0
					+844		
Na	Mg	Al	Si	P	S	Cl	Ar
−53	−21	−43	−134	−44	−200	−349	>0
					+532		
K	Ca	Ga	Ge	As	Se	Br	Kr
−48	+186	−29	−116	−78	−195	−325	>0
Rb	Sr	In	Sn	Sb	Te	I	Xe
−47	+146	−29	−116	−103	−190	−295	>0
Cs	Ba	Tl	Pb	Bi	Po	At	Rn
−46	+46	−19	−35	−91	−183	−270	>0

* Where two values are given, the first refers to the formation of X^- from the neutral atom X and the second to the formation of X^{2-} from X^-. Strictly, these values are the values of $\Delta_{eg}U$ at $T = 0$. For precise work, use $\Delta_{eg}H(T) = \Delta_{eg}U(0) - \tfrac{5}{2}RT$ with $\tfrac{5}{2}RT = 6.20\ kJ\ mol^{-1}$ at 298 K. Note that the correction cancels with the analogous one in Table 3.2.

Table 3.4 *Selected bond enthalpies, $\Delta H(AB)/(kJ\ mol^{-1})$*

Diatomic molecules

H—H	436	O=O	497	F—F	155	H—F	565
		N≡N	945	Cl—Cl	242	H—Cl	431
		O—H	428	Br—Br	193	H—Br	366
		C=O	1074	I—I	151	H—I	299

Polyatomic molecules

H—CH₃	435	H—NH₂	431	H—OH	492
H—C₆H₆	469	O₂N—NO₂	57	HO—OH	213
H₃C—CH₃	368	O=CO	799	HO—CH₃	377
H₂C=CH₂	699			Cl—CH₃	452
				Br—CH₃	293
HC≡CH	962			I—CH₃	234

in fact, two different dissociation steps. In the first step, an O—H bond is broken in an H_2O molecule:

$$H_2O(g) \rightarrow HO(g) + H(g) \qquad \Delta H = +499\ kJ$$

In the second step, the O—H bond is broken in an OH radical:

$$HO(g) \rightarrow H(g) + O(g) \qquad \Delta H = +428\ kJ$$

The sum of the two steps is the atomization of the molecule. As can be seen from this example, the O—H bonds in H_2O and HO have similar but not identical bond enthalpies.

Although accurate calculations must use bond enthalpies for the molecule in question and its successive fragments, when such data are not available there is no choice but to make estimates

Table 3.5 *Mean bond enthalpies, $\Delta H/(kJ\ mol^{-1})^*$*

	H	C	N	O	F	Cl	Br	I	S	P	Si
H	436										
C	412	348 (1)									
		612 (2)									
		518 (a)									
N	388	305 (1)	163 (1)								
		613 (2)	409 (2)								
		890 (3)	945 (3)								
O	463	360 (1)	157	146 (1)							
		743 (2)		497 (2)							
F	565	484	270	185	155						
Cl	431	338	200	203	254	242					
Br	366	276				219	193				
I	299	238				210	178	151			
S	338	259			496	250	212		264		
P	322									200	
Si	318			466							226

* Values are for single bonds except where otherwise stated (in parentheses).
(a) Denotes aromatic.

by using **mean bond enthalpies**, ΔH_B, which are the averages of bond enthalpies over a related series of compounds (Table 3.5). For example, the mean HO bond enthalpy, $\Delta H_B(H\!-\!O) = 463$ kJ mol^{-1}, is the mean of the HO bond enthalpies in H_2O and several other similar compounds, including methanol, CH_3OH.

Example 3.3 *Using mean bond enthalpies*

Estimate the enthalpy change that accompanies the reaction

$$C(s, \text{graphite}) + 2\,H_2(g) + \tfrac{1}{2}\,O_2(g) \rightarrow CH_3OH(l)$$

in which liquid methanol is formed from its elements at 25°C. Use information from Appendix 1 and bond enthalpy data from Tables 3.4 and 3.5.

Strategy In calculations of this kind, the procedure is to break the overall process down into a sequence of steps such that their sum is the chemical equation required. Always ensure, when using bond enthalpies, that all the species are in the gas phase. That may mean including the appropriate enthalpies of vaporization or sublimation. One approach is to atomize all

the reactants and then to build the products from the atoms so produced. When explicit bond enthalpies are available (that is, data are given in the tables available), use them; otherwise, use mean bond enthalpies to obtain estimates. It is often helpful to display the enthalpy changes diagrammatically.

Solution The following steps are required (Fig 3.5):

		$\Delta H/$ kJ
Atomization of graphite:	$C(s, \text{graphite}) \rightarrow C(g)$	+716.68
Dissociation of 2 mol $H_2(g)$:	$2\,H_2(g) \rightarrow 4\,H(g)$	+871.88
Dissociation of $\tfrac{1}{2}\,O_2(g)$:	$\tfrac{1}{2}\,O_2(g) \rightarrow O(g)$	+249.17

	$\Delta H/$ kJ
Overall, so far: $\quad C(s) + 2\,H_2(g) + \tfrac{1}{2}\,O_2(g)$	
$\rightarrow C(g) + 4\,H(g) + O(g)$	+1837.73

These values are accurate. In the second step, three CH bonds, one CO bond, and one OH bond are formed, and we estimate their enthalpies from mean values. The enthalpy change for bond formation (the reverse of dissociation) is the negative of the mean bond enthalpy (obtained from Table 3.5):

	$\Delta H/$ kJ
Formation of 3 C—H bonds:	−1236
Formation of 1 C—O bond:	−360
Formation of 1 O—H bond:	−463

Overall, in this step: C(g) + 4 H(g) + O(g)
$\rightarrow$ CH$_3$OH(g)   −2059

These values are estimates. The final stage of the reaction is the condensation of methanol vapour:

CH$_3$OH(g) $\rightarrow$ CH$_3$OH(l) ΔH = −38.00 kJ

The sum of the enthalpy changes is

ΔH = (+1837.73 kJ) + (−2059 kJ) + (−38.00 kJ)
= −259 kJ

The experimental value is −239.00 kJ.

Self-test 3.4

Estimate the enthalpy change for the combustion of liquid ethanol to carbon dioxide and liquid water by using the enthalpies of atomization of CO$_2$(g) and H$_2$O(g), which are 1609 kJ mol^{-1} and 920 kJ mol^{-1}, respectively, and mean bond enthalpies for atomization of the alcohol.

[*Answer*: −1348 kJ; the experimental value is −1368 kJ]

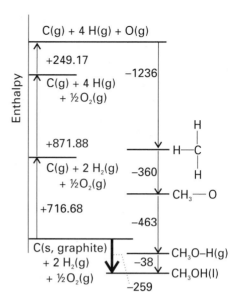

Fig 3.5 The enthalpy changes used to estimate the enthalpy change accompanying the formation of liquid methanol from its elements. The bond enthalpies on the right are mean values, so the final value is only approximate.

Chemical change

I N the remainder of this chapter we concentrate on enthalpy changes accompanying chemical reactions, such as the hydrogenation of ethene:

CH$_2$=CH$_2$(g) + H$_2$(g) $\rightarrow$ CH$_3$CH$_3$(g) ΔH = −137 kJ

The value of ΔH given here signifies that the enthalpy of the system decreases by 137 kJ (and, if the reaction takes place at constant pressure, that 137 kJ of heat is released into the surroundings) when 1 mol CH$_2$=CH$_2$ combines with 1 mol H$_2$ at 25°C.

3.3 Standard enthalpy changes

The reaction enthalpy depends on the conditions—the states of the reactants and products, the pressure,

and the temperature—under which the reaction takes place. Chemists have therefore found it convenient to report their data for a set of standard conditions at the temperature of their choice:

*The **standard state** of a substance is the pure substance at exactly 1 bar.*[11]

For example, the standard state of hydrogen gas is the pure gas at 1 bar and the standard state of solid calcium carbonate is the pure solid at 1 bar, with either the calcite or aragonite phase specified. The physical state and, when appropriate, solid phase needs to be specified because we can speak of the standard states of the solid, liquid, or vapour phases of methanol, which are the pure solid, the pure liquid, or the pure vapour, respectively, at 1 bar in each case. The temperature is not a part of the definition of a standard state, and it is possible to speak of the standard state of hydrogen gas at 100 K, 273.15 K, or any other temperature. It is conven-

---

[11] Remember that 1 bar = 10^5 Pa exactly. Solutions are a special case, and are dealt with in Section 6.5.

tional, though, for data to be reported at 298.15 K (25.00°C), and from now on, unless specified otherwise, all data will be for that temperature.

When we write $\Delta H^{\ominus}$ in a thermochemical equation, we always mean the change in enthalpy that occurs when the reactants in their standard states change into products in their standard states. For example, from the thermochemical equation

$$2\ H_2(g) + O_2(g) \rightarrow 2\ H_2O(l) \qquad \Delta H^{\ominus} = -572\ kJ$$

we know that, when 2 mol H_2 as pure hydrogen gas at 1 bar combines with 1 mol O_2 as pure oxygen gas at 1 bar to form 2 mol H_2O as pure liquid water at 1 bar, the initial and final temperatures being 25°C, then the enthalpy of the system decreases by 572 kJ, and (at constant pressure) 572 kJ of heat is released into the surroundings.

One commonly encountered reaction is **combustion**, the complete reaction of a compound, most commonly an organic compound, with oxygen,[12] as in the combustion of methane in a natural gas flame:

$$CH_4(g) + 2\ O_2(g) \rightarrow CO_2(g) + 2\ H_2O(l) \qquad \Delta H^{\ominus} = -890\ kJ$$

The **standard enthalpy of combustion**, $\Delta_c H^{\ominus}$, is the standard change in enthalpy per mole of combustible substance. In this example, we would write $\Delta_c H^{\ominus}(CH_4, g) = -890\ kJ\ mol^{-1}$. Some typical values are given in Table 3.6. Note that $\Delta_c H^{\ominus}$ is a molar quantity, and is obtained from the value of $\Delta H^{\ominus}$ by dividing by the amount of organic reactant consumed (in this case, by 1 mol CH_4).

Enthalpies of combustion are commonly measured by using a bomb calorimeter, a device in which heat is transferred at constant volume. According to the discussion in Section 2.5, the heat transfer at constant volume is equal to the change in internal energy, ΔU, not ΔH. To convert from ΔU, to ΔH we need to note that the molar enthalpy of a substance is related to its molar internal energy by $H_m = U_m + pV_m$ (eqn 2.12). For condensed phases, pV_m is so small it may be ignored. For gases, treated as

---

[12] By convention, combustion of an organic compound results in the formation of carbon dioxide gas, liquid water, and—if the compound contains nitrogen—nitrogen gas.

Table 3.6 *Standard enthalpies of combustion*

Substance	Formula	$\Delta_c H^{\ominus}/(kJ\ mol^{-1})$
Benzene	$C_6H_6(l)$	−3268
Carbon	$C(s, graphite)$	−394
Carbon monoxide	$CO(g)$	−394
Ethanol	$C_2H_5OH(l)$	−1368
Ethyne	$C_2H_2(g)$	−1300
Glucose	$C_6H_{12}O_6(s)$	−2808
Hydrogen	$H_2(g)$	−286
Methane	$CH_4(g)$	−890
Methanol	$CH_3OH(l)$	−726
Octane	$C_8H_{18}(l)$	−5471
iso-Octane*	$C_8H_{18}(l)$	−5461
Propane	$C_3H_8(g)$	−2220
Sucrose	$C_{12}H_{22}O_{11}(s)$	−5645
Toluene	$C_6H_5CH_3(l)$	−3910
Urea	$CO(NH_2)_2(s)$	−632

* 2,2,4-trimethylpentane.

perfect, pV_m may be replaced by RT. Therefore, if in the chemical equation the difference (products − reactants) in the stoichiometric coefficients of *gas-phase* species is Δv_{gas}, we can write

$$\Delta_c H = \Delta_c U + \Delta v_{gas} RT \qquad (3.3)$$

Illustration 3.1

The heat given out when glycine is burned in a bomb calorimeter is 969.6 kJ mol^{-1} at 298.15 K, so $\Delta_c U = -969.6$ kJ mol^{-1}. The chemical equation for the reaction is

$$NH_2CH_2COOH(s) + \tfrac{9}{4}O_2(g) \rightarrow 2CO_2(g) + \tfrac{5}{2}H_2O(l) + \tfrac{1}{2}N_2(g)$$

We see that $\Delta v_{gas} = (2 + \tfrac{1}{2}) - \tfrac{9}{4} = \tfrac{1}{4}$. Therefore,

$$\Delta_c H = \Delta_c U + \tfrac{1}{4}RT$$
$$= -969.6\ kJ\ mol^{-1} + \tfrac{1}{4} \times (8.3145\ J\ K^{-1}\ mol^{-1})$$
$$\times (198.15\ K) = -969.6\ kJ\ mol^{-1} + 0.62\ kJ\ mol^{-1}$$
$$= -969.0\ kJ\ mol^{-1}$$

One application of enthalpies of combustion is to judge the suitability of a fuel (Box 3.1). For example, from the value of the standard enthalpy of combustion for methane we know that for each mole of CH_4 supplied to a furnace, 890 kJ of heat can be released, whereas for each mole of iso-octane (C_8H_{18}, 2,2,4-trimethylpentane, **1**, a typical component of gasoline) supplied to an internal combustion engine, 5461 kJ of heat is released (see the data in Table 3.6). The much larger value for iso-octane is a consequence of each molecule having eight C atoms to contribute to the formation of carbon dioxide whereas methane has only one.

1 2,2,4–trimethylpentane, isooctane

3.4 The combination of reaction enthalpies

It is often the case that a reaction enthalpy is needed but is not available in tables of data. Now the fact that enthalpy is a state function comes in handy, because it implies that we can construct the required reaction enthalpy from the reaction enthalpies of known reactions. We have already seen a primitive example when we calculated the enthalpy of sublimation from the sum of the enthalpies of fusion and vaporization. The only difference is that we now apply the technique to a sequence of chemical reactions. The procedure is summarized by **Hess's law**:

The standard enthalpy of a reaction is the sum of the standard enthalpies of the reactions into which the overall reaction may be divided.

Although the procedure is given the status of a law, it hardly deserves the title because it is nothing more than a consequence of enthalpy being a state

function, which implies that an overall enthalpy change can be expressed as a sum of enthalpy changes for each step in an indirect path. The individual steps need not be actual reactions that can be carried out in the laboratory—they may be entirely hypothetical reactions, the only requirement being that their equations should balance. Each step must correspond to the same temperature.

Example 3.4 *Using Hess's law*

Given the thermochemical equations

$C_3H_6(g) + H_2(g) \rightarrow C_3H_8(g)$ $\Delta H^\ominus = -124$ kJ
$C_3H_8(g) + 5 O_2(g) \rightarrow 3 CO_2(g) + 4 H_2O(l)$ $\Delta H^\ominus = -2220$ kJ

where C_3H_6 is propene and C_3H_8 is propane, calculate the standard enthalpy of combustion of propene.

Strategy We need to add or subtract the thermochemical equations, together with any others that are needed (from Appendix 1), so as to reproduce the thermochemical equation for the reaction required. In calculations of this type, it is common to need to use the synthesis of water to balance the hydrogen or oxygen atoms in the overall equation. Once again, it may be helpful to express the changes diagrammatically.

Solution The overall reaction is

$C_3H_6(g) + \tfrac{9}{2} O_2(g) \rightarrow 3 CO_2(g) + 3 H_2O(l)$ $\Delta H^\ominus$

We can recreate this thermochemical equation from the following sum (Fig 3.6):

	$\Delta H^\ominus$/kJ
$C_3H_6(g) + H_2(g) \rightarrow C_3H_8(g)$	−124
$C_3H_8(g) + 5 O_2(g) \rightarrow 3 CO_2(g) + 4 H_2O(l)$	−2220
$H_2O(l) \rightarrow H_2(g) + \tfrac{1}{2} O_2(g)$	+286
Overall: $C_3H_6(g) + \tfrac{9}{2} O_2(g)$	
$\rightarrow 3 CO_2(g) + 3 H_2O(l)$	−2058

It follows that the standard enthalpy of combustion of propene is −2058 kJ mol^{-1}.

Self-test 3.5

Calculate the standard enthalpy of the reaction $C_6H_6(l) + 3 H_2(g) \rightarrow C_6H_{12}(l)$ from the standard enthalpies of combustion of benzene and cyclohexane.

[*Answer:* −205 kJ]

3.5 Standard enthalpies of formation

The **standard reaction enthalpy**, $\Delta_r H^{\ominus}$, is the difference between the standard molar enthalpies of the reactants and the products, with each term weighted by the stoichiometric coefficient, v (nu), in the chemical equation:

$$\Delta_r H^{\ominus} = \sum v H_m^{\ominus} \text{(products)} - \sum v H_m^{\ominus} \text{(reactants)} \tag{3.4}$$

The standard reaction enthalpy is the change in enthalpy of the system when the reactants in their standard states (pure, 1 bar) are completely converted into products in their standard states (pure, 1 bar), with the change expressed in kilojoules per mole of reaction as written. The problem with this approach, however, is that we have no way of knowing the absolute enthalpies of the substances. To avoid this problem, we can imagine the reaction as taking place by an indirect route, in which the reactants are first broken down into the elements and then the products are formed from the elements (Fig 3.7). Specifically, the **standard enthalpy of formation**, $\Delta_f H^{\ominus}$, of a substance is the standard enthalpy (per mole of the substance) for its formation from its elements in their reference states. The **reference state** of an element is its most stable form under the prevailing conditions (Table 3.7). For example, the standard enthalpy of formation of liquid water (at 25°C, as always in this text) is obtained from the thermochemical equation

$$H_2(g) + \tfrac{1}{2}O_2(g) \rightarrow H_2O(l) \qquad \Delta H^{\ominus} = -286 \text{ kJ}$$

and is $\Delta_f H^{\ominus}(H_2O, l) = -286 \text{ kJ mol}^{-1}$.[13] With the introduction of standard enthalpies of formation, we can write

$$\Delta_r H^{\ominus} = \sum v \Delta_f H^{\ominus} \text{(products)}$$
$$- \sum v \Delta_f H^{\ominus} \text{(reactants)} \tag{3.5}$$

The first term on the right is the enthalpy of formation of all the products from their elements; the second term on the right is the enthalpy of formation of all the reactants from their elements. The fact that the enthalpy is a state function means that a reaction enthalpy calculated in this way is identical to the value that would be calculated from eqn 3.4 if absolute enthalpies were available.

The values of some standard enthalpies of formation at 25°C are given in Table 3.8, and a longer list is given in Appendix 1. The standard enthalpies of formation of elements in their reference states are zero by definition (because their formation is the null reaction: element → element). Note, however, that the formation of an element in a different phase is not zero:

$$C(s, \text{graphite}) \rightarrow C(s, \text{diamond}) \quad \Delta H^{\ominus} = +1.895 \text{ kJ}$$

Therefore, although $\Delta_f H^{\ominus}(C, \text{graphite}) = 0$, $\Delta_f H^{\ominus}(C, \text{diamond}) = +1.895 \text{ kJ mol}^{-1}$.

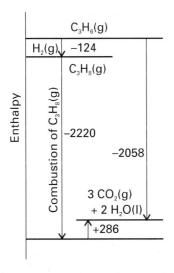

Fig 3.6 The enthalpy changes used in Example 3.4 to illustrate Hess's law.

---

[13] Note that enthalpies of formation are molar quantities, so to go from $\Delta H^{\ominus}$ in a thermochemical equation to $\Delta_f H^{\ominus}$ for that substance, divide by the amount of substance formed (in this instance, by 1 mol H_2O).

Example 3.5 *Using standard enthalpies of formation*

Calculate the standard enthalpy of combustion of liquid benzene from the standard enthalpies of formation of the reactants and products.

Strategy We write the chemical equation, identify the stoichiometric numbers of the reactants and products, and then use eqn 3.5. Note that the expression has the form 'products – reactants'. Numerical values of standard enthalpies of formation are given in Appendix 1. The standard enthalpy of combustion is the enthalpy change per mole of substance, so we need to interpret the enthalpy change accordingly.

Solution The chemical equation is

$$C_6H_6(l) + \tfrac{15}{2}O_2(g) \rightarrow 6\,CO_2(g) + 3\,H_2O(l)$$

It follows that

$$\Delta_r H^\ominus = \{6\Delta_f H^\ominus(CO_2, g) + 3\Delta_f H^\ominus(H_2O, l)\}$$

$$- \{\Delta_f H^\ominus(C_6H_6, l) + \tfrac{15}{2}\Delta_f H^\ominus - (O_2, g)\}$$

$$= \{6 \times (-393.51\,\text{kJ mol}^{-1})$$

$$+ 3 \times (-285.83\,\text{kJ mol}^{-1})\}$$

$$- \{(49.0\,\text{kJ mol}^{-1}) + 0\}$$

$$= -3268\,\text{kJ mol}^{-1}$$

Inspection of the chemical equation shows that, in this instance, the 'per mole' is per mole of C_6H_6, which is exactly what we need for an enthalpy of combustion. It follows that the standard enthalpy of combustion of liquid benzene is $-3268\,\text{kJ mol}^{-1}$.

Self-test 3.6

Use standard enthalpies of formation to calculate the enthalpy of combustion of propane gas to carbon dioxide and liquid water.

[*Answer*: $-2220\,\text{kJ mol}^{-1}$]

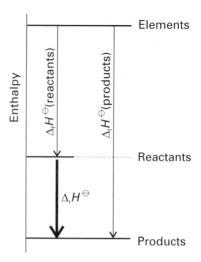

Fig 3.7 An enthalpy of reaction may be expressed as the difference between the enthalpies of formation of the products and the reactants.

Table 3.7 *Reference states of some elements at 298 K*

Element	Reference state
Arsenic	Grey arsenic
Bromine	Liquid
Carbon	Graphite
Hydrogen	Gas
Iodine	Solid
Mercury	Liquid
Nitrogen	Gas
Oxygen	Gas
Phosphorus	White phosphorus
Sulfur	Rhombic sulfur
Tin	White tin

The reference states of the elements define a thermochemical 'sea level', and enthalpies of formation can be regarded as thermochemical 'altitudes' above or below sea level (Fig 3.8). Compounds that have negative standard enthalpies of formation (such as water) are classified as **exothermic compounds**, for they lie at a lower enthalpy than their component elements (they lie below thermochemical sea level). Compounds that have positive standard enthalpies of formation (such as carbon disulfide) are classified as **endothermic compounds**, and possess a higher enthalpy than their component elements (they lie above sea level).

Box 3.1 *Food and energy reserves*

The food ingested by animals is the fuel they need for the processes of life. We shall see in Chapter 4 that the best assessment of the ability of a compound to act as a fuel to drive many of the processes occurring in the body is in terms of the 'Gibbs energy'. However, a useful guide to the resources provided by a fuel, and the only one that matters when its heat output is being considered, is the enthalpy, particularly the enthalpy of combustion. The thermochemical properties of fuels and foods are commonly discussed in terms of their *specific enthalpy*, the enthalpy of combustion per gram of material. Thus, if the standard enthalpy of combustion is $\Delta_c H^{\ominus}$ and the molar mass of the compound is M, then the specific enthalpy is $\Delta_c H^{\ominus}/M$. The table gives the specific enthalpies of a number of types of food and fuel.

A typical 18–20 y old man requires a daily input of about 12 MJ; a woman of the same age needs about 9 MJ. If the entire consumption were in the form of glucose (which has a specific enthalpy of 16 kJ g^{-1}), that would require the consumption of 750 g of glucose for a man and 560 g for a woman. In fact, digestible carbohydrates have a slightly higher specific enthalpy (17 kJ g^{-1}) than glucose itself, so a carbohydrate diet is slightly less daunting than a pure glucose diet, as well as being more appropriate in the form of fibre, the indigestible cellulose that helps move digestion products through the intestine.

The specific enthalpy of fats, which are long-chain esters like tristearin (beef fat), is much greater than that of carbohydrates, at around 38 kJ g^{-1}, slightly less than the value for the hydrocarbon oils used as fuel (48 kJ g^{-1}). The difference stems from the fact that carbohydrates can be regarded as partially oxidized versions of hydrocarbons, so less heat is released when the oxidation is completed. Fats are commonly used as an energy store, to be used only when the more readily accessible carbohydrates

have fallen into short supply. In Arctic species, the stored fat also acts as a layer of insulation; in desert species (such as the camel), the fat is also a source of water, one of its oxidation products.

Proteins are also used as a source of energy, but their components, the amino acids, are often too valuable to squander in this way, and are used to construct other proteins instead. When proteins are oxidized (to urea, $CO(NH_2)_2$), the equivalent enthalpy density is comparable to that of carbohydrates.

The heat released by the oxidation of foods needs to be discarded in order to maintain body temperature within its typical range of 35.6–37.8°C. A variety of mechanisms contribute to this aspect of homeostasis. The general uniformity of temperature throughout the body is maintained largely by the flow of blood. When heat needs to be dissipated rapidly, warm blood is allowed to flow through the capillaries of the skin, so producing flushing. Radiation is one means of discarding heat; another is evaporation and the energy demands of the enthalpy of vaporization of water. Evaporation removes about 2.4 kJ per gram of water perspired. When vigorous exercise promotes sweating (through the influence of heat selectors on the hypothalamus), 1–2 L of perspired water can be produced per hour, corresponding to a heat loss of 2.4–5.0 MJ h^{-1}.

Exercise 1 1.0 L of water is perspired by a runner in order to maintain body temperature. Assume evaporation of 1.0 L of water dissipates the heat generated by metabolism. (a) What is the change in enthalpy of the runner? (b) Assume that none of the 1.0 L of sweat evaporated and instead an increase in body temperature resulted. If the runner weighs 60 kg and has a heat capacity approximately that of water, what is the runner's body temperature?

Thermochemical properties of some fuels

Fuel	Combustion equation	$\Delta_c H^{\ominus}/(\text{kJ mol}^{-1})$	Specific enthalpy/ (kJ g^{-1})	Enthalpy density*/ (kJ L^{-1})
Hydrogen	$2\,H_2(g) + O_2(g) \rightarrow 2\,H_2O(l)$	−286	142	13
Methane	$CH_4(g) + 2\,O_2(g) \rightarrow CO_2(g) + 2\,H_2O(l)$	−890	55	40
Octane	$2\,C_8H_{18}(l) + 25\,O_2(g) \rightarrow 16\,CO_2(g) + 18\,H_2O(l)$	−5471	48	3.8×10^4
Methanol	$2\,CH_3OH(l) + 3\,O_2(g) \rightarrow 2\,CO_2(g) + 4\,H_2O(l)$	−726	23	1.8×10^4

* At atmospheric pressures and room temperature. Enthalpy density is the enthalpy of combustion divided by the molar volume.

Exercise 2 There are no dietary recommendations for consumption of carbohydrates; however, at least 65 per cent of our food calories should come from carbohydrates. A $\frac{3}{4}$-cup serving of pasta contains 40 g of carbohydrates. What percentage of the daily calorie requirement for a person on a 2200 Calorie diet (1 Cal = 1 kcal) does this represent?

Table 3.8 *Standard enthalpies of formation at 25°C*

Substance*	Formula	$\Delta_f H^{\ominus}/(\text{kJ mol}^{-1})$
Inorganic compounds		
Ammonia	$NH_3(g)$	−46.11
Ammonium nitrate	$NH_4NO_3(s)$	−365.56
Carbon monoxide	$CO(g)$	−110.53
Carbon disulfide	$CS_2(l)$	+89.70
Carbon dioxide	$CO_2(g)$	−393.51
Dinitrogen tetroxide	$N_2O_4(g)$	+9.16
Dinitrogen oxide	$N_2O(g)$	+82.05
Hydrogen chloride	$HCl(g)$	−92.31
Hydrogen fluoride	$HF(g)$	−271.1
Hydrogen sulfide	$H_2S(g)$	−20.63
Nitric acid	$HNO_3(l)$	−174.10
Nitric oxide	$NO(g)$	+90.25
Nitrogen dioxide	$NO_2(g)$	+33.18
Sodium chloride	$NaCl(s)$	−411.15
Sulfur dioxide	$SO_2(g)$	−296.83
Sulfur trioxide	$SO_3(g)$	−395.72
Sulfuric acid	$H_2SO_4(l)$	−813.99
Water	$H_2O(l)$	−285.83
	$H_2O(g)$	−241.82
Organic compounds		
Benzene	$C_6H_6(l)$	+49.0
Ethane	$C_2H_6(g)$	−84.68
Ethanol	$C_2H_5OH(l)$	−277.69
Ethene	$C_2H_4(g)$	+52.26
Ethyne	$C_2H_2(g)$	+226.73
Glucose	$C_6H_{12}O_6(s)$	−1268
Methane	$CH_4(g)$	−74.81
Methanol	$CH_3OH(l)$	−238.86
Sucrose	$C_{12}H_{22}O_{11}(s)$	−2222

* A longer list is given in Appendix 1 at the end of the book.

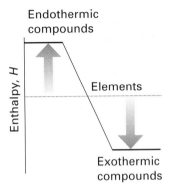

Fig 3.8 The enthalpy of formation acts as a kind of thermochemical 'altitude' of a compound with respect to the 'sea level' defined by the elements from which it is made. Endothermic compounds have positive enthalpies of formation; exothermic compounds have negative energies of formation.

3.6 The variation of reaction enthalpy with temperature

It often happens that we have data at one temperature but need it at another temperature. For example, we might want to know the enthalpy of a particular reaction at body temperature, 37°C, but may have data available for 25°C. Another type of question that might arise might be whether the oxidation of glucose is more exothermic when it takes place inside an Arctic fish that inhabits water at 0°C than when it takes place at mammalian body temperatures. Similarly, we may need to predict whether the synthesis of ammonia is more exothermic at a typical industrial temperature of 450°C than at 25°C. In precise work, every attempt would be made to measure the reaction enthalpy at the temperature of interest, but it is useful to have a 'back-of-the-envelope' way of estimating the direction of change and even a moderately reliable numerical value.

Figure 3.9 illustrates the technique we use. As we have seen, the enthalpy of a substance increases with temperature; therefore the total enthalpy of the reactants and the total enthalpy of the products increase as shown in the illustration. Provided the two total enthalpy increases are different, the standard reaction enthalpy (their difference) will change as the temperature is changed. The change in the enthalpy of a substance depends on the slope of the graph and therefore on the constant-pressure heat capacities of the substances (recall Fig 2.17). We can therefore expect the temperature dependence of the reaction enthalpy to be related to the difference in heat capacities of the products and the reactants.

As a simple example, consider the reaction

$$2 \, H_2(g) + O_2(g) \rightarrow 2 \, H_2O(l)$$

where the standard enthalpy of reaction is known at one temperature (for example, at 25°C from the tables in this book). According to eqn 3.4, we can write

$$\Delta_r H^\ominus = 2H_m^\ominus(H_2O, l) - \{2H_m^\ominus(H_2, g) + H_m^\ominus(O_2, g)\}$$

for the reaction at a temperature T. If the reaction takes place at a higher temperature T', the molar en-

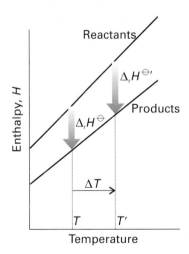

Fig 3.9 The enthalpy of a substance increases with temperature. Therefore, if the total enthalpy of the reactants increases by a different amount from that of the products, the reaction enthalpy will change with temperature. The change in reaction enthalpy depends on the relative slopes of the two lines and hence on the heat capacities of the substances.

thalpy of each substance is increased because it stores more energy and the standard reaction enthalpy becomes

$$\Delta_r H^{\ominus\prime} = 2H_m^{\ominus\prime}(H_2O, l) - \{2H_m^{\ominus\prime}(H_2, g) + H_m^{\ominus\prime}(O_2, g)\}$$

where the prime signifies the value at the new temperature. The increase in molar enthalpy of a substance is given by eqn 2.16 as $C_{p,m}\Delta T$, where $C_{p,m}$ is the molar constant-pressure heat capacity of the substance and $\Delta T = T' - T$ is the difference in temperature. For example, the molar enthalpy of water changes to

$$H_m^{\ominus\prime}(H_2O, l) = H_m^\ominus(H_2O, l) + C_{p,m}(H_2O, l) \times \Delta T$$

When we substitute terms like this into the expression above, we find

$$\Delta_r H^{\ominus\prime} = \Delta_r H^\ominus + \Delta_r C_p \times \Delta T \tag{3.6}$$

where

$$\Delta_r C_p = 2C_{p,m}(H_2O, l) - \{2C_{p,m}(H_2, g) + C_{p,m}(O_2, g)\}$$

Note that this combination has the same pattern as the reaction enthalpy, and the stoichiometric coefficients occur in the same way. In general, $\Delta_r C_p$ is the difference between the weighted sums of the molar heat capacities of the products and the reactants:[14]

$$\Delta_r C_p = \sum v \, C_{p,m}(\text{products}) - \sum v \, C_{p,m}(\text{reactants})$$

$$\tag{3.7}$$

Equation 3.6 is **Kirchhoff's law**. We see that, just as we anticipated, the standard reaction enthalpy at one temperature can be calculated from the standard reaction enthalpy at another temperature provided we know the molar constant-pressure heat capacities of all the substances. These values are given in Appendix 1. The derivation of Kirchhoff's law supposes that the heat capacities are constant over the range of temperature of interest, so the law is best restricted to small temperature differences (of no more than 100 K or so).

----

[14] Strictly, the *standard* heat capacity, the heat capacity at 1 bar, should be used.

Example 3.6 *Using Kirchhoff's law*

The standard enthalpy of formation of gaseous water at 25°C is −241.82 kJ mol^{-1}. Estimate its value at 100 °C.

Strategy First, write the chemical equation and identify the stoichiometric coefficients. Then calculate the value of $\Delta_r C_p$ from the data in Appendix 1 by using eqn 3.7 and use the result in eqn 3.6.

Solution The chemical equation is

$$H_2(g) + \tfrac{1}{2}O_2(g) \rightarrow H_2O(g)$$

and the molar constant-pressure heat capacities of $H_2O(g)$, $H_2(g)$, and $O_2(g)$ are 33.58 J K^{-1} mol^{-1}, 28.84 J K^{-1} mol^{-1}, and 29.37 J K^{-1} mol^{-1}, respectively. It follows that

$$\Delta_r C_p = C_{p,m}(H_2O, g) - \{C_{p,m}(H_2, g) + \tfrac{1}{2}C_{p,m}(O_2, g)\}$$

$$= (33.58 \text{ J K}^{-1}\text{mol}^{-1}) - \{(28.84 \text{ J K}^{-1}\text{mol}^{-1})$$

$$+ \tfrac{1}{2} \times (29.37 \text{ J K}^{-1}\text{mol}^{-1})\}$$

$$= -9.95 \text{ J K}^{-1}\text{mol}^{-1}$$

Then, because $\Delta T = +75$ K, from eqn 3.6 we find

$$\Delta_r H^\ominus \,' = (-241.82 \text{ kJ mol}^{-1})$$

$$+ (-9.95 \text{ J K}^{-1}\text{ mol}^{-1}) \times (75 \text{ K})$$

$$= (-241.82 \text{ kJ mol}^{-1}) - (0.75 \text{ kJ mol}^{-1})$$

$$= -242.57 \text{ kJ mol}^{-1}$$

The reaction is slightly more exothermic at the higher temperature.

Self-test 3.7

Estimate the standard enthalpy of formation of $NH_3(g)$ at 400 K from the data in Appendix 1.

[*Answer*: −48.4 kJ mol^{-1}]

The calculation in Example 3.6 shows that the standard reaction enthalpy at 100°C is only slightly different from that at 25°C. The reason is that the change in reaction enthalpy is proportional to the *difference* between the molar heat capacities of the products and the reactants, which is usually not very large. It is generally the case that, provided the temperature range is not too wide, enthalpies of reactions vary only slightly with temperature. A reasonable first approximation is that standard reaction enthalpies are independent of temperature.

Exercises

Assume all gases are perfect unless stated otherwise. All thermochemical data are for 298.15 K.

3.1 Liquid mixtures of sodium and potassium are used in some nuclear reactors as coolants that can survive the intense radiation inside reactor cores. Calculate the heat required to melt 224 kg of sodium metal at 371 K.

3.2 A primitive air-conditioning unit for use in places where electrical power is unavailable can be made by hanging up strips of linen soaked in water: the evaporation of the water cools the air. Calculate the heat required to evaporate 1.00 kg of water at (a) 25°C, (b) 100°C.

3.3 Isopropanol (2-propanol) is commonly used as 'rubbing alcohol' to relieve sprain injuries in sport: its action is

due to the cooling effect that accompanies its rapid evaporation when applied to the skin. In an experiment to determine its enthalpy of vaporization, a sample was brought to the boil. It was then found that when an electric current of 0.812 A from a 11.5 V supply was passed for 303 s, then 4.27 g of the alcohol was vaporized. What is the (molar) enthalpy of vaporization of isopropanol?

3.4 Refrigerators make use of the heat absorption required to vaporize a volatile liquid. A fluorocarbon liquid being investigated to replace a chlorofluorocarbon has $\Delta_{vap} H^\ominus = +26.0$ kJ mol^{-1}. Calculate q, w, ΔH, and ΔU when 1.50 mol is vaporized at 250 K and 750 Torr.

3.5 Use the information in Tables 2.1 and 3.1 to calculate the total heat required to melt 100 g of ice at 0°C, heat it to

100°C, and then vaporize it at that temperature. Sketch a graph of temperature against time on the assumption that heat is supplied to the sample at a constant rate.

3.6 The enthalpy of sublimation of calcium at 25°C is 178.2 kJ mol^{-1}. How much energy (at constant temperature and pressure) must be supplied to 10.0 g of solid calcium to produce a gas composed of Ca^{2+} ions and electrons?

3.7 The enthalpy changes accompanying the dissociation of successive bonds in $NH_3(g)$ are 460, 390, and 314 kJ mol^{-1}, respectively. (a) What is the mean enthalpy of an N—H bond? (b) Do you expect the mean bond internal energy to be larger or smaller than the mean bond enthalpy?

3.8 Use bond enthalpies and mean bond enthalpies to estimate the (a) the enthalpy of the glycolysis reaction adopted by anaerobic bacteria as a source of energy, $C_6H_{12}O_6(aq) \rightarrow 2\ CH_3CH(OH)COOH(aq)$, lactic acid, which is produced via the formation of pyruvic acid, $CH_3COCOOH$, and the action of lactate dehydrogenase and (b) the enthalpy of combustion of glucose. Ignore the contributions of enthalpies of fusion and vaporization.

3.9 The efficient design of chemical plants depends on the designer's ability to assess and use the heat output in one process to supply another process. The standard enthalpy of reaction for $N_2(g) + 3\ H_2(g) \rightarrow 2\ NH_3(g)$ is -92.22 kJ mol^{-1}. What is the change in enthalpy when (a) 1.00 mol N_2 is consumed, (b) 1.00 mol $NH_3(g)$ is formed?

3.10 Ethane is flamed off in abundance from oil wells, because it is unreactive and difficult to use commercially. But would it make a good fuel? The standard enthalpy of reaction for $2\ C_2H_6(g) + 7\ O_2(g) \rightarrow 4\ CO_2(g) + 6\ H_2O(l)$ is -3120 kJ mol^{-1}. (a) What is the standard enthalpy of combustion of ethane? (b) What is the change in enthalpy when 3.00 mol CO_2 is formed in the reaction?

3.11 Standard enthalpies of formation are widely available, but we might need a standard enthalpy of combustion instead. The standard enthalpy of formation of ethylbenzene is -12.5 kJ mol^{-1}. Calculate its standard enthalpy of combustion.

3.12 Combustion reactions are relatively easy to carry out and study, and their data can be combined to give enthalpies of other types of reaction. As an illustration, calculate the standard enthalpy of hydrogenation of cyclohexene to cyclohexane given that the standard enthalpies of combustion of the two compounds are -3752 kJ mol^{-1} (cyclohexene) and -3953 kJ mol^{-1} (cyclohexane).

3.13 Estimate the standard internal energy of formation of liquid methyl acetate (methyl ethanoate, CH_3COOCH_3) at 298 K from its standard enthalpy of formation, which is -442 kJ mol^{-1}.

3.14 The standard enthalpy of combustion of naphthalene is -5157 kJ mol^{-1}. Calculate its standard enthalpy of formation.

3.15 When 320 mg of naphthalene, $C_{10}H_8(s)$, was burned in a bomb calorimeter, the temperature rose by 3.05 K. Calculate the heat capacity of the calorimeter. By how much will the temperature rise when 100 mg of phenol, $C_6H_5OH(s)$, is burned in the calorimeter under the same conditions?

3.16 The energy resources of glucose are of major concern for the assessment of metabolic processes. When 0.3212 g of glucose was burned in a bomb calorimeter of heat capacity 641 J K^{-1} the temperature rose by 7.793 K. Calculate (a) the standard molar enthalpy of combustion, (b) the standard internal energy of combustion, and (c) the standard enthalpy of formation of glucose.

3.17 The complete combustion of fumaric acid in a bomb calorimeter released 1333 kJ per mole of $HOOCCH=CHCOOH(s)$ at 298 K. Calculate (a) the internal energy of combustion, (b) the enthalpy of combustion, (c) the enthalpy of formation of fumaric acid.

3.18 Calculate the standard enthalpy of solution of $AgBr(s)$ in water from the standard enthalpies of formation of the solid and the aqueous ions.

3.19 The standard enthalpy of decomposition of the yellow complex NH_3SO_2 into NH_3 and SO_2 is $+40$ kJ mol^{-1}. Calculate the standard enthalpy of formation of NH_3SO_2.

3.20 Given that the enthalpy of combustion of graphite is -393.5 kJ mol^{-1} and that of diamond is -395.41 kJ mol^{-1}, calculate the enthalpy of the C(s, graphite) → C(s, diamond) transition.

3.21 The pressures deep within the Earth are much greater than those on the surface, and to make use of thermochemical data in geochemical assessments we need to take the differences into account. Use the information in Exercise 3.20 together with the densities of graphite

$(2.250 \text{ g cm}^{-3})$ and diamond $(3.510 \text{ g cm}^{-3})$ to calculate the internal energy of the transition when the sample is under a pressure of 150 kbar.

3.22 The mass of a typical sugar (sucrose) cube is 1.5 g. Calculate the energy released as heat when a cube is burned in air. To what height could a person of mass 68 kg climb on the energy a cube provides assuming 20 per cent of the energy is available for work?

3.23 Camping gas is typically propane. The standard enthalpy of combustion of propane gas is $-2220 \text{ kJ mol}^{-1}$ and the standard enthalpy of vaporization of the liquid is $+15 \text{ kJ mol}^{-1}$. Calculate (a) the standard enthalpy and (b) the standard internal energy of combustion of the liquid.

3.24 Classify as endothermic or exothermic (a) a combustion reaction for which $\Delta_r H^{\ominus} = -2020 \text{ kJ mol}^{-1}$, (b) a dissolution for which $\Delta H^{\ominus} = +4.0 \text{ kJ mol}^{-1}$, (c) vaporization, (d) fusion, (e) sublimation.

3.25 Standard enthalpies of formation are of great usefulness, for they can be used to calculate the standard enthalpies of a very wide range of reactions of interest in chemistry, biology, geology, and industry. Use the data in Appendix 1 to calculate the standard enthalpies of the following reactions:

(a) $2 NO_2(g) \rightarrow N_2O_4(g)$

(b) $NO_2(g) \rightarrow \frac{1}{2} N_2O_4(g)$

(c) $3 NO_2(g) + H_2O(l) \rightarrow 2 HNO_3(aq) + NO(g)$

(d) $Cyclopropane(g) \rightarrow propene(g)$

(e) $HCl(aq) + NaOH(aq) \rightarrow NaCl(aq) + H_2O(l)$

3.26 Calculate the standard enthalpy of formation of N_2O_5 from the following data:

$2 NO(g) + O_2(g) \rightarrow 2 NO_2(g) \quad \Delta_r H^{\ominus} = -114.1 \text{ kJ mol}^{-1}$

$4 NO_2(g) + O_2(g) \rightarrow 2 N_2O_5(g) \quad \Delta_r H^{\ominus} = -110.2 \text{ kJ mol}^{-1}$

$N_2(g) + O_2(g) \rightarrow 2 NO(g) \quad \Delta_r H^{\ominus} = +180.5 \text{ kJ mol}^{-1}$

3.27 Heat capacity data can be used to estimate the reaction enthalpy at one temperature from its value at another. Use the information in Appendix 1 to predict the standard reaction enthalpy of $2 NO_2(g) \rightarrow N_2O_4(g)$ at 100°C from its value at 25°C.

3.28 It is often useful to be able to anticipate, without doing a detailed calculation, whether an increase in temperature will result in a raising or a lowering of a reaction enthalpy. The constant-pressure molar heat capacity of a gas of linear molecules is approximately $\frac{7}{2}R$ whereas that of a gas of nonlinear molecules is approximately $4R$. Decide whether the standard enthalpies of the following reactions will increase or decrease with increasing temperature:

(a) $2 H_2(g) + O_2(g) \rightarrow 2 H_2O(g)$

(b) $N_2(g) + 3 H_2(g) \rightarrow 2 NH_3(g)$

(c) $CH_4(g) + 2 O_2(g) \rightarrow CO_2(g) + 2 H_2O(g)$

3.29 The molar heat capacity of liquid water is approximately $9R$. Decide whether the standard enthalpy of the reactions (a) and (c) in Exercise 3.28 will increase or decrease with a rise in temperature if the water is produced as a liquid.

3.30 Is the standard enthalpy of combustion of glucose likely to be higher or lower at blood temperature than at 25°C?

Chapter 4

Thermodynamics: the Second Law

Contents

Entropy

The Gibbs energy

Some things happen; some things don't. A gas expands to fill the vessel it occupies; a gas that already fills a vessel does not suddenly contract into a smaller volume. A hot object cools to the temperature of its surroundings; a cool object does not suddenly become hotter than its surroundings. Hydrogen and oxygen combine explosively (once their ability to do so has been liberated by a spark) and form water; water left standing in oceans and lakes does not gradually decompose into hydrogen and oxygen. These everyday observations suggest that changes can be divided into two classes. A **spontaneous change** is a change that can occur without work being done to bring it about. A spontaneous change has a natural tendency to occur. A **non-spontaneous change** is a change that can be brought about only by doing work. A non-spontaneous change has no natural tendency to occur. Non-spontaneous changes can be *made* to occur by doing work: gas can be compressed into a smaller volume by pushing in a piston, the temperature of a cool object can be raised by forcing an electric current through a heater attached to it, and water can be decomposed by the passage of an electric current. However, in each case we need to act in some way on the system to bring the non-spontaneous change about.

There must be some feature of the world that accounts for the distinction between the two types of change. Once we have identified it, we shall be able to apply it to chemistry. Then we shall see why some reactions are spontaneous but others are not. We shall also see—and this is the principal aim of the chapter—how to predict the composition of a reaction mixture that has reached **chemical equilibrium**, when the reaction has no further tendency to form either products or reactants. A reaction at equilibrium is spontaneous in neither direction.

Throughout the chapter we shall use the terms 'spontaneous' and 'non-spontaneous' in their thermodynamic sense. That is, we use them to signify that a change does or does not have a natural *tendency* to occur. In thermodynamics the term spontaneous has nothing to do with speed. Some spontaneous changes are very fast, such as the precipitation reaction that occurs when solutions of sodium chloride and silver nitrate are mixed. However, some spontaneous changes are so slow that there may be no observable change even after millions of years. For example, although the decomposition of benzene into carbon and hydrogen is spontaneous, it does not occur at a measurable rate under normal conditions, and benzene is a common laboratory commodity with a shelf life of (in principle) millions of years. Thermodynamics deals with the tendency to change; it is silent on the rate at which that tendency is realized.

Entropy

A FEW moment's thought is all that is needed to identify the reason why some changes are spontaneous and others are not. That reason is *not* the tendency of the system to move towards lower energy. This point is easily established by identifying an example of a spontaneous change in which there is no change in energy. The isothermal expansion of a perfect gas into a vacuum is spontaneous, but the total energy of the gas does not change because the molecules continue to travel at the same average speed and so keep their same total kinetic energy. Even in a process in which the energy of a system does decrease (as in the spontaneous cooling of a block of hot metal), the First Law requires the total energy to be constant. Therefore, in this case the energy of another part of the world must increase if

the energy decreases in the part that interests us. For instance, a hot block of metal in contact with a cool block cools and loses energy; however, the second block becomes warmer and increases in energy. It is equally as valid to say that the second block moves spontaneously to higher energy as it is to say that the first block has a tendency to go to lower energy!

4.1 The direction of spontaneous change

We shall now show that *the apparent driving force of spontaneous change is the tendency of energy and matter to become disordered*. For example, the molecules of a gas may all be in one region of a container initially, but their ceaseless disorderly motion ensures that they spread rapidly throughout the entire volume of the container (Fig 4.1). Because their motion is so disorderly, there is a negligibly small probability

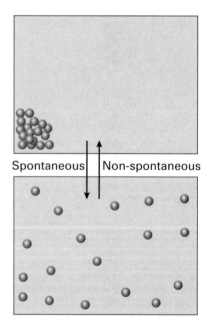

Fig 4.1 One fundamental type of spontaneous process is the chaotic dispersal of matter. This tendency accounts for the spontaneous tendency of a gas to spread into and fill the container it occupies. It is extremely unlikely that all the particles will collect into one small region of the container. (In practice, the number of particles is of the order of 10^{23}.)

that all the molecules will find their way back simultaneously into the region of the container they occupied initially. In this instance, the natural direction of change corresponds to the dispersal of matter.

A similar explanation accounts for spontaneous cooling, but now we need to consider the dispersal of energy rather than matter. In a block of hot metal, the atoms are oscillating vigorously and, the hotter the block, the more vigorous their motion. The cooler surroundings also consist of oscillating atoms, but their motion is less vigorous. The vigorously oscillating atoms of the hot block jostle their neighbours in the surroundings, and the energy of the atoms in the block is handed on to the atoms in the surroundings (Fig 4.2). The process continues until the vigour with which the atoms in the system are oscillating has fallen to that of the surroundings. The opposite flow of energy is very unlikely. It is highly improbable that there will be a net flow of energy into the system as a result of jostling from

less vigorously oscillating molecules in the surroundings. In this case, the natural direction of change corresponds to the dispersal of energy.

In summary, we have identified two basic types of spontaneous physical process:

1 Matter tends to become disordered.
2 Energy tends to become disordered.

We must now see how these two primitive types of physical change result in some chemical reactions being spontaneous and others not.

4.2 Entropy and the Second Law

To make progress, we need to make the discussion quantitative and make measurements. Measurements will help us to keep track of the degree of disorder even though it might be difficult to identify qualitatively. A change in disorder may be hard to identify, for instance, when one substance changes into another in the course of a chemical reaction.

The measure of the disorder of matter and energy used in thermodynamics is called the **entropy**, S. Initially, we can take entropy to be a synonym for the extent of disorder, but shortly we shall see that it can be defined precisely and quantitatively, measured, and then applied to chemical reactions. At this point, all we need know is that, when matter and energy become disordered, the entropy increases. That being so, we can combine the two remarks above into a single statement known as the **Second Law of thermodynamics**:

The entropy of the universe tends to increase.

The remarkable feature of this law is that it accounts for change in all its forms—for chemical reactions as well as the physical changes we have already considered.

To make progress and turn the Second Law into a quantitatively useful statement, we need to define entropy precisely. We shall use the following definition of a *change* in entropy:

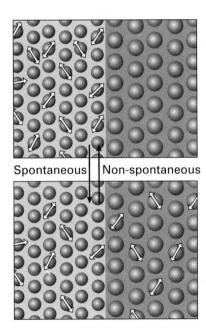

Fig 4.2 Another fundamental type of spontaneous process is the chaotic dispersal of energy (represented by the small arrows). In these diagrams, the small spheres represent the system and the large spheres represent the surroundings. The double-headed arrows represent the thermal motion of the atoms.

Spontaneous | Non-spontaneous

$$\Delta S = \frac{q_{rev}}{T} \qquad (4.1)$$

That is, the change in entropy of a substance is equal to the energy transferred as heat to it *reversibly* divided by the temperature at which the transfer takes place. The formal derivation of this expression is based on a special kind of process called a 'Carnot cycle' that was originally devised to assess the efficiency of steam engines. However, we shall not go into that formal justification. Instead, we shall show that eqn 4.1 is a plausible formula for the change in entropy, and then show how to use it to obtain numerical values for a range of processes.

There are three points we need to understand about the definition in eqn 4.1: the significance of the term 'reversible', why heat (not work) appears in the definition, and why the entropy change depends on the temperature at which the transfer takes place.

We met the concept of reversibility in Section 2.3, where we saw that it refers to the ability of an infinitesimal change in a variable to change the direction of a process. Mechanical reversibility refers to the equality of pressure acting on either side of a movable wall. Thermal reversibility, the type involved in eqn 4.1, refers to the equality of temperature on either side of a thermally conducting wall. Reversible transfer of heat is smooth, careful, restrained transfer between two bodies at the same temperature. By making the transfer reversible we ensure that there are no hot spots generated in the object that later disperse spontaneously and hence add to the entropy.

Now consider why heat and not work appears in eqn 4.1. Recall from Section 2.2 that to transfer energy as heat we make use of the disorderly motion of molecules, whereas to transfer energy as work we make use of orderly motion. It should be plausible that the change in entropy—the change in the degree of disorder—is proportional to the energy transfer that takes place by making use of disorderly motion rather than orderly motion.

Finally, the presence of the temperature in the denominator in eqn 4.1 takes into account the disorder that is already present. If a given quantity of energy is transferred as heat to a hot object (one in which the atoms have a lot of disorderly thermal motion), then the additional disorder generated is less significant than if the same quantity of energy is transferred as heat to a cold object in which the

atoms have less thermal motion. The difference is like sneezing in a busy street and sneezing in a quiet library.

Illustration 4.1

Transferring 100 kJ of heat to a large mass of water at 0°C (273 K) results in a change in entropy of [1]

$$\Delta S = \frac{q_{rev}}{T} = \frac{100 \times 10^3\,\text{J}}{273\,\text{K}} = +366\,\text{J K}^{-1}$$

whereas the same transfer at 100°C (373 K) results in

$$\Delta S = \frac{100 \times 10^3\,\text{J}}{373\,\text{K}} = +268\,\text{J K}^{-1}$$

The change in entropy is greater at the lower temperature. Notice that the units of entropy are joules per kelvin (J K^{-1}). Entropy is an extensive property. When we deal with molar entropy, an intensive property, the units will be joules per kelvin per mole ($\text{J K}^{-1}\,\text{mol}^{-1}$).

Entropy is a state function, a property with a value that depends only on the present state of the system. The entropy is a measure of the current state of disorder of the system, and how that disorder was achieved is not relevant to its current value. A sample of liquid water of mass 100 g at 60°C and 98 kPa has exactly the same degree of molecular disorder—the same entropy—regardless of what has happened to it in the past. The implication of entropy being a state function is that a change in its value when a system undergoes a change of state is independent of how the change of state is brought about.

4.3 Entropy changes for typical processes

We can often rely on intuition to judge whether the entropy increases or decreases when a substance undergoes a physical change. For instance, the entropy of a sample of gas increases as it expands because the molecules get to move in a greater volume and so have a greater degree of disorder. However, the ad-

---

[1] We use a large mass of water to ensure that the temperature of the sample does not change as heat is transferred.

vantage of eqn 4.1 is that it lets us express the increase *quantitatively* and make numerical calculations. For instance, we can use the definition to calculate the change in entropy when a perfect gas expands isothermally from a volume V_i to a volume V_f.

Derivation 4.1 *The variation of entropy with volume*

To calculate the entropy change from eqn 4.1, we need to know q_{rev}, the energy transferred as heat in the course of a *reversible* change at the temperature T. We saw in Section 2.4 (eqn 2.7) that the heat transferred to a perfect gas when it undergoes reversible isothermal expansion from a volume V_i to a volume V_f at a temperature T is

$$q_{rev} = nRT \ln \frac{V_f}{V_i}$$

It follows that

$$\Delta S = \underbrace{\frac{q_{rev}}{T}}_{\text{Definition}} = \underbrace{nR \ln \frac{V_f}{V_i}}_{\substack{\text{Isothermal,} \\ \text{reversible,} \\ \text{perfect gas}}} \qquad (4.2)$$

We have already stressed the importance of reading equations for their physical content. In this case we see that, if $V_f > V_i$, as in an expansion, then $V_f/V_i > 1$ and the logarithm is positive. Consequently, eqn 4.2 predicts a positive value for ΔS, corresponding to an increase in entropy, just as we anticipated (Fig 4.3). Perhaps surprisingly, though, the equation shows that the change in entropy is independent of the temperature at which expansion occurs. The explanation is that more work is done if the temperature is high (because the external pressure must be matched to a higher value of the pressure of the gas), so more heat must be supplied to maintain the temperature. The temperature in the denominator of eqn 4.1 is higher, but the 'sneeze' (in terms of the analogy introduced earlier) is greater too, and the two effects cancel.

Self-test 4.1

Calculate the change in molar entropy when a sample of hydrogen gas expands isothermally to twice its initial volume.

[*Answer:* $+5.8\ \text{J K}^{-1}\ \text{mol}^{-1}$]

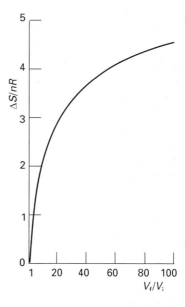

Fig 4.3 The entropy of a perfect gas increases logarithmically (as ln V) as the volume is increased.

The second type of change we consider is raising the temperature. We should expect the entropy of a sample to increase as the temperature is raised, because the thermal disorder of the system is greater at the higher temperature, when the molecules move more vigorously. Once again, to calculate the value of the change in entropy, we go back to the definition in eqn 4.1.

Derivation 4.2 *The variation of entropy with temperature*

Equation 4.1 refers to the transfer of heat to a system at a temperature T. In general, the temperature changes as we heat a system, so we cannot use eqn 4.1 directly. Suppose, however, that we transfer only an infinitesimal energy, dq, to the system, then there is only an infinitesimal change in temperature and we introduce negligible error if we keep the temperature in the denominator of eqn 4.1 equal to T during that transfer. As a result, the entropy increases by an infinitesimal amount dS given by

$$dS = \frac{dq_{rev}}{T}$$

Although we can neglect the change in temperature in the denominator of eqn 4.1, the transfer of heat to a system at constant volume does result in an infinitesimal increase in temperature dT, where $dq = C_V dT$, with C_V the heat capacity at constant volume (Section 2.4). This relation also applies when the transfer is carried out reversibly, so

$$dS = \frac{C_V dT}{T}$$

The total change in entropy, ΔS, when the temperature changes from T_i to T_f is the sum (integral) of all such infinitesimal terms:

$$\Delta S = \int_{T_i}^{T_f} \frac{C_V dT}{T} \tag{4.3}$$

Now recall that the integral of a function $f(x)$ between two values of x, is the area under the curve of $f(x)$ against x between the two limits. We see that to determine an entropy change experimentally, we need to measure the area under a plot of C_V/T between the two temperatures (Fig 4.4).

For many substances and for small temperature ranges we may take C_V to be constant. That is strictly true for a monatomic perfect gas. Then C_V may be taken outside the integral, and the latter evaluated as follows:

$$\Delta S = \int_{T_i}^{T_f} \frac{C_V dT}{T} \overbrace{=}^{\text{Constant heat capacity}} C_V \int_{T_i}^{T_f} \frac{dT}{T} = C_V \ln\frac{T_f}{T_i} \tag{4.4}$$

(We have used the same standard integral as in Derivation 2.2.) The same expression, except for the replacement of C_V by C_p, is obtained when the heating takes place at constant pressure.

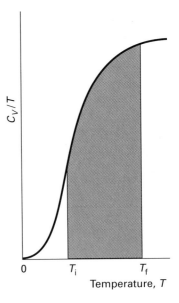

Fig 4.4 The experimental determination of the change in entropy of a sample that has a heat capacity that varies with temperature involves measuring the heat capacity over the range of temperatures of interest, then plotting C_V/T against T and determining the area under the curve (the tinted area shown here). The heat capacity of all solids decreases towards zero as the temperature is reduced.

Self-test 4.2

Calculate the change in molar entropy when hydrogen gas is heated from 20°C to 30°C at constant volume. ($C_{V,m} = 22.44\ \text{J K}^{-1}\ \text{mol}^{-1}$.)

[*Answer:* $+0.75\ \text{J K}^{-1}\ \text{mol}^{-1}$]

Equation 4.4 is in line with what we expect. When $T_f > T_i$, $T_f/T_i > 1$, which implies that the logarithm is positive, that $\Delta S > 0$, and therefore that the entropy increases (Fig 4.5). Note that the relation also shows a less obvious point, that the higher the heat capacity of the substance is, the greater the change in entropy for a given rise in temperature. A moment's thought shows that that is reasonable too. A high heat capacity implies that a lot of heat is required to produce a given change in temperature, so the 'sneeze' must be more powerful than for the case when the heat capacity is low, and the entropy increase is correspondingly high.

The third common process to consider is a phase transition, as in melting or boiling. We can suspect that the entropy of a substance increases when it melts and when it boils because its molecules become more disordered as it changes from solid to liquid and from liquid to vapour.

Suppose a solid is at its melting temperature. Transfer of energy as heat then occurs reversibly. If the temperature of the surroundings is infinitesimally lower than that of the system, then heat flows out of the system and the substance freezes. If the temperature is infinitesimally higher, then heat flows into the system and the substance melts. Moreover, because the transition occurs at constant pressure, we can identify the heat transferred per

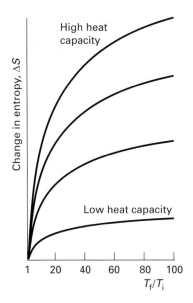

Fig 4.5 The entropy of a sample with a heat capacity that is independent of temperature, such as a monatomic perfect gas, increases logarithmically (as $\ln T$) as the temperature is increased. The increase is proportional to the heat capacity of the sample.

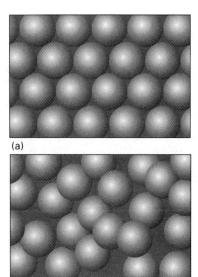

Fig 4.6 When a solid, depicted by the orderly array of spheres (a), melts, the molecules form a more chaotic liquid, the disorderly array of spheres (b). As a result, the entropy of the sample increases.

mole of substance with the enthalpy of fusion (melting). Therefore, the **entropy of fusion**, $\Delta_{fus}S$, the change of entropy per mole of substance, at the melting temperature, T_f, is

At the melting temperature: $\Delta_{fus} S = \dfrac{\Delta_{fus} H}{T_f}$ (4.5)

All enthalpies of fusion are positive (melting is endothermic: it requires heat), so all entropies of fusion are positive too: disorder increases on melting. The entropy of water, for example, increases when it melts because the orderly structure of ice collapses as the liquid forms (Fig 4.6).

Self-test 4.3

Calculate the entropy of fusion of ice at 0°C from the information in Table 3.1.

[*Answer:* +22 J K^{-1} mol^{-1}]

The entropy of other types of transition may be discussed similarly. Thus, the **entropy of vaporization**, $\Delta_{vap}S$, at the boiling temperature, T_b, of a liquid is related to its enthalpy of vaporization at that temperature by

At the boiling temperature: $\Delta_{vap} S = \dfrac{\Delta_{vap} H}{T_b}$ (4.6)

Because vaporization is endothermic for all substances, all entropies of vaporization are positive. The increase in entropy on vaporization is in line with what we should expect when a compact liquid turns into a gas.

Self-test 4.4

Calculate the entropy of vaporization of water at 100°C.

[*Answer:* +109 J K^{-1} mol^{-1}]

Entropies of vaporization shed light on an empirical relation known as **Trouton's rule**. Trouton noticed that $\Delta_{vap}H/T_b$ is approximately the same (and equal to about 85 J K^{-1} mol^{-1}) for all liquids except when hydrogen bonding or some other kind of specific bonding is present (see Table 4.1). We know that the quantity $\Delta_{vap}H/T_b$, however, is the entropy of vaporization of the liquid at its boiling point, so Trouton's rule is explained if all liquids have approximately the same entropy of vaporization. This near equality is to be expected because, when a

Table 4.1 *Entropies of vaporization at 1 atm and the normal boiling point*

	$\Delta_{vap}S/(J\ K^{-1}\ mol^{-1})$
Bromine, Br_2	88.6
Benzene, C_6H_6	87.2
Carbon tetrachloride, CCl_4	85.9
Cyclohexane, C_6H_{12}	85.1
Hydrogen sulfide, H_2S	87.9
Ammonia, NH_3	97.4
Water, H_2O	109.1
Mercury	94.2

liquid vaporizes, the compact condensed phase changes into a widely dispersed gas that occupies approximately the same volume whatever its identity. To a good approximation, therefore, we expect the increase in disorder, and therefore the entropy of vaporization, to be almost the same for all liquids at their boiling temperatures.

Illustration 4.2

We can estimate the enthalpy of vaporization of liquid bromine from its boiling temperature, 59.2°C. No hydrogen bonding or other kind of special interaction is present, so we use Trouton's rule after converting the boiling point to 332.4 K:

$$\Delta_{vap}H \approx (332.4\ \text{K}) \times (85\ \text{J K}^{-1}\ \text{mol}^{-1}) = 28\ \text{kJ mol}^{-1}$$

The experimental value is 29 kJ mol^{-1}.

Self-test 4.5

Estimate the enthalpy of vaporization of ethane from its boiling point, which is −88.6°C.

[*Answer*: 16 kJ mol^{-1}]

The exceptions to Trouton's rule include liquids in which the interactions between molecules result in the liquid being less disordered than a random jumble of molecules. For example, the high value for water implies that the H_2O molecules are linked together in some kind of ordered structure by hydrogen bonding (see Section 16.6), with the result that the entropy change is greater when this rela-

tively ordered liquid forms a disordered gas. The high value for mercury has a similar explanation but stems from the presence of metallic bonding in the liquid, which organizes the atoms into more definite patterns than would be the case if such bonding were absent.

4.4 Entropy changes in the surroundings

We can use the definition of entropy in eqn 4.1 to calculate the entropy change of the surroundings in contact with the system at the temperature T:

$$\Delta S_{sur} = \frac{q_{sur,rev}}{T}$$

The surroundings are so extensive that, by the time the heat supplied has spread throughout them, the transfer of heat is effectively reversible and we can write[2]

$$\Delta S_{sur} = \frac{q_{sur}}{T} \tag{4.7}$$

We can use this formula to calculate the entropy change of the surroundings regardless of whether the change in the system is reversible or not.

Example 4.1 *Estimating the entropy change of the surroundings*

A typical resting person generates about 100 W of heat. Estimate the entropy they generate in the surroundings in the course of a day at 20°C.

Strategy We can estimate the approximate change in entropy from eqn 4.7 once we have calculated the energy transferred as heat. To find this quantity, we use the facts that 1 W = 1 J s^{-1} and there are 86 400 s in a day. Convert the temperature to kelvins.

Solution The heat transferred to the surroundings in the course of a day is

$$q_{sur} = (86\ 400\ \text{s}) \times (100\ \text{J s}^{-1}) = 86\ 400 \times 100\ \text{J}$$

---

[2] More formally, the surroundings remain at constant pressure regardless of any events taking place in the system, so $q_{sur,rev} = \Delta H_{sur}$. The enthalpy is a state function, so a change in its value is independent of the path and we get the same value of ΔH_{sur} regardless of how the heat is transferred. Therefore, we can drop the label 'rev' from q.

The increase in entropy of the surroundings is therefore

$$\Delta S_{sur} = \frac{q_{sur}}{T} = \frac{86400 \times 100 \text{ J}}{293 \text{ K}} = +2.95 \times 10^4 \text{ J K}^{-1}$$

That is, the entropy production is about 30 kJ K^{-1}. Just to stay alive, each person on the planet contributes about 30 kJ K^{-1} each day to the ever-increasing entropy of their surroundings. The use of transport, machinery, and communications generates far more in addition.

Self-test 4.6

Suppose a small reptile operates at 0.50 W. What entropy does it generate in the course of a day in the water in the lake that it inhabits, where the temperature is 15°C?

[*Answer:* +150 J K^{-1}]

Equation 4.7 is expressed in terms of the heat supplied to the surroundings, q_{sur}. Normally, we have information about the heat supplied to or escaping from the system, q. The two quantities are related by $q_{sur} = -q$. (For instance, if $q = +100$ J, then $q_{sur} = -100$ J.) Therefore, at this stage we can write

$$\Delta S_{sur} = -\frac{q}{T} \qquad (4.8)$$

This expression is in terms of the properties of the system. Moreover, it applies whether or not the process taking place in the system is reversible.

As an illustration, suppose a perfect gas expands isothermally and reversibly from V_i to V_f. The entropy change of the gas itself (the system) is given by eqn 4.2. To calculate the entropy change in the surroundings, we note that q, the heat required to keep the temperature constant, is given in Derivation 4.1. Therefore,

$$\Delta S_{sur} = -\frac{q}{T} = -nR \ln\frac{V_f}{V_i}$$

The change of entropy in the surroundings is therefore the negative of the change in entropy of the system, and the total entropy change is zero:

$$\Delta S_{total} = \Delta S + \Delta S_{sur} = 0$$

Now suppose that the gas expands isothermally but freely ($p_{ex} = 0$) between the same two volumes. The change in entropy of the system is the same, because entropy is a state function. However, because no work is done, no heat is taken in from the surroundings. Because $q = 0$, it follows from eqn 4.8 (which, remember, can be used for either reversible or irreversible heat transfers), that $\Delta S_{sur} = 0$. The total change in entropy is therefore equal to the change in entropy of the system, which is positive. We see that, for this irreversible process, the entropy of the universe has increased, in accord with the Second Law.

Finally, suppose the process taking place in the system is at constant pressure, such as a chemical reaction or a phase transition. Then we can identify q with the change in enthalpy of the system, and obtain

For a process at constant pressure: $\Delta S_{sur} = -\frac{\Delta H}{T}$ (4.9)

This enormously important expression will lie at the heart of our discussion of chemical equilibria. We see that it is consistent with common sense: if the process is exothermic, ΔH is negative and therefore ΔS_{sur} is positive. The entropy of the surroundings increases if heat is released into them. If the process is endothermic ($\Delta H > 0$), then the entropy of the surroundings decreases.

4.5 Absolute entropies and the Third Law of thermodynamics

The graphical procedure summarized by Fig 4.4 and eqn 4.3 for the determination of the difference in entropy of a substance at two temperatures has a very important application. If $T_i = 0$, then the area under the graph between $T = 0$ and some temperature T gives us the value of $\Delta S = S(T) - S(0)$.[3] However, at $T = 0$, all the motion of the atoms has been eliminated, and there is no thermal disorder. Moreover,

---

[3] We are supposing that there are no phase transitions below the temperature T. If there are any phase transitions (for example, melting) in the temperature range of interest, then the entropy of each transition at the transition temperature is calculated like that in eqn 4.5.

if the substance is perfectly crystalline, with every atom in a well-defined location, then there is no spatial disorder either. We can therefore suspect that, at $T = 0$, the entropy is zero.

The thermodynamic evidence for this conclusion is as follows. Sulfur undergoes a phase transition from rhombic to monoclinic at 96°C (369 K) and the enthalpy of transition is +402 J mol^{-1}. The entropy of transition is therefore +1.09 J K^{-1} mol^{-1} at this temperature. We can also measure the molar entropy of each phase relative to its value at $T = 0$ by determining the heat capacity from $T = 0$ up to the transition temperature (Fig 4.7). At this stage, we do not know the values of the entropies at $T = 0$. However, as we see from the illustration, to match the observed entropy of transition at 369 K, the molar entropies of the two crystalline forms must be the same at $T = 0$. We cannot say that the entropies are zero at $T = 0$, but from the experimental data we do know that they are the same. This observation is generalized into the **Third Law of thermodynamics**:

The entropies of all perfectly crystalline substances are the same at $T = 0$.

For convenience (and in accord with our understanding of entropy as a measure of disorder), we take this common value to be zero. Then, with this convention, according to the Third Law, $S(0) = 0$ for all perfectly ordered crystalline materials.

The **Third-Law entropy** at any temperature, $S(T)$, is equal to the area under the graph of C_V/T between $T = 0$ and the temperature T (Fig 4.8). Its value, which is commonly called simply 'the entropy', of a substance depends on the pressure. A high pressure, for instance, would confine a gas to a smaller

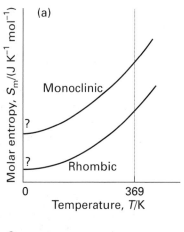

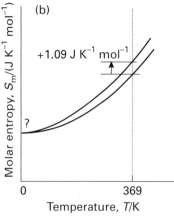

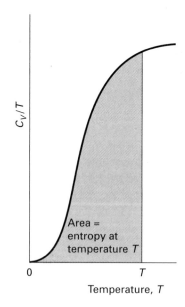

Fig 4.7 (a) The molar entropies of monoclinic and rhombic sulfur vary with temperature as shown here. At this stage we do not know their values at $T = 0$. (b) When we slide the two curves together by matching their separation to the measured entropy of transition at the transition temperature, we find that the entropies of the two forms are the same at $T = 0$.

Fig 4.8 The absolute entropy (or Third-Law entropy) of a substance is calculated by extending the measurement of heat capacities down to $T = 0$ (or as close to that value as possible), and then determining the area of the graph of C_V/T against T up to the temperature of interest. The area is equal to the absolute entropy at the temperature T.

volume and so reduce its entropy. We therefore select a standard pressure and report the **standard molar entropy**, $S_m^{\ominus}$, the molar entropy of a substance in its standard state (as specified in Section 3.3: pure, 1 bar) at the temperature of interest. Some values at 25°C (the conventional temperature for reporting data) are given in Table 4.2.

It is worth spending a moment to look at the values in Table 4.2 to see that they are consistent with our understanding of entropy. All standard molar entropies are positive, because raising the temperature of a sample above $T = 0$ invariably increases its entropy above the value $S(0) = 0$. Another feature is that the standard molar entropy of diamond (2.4 J K^{-1} mol^{-1}) is lower than that of graphite (5.70 J K^{-1}mol^{-1}). This difference is consistent with the atoms being linked less rigidly in graphite than in diamond and their thermal disorder being greater. The standard molar entropies of ice, water, and water vapour at 25°C are, respectively, 45, 70, and 189 J K^{-1} mol^{-1}, and the increase in values corresponds to the increasing disorder on going from a solid to a liquid to a gas.

Heat capacities can be measured only with great difficulty at very low temperatures, particularly close to $T = 0$. However, it has been found that many nonmetallic substances have a heat capacity that obeys the **Debye T^3-law**:

$$\text{At temperatures close to } T = 0, \; C_{V,m} = aT^3 \qquad (4.10)$$

where a is an empirical constant that depends on the substance and is found by fitting eqn 4.10 to a series of measurements close to $T = 0$. With a determined, it is easy to deduce the molar entropy at low temperatures.

Table 4.2 *Standard molar entropies of some substances at 25°C*

Substance	$S_m^{\ominus}$ /(J K^{-1} mol^{-1})
Gases	
Ammonia, NH$_3$	192.5
Carbon dioxide, CO$_2$	213.7
Helium, He	126.2
Hydrogen, H$_2$	130.7
Neon, Ne	146.3
Nitrogen, N$_2$	191.6
Oxygen, O$_2$	205.1
Water vapour, H$_2$O	188.8
Liquids	
Benzene, C$_6$H$_6$	173.3
Ethanol, CH$_3$CH$_2$OH	160.7
Water, H$_2$O	69.9
Solids	
Calcium oxide, CaO	39.8
Calcium carbonate, CaCO$_3$	92.9
Copper, Cu	33.2
Diamond, C	2.4
Graphite, C	5.7
Lead, Pb	64.8
Magnesium carbonate, MgCO$_3$	65.7
Magnesium oxide, MgO	26.9
Sodium chloride, NaCl	72.1
Sucrose, C$_{12}$H$_{22}$O$_{11}$	360.2
Tin, Sn (white)	51.6
Sn (grey)	44.1

See Appendix 1 for more values.

Derivation 4.3 *Entropies close to T = 0*

Suppose we have determined the constant a for a substance; then we can combine eqns 4.3 and 4.10 and obtain

$$S_m(T) - \underset{\substack{\text{0, by the}\\\text{Third Law}}}{\underline{S_m(0)}} = \int_0^T \frac{C_{V,m}\,dT}{T} \overset{\substack{\text{Debye}\\\text{law}}}{=} \int_0^T \frac{aT^3\,dT}{T}$$

$$= a\int_0^T T^2\,dT = \tfrac{1}{3}aT^3 \overset{\substack{\text{Debye}\\\text{law}}}{=} \underset{\substack{\text{Heat capacity}\\\text{at the temperature } T}}{\underline{\tfrac{1}{3}C_{V,m}(T)}}$$

The standard integral we have used is

$$\int x^n\,dx = \frac{x^{n+1}}{n+1} + \text{constant}$$

This calculation tells us that, to measure the molar entropy at a temperature T, $S_m(T)$, all we need do is to determine the molar heat capacity at that temperature and divide by 3. This conclusion is valid only for temperatures close to $T = 0$, where the Debye law is valid.

4.6 **The standard reaction entropy**

Now we move into the arena of chemistry, where re-actants are transformed into products. When there is net formation of a gas in a reaction, as in a combustion, we can usually anticipate that the entropy increases. When there is net consumption of gas, as in photosynthesis, it is usually safe to predict that the entropy decreases. However, for a quantitative value of the change in entropy, and to predict the sign of the change when no gases are involved, we need to carry out an explicit calculation.

The difference in molar entropy between the products and the reactants in their standard states is called the **standard reaction entropy**, $\Delta_r S^{\ominus}$. It can be expressed in terms of the molar entropies of the substances in much the same way as that we have already used for the standard reaction enthalpy:

$$\Delta_r S^{\ominus} = \sum v S_m^{\ominus}(\text{products}) - \sum v S_m^{\ominus}(\text{reactants})$$

(4.11)

where the vs are the stoichiometric coefficients in the chemical equation.

Illustration 4.3

For the reaction $2\,H_2(g) + O_2(g) \rightarrow 2\,H_2O(l)$ we expect a negative entropy of reaction as gases are consumed. To find the explicit value we use the values in Appendix 1 to write

$\Delta S^{\ominus} = 2 S_m^{\ominus}(H_2O, l) - \{2 S_m^{\ominus}(H_2, g) + S_m^{\ominus}(O_2, g)\}$

$= 2 \times (70\ J\ K^{-1}\ mol^{-1})$

$\quad - \{2 \times (131\ J\ K^{-1}\ mol^{-1}) + (205\ J\ K^{-1}\ mol^{-1})\}$

$= -327\ J\ K^{-1}\ mol^{-1}$

Do not make the mistake of setting the standard molar entropies of elements equal to zero: they have nonzero values (provided $T > 0$), as we have already discussed.

Self-test 4.7

(a) Calculate the standard reaction entropy for $N_2(g) + 3\,H_2(g) \rightarrow 2\,NH_3(g)$ at 25°C. (b) What is the change in entropy when 2 mol H_2 reacts?

[*Answer*: (a) $-198.76\ J\ K^{-1}\ mol^{-1}$; (b) $-132.51\ J\ K^{-1}$]

4.7 **The spontaneity of chemical reactions**

The result of the H_2O calculation in Illustration 4.3 should be rather surprising at first sight. We know that the reaction between hydrogen and oxygen is spontaneous and that, once initiated, it proceeds with explosive violence. Nevertheless, the entropy change that accompanies it is negative: the reaction results in less disorder, yet it is spontaneous!

The resolution of this apparent paradox under-scores a feature of entropy that recurs throughout chemistry: it is essential to consider the entropy of both the system and its surroundings when deciding whether a process is spontaneous or not. The reduction in entropy by $327\ J\ K^{-1}\ mol^{-1}$ relates only to the system, the reaction mixture. To apply the Second Law correctly, we need to calculate the total entropy, the sum of the changes in the system and the surroundings. It may well be the case that the entropy of the system decreases when a change takes place, but there may be a more than compensating increase in entropy of the surroundings so that overall the entropy change is positive. The opposite may also be true: a large decrease in entropy of the surroundings may occur when the entropy of the system increases. In that case we would be wrong to conclude from the increase of the system alone that the change is spontaneous. *Whenever considering the implications of entropy, we must always consider the total change of the system and its surroundings* (Box 4.1).

To calculate the entropy change in the surroundings when a reaction takes place at constant pressure, we use eqn 4.9, interpreting the ΔH in that expression as the reaction enthalpy. For example, for the water formation reaction in Illustration 4.3, with $\Delta_r H^{\ominus} = +572\ kJ\ mol^{-1}$, the change in entropy of the surroundings (which are maintained at 25°C, the same temperature as the reaction mixture) is

$$\Delta_r S_{sur} = -\frac{\Delta_r H}{T} = -\frac{(-572\ kJ\ mol^{-1})}{298\ K}$$
$$= +1.92 \times 10^3\ J\ K^{-1}\ mol^{-1}$$

Now we can see that the total entropy change is positive:

$$\Delta_r S_{total} = (-327\ J\ K^{-1}\ mol^{-1}) + (1.92 \times 10^3\ J\ K^{-1}\ mol^{-1})$$
$$= +1.59 \times 10^3\ J\ K^{-1}\ mol^{-1}$$

Box 4.1 *The hydrophobic effect*

Many of the side chains of the amino acids that are used to form the polypeptide chains of proteins are hydrophobic. That is, they have a tendency to avoid water in their immediate environment and to cluster together. This effect is a major contribution to the tertiary structure of polypeptides (the way the helical and sheet-like parts of the chain stack together to form, for instance, a globular unit). But what is meant by the tendency of a hydrocarbon group to avoid water?

Whenever we think about a tendency for an event to occur, we have to consider the total change in entropy of the system and its surroundings, not the system alone. The clustering together of hydrophobic groups results in a negative contribution to the change in entropy of the system because the clustering corresponds to a decrease in the disorder of the system. This tendency favours a random coil over a well-organized arrangement of the peptide residues. However, we must not forget the role of the solvent and the effect on the entropy of its reorganization when a random coil adopts a specific structure.

Consider two hydrocarbon groups, –R, surrounded by water. To accommodate the two hydrophobic groups, the water molecules need to adjust their hydrogen-bonded arrangement and to form a cavity (see the illustration). The net effect is the reduction in the disorder of the system as the water molecules form a cage around the hydrocarbons. Measurements on small hydrocarbon molecules (such as propane) suggest that the change in entropy is about -80 J K^{-1} mol^{-1} at room temperature. Now consider the arrangement when the two hydrocarbon groups are very close together. They occupy a single bigger cavity, but fewer water molecules need to become organized into the cage-like structure, which has a lower surface area than the sum of the surface areas of the original two cavities, so there is a net increase in entropy when the two groups come together. If we disregard any enthalpy effects, there is a thermodynamic tendency for the hydrophobic groups to move together.

One consequence of the hydrophobic effect is that lowering the temperature of the system favours a more disorganized arrangement. To see why, we have to think about the entropy change in the surroundings. For a

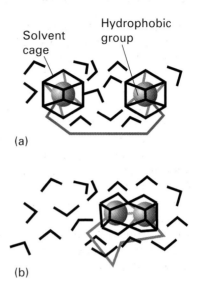

(a)

(b)

(a) When two hydrophobic groups (the spheres) are well apart, the water molecules form two structured cages around them. (b) When the two groups come together, they can be accommodated in a single cage. The entropy of the solvent is greater in the latter case as it has less structure overall.

given transfer of heat into them, the change in their entropy increases as the temperature is decreased (eqn 4.8). Therefore, the entropy changes in the system become relatively less important, the system tends to change in its exothermic direction (the direction corresponding to an increase in entropy of the surroundings), and the hydrophobic effect becomes less important. This is the reason why some proteins dissociate into their individual subunits as the temperature is lowered to 0°C.

Exercise 1 Explain the statement 'There is no such thing as a hydrophobic bond'.

Exercise 2 Two long-chain hydrophobic polypeptides can associate end-to-end so that only the ends meet, or side-to-side so that the entire chains are in contact. Which arrangement would produce a larger entropy change when they come together?

This calculation confirms, as we know from experience, that the reaction is strongly spontaneous. In this case, the spontaneity is a result of the considerable disorder that the reaction generates in the surroundings: water is dragged into existence, even though $H_2O(l)$ has a lower entropy than the gaseous reactants, by the tendency of energy to disperse into the surroundings.

The Gibbs energy

O NE of the problems with entropy calculations is already apparent: we have to work out two entropy changes, the change in the system and the change in the surroundings, and then consider the sign of their sum. The great American theoretician J.W. Gibbs (1839–1903), who laid the foundations of chemical thermodynamics towards the end of the nineteenth century, discovered how to combine the two calculations into one. The combination of the two procedures in fact turns out to be of much greater relevance than just saving a little labour, and throughout this text we shall see consequences of the procedure he developed.

4.8 Focusing on the system

The total entropy change that accompanies a process is

$$\Delta S_{total} = \Delta S + \Delta S_{sur}$$

where ΔS is the entropy change for the system; for a spontaneous change, $\Delta S_{total} > 0$. If the process occurs at constant pressure and temperature, we can use eqn 4.9 to express the change in entropy of the surroundings in terms of the enthalpy change of the system, ΔH. When the resulting expression is inserted into this one, we obtain

At constant temperature and pressure:

$$\Delta S_{total} = \Delta S - \frac{\Delta H}{T} \qquad (4.12)$$

The great advantage of this formula is that it expresses the total entropy change of the system and its surroundings in terms of properties of the system alone. The only restriction is to changes at constant pressure and temperature.

Now we take a very important step. First, we introduce the **Gibbs energy**, G, which is defined as[4]

$$G = H - TS \qquad (4.13)$$

Because H, T, and S are state functions, G is a state function too. A change in Gibbs energy, ΔG, at constant temperature arises from changes in enthalpy and entropy, and is

At constant temperature: $\Delta G = \Delta H - T\Delta S$ (4.14)

By comparing eqns 4.12 and 4.14 we see that

At constant temperature and pressure: $\Delta G = -T\Delta S_{total}$
$$(4.15)$$

We see that, at constant temperature and pressure, the change in Gibbs energy of a system is proportional to the overall change in entropy of the system plus its surroundings.

4.9 Properties of the Gibbs energy

The difference in sign between ΔG and ΔS_{total} implies that the condition for a process being spontaneous changes from $\Delta S_{total} > 0$ in terms of the total entropy (which is universally true) to $\Delta G < 0$ in terms of the Gibbs energy (for processes occurring at constant temperature and pressure). That is, *in a spontaneous change at constant temperature and pressure, the Gibbs energy decreases* (Fig 4.9).

It may seem more natural to think of a system as falling to a lower value of some property. However, it must never be forgotten that to say that a system tends to fall toward lower Gibbs energy is only a modified way of saying that a system and its surroundings jointly tend towards a greater total entropy. The only criterion of spontaneous change is the total entropy of the system and its surroundings; the Gibbs energy merely contrives a way of expressing that total change in terms of the properties

----

[4] The Gibbs energy is commonly referred to as the 'free energy'.

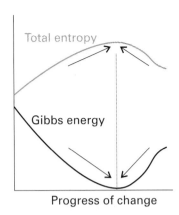

Fig 4.9 The criterion of spontaneous change is the increase in total entropy of the system and its surroundings. Provided we accept the limitation of working at constant pressure and temperature, we can focus entirely on properties of the system, and express the criterion as a tendency to move to lower Gibbs energy.

of the system alone, and is valid only for processes that occur at constant temperature and pressure. All chemical reactions that are spontaneous under conditions of constant temperature and pressure, including those that drive the processes of growth, learning, and reproduction, are reactions that change in the direction of lower Gibbs energy.

A second feature of the Gibbs energy is that *the value of ΔG for a process gives the maximum amount of non-expansion work that can be extracted from the process at constant temperature and pressure*. By **non-expansion work**, w', we mean any work other than that arising from the expansion of the system. It may include electrical work, if the process takes place inside an electrochemical or biological cell, or other kinds of mechanical work, such as the winding of a spring or the contraction of a muscle. To demonstrate this property, we need to combine the First and Second Laws.

Derivation 4.4 *Maximum non-expansion work*

We need to consider infinitesimal changes because dealing with reversible processes is then much easier. Our aim is to derive the relation between the infinitesimal change in Gibbs energy, dG, accompanying a process and the maximum amount of non-expansion work that the process can do, dw'. We start with the infinitesimal form of eqn 4.14,

At constant temperature: $dG = dH - TdS$

where, as usual, d denotes an infinitesimal difference. A good rule for the manipulation of thermodynamic expressions is to feed in definitions of the terms that appear. We do this twice. First, we use the expression for the change in enthalpy at constant pressure (eqn 2.14; $dH = dU + pdV$), and obtain

At constant temperature and pressure:
$$dG = dU + pdV - TdS$$

Then we replace dU in terms of infinitesimal contributions from work and heat ($dU = dw + dq$):

$$dG = dw + dq + pdV - TdS$$

The work done on the system consists of expansion work, $-p_{ex}dV$, and non-expansion work, dw'. Therefore,

$$dG = -p_{ex}dV + dw' + dq + pdV - TdS$$

This derivation is valid for any process taking place at constant temperature and pressure.

Now we specialize to a reversible change. For expansion work to be reversible, we need to match p and p_{ex}, in which case the first and fourth terms on the right cancel. Moreover, because the heat transfer is also reversible, we can replace dq by TdS, in which case the third and fifth terms also cancel. We are left with

At constant temperature and pressure, for a reversible process: $dG = dw'_{rev}$

Maximum work is done during a reversible change, so another way of writing this expression is

At constant temperature and pressure: $dG = dw'_{max}$

$$(4.16)$$

Because this relation holds for each infinitesimal step between the specified initial and final states, it applies to the overall change too. Therefore, we can write

At constant temperature and pressure: $\Delta G = w'_{max}$

$$(4.17)$$

Illustration 4.4

Experiments show that for the formation of 1 mol $H_2O(l)$ at 25°C and 1 bar, $\Delta G = -237$ kJ, so up to 237 kJ of non-expansion work can be extracted from the reaction between hydrogen and oxygen to produce 1 mol $H_2O(l)$ at 25°C. If the reaction takes place in a fuel cell—a device for using a chemical reaction to produce an electric current, like those used on the space shuttle—then up to 237 kJ of electrical energy can be generated for each mole of H_2O produced. If no attempt is made to extract any energy as work, then 286 kJ (in general, $-\Delta H$) of heat will be produced. If some of the energy released is used to do work, then up to 237 kJ (in general, $-\Delta G$) of non-expansion work can be obtained.

The great importance of the Gibbs energy in chemistry should be beginning to become apparent. At this stage, we see that it is a measure of the non-expansion work resources of chemical reactions: if we know ΔG, then we know the maximum non-expansion work that we can do by harnessing the reaction in some way. In some cases, the non-expansion work is extracted as electrical energy. This is the case when the reaction takes place in an electrochemical cell, of which a fuel cell is a special case, as we see in Chapter 9. In other cases, the reaction may be used to build other molecules. This is the case in biological cells, where the Gibbs energy available from the hydrolysis of ATP (adenosine triphosphate) to ADP is used to build proteins from amino acids, to power muscular contraction, and to drive the neuronal circuits in our brains.

Example 4.2 *Estimating a change in Gibbs energy*

Suppose a certain small bird has a mass of 30 g. What is the minimum mass of glucose that it must consume to fly to a branch 10 m above the ground? The change in Gibbs energy that accompanies the oxidation of 1.0 mol $C_6H_{12}O_6(s)$ to carbon dioxide and water vapour at 25°C is −2828 kJ.

Strategy First, we need to calculate the work needed to raise a mass m through a height h on the surface of the Earth. As we saw in eqn 2.1, this work is equal to mgh, where g is the acceleration of free fall. This work, which is non-expansion work, can be identified with ΔG. We need to determine the amount of substance that corresponds to the required change in Gibbs energy, and then convert that amount to a mass by using the molar mass of glucose.

Solution The work to be done is

$$w' = (30 \times 10^{-3} \text{ kg}) \times (9.81 \text{ m s}^{-2}) \times (10 \text{ m})$$
$$= 3.0 \times 9.81 \times 1.0 \times 10^{-1} \text{ J}$$

(because 1 kg m^2 s^{-2} = 1 J). The amount, n, of glucose molecules required for oxidation to give a change in Gibbs energy of this value given that 1 mol provides 2828 kJ is

$$n = \frac{3.0 \times 9.81 \times 1.0 \times 10^{-1} \text{ J}}{2.828 \times 10^6 \text{ J mol}^{-1}} = \frac{3.0 \times 9.81 \times 1.0 \times 10^{-7}}{2.828} \text{ mol}$$

Therefore, because the molar mass, M, of glucose is 180 g mol^{-1}, the mass, m, of glucose that must be oxidized is

$$m = nM = \left(\frac{3.0 \times 9.81 \times 1.0 \times 10^{-7}}{2.828} \text{ mol} \right) \times (180 \text{ g mol}^{-1})$$

$$= 1.9 \times 10^{-4} \text{ g}$$

That is, the bird must consume at least 0.19 mg of glucose for the mechanical effort (and more if it thinks about it).

Self-test 4.8

A hard-working human brain, perhaps one that is grappling with physical chemistry, operates at about 25 W. What mass of glucose must be consumed to sustain that power output for an hour?

[*Answer*: 5.7 g]

Some insight into the significance of G comes from its definition as $H - TS$. The enthalpy is a measure of the energy that can be obtained from the system as heat. The term TS is a measure of the quantity of energy stored in the random motion of the molecules making up the sample. Work, as we have seen, is energy transferred in an orderly way, so we cannot expect to obtain work from the energy stored randomly. The difference between the total stored energy and the energy stored randomly, $H - TS$, is available for doing work, and we recognize that difference as the Gibbs energy. In other words, the Gibbs energy is the energy stored in the orderly motion and arrangement of the molecules in the system.

Exercises

4.1 A goldfish swims in a bowl of water at 20°C. Over a period of time, the fish transfers 120 J to the water as a result of its metabolism. What is the change in entropy of the water?

4.2 Suppose you put a cube of ice of mass 100 g into a glass of water at just above 0°C. When the ice melts, about 33 kJ of energy is absorbed from the surroundings as heat. What is the change in entropy of (a) the sample (the ice), (b) the surroundings (the glass of water)?

4.3 A sample of aluminium of mass 1.25 kg is cooled at constant pressure from 300 K to 260 K. Calculate the energy that must be removed as heat and the change in entropy of the sample. The molar heat capacity of aluminum is 24.35 J K^{-1} mol^{-1}.

4.4 Calculate the change in entropy of 100 g of ice at 0°C as it is melted, heated to 100°C, and then vaporized at that temperature. Suppose that the changes are brought about by a heater that supplies heat at a constant rate, and sketch a graph showing (a) the change in temperature of the system, (b) the enthalpy of the system, (c) the entropy of the system as a function of time.

4.5 Calculate the change in molar entropy when carbon dioxide expands isothermally from 1.5 L to 4.5 L.

4.6 A sample of carbon dioxide that initially occupies 15.0 L at 250 K and 1.00 atm is compressed isothermally. Into what volume must the gas be compressed to reduce its entropy by 10.0 J K^{-1}?

4.7 Whenever a gas expands—when we exhale, when a flask is opened, and so on—the gas undergoes an increase in entropy. A sample of methane gas of mass 25 g at 250 K and 185 kPa expands isothermally and (a) reversibly, (b) irreversibly until its pressure is 2.5 kPa. Calculate the change in entropy of the gas.

4.8 What is the change in entropy of 100 g of water when it is heated from room temperature (20°C) to body temperature (37°C)? Use $C_{p,m} = 75.5$ J K^{-1} mol^{-1}.

4.9 Calculate the change in molar entropy when a sample of argon is compressed from 2.0 L to 500 mL and simultaneously heated from 300 K to 400 K. Take $C_{V,m} = \frac{3}{2}R$.

4.10 A monatomic perfect gas at a temperature T_i is expanded isothermally to twice its initial volume. To what temperature should it be cooled to restore its entropy to its initial value? Take $C_{V,m} = \frac{3}{2}R$.

4.11 In a certain cyclic engine (technically, a Carnot cycle), a perfect gas expands isothermally and reversibly, then adiabatically ($q = 0$) and reversibly. In the adiabatic expansion step the temperature falls. At the end of the expansion stage, the sample is compressed reversibly first isothermally and then adiabatically in such a way as to end up at the starting volume and temperature. Draw a graph of entropy against temperature for the entire cycle.

4.12 Calculate the change in entropy when 100 g of water at 80°C is poured into 100 g of water at 10°C in an insulated vessel given that $C_{p,m} = 75.5$ J K^{-1} mol^{-1}.

4.13 The enthalpy of the graphite → diamond phase transition, which under 100 kbar occurs at 2000 K, is +1.9 kJ mol^{-1}. Calculate the entropy change of the transition.

4.14 The enthalpy of vaporization of chloroform (trichloromethane), $CHCl_3$, is 29.4 kJ mol^{-1} at its normal boiling point of 334.88 K. (a) Calculate the entropy of vaporization of chloroform at this temperature. (b) What is the entropy change in the surroundings?

4.15 Octane is typical of the components of gasoline. Estimate (a) the entropy of vaporization, (b) the enthalpy of vaporization of octane, which boils at 126°C.

4.16 Estimate the molar entropy of potassium chloride at 5.0 K given that its molar heat capacity at that temperature is 1.2 mJ K^{-1} mol^{-1}.

4.17 Without performing a calculation, estimate whether the standard entropies of the following reactions are positive or negative:

(a) Ala–Ser–Thr–Lys–Gly–Arg–Ser $\xrightarrow{\text{trypsin}}$
$$\text{Ala–Ser–Thr–Lys} + \text{Gly–Arg}$$
(b) $N_2(g) + 3\,H_2(g) \rightarrow 2\,NH_3(g)$
(c) $ATP^{4-}(aq) + 2\,H_2O(l) \rightarrow ADP^{3-}(aq) + HPO_4^{2-}(aq) + H_3O^+(aq)$

4.18 Calculate the standard reaction entropy at 298 K of

(a) $2\,CH_3CHO(g) + O_2(g) \rightarrow 2\,CH_3COOH(l)$

(b) $2 AgCl(s) + Br_2(l) \rightarrow 2 AgBr(s) + Cl_2(g)$

(c) $Hg(l) + Cl_2(g) \rightarrow HgCl_2(s)$

(d) $Zn(s) + Cu^{2+}(aq) \rightarrow Zn^{2+}(aq) + Cu(s)$

(e) $C_{12}H_{22}O_{11}(s) + 12 O_2(g) \rightarrow 12 CO_2(g) + 11 H_2O(l)$

4.19 The constant-pressure molar heat capacities of linear gaseous molecules are approximately $\frac{7}{2}R$ and those of nonlinear gaseous molecules are approximately $4R$. Estimate the change in standard reaction entropy of the following two reactions when the temperature is increased by 10 K from 273 K at constant pressure:

(a) $2 H_2(g) + O_2(g) \rightarrow 2 H_2O(g)$

(b) $CH_4(g) + 2 O_2(g) \rightarrow CO_2(g) + 2 H_2O(g)$

4.20 Suppose that, when you exercise, you consume 100 g of glucose and that all the energy released as heat remains in your body at 37°C. What is the change in entropy of your body? (Use $\Delta_c H = -2808$ kJ mol^{-1}.)

4.21 In a particular biological reaction taking place in the body at 37°C, the change in enthalpy was -125 kJ mol^{-1} and the change in entropy was -126 J K^{-1} mol^{-1}. (a) Calculate the change in Gibbs energy. (b) Is the reaction spontaneous? (c) Calculate the total change in entropy of the system and the surroundings.

4.22 The change in Gibbs energy that accompanies the oxidation of $C_6H_{12}O_6(s)$ to carbon dioxide and water vapour at 25°C is -2828 kJ mol^{-1}. How much glucose does a person of mass 65 kg need to consume to climb through 10 m?

4.23 The formation of glutamine from glutamate and ammonium ions requires 14.2 kJ mol^{-1} of energy input. It is driven by the hydrolysis of ATP to ADP mediated by the enzyme glutamine synthetase. (a) Given that the change in Gibbs energy for the hydrolysis of ATP corresponds to $\Delta G = -31$ kJ mol^{-1} under the conditions prevailing in a typical cell, can the hydrolysis drive the formation of glutamine? (b) How many moles of ATP must be hydrolysed to form 1 mol glutamine?

4.24 The hydrolysis of acetyl phosphate has $\Delta G = -42$ kJ mol^{-1} under typical biological conditions. If acetyl phosphate were to be synthesized by coupling to the hydrolysis of ATP, what is the minimum number of ATP molecules that would need to be involved?

4.25 Suppose that the radius of a typical cell is 10 μm and that inside it 10^6 ATP molecules are hydrolysed each second. What is the power density of the cell in watts per metre cubed (1 W = 1 J s^{-1}). A computer battery delivers about 15 W and has a volume of 100 cm^3. Which has the greater power density, the cell or the battery? (For data, see Exercise 4.23.)

Chapter 5

Phase equilibria: pure substances

Contents

The thermodynamics of transition

Phase diagrams

BOILING, freezing, and the conversion of graphite to diamond are all examples of **phase transitions**, or changes of phase without change of chemical composition. Many phase changes are common everyday phenomena and their description is an important part of physical chemistry. They occur whenever a solid changes into a liquid, as in the melting of ice, or a liquid changes into a vapour, as in the vaporization of water in our lungs. They also occur when one solid phase changes into another, as in the conversion of graphite into diamond under high pressure, or the conversion of one phase of iron into another as it is heated in the process of steel-making. Phase changes are important geologically too, for calcium carbonate is typically deposited as aragonite, but then gradually changes into another crystal form, calcite.

The thermodynamics of transition

THE Gibbs energy of a substance will be at centre stage in all that follows. We need to know how its value depends on the pressure and temperature. As we work out these dependencies, we shall acquire deep insight into the thermodynamic properties of matter.

5.1 The condition of stability

First, we need to establish the importance of the *molar* Gibbs energy in the discussion of phase transitions. The Gibbs energy of a sample of substance, G, is equal to nG_m, where n is the amount of substance in the sample and G_m is its molar Gibbs energy. The molar Gibbs energy depends on the phase of the substance. For instance, the molar Gibbs energy of liquid water is in general different from that of water vapour at the same temperature and pressure. Therefore, when an amount n of the substance changes from phase 1 (for instance, liquid), with molar Gibbs energy $G_m(1)$ to phase 2 (for instance, vapour) with molar Gibbs energy $G_m(2)$, the change in Gibbs energy is

$$\Delta G = nG_m(2) - nG_m(1) = n\{G_m(2) - G_m(1)\}$$

We know that a spontaneous change at constant temperature and pressure is signalled by a negative value of ΔG. This expression shows, therefore, that a change from phase 1 to phase 2 is spontaneous if the molar Gibbs energy of phase 2 is lower than that of phase 1. In other words, *a substance has a spontaneous tendency to change into the phase with the lowest molar Gibbs energy.*

If at a certain temperature and pressure the solid phase of a substance has a lower molar Gibbs energy than its liquid phase, then the solid phase is thermodynamically more stable and the liquid will freeze. If the opposite is true, the liquid phase is thermodynamically more stable and the solid will melt. For example, at 1 atm, ice has a lower molar Gibbs energy than liquid water when the temperature is below 0°C, and under these conditions water converts spontaneously to ice.

Self-test 5.1

The standard Gibbs energy of formation of metallic white tin (α-Sn) is 0 at 25°C and that of the nonmetallic grey tin (β-Sn) is +0.13 kJ mol^{-1} at the same temperature. Which is the thermodynamically stable phase at 25°C?

[*Answer*: white tin]

5.2 The variation of Gibbs energy with pressure

The general equation for the change in Gibbs energy when the pressure and temperature are changed by the infinitesimal amounts dp and dT is

$$dG = Vdp - SdT \qquad (5.1a)$$

The derivation of this important equation can be found in *Further information* 5. It follows that the change in *molar* Gibbs energy of a substance is

$$dG_m = V_m dp - S_m dT \qquad (5.1b)$$

where V_m is the *molar* volume of the substance and S_m is its *molar* entropy.

For a change in the pressure of a system at constant temperature we can set dT = 0 in eqn 5.1b, which leaves

$$dG_m = V_m dp \qquad (5.2)$$

This expression tells us that, because all molar volumes are positive, *the molar Gibbs energy increases* (dG_m > 0) *when the pressure increases* (dp > 0). We also see that, for a given change in pressure, the resulting change in molar Gibbs energy is greatest for substances with large molar volumes. Therefore, because the molar volume of a gas is much larger than that of a condensed phase (a liquid or a solid), the slope of a graph of G_m against p is much steeper for a gas than for a condensed phase. For most substances, the molar volume of the liquid phase is greater than that of the solid phase. Therefore, for most substances, the slope of G_m is greater for a liquid than for a solid. These characteristics are illustrated in Fig 5.1.

As we see from Fig 5.1, when we increase the pressure on a substance, the molar Gibbs energy of the gas phase rises above that of the liquid, then the molar Gibbs energy of the liquid rises above that of the solid. Because the system has a tendency to convert into the state of lowest molar Gibbs energy, the graphs show that at low pressures the gas phase is the most stable, then at higher pressures the liquid phase becomes the most stable, followed by solid phase. In other words, under pressure the

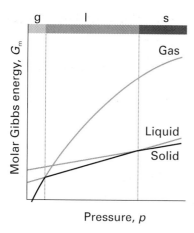

Fig 5.1 The variation of molar Gibbs energy with pressure. The region where the molar Gibbs energy of a particular phase is least is shown by a dark line and the corresponding region of stability of each phase is indicated in the band at the top of the illustration.

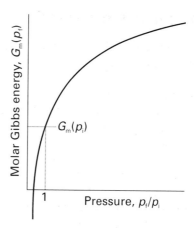

Fig 5.2 The variation of the molar Gibbs energy of a perfect gas with pressure.

substance condenses to a liquid, and then further pressure can result in the formation of a solid.

So far, we have argued qualitatively about the variation of G with pressure. In fact, eqn 5.2 lets us go further, for we can use it to predict the actual shape of graphs like those in Fig 5.1.

Derivation 5.1 *The pressure variation of Gibbs energy*

First, suppose the substance is incompressible, so that V is independent of p. This is a good approximation for liquids and solids. Then for a change of pressure from p_i to p_f, the change in molar Gibbs energy is

$$\Delta G_m = \int_{p_i}^{p_f} V_m \, dp \overbrace{=}^{\text{Constant volume}} V_m \int_{p_i}^{p_f} dp = V_m \Delta p \quad (5.3)$$

Therefore, as the pressure is increased, the molar Gibbs energy increases in proportion to the change in pressure and the molar volume of the sample. Graphs of G_m against p are therefore straight lines. Because the molar volumes of liquids and solids are small, the graphs have only very shallow slopes. It follows that, for small variations in pressure, we can make the approximation that the molar Gibbs energies of liquids and solids are independent of pressure.

If the substance is a gas, the change in Gibbs energy when the pressure of the gas changes from p_i to p_f

---

must allow for the fact that V_m decreases with pressure. The easiest case to consider is a perfect gas, for which $V_m = RT/p$. Then

$$\Delta G_m = \int_{p_i}^{p_f} V_m \, dp \overbrace{=}^{\substack{\text{Perfect} \\ \text{gas}}} \int_{p_i}^{p_f} \frac{RT}{p} \, dp \overbrace{=}^{\substack{\text{Constant} \\ \text{temperature}}} RT \int_{p_i}^{p_f} \frac{dp}{p}$$

$$= RT \ln \frac{p_f}{p_i} \quad (5.4)$$

That is, with $\Delta G_m = G_m(p_f) - G_m(p_i)$,

$$G_m(p_f) = G_m(p_i) + RT \ln \frac{p_f}{p_i} \quad (5.5)$$

The molar Gibbs energy of a sample of gas therefore increases logarithmically (as $\ln p$) with the pressure (Fig 5.2). The flattening of the variation at high pressures reflects the fact that, as V_m gets smaller, G_m becomes less responsive to pressure.

5.3 The variation of Gibbs energy with temperature

Equation 5.1b is also our starting point for finding out how the molar Gibbs energy varies with temperature. This time we keep the pressure constant ($dp = 0$), so

$$dG_m = -S_m dT \quad (5.6)$$

This simple expression tells us that, because molar entropy is positive, *an increase in temperature (dT > 0) results in a decrease in G_m (dG_m < 0)*. We see that, for a given change of temperature, the change in molar Gibbs energy is proportional to the molar entropy. For a given substance, the molar entropy of the gas phase is greater than that for a condensed phase, so the molar Gibbs energy falls more steeply with temperature for a gas than for a condensed phase. The molar entropy of the liquid phase of a substance is greater than that of its solid phase, so the slope is least steep for a solid. Figure 5.3 summarizes these characteristics.

Figure 5.3 also reveals the thermodynamic reason why substances melt and vaporize as the temperature is raised. At low temperatures, the solid phase has the lowest molar Gibbs energy and is therefore the most stable. However, as the temperature is raised, the molar Gibbs energy of the liquid phase falls below that of the solid phase, and the substance melts. At even higher temperatures, the molar Gibbs energy of the gas phase plunges down below that of the liquid phase, and the gas becomes the most stable phase. In other words, above a certain temperature, the liquid vaporizes to a gas.

We can also start to understand why some substances, such as carbon dioxide, sublime to a vapour without first forming a liquid. There is no requirement for the three lines to lie exactly in the positions in which we have drawn them in Fig 5.3: the liquid line, for instance, could lie where we have drawn it in Fig 5.4. Now we see that at no temperature (at the given pressure) does the liquid phase have the lowest molar Gibbs energy. Such a substance converts spontaneously directly from the solid to the vapour. That is, the substance sublimes.

The **transition temperature** between two phases, such as between liquid and solid, is the temperature, at a given pressure, at which the molar Gibbs energies of the two phases are equal. Above the transition temperature the liquid phase (for instance) is thermodynamically more stable; below it, the solid phase is more stable. For example, at 1 atm, the transition temperature for ice and liquid water is 0°C and that for grey and white tin is 13°C. At the transition temperature itself, the molar Gibbs energies of the two phases are identical and there is no tendency for either phase to change into the other. At this temperature, therefore, the two phases are in equilibrium. At 1 atm, ice and liquid water are in equilibrium at 0°C and the two allotropes of tin are in equilibrium at 13°C.

As always when using thermodynamic arguments, it is important to keep in mind the distinc-

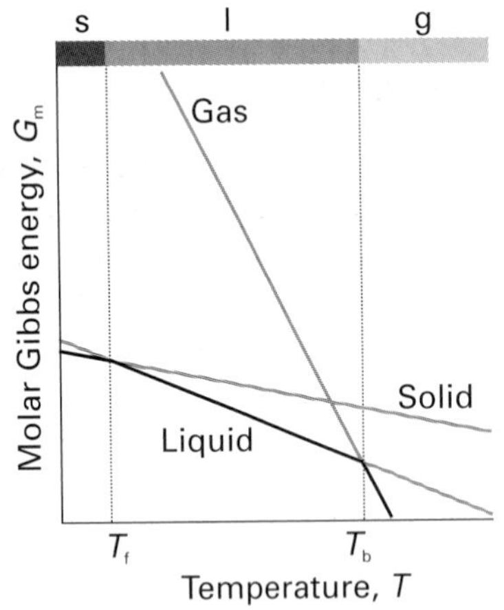

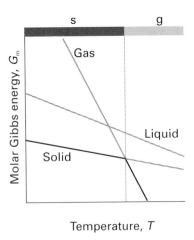

Fig 5.3 The variation of molar Gibbs energy with temperature. All molar Gibbs energies decrease with increasing temperature. The regions of temperature over which the solid, liquid, and gaseous forms of a substance have the lowest molar Gibbs energy are indicated in the band at the top of the illustration.

Fig 5.4 If the line for the Gibbs energy of the liquid phase does not cut through the line for the solid phase (at a given pressure) before the line for the gas phase cuts through the line for the solid, the liquid is not stable at any temperature at that pressure. Such a substance sublimes.

tion between the spontaneity of a phase transition and its rate. *Spontaneity is a tendency, not necessarily an actuality.* A phase transition predicted to be spontaneous may occur so slowly as to be unimportant in practice. For instance, at normal temperatures and pressures the molar Gibbs energy of graphite is 3 kJ mol^{-1} lower than that of diamond, so there is a thermodynamic tendency for diamond to convert into graphite. However, for this transition to take place, the C atoms of diamond must change their locations. Because the bonds between the atoms are so strong and large numbers of bonds must change simultaneously, this process is unmeasurably slow except at high temperatures. In gases and liquids the mobilities of the molecules normally allow phase transitions to occur rapidly, but in solids thermodynamic instability may be frozen in and a thermodynamically unstable phase may persist for thousands of years.

Phase diagrams

The **phase diagram** of a substance is a map that shows the conditions of temperature and pressure at which its various phases are thermodynamically most stable (Fig 5.5). For example, at point A in the illustration, the vapour phase of the substance is thermodynamically the most stable, but at B the liquid phase is the most stable.

The boundaries between regions, which are called **phase boundaries**, show the values of p and T at which the two neighbouring phases are in equilibrium. For example, if the system is arranged to have a pressure and temperature represented by point C, then the liquid and its vapour are in equilibrium (like liquid water and water vapour at 1 atm and 100°C). If the temperature is reduced at constant pressure, the system moves to point B where the liquid is stable (like water at 1 atm and at temperatures between 0°C and 100°C). If the temperature is reduced still further to D, then the solid and the liquid phases are in equilibrium (like ice and water at 1 atm and 0°C). A further reduction in tem-

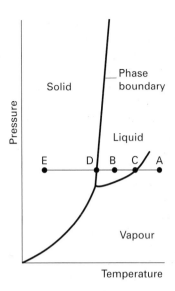

Fig 5.5 A typical phase diagram, showing the regions of pressure and temperature at which each phase is the most stable. The phase boundaries (three are shown here) show the values of pressure and temperature at which the two phases separated by the line are in equilibrium. The significance of the letters A, B, C, D, and E is explained in the text and they are referred to in Fig 5.8.

perature takes the system into the region where the solid is the stable phase.

5.4 Phase boundaries

The pressure of the vapour in equilibrium with its condensed phase is called the **vapour pressure** of the substance. Vapour pressure increases with temperature because, as the temperature is raised, more molecules have sufficient energy to leave their neighbours in the liquid.

The liquid–vapour boundary in a phase diagram is a plot of the vapour pressure against temperature. To determine the boundary, we can introduce a liquid into the near-vacuum at the top of a mercury barometer and measure how much the column is depressed (Fig 5.6). To ensure that the pressure exerted by the vapour is truly the vapour pressure, we have to add enough liquid for some to remain after the vapour forms, for only then are the liquid and vapour phases in equilibrium. We can

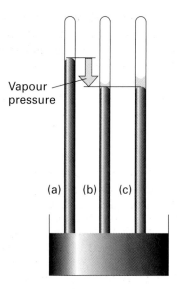

Vapour pressure

(a)   (b)   (c)

Fig 5.6 When a small volume of liquid is introduced into the vacuum above the mercury in a barometer (a), the mercury is depressed (b) by an amount that is proportional to the vapour pressure of the liquid. (c) The same pressure is observed however much liquid is present (provided some is present).

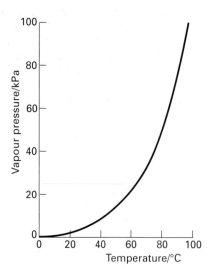

Fig 5.7 The variation of the vapour pressure of water with temperature.

change the temperature and determine another point on the curve, and so on (Fig 5.7).

Suppose we use a piston to apply a pressure greater than the vapour pressure of the liquid. Then the vapour is eliminated, the piston rests on the sur-

face of the liquid, and the system moves to one of the points in the 'liquid' region of the phase diagram. Only a single phase is present. If instead we reduce the pressure on the system to a value below the vapour pressure, the system moves to one of the points in the 'vapour' region of the diagram. Reducing the pressure will involve pulling out the piston a long way, so that all the liquid evaporates; while any liquid is present, the pressure in the system remains constant at the vapour pressure of the liquid.

Self-test 5.2

What would be observed when a pressure of 50 Torr is applied to a sample of water in equilibrium with its vapour at 25°C, when its vapour pressure is 23.8 Torr?

[*Answer*: entirely condense to liquid]

The same approach can be used to plot the solid–vapour boundary, which is a graph of the vapour pressure of the solid against temperature. The **sublimation vapour pressure** of a solid, the pressure of the vapour in equilibrium with a solid at a particular temperature, is usually much lower than that of a liquid.

A more sophisticated procedure is needed to determine the locations of solid–solid phase boundaries like that between calcite and aragonite, for instance, because the transition between two solid phases is more difficult to detect. One approach is to use **thermal analysis**, which takes advantage of the heat released during a transition. In a typical thermal analysis experiment, a sample is allowed to cool and its temperature is monitored. When the transition occurs, heat is evolved and the cooling stops until the transition is complete (Fig 5.8). The transition temperature is obvious from the shape of the graph and is used to mark a point on the phase diagram. The pressure can then be changed, and the corresponding transition temperature determined.

Any point lying on a phase boundary represents a pressure and temperature at which there is a 'dynamic equilibrium' between the two adjacent phases. A state of **dynamic equilibrium** is one in which a reverse process is taking place at the same rate as the forward process. Although there may be a great deal of activity at a molecular level, there is

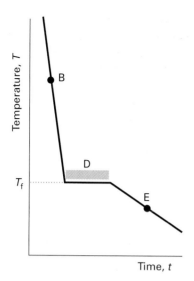

Fig 5.8 The cooling curve for the B–E section of the horizontal line in Fig 5.5. The halt at D corresponds to the pause in cooling while the liquid freezes and releases its enthalpy of transition. The halt lets us locate T_f even if the transition cannot be observed visually.

no net change in the bulk properties or appearance of the sample. For example, any point on the liquid–vapour boundary represents a state of dynamic equilibrium in which vaporization and condensation continue at matching rates. Molecules are leaving the surface of the liquid at a certain rate, and molecules already in the gas phase are returning to the liquid at the same rate; as a result, there is no net change in the number of molecules in the vapour and hence no net change in its pressure. Similarly, a point on the solid–liquid curve represents conditions of pressure and temperature at which molecules are ceaselessly breaking away from the surface of the solid and contributing to the liquid. However, they are doing so at a rate that exactly matches that at which molecules already in the liquid are settling on to the surface of the solid and contributing to the solid phase.

5.5 The location of phase boundaries

Thermodynamics provides us with a way of predicting the location of the phase boundaries. Suppose

two phases are in equilibrium at a given pressure and temperature. Then, if we change the pressure, we must adjust the temperature to a different value to ensure that the two phases remain in equilibrium. In other words, there must be a relation between the change in pressure that we exert and the change in temperature we must make to ensure that the two phases remain in equilibrium.

Derivation 5.2 *The Clapeyron equation*

Consider two phases 1 (for instance, a liquid) and 2 (a vapour). At a certain pressure and temperature the two phases are in equilibrium and $G_m(1) = G_m(2)$, where $G_m(1)$ is the molar Gibbs energy of phase 1 and $G_m(2)$ that of phase 2 (Fig 5.9). Now change the pressure by an infinitesimal amount dp and the temperature by dT. According to eqn 5.1, the molar Gibbs energies of each phase change as follows:

$$dG_m(1) = V_m(1)dp - S_m(1)dT$$

$$dG_m(2) = V_m(2)dp - S_m(2)dT$$

where $V_m(1)$ and $S_m(1)$ are the molar volume and molar entropy of phase 1 and $V_m(2)$ and $S_m(2)$ those of phase 2. The two phases were in equilibrium before the change, so the two molar Gibbs energies were equal. The two phases are still in equilibrium after the pressure and temperature are changed, so their two molar Gibbs energies are still equal. Therefore, the two *changes* in molar Gibbs energy must be equal, and we can write

$$V_m(1)dp - S_m(1)dT = V_m(2)dp - S_m(2)dT$$

This equation can be rearranged to

$$\{S_m(2) - S_m(1)\}dT = \{V_m(2) - V_m(1)\}dp$$

The entropy of transition, $\Delta_{trs}S$, is the difference between the two molar entropies, and the volume of transition, $\Delta_{trs}V$, is the difference between the molar volumes of the two phases:

$$\Delta_{trs}S = S_m(2) - S_m(1) \qquad \Delta_{trs}V = V_m(2) - V_m(1)$$

We can therefore write

$$\Delta_{trs}S \times dT = \Delta_{trs}V \times dp$$

and rearrange this expression into the **Clapeyron equation**:

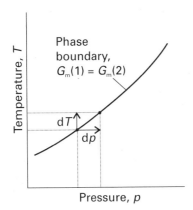

Fig 5.9 At equilibrium, two phases have the same molar Gibbs energy. When the temperature is changed, for the two phases to remain in equilibrium, the pressure must be changed so that the Gibbs energies of the two phases remain equal.

$$\frac{dp}{dT} = \frac{\Delta_{trs} S}{\Delta_{trs} V} \tag{5.7}$$

The Clapeyron equation tells us the slope (the value of dp/dT) of any phase boundary in terms of the entropy and volume of transition.

Now we put the Clapeyron equation, eqn 5.7, to work. Let's use it to predict how the vapour pressure of a substance varies with temperature.

Derivation 5.3 *The Clausius–Clapeyron equation*

For the liquid–vapour boundary the 'trs' label in eqn 5.7 becomes 'vap'. Then we note that the entropy of vaporization is related to the enthalpy of vaporization by $\Delta_{vap} S = \Delta_{vap} H/T$. Therefore, for the liquid–vapour phase boundary, eqn 5.7 becomes

$$\frac{dp}{dT} = \frac{\Delta_{vap} H}{T \Delta_{vap} V}$$

Next, we note that, because the molar volume of a gas is much larger than the molar volume of a liquid, the volume of vaporization, $\Delta_{vap} V = V_m(g) - V_m(l)$, is approximately equal to the molar volume of the gas itself. Therefore,

$$\frac{dp}{dT} = \frac{\Delta_{vap} H}{T V_m(g)}$$

To a good approximation, we can treat the vapour as a perfect gas and write its molar volume as $V_m = RT/p$. Then

$$\frac{dp}{dT} = \frac{\Delta_{vap} H}{T(RT/p)} = \frac{p \Delta_{vap} H}{RT^2}$$

A standard mathematical relation in calculus is $dp/p = d \ln p$ so, after dividing both sides by p, we obtain the **Clausius–Clapeyron equation**:

$$\frac{d \ln p}{dT} = \frac{\Delta_{vap} H}{RT^2} \tag{5.8}$$

The Clausius–Clapeyron equation is an approximate equation for the slope of a plot of the logarithm of the vapour pressure against the temperature (Fig 5.10). We can turn it into an expression for the vapour pressure itself by more mathematical manipulation.

First, rearrange eqn 5.8 so that $d \ln p$ occurs on the left and dT occurs on the right:

$$d \ln p = \frac{\Delta_{vap} H}{RT^2} dT$$

Now integrate both sides. Let's suppose that the vapour pressure is p at a temperature T and that it becomes p' when the temperature has been changed to T'. Then

$$\int_p^{p'} d \ln p = \int_T^{T'} \frac{\Delta_{vap} H}{RT^2} dT$$

The integral on the left evaluates to $\ln(p'/p)$. To evaluate the integral on the right, we suppose that the enthalpy of vaporization is constant over the temperature range of interest, so together with R it can be taken outside the integral:

$$\ln \frac{p'}{p} = \frac{\Delta_{vap} H}{R} \int_T^{T'} \frac{1}{T^2} dT = \frac{\Delta_{vap} H}{R} \left(\frac{1}{T} - \frac{1}{T'} \right) \tag{5.9}$$

To obtain this result we have used the standard integral

$$\int \frac{dx}{x^2} = -\frac{1}{x} + \text{constant}$$

Equation 5.9 lets us calculate the vapour pressure at one temperature provided we know it at another temperature. The equation tells us that, for a given change in temperature, the larger the enthalpy of vaporization, the greater the change in vapour pressure. The vapour pressure of water, for instance,

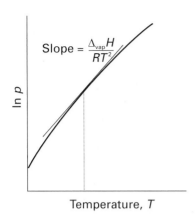

Fig 5.10 The Clausius–Clapeyron equation gives the slope of a plot of the logarithm of the vapour pressure of a substance against the temperature. That slope at a given temperature is proportional to the enthalpy of vaporization of the substance.

Table 5.1 *Vapour pressure**

Substance	A	B/K	Temperature range/°C
P_4(s, white)	9.6511	3297	20 to 44
SO_3(l)	10.022	2269	24 to 48
CH_3OH(l)	8.8017	2002	−10 to +80
C_6H_6(l)	7.9622	1785	0 to +42
CCl_4(l)	8.004	1771	−19 to +20
C_6H_{14}(l)	7.724	1655	−10 to +90
$C_6H_5CH_3$(l)	8.330	2047	−92 to +15

*A and B are the constants in the expression $\log(p/\text{Torr}) = A - B/T$.

responds more sharply to a change in temperature than benzene does. Note too that we can rearrange eqn 5.9 into the form[1]

$$\log\ p = A - \frac{B}{T} \tag{5.10}$$

where A and B are constants. This is the form in which vapour pressures are commonly reported (Table 5.1 and Fig 5.11).

5.6 Characteristic points

As we have seen, when a liquid is heated, its vapour pressure increases. First, consider what we would observe when we heat a liquid in an open vessel. At a certain temperature, the vapour pressure becomes equal to the external pressure. At this temperature, the vapour can drive back the surrounding atmosphere and expand indefinitely. Moreover, because there is no constraint on expansion, bubbles of vapour can form throughout the body of the liquid. This condition is known as **boiling**. The temperature at which the vapour pressure of a liquid is equal to the external pressure is called

the **boiling temperature**. When the external pressure is 1 atm, the boiling temperature is called the **normal boiling point**, T_b. It follows that we can predict the normal boiling point of a liquid by noting the temperature on the phase diagram at which its vapour pressure is 1 atm.

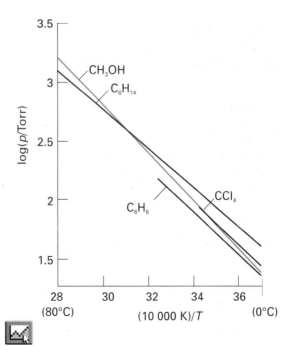

Fig 5.11 The vapour pressures of some substances based on the data in Table 5.1.

---

[1] To derive this expression, use $\ln x = \ln 10 \times \log x$. You will find that $B = \Delta_{vap}H/(R \ln 10)$. The value of A depends on the units adopted for p.

Table 5.2 *Critical constants* *

	p_c/atm	V_c/(cm³ mol⁻¹)	T_c/K
Ammonia, NH_3	111	73	406
Argon, Ar	48	75	151
Benzene, C_6H_6	49	260	563
Bromine, Br_2	102	135	584
Carbon dioxide, CO_2	73	94	304
Chlorine, Cl_2	76	124	417
Ethane, C_2H_6	48	148	305
Ethene, C_2H_4	51	124	283
Hydrogen, H_2	13	65	33
Methane, CH_4	46	99	191
Oxygen, O_2	50	78	155
Water, H_2O	218	55	647

* The critical volume, V_c, is the molar volume at the critical pressure and critical volume.

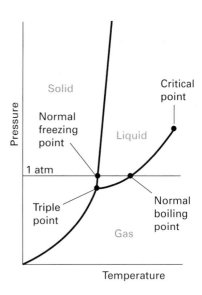

Fig 5.12 The significant points of a phase diagram. The liquid–vapour phase boundary terminates at the *critical point*. At the *triple point*, solid, liquid, and vapour are in dynamic equilibrium. The *normal freezing point* is the temperature at which the liquid freezes when the pressure is 1 atm; the *normal boiling point* is the temperature at which the vapour pressure of the liquid is 1 atm.

Now consider what happens when we heat the liquid in a closed vessel. Because the vapour cannot escape, its density increases as the vapour pressure rises and in due course the density of the vapour becomes equal to that of the remaining liquid. At this stage the surface between the two phases disappears, as was depicted in Fig 1.13. The temperature at which the surface disappears is the **critical temperature**, T_c, which we first encountered in Section 1.10. The vapour pressure at the critical temperature is called the **critical pressure**, p_c, and the critical temperature and critical pressure together identify the **critical point** of the substance (see Table 5.2). If we exert pressure on a sample that is above its critical temperature, we produce a denser fluid. However, no surface appears to separate the two parts of the sample and a single uniform phase continues to fill the container. That is, we have to conclude that *a liquid cannot be produced by the application of pressure to a substance if it is at or above its critical temperature.* That is why the liquid–vapour boundary in a phase diagram terminates at the critical point (Fig 5.12).

The temperature at which the liquid and solid phases of a substance coexist in equilibrium at a specified pressure is called the **melting temperature** of the substance. Because a substance melts at the same temperature as it freezes, the melting

temperature is the same as the **freezing temperature**. The solid–liquid boundary therefore shows how the melting temperature of a solid varies with pressure. The melting temperature when the pressure on the sample is 1 atm is called the **normal melting point** or the **normal freezing point**, T_f. A liquid freezes when the energy of the molecules in the liquid is so low that they cannot escape from the attractive forces of their neighbours and lose their mobility.

There is a set of conditions under which three different phases (typically solid, liquid, and vapour) all simultaneously coexist in equilibrium. It is represented by the **triple point**, where the three phase boundaries meet. The triple point of a pure substance is a characteristic, unchangeable physical property of the substance. For water the triple point lies at 273.16 K and 611 Pa, and ice, liquid water, and water vapour coexist in equilibrium at no other combination of pressure and temperature.[2] At the

---

[2] The triple point of water is used to define the Kelvin scale of temperature.

triple point, the rates of each forward and reverse process are equal (but the three individual rates are not necessarily the same).

The triple point and the critical point are important features of a substance because they act as frontier posts for the existence of the liquid phase. As we see from Fig 5.13a, if the slope of the solid–liquid phase boundary is as shown in the diagram:

The triple point marks the lowest temperature at which the liquid can exist.

The critical point marks the highest temperature at which the liquid can exist.

We shall see in the following section that for a few materials (most notably water) the solid–liquid phase boundary slopes in the opposite direction, and then only the second of these conclusions is relevant (see Fig 5.13b).

5.7 The phase rule

We might wonder whether *four* phases of a single substance could ever be in equilibrium (such as the two solid forms of tin, liquid tin, and tin vapour). To explore this question we think about the thermodynamic criterion for four phases to be in equilibrium. For equilibrium, the four molar Gibbs energies would all have to be equal and we could write

$$G_m(1) = G_m(2) \qquad G_m(2) = G_m(3) \qquad G_m(3) = G_m(4)$$

Each Gibbs energy is a function of the pressure and temperature, so we should think of these three relations as three equations for the two unknowns p and T. In general, three equations for two unknowns have no solution.[3] Therefore, we have to conclude that the four molar Gibbs energies cannot all be equal. In other words, *four phases of a single substance cannot coexist in mutual equilibrium.*

The conclusion we have reached is a special case of one of the most elegant results of chemical thermodynamics. The **phase rule** was derived by Gibbs and states that, for a system at equilibrium,

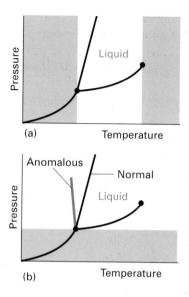

Fig 5.13 (a) For substances that have phase diagrams resembling the one shown here (which is common for most substances, with the important exception of water), the triple point and the critical point mark the range of temperatures over which the substance may exist as a liquid. The shaded areas show the regions of temperature in which a liquid cannot exist as a stable phase. (b) A liquid cannot exist as a stable phase if the pressure is below that of the triple point for normal or anomalous liquids.

$$F = C - P + 2 \tag{5.11}$$

Here F is the number of degrees of freedom, C is the number of components, and P is the number of phases. The **number of components**, C, in a system is the minimum number of independent species necessary to define the composition of all the phases present in the system. The definition is easy to apply when the species present in a system do not react, for then we simply count their number. For instance, pure water is a one-component system ($C = 1$) and a mixture of ethanol and water is a two-component system ($C = 2$). The **number of degrees of freedom**, F, of a system is the number of intensive variables (such as the pressure, temperature, or mole fractions) that can be changed independently without disturbing the number of phases in equilibrium.[4]

---

[3] For instance, the three equations $5x + 3y = 4$, $2x + 6y = 5$, and $x + y = 1$ have no solutions. Try it.

[4] An intensive property, recall from Section 2.4, is one that is independent of the size of the sample.

For a one-component system, such as pure water, we set $C = 1$ and the phase rule simplifies to $F = 3 - P$. When only one phase is present, $F = 2$, which implies that p and T can be varied independently. In other words, a single phase is represented by an *area* on a phase diagram. When two phases are in equilibrium $F = 1$, which implies that pressure is not freely variable if we have set the temperature. That is, the equilibrium of two phases is represented by a *line* in a phase diagram: a line in a graph shows how one variable must change if another variable is varied (Fig 5.14). Instead of selecting the temperature, we can select the pressure, but having done so the two phases come into equilibrium at a single definite temperature. Therefore, freezing (or any other phase transition) occurs at a definite temperature at a given pressure. When three phases are in equilibrium $F = 0$. This special 'invariant condition' can therefore be established only at a definite temperature and pressure. The equilibrium of three phases is therefore represented by a *point*, the triple point, on the phase diagram. If we set $P = 4$, we get the absurd result that F is negative; that result is in accord with the conclusion at the start of this section that four phases cannot be in equilibrium in a one-component system.

5.8 Phase diagrams of typical materials

We shall now see how these general features appear in the phase diagrams of a selection of pure substances.

Figure 5.15 is the phase diagram for water. The liquid–vapour phase boundary shows how the vapour pressure of liquid water varies with temperature. We can use this curve, which is shown in more detail in Fig 5.7, to decide how the boiling temperature varies with changing external pressure. For example, when the external pressure is 149 Torr (at an altitude of 12 km), water boils at 60°C because that is the temperature at which the vapour pressure is 149 Torr (19.9 kPa).

Self-test 5.3

What is the minimum pressure at which liquid is the thermodynamically stable phase of water at 25°C? Use eqn 5.9.

[*Answer*: 21.4 Torr, 2.85 kPa]

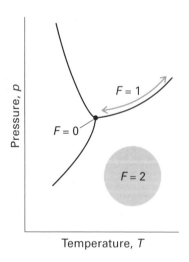

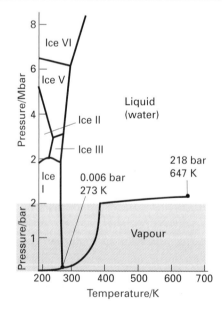

Fig 5.14 The features of a phase diagram represent different degrees of freedom. When only one phase is present, $F = 2$ and the pressure and temperature can be varied at will. When two phases are present in equilibrium, $F = 1$: now, if the temperature is changed, the pressure must be changed by a specific amount. When three phases are present in equilibrium, $F = 0$ and there is no freedom to change either variable.

Fig 5.15 The phase diagram for water showing the different solid phases. Note the change in the vertical scale at 2 bar.

The solid–liquid boundary line in Fig 5.15, which is shown in more detail in Fig 5.16, shows how the melting temperature of water depends on the pressure. For example, although ice melts at 0°C at 1 atm, it melts at −1°C when the pressure is 130 atm. The very steep slope of the boundary indicates that enormous pressures are needed to bring about significant changes. Notice that the line slopes down from left to right, which means that the melting temperature of ice falls as the pressure is raised. We can trace the reason for this unusual behaviour to the decrease in volume that occurs when ice melts: it is favourable for the solid to transform into the denser liquid as the pressure is raised. The decrease in volume is a result of the very open structure of the crystal structure of ice: the water molecules are held apart, as well as together, by the hydrogen bonds between them but the structure partially collapses on melting and the liquid is denser than the solid. Mathematically, $\Delta_{fus}V < 0$, so the slope of the phase boundary as given by the Clapeyron equation (eqn 5.7) is negative (down from left to right).

Figure 5.15 shows that water has many different solid phases other than ordinary ice ('ice I'). These solid phases differ in the arrangement of the water molecules: under the influence of very high pressures, hydrogen bonds buckle and the H_2O molecules adopt different arrangements. These **polymorphs**, or different forms, of ice may be responsible for the advance of glaciers, for ice at the bottom of glaciers experiences very high pressures where it rests on jagged rocks. The sudden apparent explosion of Halley's comet in 1991 may have been due to the conversion of one form of ice into another in its interior.

Figure 5.17 shows the phase diagram for carbon dioxide. The features to notice include the slope of the solid–liquid boundary: this positive slope is typical of almost all substances. The slope indicates that the melting temperature of solid carbon dioxide rises as the pressure is increased. As the triple point (217 K, 5.11 bar) lies well above ordinary atmospheric pressure, liquid carbon dioxide does not exist at normal atmospheric pressures whatever the temperature, and the solid sublimes when left in the open (hence the name 'dry ice'). To obtain liquid carbon dioxide, it is necessary to exert a pressure of at least 5.11 bar.

Cylinders of carbon dioxide generally contain the liquid or compressed gas; if both gas and liquid are present inside the cylinder, then at 20°C the pressure must be about 65 atm. When the gas squirts

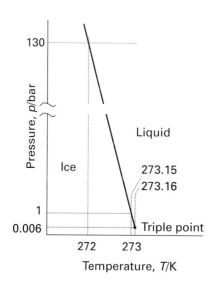

Fig 5.16 The solid–liquid boundary of water in more detail. The graph is schematic, and not to scale.

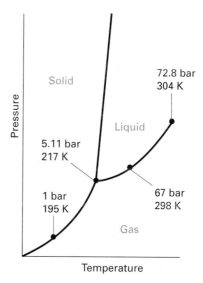

Fig 5.17 The phase diagram of carbon dioxide. Note that, as the triple point lies well above atmospheric pressure, liquid carbon dioxide does not exist under normal conditions (a pressure of at least 5.11 bar must be applied).

through the throttle it cools by the Joule–Thomson effect so, when it emerges into a region where the pressure is only 1 atm, it condenses into a finely divided snow-like solid.

Figure 5.18 shows the phase diagram of helium. Helium behaves unusually at low temperatures. For instance, the solid and gas phases of helium are never in equilibrium however low the temperature: the atoms are so light that they vibrate with a large-amplitude motion even at very low temperatures and the solid simply shakes itself apart. Solid helium can be obtained, but only by holding the atoms together by applying pressure. A second unique feature of helium is that pure helium-4 has two liquid phases. The phase marked He-I in the diagram behaves like a normal liquid; the other phase, He-II, is a **superfluid**; it is so called because it flows without viscosity. Helium is the only known substance with a liquid–liquid boundary in its phase diagram.[5]

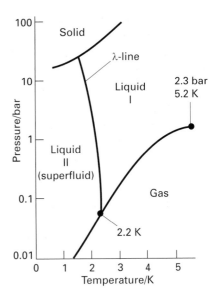

Fig 5.18 The phase diagram for helium-4. The 'λ-line' marks the conditions under which the two liquid phases are in equilibrium. Helium-I is a conventional liquid and helium-II is a superfluid. Note that a pressure of at least 20 bar must be exerted before solid helium can be obtained.

Exercises

5.1 The standard Gibbs energy of formation of rhombic sulfur is 0 and that of monoclinic sulfur is +0.33 kJ mol^{-1} at 25°C. Which polymorph is the more stable at that temperature?

5.2 The density of rhombic sulfur is 2.070 g cm^{-3} and that of monoclinic sulfur is 1.957 g cm^{-3}. Can the application of pressure be expected to make monoclinic sulfur more stable than rhombic sulfur? (See Exercise 5.1.)

5.3 What is the difference in molar Gibbs energy due to pressure alone of (a) water (density 1.03 g cm^{-3}), at the ocean surface and in the Mindañao trench (depth 11.5 km), (b) mercury (density 13.6 g cm^{-3}), at the top and bottom of the column in a barometer? (*Hint*. At the

very top, the pressure on the mercury is equal to the vapour pressure of mercury, which at 20°C is 160 mPa.)

5.4 The density of the fat tristearin is 0.95 g cm^{-3}. Calculate the change in molar Gibbs energy of tristearin when a deep-sea creature is brought to the surface (p = 1.0 atm) from a depth of 2.0 km. To calculate the hydrostatic pressure, take the mean density of water to be 1.03 g cm^{-3}.

5.5 Calculate the change in molar Gibbs energy of carbon dioxide (treated as a perfect gas) at 20°C when its pressure is changed isothermally from 1.0 bar to (a) 2.0 bar, (b) 0.00027 atm, its partial pressure in air.

5.6 A sample of water vapour at 200°C is compressed isothermally from 300 mL to 100 mL. What is the change in its molar Gibbs energy?

---

[5] Recent work has suggested that water may also have a superfluid liquid phase.

5.7 Without doing a calculation, decide whether the presence of (a) attractive, (b) repulsive interactions between gas molecules will raise or lower the molar Gibbs energy of a gas relative to its 'perfect' value.

5.8 Suppose that a gas obeys the van der Waals equation of state with the repulsive effects much greater than the attractive effects (that is, neglect the parameter a). (a) Find an expression for the change in molar Gibbs energy when the pressure is changed from p_i to p_f. (b) Is the change greater or smaller than for a perfect gas? (c) Estimate the percentage difference between the van der Waals and perfect gas calculations for carbon dioxide undergoing a change from 1.0 atm to 10.0 atm at 298.15 K. (*Hint.* For the first part, use calculus as in Derivation 5.1.)

5.9 The standard molar entropy of rhombic sulfur is 31.80 J K^{-1} mol^{-1} and that of monoclinic sulfur is 32.6 J K^{-1} mol^{-1}. (a) Can an increase in temperature be expected to make monoclinic sulfur more stable than rhombic sulfur? (b) If so, at what temperature will the transition occur at 1 bar? (See Exercise 5.1 for data.)

5.10 The standard molar entropy of benzene is 173.3 J K^{-1} mol^{-1}. Calculate the change in its standard molar Gibbs energy when it is heated from 20°C to 50°C.

5.11 The standard molar entropies of water ice, liquid, and vapour are 37.99, 69.91, and 188.83 J K^{-1} mol^{-1}, respectively. On a single graph, show how the Gibbs energies of each of these phases varies with temperature.

5.12 An open vessel containing (a) water, (b) benzene, (c) mercury stands in a laboratory measuring 6.0 m × 5.3 m × 3.2 m at 25°C. What mass of each substance will be found in the air if there is no ventilation? (The vapour pressures are (a) 24 Torr, (b) 98 Torr, (c) 1.7 mTorr.)

5.13 (a) Use the Clapeyron equation to estimate the slope of the solid–liquid phase boundary of water given that the enthalpy of fusion is 6.008 kJ mol^{-1} and the densities of ice and water at 0°C are 0.916 71 and 0.999 84 g cm^{-3}, respectively. (*Hint*: Express the entropy of fusion in terms of the enthalpy of fusion and the melting point of ice.) (b) Estimate the pressure required to lower the melting point of ice by 1°C.

5.14 Given the parametrization of the vapour pressure in eqn 5.10 and Table 5.1, what is the enthalpy of vaporization of hexane?

5.15 The vapour pressure of mercury at 20°C is 160 mPa; what is its vapour pressure at 50°C given that its enthalpy of vaporization is 59.30 kJ mol^{-1}?

5.16 The vapour pressure of pyridine is 50.0 kPa at 365.7 K and the normal boiling point is 388.4 K. What is the enthalpy of vaporization of pyridine?

5.17 Estimate the boiling point of benzene given that its vapour pressure is 20 kPa at 35°C and 50.0 kPa at 58.8°C.

5.18 On a cold, dry morning after a frost, the temperature was −5°C and the partial pressure of water in the atmosphere fell to 2 Torr. Will the frost sublime? What partial pressure of water would ensure that the frost remained?

5.19 (a) Refer to Fig 5.15 and describe the changes that would be observed when water vapour at 1.0 bar and 400 K is cooled at constant pressure to 260 K. (b) Suggest the appearance of a plot of temperature against time if energy is removed at a constant rate. To judge the relative slopes of the cooling curves, you need to know that the constant-pressure molar heat capacities of water vapour, liquid, and solid are approximately $4R$, $9R$, and $4.5R$; the enthalpies of transition are given in Table 3.1.

5.20 Refer to Fig 5.15 and describe the changes that would be observed when cooling takes place at the pressure of the triple point.

5.21 Use the phase diagram in Fig 5.17 to state what would be observed when a sample of carbon dioxide, initially at 1.0 atm and 298 K, is subjected to the following cycle: (a) constant-pressure heating to 320 K, (b) isothermal compression to 100 atm, (c) constant-pressure cooling to 210 K, (d) isothermal decompression to 1.0 atm, (e) constant-pressure heating to 298 K.

5.22 Infer from the phase diagram for helium in Fig 5.18 whether helium-I is more or less dense than helium-II given that the molar entropy of He-I is greater than that of He-II.

The properties of mixtures

Contents

W<small>E</small> now leave pure materials and the limited but important changes they can undergo, and examine mixtures of substances. Here we consider mainly **nonelectrolyte solutions**, where the solute is not present as ions. Examples are sucrose dissolved in water and sulfur dissolved in carbon disulfide. We delay until Chapter 9 the special problems of **electrolyte solutions**, in which the solute consists of ions that interact strongly with one another.

The thermodynamic description of mixtures

W<small>E</small> need a set of concepts that enable us to apply thermodynamics to mixtures of variable composition. We have already seen how to use the partial pressure, the contribution of one component in a gaseous mixture to the total pressure, to discuss the properties of mixtures of gases. For a more general description of the thermodynamics of mixtures we have to introduce other 'partial' properties, each one being the contribution that a particular component makes to the mixture.

6.1 Measures of concentration

There are essentially three measures of concentration that we shall employ. One, the *molar concentration*, is used when we need to know the amount of solute in a sample of known volume. The other two, the *molality* and the *mole fraction*, are used when we need to know the relative numbers of solute and solvent molecules in a sample.

The **molar concentration**, [J] or c_J, of a solute J in a solution is the chemical amount of J divided by the volume it occupies:[1]

$$[J] = \frac{\text{amount of solute}}{\text{volume of solution}} = \frac{n_J}{V} \qquad (6.1)$$

Molar concentration is typically reported in moles per litre (mol L^{-1}; more formally, as mol dm^{-3}). The unit 1 mol L^{-1} is commonly denoted 1 M (and read 'molar'). In practice, a solution containing a solute of given molar concentration is prepared by measuring out a known mass of solute into a volumetric flask, dissolving the solute in a little solvent, and then adding enough solvent to produce the desired volume. For instance, to prepare 1.00 M $C_6H_{12}O_6$(aq), we dissolve 180 g of glucose in enough water to produce 1.00 L of solution.[2] Once we know the molar concentration of a solute, we can calculate the amount, n_J, of that substance in a given volume, V, of solution by writing

$$n_J = [J]V \qquad (6.2)$$

Self-test 6.1

What mass of glycine, NH_2CH_2COOH, should be used to make 250 mL of 0.015 M NH_2CH_2COOH(aq)?

[*Answer:* 0.282 g]

The **molality**, b_J, of a solute J in a solution is the amount of substance divided by the mass of solvent used to prepare the solution:

$$b_J = \frac{\text{amount of solute}}{\text{mass of solvent}} = \frac{n_J}{m_{\text{solvent}}} \qquad (6.3)$$

---

[1] Molar concentration is still widely called 'molarity'.
[2] Note that the solute is *not* dissolved in 1.00 L of water.

Molality is typically reported in moles of solute per kilogram of solvent (mol kg^{-1}). This unit is sometimes (but unofficially) denoted m, with 1 m = 1 mol kg^{-1}. An important distinction between molar concentration and molality is that, whereas the former is defined in terms of the volume of the solution, the molality is defined in terms of the mass of solvent used to prepare the solution. Thus, to prepare 1.0 m $C_6H_{12}O_6$(aq) we dissolve 180 g of glucose in 1.0 kg of water. A distinction between molar concentration and molality is that the former varies with temperature as the solution expands and contracts, but the molality does not.

As we have indicated, we use molality when we need to emphasize the relative amounts of solute and solvent molecules. To see why this is so, we note that the mass of solvent is proportional to the amount of solvent molecules present, so from eqn 6.3 we see that the molality is proportional to the ratio of the amounts of solute and solvent molecules. For example, any 1.0 m aqueous nonelectrolyte solution contains 1.0 mol solute particles per 55.5 mol H_2O molecules, so in each case there is 1 solute molecule per 55.5 solvent molecules.

Closely related to the molality of a solute is the **mole fraction**, x_J, which was introduced in Chapter 1 in connection with mixtures of gases:

$$x_J = \frac{\text{amount of J}}{\text{total amount of molecules}} = \frac{n_J}{n} \qquad (6.4)$$

Here n_J is the amount (in moles) of J and n is the total amount of species (the total number of moles) in the sample.

Example 6.1 *Relating mole fraction and molality*

What is the mole fraction of glycine molecules in 0.140 m NH_2CH_2COOH(aq)?

Strategy We consider a sample that contains (exactly) 1 kg of solvent, and hence an amount $n_J = b_J \times (1$ kg) of solute molecules. The amount of solvent molecules in 1 kg of solvent is

$$n_{\text{solvent}} = \frac{1 \text{ kg}}{M}$$

where M is the molar mass of the solvent. Once these two amounts are available, we can calculate the mole fraction by using eqn 6.4 with $n = n_J + n_{solvent}$.

Solution It follows from the discussion in the *Strategy* that the amount of glycine (gly) molecules in 1 kg of solvent is

$$n_{gly} = (0.140 \text{ mol kg}^{-1}) \times (1 \text{ kg}) = 0.140 \text{ mol}$$

The amount of water molecules in 1 kg of water is

$$n_{water} = \frac{10^3 \text{ g}}{18.02 \text{ g mol}^{-1}} = \frac{10^3}{18.02} \text{ mol}$$

The total amount of molecules present is

$$n = 0.140 + \frac{10^3}{18.02} \text{ mol}$$

The mole fraction of glycine is therefore

$$x_{gly} = \frac{0.140 \text{ mol}}{0.140 + (10^3/18.02) \text{ mol}} = 2.52 \times 10^{-3}$$

Self-test 6.2

Calculate the mole fraction of sucrose molecules in 1.22 m $C_{12}H_{22}O_{11}$(aq).

[*Answer*: 2.15×10^{-2}]

6.2 Partial molar properties

A **partial molar property** is the contribution (per mole) that a substance makes to an overall property of a mixture. The easiest partial molar property to visualize is the **partial molar volume**, V_J, of a substance J, the contribution J makes to the total volume of a mixture.[3] We have to be alert to the fact that, although 1 mol of a substance has a characteristic volume when it is pure, 1 mol of a substance can make different contributions to the total volume of a mixture because molecules pack together in different ways in the pure substances and in mixtures.

---

[3] Partial molar quantities are also commonly denoted by a bar over the symbol, as in $\bar{V}_J$.

Illustration 6.1

To grasp the meaning of the concept of partial molar volume, imagine a huge volume of pure water. When a further 1 mol H_2O is added, the volume increases by 18 cm^3. However, when we add 1 mol H_2O to a huge volume of pure ethanol, the volume increases by only 14 cm^3. The quantity 18 cm^3 mol^{-1} is the volume occupied per mole of water molecules in pure water; 14 cm^3 mol^{-1} is the volume occupied per mole of water molecules in virtually pure ethanol. In other words, the partial molar volume of water in pure water is 18 cm^3 mol^{-1} and the partial molar volume of water in pure ethanol is 14 cm^3 mol^{-1}. In the latter case there is so much ethanol present that each H_2O molecule is surrounded by ethanol molecules and the packing of the molecules results in the water molecules occupying only 14 cm^3.

The partial molar volume at an intermediate composition of the water/ethanol mixture is an indication of the volume the H_2O molecules occupy when they are surrounded by a mixture of molecules representative of the overall composition (half water, half ethanol, for instance, when the mole fractions are both 0.5). The partial molar volume of ethanol varies as the composition of the mixture is changed, because the environment of an ethanol molecule changes from pure ethanol to pure water as the proportion of water increases and the volume occupied by the ethanol molecules varies accordingly. Figure 6.1 shows the variation of the two partial molar volumes across the full composition range at 25°C.

Once we know the partial molar volumes V_A and V_B of the two components A and B of a mixture at the composition (and temperature) of interest we can state the total volume V of the mixture by using

$$V = n_A V_A + n_B V_B \tag{6.5}$$

To verify this relation, consider a very large sample of the mixture of the specified composition. Then, when an amount n_A of A is added, the composition remains virtually unchanged but the volume of the sample increases by $n_A V_A$. Similarly, when an amount n_B of B is added, the volume increases by $n_B V_B$. The total increase in volume is $n_A V_A + n_B V_B$. The mixture now occupies a larger volume but the proportions of the components are still the same. Next,

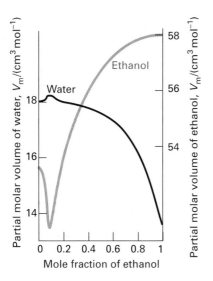

Fig 6.1 The partial molar volumes of water and ethanol at 25°C. Note the different scales (water on the left, ethanol on the right).

scoop out of this enlarged volume a sample containing n_A of A and n_B of B. Its volume is $n_A V_A + n_B V_B$. Because volume is a state function, the same sample could have been prepared simply by mixing the appropriate amounts of A and B.

Example 6.2 *Using partial molar volumes*

What is the total volume of a mixture of 50.0 g of ethanol and 50.0 g of water at 25°C?

Strategy To use eqn 6.5, we need the mole fractions of each substance and the corresponding partial molar volumes. We calculate the mole fractions in the same way as in Section 1.3, by using the molar masses of the components to calculate the amounts. We can then find the partial molar volumes corresponding to these mole fractions by referring to Fig 6.1.

Solution We find that $n_{ethanol} = 1.09$ mol and $n_{water} = 2.77$ mol, and hence that $x_{ethanol} = 0.282$ and $x_{water} = 0.718$. According to Fig 6.1, the partial molar volumes of the two substances in a mixture of this composition are 55 cm^3 mol^{-1} and 18 cm^3 mol^{-1}, respectively, so from eqn 6.5 the total volume of the mixture is

---

$$V = (1.09 \text{ mol}) \times (55 \text{ cm}^3 \text{ mol}^{-1})$$
$$+ (2.77 \text{ mol}) \times (18 \text{ cm}^3 \text{ mol}^{-1})$$
$$= 1.09 \times 55 + 2.77 \times 18 \text{ cm}^3 = 110 \text{ cm}^3$$

Self-test 6.3

Use Fig 6.1 to calculate the mass density of a mixture of 20 g of water and 100 g of ethanol.

[*Answer*: 0.84 g cm^{-3}]

Now we extend the concept of a partial molar quantity to other state functions. The most important for our purposes is the **partial molar Gibbs energy**, G_J, of a substance J, which is the contribution of J (per mole of J) to the total Gibbs energy of a mixture. It follows in the same way as for volume that, if we know the partial molar Gibbs energies of two substances A and B in a mixture of a given composition, then we can calculate the total Gibbs energy of the mixture by using an expression like eqn 6.5:

$$G = n_A G_A + n_B G_B \qquad (6.6)$$

The partial molar Gibbs energy has exactly the same significance as the partial molar volume. For instance, ethanol has a particular partial molar Gibbs energy when it is pure (and every molecule is surrounded by other ethanol molecules), and it has a different partial molar Gibbs energy when it is in an aqueous solution of a certain composition (because then each ethanol molecule is surrounded by a mixture of ethanol and water molecules).

The partial molar Gibbs energy is so important in chemistry that it is given a special name and symbol. From now on, we shall call it the **chemical potential** and denote it μ (mu). Then, eqn 6.6 becomes

$$G = n_A \mu_A + n_B \mu_B \qquad (6.7)$$

where μ_A is the chemical potential of A in the mixture and μ_B is the chemical potential of B. In the course of this chapter and the next we shall see that the name 'chemical potential' is very appropriate, for it will become clear that μ_J is a measure of the ability of J to bring about physical and chemical change. A substance with a high chemical potential has a high ability, in a sense we shall explore, to drive a reaction or some other physical process forward.

To make progress, we need an explicit formula for the variation of the chemical potential of a substance with the composition of the mixture. Our starting point is eqn 5.5, which shows how the molar Gibbs energy of a perfect gas depends on pressure:

$$G_m(p_f) = G_m(p_i) + RT \ln \frac{p_f}{p_i}$$

First, we set $p_f = p$, the pressure of interest, and $p_i = p^{\ominus}$, the standard pressure (1 bar). At the latter pressure, the molar Gibbs energy has its standard value, $G_m^{\ominus}$, so we can write

$$G_m(p) = G_m^{\ominus} + RT \ln \frac{p}{p^{\ominus}} \tag{6.8}$$

Next, for a *mixture* of perfect gases, we interpret p as the *partial* pressure of the gas, and the G_m is the *partial* molar Gibbs energy, the chemical potential. Therefore, for a mixture of perfect gases, for each component J,

$$\mu_J = \mu_J^{\ominus} + RT \ln \frac{p_J}{p^{\ominus}} \tag{6.9a}$$

In this expression, $\mu_J^{\ominus}$ is the **standard chemical potential** of the gas J, which is identical to its standard molar Gibbs energy, the value of G_m for the pure gas at 1 bar. If we adopt the convention that, whenever p_J appears in a formula it is to be interpreted as $p_J/p^{\ominus}$ (so, if the pressure is 2.0 bar, $p_J = 2.0$), we can write eqn 6.9a more simply as

$$\mu_J = \mu_J^{\ominus} + RT \ln p_J \tag{6.9b}$$

Figure 6.2 illustrates the pressure dependence of the chemical potential of a perfect gas predicted by this equation. Note that the chemical potential becomes negatively infinite as the pressure tends to zero, rises to its standard value at 1 bar, and then increases slowly (logarithmically, as ln p) as the pressure is increased further.

As always, we can become familiar with an equation by listening to what it tells us. In this case, we note that, as p_J increases, so does ln p_J. Therefore,

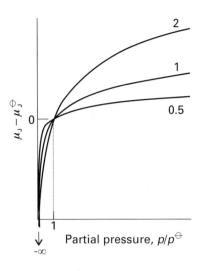

Fig 6.2 The variation with partial pressure of the chemical potential of a perfect gas at three different temperatures (in the ratios 0.5:1:2). Note that the chemical potential increases with pressure and, at a given pressure, with temperature.

eqn 6.9 tells us that, *the higher the partial pressure of a gas, the higher its chemical potential.* This conclusion is consistent with the interpretation of the chemical potential as an indication of the potential of a substance to be active chemically: the higher the partial pressure, the more active chemically the species. In this instance the chemical potential represents the tendency of the substance to react when it is in its standard state (the significance of the term $\mu^{\ominus}$) plus an additional tendency that reflects whether it is at a different pressure. A higher pressure gives a substance more chemical 'punch', just like winding a spring gives a spring more physical punch (that is, enables it to do more work).

Self-test 6.4

Suppose that the partial pressure of a perfect gas falls from 1.00 bar to 0.50 bar as it is consumed in a reaction at 25°C. What is the change in chemical potential of the substance?

[*Answer*: –1.7 kJ mol^{-1}]

We saw in Section 5.1 that the molar Gibbs energy of a pure substance is the same in all the phases at equilibrium. We can use the same argument to show that *a system is at equilibrium when the chemical potential of each substance has the same value in every phase in which it occurs.* We can think of the chemical potential as the pushing power of each substance, and equilibrium is reached only when each substance pushes with the same strength in any phase it occupies.

Derivation 6.1 *The uniformity of chemical potential*

Suppose a substance J occurs in different phases in different regions of a system. For instance, we might have a liquid mixture of ethanol and water and a mixture of their vapours. Let the substance J have chemical potential $\mu_J(l)$ in the liquid mixture and $\mu_J(g)$ in the vapour. We could imagine an infinitesimal amount, dn_J, of J migrating from the liquid to the vapour. As a result, the Gibbs energy of the liquid phase falls by $\mu_J(l)dn_J$ and that of the vapour rises by $\mu_J(g)dn_J$. The net change in Gibbs energy is

$$dG = \mu_J(g)dn_J - \mu_J(l)dn_J = \{\mu_J(g) - \mu_J(l)\}dn_J$$

There is no tendency for this migration to occur (that is, the system is at equilibrium) if $dG = 0$, which requires that $\mu_J(g) = \mu_J(l)$. The argument applies to each substance in the system. Therefore, *for a substance to be at equilibrium throughout the system, its chemical potential must be the same everywhere.*

6.3 Spontaneous mixing

All gases mix spontaneously with one another because the molecules of one gas can mingle with the molecules of the other gas. But how can we show thermodynamically that mixing is spontaneous? At constant temperature and pressure, we need to show that $\Delta G < 0$. The calculation is also a good illustration of how to use the chemical potential.

Derivation 6.2 *The Gibbs energy of mixing*

Suppose we have an amount n_A of a perfect gas A at a certain temperature T and pressure p, and an amount n_B of a perfect gas B at the same temperature and pressure. The two gases are in separate compartments initially (Fig 6.3). The Gibbs energy of the system (the two unmixed gases) is the sum of their individual Gibbs energies:

$$G_i = n_A\mu_A + n_B\mu_B$$
$$= n_A\{\mu_A^\ominus + RT \ln p\} + n_B\{\mu_B^\ominus + RT \ln p\}$$

The chemical potentials are those for the two gases each at a pressure p. When the partition is removed, the total pressure remains the same, but, according to Dalton's law (Section 1.3), the partial pressures fall to $x_A p$ and $x_B p$, where the x_J are the mole fractions of the two gases in the mixture ($x_J = n_J/n$, with $n = n_A + n_B$). The final Gibbs energy of the system is therefore

$$G_f = n_A\{\mu_A^\ominus + RT \ln x_A p\} + n_B\{\mu_B^\ominus + RT \ln x_B p\}$$

The difference $G_f - G_i$ is the change in Gibbs energy that accompanies mixing. The standard chemical potentials cancel, and by making use of the relation

$$\ln xp - \ln p = \ln \frac{xp}{p} = \ln x$$

for each gas, we obtain

$$\Delta G = RT\{n_A \ln x_A + n_B \ln x_B\}$$
$$= nRT\{x_A \ln x_A + x_B \ln x_B\} \qquad (6.10)$$

Figure 6.4 shows the variation of Gibbs energy with composition of the mixture predicted by this equation.

Equation 6.10 tells us the change in Gibbs energy when two gases mix at constant temperature and pressure. The crucial feature is that, because x_A and x_B are both less than 1, the two logarithms are negative ($\ln x < 0$ if $x < 1$), so $\Delta G < 0$ at all compositions. Therefore, *perfect gases mix spontaneously in all proportions.* Furthermore, if we compare eqn 6.10 with $\Delta G = \Delta H - T\Delta S$, we can conclude that

$$\Delta H = 0 \text{ and } \Delta S = -nR\{x_A \ln x_A + x_B \ln x_B\} \qquad (6.11)$$

That is, there is no change in enthalpy when two perfect gases mix, which reflects the fact that there are no interactions between the molecules. There is an increase in entropy, because the mixed gas is

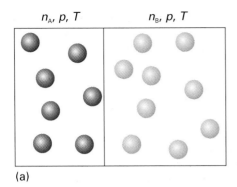

n_A, p, T n_B, p, T

(a)

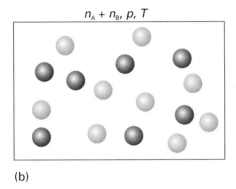

$n_A + n_B, p, T$

(b)

Fig 6.3 The (a) initial and (b) final states of a system in which two perfect gases mix. The molecules do not interact, so the enthalpy of mixing is zero. However, because the final state is more disordered than the initial state, there is an increase in entropy.

more disordered than the unmixed gases (Fig 6.5). This increase in entropy of the system is the 'driving force' of the mixing.[4]

6.4 Ideal solutions

In chemistry we are concerned with liquids as well as gases, so we need an expression for the chemical potential of a substance in a liquid solution. We can anticipate that the chemical potential of a species ought to increase with concentration, because the higher its concentration the greater its chemical 'punch'.

---

[4] The entropy of the surroundings is unchanged because the enthalpy of the system is constant, so no heat escapes into the surroundings.

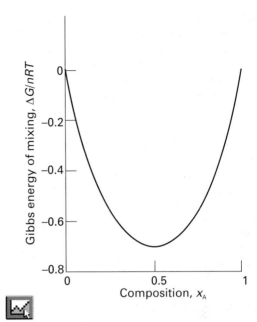

Fig 6.4 The variation of the Gibbs energy of mixing with composition for two perfect gases at constant temperature and pressure. Note that $\Delta G < 0$ for all compositions, which indicates that two gases mix spontaneously in all proportions.

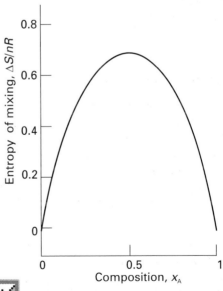

Fig 6.5 The variation of the entropy of mixing with composition for two perfect gases at constant temperature and pressure.

The key to setting up an expression for the chemical potential of a solute is the work done by the French chemist François Raoult (1830–1901), who spent most of his life measuring the vapour pressures of solutions. He measured the **partial vapour pressure**, p_J, of each component in the mixture, the partial pressure of the vapour of each component in dynamic equilibrium with the solution, and established what is now called **Raoult's law**:

The partial vapour pressure of a substance in a mixture is proportional to its mole fraction in the solution and its vapour pressure when pure:

$$p_J = x_J p_J^* \qquad (6.12)$$

In this expression, p_J^* is the vapour pressure of the pure substance. For example, when the mole fraction of water in an aqueous solution is 0.90, then, provided Raoult's law is obeyed, the partial vapour pressure of the water in the solution is 90 per cent that of pure water. This conclusion is approximately true whatever the identity of the solute and the solvent (Fig 6.6).

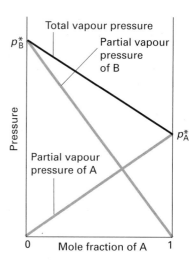

Fig 6.6 The partial vapour pressures of the two components of an ideal binary mixture are proportional to the mole fractions of the components in the liquid. The total pressure of the vapour is the sum of the two partial vapour pressures.

Self-test 6.5

A solution is prepared by dissolving 1.5 mol $C_{10}H_8$ (naphthalene) in 1.00 kg of benzene. The vapour pressure of pure benzene is 94.6 Torr at 25°C. What is the partial vapour pressure of benzene in the solution?

[*Answer*: 85 Torr]

The molecular origin of Raoult's law is the effect of the solute on the entropy of the solution. In the pure solvent, the molecules have a certain disorder and a corresponding entropy; the vapour pressure then represents the tendency of the system and its surroundings to reach a higher entropy. When a solute is present, the solution has a greater disorder than the pure solvent because we cannot be sure that a molecule chosen at random will be a solvent molecule (Fig 6.7). Because the entropy of the solution is higher than that of the pure solvent, the solution has a lower tendency to acquire an even higher entropy by the solvent vaporizing. In other words, the vapour pressure of the solvent in the solution is lower than that of the pure solvent.

A hypothetical solution that obeys Raoult's law

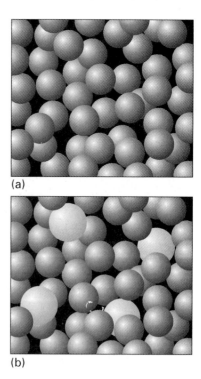

Fig 6.7 (a) In a pure liquid, we can be confident that any molecule selected from the sample is a solvent molecule. (b) When a solute is present, we cannot be sure that blind selection will give a solvent molecule, so the entropy of the system is greater than in the absence of the solute.

throughout the composition range from pure A to pure B is called an **ideal solution**. The law is most reliable when the components of a mixture have similar molecular shapes and are held together in the liquid by similar types and strengths of intermolecular forces. An example is a mixture of two structurally similar hydrocarbons. A mixture of benzene and methylbenzene (toluene) is a good approximation to an ideal solution, for the partial vapour pressure of each component satisfies Raoult's law reasonably well throughout the composition range from pure benzene to pure toluene (Fig 6.8).

No mixture is perfectly ideal and all real mixtures show deviations from Raoult's law. However, the deviations are small for the component of the mixture that is in large excess (the solvent) and become smaller as the concentration of solute decreases (Fig 6.9). We can usually be confident that Raoult's law is reliable for the solvent when the solution is very dilute. More formally, Raoult's law is a *limiting law* (like the perfect gas law), and is strictly valid only in the limit of zero concentration.

The theoretical importance of Raoult's law is that, because it relates vapour pressure to composition, and we know how to relate pressure to chemical potential, we can use the law to relate chemical potential to the composition of a solution.

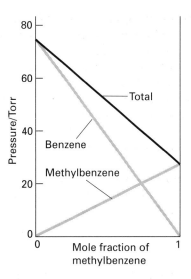

Fig 6.8 Two similar substances, in this case benzene and methylbenzene (toluene), behave almost ideally and have vapour pressures that closely resemble those for the ideal case depicted in Fig 6.6.

Derivation 6.3 *The chemical potential of a solvent*

We use J to denote a substance in general, A to denote a solvent, and B the solute. We have seen that, when a liquid A in a mixture is in equilibrium with its vapour at a partial pressure p_A, the chemical potentials of the two phases are equal (Fig 6.10):

$$\mu_A(l) = \mu_A(g)$$

However, we already have an expression for the chemical potential of a vapour, eqn 6.9; so at equilibrium

$$\mu_A(l) = \mu_A^{\ominus}(g) + RT \ln p_A$$

According to Raoult's law, $p_A = x_A p_A^*$, so we can write

$$\mu_A(l) = \mu_A^{\ominus}(g) + RT \ln x_A p_A^*$$
$$= \mu_A^{\ominus}(g) + RT \ln p_A^* + RT \ln x_A$$

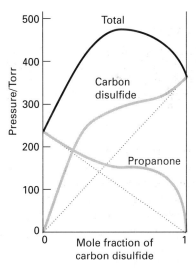

Fig 6.9 Strong deviations from ideality are shown by dissimilar substances, in this case carbon disulfide and acetone (propanone). Note, however, that Raoult's law is obeyed by propanone when only a small amount of carbon disulfide is present (on the left) and by carbon disulfide when only a small amount of propanone is present (on the right).

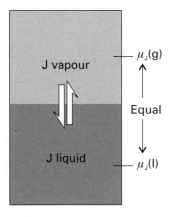

Fig 6.10 At equilibrium, the chemical potential of a substance in its liquid phase is equal to the chemical potential of the substance in its vapour phase.

The first two terms on the right, $\mu_A^{\ominus}(g)$ and $RT \ln p_A^*$, are independent of the composition of the mixture. We can write them as the constant $\mu_A^{\ominus}$, the standard chemical potential of the liquid. Then

$$\mu_A = \mu_A^{\ominus} + RT \ln x_A \tag{6.13}$$

Figure 6.11 shows the variation of chemical potential of the solvent predicted by this expression. Note that the chemical potential has its standard value at $x_A = 1$ (when it is pure).

The essential feature of eqn 6.13 is that, because $x_A < 1$ implies that $\ln x_A < 0$, *the chemical potential of a solvent is lower in a solution than when it is pure*. A solvent in which a solute is present has less chemical 'punch' (including a lower ability to generate a vapour pressure) than when it is pure.

Self-test 6.6

By how much is the chemical potential of benzene reduced at 25°C by a solute that is present at a mole fraction of 0.1.

[*Answer*: −0.26 kJ mol⁻¹]

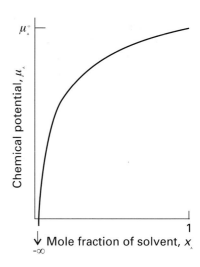

Fig 6.11 The variation of the chemical potential of the solvent with the composition of the solution. Note that the chemical potential of the solvent is lower in the mixture than for the pure liquid (for an ideal system). This behaviour is likely to be shown by a dilute solution in which the solvent is almost pure (and obeys Raoult's law).

6.5 Ideal-dilute solutions

Raoult's law provides a good description of the vapour pressure of the *solvent* in a very dilute solution. However, we cannot expect it to be a good description of the vapour pressure of the solute because a solute in dilute solution is very far from being pure. In a dilute solution, each solute molecule is surrounded by nearly pure solvent, so its environment is quite unlike that in the pure solute and it is very unlikely that its vapour pressure will be related to that of the pure solute. However, it is found experimentally that in dilute solutions the vapour pressure of the solute is in fact proportional to its mole fraction, just as for the solvent. Unlike the solvent, though, the constant of proportionality is not the vapour pressure of the pure solute. This linear but different dependence was discovered by the English chemist William Henry (1775–1836), and is summarized as **Henry's law**:

The vapour pressure of a volatile solute B is proportional to its mole fraction in a solution:

$$p_B = x_B K_B \tag{6.14}$$

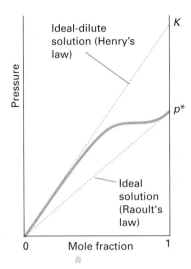

Fig 6.12 When a component (the solvent) is almost pure, it behaves in accord with Raoult's law and has a vapour pressure that is proportional to the mole fraction in the liquid mixture, and a slope p^*, the vapour pressure of the pure substance. When the same substance is the minor component (the solute), its vapour pressure is still proportional to its mole fraction, but the constant of proportionality is now K.

Here K_B, which is called **Henry's law constant**, is characteristic of the solute and chosen so that the straight line predicted by eqn 6.14 is tangent to the experimental curve at $x_B = 0$ (Fig 6.12).

Henry's law is usually obeyed only at low concentrations of the solute (close to $x_B = 0$). Solutions that are dilute enough for the solute to obey Henry's law are called **ideal-dilute solutions**.

Example 6.3 *Verifying Raoult's and Henry's laws*

The partial vapour pressures of each component in a mixture of propanone (acetone, A) and trichloromethane (chloroform, C) were measured at 35°C with the following results:

x_C	0	0.20	0.40	0.60	0.80	1
p_C/Torr	0	35	82	142	219	293
p_A/Torr	347	270	185	102	37	0

Confirm that the mixture conforms to Raoult's law for the component in large excess and to Henry's law for the minor component. Find the Henry's law constants.

Strategy We need to plot the partial vapour pressures against mole fraction. To verify Raoult's law, we com-

pare the data to the straight line $p_J = x_J p_J^*$ for each component in the region in which it is in excess and therefore acting as the solvent. We verify Henry's law by finding a straight line $p_J = x_J K_J$ that is tangent to each partial vapour pressure at low x_J where the component can be treated as the solute.[5]

Solution The data are plotted in Fig 6.13 together with the Raoult's law lines. Henry's law requires $K_A = 175$ Torr and $K_C = 165$ Torr. Notice how the data deviate from both Raoult's and Henry's laws for even quite small departures from $x = 1$ and $x = 0$, respectively.

Self-test 6.7

The vapour pressure of chloromethane at various mole fractions in a mixture at 25°C was found to be as follows:

x	0.005	0.009	0.0019	0.0024
p/Torr	205	363	756	946

Estimate Henry's law constant.

[*Answer*: 4×10^5 Torr]

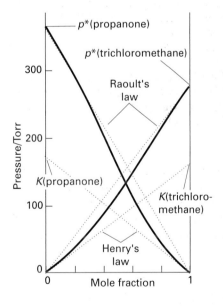

Fig 6.13 The experimental partial vapour pressures of a mixture of trichloromethane, $CHCl_3$, and propanone, CH_3COCH_3 (acetone), based on the data in Example 6.3.

---

[5] In a more professional approach, the data are fitted to a polynomial function (using a computer) and then the tangent is calculated by evaluating the first derivative of the polynomial at $x_J = 0$.

The Henry's law constants of some gases are listed in Table 6.1. They are often used in calculations relating to gas solubilities, as in the estimation of the concentration of O_2 in natural waters or the concentration of carbon dioxide in blood plasma. To apply Henry's law to this kind of problem, we treat the gas as the solute and use its partial pressure above the solvent to calculate the mole fraction in the solution by rearranging eqn 6.14 into

$$x_B = \frac{p_B}{K_B} \qquad (6.15)$$

A knowledge of Henry's law constants for gases in fats and lipids is important for the discussion of respiration, especially when the partial pressure of oxygen is abnormal, as in diving and mountaineering (Box 6.1).

Box 6.1 *Gas solubility and breathing*

We inhale about 500 mL of air with each breath we take. The influx of air is a result of changes in volume of the lungs as the diaphragm is depressed and the chest expands, which results in a decrease in pressure of about 1 Torr relative to atmospheric pressure. Expiration occurs as the diaphragm rises and the chest contracts, and gives rise to a differential pressure of about 1 Torr above atmospheric pressure. The total volume of air in the lungs is about 6 L, and the additional volume of air that can be exhaled forcefully after normal expiration is about 1.5 L. Some air remains in the lungs at all times to prevent the collapse of the alveoli.

The effect of gas exchange between blood and the air inside the alveoli of the lungs means that the composition of the air in the lungs changes throughout the breathing cycle. Alveolar gas is, in fact, a mixture of newly inhaled air and air about to be exhaled. The concentration of oxygen present in arterial blood is equivalent to a partial pressure of about 40 Torr, whereas the partial pressure of freshly inhaled air is about 104 Torr. Arterial blood remains in the capillary passing through the wall of an alveolus for about 0.75 s, but such is the steepness of the pressure gradient that it becomes fully saturated with oxygen in about 0.25 s. If the lungs collect fluids (as in pneumonia), the respiratory membrane thickens, diffusion is greatly slowed, and body tissues begin to suffer from oxygen starvation. Carbon dioxide moves in the opposite direction across the respiratory tissue, but the partial pressure gradient is much less, corresponding to about 5 Torr in blood and 40 Torr in air at equilibrium. However, because carbon dioxide is much more soluble in the alveolar fluid than oxygen is, equal amounts of oxygen and carbon dioxide are exchanged in each breath.

A hyperbaric oxygen chamber, in which oxygen is at an elevated partial pressure, is used to treat certain types of disease. Carbon monoxide poisoning can be treated in this way as can the consequences of shock. Diseases that are caused by anaerobic bacteria, such as gas gangrene

and tetanus, can also be treated because the bacteria cannot thrive in high oxygen concentrations.

In scuba diving (where *scuba* is an acronym formed from 'self-contained underwater breathing apparatus'), air is supplied at a higher pressure, so that the pressure within the diver's chest matches the pressure exerted by the surrounding water. The latter increases by about 1 atm for each 10 m of descent. One unfortunate consequence of breathing air at high pressures is that nitrogen is much more soluble in fatty tissues than in water, so it tends to dissolve in the central nervous system, bone marrow, and fat reserves. The result is *nitrogen narcosis*, with symptoms like intoxication. If the diver rises too rapidly to the surface, the nitrogen comes out of its lipid solution as bubbles, which causes the painful and sometimes fatal condition known as *the bends*. Many cases of scuba drowning appear to be consequences of arterial embolisms and loss of consciousness as the air bubbles rise into the head.

Exercise 1   Haemoglobin, the red blood pigment responsible for oxygen transport, binds about 1.34 mL of oxygen per gram. Normal blood has a haemoglobin concentration of 15 g/100 mL. Haemoglobin in the lungs is about 97 per cent saturated, but in the capillary is only about 75 per cent saturated. What volume of oxygen is given up by 100 mL of blood flowing from the lungs to the capillary?

Exercise 2   Breathing air at high pressures, such as in scuba driving, results in increased concentrations of nitrogen. The Henry's law constant for the solubility of nitrogen is 1.8×10^{-4} mg/(g H_2O atm). What mass of nitrogen is dissolved in 100 g of water saturated with air at 4.0 atm and 20°C? Compare your answer to that for 100 g of water saturated with air at 1.0 atm. (Air is 78.08 mole per cent nitrogen.) If nitrogen is four times as soluble in fatty tissues as in water, what is the increase in nitrogen concentration in fatty tissue in going from 1 atm to 4 atm?

Example 6.4 *Determining whether a natural water can support aquatic life*

The concentration of O_2 in water required to support aquatic life is about 4 mg L^{-1}. What is the minimum partial pressure of oxygen in the atmosphere that can achieve this concentration and 25°C?

Strategy The strategy of the calculation is to determine the partial pressure of oxygen that, according to Henry's law, corresponds to the concentration specified. To use eqn 6.15 we need to convert the stated mass concentration to a mole fraction of the solute. For that, we consider (exactly) 1 L of solution, calculate the mass of solute present, and then use its molar mass to convert to an amount in moles. Then we convert the mass of solvent present to an amount of solvent molecules in moles. To do so, we make the approximation that the solution is so dilute that the solvent is almost pure water, and use the fact that the density of water is approximately 1 kg L^{-1}. At that stage we can find the mole fractions, and use Henry's law to calculate the required partial pressure.

Solution Because 1 L of solution contains 4 mg of oxygen, the amount of O_2 present in 1 L of solution is

$$n_{O_2} = \frac{4 \times 10^{-3}\ \text{g}}{32.00\ \text{g mol}^{-1}} = \frac{4 \times 10^{-3}}{32.00}\ \text{mol}$$

The amount of H_2O present in 1 L of solution, which we take to correspond to 1 kg of water, is similarly

$$n_{H_2O} = \frac{1 \times 10^3\ \text{g}}{18.02\ \text{g mol}^{-1}} = \frac{1 \times 10^3}{18.02}\ \text{mol}$$

Therefore, the mole fraction of O_2 is

$$x_{O_2} = \frac{(4 \times 10^{-3}/32.00)\ \text{mol}}{(4 \times 10^{-3}/32.00)\ \text{mol} + (1 \times 10^3/18.02)\ \text{mol}}$$

$$= \frac{(4 \times 10^{-3}/32.00)}{(4 \times 10^{-3}/32.00) + (1 \times 10^3/18.02)}$$

(This expression evaluates to 2.3×10^{-6}.) Because the Henry's law constant for oxygen in water at 25°C is 3.3×10^7 Torr, the partial pressure required to achieve the mole fraction is

$$p_{O_2} = x_{O_2} K$$

$$= \frac{(4 \times 10^{-3}/32.00)}{(4 \times 10^{-3}/32.00) + (1 \times 10^3/18.02)}$$

$$\times (3.3 \times 10^7\ \text{Torr})$$

$$= 7 \times 10\ \text{Torr}$$

The partial pressure of oxygen in air at sea level is 0.21×760 Torr $= 1.6 \times 10^2$ Torr, which is greater than 70 Torr, so the required concentration can be maintained under normal conditions.

Self-test 6.8

What partial pressure of methane is needed to achieve 21 mg of methane in 100 g of benzene at 25°C?

[*Answer*: 4.3×10^2 Torr]

Henry's law lets us write an expression for the chemical potential of a solute in a solution. By exactly the same reasoning as in *Derivation 6.3*, but with the empirical constant K_B used in place of the vapour pressure of the pure solute, p_B^*, the chemical potential of the solute when it is present at a mole fraction x_B is

$$\mu_B = \mu_B^{\ominus} + RT \ln x_B \tag{6.16}$$

This expression, which is illustrated in Fig 6.14, applies when Henry's law is valid, in very dilute solutions. The chemical potential of the solute has its standard value when it is pure ($x_B = 1$) and a smaller value when dissolved ($x_B < 1$).

We often express the composition of a solution in terms of the molar concentration of the solute, [B], rather than as a mole fraction. The mole fraction and the molar concentration are proportional to each other in dilute solutions, so we write $x_B = $ constant $\times$ [B]. To avoid complications with units, we shall interpret [B] wherever it appears as the

Table 6.1 *Henry's law constants for gases at 25°C, K/Torr*

	Solvent	
	Water	Benzene
Methane, CH_4	3.14×10^5	4.27×10^5
Carbon dioxide, CO_2	1.25×10^6	8.57×10^4
Hydrogen, H_2	5.34×10^7	2.75×10^6
Nitrogen, N_2	6.51×10^7	1.79×10^6
Oxygen, O_2	3.30×10^7	

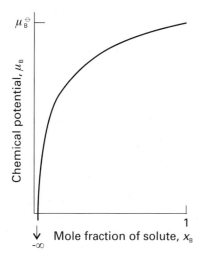

Fig 6.14 The variation of the chemical potential of the solute with the composition of the solution expressed in terms of the mole fraction of solute. Note that the chemical potential of the solvent is lower in the mixture than for the pure solute (for an ideal system). This behaviour is likely to be shown by a dilute solution in which the solvent is almost pure and the solute obeys Henry's law.

numerical value of the molar concentration in moles per litre.[6] Then eqn 6.16 becomes

$$\mu_B = \mu_B^{\ominus} + RT \ln(\text{constant}) + RT \ln [B]$$

We can combine the first two terms into a single constant, which we shall also denote $\mu_B^{\ominus}$, and write this relation as

$$\mu_B = \mu_B^{\ominus} + RT \ln [B] \tag{6.17}$$

Figure 6.15 illustrates the variation of chemical potential with concentration predicted by this equation. The chemical potential of the solute has its standard value when the molar concentration of the solute is 1 mol L^{-1}.

6.6 Real solutions: activities

No actual solutions are ideal, and many solutions deviate from ideal-dilute behaviour as soon as the concentration of solute rises above a small value. In thermodynamics we try to preserve the form of equations developed for ideal systems so that it becomes easy to step between the two types of system.[7] This is the thought behind the introduction of the **activity**, a_J, of a substance, which is a kind of effective concentration. The activity is defined so that the expression

$$\mu_J = \mu_J^{\ominus} + RT \ln a_J \tag{6.18}$$

is *always* true at any concentration and for both the solvent and the solute.

For ideal solutions, $a_A = x_A$, and the activity of each component is equal to its mole fraction. For ideal-dilute solutions using the definition in eqn 6.17, $a_B = [B]$, and the activity of the solute is equal to the numerical value of its molar concentration. For *non*-ideal solutions we write

$$a_A = \gamma_A x_A \qquad a_B = \gamma_B [B] \tag{6.19}$$

where the γ in each case is the **activity coefficient**. Note that:

Because the solvent behaves more in accord with Raoult's law as it becomes pure, $\gamma_A \to 1$ as $x_A \to 1$.

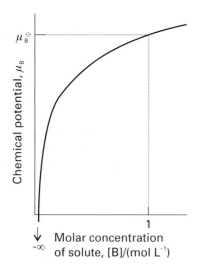

Fig 6.15 The variation of the chemical potential of the solute with the composition of the solution that obeys Henry's law expressed in terms of the molar concentration of solute. The chemical potential has its standard value at [B] = 1 mol L^{-1}.

---

[6] Thus, if the molar concentration of B is 1.0 mol L^{-1}, then [B] = 1.0.

[7] An added advantage is that there are fewer equations to remember!

Table 6.2 *Activities and standard states*

Substance	Standard state	Activity*
Solid	Pure solid, 1 bar	1
Liquid	Pure liquid, 1 bar	1
Gas	Pure gas, 1 bar	p_J (strictly, $p_J/p^{\ominus}$, $p^{\ominus}$ = 1 bar)
Solute	Molar concentration of 1 mol L^{-1}	[J] (strictly, [J]/mol L^{-1})

*For perfect gases and ideal-dilute solutions. All activities are dimensionless.

Because the solute behaves more in accord with Henry's law as the solution becomes very dilute, $\gamma_B \to 1$ as [B] $\to 0$.

Because a pure liquid or solid is in its standard state, the activity of a pure liquid or solid is 1, and we get $\mu_J = \mu_J^{\ominus}$ by setting $a_J = 1$ in eqn 6.18.

Conventions and relations about activities and standard states are summarized in Table 6.2.

Activities and activity coefficients are often branded as 'fudge factors'. To some extent that is true. However, their introduction does allow us to derive thermodynamically exact expressions for the properties of non-ideal solutions. A good example of this application is the use of pH as a measure of hydrogen ion activity (Chapter 8). A second point is that it is possible in a number of cases to calculate or measure the activity coefficient of a species in solution. In this text we shall normally derive thermodynamic relations in terms of activities but, when we want to make contact with actual measurements, we shall set the activities equal to the 'ideal' values in Table 6.2.

Colligative properties

A^{N} ideal solute has no effect on the enthalpy of a solution. However, it does affect the entropy by introducing a degree of disorder that is not present in the pure solvent. We can therefore expect a solute to modify the physical properties of the solution. Apart from lowering the vapour pressure of the solvent, which we have already considered, a nonvolatile solute has three main effects: it raises the boiling point of a solution, it lowers the freezing point, and it gives rise to an osmotic pressure. (The meaning of the last will be explained shortly.) Because these properties all stem from changes in the disorder of the solvent, and the increase in disorder is independent of the identity of the species we use to bring it about, all of them depend only on the number of solute particles present, not their chemical identity. For this reason they are called **colligative properties**.[8] Thus, a 0.01 mol kg^{-1} aqueous solution of any nonelectrolyte should have the same boiling point, freezing point, and osmotic pressure.

6.7 The modification of boiling and freezing points

As indicated above, the effect of a solute is to raise the boiling point of a solvent and to lower its freezing point. It is found empirically, and can be justified thermodynamically, that the **elevation of boiling point**, ΔT_B, and the **depression of freezing point**, ΔT_f, are both proportional to the molality, b_B, of the solute:

$$\Delta T_B = K_B b_B \qquad \Delta T_f = K_f b_B \qquad (6.20)$$

K_B is the **ebullioscopic constant** and K_f is the **cryoscopic constant** of the solvent.[9] The two constants can be estimated from other properties of the

---

[8] Colligative denotes 'depending on the collection'.
[9] They are also called the 'boiling-point constant' and the 'freezing-point constant', respectively.

Table 6.3 *Cryoscopic and ebullioscopic constants*

Solvent	$K_f/(\text{K kg mol}^{-1})$	$K_b/(\text{K kg mol}^{-1})$
Acetic acid	3.90	3.07
Benzene	5.12	2.53
Camphor	40	
Carbon disulfide	3.8	2.37
Carbon tetrachloride	30	4.95
Naphthalene	6.94	5.8
Phenol	7.27	3.04
Water	1.86	0.51

solvent, but both are best treated as empirical constants (Table 6.3).

Self-test 6.9

Estimate the lowering of the freezing point of the solution made by dissolving 3.0 g (about one cube) of sucrose in 100 g of water.

[*Answer*: −0.16 K]

To understand the origin of these effects we shall make two simplifying assumptions:

1 The solute is not volatile, and therefore does not appear in the vapour phase.

2 The solute is insoluble in the solid solvent, and therefore does not appear in the solid phase.

For example, a solution of sucrose in water consists of a solute (sucrose, $C_{12}H_{22}O_{11}$) that is not volatile and therefore never appears in the vapour, which is therefore pure water vapour. The sucrose is also left behind in the liquid solvent when ice begins to form, so the ice remains pure.

The origin of colligative properties is the lowering of chemical potential of the solvent by the presence of a solute, as expressed by eqn 6.13. We saw in Section 5.3 that the freezing and boiling points correspond to the temperatures at which the graph of the molar Gibbs energy of the liquid intersects the graphs of the molar Gibbs energy of the solid and gas phases, respectively. Because we are now dealing with mixtures, we have to think about the *partial* molar Gibbs energy (the chemical potential) of the solvent. The presence of a solute lowers the chemical potential of the liquid but, because the

vapour and solid remain pure, their chemical potentials remain unchanged. As a result, we see from Fig 6.16 that the freezing point moves to lower values; likewise, from Fig 6.17 we see that the boiling point moves to higher values. In other words, the freezing point is depressed, the boiling point is elevated, and the liquid phase exists over a wider range of temperatures.

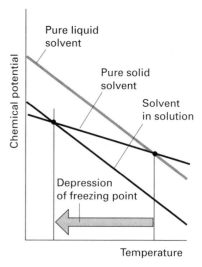

Fig 6.16 The chemical potentials of pure solid solvent and pure liquid solvent also decrease with temperature, and the point of intersection, where the chemical potential of the liquid rises above that of the solid, marks the freezing point of the pure solvent. A solute lowers the chemical potential of the solvent but leaves that of the solid unchanged. As a result, the intersection point lies further to the left and the freezing point is therefore lowered.

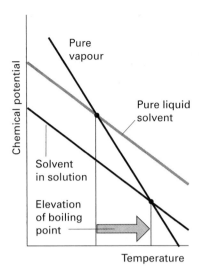

Fig 6.17 The chemical potentials of pure solvent vapour and pure liquid solvent decrease with temperature, and the point of intersection, where the chemical potential of the vapour falls below that of the liquid, marks the boiling point of the pure solvent. A solute lowers the chemical potential of the solvent but leaves that of the vapour unchanged. As a result, the intersection point lies further to the right, and the boiling point is therefore raised.

The elevation of boiling point is too small to have any practical significance. A practical consequence of the lowering of freezing point, and hence the lowering of the melting point of the pure solid, is its employment in organic chemistry to judge the purity of a sample, for any impurity lowers the melting point of a substance from its accepted value. The salt water of the oceans freezes at temperatures lower than that of fresh water, and salt is spread on highways to delay the onset of freezing. The addition of 'antifreeze' to car engines and, by natural processes, to arctic fish, is commonly held up as an example of the lowering of freezing point, but the concentrations are far too high for the arguments we have used here to be applicable. The 1,2-ethandiol ('glycol') used as antifreeze and the proteins present in fish body fluids probably simply interfere with bonding between water molecules.

6.8 Osmosis

The phenomenon of **osmosis** is the passage of a pure solvent into a solution separated from it by a

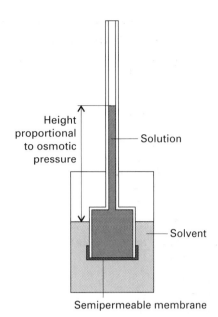

Semipermeable membrane

Fig 6.18 In a simple osmosis experiment, a solution is separated from the pure solvent by a semipermeable membrane. Pure solvent passes through the membrane and the solution rises in the inner tube. The net flow ceases when the pressure exerted by the column of liquid is equal to the osmotic pressure of the solution.

semipermeable membrane.[10] A **semipermeable membrane** is a membrane that is permeable to the solvent but not to the solute (Fig 6.18). The membrane might have microscopic holes that are large enough to allow water molecules to pass through, but not ions or carbohydrate molecules with their bulky coating of hydrating water molecules. The **osmotic pressure**, Π (uppercase pi), is the pressure that must be applied to the solution to stop the inward flow of solvent. One of the most important examples of osmosis is transport of fluids through cell membranes, but osmosis is also the basis of the technique called **osmometry**, the determination of molar mass by measurement of osmotic pressure, especially of macromolecules. The migration of species through membranes is also used to study the binding of small molecules to proteins (Box 6.2).

In the simple arrangement shown in Fig 6.18, the pressure opposing the passage of solvent into the

---

[10] The name *osmosis* is derived from the Greek word for 'push'.

Box 6.2 *Dialysis and protein building*

In a dialysis experiment, a solution of macromolecules and smaller ions is placed in a bag made of a material that acts a semipermeable membrane and the whole is immersed in a solvent. The membrane permits the passage of the small ions but not the macromolecules, so the former migrate through the membrane, leaving the macromolecule behind. Dialysis is used to study the binding of small molecules to macromolecules, such as an inhibitor to an enzyme, an antibiotic to DNA, and any other instance of cooperation or inhibition by small molecules attaching to large ones.

Suppose the molar concentration of the macromolecule M is [M] and the total concentration of small molecule A in the compartment containing the macromolecule is $[A]_{in}$. This total concentration is the sum of the concentrations of free A and bound A, which we write $[A]_{free}$ and $[A]_{bound}$, respectively. At equilibrium, the chemical potential of free A in the macromolecule solution is equal to the chemical potential of A in the solution on the other side of the membrane, where its concentration is $[A]_{out}$. The equality $\mu_{A,free} = \mu_{A,out}$ implies that $[A]_{free} = [A]_{out}$, provided the activity coefficient of A is the same in both solutions. Therefore, by measuring the concentration of A in the 'outside' solution, we can find the concentration of unbound A in the macromolecule solution and, from the difference $[A]_{in} - [A]_{free}$, which is equal to $[A]_{in} - [A]_{out}$, the concentration of bound A. The average number of A molecules bound to M molecules, v, is then the ratio

$$v = \frac{[A]_{bound}}{[M]} = \frac{[A]_{in} - [A]_{out}}{[M]}$$

The bound and unbound A molecules are in equilibrium, $M + A \rightleftharpoons MA$, so their concentrations satisfy an equilibrium constant K (see Chapter 8), where

$$K = \frac{[MA]}{[M]_{free}[A]_{free}} = \frac{[A]_{bound}}{([M] - [A]_{bound})[A]_{free}}$$

On division of both numerator and denominator by [M], and replacement of $[A]_{free}$ by $[A]_{out}$, this expression becomes

$$K = \frac{v}{(1 - v)[A]_{out}}$$

If there are N identical and independent binding sites on each macromolecule, each macromolecule behaves like N separate smaller macromolecules, with the same value of K for each site. The average number of A molecules per macromolecule is v/N, so the last equation becomes

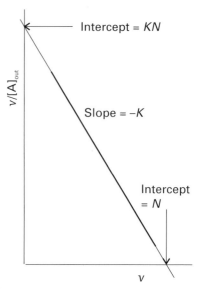

Fig B1 A Scatchard plot for the determination of the binding constant K and the number of binding sites v.

$$K = \frac{v/N}{\left(1 - \dfrac{v}{N}\right)[A]_{out}}$$

It then follows that

$$\frac{v}{[A]_{out}} = KN - Kv$$

This expression is the *Scatchard equation*. It implies that a plot of $v/[A]_{out}$ against v should be a straight line of slope $-K$ and intercept KN at $v = 0$ (see the illustration). From these two quantities, we can find the equilibrium constant for binding, the enthalpy of binding (from the temperature dependence of K and the van't Hoff equation, eqn 7.14), and the number of binding sites on each macromolecule. If a straight line is not obtained we can conclude that the binding sites are not equivalent or independent.

Exercise 1 Ethidium bromide binds to DNA by a process called *intercalation*, in which the aromatic ethidium bromide fits between two adjacent DNA bases. An equilibrium dialysis experiment was used to study the ethidium bromide (EB), binding to a short piece of DNA. A 1.00×10^{-6} M aqueous solution of DNA oligonucleotide was dialysed against an excess of EB. The following data were obtained for the total concentration of EB:

[EB]/μmol L^{-1}	
Side without DNA	**Side with DNA**
0.042	0.292
0.092	0.590
0.204	1.204
0.526	2.531
1.150	4.150

From these data, make a Scatchard plot and evaluate the intrinsic equilibrium constant, K, and total number of sites per DNA oligonucleotide. Is the identical and independent sites model for binding applicable?

Exercise 2 For nonidentical independent binding sites, the Scatchard equation is

$$\frac{v}{[A]_{out}} = \sum_i \frac{N_i K_i}{1 + K_i [A]_{out}}$$

Plot $v/[A]$ for the following cases. (a) There are four independent sites on an enzyme molecule and the intrinsic binding constant is $K = 1.0 \times 10^7$. (b) There are a total of six sites per polymer. Four of the sites are identical and have an intrinsic binding constant of 1×10^5. The binding constants for the other two sites are 2×10^6.

solution arises from the hydrostatic pressure of the column of solution that the osmosis itself produces. This column is formed when the pure solvent flows through the membrane into the solution and pushes the column of solution higher up the tube. Equilibrium is reached when the downward pressure exerted by the column of solution is equal to the upward osmotic pressure. A complication of this arrangement is that the entry of solvent into the solution results in dilution of the latter, so it is more difficult to treat than an arrangement in which an externally applied pressure opposes any flow of solvent into the solution.

The osmotic pressure of a solution is proportional to the concentration of solute. In fact, we can show that the expression for the osmotic pressure of an ideal solution bears an uncanny resemblance to the expression for the pressure of a perfect gas.

Derivation 6.4 *The van't Hoff equation*

The thermodynamic treatment of osmosis makes use of the fact that, at equilibrium, the chemical potential of the solvent A is the same on each side of the membrane (Fig 6.19). The starting relation is therefore

μ_A(solvent in the solution at pressure $p + \Pi$)
$= \mu_A$(pure solvent at pressure p)

The pure solvent is at atmospheric pressure, p, and the solution is at a pressure $p + \Pi$ on account of the additional pressure, Π, that has to be exerted on the solution to establish equilibrium. We shall write

the chemical potential of the pure solvent at the pressure p as $\mu_A^*(p)$. The chemical potential of the solvent in the solution is lowered by the solute but it is raised on account of the greater pressure, $p + \Pi$, acting on the solution. We shall denote this chemical potential by $\mu_A(x_A, p + \Pi)$. Our task is to find the extra pressure needed to balance the lowering of chemical potential caused by the solute.

The condition for equilibrium is

$$\mu_A^*(p) = \mu_A(x_A, p + \Pi)$$

We take the effect of the solute into account by using eqn 6.13:

$$\mu_A(x_A, p + \Pi) = \mu_A^*(p + \Pi) + RT \ln x_A$$

The effect of pressure on an (assumed incompressible) liquid is given by eqn 5.3 ($\Delta G_m = V_m \Delta p$) but now expressed in terms of the chemical potential:

$$\mu_A^*(p + \Pi) = \mu_A^*(p) + V_A \Delta p$$

At this point we identify the difference in pressure Δp as Π. When the last three equations are combined we get

$$-RT \ln x_A = \Pi V_A$$

The mole fraction of the solvent is equal to $1 - x_B$, where x_B is the mole fraction of solute molecules. In dilute solution, $\ln(1 - x_B)$ is approximately equal to $-x_B$ (for example, $\ln(1 - 0.01) = \ln 0.99 = -0.010050$), so this equation becomes

$$RTx_B \approx \Pi V_A$$

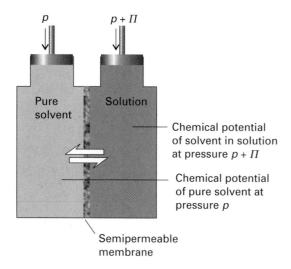

Fig 6.19 The basis of the calculation of osmotic pressure. The presence of a solute lowers the chemical potential of the solvent in the right-hand compartment, but the application of pressure raises it. The osmotic pressure is the pressure needed to equalize the chemical potential of the solvent in the two compartments.

When the solution is dilute, $x_B = n_B/n \approx n_B/n_A$. Moreover, because $n_A V_A \approx V$, the total volume of the solution, this equation becomes

$$n_B RT \approx \Pi V \qquad (6.21)$$

To remember this expression, note its resemblance to the perfect gas law, $nRT = pV$.

Equation 6.21 is called the **van't Hoff equation** for the osmotic pressure. Because $n_B/V = [B]$, the molar concentration of the solute, a simpler form is

$$\Pi \approx [B]RT \qquad (6.22)$$

This equation applies only to solutions that are sufficiently dilute to behave ideally.

One of the most common applications of osmosis is **osmometry**, the measurement of molar masses of proteins and synthetic polymers from the osmotic pressure of their solutions. As these huge molecules dissolve to produce solutions

that are far from ideal, we assume that the van't Hoff equation is only the first term of an expansion:

$$\Pi = [B]RT\{1 + B[B] + \cdots\} \qquad (6.23)$$

Exactly the same expansion was used in Section 1.12 to extend the perfect gas equation to real gases and led to the virial equation of state. The empirical parameter B in this expression is called the **osmotic virial coefficient**. To use eqn 6.23, we rearrange it into a form that gives a straight line by dividing both sides by $[B]$:

$$\frac{\Pi}{[B]} = \overbrace{RT}^{\text{Intercept}} + \overbrace{BRT}^{\text{Slope}}[B] + \cdots \qquad (6.24)$$

As we illustrate in the following example, we can find the molar mass of the solute B by measuring the osmotic pressure at a series of mass concentrations and making a plot of $\Pi/[B]$ against $[B]$ (Fig 6.20).

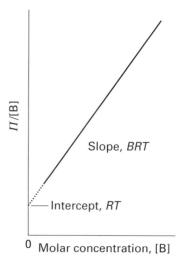

Fig 6.20 The plot and extrapolation made to analyse the results of an osmometry experiment.

Example 6.5 *Using osmometry to determine molar mass*

The osmotic pressures of solutions of an enzyme in water at 298 K are given below. The pressures are expressed in terms of the heights of solution (of density ρ = 0.9998 g cm^{-3}) in balance with the osmotic pressure. Find the molar mass of the enzyme.

c/(g dm^{-3})	1.00	2.00	4.00	7.00	9.00
h/cm	0.28	0.71	2.01	5.17	8.00

Strategy First, we need to express eqn 6.24 in terms of the mass concentration, c, and the height of solution, h, so that we can use the data. The osmotic pressure is related to the height h of the solution above the level of the solvent by[11] $\Pi = \rho g h$, where ρ is the mass density of the solution and g is the acceleration of free fall (9.81 m s^{-2}). The molar concentration [B] of the solute is related to the mass concentration c (in grams per litre) by

$$c = \frac{mass}{volume} = \frac{mass}{amount} \times \frac{amount}{volume} = M \times [B]$$

where M is the molar mass of the solute, so $[B] = c/M$. With these substitutions, eqn 6.24 becomes

$$\frac{\rho g h}{c/M} = RT + \frac{BRTc}{M} + + \cdots$$

Division through by $\rho g M$ gives

$$\frac{h}{c} = \frac{RT}{\rho g M} + \left(\frac{RTB}{\rho g M^2}\right) c + \cdots$$

It follows that, by plotting h/c against c, the results should fall on a straight line with intercept $RT/\rho g M$ on the vertical axis at c = 0. Therefore, by locating the intercept by extrapolation of the data to c = 0, we can find the molar mass of the solute.

Solution The following values of h/c can be calculated from the data:

c/(g dm^{-3})	1.00	2.00	4.00	7.00	9.00
$(h/\text{cm})/$ $(c/\text{g dm}^{-3})$	0.28	0.36	0.503	0.739	0.889

The points are plotted in Fig 6.21. The intercept with the vertical axis at c = 0 is at $(h/\text{cm})/(c/\text{g dm}^{-3})$ = 0.21, corresponding to h/c = 0.21 cm g^{-1} dm^3. Now note that

$$0.21 \ \frac{\text{cm dm}^3}{\text{g}} = 0.21 \times \frac{10^{-2} \ \text{m} \times 10^{-3} \ \text{m}^3}{10^{-3} \ \text{kg}}$$

$$= 2.1 \times 10^{-3} \ \text{m}^4 \ \text{kg}^{-1}$$

Therefore, because this intercept is equal to $RT/\rho g M$, we can write

$$M = \frac{RT/\rho g}{2.1 \times 10^{-3} \ \text{m}^4 \ \text{kg}^{-1}}$$

It follows that

$$M = \frac{(8.31451 \ \text{J K}^{-1} \text{mol}^{-1}) \times (298 \ \text{K})}{(999.8 \ \text{kg m}^{-3}) \times (9.81 \ \text{m s}^{-2}) \times (2.1 \times 10^{-3} \ \text{m}^4 \text{kg}^{-1})}$$

$$= 1.2 \times 10^2 \ \text{kg mol}^{-1}$$

The molar mass of the enzyme is therefore close to 120 kDa.

Self-test 6.10

The heights of the solution in an osmometry experiment on a solution of poly(vinyl chloride), PVC, in dioxane at 25°C were as follows:

c/(g dm^{-3})	0.50	1.00	1.50	2.00	2.50
h/cm	0.18	0.35	0.53	0.71	0.90

The density of the solution is 0.980 g cm^{-3}. What is the molar mass of the polymer?

[*Answer:* 72 kg mol^{-1}]

[11] This expression for the pressure at the foot of a column of liquid was calculated in *Derivation 0.1*.

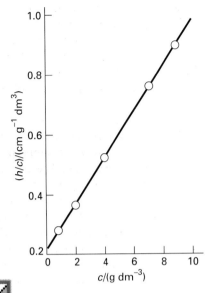

Fig 6.21 The plot of the data in Example 6.5. The molar mass is determined from the intercept at $c = 0$.

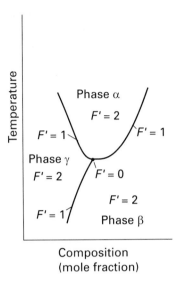

Fig 6.22 The interpretation of a temperature–composition phase diagram at constant pressure. In a region, where only one phase is present, $F' = 2$ and both composition and temperature can be varied. On a phase boundary, where two phases are in equilibrium, $F' = 1$ and only one variable can be changed independently. At a point where three phases are present in equilibrium, $F' = 0$, and the temperature and composition are fixed.

Phase diagrams of mixtures

As in the discussion of pure substances, the phase diagram of a mixture shows which phase is most stable for the given conditions. However, composition is now a variable in addition to the pressure and temperature.

It will be useful to keep in mind the implications of the phase rule ($F = C - P + 2$, Section 5.7). We shall consider only **binary mixtures**, which are mixtures of two components (such as ethanol and water) and may therefore set $C = 2$. Then $F = 4 - P$. For simplicity we keep the pressure constant (at 1 atm, for instance), which uses up one of the degrees of freedom, and write $F' = 3 - P$ for the number of degrees of freedom remaining. One of these degrees of freedom is the temperature; the other is the composition. Hence we should be able to depict the phase equilibria of the system on a **temperature–composition diagram** in which one axis is the temperature and the other axis is the mole fraction. In a region where there is only one phase, $F' = 2$ and

both the temperature and the composition can be varied (Fig 6.22). If two phases are present at equilibrium, $F' = 1$, and only one of the two variables may be changed at will. For example, if we change the composition, then to maintain equilibrium between the two phases we have to adjust the temperature too. Such two-phase equilibria therefore define a line in the phase diagram. If three phases are present, $F' = 0$ and there is no degree of freedom for the system. To achieve equilibrium between three phases we must adopt a specific temperature and composition. Such a condition is therefore represented by a point on the phase diagram.

6.9 Mixtures of volatile liquids

First, we consider the phase diagram of a binary mixture of two volatile components. This kind of system is important for understanding fractional distillation, which is a widely used technique in industry and the laboratory. Intuitively, we might expect the boiling point of a mixture of two volatile liquids to vary smoothly from the boiling point of one pure component when only that liquid is present to the boiling point of the other pure component when only that liquid is present. This expectation is often borne out in practice, and Fig 6.23 shows a typical plot of boiling point against composition (the lower curve).

The vapour in equilibrium with the boiling mixture is also a mixture of the two components. We should expect the vapour to be richer than the liquid mixture in the more volatile of the two substances. This difference is also often found in practice, and the upper curve in the illustration shows the composition of the vapour in equilibrium with the boiling liquid. To identify the composition of the vapour, we note the boiling point of the liquid mixture (point a, for instance) and draw a horizontal **tie line**, a line joining two phases that are in equilibrium with each other, across to the upper curve. Its point of intersection (a') gives the composition of the vapour. In this example, we see that the mole fraction of A in the vapour is about 0.6. As expected, the vapour is richer than the liquid in the more volatile component.

Graphs like these are determined empirically, by measuring the boiling points of a series of mixtures (to plot the lower curve of boiling point against composition), and measuring the composition of the vapour in equilibrium with each boiling mixture (to plot the corresponding points of the vapour-composition curve).

We can follow the changes that occur during the fractional distillation of a mixture of volatile liquids by following what happens when a mixture of composition a_1 is heated (Fig 6.24). The mixture boils at a_2 and its vapour has composition a_2'. This vapour condenses to a liquid of the same composition when it has risen to a cooler part of the 'fractionating column', a vertical column packed with glass rings or beads to give a large surface area. This condensate boils at the temperature corresponding to the point a_3 and yields a vapour of composition a_3'. This vapour is even richer in the

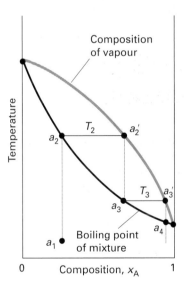

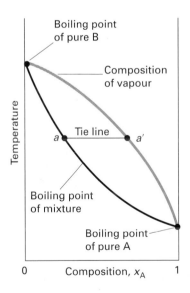

Fig 6.23 A temperature–composition diagram for a binary mixture of volatile liquids. The tie line connects the points that represent the compositions of liquid and vapour that are in equilibrium at each temperature. The lower curve is a plot of the boiling point of the mixture against composition.

Fig 6.24 The process of fractional distillation can be represented by a series of steps on a temperature–composition diagram like that in Fig 6.23. The initial liquid mixture may be at a temperature and have a composition like that represented by point a_1. It boils at the temperature T_2, and the vapour in equilibrium with the boiling liquid has composition a_2'. If that vapour is condensed (to a_3 or below), the resulting condensate boils at T_3 and gives rise to a vapour of composition represented by a_3'. As the succession of vaporizations and condensations is continued, the composition of the distillate moves towards pure A (the more volatile component).

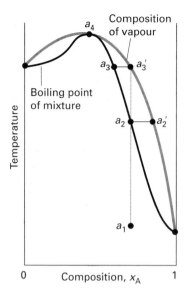

Fig 6.25 The temperature–composition diagram for a high-boiling azeotrope. As fractional distillation proceeds, the composition of the remaining liquid moves towards a_4; however, once there, the vapour in equilibrium with that liquid has the same composition, so the mixture evaporates with an unchanged composition and no further separation can be achieved.

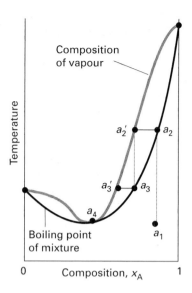

Fig 6.26 The temperature–composition diagram for a low-boiling azeotrope. As fractional distillation proceeds, the composition of the vapour moves towards a_4; however, once there, the vapour in equilibrium with that liquid has the same composition, so no further separation of the distillate can be achieved.

more volatile component. That vapour condenses to a liquid which boils at the temperature corresponding to the point a_4. The cycle is repeated until almost pure A emerges from the top of the column.

Whereas many binary liquid mixtures do have temperature–composition diagrams resembling that shown in Fig 6.24, in a number of important cases there are marked differences. For example, a maximum in the boiling point curve is sometimes found (Fig 6.25). This behaviour is a sign that favourable interactions between the molecules of the two components reduce the vapour pressure of the mixture below the ideal value. Examples of this behaviour include trichloromethane/propanone and nitric acid/water mixtures. Temperature–composition curves are also found that pass through a minimum (Fig 6.26). This behaviour indicates that the (A,B) interactions are unfavourable and hence that the mixture is more volatile than expected on the basis of simple mingling of the two species. Examples include dioxane/water and ethanol/water.

There are important consequences for distillation when the temperature–composition diagram has a maximum or a minimum. Consider a liquid of composition a_1 on the right of the maximum in Fig 6.25. It boils at a_2 and its vapour (of composition a_2') is richer in the more volatile component A. If that vapour is removed, the composition of the remaining liquid moves towards a_3. The vapour in equilibrium with this boiling liquid has composition a_3': note that the two compositions are more similar than the original pair (a_3 and a_3' are closer together than a_2 and a_2'). If that vapour is removed, the composition of the boiling liquid shifts towards a_4 and the vapour of that boiling mixture has a composition identical to that of the liquid. At this stage, evaporation occurs without change of composition. The mixture is said to form an **azeotrope**.[12] When the azeotropic composition has been reached, distillation cannot separate the two liquids because the condensate retains the composition of the liquid. One example of azeotrope formation is hydrochloric

---

[12] The name comes from the Greek words for 'boiling without changing'.

acid/water, which is azeotropic at 80 per cent water (by mass) and boils unchanged at 108.6°C.

The system shown in Fig 6.26 is also azeotropic, but shows its character in a different way. Suppose we start with a mixture of composition a_1 and follow the changes in the composition of the vapour that rises through a fractionating column. The mixture boils at a_2 to give a vapour of composition a_2'. This vapour condenses in the column to a liquid of the same composition (now marked a_3). That liquid reaches equilibrium with its vapour at a_3', which condenses higher up the tube to give a liquid of the same composition. The fractionation therefore shifts the vapour towards the azeotropic composition at a_4, but the composition cannot move beyond a_4 because now the vapour and the liquid have the same composition. Consequently, the azeotropic vapour emerges from the top of the column. An example is ethanol/water, which boils unchanged when the water content is 4 per cent and the temperature is 78°C.

6.10 Liquid–liquid phase diagrams

Partially miscible liquids are liquids that do not mix together in all proportions. An example is a mixture of hexane and nitrobenzene: when the two liquids are shaken together, the liquid consists of two liquid phases: one is a saturated solution of hexane in nitrobenzene and the other is a saturated solution of nitrobenzene in hexane. Because the two solubilities vary with temperature, the compositions and proportions of the two phases change as the temperature is changed. We can use a temperature–composition diagram to display the composition of the system at each temperature.

Suppose we add a small amount of nitrobenzene to hexane at a temperature T'. The nitrobenzene dissolves completely; however, as more nitrobenzene is added, a stage comes when no more dissolves. The sample now consists of two phases in equilibrium with each other, the more abundant one consisting of hexane saturated with nitrobenzene, the less abundant one a trace of nitrobenzene saturated with hexane. In the temperature–composition diagram drawn in Fig 6.27, the composition of the former is represented by the point a' and that

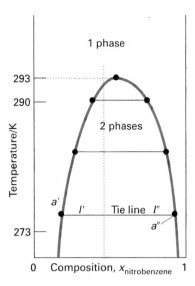

Fig 6.27 The temperature–composition diagram for hexane and nitrobenzene at 1 atm. The upper critical solution temperature, T_{uc}, is the temperature above which no phase separation occurs. For this system it lies at 293 K (when the pressure is 1 atm).

of the latter by the point a''. The relative abundances of the two phases are given by the **lever rule** (Fig 6.28), which we can derive as follows.

Derivation 6.5 *The lever rule*

We write $n = n' + n''$, where n' is the total amount of molecules in the one phase, n'' is the total amount in the other phase, and n is the total amount of molecules in the sample. The total amount of A in the sample is nx_A, where x_A is the overall mole fraction of A in the sample (this is the quantity plotted along the horizontal axis). The overall amount of A is also the sum of its amounts in the two phases, where it has the mole fractions x_A' and x_A'', respectively:

$$nx_A = n'x_A' + n''x_A''$$

We can also multiply each side of the relation $n = n' + n''$ by x_A and obtain

$$nx_A = n'x_A + n''x_A$$

Then, by equating these two expressions and rearranging them slightly, it follows that

$$n'(x_A' - x_A) = n''(x_A - x_A'')$$

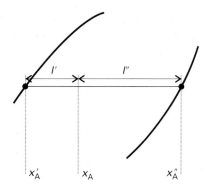

Fig 6.28 The coordinates and compositions referred to by the lever rule.

or (as can be seen by referring to Fig 6.28)

$$n'l' = n''l''$$

In other words,

$$\frac{\text{Amount of phase of composition } a''}{\text{Amount of phase of composition } a'} = \frac{l'}{l''} \quad (6.25)$$

Example 6.6 *Interpreting a liquid–liquid phase diagram*

A mixture of 50 g (0.59 mol) of hexane and 50 g (0.41 mol) of nitrobenzene was prepared at 290 K. What are the compositions of the phases, and in what proportions do they occur? To what temperature must the sample be heated in order to obtain a single phase at 1 atm?

Strategy The answer is based on Fig 6.27. First, we need to identify the tie line corresponding to the temperature specified: the points at its two ends give the compositions of the two phases in equilibrium. Next, we identify the location on the horizontal axis corresponding to the overall composition of the system and draw a vertical line. Where that line cuts the tie line it divides it into the two lengths needed to use the lever rule, eqn 6.25. For the final part, we note the temperature at which the same vertical line cuts through the phase boundary: at that temperature and above, the system consists of a single phase.

Solution We denote hexane by H and nitrobenzene by N. The horizontal tie line at 290 K cuts the phase

boundary at $x_N = 0.38$ and at $x_N = 0.74$, so those mole fractions are the compositions of the two phases. The overall composition of the system corresponds to $x_N = 0.41$, so we draw a vertical line at that mole fraction. The lever rule then gives the ratio of amounts of each phase as

$$\frac{l'}{l''} = \frac{0.41 - 0.38}{0.74 - 0.41} = \frac{0.03}{0.33} = 0.1$$

We conclude that the hexane-rich phase is ten times more abundant than the nitrobenzene-rich phase at this temperature. Heating the sample to 292 K takes it into the single-phase region.

Self-test 6.11

Repeat the problem for 50 g hexane and 100 g nitrobenzene at 273 K.

[*Answer:* $x_N = 0.09$ and 0.95 in the ratio 1:1.3; 290 K]

When more nitrobenzene is added to the two-phase mixture at the temperature T', hexane dissolves in it slightly. The overall composition moves to the right in the phase diagram, but the compositions of the two phases in equilibrium remain a' and a''. The difference is that the amount of the second phase increases at the expense of the first. A stage is reached when so much nitrobenzene is present that it can dissolve all the hexane, and the system reverts to a single phase. Now the point representing the overall composition and temperature lies to the right of the phase boundary in the illustration and the system is a single phase.

The **upper critical solution temperature**, T_{uc}, is the upper limit of temperatures at which phase separation occurs.[13] Above the upper critical solution temperature the two components are fully miscible. In molecular terms, this temperature exists because the greater thermal motion of the molecules leads to greater miscibility of the two components. In thermodynamic terms, the Gibbs energy of mixing becomes negative above a certain temperature, regardless of the composition.

Some systems show a **lower critical solution temperature**, T_{lc}, below which they mix in all

[13] The upper critical solution temperature is also called the *upper consolute temperature*.

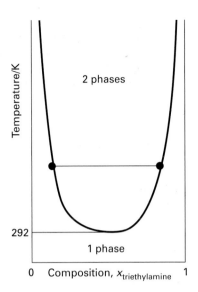

Fig 6.29 The temperature–composition diagram for water and triethylamine. The lower critical solution temperature, T_{lc}, is the temperature below which no phase separation occurs. For this system it lies at 292 K (when the pressure is 1 atm).

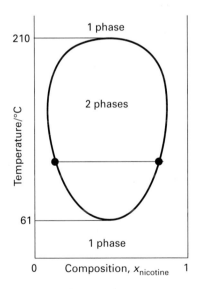

Fig 6.30 The temperature–composition diagram for water and nicotine, which has both upper and lower critical solution temperatures. Note the high temperatures on the graph: the diagram corresponds to a sample under pressure.

proportions and above which they form two phases.[14] An example is water and triethylamine (Fig 6.29). In this case, at low temperatures the two components are more miscible because they form a weak complex; at higher temperatures the complexes break up and the two components are less miscible.

A few systems have both upper and lower critical temperatures. The reason can be traced to the fact that, after the weak complexes have been disrupted, leading to partial miscibility, the thermal motion at higher temperatures homogenizes the mixture again, just as in the case of ordinary partially miscible liquids. One example is nicotine and water, which are partially miscible between 61°C and 210°C (Fig 6.30).

6.11 Liquid–solid phase diagrams

Phase diagrams are also used to show the regions of temperature and composition at which solids and liquids exist in binary systems. Such diagrams are useful for discussing the techniques that are used to prepare the high-purity materials used in the electronics industry and are also of great importance in metallurgy.

Figure 6.31 shows the phase diagram for a system composed of two metals that are almost completely immiscible right up to their melting points (such as antimony and bismuth). Consider the molten liquid of composition a_1. When the liquid is cooled to a_2, the system enters the two-phase region labelled 'Liquid + A'. Almost pure solid A begins to come out of solution and the remaining liquid becomes richer in B. On cooling to a_3, more of the solid forms, and the relative amounts of the solid and liquid (which are in equilibrium) are given by the lever rule: at this stage there are nearly equal amounts of each. The liquid phase is richer in B than before (its composition is given by b_3) because A has been deposited. At a_4 there is less liquid than at a_3 and its composition is given by e. This liquid now freezes to give a two-phase system of almost pure A and

---

[14] The lower critical solution temperature is also called the *lower consolute temperature*.

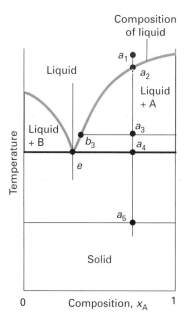

Fig 6.31 The temperature–composition diagram for two almost immiscible solids and their completely immiscible liquids. The vertical line through *e* corresponds to the eutectic composition, the mixture with lowest melting point.

almost pure B and cooling down to a_5 leads to no further change in composition.

The vertical line through *e* in Fig 6.31 corresponds to the **eutectic composition**.[15] A solid with the eutectic composition melts, without change of composition, at the lowest temperature of any mixture. Solutions of composition to the right of *e* deposit A as they cool, and solutions to the left deposit B: only the eutectic mixture (apart from pure A or pure B) solidifies at a single definite temperature without gradually unloading one or other of the components from the liquid.

One technologically important eutectic is solder, which consists of 67 per cent tin and 33 per cent lead by mass and melts at 183°C. Eutectic formation occurs in the great majority of binary alloy systems. It is of great importance for the microstructure of solid materials for, although a eutectic solid is a two-phase system, it crystallizes out in a nearly homogeneous mixture of microcrystals. The two microcrystalline phases can be distinguished by

---

[15] The name comes from the Greek words for 'easily melted'.

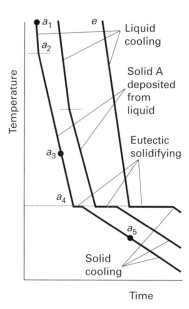

Fig 6.32 The cooling curves for the system shown in Fig 6.31. For a sample of composition represented by the vertical line through a_1 to a_5, the rate of cooling decreases at a_2 because solid A comes out of solution. The second cooling curve is for a sample of intermediate composition (between the vertical lines through *a* and *e*). If the experiment is repeated using a sample of composition represented by the vertical line through *e*, then there is a complete halt at *e* when the eutectic solidifies without change of composition. The halt is longest for the mixture of eutectic composition. The cooling curves can be used to construct the phase diagram.

microscopy and structural techniques such as X-ray diffraction.

Thermal analysis is a very useful practical way of detecting eutectics. We can see how it is used by considering the rate of cooling down the vertical line at a_1 in Fig 6.31. The liquid cools steadily until it reaches a_2, when A begins to be deposited. Cooling is now slower because the solidification of A is exothermic and retards the cooling (Fig 6.32). When the remaining liquid reaches the eutectic composition, the temperature remains constant until the whole sample has solidified: this pause in the decrease in temperature is known as the **eutectic halt**. If the liquid has the eutectic composition *e* initially, then the liquid cools steadily down to the freezing temperature of the eutectic, when there is a long eutectic halt as the entire sample solidifies just like the freezing of a pure liquid.

Monitoring the cooling curves at different overall compositions gives a clear indication of the structure of the phase diagram. The solid–liquid boundary is given by the points at which the rate of cooling changes. The longest eutectic halt gives the location of the eutectic composition and its melting temperature.

6.12 **Ultrapurity and controlled impurity**

Advances in technology have called for materials of extreme purity. For example, semiconductor devices consist of almost perfectly pure silicon or germanium doped to a precisely controlled extent. For these materials to operate successfully, the impurity level must be kept down to less than 1 in 10^9. The technique of **zone refining** makes use of the nonequilibrium properties of mixtures. It relies on the impurities being more soluble in the molten sample than in the solid, and sweeps them up by passing a molten zone repeatedly from one end to the other along a sample (Fig 6.33). In practice, a train of hot and cold zones are swept repeatedly from one end to the other. The zone at the end of the sample is the impurity dump: when the heater has gone by, it cools to a dirty solid that can be discarded.

We can use a phase diagram to discuss zone refining, but we have to allow for the fact that the molten zone moves along the sample and the sample is uniform in neither temperature nor composition. Consider a liquid (which represents the molten zone) on the vertical line at a_1 in Fig 6.34, and let it cool without the entire sample coming to overall equilibrium. If the temperature falls to a_2, a solid of composition b_2 is deposited and the remaining liquid (the zone where the heater has moved on) is at a_2'. Cooling that liquid down a vertical line passing through a_2' deposits solid of composition b_3 and leaves liquid at a_3'. The process continues until the last drop of liquid to solidify is heavily contaminated with A. There is plenty of everyday evidence that impure liquids freeze in this way. For example, an ice cube is clear near the surface but misty in the core. The water used to make ice normally contains dissolved air; freezing proceeds from the outside, and air is accumulated in the retreating liquid phase. The air cannot escape from the interior of the cube, so when that freezes the air is trapped in a mist of tiny bubbles.

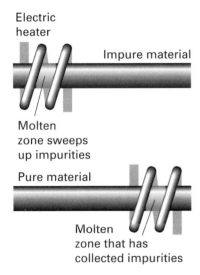

Fig 6.33 In the zone refining procedure, a heater is used to melt a small region of a long cylindrical sample of the impure solid, and that zone is swept to the other end of the rod. As it moves, it collects impurities. If a series of passes is made, the impurities accumulate at one end of the rod and can be discarded.

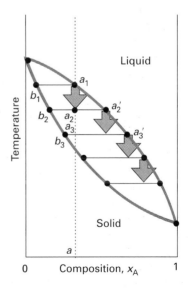

Fig 6.34 A binary temperature–composition diagram can be used to discuss zone refining, as explained in the text.

A modification of zone refining is **zone levelling**. This technique is used to introduce controlled amounts of impurity (for example, of indium into germanium). A sample rich in the required dopant is put at the head of the main sample, and made molten. The zone is then dragged repeatedly in alternate directions through the sample, where it deposits a uniform distribution of the impurity.

Exercises

6.1 What mass of glucose should you use to prepare 250.0 mL of 0.112 M $C_6H_{12}O_6(aq)$?

6.2 What mass of glucose should you use to prepare 0.112 m $C_6H_{12}O_6(aq)$ using 250.0 g of water?

6.3 What is the mass of glycine in 25.00 mL of 0.245 M $NH_2CH_2COOH(aq)$?

6.4 What is the mole fraction of alanine in 0.134 m $CH_3CH(NH_2)COOH(aq)$?

6.5 What mass of sucrose, $C_{12}H_{22}O_{11}$, should you dissolve in 100.0 g of water to obtain a solution in which the mole fraction of $C_{12}H_{22}O_{11}$ is 0.124?

6.6 A mixture was prepared consisting of 50.0 g of 1-propanol and 50.0 g of 2-propanol. What are the mole fractions of the two alcohols?

6.7 A mixture was prepared that consists of 40.0 g of 1-propanol and 60.0 g of 1-butanol. Calculate the mole fractions of the two components.

6.8 The partial molar volumes of propanone and trichloromethane in a mixture in which the mole fraction of $CHCl_3$ is 0.4693 are 74.166 and 80.235 $cm^3\ mol^{-1}$, respectively. What is the volume of a solution of total mass 1.000 kg?

6.9 Use Fig 6.1 to estimate the total volume of a solution formed by mixing 50.0 mL of ethanol with 50.0 mL of water. The densities of the two liquids are 0.789 and 1.000 g cm^{-3}, respectively.

6.10 The partial molar volume of ethanol in a mixture with water at 25°C is

$$V_{ethanol}/(mL\ mol^{-1}) = 54.6664 - 0.72788b + 0.084768b^2$$

where b is the numerical value of the molality of ethanol. Plot this quantity as a function of b and identify the com-

position at which the partial molar volume is a minimum (*Hint.* A more accurate value is found by using calculus.) Express that composition as a mole fraction.

6.11 The total volume of a water–ethanol mixture at 25°C fits the expression

$$V/mL = 1002.93 + 54.6664b - 0.36394b^2 + 0.028256b^3$$

where b is the numerical value of the molality of ethanol. With the information in Exercise 6.10, find an expression for the partial molar volume of water. Plot the curve. Show that the partial molar volume of water has a maximum value where the partial molar volume of ethanol is a minimum.

6.12 Calculate (a) the (molar) Gibbs energy of mixing, (b) the (molar) entropy of mixing when the two major components of air (nitrogen and oxygen) are mixed to form air at 298 K. The mole fractions of N_2 and O_2 are 0.78 and 0.22, respectively. Is the mixing spontaneous?

6.13 Suppose now that argon is added to the mixture in Exercise 6.12 to bring the composition closer to real air, with mole fractions 0.780, 0.210, and 0.0096, respectively. What is the additional change in molar Gibbs energy and entropy? Is the mixing spontaneous?

6.14 A solution is prepared by dissolving 1.23 g of C_{60} (buckminsterfullerene) in 100.0 g of toluene (methylbenzene). Given that the vapour pressure of pure toluene is 5.00 kPa at 30°C, what is the vapour pressure of toluene in the solution?

6.15 Estimate the vapour pressure of sea water at 20°C given that the vapour pressure of pure water is 2.338 kPa at that temperature and the solute is largely Na^+ and Cl^- ions, each present at about 0.50 mol L^{-1}.

6.16 At 300 K, the vapour pressure of dilute solutions of HCl in liquid $GeCl_4$ are as follows:

$x(HCl)$	0.005	0.012	0.019
p/kPa	32.0	76.9	121.8

Show that the solution obeys Henry's law in this range of mole fractions and calculate the Henry's law constant at 300 K.

6.17 Calculate the concentration of carbon dioxide in fat given that the Henry's law constant is 8.6×10^4 Torr and the partial pressure of carbon dioxide is 55 kPa.

6.18 The rise in atmospheric carbon dioxide results in higher concentrations of dissolved carbon dioxide in natural waters. Use Henry's law and the data in Table 6.1 to calculate the solubility of CO_2 in water at 25°C when its partial pressure is (a) 4.0 kPa, (b) 100 kPa.

6.19 The mole fractions of N_2 and O_2 in air at sea level are approximately 0.78 and 0.21. Calculate the molalities of the solution formed in an open flask of water at 25°C.

6.20 A water-carbonating plant is available for use in the home and operates by providing carbon dioxide at 3.0 atm. Estimate the molar concentration of the CO_2 in the soda water it produces.

6.21 At 90°C the vapour pressure of toluene (methylbenzene) is 400 Torr and that of o-xylene (1,2-dimethylbenzene) is 150 Torr. What is the composition of the liquid mixture that boils at 25°C when the pressure is 0.50 atm? What is the composition of the vapour produced?

6.22 The vapour pressure of a sample of benzene is 400 Torr at 60.6°C, but it fell to 386 Torr when 0.125 g of an organic compound was dissolved in 5.00 g of the solvent. Calculate the molar mass of the compound.

6.23 Estimate the freezing point of 150 cm^3 of water sweetened with 7.5 g of sucrose.

6.24 The addition of 28.0 g of a compound to 750 g of tetrachloromethane, CCl_4, lowered the freezing point of the solvent by 5.40 K. Calculate the molar mass of the compound.

6.25 A compound A existed in equilibrium with its dimer, A_2, in propanone solution. Derive an expression for the equilibrium constant $K = [A_2]/[A]^2$ in terms of the depression in vapour pressure caused by a given concentration of compound. (Hint. Suppose that a fraction f of the A molecules are present as the dimer. The depression of vapour pressure is proportional to the total concen-

tration of A and A_2 molecules regardless of their chemical identities.)

6.26 The osmotic pressure of an aqueous solution of urea at 300 K is 120 kPa. Calculate the freezing point of the same solution.

6.27 The osmotic pressure of a solution of polystyrene in toluene was measured at 25°C and the pressure was expressed in terms of the height of the solvent of density 1.004 g cm^{-3}:

$c/(g\,L^{-1})$	2.042	6.613	9.521	12.602
h/cm	0.592	1.910	2.750	3.600

Calculate the molar mass of the polymer.

6.28 The molar mass of an enzyme was determined by dissolving it in water, measuring the osmotic pressure at 20°C, and extrapolating the data to zero concentration. The following data were used:

$c/(mg\,cm^{-3})$	3.221	4.618	5.112	6.722
h/cm	5.746	8.238	9.119	11.990

Calculate the molar mass of the enzyme.

6.29 The following temperature/composition data were obtained for a mixture of octane (O) and toluene (T) at 760 Torr, where x is the mole fraction in the liquid and y the mole fraction in the vapour at equilibrium.

$\theta/°C$	110.9	112.0	114.0	115.8	117.3	119.0	120.0	123.0
x_T	0.908	0.795	0.615	0.527	0.408	0.300	0.203	0.097
y_T	0.923	0.836	0.698	0.624	0.527	0.410	0.297	0.164

The boiling points are 110.6°C for toluene and 125.6°C for octane. Plot the temperature–composition diagram of the mixture. What is the composition of the vapour in equilibrium with the liquid of composition (a) $x_T = 0.250$ and (b) $x_O = 0.250$?

6.30 Sketch the phase diagram of the system NH_3/N_2H_4 given that the two substances do not form a compound with each other, that NH_3 freezes at −78°C and N_2H_4 freezes at +2°C, and that a eutectic is formed when the mole fraction of N_2H_4 is 0.07 and that the eutectic melts at −80°C.

6.31 Figure 6.35 shows the phase diagram for two partially miscible liquids, which can be taken to be that for water (A) and 2-methyl-1-propanol (B). Describe what will be observed when a mixture of composition b_3 is heated, at each stage giving the number, composition, and relative amounts of the phases present.

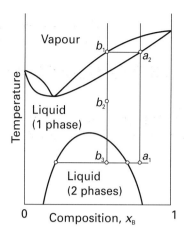

Fig 6.35

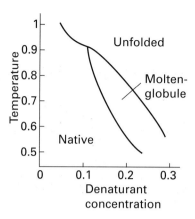

Fig 6.37

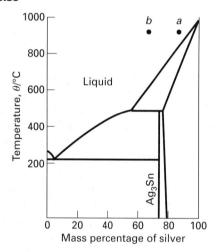

Fig 6.36

phase diagram. Describe the phase changes that occur when perfluorohexane is added to a fixed amount of hexane at (a) 23°C, (b) 25°C.

6.36 In a theoretical study of protein-like polymers, the phase diagram shown in Fig 6.37 was obtained. It shows three structural regions: the native form, the unfolded form, and a 'molten globule' form. (a) Is the molten-globule form ever stable when the denaturant concentration is below 0.1? (b) Describe what happens to the polymer as the native form is heated in the presence of denaturant at concentration 0.15.

6.37 In an experimental study of membrane-like assemblies of synthetic materials, a phase diagram like that shown in Fig 6.38 was obtained. The two components are dielaidoylphosphatidylcholine (DEL) and dipalmitoylphosphatidylcholine (DPL). Explain what happens as a liquid mixture of composition $x_{DEL} = 0.5$ is cooled from 45°C.

6.32 Figure 6.36 is the phase diagram for silver/tin. Label the regions, and describe what will be observed when liquids of compositions a and b are cooled to 200°C.

6.33 Sketch the cooling curves for the compositions a and b in Fig 6.36.

6.34 Use the phase diagram in Fig 6.36 to determine (a) the solubility of silver in tin at 800°C, (b) the solubility of Ag_3Sn in silver at 460°C, and (c) the solubility of Ag_3Sn in silver at 300°C.

6.35 Hexane and perfluorohexane (C_6F_{14}) show partial miscibility below 22.70°C. The critical concentration at the upper critical temperature is $x = 0.355$, where x is the mole fraction of C_6F_{14}. At 22.0°C the two solutions in equilibrium have $x = 0.24$ and $x = 0.48$ respectively, and at 21.5°C the mole fractions are 0.22 and 0.51. Sketch the

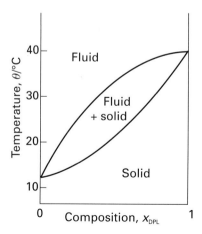

Fig 6.38

Principles of chemical equilibrium

Contents

CHEMICAL thermodynamics is used to predict whether a mixture of reactants has a spontaneous tendency to change into products, to predict the composition of the reaction mixture at equilibrium, and to predict how that composition will be modified by changing the conditions. Although reactions in industry are rarely allowed to reach equilibrium, knowing whether equilibrium lies in favour of reactants or products under certain conditions is a good indication of the feasibility of a process. Much the same is true of biochemical reactions, where the avoidance of equilibrium is life and the attainment of equilibrium is death. Nevertheless, the material we cover in this chapter is of crucial importance for understanding the processes of metabolism, respiration, and all the processes going on inside organisms.

Thermodynamic background

THE thermodynamic criterion for spontaneous change at constant temperature and pressure is $\Delta G < 0$. The principal idea behind this chapter, therefore, is that, *at constant temperature and pressure, a reaction mixture tends to adjust its composition until its Gibbs energy is a minimum.* If the Gibbs energy of a mixture varies as shown in Fig 7.1a, very little of the

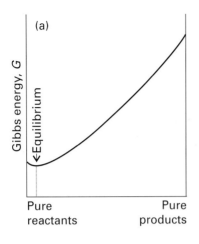

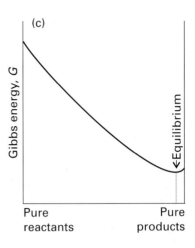

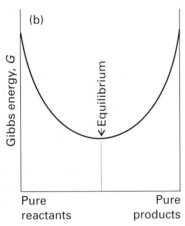

Fig 7.1 The variation of Gibbs energy of a reaction mixture with progress of the reaction, pure reactants on the left and pure products on the right. (a) This reaction 'does not go': the minimum in the Gibbs energy occurs very close to reactants. (b) This reaction reaches equilibrium with approximately equal amounts of reactants and products present in the mixture. (c) This reaction goes almost to completion, as the minimum in Gibbs energy lies very close to pure products.

reactants convert into products before G has reached its minimum value and the reaction 'does not go'. If G varies as shown in Fig 7.1c, then a high proportion of products must form before G reaches its minimum and the reaction 'goes'. Many reactions have a Gibbs energy that varies as shown in Fig 7.1b, and at equilibrium the reaction mixture contains substantial amounts of both reactants and products. One of our tasks is to see how to use thermodynamic data to predict the equilibrium composition in all three cases and to see how that composition depends on the conditions.

7.1 The reaction Gibbs energy

To keep our ideas in focus, we consider two reactions that are important to the survival of civilization. One is the isomerism of glucose-6-phosphate (**1**, G6P) to fructose-6-phosphate (**2**, F6P), which is

the second step in the glycolytic pathway in which 6-carbon molecules are broken down into 3-carbon molecules in the course of metabolism of carbohydrates:

$$G6P(aq) \rightarrow F6P(aq) \qquad (A)$$

This reaction takes place in the aqueous environment of the cell. Of great importance in industry is the synthesis of ammonia:

$$N_2(g) + 3 H_2(g) \rightarrow 2 NH_3(g) \qquad (B)$$

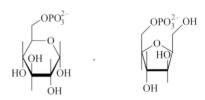

1 Glucose-6-phosphate   **2** Fructose-6-phosphate

These two reactions are specific examples of a general reaction of the form

$$a\,A + b\,B \rightarrow c\,C + d\,D \tag{C}$$

with arbitrary physical states.

First, consider reaction A. Suppose that, in a short interval while the reaction is in progress, the amount of G6P changes by $-\Delta n$. As a result of this change in amount, the contribution of G6P to the total Gibbs energy of the system changes by $-\mu_{G6P}\Delta n$, where μ_{G6P} is the chemical potential (the partial molar Gibbs energy) of G6P in the reaction mixture. In the same interval, the amount of F6P changes by $+\Delta n$, so its contribution to the total Gibbs energy changes by $+\mu_{F6P}\Delta n$, where μ_{F6P} is the chemical potential of F6P. Provided Δn is small enough to leave the composition virtually unchanged, the net change in Gibbs energy of the system is

$$\Delta G = \mu_{F6P} \times \Delta n - \mu_{G6P} \times \Delta n$$

If we divide through by Δn, we obtain the **reaction Gibbs energy**, $\Delta_r G$:

$$\Delta_r G = \frac{\Delta G}{\Delta n} = \mu_{F6P} - \mu_{G6P} \tag{7.1a}$$

There are two ways to interpret $\Delta_r G$. First, it is the difference of the chemical potentials of the products and reactants *at the composition of the reaction mixture*. Second, because $\Delta_r G$ is the change in G divided by the change in composition, we can think of it as the slope of the graph of G plotted against the changing composition of the system (Fig 7.2).

The synthesis of ammonia provides a slightly more complicated example. If the amount of N_2 changes by $-\Delta n$, then from the reaction stoichiometry we know that the change in the amount of H_2 will be $-3\Delta n$ and the change in the amount of NH_3 will be $+2\Delta n$. Each change contributes to the change in the total Gibbs energy of the mixture, and the overall change is

$$\Delta G = \mu_{NH_3} \times 2\Delta n - \mu_{N_2} \times \Delta n - \mu_{H_2} \times 3\Delta n$$

$$= (2\mu_{NH_3} - \mu_{N_2} - 3\mu_{H_2})\Delta n$$

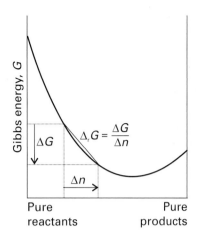

Fig 7.2 The variation of Gibbs energy with progress of reaction showing how the reaction Gibbs energy, $\Delta_r G$, is related to the slope of the curve at a given composition.

where the μ_J are the chemical potentials of the species in the reaction mixture. In this case, therefore, the reaction Gibbs energy is

$$\Delta_r G = \frac{\Delta G}{\Delta n} = 2\mu_{NH_3} - (\mu_{N_2} + 3\mu_{H_2}) \tag{7.1b}$$

Note that each chemical potential is multiplied by the corresponding stoichiometric coefficient and that reactants are subtracted from products. For the general reaction C,

$$\Delta_r G = (c\mu_C + d\mu_D) - (a\mu_A - b\mu_B) \tag{7.1c}$$

The chemical potential of a substance depends on the composition of the mixture in which it is present, and is high when its concentration or partial pressure is high. Therefore, $\Delta_r G$ changes as the composition changes (Fig 7.3). Remember that $\Delta_r G$ is the *slope* of G plotted against composition. We see that $\Delta_r G < 0$ and the slope of G is negative (down from left to right) when the mixture is rich in the reactants A and B because μ_A and μ_B are then high. Conversely, $\Delta_r G > 0$ and the slope of G is positive (up from left to right) when the mixture is rich in the products C and D because μ_C and μ_D are then high. At compositions corresponding to $\Delta_r G < 0$ the reaction tends to form more products; where $\Delta_r G > 0$, the *reverse* reaction is spontaneous, and the products tend to decompose into reactants. Where $\Delta_r G = 0$ (at the

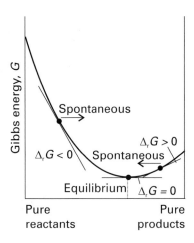

Fig 7.3 At the minimum of the curve, corresponding to equilibrium, $\Delta_r G = 0$. To the left of the minimum, $\Delta_r G < 0$, and the forward reaction is spontaneous. To the right of the minimum, $\Delta_r G > 0$ and the reverse reaction is spontaneous.

minimum of the graph), the reaction has no tendency to form either products or reactants. In other words, the reaction is at equilibrium. That is, *the criterion for chemical equilibrium* is

At constant temperature and pressure: $\Delta_r G = 0$ (7.2)

7.2 The variation of $\Delta_r G$ with composition

Our next step is to find how $\Delta_r G$ varies with the composition of the system. Once we know that, we shall be able to identify the composition corresponding to $\Delta_r G = 0$. Our starting point is the general expression for the composition-dependence of the chemical potential derived in Section 6.6:

$$\mu_J = \mu_J^{\ominus} + RT \ln a_J \qquad (7.3)$$

where a_J is the activity of the species J. When we are dealing with ideal systems, which will be the case in this chapter, we use the identifications given in Table 6.2 and replace the activities of gases by partial pressures and the activities of solutes by molar concentrations. Substitution of eqn 7.3 into eqn 7.1c gives

$$\Delta_r G = \{c(\mu_C^{\ominus} + RT \ln a_C^{\ominus}) + d(\mu_D^{\ominus} + RT \ln a_D^{\ominus})\}$$
$$- \{a(\mu_A^{\ominus} + RT \ln a_A^{\ominus}) + b(\mu_B^{\ominus} + RT \ln a_B^{\ominus})\}$$
$$= \{(c\mu_C^{\ominus} + d\mu_D^{\ominus}) - (c\mu_A^{\ominus} + dc\mu_B^{\ominus})\}$$
$$+ RT\{c \ln a_C + d \ln a_D - a \ln a_A - b \ln a_B\}$$

The first term on the right in the second line is the **standard reaction Gibbs energy**, $\Delta_r G^{\ominus}$:

$$\Delta_r G^{\ominus} = \{c\mu_C^{\ominus} + d\mu_D^{\ominus}\} - \{a\mu_A^{\ominus} + b\mu_B^{\ominus}\} \qquad (7.4a)$$

Because the standard states refer to the pure materials, the standard chemical potentials in this expression are the standard molar Gibbs energies of the (pure) species. Therefore, eqn 7.4a is the same as

$$\Delta_r G^{\ominus} = \{cG_m^{\ominus}(C) + dG_m^{\ominus}(D)\} - \{aG_m^{\ominus}(A) + bG_m^{\ominus}(B)\}$$

(7.4b)

We consider this important quantity in more detail shortly. We can rearrange the remaining term in the expression for $\Delta_r G$ as follows:[1]

$$c \ln a_C + d \ln a_D - a \ln a_A - b \ln a_B$$
$$= \ln a_C^c + \ln a_D^d - \ln a_A^a - \ln a_B^b$$
$$= \ln a_C^c a_D^d - \ln a_A^a a_B^b = \ln \frac{a_C^c a_D^d}{a_A^a a_B^b}$$

To simplify the appearance of this expression we introduce the (dimensionless) **reaction quotient**, Q, for reaction C:

$$Q = \frac{a_C^c a_D^d}{a_A^a a_B^b} \qquad (7.5)$$

Note that Q has the form of products divided by reactants, with each species raised to a power equal to its stoichiometric coefficient in the reaction.

Illustration 7.1

The reaction quotient for reaction A is

$$Q = \frac{a_{F6P}}{a_{G6P}} = \frac{[F6P]}{[G6P]}$$

---

[1] We use $a \ln x = \ln x^a$, $\ln x + \ln y = \ln xy$, and $\ln x - \ln y = \ln(x/y)$.

with [J] the numerical value of the molar concentration of J. For reaction B, the synthesis of ammonia, the reaction quotient is

$$Q = \frac{p_{NH_3}^2}{p_{N_2}\, p_{H_2}^3}$$

with p_J the numerical value of the partial pressure of J in bar.

Self-test 7.1

Write the reaction quotient for an esterification reaction of the form $CH_3COOH + C_2H_5OH \rightleftharpoons CH_3COOC_2H_5 + H_2O$. (All four components are present in the reaction mixture as liquids: the mixture is not an aqueous solution.)

[*Answer*: $Q \approx [CH_3COOC_2H_5][H_2O]/[CH_3COOH][C_2H_5OH]]$

At this stage, we can write the overall expression for the reaction Gibbs energy at any composition of the reaction mixture as

$$\Delta_r G = \Delta_r G^\ominus + RT \ln Q \tag{7.6}$$

This important equation will occur several times in different disguises.

Example 7.1 *Calculating the reaction Gibbs energy at a specified composition*

The standard reaction Gibbs energy for the hydrolysis of ATP in the reaction $ATP(aq) \rightarrow ADP(aq) + P_i(aq)$, where P_i is inorganic phosphate, is -31 kJ mol^{-1} at 37°C. In a typical bacterial cell the concentrations of ATP, ADP, and P_i are 8 mmol L^{-1}, 1 mmol L^{-1}, and 8 mmol L^{-1}, respectively. What is the reaction Gibbs energy under these conditions?

Strategy First, write down the expression for Q for the reaction, and substitute the data, remembering that activities are the numerical values of the molar concentrations expressed in moles per litre.

Solution For the reaction,

$$Q = \frac{a_{ADP}\, a_{P_i}}{a_{ATP}} \approx \frac{[ADP]\,[P_i]}{[ATP]} = \frac{(1 \times 10^{-3}) \times (8 \times 10^{-3})}{8 \times 10^{-3}}$$

$$= 1 \times 10^{-3}$$

It then follows from eqn 7.6 that

$$\Delta_r G = -31 \text{ kJ mol}^{-1}$$
$$+ (8.3145 \text{ J K}^{-1} \text{ mol}^{-1}) \times (310 \text{ K}) \times \ln(1 \times 10^{-3})$$
$$= -31 - 18 \text{ kJ mol}^{-1} = -49 \text{ kJ mol}^{-1}$$

We see that, under the conditions typical of a bacterial cell, the reaction has an even greater driving power towards equilibrium ($\Delta_r G$ is more negative) than under standard conditions.

Self-test 7.2

Calculate the reaction Gibbs energy for $N_2(g) + 3 H_2(g) \rightarrow 2 NH_3(g)$ at 25°C when the partial pressures of nitrogen, hydrogen, and ammonia are 0.20 bar, 0.42 bar, and 0.61 bar, respectively. In which direction is the reaction spontaneous under these conditions? Use $\Delta_r G^\ominus = -32.90$ kJ mol^{-1}.

[*Answer*: -25 kJ mol^{-1}, forward]

7.3 Reactions at equilibrium

When the reaction has reached equilibrium, the composition has no further tendency to change because $\Delta_r G = 0$ and the reaction is spontaneous in neither direction. At equilibrium, the reaction quotient has a certain value called the **equilibrium constant**, K, of the reaction:

$$K = \left(\frac{a_C^c\, a_D^d}{a_A^a\, a_B^b} \right)_{equilibrium} \tag{7.7}$$

We shall not normally write 'equilibrium'; the context will always make it clear that Q refers to an *arbitrary* stage of the reaction whereas K, the value of Q at equilibrium, is calculated from the equilibrium composition. It now follows from eqn 7.6, that at equilibrium

$$0 = \Delta_r G^\ominus + RT \ln K$$

and therefore that

$$\Delta_r G^\ominus = -RT \ln K \tag{7.8}$$

This is one of the most important equations in the whole of chemical thermodynamics. Its principal use is to predict the value of the equilibrium constant of any reaction from tables of thermodynamic data, like those in Appendix 1. Alternatively, we can use it to determine $\Delta_r G^{\ominus}$ by measuring the equilibrium constant of a reaction.

Illustration 7.2

Suppose we know that $\Delta_r G^{\ominus} = +3.40$ kJ mol^{-1} for the reaction $H_2(g) + I_2(s) \rightarrow 2$ HI(g) at 25°C; then to calculate the equilibrium constant we write

$$\ln K = -\frac{\Delta_r G^{\ominus}}{RT} = -\frac{3.40 \times 10^3 \text{ J mol}^{-1}}{(8.314\,51 \text{ J K}^{-1}\text{mol}^{-1}) \times (298 \text{ K})}$$

$$= -\frac{3.40 \times 10^3}{8.314\,51 \times 298}$$

On taking the natural logarithm (e^x) of the term on the right, we find $K = 0.25$.

Self-test 7.3

Calculate the equilibrium constant of the reaction $N_2(g) + 3 H_2(g) \rightleftharpoons 2$ NH$_3$(g) at 25°C, given that $\Delta_r G^{\ominus} = -32.90$ kJ mol^{-1}.

[Answer: 5.8×10^5]

An important feature of eqn 7.8 is that it tells us that $K > 1$ if $\Delta_r G^{\ominus} < 0$. Broadly speaking, $K > 1$ implies that products are dominant at equilibrium, so we can conclude that *a reaction is thermodynamically feasible if* $\Delta_r G^{\ominus} < 0$ (Fig 7.4). Conversely, because eqn 7.8 tells us that $K < 1$ if $\Delta_r G^{\ominus} > 0$, then we know that the reactants will be dominant in a reaction mixture at equilibrium if $\Delta_r G^{\ominus} > 0$. In other words, *a reaction with $\Delta_r G^{\ominus} > 0$ is not thermodynamically feasible*. Some care must be exercised with these rules, however, because the products will be significantly more abundant than reactants only if $K \gg 1$ (more than about 10^3) and even a reaction with $K < 1$ may have a reasonable abundance of products at equilibrium.

Table 7.1 summarizes the conditions under which $\Delta_r G^{\ominus} < 0$ and $K > 1$. Because $\Delta_r G^{\ominus} = \Delta_r H^{\ominus} - T\Delta_r S^{\ominus}$, the standard reaction Gibbs energy is certainly negative if both $\Delta_r H^{\ominus} < 0$ (an exothermic reaction) and $\Delta_r S^{\ominus} > 0$ (a reaction system that becomes more dis-

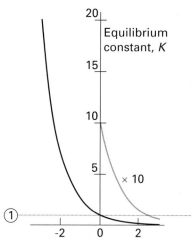

Fig 7.4 The relation between standard reaction Gibbs energy and the equilibrium constant of the reaction. The grey curve is magnified by a factor of 10.

Table 7.1 *Thermodynamic criteria of spontaneity*

---

1 If the reaction is exothermic ($\Delta_r H^{\ominus} < 0$) and $\Delta_r S^{\ominus} > 0$

   $\Delta_r G^{\ominus} < 0$ and $K > 1$ at all temperatures

2 If the reaction is exothermic ($\Delta_r H^{\ominus} < 0$) and $\Delta_r S^{\ominus} < 0$

   $\Delta_r G^{\ominus} < 0$ and $K > 1$ provided that $T < \Delta_r H^{\ominus}/\Delta_r S^{\ominus}$

3 If the reaction is endothermic ($\Delta_r H^{\ominus} > 0$) and $\Delta_r S^{\ominus} > 0$

   $\Delta_r G^{\ominus} < 0$ and $K > 1$ provided that $T > \Delta_r H^{\ominus}/\Delta_r S^{\ominus}$

4 If the reaction is endothermic ($\Delta_r H^{\ominus} > 0$) and $\Delta_r S^{\ominus} < 0$

   $\Delta_r G^{\ominus} < 0$ and $K > 1$ at no temperature

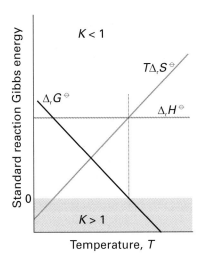

Fig 7.5 An endothermic reaction may have $K > 1$ provided the temperature is high enough for $T\Delta_r S^{\ominus}$ to be large enough that, when subtracted from $\Delta_r H^{\ominus}$ the result is negative.

orderly, such as by forming a gas). The standard reaction Gibbs energy is also negative if the reaction is endothermic ($\Delta_r H^{\ominus} > 0$) and $T\Delta_r S^{\ominus}$ is sufficiently large and positive. Note that, for an endothermic reaction to have $\Delta_r G^{\ominus} < 0$, its standard reaction entropy *must* be positive. Moreover, the temperature must be high enough for $T\Delta_r S^{\ominus}$ to be greater than $\Delta_r H^{\ominus}$ (Fig 7.5). The switch of $\Delta_r G^{\ominus}$ from positive to negative, corresponding to the switch from $K < 1$ (the reaction 'does not go') to $K > 1$ (the reaction 'goes'), occurs at a temperature given by equating $\Delta_r H^{\ominus} - T\Delta_r S^{\ominus}$ to 0, which gives:

$$T = \frac{\Delta_r H^{\ominus}}{\Delta_r S^{\ominus}} \qquad (7.9)$$

Illustration 7.3

Consider the (endothermic) thermal decomposition of calcium carbonate:

$$CaCO_3(s) \rightarrow CaO(s) + CO_2(g)$$

For this reaction $\Delta_r H^{\ominus} = +178 \text{ kJ mol}^{-1}$ and $\Delta_r S^{\ominus} = +161 \text{ J K}^{-1} \text{ mol}^{-1}$. The decomposition temperature, the temperature at which the decomposition becomes spontaneous, is

$$T = \frac{1.78 \times 10^5 \text{ J mol}^{-1}}{161 \text{ J K}^{-1} \text{ mol}^{-1}} = 1.11 \times 10^3 \text{ K}$$

---

or about 837°C. Because the entropy of decomposition is similar for all such reactions (they all involve the decomposition of a solid into a gas), we can conclude that the decomposition temperature of solids increases as their enthalpy of decomposition increases, so substances with high lattice enthalpies can be expected to have high decomposition temperatures.

7.4 The standard reaction Gibbs energy

The standard reaction Gibbs energy, $\Delta_r G^{\ominus}$, is central to the discussion of chemical equilibria and the calculation of equilibrium constants. We have seen that it is defined as the difference in standard molar Gibbs energies of the products and the reactants weighted by the stoichiometric coefficients, v, in the chemical equation:

$$\Delta_r G^{\ominus} = \sum v\, G_m^{\ominus}(\text{products}) - \sum v\, G_m^{\ominus}(\text{reactants}) \quad (7.10)$$

For example, the standard reaction Gibbs energy for reaction A is the difference between the molar Gibbs energies of fructose-6-phosphate and glucose-6-phosphate in solution at 1 mol L^{-1} and 1 bar.

We cannot calculate $\Delta_r G^{\ominus}$ from the standard molar Gibbs energies themselves, because these quantities are not known. One practical approach is to calculate the standard reaction enthalpy from standard enthalpies of formation (Section 3.5), the standard reaction entropy from Third-Law entropies (Section 4.6), and then to combine the two quantities by using

$$\Delta_r G^{\ominus} = \Delta_r H^{\ominus} - T\Delta_r S^{\ominus} \qquad (7.11)$$

Example 7.2 *Evaluating the standard reaction Gibbs energy*

Evaluate the standard reaction Gibbs energy at 25°C for the reaction $H_2(g) + \frac{1}{2} O_2(g) \rightarrow H_2O(l)$.

Strategy To use eqn 7.11, we need two pieces of information. One is the standard reaction enthalpy, which we can find from tables of enthalpies of formation. The second is the standard reaction entropy, which we can find from the standard entropies of the reactants and products.

Solution The standard reaction enthalpy is

$$\Delta_r H^\ominus = \Delta_f H^\ominus(H_2O, l) = -285.83 \text{ kJ mol}^{-1}$$

The standard reaction entropy, calculated as in Illustration 4.3, is

$$\Delta_r S^\ominus = -163.34 \text{ J K}^{-1} \text{ mol}^{-1}$$

which corresponds to $-0.163\,34 \text{ kJ K}^{-1} \text{ mol}^{-1}$. Therefore, from eqn 7.11,

$$\Delta_r G^\ominus = (-285.83 \text{ kJ mol}^{-1}) - $$
$$(298.15 \text{ K}) \times (-0.163\,34 \text{ kJ K}^{-1} \text{ mol}^{-1})$$
$$= -237.13 \text{ kJ mol}^{-1}$$

Self-test 7.4

Use the information in Appendix 1 to determine the standard reaction Gibbs energy for $3\,O_2(g) \rightarrow 2\,O_3(g)$ from standard enthalpies of formation and standard entropies.

[*Answer:* +327.4 kJ mol^{-1}]

We already know how to use standard enthalpies of formation of substances to calculate standard reaction enthalpies. We can use the same technique for standard reaction Gibbs energies. To do so, we list the **standard Gibbs energy of formation**, $\Delta_f G^\ominus$, of a substance, which is the standard reaction Gibbs energy (per mole of the species) for its formation from the elements in their reference states. The concept of reference state was introduced in Section 3.5; the temperature is arbitrary, but we shall almost always take it to be 25°C (298 K). For example, the standard Gibbs energy of formation of liquid water, $\Delta_f G^\ominus(H_2O, l)$, is the standard reaction Gibbs energy for

$$H_2(g) + \tfrac{1}{2}O_2(g) \rightarrow H_2O(l)$$

and is -237 kJ mol^{-1} at 298 K. Some standard Gibbs energies of formation are listed in Table 7.2 and more can be found in Appendix 1. It follows from the definition that the standard Gibbs energy of formation of an element in its reference state is zero because reactions such as

Table 7.2 *Standard Gibbs energies of formation at 25°C*

Substance	$\Delta_f G^\ominus/(\text{kJ mol}^{-1})$
Gases	
Ammonia, NH_3	−16.45
Carbon dioxide, CO_2	−394.36
Dinitrogen tetroxide, N_2O_4	+97.89
Hydrogen iodide, HI	+1.70
Nitrogen dioxide, NO_2	+51.31
Sulfur dioxide, SO_2	−300.19
Water, H_2O	−228.57
Liquids	
Benzene, C_6H_6	+124.3
Ethanol, CH_3CH_2OH	−174.78
Water, H_2O	−237.13
Solids	
Calcium carbonate, $CaCO_3$	−1128.8
Iron(III) oxide, Fe_2O_3	−742.2
Silver bromide, AgBr	−96.90
Silver chloride, AgCl	−109.79

Additional values are given in Appendix 1.

$$C(s, \text{graphite}) \rightarrow C(s, \text{graphite})$$

are null (that is, nothing happens). The standard Gibbs energy of formation of an element in a phase different from its reference state is nonzero:

$$C(s, \text{graphite}) \rightarrow C(s, \text{diamond})$$
$$\Delta_f G^\ominus(C, \text{diamond}) = +2.90 \text{ kJ mol}^{-1}$$

Many of the values in the tables have been compiled by combining the standard enthalpy of formation of the species with the standard entropies of the compound and the elements, as illustrated above, but there are other sources of data and we encounter some of them later.

Standard Gibbs energies of formation can be combined to obtain the standard Gibbs energy of almost any reaction. We use the now familiar expression

$$\Delta_r G^\ominus = \sum v\,\Delta_f G^\ominus(\text{products}) - \sum v\,\Delta_f G^\ominus(\text{reactants}) \quad (7.12)$$

Illustration 7.4

To determine the standard reaction Gibbs energy for

$$2\,CO(g) + O_2(g) \rightarrow 2\,CO_2(g)$$

we carry out the following calculation:

$$\Delta_r G^\ominus = 2\Delta_f G^\ominus(CO_2, g) - \{2\Delta_f G^\ominus(CO, g) + \Delta_f G^\ominus(O_2, g)\}$$
$$= 2 \times (-394\ \text{kJ mol}^{-1}) - \{2 \times (-137\ \text{kJ mol}^{-1}) + 0\}$$
$$= -514\ \text{kJ mol}^{-1}$$

Self-test 7.5

Calculate the standard reaction Gibbs energy of the oxidation of ammonia to nitric oxide according to the equation $4\,NH_3(g) + 5\,O_2(g) \rightarrow 4\,NO(g) + 6\,H_2O(g)$.

[*Answer:* −959.42 kJ mol⁻¹]

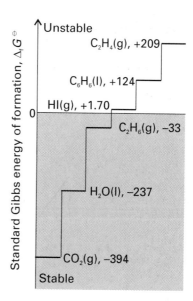

Fig 7.6 The standard Gibbs energy of formation of a compound is like a measure of the compound's altitude above (or below) sea level: compounds that lie above sea level have a spontaneous tendency to decompose into the elements (and to revert to sea level). Compounds that lie below sea level are stable with respect to decomposition into the elements.

Standard Gibbs energies of formation of compounds have their own significance as well as being useful in calculations of K. They are a measure of the 'thermodynamic altitude' of a compound above or below a 'sea level' of stability represented by the elements in their reference states (Fig 7.6). If the standard Gibbs energy of formation is positive and the compound lies above 'sea level', then the compound has a spontaneous tendency to sink towards thermodynamic sea level and decompose into the elements. That is, $K < 1$ for their formation reaction. We say that a compound with $\Delta_f G^\ominus > 0$ is **thermodynamically unstable** with respect to its elements or that it is **endergonic**. Thus, the endergonic substance ozone, for which $\Delta_f G^\ominus = +163$ kJ mol⁻¹, has a spontaneous tendency to decompose into oxygen under standard conditions at 25°C. More precisely, the equilibrium constant for the reaction $\frac{3}{2}O_2(g) \rightleftharpoons O_3(g)$ is less than 1 (in fact, $K = 2.7 \times 10^{-29}$). However, although ozone is thermodynamically unstable, it can survive if the reactions that convert it into oxygen are slow. That is the case in the upper atmosphere, and the O_3 molecules in the ozone layer survive for long periods. Benzene ($\Delta_f G^\ominus = +124$ kJ mol⁻¹) is also thermodynamically unstable with respect to its elements ($K = 1.8 \times 10^{-22}$). However, the fact that bottles of benzene are everyday laboratory commodities also reminds us that *spontaneity is a thermodynamic tendency that might not be realized at a significant rate in practice*.

Another useful point that can be made about standard Gibbs energies of formation is that there is no point in searching for *direct* syntheses of a thermodynamically unstable compound from its elements (under standard conditions, at the temperature to which the data apply), because the reaction does not occur in the required direction: the *reverse* reaction, decomposition, is spontaneous. Endergonic compounds must be synthesized by alternative routes or under conditions for which their Gibbs energy of formation is negative and they lie beneath thermodynamic sea level.

Compounds with $\Delta_f G^\ominus < 0$ (corresponding to $K > 1$ for their formation reactions) are said to be **thermodynamically stable** with respect to their elements or **exergonic**. Exergonic compounds lie below the thermodynamic sea level of the elements (under standard conditions). An example is the exergonic compound ethane, with $\Delta_f G^\ominus = -33$ kJ mol⁻¹: the negative sign shows that the formation of ethane gas is spontaneous in the sense that $K > 1$ (in fact, $K = 6.1 \times 10^5$ at 25°C).

7.5 Coupled reactions

A reaction that is not spontaneous may be driven forward by a reaction that is spontaneous. A simple mechanical analogy is a pair of weights joined by a string (Fig 7.7): the lighter of the pair of weights will be pulled up as the heavier weight falls. Although the lighter weight has a natural tendency to move downwards, its coupling to the heavier weight results in it being raised. The thermodynamic analogue is an **endergonic reaction**, a reaction with a positive Gibbs energy, $\Delta_r G$, (the analogue of the lighter weight) being forced to occur by coupling it to an **exergonic reaction**, a reaction with a negative Gibbs energy, $\Delta_r G'$, (the analogue of the heavier weight falling to the ground). The overall reaction is spontaneous because the sum $\Delta_r G + \Delta_r G'$ is negative. The whole of life's activities depend on coupling of this kind, for the oxidation reactions of food act as the heavy weights that drive other reactions forward and result in the formation of proteins from amino acids, the actions of muscles for propulsion, and even the activities of the brain for reflection, learning, and imagination.

Fig 7.7 If two weights are coupled as shown here, then the heavier weight will move the lighter weight in its non-spontaneous direction: overall, the process is still spontaneous. The weights are the analogues of two chemical reactions: a reaction with a large negative ΔG can force another reaction with a smaller $|\Delta G|$ to run in its non-spontaneous direction.

3 ATP, adenosine triphosphate

4 ADP, adenosine diphosphate

The function of adenosine triphosphate, ATP (**3**), for instance, is to store the energy made available when food is metabolized and then to supply it on demand to a wide variety of processes, including muscular contraction, reproduction, and vision. The essence of ATP's action is its ability to lose its terminal phosphate group by hydrolysis and to form adenosine diphosphate, ADP (**4**):[2]

$$ATP(aq) + H_2O(l) \rightarrow ADP(aq) + P_i^-(aq) + H^+(aq)$$

This reaction is exergonic under the conditions prevailing in cells and can drive an endergonic reaction forward if suitable enzymes are available to couple the reactions.

Before discussing the hydrolysis of ATP quantitatively, we need to note that the conventional standard state of hydrogen ions ($a = 1$, corresponding to pH = 0, a strongly acidic solution)[3] is not appropriate to normal biological conditions inside cells, where the pH is close to 7. Therefore, in biochemistry it is common to adopt the **biological standard state**, in

---

[2] We have already met the symbol P_i^-, which denotes an inorganic phosphate group, such as $H_2PO_4^-$.

[3] The pH scale is discussed in Chapter 8.

which pH = 7, a neutral solution. We shall adopt this convention in this section, and label the corresponding standard quantities as $G^\oplus$, $H^\oplus$, and $S^\oplus$.[4]

Example 7.3 *Converting between thermodynamic and biological standard states*

The standard reaction Gibbs energy for the hydrolysis of ATP (given above) is +10 kJ mol^{-1} at 298 K. What is the biological standard state value?

Strategy Because protons occur as products, lowering their concentration (from 1 mol L^{-1} to 10^{-7} mol L^{-1}) suggests that the reaction will have a higher tendency to form products. Therefore, we expect a more negative value of the reaction Gibbs energy for the biological standard than for the thermodynamic standard. The two types of standard are related by eqn 7.6, with the activity of hydrogen ions set equal to 10^{-7} in place of 1.

Solution The reaction quotient for the hydrolysis reaction when all the species are in their standard states except the hydrogen ions, which are present at 10^{-7} mol L^{-1}, is

$$Q = \frac{a_{ADP}\, a_{P_i}\, a_{H^+}}{a_{ATP}\, a_{H_2O}} = \frac{1 \times 1 \times 10^{-7}}{1 \times 1} = 1 \times 10^{-7}$$

The thermodynamic and biological standard values are therefore related by eqn 7.6 in the form

$$\Delta_r G^\oplus = \Delta_r G^\ominus$$

$$+ (8.31451\ J\ K^{-1}\ mol^{-1}) \times (298\ K) \times \ln(1 \times 10^{-7})$$

$$= 10 - 40\ kJ\ mol^{-1} = -30\ kJ\ mol^{-1}$$

Note how the large change in pH changes the sign of the standard reaction Gibbs energy.

Self-test 7.6

The overall reaction for the glycolysis reaction is
$C_6H_{12}O_6(aq) + 2\ NAD^+(aq) + 2\ ADP(aq) + 2\ P_i \rightarrow 2\ CH_3COCO_2^-(aq) + 2\ NADH(aq) + 2\ ATP(aq) + 2\ H_3O^+(aq)$. For this reaction, $\Delta_r G^\ominus = -80.6$ kJ mol^{-1} at 298 K. What is the value of $\Delta_r G^\oplus$?

[*Answer:* −0.7 kJ mol^{-1}]

The biological standard values for ATP hydrolysis at 37°C (310 K, body temperature) are

$$\Delta_r G^\oplus = -31\ kJ\ mol^{-1} \quad \Delta_r H^\oplus = -20\ kJ\ mol^{-1}$$
$$\Delta_r S^\oplus = +34\ J\ K^{-1}\ mol^{-1}$$

The hydrolysis is therefore exergonic ($\Delta_r G < 0$) under these conditions, and 31 kJ mol^{-1} is available for driving other reactions. On account of its exergonic character, the ADP–phosphate bond has been called a 'high-energy phosphate bond'. The name is intended to signify a high tendency to undergo reaction and should not be confused with 'strong' bond in its normal chemical sense (that of a high bond enthalpy). In fact, even in the biological sense it is not of very 'high energy'. The action of ATP depends on the bond being intermediate in strength. Thus ATP acts as a phosphate donor to a number of acceptors (such as glucose), but is recharged with a new phosphate group by more powerful phosphate donors in the phosphorylation steps in the respiration cycle (Box 7.1).

Adenosine triphosphate is not the only phosphate species capable of driving other less exergonic reactions. For instance, creatine phosphate (5) can release its phosphate group in a hydrolysis reaction, and $\Delta_r G^\oplus = -43$ kJ mol^{-1}. These different exergonicities give rise to the concept of **transfer potential**, which is the negative of the value of $\Delta_r G^\oplus$ for the hydrolysis reaction. Thus, the transfer potential of creatine phosphate is 43 kJ mol^{-1}. Just as one exergonic reaction can drive a less exergonic reaction, so the hydrolysis of a species with a high transfer potential can drive the phosphorylation of a species with a lower transfer potential (Table 7.3).

5 Creatine phosphate

---

[4] Another convention to denote the biological standard state is to write $X^{\circ\prime}$ or $X^{\ominus\prime}$. There is no difference between thermodynamic and biological standard values if hydrogen ions are not involved in the reaction.

Box 7.1 *Anaerobic and aerobic metabolism*

The energy source of anaerobic cells is *glycolysis*, the partial oxidation of glucose to pyruvate ions, $CH_3COCO_2^-$, and at blood temperature $\Delta_r G^\ominus = -147$ kJ mol^{-1}. The glycolysis (the 'heavy weight' spontaneous driving reaction) is coupled to a reaction (the 'light weight' non-spontaneous reaction) in which two ADP molecules are converted into two ATP molecules:

$$C_6H_{12}O_6 + 2\ P_i^- + 2\ ADP \rightarrow 2\ CH_3COCO_2^- + 2\ ATP + 2\ H_2O$$

The biological standard Gibbs energy for this reaction is $(-147) - 2(-31)$ kJ mol$^{-1} = -85$ kJ mol^{-1}. The reaction is exergonic, and therefore spontaneous: the metabolism of the food has been used to 'recharge' the ATP.

Metabolism by aerobic respiration is much more efficient, as is indicated by the standard Gibbs energy of combustion of glucose, which is -2880 kJ mol^{-1}. Therefore, terminating the oxidation of glucose at lactic acid is a poor use of resources. In aerobic respiration the oxidation is carried out to completion and a highly complex set of reactions preserves as much of the energy released as possible. In the overall reaction, 38 ATP molecules are generated for each glucose molecule consumed. Each mole of ATP molecules extracts 30 kJ from the 2880 kJ supplied by 1 mol $C_6H_{12}O_6$ (180 g of glucose), and so 1140 kJ per mole of glucose molecules has been stored for later use.

Each ATP molecule can be used to drive an endergonic reaction for which $\Delta_r G^\ominus$ does not exceed 30 kJ mol^{-1}. For example, the biosynthesis of sucrose from glucose and fructose can be driven, if a suitable enzyme system is available, because the reaction is endergonic to the extent $\Delta_r G^\ominus = +23$ kJ mol^{-1}. The biosynthesis of proteins is strongly endergonic, not only on account of the enthalpy change but also on account of the large decrease in entropy that occurs when many amino acids are assembled into a precisely determined sequence. For instance, the formation of a peptide link is endergonic, with $\Delta_r G^\ominus = +17$ kJ mol^{-1}, but the biosynthesis occurs indirectly and is equivalent to the consumption of three ATP molecules for each link. In a moderately small protein like myoglobin, with about 150 peptide links, the construction alone requires 450 ATP molecules, and therefore about 12 mol of glucose molecules for 1 mol of protein molecules.

Exercise 1 A 'food calorie', 1 Cal, is equivalent to 1 thermochemical kilocalorie (1 kcal). A manufacturer claims that a teaspoon of sucrose (about 5.0 g) provides only 20 Cal. Is this correct?

Exercise 2 Fats yield almost twice as much energy per gram as carbohydrates. How many grams of fat would need to be metabolized to synthesize 1.0 mol of myoglobin molecules?

Table 7.3 *Transfer potentials*

Substance	Transfer potential, $-\Delta_r G^\ominus/(\text{kJ mol}^{-1})$
Phosphoenolpyruvate	62
1,3-bis(phospho)glycerate	49
Creatine phosphate	43
Pyrophosphate, $HP_2O_7^{3-}$	33
ATP, ADP	31
AMP	14
Glucose-6-phosphate	14
Glycerol-1-phosphate	10

7.6 **The equilibrium composition**

The magnitude of an equilibrium constant is a good *qualitative* indication of the feasibility of a reaction regardless of whether the system is ideal or not. Broadly speaking, if $K \gg 1$ (typically $K > 10^3$, corresponding to $\Delta_r G^\ominus < -17$ kJ mol^{-1} at 25°C), the reaction has a strong tendency to form products. If $K \ll 1$ (that is, for $K < 10^{-3}$, corresponding to $\Delta_r G^\ominus > +17$ kJ mol^{-1} at 25°C), the equilibrium composition will consist of largely unchanged reactants. If K is comparable to 1 (typically lying between 10^{-3} and 10^3), then significant amounts of both reactants and products will be present at equilibrium.

An equilibrium constant expresses the composition of an equilibrium mixture as a ratio of products of activities. Even if we confine our attention to ideal systems it is still necessary to do some work to extract the actual equilibrium concentrations or partial pressures of the reactants and products given their initial values.

Example 7.4 *Calculating an equilibrium composition 1*

Estimate the composition of a solution in which G6P and F6P are in equilibrium at 25°C, and draw a graph to show how the spontaneity of the reaction varies with composition. For the reaction, $\Delta_r G^\ominus =$ +1.7 kJ mol^{-1}.

Strategy Calculate the equilibrium constant by using eqn 7.8. Express the composition in terms of the fraction, f, of F6P, and solve for f. For the second part of the question, express $\Delta_r G$ in terms of f, and plot $\Delta_r G$ against f. Where $\Delta_r G < 0$, the formation of F6P is spontaneous; where $\Delta_r G > 0$, the formation of G6P is spontaneous.

Solution We find the equilibrium constant by rearranging eqn 7.8 into

$$K = e^{-\Delta_r G^\ominus /RT}$$

First, note that

$$\frac{\Delta_r G^\ominus}{RT} = \frac{1.7 \times 10^3 \text{ J mol}^{-1}}{(8.3145 \text{ J K}^{-1} \text{mol}^{-1}) \times (298 \text{ K})} = \frac{1.7 \times 10^3}{8.3145 \times 298}$$

Therefore, $K = 0.50$. Next, note that

$$K = \frac{[\text{F6P}]}{[\text{G6P}]} = \frac{[\text{F6P}]/([\text{G6P}] + [\text{F6P}])}{[\text{G6P}]/([\text{G6P}] + [\text{F6P}])} = \frac{f_{eq}}{1 - f_{eq}}$$

where f_{eq} is the fraction of F6P in the solution at equilibrium (and $1 - f_{eq}$ is the fraction of G6P). Therefore,

$$f_{eq} = \frac{K}{1 + K}$$

which evaluates to $f_{eq} = 0.33$. That is, at equilibrium, 33 per cent of the solute is F6P and 67 per cent is G6P. For the second part of the problem, we use eqn 7.6 in the form

$$\Delta_r G = \Delta_r G^\ominus + RT \ln Q = \Delta_r G^\ominus + RT \ln\left(\frac{f}{1-f}\right)$$

where f is the fraction of F6P in the solution (not necessarily at equilibrium). Insertion of the data gives

$$\Delta_r G/(\text{kJ mol}^{-1}) = 1.7 + 2.48 \ln\left(\frac{f}{1-f}\right)$$

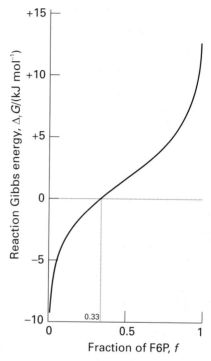

Fig 7.8 The variation of reaction Gibbs energy with fraction, f, of fructose-6-phosphate (F6P) in an aqueous solution of glucose-6-phosphate (G6P). Equilibrium corresponds to $f = 0.33$.

This function is plotted in Fig 7.8. Notice how the formation of F6P is spontaneous when its fraction present is less than $f_{eq} = 0.33$.

Self-test 7.7

Estimate the composition of a solution in which two isomers are in equilibrium at 37°C and $\Delta_r G^\ominus =$ −2.2 kJ mol^{-1}.

[*Answer: f = 0.70*]

In more complicated cases it is best to organize the necessary work into a systematic procedure resembling a spreadsheet by constructing a table with columns headed by the species and, in successive rows:

1 The initial molar concentrations or partial pressures of the species.

2 The changes in these quantities that must take place for the system to reach equilibrium.

3 The resulting equilibrium values.

Box 7.2 *Myoglobin and haemoglobin*

Myoglobin (Mb) stores oxygen; haemoglobin[5] (Hb) transports oxygen. These two proteins are closely related, for haemoglobin can be regarded, as a first approximation, as a tetramer of myoglobin. There are, in fact, slight differences in the peptide sequence of the myoglobin-like components of haemoglobin, but we can ignore them at this stage. In each protein, the O_2 molecule attaches to an iron atom in a haem group (see Box 13.1 for more information). For our purposes here, we are concerned with the different equilibrium characteristics for the uptake of oxygen by myoglobin and haemoglobin.

First, consider the equilibrium between Mb and O_2:

$$Mb(aq) + O_2(g) \rightleftharpoons MbO_2(aq) \qquad K = \frac{[MbO_2]}{p[Mb]}$$

where p is the numerical value of the partial pressure of oxygen. It follows that the *fractional saturation*, s, the fraction of Mb molecules that are oxygenated, is

$$s = \frac{[MbO_2]}{[Mb]_{total}} = \frac{[MbO_2]}{[Mb] + [MbO_2]} = \frac{Kp}{1 + Kp}$$

The dependence of s on p is shown in the illustration.

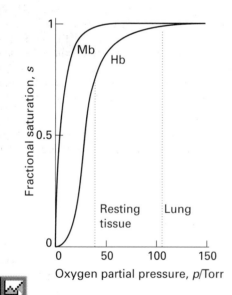

Fig B1 The variation of the fractional saturation of myoglobin and haemoglobin molecules with the partial pressure of oxygen. The different shapes of the curves account for the different biological functions of the two proteins.

-----

[5] Haemoglobin and haem in British English; hemoglobin and heme in American English.

Now consider the equilibrium between Hb and O_2:

$$Hb(aq) + O_2(g) \rightleftharpoons HbO_2(aq) \qquad K_1 = \frac{[HbO_2]}{p[Hb]}$$

$$HbO_2(aq) + O_2(g) \rightleftharpoons Hb(O_2)_2(aq) \qquad K_2 = \frac{[Hb(O_2)_2]}{p[HbO_2]}$$

and so on for K_3 and K_4. To develop an expression for s, we express $[Hb(O_2)_2]$ in terms of $[HbO_2]$ by using K_2, then express $[HbO_2]$ in terms of $[Hb]$ by using K_1, and likewise for all the other concentrations of oxygenated material. For instance,

$$[Hb(O_2)_4] = K_1 K_2 K_3 K_4 p^4 [Hb]$$

Then, the total concentration of bound oxygen is

$$[bound\ O_2] = [HbO_2] + 2[Hb(O_2)_2] + 3[Hb(O_2)_3] + 4[Hb(O_2)_4]$$

$$= (1 + 2K_2 p + 3K_2 K_3 p^2 + 4K_2 K_3 K_4 p^3) K_1 p [Hb]$$

and the total concentration of haemoglobin is

$$[Hb]_{total} = (1 + K_1 p + K_1 K_2 p^2 + K_1 K_2 K_3 p^3 + K_1 K_2 K_3 K_4 p^4)[Hb]$$

Because each Hb molecule has four sites at which O_2 can attach, the fractional saturation is

$$s = \frac{[bound\ O_2]}{4[Hb]_{total}}$$

$$= \frac{(1 + 2K_2 p + 3K_2 K_3 p^2 + 4K_2 K_3 K_4 p^3) K_1 p}{4(1 + K_1 p + K_1 K_2 p^2 + K_1 K_2 K_3 p^3 + K_1 K_2 K_3 K_4 p^4)}$$

A reasonable fit can be obtained with $K_1 = 0.01$, $K_2 = 0.02$, $K_3 = 0.04$, and $K_4 = 0.08$ when p is expressed in torr, and the resulting function is also plotted in the illustration. We see that, unlike the myoglobin saturation curve, the graph is *sigmoidal* (S-shaped). Box 13.1 explains the origin of this cooperativity in terms of the *allosteric effect*, in which an adjustment of the conformation of a molecule when one substrate species binds affects the ease with which a subsequent substrate molecule binds.

The differing shapes of the two curves have important consequences for the way oxygen is made available in the body: in particular, the greater sharpness of the Hb saturation curve means that Hb can load O_2 more fully in the lungs and unload it more fully in different regions of the organism. In the lungs, where $p \approx 105$ Torr, $s \approx 0.98$, representing almost complete saturation. In resting muscular tissue, p is equivalent to about 40 Torr, corresponding to $s \approx 0.75$. That figure implies that sufficient oxygen is still available should a sudden surge of activity take place. If the local partial pressure falls to 20 Torr, s falls to about 0.1. Note that the steepest part of the curve falls in the range of typical tissue oxygen partial pressure. Myoglobin, on the other hand, begins to release O_2 only when p has fallen below about 20 Torr, so it acts as a reserve to be drawn on only when the Hb oxygen has been used up.

Exercise 1 The Scatchard equation applies to identical and non-interacting sites. This is not the case for haemoglobin, the oxygen transport protein. The Hill coefficient, ν, varies from 1, for no cooperativity, to N for all-or-none binding of N ligands. The Hill coefficient is determined from a plot of

$$\log \frac{s}{1-s} = \nu \log p_{O_2} - \nu \log K$$

where K is a constant, not the binding constant for one ligand, and s is the saturation. The Hill coefficient for myoglobin is 1, and for haemoglobin 2.8. Determine the constant K for both Mb and Hb from the graph of fractional saturation (at $s = 0.5$) and then calculate the fractional saturation of Mb and Hb for the following values of p_{O_2}/Torr: 5, 10, 20, 30, and 60.

Exercise 2 Using the information from the first exercise, calculate the value of s at the same p_{O_2} values assuming ν has the theoretical maximum value of 4.

In most cases, we do not know the change that must occur for the system to reach equilibrium, so the change in the concentration or partial pressure of one species is written as x and the reaction stoichiometry is used to write the corresponding changes in the other species. When the values at equilibrium (the last row of the table) are substituted into the expression for the equilibrium constant, we obtain an equation for x in terms of K. This equation can be solved for x, and hence the concentrations of all the species at equilibrium may be found.

Example 7.5 *Calculating an equilibrium composition 2*

Suppose that in an industrial process, N_2 at 1.00 bar is mixed with H_2 at 3.00 bar and the two gases are allowed to come to equilibrium with the product ammonia (in the presence of a catalyst) in a reactor of constant volume. At the temperature of the reaction, it has been determined experimentally that $K = 977$. What are the equilibrium partial pressures of the three gases?

Strategy Proceed as set out above. Write down the chemical equation of the reaction and the expression for K. Set up the equilibrium table, express K in terms of x, and solve the equation for x. Because the volume of the reaction vessel is constant, each partial pressure is proportional to the amount of its molecules present ($p_J = n_J RT/V$), so the stoichiometric relations apply to the partial pressures directly. In general, solution of the equation for x results in several mathematically possible values of x; select the chemically acceptable solution by considering the signs of the predicted concentrations or partial pressures: they must be positive. In some cases, the equation for x is quadratic:

$$ax^2 + bx + c = 0 \text{ with the roots } \quad x = \frac{-b \pm \sqrt{b^2 - 4ac}}{2a}$$

More complex equations must be solved graphically or by using mathematical software; in some cases, approximations may be used.[6] Confirm the accuracy of the calculation by substituting the calculated equilibrium partial pressures into the expression for the equilibrium constant to verify that the value so calculated is equal to the experimental value used in the calculation.

Solution The chemical equation is reaction B ($N_2(g)$ + 3 $H_2(g) \rightarrow$ 2 $NH_3(g)$), and the equilibrium constant is

$$K = \frac{p_{NH_3}^2}{p_{N_2}\, p_{H_2}^3}$$

with the partial pressures those at equilibrium. The equilibrium table is

Species:	N_2	H_2	NH_3
Initial partial pressure/bar	1.00	3.00	0
Change to reach equilibrium/bar	$-x$	$-3x$	$+2x$
Equilibrium partial pressure/bar	$1.00 - x$	$3.00 - 3x$	$2x$

The equilibrium constant for the reaction is therefore

---

[6] For example, that x is very small when $K \ll 1$; examples are given in Section 8.2.

$$K = \frac{(2x)^2}{(1.00 - x) \times (3.00 - 3x)^3}$$

Then, with $K = 977$, this equation rearranges first to

$$977 = \frac{4}{27}\left(\frac{x}{(1.00 - x)^2}\right)^2$$

and then, after taking the square root of both sides, to

$$\sqrt{\tfrac{1}{4} \times 27 \times 977} = \frac{x}{(1.00 - x)^2}$$

We now rearrange this expression into the quadratic equation

$gx^2 - (2.00g + 1)x + 1.00g = 0$ with $g = (\tfrac{1}{4} \times 27 \times 977)^{1/2}$

The quadratic formula gives the solutions $x = 1.12$ and $x = 0.895$. Because p_{N_2} cannot be negative, and $p_{N_2} = 1.00 - x$ (from the equilibrium table), we know that x cannot be greater than 1.00; therefore, we select $x = 0.895$ as the acceptable solution. It then follows from the last line of the equilibrium table that (with the units bar restored):

$p_{N_2} = 0.10$ bar $\qquad p_{H_2} = 0.31$ bar $\qquad p_{NH_3} = 1.8$ bar

This is the composition of the reaction mixture at equilibrium. Note that, because K is large (of the order of 10^3), the products dominate. To verify the result, we calculate

$$\frac{p_{NH_3}^2}{p_{N_2}\,p_{H_2}^3} = \frac{(1.8)^2}{(0.10) \times (0.32)^3} = 9.9 \times 10^2$$

The result is close to the experimental value (the discrepancy stems from rounding errors).

Self-test 7.8

In an experiment to study the formation of nitrogen oxides in jet exhausts, N_2 at 0.100 bar is mixed with O_2 at 0.200 bar and the two gases are allowed to come to equilibrium with the product NO in a reactor of constant volume. Take $K = 3.4 \times 10^{-21}$ at 800 K. What is the equilibrium partial pressure of NO?

[*Answer*: 8.2 pbar]

The response of equilibria to the conditions

IN introductory chemistry, we meet the empirical rule of thumb known as **Le Chatelier's principle**:

When a system at equilibrium is subjected to a disturbance, the composition of the system adjusts so as to tend to minimize the effect of the disturbance.

For instance, if a system is compressed, then the equilibrium position can be expected to shift in the direction that leads to a reduction in the number of molecules in the gas phase, for that tends to minimize the effect of compression. Le Chatelier's principle, though, is only a rule of thumb, and to understand why reactions respond as they do, and to calculate the new equilibrium composition, we need to use thermodynamics. In large part, that depends on judging how the conditions affect the value of $\Delta_r G^\ominus$.

7.7 The presence of a catalyst

A catalyst is a substance that accelerates a reaction without itself appearing in the overall chemical equation. Enzymes are biological versions of catalysts. We study the action of catalysts in Section 10.12, and at this stage do not need to know in detail how they work other than that they provide an alternative, faster route from reactants to products. Although the new route is faster, the initial reactants and the final products are the same. We know that G is a state function, so $\Delta_r G^\ominus$ has the same value however the reaction is brought about. Therefore, an alternative pathway between reactants and products leaves $\Delta_r G^\ominus$ and therefore K unchanged. That is, *the presence of a catalyst does not change the equilibrium constant of a reaction.*

7.8 **The effect of temperature**

The equilibrium constant of a reaction changes when the temperature is changed. According to Le Chatelier's principle, we can expect a reaction to respond to a lowering of temperature by releasing heat and to respond to an increase of temperature by absorbing heat. That is, when the temperature is increased:

> The equilibrium composition of an exothermic reaction will shift towards reactants.

> The equilibrium composition of an endothermic reaction will shift towards products.

In each case, the response tends to minimize the effect of raising the temperature. But *why* do reactions at equilibrium respond in this way? Le Chatelier's principle is only a rule of thumb, and gives no clue to the reason for this behaviour. As we shall now see, the origin of the effect is the dependence of $\Delta_r G^{\ominus}$, and therefore of K, on the temperature.

First, we consider the effect of temperature on $\Delta_r G^{\ominus}$. We use the relation $\Delta_r G^{\ominus} = \Delta_r H^{\ominus} - T \Delta_r S^{\ominus}$ and make the assumption that neither the reaction enthalpy nor the reaction entropy varies much with temperature (over small ranges, at least). It follows that

$$\text{Change in } \Delta_r G^{\ominus} = -(\text{change in } T) \times \Delta_r S^{\ominus} \qquad (7.13)$$

This expression is easy to apply when there is a consumption or formation of gas because, as we have seen (Section 4.6), gas formation dominates the sign of the reaction entropy.

Illustration 7.5

Consider the three reactions

(i) $\frac{1}{2}C(s) + \frac{1}{2}O_2(g) \to \frac{1}{2}CO_2(g)$
(ii) $C(s) + \frac{1}{2}O_2(g) \to CO(g)$
(iii) $CO(g) + \frac{1}{2}O_2(g) \to CO_2(g)$

all of which are important in the discussion of the extraction of metals from their ores. In reaction (i), the amount of gas is constant, so the entropy change is small and $\Delta_r G^{\ominus}$ for this reaction changes

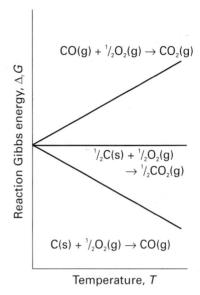

Fig 7.9 The variation of reaction Gibbs energy with temperature depends on the reaction entropy and therefore on the net production or consumption of gas in a reaction. The Gibbs energy of a reaction that produces gas decreases with increasing temperature. The Gibbs energy of a reaction that results in a net consumption of gas increases with temperature.

only slightly with temperature.[7] Because in reaction (ii) there is a net increase in the amount of gas molecules, from $\frac{1}{2}$ mol to 1 mol, the standard reaction entropy is large and positive; therefore, $\Delta_r G^{\ominus}$ for this reaction decreases sharply with increasing temperature. In reaction (iii), there is a similar net decrease in the amount of gas molecule, from $\frac{3}{2}$ mol to 1 mol, so $\Delta_r G^{\ominus}$ for this reaction increases sharply with increasing temperature. These remarks are summarized in Fig 7.9.

Now consider the effect of temperature on K itself. At first, this problem looks troublesome, because both T and $\Delta_r G^{\ominus}$ appear in the expression for K. However, in fact, the effect of temperature can be expressed very simply.

---

[7] Note, however, that K changes, because $\Delta_r G^{\ominus}/RT$ becomes less negative as T increases, so K decreases.

Derivation 7.1 *The van't Hoff equation*

As before, we use the approximation that the standard reaction enthalpy and entropy are independent of temperature over the range of interest, so the entire temperature dependence of $\Delta_r G^{\ominus}$ stems from the T in $\Delta_r G^{\ominus} = \Delta_r H^{\ominus} - T\Delta_r S^{\ominus}$. At a temperature T,

$$\ln K = -\frac{\Delta_r G^{\ominus}}{RT} \overset{\substack{\text{Substitute} \\ \Delta_r G^{\ominus} = \Delta_r H^{\ominus} - T\Delta_r S^{\ominus}}}{=} -\frac{\Delta_r H^{\ominus}}{RT} + \frac{\Delta_r S^{\ominus}}{R}$$

At another temperature T', when $\Delta_r G^{\ominus'} = \Delta_r H^{\ominus} - T'\Delta_r S^{\ominus}$ and the equilibrium constant is K', a similar expression holds:

$$\ln K' = -\frac{\Delta_r H^{\ominus}}{RT'} + \frac{\Delta_r S^{\ominus}}{R}$$

The difference between the two is

$$\ln K' - \ln K = \frac{\Delta_r H^{\ominus}}{R}\left(\frac{1}{T} - \frac{1}{T'}\right) \qquad (7.14)$$

This formula is a version of the **van't Hoff equation**.[8] All we need to know to calculate the temperature dependence of an equilibrium constant, therefore, is the standard reaction enthalpy.

Let's explore the information in the van't Hoff equation. Consider the case when $T' > T$. Then the term in parentheses in eqn 7.14 is positive. If $\Delta_r H^{\ominus} > 0$, corresponding to an endothermic reaction, the entire term on the right is positive. In this case, therefore, $\ln K' > \ln K$. That being so, we conclude that $K' > K$ for an endothermic reaction. In general, *the equilibrium constant of an endothermic reaction increases with temperature.* The opposite is true when $\Delta_r H^{\ominus} < 0$, so we can conclude that *the equilibrium constant of an exothermic reaction decreases with an increase in temperature.*

The conclusions we have outlined are of considerable commercial and environmental significance. For example, the synthesis of ammonia is exothermic, so its equilibrium constant decreases as the temperature is increased; in fact, K falls below 1 when the temperature is raised to above 200°C. Unfortunately, the reaction is slow at low temperatures and is commercially feasible only if the

temperature exceeds about 750°C even in the presence of a catalyst; but then K is very small. We shall see shortly how Fritz Haber, the inventor of the Haber process for the commercial synthesis of ammonia, was able to overcome this difficulty. Another example is the oxidation of nitrogen:

$$N_2(g) + O_2(g) \rightarrow 2\,NO(g)$$

This reaction is endothermic ($\Delta_r H^{\ominus} = +180\text{ kJ mol}^{-1}$) largely as a consequence of the very high bond enthalpy of N_2, so its equilibrium constant increases with temperature. It is for this reason that nitrogen monoxide (nitric oxide) is formed in significant quantities in the hot exhausts of jet engines and in the hot exhaust manifolds of internal combustion engines, and then goes on to contribute to the problems caused by acid rain.

7.9 The effect of compression

We have seen that Le Chatelier's principle suggests that the effect of compression (decrease in volume) on a gas-phase reaction at equilibrium is as follows:

When a system at equilibrium is compressed, the composition of a gas-phase equilibrium adjusts so as to reduce the number of molecules in the gas phase.

For example, in the synthesis of ammonia, reaction B, four reactant molecules give two product molecules, so compression favours the formation of ammonia. Indeed, this is the key to resolving Haber's dilemma, for by working with highly compressed gases he was able to increase the yield of ammonia.

Self-test 7.9

Is the formation of products in the reaction $4\,NH_3(g) + 5\,O_2(g) \rightleftharpoons 4\,NO(g) + 6\,H_2O(g)$ favoured by compression or expansion of the reaction vessel?

[*Answer*: expansion]

Let's explore the thermodynamic basis of this dependence. First, we note that $\Delta_r G^{\ominus}$ is defined as the difference between the Gibbs energies of sub-

---

[8] There are several 'van't Hoff equations'. To distinguish them, this one is sometimes called the *van't Hoff isochore*.

However, we have seen that compression leaves K unchanged. Therefore, the two partial pressures must adjust by different amounts. In this instance, K' will remain equal to K if the partial pressure of HI changes by a factor of less than 2 and the partial pressure of H_2 increases by more than a factor of 2. In other words, the equilibrium composition must shift in the direction of the reactants in order to preserve the equilibrium constant.

Compression has no effect on the composition when the number of gas-phase molecules is the same in the reactants as in the products. An example is the synthesis of hydrogen iodide in which all three substances are present in the gas phase and the chemical equation is $H_2(g) + I_2(g) \rightleftharpoons 2\ HI(g)$.

A more subtle example is the effect of the addition of an inert gas to a reaction mixture contained inside a vessel of constant volume. The overall pressure increases as the gas (such as argon) is added, but the addition of a foreign gas does not affect the *partial* pressures of the other gases present.[9] Therefore, under these circumstances, not only does the equilibrium constant remain unchanged, but the partial pressures of the reactants and products remain the same whatever the stoichiometry of the reaction.

stances in their standard states and therefore at 1 bar. It follows that $\Delta_r G^\ominus$ has the same value whatever the actual pressure used for the reaction. Therefore, because ln K is proportional to $\Delta_r G^\ominus$, K is *independent of pressure*. Thus, if the reaction mixture in which ammonia is being synthesized is compressed isothermally, the equilibrium constant remains unchanged.

This rather startling conclusion should not be misinterpreted. The value of K is independent of the pressure to which the system is subjected but, because partial pressures occur in the expression for K in a rather complicated way, that does not mean that the *individual* partial pressures or concentrations are unchanged. Suppose, for example, the volume of the reaction vessel in which the reaction $H_2(g) + I_2(s) \rightleftharpoons 2\ HI(g)$ has reached equilibrium is reduced by a factor of 2 and the system is allowed to reach equilibrium again. If the partial pressures were simply to double (that is, there is no adjustment of composition by further reaction), the equilibrium constant would double from

$$K = \frac{p_{HI}^2}{p_{H_2}} \quad \text{to} \quad K' = \frac{(2\,p_{HI})^2}{2\,p_{H_2}} = 2\,K$$

Exercises

7.1 Write the reaction quotients for the following reactions making the approximation of replacing activities by molar concentrations or partial pressures:

(a) $G6P(aq) + H_2O(l) \rightarrow G(aq) + P_i(aq)$

where G6P is glucose 6-phosphate, G is glucose, and P_i is inorganic phosphate.

(b) $Gly(aq) + Ala(aq) \rightarrow Gly-Ala(aq) + H_2O(l)$

(c) $Mg^{2+}(aq) + ATP^{4-}(aq) \rightarrow MgATP^{2-}(aq)$

(d) $2\ CH_3COCOOH(aq) + 5\ O_2(g) \rightarrow 6\ CO_2(g) + 4\ H_2O(l)$

7.2 The concentration of fructose 6-phosphate (F6P) and fructose 1,6-biphosphate (F16bP), ATP, and ADP in muscle tissue were measured as 0.089, 0.012, 12.0,

and 1.2 mmol L^{-1}. For the reaction F6P(aq) + ATP(aq) $\rightarrow$ F16bP(aq) + ADP(aq) at 37°C, the standard Gibbs energy (at pH = 7) is -18.3 kJ mol^{-1}. (a) Calculate the reaction quotient and the reaction Gibbs energy for the reaction in the muscle tissue environment. (b) Is the reaction spontaneous?

7.3 One of the most extensively studied reactions of industrial chemistry is the synthesis of ammonia, for its successful operation helps to govern the efficiency of the entire economy. The standard Gibbs energy of formation

---

[9] The partial pressure of a perfect gas (Section 1.3), is the pressure a gas would exert if it alone occupied the vessel, so it is independent of the presence or absence of any other gases.

of $NH_3(g)$ is -16.5 kJ mol^{-1} at 298 K. What is the reaction Gibbs energy when the partial pressures of the N_2, H_2, and NH_3 (treated as perfect gases) are 3.0 bar, 1.0 bar, and 4.0 bar, respectively? What is the spontaneous direction of the reaction in this case?

7.4 Write the expressions for the equilibrium constants of the following reactions:

(a) $CO(g) + Cl_2(g) \rightleftharpoons COCl(g) + Cl(g)$
(b) $2 SO_2(g) + O_2(g) \rightleftharpoons 2 SO_3(g)$
(c) $H_2(g) + Br_2(g) \rightleftharpoons 2 HBr(g)$
(d) $2 O_3(g) \rightleftharpoons 3 O_2(g)$

7.5 If the equilibrium constant for the reaction $A + B \rightleftharpoons C$ is reported as 0.224, what would be the equilibrium constant for the reaction written as $C \rightleftharpoons A + B$?

7.6 The equilibrium constant for the reaction $A + B \rightleftharpoons 2 C$ is reported as 3.4×10^4. What would it be for the reaction written as (a) $2 A + 2 B \rightleftharpoons 4 C$, (b) $\frac{1}{2} A + \frac{1}{2} B \rightleftharpoons C$?

7.7 The equilibrium constant for the isomerization of *cis*-2-butene to *trans*-2-butene is $K = 2.07$ at 400 K. Calculate the standard reaction Gibbs energy for the isomerization.

7.8 The standard reaction Gibbs energy of the isomerization of *cis*-2-pentene to *trans*-2-pentene at 400 K is -3.67 kJ mol^{-1}. Calculate the equilibrium constant of the isomerization.

7.9 One biochemical reaction has a standard Gibbs energy of -200 kJ mol^{-1} and a second biochemical reaction has a standard Gibbs energy of -100 kJ mol^{-1}. What is the ratio of their equilibrium constants at 310 K?

7.10 One enzyme-catalysed reaction in a biochemical cycle has an equilibrium constant that is 10 times the equilibrium constant of a second reaction. If the standard Gibbs energy of the former reaction is -300 kJ mol^{-1}, what is the standard reaction Gibbs energy of the second reaction?

7.11 What is the value of the equilibrium constant of a reaction for which $\Delta_r G^{\ominus} = 0$?

7.12 The standard reaction Gibbs energies (at pH = 7) for the hydrolysis of glucose-1-phosphate, glucose-6-phosphate, and glucose-3-phosphate are -21, -14, and -9.2 kJ mol^{-1}. Calculate the equilibrium constants for the hydrolyses at 37°C.

7.13 The biological standard Gibbs energy for the hydrolysis of ATP to ADP is -30.5 kJ mol^{-1}; what is the

Gibbs energy of reaction in an environment at 37°C in which the ATP, ADP, and P_i concentrations are all (a) 1.0 mmol L^{-1}, (b) 1.0 μmol L^{-1}?

7.14 Calculate the equilibrium constant of the reaction that results in the formation of glutamine (G, **6**) from glutamate (G′, **7**) in living cells, which is facilitated by the enzyme glutamine synthetase, which is a means of transporting ammonia from the kidneys to other cells. Base your answer on the fact that the standard reaction Gibbs energy for $G'(aq) + NH_4^+(aq) \rightarrow G(aq)$ is $+15.7$ kJ mol^{-1} at blood temperature 37°C. The reaction is in fact driven by coupling with $ATP(aq) \rightarrow ADP(aq) + P_i(aq)$, for which the standard Gibbs energy is -31.0 kJ mol^{-1}. What is the equilibrium constant for the overall reaction $G'(aq) + NH_4^+ + ATP(aq) \rightarrow G(aq) + ADP(aq) + P_i(aq)$?

7.15 The distribution of Na^+ ions across a typical biological membrane is 10 mmol L^{-1} inside the cell and 140 mmol L^{-1} outside the cell. At equilibrium the concentrations are equal. What is the Gibbs energy difference across the membrane at 37°C? The difference in concentration must be sustained by coupling to reactions that have at least that difference of Gibbs energy.

7.16 Use the information in Appendix 1 to estimate the temperature at which (a) $CaCO_3$ decomposes spontaneously and (b) $CuSO_4 \cdot 5H_2O$ undergoes dehydration.

7.17 The standard reaction enthalpy of $Zn(s) + H_2O(g) \rightarrow ZnO(s) + H_2(g)$ is approximately constant at $+224$ kJ mol^{-1} from 920 K up to 1280 K. The standard reaction Gibbs energy is $+33$ kJ mol^{-1} at 1600 K. Assuming that both quantities remain constant, estimate the temperature at which the equilibrium constant becomes greater than 1.

7.18 Creatine (**8**) is phosphorylated to creatine phosphate (**9**) by ATP in conjunction with the enzyme creatine phosphokinase. The standard Gibbs energy for the hydrolysis of creatine phosphate is -43 kJ mol^{-1} (at pH = 7) and that of ATP is -31 kJ mol^{-1}. Calculate the standard reaction Gibbs energy for the phosphorylation and the equilibrium constant at 37°C.

6 Glutamine   **7** Glutamate

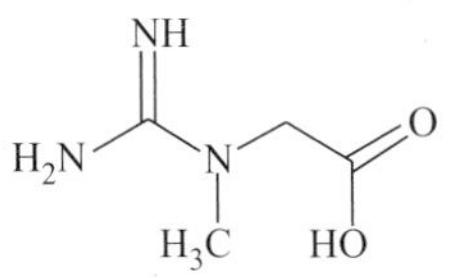

8 Creatine

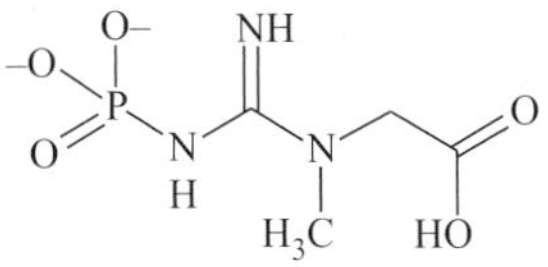

9 Creatine phosphate

7.19 Classify the following compounds as endergonic or exergonic: (a) glucose, (b) methylamine, (c) octane, (d) ethanol.

7.20 Combine the reaction entropies calculated in the following reactions with the reaction enthalpies and calculate the standard reaction Gibbs energies at 298 K:

(a) $HCl(g) + NH_3(g) \rightarrow NH_4Cl(s)$
(b) $2 Al_2O_3(s) + 3 Si(s) \rightarrow 3 SiO_2(s) + 4 Al(s)$
(c) $Fe(s) + H_2S(g) \rightarrow FeS(s) + H_2(g)$
(d) $FeS_2(s) + 2 H_2(g) \rightarrow Fe(s) + 2 H_2S(g)$
(e) $2 H_2O_2(l) + H_2S(g) \rightarrow H_2SO_4(l) + 2 H_2(g)$

7.21 Use the Gibbs energies of formation in Appendix 1 to decide which of the following reactions have $K > 1$ at 298 K.

(a) $2 CH_3CHO(g) + O_2(g) \rightleftharpoons 2 CH_3COOH(l)$
(b) $2 AgCl(s) + Br_2(l) \rightleftharpoons 2 AgBr(s) + Cl_2(g)$
(c) $Hg(l) + Cl_2(g) \rightleftharpoons HgCl_2(s)$
(d) $Zn(s) + Cu^{2+}(aq) \rightleftharpoons Zn^{2+}(aq) + Cu(s)$
(e) $C_{12}H_{22}O_{11}(s) + 12 O_2(g) \rightleftharpoons 12 CO_2(g) + 11 H_2O(l)$

7.22 Recall from Chapter 4 that the change in Gibbs energy can be identified with the maximum non-expansion work that can be extracted from a process. What is the maximum energy that can be extracted as (a) heat, (b) non-expansion work when 1.0 kg of natural gas (taken to be pure methane) is burned under standard conditions at 25°C? Take the reaction to be $CH_4(g) + 2 O_2(g) \rightarrow CO_2(g) + 2 H_2O(l)$.

7.23 Could the synthesis of ammonia be used as the basis of a fuel cell? What is the maximum electrical energy output for the consumption of 100 g of nitrogen?

7.24 In assessing metabolic processes we are usually more interested in the work that may be performed for the consumption of a given mass of compound than the heat it can produce (which merely keeps the body warm). What is the maximum energy that can be extracted as (a) heat, (b) non-expansion work when 1.0 kg of glucose is burned under standard conditions at 25°C with the production of water vapour? The reaction is $C_6H_{12}O_6(s) + 6 O_2(g) \rightarrow 6 CO_2(g) + 6 H_2O(g)$.

7.25 Is it more energy-effective to ingest sucrose or glucose? Calculate the non-expansion work, the expansion work, and the total work that can be obtained from the combustion of 1.0 kg of sucrose under standard conditions at 25°C when the product includes (a) water vapour, (b) liquid water.

7.26 A person of mass 65 kg has a daily food intake of about 9.0 MJ. Suppose that the food is used to phosphorylate ADP, which then supplies energy via hydrolysis, with a reaction Gibbs energy of about 50 kJ mol^{-1} in the prevailing conditions of cell composition and an efficiency of 50 per cent. What mass of disodium ATP (of molar mass 551 g mol^{-1}) is synthesized each day?

7.27 The standard enthalpy of combustion of solid phenol, C_6H_5OH, is -3054 kJ mol^{-1} at 298 K and its standard molar entropy is 144.0 J $K^{-1} mol^{-1}$. Calculate the standard Gibbs energy of formation of phenol at 298 K.

7.28 Calculate the maximum non-expansion work per mole that may be obtained from a fuel cell in which the chemical reaction is the combustion of methane at 298 K.

7.29 The oxidation of glucose in the mitochondria in energy-hungry brain cells leads to the formation of pyruvate ions, $CH_3COCO_2^-$, which are then decarboxylated to ethanal (acetaldehyde, CH_3CHO) in the course of the ultimate formation of carbon dioxide. The standard Gibbs energies of formation of pyruvate ions in aqueous solution and gaseous ethanal are -474 and -133 kJ mol^{-1}, respectively. Calculate the Gibbs energy of the reaction in which pyruvate ions are converted to ethanal by the action of pyruvate decarboxylase with the release of carbon dioxide.

7.30 Acetaldehyde is soluble in water. Would you expect the standard Gibbs energy of the enzyme-catalysed reaction in which pyruvate ions are decarboxylated to ethanal in solution to be larger or smaller than the value for the production of gaseous ethanal? (See the preceding exercise.)

7.31 Pyruvic acid is a weak acid with $pK_a = 2.49$. Would you expect the standard Gibbs energy of the enzyme-catalysed reaction in which pyruvate ions are decarboxylated to ethanal in solution to be larger or smaller than the value for the decarboxylation of pyruvate ions in the gas phase?

7.32 Calculate the standard biological Gibbs energy for the reaction

$$\text{Pyruvate}^- + \text{NADH} + \text{H}^+ \rightarrow \text{lactate}^- + \text{NAD}^+$$

at 310 K given that $\Delta_r G^{\ominus} = -66.6$ kJ mol^{-1}. This reaction occurs in muscle cells during strenuous exercise and can lead to cramp.

7.33 The standard biological reaction Gibbs energy for the removal of the phosphate group from adenosine monophosphate is -14 kJ mol^{-1} at 298 K. What is the value of the thermodynamic standard reaction Gibbs energy?

7.34 The equilibrium constant of the reaction $2\ \text{C}_3\text{H}_6(g) \rightleftharpoons \text{C}_2\text{H}_4(g) + \text{C}_4\text{H}_8(g)$ is found to fit the expression

$$\ln K = -1.04 - \frac{1088\ \text{K}}{T} + \frac{1.51 \times 10^5\ \text{K}^2}{T^2}$$

between 300 K and 600 K. Calculate the standard reaction enthalpy and standard reaction entropy at 400 K. *Hint.* Begin by calculating ln K at 390 K and 410 K; then use eqn 7.14.

7.35 Borneol is a pungent compound obtained from the camphorwood tree of Borneo and Sumatra. The standard reaction Gibbs energy of the isomerization of borneol (**10**) to isoborneol (**11**) in the gas phase at 503 K is $+9.4$ kJ mol^{-1}. Calculate the reaction Gibbs energy in a mixture consisting of 0.15 mol of borneol and 0.30 mol of isoborneol when the total pressure is 600 Torr.

7.36 The equilibrium constant for the gas-phase isomerization of borneol, $\text{C}_{10}\text{H}_{17}\text{OH}$, to isoborneol (see Exercise 7.35) at 503 K is 0.106. A mixture consisting of 7.50 g of borneol and 14.0 g of isoborneol in a container of volume 5.0 L is heated to 503 K and allowed to come to equilibrium. Calculate the mole fractions of the two substances at equilibrium.

10 d-Borneol

11 d-Isoborneol

7.37 Calculate the composition of a system in which nitrogen and hydrogen are mixed at partial pressures of 1.00 bar and 4.00 bar and allowed to reach equilibrium with their product, ammonia, under conditions when $K = 89.8$.

7.38 The bond in molecular iodine is quite weak, and hot iodine vapour contains a proportion of atoms. When 1.00 g of I_2 is heated to 1000 K in a sealed container of volume 1.00 L, the resulting equilibrium mixture contains 0.830 g of I_2. Calculate K for the dissociation equilibrium $\text{I}_2(g) \rightleftharpoons 2\ \text{I}(g)$.

7.39 In a gas-phase equilibrium mixture of SbCl_5, SbCl_3, and Cl_2 at 500 K, $p_{\text{SbCl}_5} = 0.15$ bar and $p_{\text{SbCl}_3} = 0.20$ bar. Calculate the equilibrium partial pressure of Cl_2 given that $K = 3.5 \times 10^{-4}$ for the reaction $\text{SbCl}_5(g) \rightleftharpoons \text{SbCl}_3(g) + \text{Cl}_2(g)$.

7.40 The equilibrium constant $K = 0.36$ for the reaction $\text{PCl}_5(g) \rightleftharpoons \text{PCl}_3(g) + \text{Cl}_2(g)$ at 400 K. (a) Given that 2.0 g of PCl_5 was initially placed in a reaction vessel of volume 250 mL, determine the molar concentrations in the mixture at equilibrium. (b) What is the percentage of PCl_5 decomposed at 400 K?

7.41 In the Haber process for ammonia, $K = 0.036$ for the reaction $\text{N}_2(g) + 3\ \text{H}_2(g) \rightleftharpoons 2\ \text{NH}_3(g)$ at 500 K. If a reactor is charged with partial pressures of 0.020 bar of N_2 and 0.020 bar of H_2, what will be the equilibrium partial pressure of the components?

7.42 Express the equilibrium constant for $\text{N}_2\text{O}_4(g) \rightleftharpoons 2\ \text{NO}_2(g)$ in terms of the fraction α of N_2O_4 that has dissociated and the total pressure p of the reaction mixture, and show that, when the extent of dissociation is small ($\alpha \ll 1$), α is inversely proportional to the square root of the total pressure ($\alpha \propto p^{-1/2}$).

7.43 The equilibrium pressure of H_2 over a mixture of solid uranium and solid uranium hydride at 500 K is 1.04 Torr. Calculate the standard Gibbs energy of formation of $\text{UH}_3(s)$ at 500 K.

7.44 Which of the products in Exercise 7.4 are favoured by a rise in temperature at constant pressure (in the sense that K increases)?

7.45 What is the standard enthalpy of a reaction for which the equilibrium constant is (a) doubled, (b) halved when the temperature is increased by 10 K at 298 K?

7.46 The dissociation vapour pressure (the pressure of gaseous products in equilibrium with the solid reactant) of NH_4Cl at 427°C is 608 kPa but at 459°C it has risen to

1115 kPa. Calculate (a) the equilibrium constant, (b) the standard reaction Gibbs energy, (c) the standard enthalpy, (d) the standard entropy of dissociation, all at 427°C. Assume that the vapour behaves as a perfect gas and that $\Delta H^{\ominus}$ and $\Delta S^{\ominus}$ are independent of temperature in the range given.

7.47 One of the key steps in the tricarboxylic acid cycle is the hydrolysis of fumarate ions to malate ions: fumarate^{2-} + H$_2$O $\rightleftharpoons$ malate^{2-}. The standard reaction Gibbs energy and enthalpy are -880 kcal mol^{-1} and $+3560$ kcal mol^{-1} at 25°C (note the units). What is the equilibrium constant for the reaction at 37°C?

7.48 The native (active) form of an enzyme is in equilibrium with its denatured (inactive) form, and their relative abundances change with temperature. In a study of ribonuclease, the ratio of concentrations (active over inactive) was found to be 390 at 50°C and 6.2 at 100°C. Estimate the enthalpy of denaturation.

Chapter 8

Consequences of equilibrium

Contents

I**N** this chapter we examine two consequences of dynamic chemical equilibria. We concentrate on the equilibria that exist in solutions of acids, bases, and their salts in water, where rapid proton transfer between species ensures that equilibrium is maintained at all times. The link between Chapter 7 and the discussion here is that, provided the temperature is held constant, *an equilibrium constant retains its value even though the individual activities may change.* So, if one substance is added to a mixture at equilibrium, the other substances adjust their abundances to restore the value of K.

Proton transfer equilibria

P**ROTON** transfer equilibrium is maintained in living cells and helps to keep proteins viable. Even small drifts in the equilibrium concentration of hydrogen ions can result in disease, cell damage, and death. We shall see the basis of our bodies' mechanism for maintaining homeostasis.

8.1 Brønsted–Lowry theory

We assume that the general ideas of the **Brønsted–Lowry theory** of acids and bases are familiar. In that theory, an **acid** is a proton donor and a **base** is a proton acceptor. The proton, which in this context means a hydrogen ion, H^+, is highly mobile and acids and bases in water are always in equilibrium with their conjugate bases and acids and hydronium ions (H_3O^+). Thus, an acid HA, such as HCN, immediately establishes the equilibrium

$$HA(aq) + H_2O(aq) \rightleftharpoons H_3O^+(aq) + A^-(aq) \quad K = \frac{a_{H_3O^+} a_{A^-}}{a_{HA} a_{H_2O}}$$

A base B (such as NH_3) immediately establishes the equilibrium

$$B(aq) + H_2O(aq) \rightleftharpoons BH^+(aq) + OH^-(aq) \quad K = \frac{a_{BH^+} a_{OH^-}}{a_B a_{H_2O}}$$

In these equilibria, A^- is the **conjugate base** of the acid HA, and BH^+ is the **conjugate acid** of the base B. Even in the absence of added acids and bases, proton transfer occurs between water molecules and the **autoprotolysis equilibrium**[1]

$$2 H_2O(l) \rightleftharpoons H_3O^+(aq) + OH^-(aq) \quad K = \frac{a_{H_3O^+} a_{OH^-}}{a_{H_2O}^2}$$

is always present.

As will be familiar from introductory chemistry, the hydronium ion concentration is commonly expressed in terms of the pH, which is defined formally as

$$pH = -\log a_{H_3O^+} \quad (8.1)$$

In elementary work, the hydronium ion activity is replaced by the numerical value of its molar concentration, $[H_3O^+]$. For example, if the molar concentration of H_3O^+ is 2.0 mmol L^{-1} then

$$pH \approx -\log(2.0 \times 10^{-3}) = 2.70$$

If the molar concentration were ten times less, at 0.20 mmol L^{-1}, then the pH would be 3.70. Notice that, *the higher the pH, the lower the concentration of*

hydronium ions in the solution and that a change in pH by 1 unit corresponds to a tenfold change in their molar concentration. However, it should never be forgotten that the replacement of activities by molar concentration is invariably hazardous. Because ions interact over long distances, the replacement is unreliable for all but the most dilute solutions.

Self-test 8.1

Death is likely if the pH of human blood plasma changes by more than ±0.4 from its normal value of 7.4. What is the approximate range of molar concentrations of hydrogen ions for which life can be sustained?

[*Answer*: 16 nmol L^{-1} to 100 nmol L^{-1} (1 nmol = 10^{-9} mol)]

8.2 Protonation and deprotonation

All the solutions we consider are so dilute that we can regard the water present as being a nearly pure liquid. When we set $a_{H_2O} = 1$ for all the solutions we consider, the resulting equilibrium constant is called the **acidity constant**, K_a, of the acid HA:[2]

$$K_a = \frac{a_{H_3O^+} a_{A^-}}{a_{HA}} \approx \frac{[H_3O^+][A^-]}{[HA]} \quad (8.2)$$

Data are widely reported in terms of the negative common logarithm of this quantity (Table 8.1):

$$pK_a = -\log K_a \quad (8.3)$$

It follows from eqn 7.8 ($\Delta_r G^{\ominus} = -RT \ln K$) that pK_a is proportional to $\Delta_r G^{\ominus}$ for the proton transfer reaction. Therefore, manipulations of pK_a and related quantities are actually manipulations of standard reaction Gibbs energies in disguise.

The value of the acidity constant indicates the extent to which proton transfer has occurred. The smaller the value of K_a, and therefore the larger the value of pK_a, the lower is the concentration of deprotonated molecules. Most acids

---

[1] Autoprotolysis is also called *autoionization*.

[2] Acidity constants are also called *acid ionization constants* and, less appropriately, *dissociation constants*.

Table 8.1 *Acidity and basicity constants* at 25 °C*

Acid/base	K_b	pK_b	K_a	pK_a
Strongest weak acids				
Trichloroacetic acid, CCl_3COOH	3.3×10^{-14}	13.48	3.0×10^{-1}	0.52
Benzenesulfonic acid, $C_6H_5SO_3H$	5.0×10^{-14}	13.30	2×10^{-1}	0.70
Iodic acid, HIO_3	5.9×10^{-14}	13.23	1.7×10^{-1}	0.77
Sulfurous acid, H_2SO_3	6.3×10^{-13}	12.19	1.6×10^{-2}	1.81
Chlorous acid, $HClO_2$	1.0×10^{-12}	12.00	1.0×10^{-2}	2.00
Phosphoric acid, H_3PO_4	1.3×10^{-12}	11.88	7.6×10^{-3}	2.12
Chloroacetic acid, $CH_2ClCOOH$	7.1×10^{-12}	11.15	1.4×10^{-3}	2.85
Lactic acid, $CH_3CH(OH)COOH$	1.2×10^{-11}	10.92	8.4×10^{-4}	3.08
Nitrous acid, HNO_2	2.3×10^{-11}	10.63	4.3×10^{-4}	3.37
Hydrofluoric acid, HF	2.9×10^{-11}	10.55	3.5×10^{-4}	3.45
Formic acid, HCOOH	5.6×10^{-11}	10.25	1.8×10^{-4}	3.75
Benzoic acid, C_6H_5COOH	1.5×10^{-10}	9.81	6.5×10^{-5}	4.19
Acetic acid, CH_3COOH	5.6×10^{-10}	9.25	1.8×10^{-5}	4.75
Carbonic acid, H_2CO_3	2.3×10^{-8}	7.63	4.3×10^{-7}	6.37
Hypochlorous acid, HClO	3.3×10^{-7}	6.47	3.0×10^{-8}	7.53
Hypobromous acid, HBrO	5.0×10^{-6}	5.31	2.0×10^{-9}	8.69
Boric acid, $B(OH)_3$†	1.4×10^{-5}	4.86	7.2×10^{-10}	9.14
Hydrocyanic acid, HCN	2.0×10^{-5}	4.69	4.9×10^{-10}	9.31
Phenol, C_6H_5OH	7.7×10^{-5}	4.11	1.3×10^{-10}	9.98
Hypoiodous acid, HIO	4.3×10^{-4}	3.36	2.3×10^{-11}	10.64
Weakest weak acids				
Weakest weak bases				
Urea, $CO(NH_2)_2$	1.3×10^{-14}	13.90	7.7×10^{-1}	0.10
Aniline, $C_6H_5NH_2$	4.3×10^{-10}	9.37	2.3×10^{-5}	4.63
Pyridine, C_5H_5N	1.8×10^{-9}	8.75	5.6×10^{-6}	5.35
Hydroxylamine, NH_2OH	1.1×10^{-8}	7.97	9.1×10^{-7}	6.03
Nicotine, $C_{10}H_{11}N_2$	1.0×10^{-6}	5.98	1.0×10^{-8}	8.02
Morphine, $C_{17}H_{19}O_3N$	1.6×10^{-6}	5.79	6.3×10^{-9}	8.21
Hydrazine, NH_2NH_2	1.7×10^{-6}	5.77	5.9×10^{-9}	8.23
Ammonia, NH_3	1.8×10^{-5}	4.75	56×10^{-10}	9.25
Trimethylamine, $(CH_3)_3N$	6.5×10^{-5}	4.19	1.5×10^{-10}	9.81
Methylamine, CH_3NH_2	3.6×10^{-4}	3.44	2.8×10^{-11}	10.56
Dimethylamine, $(CH_3)_2NH$	5.4×10^{-4}	3.27	1.9×10^{-11}	10.73
Ethylamine, $C_2H_5NH_2$	6.5×10^{-4}	3.19	1.5×10^{-11}	10.81
Triethylamine, $(C_2H_5)_3N$	1.0×10^{-3}	2.99	1.0×10^{-11}	11.01
Strongest weak bases				

*Values for polyprotic acids—those capable of donating more than one proton—refer to the first deprotonation.
†The proton transfer equilibrium is $B(OH)_3(aq) + 2 H_2O(l) = H_3O^+(aq) + B(OH)_4^-(aq)$.

have $K_a < 1$ (and usually much less than 1), with $pK_a > 0$, indicating only a small extent of deprotonation in water. These acids are classified as **weak acids**. A few acids, most notably, in aqueous solution, HCl, HBr, HI, HNO_3, H_2SO_4, and $HClO_4$, have $K_a > 1$ and $pK_a < 0$. These acids are classified as **strong acids**, and are commonly regarded as being completely deprotonated in aqueous solution.[3]

The corresponding expression for a base is called the **basicity constant**, K_b, and is

$$K_b = \frac{a_{BH^+}\, a_{OH^-}}{a_B} \approx \frac{[BH^+][OH^-]}{[B]} \qquad pK_b = -\log K_b \quad (8.4)$$

A **strong base** is fully protonated in solution in the sense that $K_b > 1$. One example is the oxide ion, which cannot survive in water but is immediately and fully converted into its conjugate acid OH^-. A **weak base** is not fully protonated in water in the sense that $K_b < 1$ (and usually much less than 1). Ammonia, NH_3, and its organic derivatives the amines are all weak bases in water, and only a small proportion of their molecules exist as the conjugate acid (NH_4^+ or RNH_3^+).

The **autoprotolysis constant** for water, K_w, is

$$K_w = a_{H_3O^+}\, a_{OH^-} \qquad pK_w = -\log K_w \quad (8.5)$$

with $K_w = 1.00 \times 10^{-14}$ and $pK_w = 14.00$ at 25°C. The acidity constant of the conjugate acid, BH^+, of a base B is related to the basicity constant of B by

$$K_a \times K_b = K_w \qquad (8.6a)$$

as may be confirmed by multiplying the two constants together. The implication of this relation is that K_a increases as K_b decreases to maintain a product equal to the constant K_w. That is, *as the strength of a base decreases, the strength of its conjugate acid increases*, and vice versa. On taking the negative logarithm of both sides of eqn 8.6a, we obtain

$$pK_a + pK_b = pK_w \qquad (8.6b)$$

---

[3] Sulfuric acid, H_2SO_4, is strong with respect only to its first deprotonation; HSO_4^- is weak.

The great advantage of this relation is that the pK_b values of bases may be expressed as the pK_a of their conjugate acids, so the strengths of all acids and bases may be listed in a single table.

Another useful relation is obtained by taking the negative logarithm of both sides of the definition of K_w in eqn 8.5, which gives

$$pH + pOH = pK_w \qquad (8.7)$$

where $pOH = -\log a_{OH^-}$. This enormously important relation means that the activities (in elementary work, the molar concentrations) of hydronium and hydroxide ions are related by a seesaw relation: as one goes up, the other goes down to preserve the value of pK_w.

Self-test 8.2

The molar concentration of OH^- ions in a certain solution is $0.010\,\text{mmol L}^{-1}$. What is the pH of the solution?

[*Answer*: 9.00]

The extent of deprotonation of a weak acid in solution depends on the acidity constant and the initial concentration of the acid, its concentration as prepared. The **fraction deprotonated**, the fraction of acid molecules that have donated a proton, is

$$\text{Fraction deprotonated} =$$

$$\frac{\text{molar concentration of conjugate base}}{\text{initial molar concentration of base}} \qquad (8.8)$$

The extent to which a weak base is protonated is reported in terms of the **fraction protonated**:

$$\text{Fraction protonated} =$$

$$\frac{\text{molar concentration of conjugate acid}}{\text{initial molar concentration of acid}} \qquad (8.9)$$

We can estimate the pH of a solution of a weak acid or a weak base and calculate either of these fractions by using the equilibrium-table technique described in Section 7.6.

Example 8.1 *Assessing the deprotonation of a weak acid*

Estimate the pH and the fraction of CH_3COOH molecules deprotonated in 0.15 M CH_3COOH(aq).

Strategy The aim is to calculate the equilibrium composition of the solution. To do so, we use the technique illustrated in Example 7.5, with x the change in molar concentration of H_3O^+ ions required to reach equilibrium. We ignore the tiny concentration of hydronium ions present in pure water. Once x has been found, calculate pH = $-\log x$. Because we can anticipate that the extent of deprotonation is small (the acid is weak), use the approximation that x is very small to simplify the equations.

Solution We draw up the following equilibrium table:

Species:	CH_3COOH	H_3O^+	$CH_3CO_2^-$
Initial concentration/ (mol L^{-1})	0.15	0	0
Change to reach equilibrium/ (mol L^{-1})	$-x$	$+x$	$+x$
Equilibrium concentration/ (mol L^{-1})	$0.15-x$	x	x

The value of x is found by inserting the equilibrium concentrations into the expression for the acidity constant:

$$K_a = \frac{[H_3O^+][CH_3CO_2^-]}{[CH_3COOH]} = \frac{x \times x}{0.15-x}$$

We could arrange the expression into a quadratic equation and use the solution in Example 7.5. However, it is more instructive to make use of the smallness of x to replace $0.15 - x$ by 0.15 (this approximation is valid if $x \ll 0.15$). Then the simplified equation rearranges first to $0.15 \times K_a = x^2$ and then to

$$x = (0.15 \times K_a)^{1/2} = (0.15 \times 1.8 \times 10^{-5})^{1/2} = 1.6 \times 10^{-3}$$

Therefore, pH = 2.80. Calculations of this kind are rarely accurate to more than one decimal place in the pH (and even that may be overly optimistic) because the effects of ion–ion interactions have been ignored, so this answer would be reported as pH = 2.8. The fraction deprotonated, f, is

$$f = \frac{[CH_3CO_2^-]_{equilibrium}}{[CH_3COOH]_{added}} = \frac{x}{0.15} = \frac{1.6 \times 10^{-3}}{0.15} = 0.011$$

That is, only 1.1 per cent of the acetic acid molecules have donated a proton.

Self-test 8.3

Estimate the pH of 0.010 M $CH_3CH(OH)COOH$(aq) (lactic acid) from the data in Table 8.1. Before carrying out the numerical calculation, decide whether you expect the pH to be higher or lower than that calculated for the same concentration of acetic acid.

[*Answer:* 2.6]

Example 8.2 *Assessing the protonation of a weak base*

The base quinoline (**1**) has pK_a = 4.88. Estimate the pH and the fraction of molecules protonated in an 0.010 M aqueous solution of quinoline.

Strategy The calculation of the pH of a solution of a base involves one more step than that for the pH of a solution of an acid. The first step is to calculate the concentration of OH^- ions in the solution by using the equilibrium-table technique, and to express it as the pOH of the solution. The additional step is to convert that pOH into a pH by using the water autoprotolysis equilibrium, eqn 8.7, in the form pH = pK_w − pOH, with pK_w = 14.00 at 25°C. We also need to compute $pK_b = pK_w − pK_a$.

Solution First, we write

$$pK_b = 14.00 - 4.88 = 9.12,$$
corresponding to $K_b = 10^{-9.12} = 7.6 \times 10^{-10}$

Now draw up the following equilibrium table, denoting quinoline by Q and its conjugate acid by QH^+:

Species:	Q	OH^-	QH^+
Initial concentration/ (mol L^{-1})	0.010	0	0
Change to reach equilibrium/ (mol L^{-1})	$-x$	$+x$	$+x$
Equilibrium concentration/ (mol L^{-1})	$0.010-x$	x	x

1 Quinoline

The value of x is found by inserting the equilibrium concentrations into the expression for the basicity constant:

$$K_b = \frac{[OH^-][QH^+]}{[Q]} = \frac{x \times x}{0.010 - x}$$

We suppose that $x \ll 0.010$. Then the simplified equation rearranges to

$$x = (0.010 \times K_b)^{1/2} = (0.010 \times 7.6 \times 10^{-10})^{1/2} = 2.8 \times 10^{-6}$$

Therefore, pOH = 5.55, and consequently pH = 14.00 − 5.55 = 8.45, or about 8.4. The fraction protonated, f, is

$$f = \frac{[QH^+]_{equilibrium}}{[Q]_{added}} = \frac{x}{0.010} = \frac{2.8 \times 10^{-6}}{0.010} = 2.8 \times 10^{-4}$$

or 1 molecule in about 3500.

Self-test 8.4

The pK_a for the first protonation of nicotine (**2**) is 8.02. What is the pH and the fraction of molecules protonated in a 0.015 M aqueous solution of nicotine?

[*Answer*: 10.08; 1/125]

2 Nicotine

8.3 Polyprotic acids

A **polyprotic acid** is a molecular compound that can donate more than one proton. Two examples are sulfuric acid, H_2SO_4, which can donate up to two protons, and phosphoric acid, H_3PO_4, which can donate up to three. A polyprotic acid is best considered to be a molecular species that can give rise to a series of Brønsted acids as it donates its succession of protons. Thus, sulfuric acid is the parent of two Brønsted acids, H_2SO_4 itself and HSO_4^-, and phosphoric acid is the parent of three Brønsted acids, namely H_3PO_4, $H_2PO_4^-$, and HPO_4^{2-}.

For a species H_2A with two acidic protons (such as H_2SO_4), the successive equilibria we need to consider are

$$H_2A(aq) + H_2O(l) \rightleftharpoons H_3O^+(aq) + HA^-(aq) \quad K_{a1} = \frac{a_{H_3O^+} a_{HA^-}}{a_{H_2A}}$$

$$HA^-(aq) + H_2O(l) \rightleftharpoons H_3O^+(aq) + A^{2-}(aq) \quad K_{a2} = \frac{a_{H_3O^+} a_{A^{2-}}}{a_{HA^-}}$$

In the first of these equilibria, HA^- is the conjugate base of H_2A. In the second, HA^- acts as the acid and A^{2-} is its conjugate base. Values are given in Table 8.2. In all cases, K_{a2} is smaller than K_{a1}, typically by three orders of magnitude for small molecular species, because the second proton is more difficult to remove, partly on account of the negative charge on HA^-. Enzymes are polyprotic acids, for they possess many protons that can be donated to a substrate molecule or to the surrounding aqueous medium of the cell. For them, successive acid-

Table 8.2 *Successive acidity constants of polyprotic acids*

Acid	K_{a1}	pK_{a1}	K_{a2}	pK_{a2}	K_{a3}	pK_{a3}
Carbonic acid, H_2CO_3	4.3×10^{-7}	6.37	5.6×10^{-11}	10.25		
Hydrosulfuric acid, H_2S	1.3×10^{-7}	6.88	7.1×10^{-15}	14.15		
Oxalic acid, $(COOH)_2$	5.9×10^{-2}	1.23	6.5×10^{-5}	4.19		
Phosphoric acid, H_3PO_4	7.6×10^{-3}	2.12	6.2×10^{-8}	7.21	2.1×10^{-13}	12.67
Phosphorous acid, H_2PO_3	1.0×10^{-2}	2.00	2.6×10^{-7}	6.59		
Sulfuric acid, H_2SO_4	Strong		1.2×10^{-2}	1.92		
Sulfurous acid, H_2SO_3	1.5×10^{-2}	1.81	1.2×10^{-7}	6.91		
Tartaric acid, $C_2H_4O_2(COOH)_2$	6.0×10^{-4}	3.22	1.5×10^{-5}	4.82		

ity constants vary much less because the molecules are so large that the loss of a proton from one part of the molecule has little effect on the ease with which another some distance away may be lost.

Example 8.3 *Calculating the concentration of carbonate ion in carbonic acid*

Ground water contains dissolved carbon dioxide, carbonic acid, hydrogencarbonate ions, and a very low concentration of carbonate ions. Estimate the molar concentration of CO_3^{2-} ions in a saturated solution.

Strategy Start with the equilibrium that produces the ion of interest (such as A^{3-}) and write its activity in terms of the acidity constant for its formation (K_{a3}). That expression will contain the activity of the conjugate acid (HA^{2-}), which we can express in terms of the activity of *its* conjugate acid (H_2A^-) by using the appropriate acidity constant (K_{a2}), and so on. In the last step, we arrive at the first proton loss (K_{a1}) from the overall parent acid (H_3A, if the acid is triprotic). This equilibrium dominates all the rest provided the molecule is small and there are marked differences between its acidity constants, so it may be possible to make an approximation at this stage.

Solution The CO_3^{2-} ion, the conjugate base of the acid HCO_3^-, is produced in the equilibrium

$$HCO_3^-(aq) + H_2O(l) \rightleftharpoons H_3O^+(aq) + CO_3^{2-}(aq)$$

$$K_{a2} = \frac{a_{H_3O^+}\, a_{CO_3^{2-}}}{a_{HCO_3^-}}$$

Hence,

$$a_{CO_3^{2-}} = \frac{a_{HCO_3^-}\, K_{a2}}{a_{H_3O^+}} \approx \frac{[HCO_3^-]\,K_{a2}}{[H_3O^+]}$$

The HCO_3^- ions are produced in the equilibrium

$$H_2CO_3(aq) + H_2O(l) \rightleftharpoons H_3O^+(aq) + HCO_3^-(aq)$$

This equilibrium dominates the second deprotonation, so it also dominates the molar concentrations of HCO_3^- and H_3O^+. These two concentrations are not exactly the same, because a little HCO_3^- has been lost in the second deprotonation and the amount of H_3O^+ has been increased by it; however, those secondary changes can safely be ignored in an approximate calculation. Because the molar concentrations are approximately the same, we can suppose that the activities are also approximately the same, and

set $a_{HCO_3^-} \approx a_{H_3O^+}$. When this equality is substituted into the preceding expression, we obtain

$$[CO_3^{2-}] \approx K_{a2}$$

Because we know from Table 8.2 that $pK_{a2} = 10.25$, it follows that $[CO_3^{2-}] = 5.6 \times 10^{-11}$, and therefore that the molar concentration of CO_3^{2-} ions is 5.6×10^{-11} mol L^{-1}.

Self-test 8.5

Calculate the molar concentration of S^{2-} ions in H_2S(aq).

[*Answer*: 7.1×10^{-15} mol L^{-1}]

Example 8.4 *Calculating the fractional composition of a solution*

The amino acid lysine (Lys, **3**) can accept two protons on its nitrogen atoms and donate one from its carboxyl group. Show how the composition of an aqueous solution that contains 0.010 mol L^{-1} of lysine varies with pH. The pK_a values of amino acids are given in Table 8.3.

Strategy We expect the fully protonated species (H_3Lys^{2+}) at low pH, the partially protonated species (H_2Lys^+ and HLys) at intermediate pH, and the fully deprotonated species (Lys$^-$) at high pH. Set up the expressions for the three acidity constants, treating H_3Lys^{2+} as the parent acid, and an expression for the total concentration of lysine, L. Solve the resulting expressions for the fraction of each species in terms of the hydronium ion concentration, H.

Solution The three acidity constants we require are

$$H_3Lys^{2+}(aq) + H_2O(l) \rightleftharpoons H_3O^+(aq) + H_2Lys^+(aq)$$

$$K_{a1} = \frac{[H_3O^+][H_2Lys^+]}{[H_3Lys^{2+}]} = \frac{H[H_2Lys^+]}{[H_3Lys^{2+}]}$$

3 Lysine (Lys)

Table 8.3 *Acidity constants of amino acids* *

Acid	pK_{a1}	pK_{a2}	pK_{a3}
Ala	2.35	9.87	
Arg	2.01	9.04	12.48
Asn	2.02	8.80	
Asp	2.10	3.86	9.82
Cys	1.65	2.26	7.85
Glu	2.19	4.25	9.67
Gly	2.35	9.60	
His	1.77	6.10	9.18
Ile	2.32	9.76	
Leu	2.33	9.74	
Lys	2.18	8.95	10.53
Met	2.28	9.21	
Phe	2.58	9.24	
Pro	1.99	10.6	
Ser	2.21	9.15	
Thr	2.09	9.10	
Trp	2.38	9.39	
Tyr	2.20	9.11	10.07
Val	2.29	9.72	

* For the identities of the acids, see Appendix 3. The acidity constants refer, respectively, to the most highly protonated form, the next most, and so on. So the values for Lys, for instance, refer to H_3Lys^{2+}, H_2Lys^+, and HLys (the electrically neutral molecule).

$$H_2Lys^+(aq) + H_2O(l) \rightleftharpoons H_3O^+(aq) + HLys(aq)$$

$$K_{a2} = \frac{[H_3O^+][HLys]}{[H_2Lys^+]} = \frac{H[HLys]}{[H_2Lys^+]}$$

$$HLys(aq) + H_2O(l) \rightleftharpoons H_3O^+(aq) + Lys^-(aq)$$

$$K_{a2} = \frac{[H_3O^+][Lys^-]}{[HLys]} = \frac{H[Lys^-]}{[HLys]}$$

We also know that the total concentration of lysine in all its forms is

$$[H_3Lys^{2+}] + [H_2Lys^+] + [HLys] + [Lys^-] = L$$

We now have four equations for four unknown concentrations. To solve the equations, we proceed systematically by using K_{a3} to express $[Lys^-]$ in terms of [HLys], then K_{a2} to express [HLys] in terms of $[H_2Lys^+]$, and so on. For instance,

$$[Lys^-] = \frac{K_{a3}[HLys]}{H} = \frac{K_{a3}K_{a2}[H_2Lys^+]}{H^2}$$

$$= \frac{K_{a3}K_{a2}K_{a1}[H_3Lys^{2+}]}{H^3}$$

Then the expression for the total concentration can be expressed in terms of $[H_3Lys^{2+}]$, H, and L. If we write

$$K = H^3 + H^2K_{a1} + HK_{a1}K_{a2} + K_{a1}K_{a2}K_{a3}$$

then for the fractions of each species present in the solution we find

$$f(H_3Lys^{2+}) = \frac{[H_3Lys^{2+}]}{L} = \frac{H^3}{K}$$

$$f(H_2Lys^+) = \frac{[H_2Lys^+]}{L} = \frac{H^2K_{a1}}{K}$$

$$f(HLys) = \frac{[HLys]}{L} = \frac{HK_{a1}K_{a2}}{K}$$

$$f(Lys^-) = \frac{[HLys^-]}{L} = \frac{K_{a1}K_{a2}K_{a3}}{K}$$

These fractions are plotted against pH = $-\log H$ in Fig 8.1. Note how H_3Lys^{2+} is dominant for pH < pK_{a3}, that H_3Lys^{2+} and H_3Lys^+ have the same concentration at pH = pK_{a3}, and that H_2Lys^+ is dominant for pH > pK_{a3}, until HLys becomes dominant, and so on.

Self-test 8.6

Construct the diagram for the fraction of protonated species in an aqueous solution of histidine (**4**).

[*Answer*: Fig 8.2]

4 Histidine (His)

We can summarize the behaviour found in Example 8.4 and illustrated in Figs 8.1 and 8.2 as follows. Consider each conjugate acid–base pair, with acidity constant K; then:

The acid form is dominant for pH < pK.

The conjugate pair have equal concentrations at pH = pK.

The base form is dominant for pH > pK.

In each case, the other possible forms of a polyprotic system can be ignored, provided the pK values are not too close together.

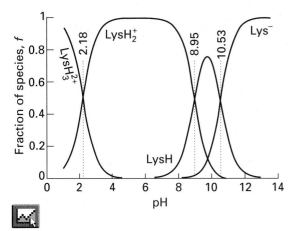

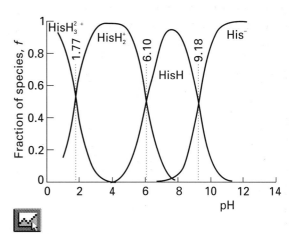

Fig 8.1 The fractional composition of the protonated and de-protonated forms of lysine (Lys) in aqueous solution as a function of pH. Note that conjugate pairs are present at equal concentrations when the pH is equal to the pK_a of the acid member of the pair.

Fig 8.2 The fractional composition of the protonated and de-protonated forms of histidine (His) in aqueous solution as a function of pH.

8.4 Amphiprotic systems

An **amphiprotic** species is a molecule or ion that can both accept and donate protons. For instance, $H_2PO_4^-$ can act as an acid (to form HPO_4^{2-}) and as a base (to form H_3PO_4). Among the most important amphiprotic compounds are the amino acids, which can act as proton donors by virtue of their carboxyl groups and as bases by virtue of their amino groups. Indeed, in solution, amino acids are present largely in their **zwitterionic** ('double ion') form, in which the amino group is protonated and the carboxyl group is deprotonated. The zwitterionic form of glycine, NH_2CH_2COOH, for instance, is $^+H_3NCH_2CO_2^-$.

The question we need to tackle is the pH of a solution of a salt with an amphiprotic anion, such as a solution of $NaHCO_3$. Is the solution acidic on account of the acid character of HCO_3^-, or is it basic on account of the anion's basic character? Let's suppose that we make up a solution of the salt MHA, where HA^- is the amphiprotic anion (such as HCO_3^-) and M^+ is a cation (such as Na^+). The equilibrium table is as follows:

Species:	H_2A	HA^-	A^{2-}	H_3O^+
Initially	0	A	0	0
Change to reach equilibrium	$+x$	$-(x+y)$	$+y$	$+(y-x)$
Equilibrium	x	$A-x-y$	y	$y-x$

The two acidity constants are

$$K_{a1} = \frac{[H_3O^+][HA^-]}{[H_2A]} = \frac{(y-x)(A-x-y)}{x}$$

$$K_{a2} = \frac{[H_3O^+][A^{2-}]}{[HA^-]} = \frac{(y-x)y}{A-x-y}$$

The first of these two expressions expands to

$$xK_{a2} = Ay - y^2 - Ax + x^2$$

If we accept that xK_{a1}, x^2, and y^2 are all very small compared with terms that have A in them, we conclude that $x \approx y$. Now multiply the two expressions for the acidity constants together, and find

$$K_{a1}K_{a2} = \frac{(y-x)^2 y}{x} \approx (y-x)^2 = [H_3O^+]^2$$

It follows that $[H_3O^+] = (K_{a1}K_{a2})^{1/2}$ and, on taking the negative common logarithms of both sides, that

$$pH = \tfrac{1}{2}(pK_{a1} + pK_{a2}) \qquad (8.10)$$

Illustration 8.1

If the dissolved salt is sodium hydrogencarbonate, we can immediately conclude that the pH of the solution *of any concentration* is

$$pH = \tfrac{1}{2}(6.37 + 10.25) = 8.31$$

The solution is basic. We can treat a solution of potassium hydrogenphosphate in the same way, taking into account the second and third acidity constants of H_3PO_4:

$$pH = \tfrac{1}{2}(7.21 + 12.67) = 9.94$$

Salts in water

THE ions present when a salt is added to water may themselves be either acids or bases and consequently affect the pH of the solution. For example, when ammonium chloride is added to water, it provides both an acid (NH_4^+) and a base (Cl^-). The solution consists of a weak acid (NH_4^+) and a very weak base (Cl^-). The net effect is that the solution is acidic. Similarly, a solution of sodium acetate consists of a neutral ion (the Na^+ ion) and a base ($CH_3CO_2^-$). The net effect is that the solution is basic, and its pH is greater than 8.

Self-test 8.7

Is an aqueous solution of potassium tartrate likely to be acidic or basic?

[*Answer*: basic]

To estimate the pH of the solution, we proceed in exactly the same way as for the addition of a 'conventional' acid or base, for in the Brønsted–Lowry theory, there is no distinction between 'conventional' acids like acetic acid and the conjugate acids of bases (like NH_4^+). For example, to calculate the pH of 0.010 M $NH_4Cl(aq)$ at 25°C, we proceed exactly as in Example 8.1, taking the initial concentration of the acid (NH_4^+) to be 0.010 mol L^{-1}. The K_a to use is the acidity constant of the acid NH_4^+, which is listed

in Table 8.1. Alternatively, we use K_b for the conjugate base (NH_3) of the acid and convert that quantity to K_a by using eqn 8.6 ($K_aK_b = K_w$). We find pH = 5.63, which is on the acid side of neutral. Exactly the same procedure is used to find the pH of a solution of a salt of a weak acid, such as sodium acetate. The equilibrium table is set up by treating the anion $CH_3CO_2^-$ as a base (which it is), and using for K_b the value obtained from the value of K_a for its conjugate acid (CH_3COOH).

Self-test 8.8

Estimate the pH of 0.0025 M $NH(CH_3)_3Cl(aq)$ at 25°C.

[*Answer*: 6.2]

8.5 Acid–base titrations

Acidity constants play an important role in acid–base titrations, for we can use them to decide the value of the pH that signals the **stoichiometric point**, the stage at which a stoichiometrically equivalent amount of acid has been added to a given amount of base.[4] The plot of the pH of the **analyte**, the solution being analysed, against the volume of **titrant**, the solution in the burette, added is called the **pH curve**. It shows a number of features that are still of interest even nowadays when many titrations are carried out in automatic titrators with the pH monitored electronically: automatic titration equipment is built to make use of the concepts we describe here.

First, consider the titration of a strong acid with a strong base, such as the titration of hydrochloric acid with sodium hydroxide. The reaction is

$$HCl(aq) + NaOH(aq) \rightarrow NaCl(aq) + H_2O(l)$$

Initially, the analyte (hydrochloric acid) has a low pH. The ions present at the stoichiometric point (the Na^+ ions from the strong base and the Cl^- ions from the strong acid) barely affect the pH, so the pH is that of almost pure water, namely, pH = 7. After the stoichiometric point, when base is added

---

[4] The stoichiometric point is widely called the *equivalence point* of a titration. The meaning of *end point* is explained in Section 8.7.

to a neutral solution, the pH rises sharply to a high value. The pH curve for such a titration is shown in Fig 8.3.

Figure 8.4 shows the pH curve for the titration of a weak acid (such as CH_3COOH) with a strong base (NaOH). At the stoichiometric point the solution contains $CH_3CO_2^-$ ions and Na^+ ions together with any ions stemming from autoprotolysis. The presence of the Brønsted base $CH_3CO_2^-$ in the solution means that we can expect pH > 7. In a titration of a weak base (such as NH_3) and a strong acid (HCl), the solution contains NH_4^+ ions and Cl^- ions at the stoichiometric point. Because Cl^- is only a very weak Brønsted base and NH_4^+ is a weak Brønsted acid the solution is acidic and its pH will be less than 7.

Now we consider the shape of the pH curve in terms of the acidity constants of the species involved. The approximations we make are based on the fact that the acid is weak, and therefore that HA is more abundant than any A^- ions in the solution. Furthermore, when HA is present, it provides so many H_3O^+ ions, even though it is a weak acid, that they greatly outnumber any H_3O^+ ions that come from the very feeble autoprotolysis of water. Finally, when excess base is present after the

stoichiometric point has been passed, the OH^- ions it provides dominate any that come from the water autoprotolysis.

To be definite, let's suppose that we are titrating 25.00 mL of 0.100 M HClO(aq) with 0.200 M NaOH(aq) at 25°C. We can calculate the pH at the start of a titration of a weak acid with a strong base as explained in Example 8.1, and find pH = 4.3. The addition of titrant converts some of the acid to its conjugate base in the reaction

$$HClO(aq) + OH^-(aq) \rightarrow H_2O(l) + ClO^-(aq)$$

Suppose we add enough titrant to produce a concentration [base] of the conjugate base and simultaneously reduce the concentration of acid to [acid]. Then, because the solution remains at equilibrium,

$$K_a = \frac{a_{H_3O^+} a_{ClO^-}}{a_{HClO}} \approx \frac{a_{H_3O^+}[\text{base}]}{[\text{acid}]}$$

which rearranges first to

$$a_{H_3O^+} \approx \frac{K_a[\text{acid}]}{[\text{base}]}$$

Fig 8.3 The pH curve for the titration of a strong acid (the analyte) with a strong base (the titrant). There is an abrupt change in pH near the stoichiometric point at pH = 7. The final pH of the medium approaches that of the titrant.

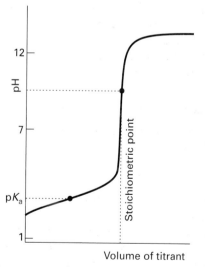

Fig 8.4 The pH curve for the titration of a weak acid (the analyte) with a strong base (the titrant). Note that the stoichiometric point occurs at pH > 7 and that the change in pH near the stoichiometric point is less abrupt than in Fig 8.3. The pK_a of the acid is equal to the pH halfway to the stoichiometric point.

and then, by taking negative common logarithms, to the **Henderson–Hasselbalch equation**

$$pH \approx pK_a - \log\frac{[\text{acid}]}{[\text{base}]} \qquad (8.11)$$

Example 8.5 *Estimating the pH at an intermediate stage in a titration*

Calculate the pH of the solution after the addition of 5.00 mL of the titrant to the analyte in the titration described above.

Strategy The first step involves deciding the amount of OH^- ions added in the titrant, and then to use that amount to calculate the amount of HClO remaining. Notice that, because the ratio of acid to base molar concentrations occurs in eqn 8.11, the volume of solution cancels, and we can equate the ratio of concentrations to the ratio of amounts.

Solution The addition of 5.00 mL of titrant corresponds to the addition of

$$n_{OH^-} = (5.00 \times 10^{-3}\,L) \times (0.200\,mol\,L^{-1})$$
$$= 1.00 \times 10^{-3}\,mol$$

This amount of OH^- converts 1.00 mmol HClO to the base ClO^-. The initial amount of HClO in the analyte is

$$n_{HClO} = (25.00 \times 10^{-3}\,L) \times (0.200\,mol\,L^{-1})$$
$$= 2.50 \times 10^{-3}\,mol$$

so the amount remaining after the addition of titrant is 1.50 mmol. It then follows from the Henderson–Hasselbalch equation that

$$pH \approx 7.53 - \log\frac{1.50 \times 10^{-3}}{1.00 \times 10^{-3}} = 7.4$$

As expected, the addition of base has resulted in an increase in pH from 4.3. You should not take the value 7.4 too seriously, but should note that the pH has increased from its initial acidic value.

Self-test 8.9

Calculate the pH after the addition of a further 5.00 mL of titrant.

[*Answer*: 8.1]

Halfway to the stoichiometric point, when enough base has been added to neutralize half the acid, the concentrations of acid and base are equal

and because log 1 = 0 the Henderson–Hasselbalch equation gives

$$pH \approx pK_a \qquad (8.12)$$

In the present titration, we see that, at this stage of the titration, pH ≈ 7.5. Note from the pH curve in Fig 8.4 how much more slowly the pH is changing compared with initially: this point will prove important shortly. Equation 8.12 implies that we can determine the pK_a of the acid directly from the pH of the mixture. In practice this is done by recording the pH during a titration and then examining the record for the pH halfway to the stoichiometric point.

At the stoichiometric point, enough base has been added to convert all the acid to its base, and so the solution consists only of ClO^- ions. These ions are Brønsted bases, so we can expect the solution to be basic with a pH of well above 7. We have already seen how to estimate the pH of a solution of a weak base in terms of its concentration B (Example 8.2), so all that remains to be done is to calculate the concentration of ClO^- at the stoichiometric point.

Illustration 8.2

Because the analyte initially contained 2.50 mmol HClO, the volume of titrant needed to neutralize it is the volume that contains the same amount of base:

$$V_{\text{base}} = \frac{2.50 \times 10^{-3}\,mol}{0.200\,mol\,L^{-1}} = 1.25 \times 10^{-2}\,L$$

or 12.5 mL. The total volume of the solution at this stage is therefore 37.5 mL, so the concentration of base is

$$[ClO^-] = \frac{2.50 \times 10^{-3}\,mol}{37.5 \times 10^{-3}\,L} = 6.67 \times 10^{-2}\,mol\,L^{-1}$$

It then follows from Example 8.2 that the pH of the solution at the stoichiometric point is 10.2.

It is very important to note that the pH at *the stoichiometric point of a weak-acid–strong-base titration is on the basic side of neutrality* (pH > 7). At the stoichiometric point, the solution consists of a weak base (the conjugate base of the weak acid, here the ClO^- ions) and neutral cations (the Na^+ ions from the titrant).

The general form of the pH curve throughout a weak-acid–strong-base titration is illustrated in Fig 8.4. The pH rises slowly from its initial value, passing through the values given by the Henderson–Hasselbalch equation when the acid and its conjugate base are both present, until the stoichiometric point is approached. It then changes rapidly to and through the value characteristic of a solution of a salt, which takes into account the effect on the pH of a solution of a weak base, the conjugate base of the original acid. The pH then climbs less rapidly towards the value corresponding to a solution consisting of excess base, and finally approaches the pH of the original base solution when so much titrant has been added that the solution is virtually the same as the titrant itself. The stoichiometric point is detected by observing where the pH changes rapidly through the value calculated in Illustration 8.2 as described at the beginning of this section.

A similar sequence of changes occurs when the analyte is a weak base (such as ammonia) and the titrant is a strong acid (such as hydrochloric acid). In this case the pH curve is like that shown in Fig 8.5: the pH falls as acid is added, plunges through the pH corresponding to a solution of a weak acid (the conjugate acid of the original base, such as NH_4^+), and then slowly approaches the pH of the original strong acid. The pH of the stoichiometric point is that of a solution of a weak acid, and is calculated as illustrated in Example 8.1.

8.6 **Buffer action**

The slow variation of the pH when the concentrations of the conjugate acid and base are equal, when pH = pK_a, is the basis of **buffer action**, the ability of a solution to oppose changes in pH when small amounts of strong acids and bases are added (Fig 8.6). An **acid buffer** solution, one that stabilizes the solution at a pH below 7, is typically prepared by making a solution of a weak acid (such as acetic acid) and a salt that supplies its conjugate base (such as sodium acetate). A **base buffer**, one that stabilizes a solution at a pH above 7, is prepared by making a solution of a weak base (such as ammonia) and a salt that supplies its conjugate acid (such as ammonium chloride).

An acid buffer stabilizes the pH of a solution because the abundant supply of A^- ions (from the salt) can remove any H_3O^+ ions brought by additional acid; furthermore, the abundant supply of HA molecules can provide H_3O^+ ions to react with any base

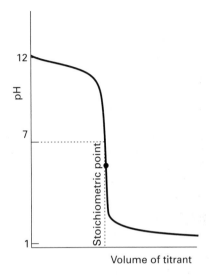

Fig 8.5 The pH curve for the titration of a weak base (the analyte) with a strong acid (the titrant). The stoichiometric point occurs at pH < 7. The final pH of the solution approaches that of the titrant.

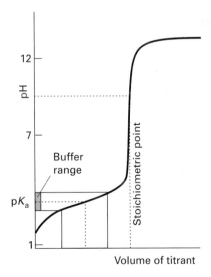

Fig 8.6 The pH of a solution changes only slowly in the region of halfway to the stoichiometric point. In this region the solution is buffered to a pH close to pK_a.

that is added. Similarly, in a base buffer the weak base B can accept protons when an acid is added and its conjugate acid BH^+ can supply protons if a base is added.

Illustration 8.3

Suppose we need to estimate the pH of a buffer formed from equal amounts of $KH_2PO_4(aq)$ and $K_2HPO_4(aq)$. We note that the two anions present are $H_2PO_4^-$ and HPO_4^{2-}. The former is the conjugate acid of the latter:

$$H_2PO_4^-(aq) + H_2O(l) \rightleftharpoons H_3O^+(aq) + HPO_4^{2-}(aq)$$

so we need its pK_a. In this case we recognize it as the pK_{a2} of phosphoric acid, and take it from Table 8.2. In either case, $pK_a = 7.21$. Hence, the solution should buffer close to pH = 7.

Self-test 8.10

Calculate the pH of an aqueous buffer solution that contains equal amounts of NH_3 and NH_4Cl.

[*Answer*: 9.25; more realistically: 9]

8.7 Indicators

The rapid change of pH near the stoichiometric point of an acid–base titration is the basis of indicator detection. An **acid–base indicator** is a large, water-soluble, organic molecule with acid (HIn) and conjugate base (In^-) forms that differ in colour. The two forms are in equilibrium in solution:

$$HIn(aq) + H_2O(l) \rightleftharpoons H_3O^+(aq) + In^-(aq) \qquad K_{In} = \frac{a_{H_3O^+} a_{In^-}}{a_{HIn}}$$

The pK_{In} of some indicators are listed in Table 8.4. The ratio of the concentrations of the conjugate acid and base forms of the indicator is

$$\frac{[In^-]}{[HIn]} \approx \frac{K_{In}}{a_{H_3O^+}}$$

This expression can be rearranged (after taking logarithms) to

$$\log \frac{[In^-]}{[HIn]} \approx pH - pK_{In} \qquad (8.13)$$

We see that, as the pH swings from higher than pK_{In} to lower than pK_{In} as acid is added to the solution, the ratio of In^- to HIn swings from well above 1 to well below 1 (Fig 8.7).

Self-test 8.11

What is the ratio of the yellow and blue forms of bromocresol green in solution of pH (a) 3.7, (b) 4.7, and (c) 5.7?

[*Answer*: (a) 10:1, (b) 1:1, (c) 1:10]

At the stoichiometric point, the pH changes sharply through several pH units, so the molar concentration of H_3O^+ changes through several orders of magnitude. The indicator equilibrium changes so as to accommodate the change of pH, with HIn the dominant species on the acid side of the stoichiometric point, when H_3O^+ ions are abundant, and In^- dominant on the basic side, when the base can remove protons from HIn. The accompanying colour change signals the stoichiometric point of the titration. The colour in fact changes over a range of pH, typically from pH $\approx pK_{In} - 1$, when HIn is ten times as abundant as In^-, to pH $\approx pK_{In} + 1$, when In^- is ten times as abundant as HIn. The pH halfway through a colour change, when pH $\approx pK_{In}$ and the two forms, HIn and In^-, are in equal abundance, is the **end point** of the indicator. With a well-chosen indicator, the end point of the indicator coincides with the stoichiometric point of the titration.

Care must be taken to use an indicator that changes colour at the pH appropriate to the type of titration. Specifically, we need to match the end point to the stoichiometric point, and therefore select an indicator for which pK_{In} is close to the pH of the stoichiometric point. Thus, in a weak-acid–strong-base titration, the stoichiometric point lies at pH > 7, and we should select an indicator that changes at that pH (Fig 8.8). Similarly, in a strong-acid–weak-base titration, we need to select an indicator with an end point at pH < 7. Qualitatively, we should choose an indicator with $pK_{In} \approx 7$ for strong-acid–strong-base titrations, one with $pK_{In} < 7$ for strong-acid–weak-base titrations, and one with $pK_{In} > 7$ for weak-acid–strong-base titrations.

Table 8.4 *Indicator colour changes*

Indicator	Acid colour	pH range of colour change	pK_{In}	Base colour
Thymol blue	Red	1.2 to 2.8	1.7	Yellow
Methyl orange	Red	3.2 to 4.4	3.4	Yellow
Bromophenol blue	Yellow	3.0 to 4.6	3.9	Blue
Bromocresol green	Yellow	4.0 to 5.6	4.7	Blue
Methyl red	Red	4.8 to 6.0	5.0	Yellow
Bromothymol blue	Yellow	6.0 to 7.6	7.1	Blue
Litmus	Red	5.0 to 8.0	6.5	Blue
Phenol red	Yellow	6.6 to 8.0	7.9	Red
Thymol blue	Yellow	9.0 to 9.6	8.9	Blue
Phenolphthalein	Colourless	8.2 to 10.0	9.4	Pink
Alizarin yellow	Yellow	10.1 to 12.0	11.2	Red
Alizarin	Red	11.0 to 12.4	11.7	Purple

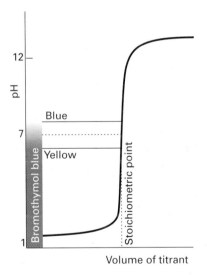

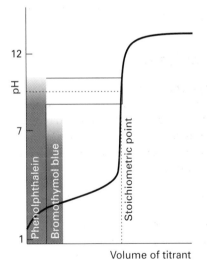

Fig 8.7 The range of pH over which an indicator changes colour is depicted by the tinted band. For a strong-acid–strong-base titration, the stoichiometric point is indicated accurately by an indicator that changes colour at pH = 7 (such as bromothymol blue). However, the change in pH is so sharp that accurate results are also obtained even if the indicator changes colour in neighbouring values. Thus, phenolphthalein (which has $pK_{In} = 9.4$, see Table 8.4) is also often used.

Fig 8.8 In a weak-acid–strong-base titration, an indicator with $pK_{In} \approx 7$ (the lower band, like bromothymol blue) would give a false indication of the stoichiometric point; it is necessary to use an indicator that changes colour close to the pH of the stoichiometric point. If that lies at about pH = 9, then phenolphthalein would be appropriate.

Solubility equilibria

A SOLID dissolves in a solvent until the solution and the solid solute are in equilibrium. At this stage, the solution is said to be **saturated**, and its molar concentration is the **molar solubility** of the solid. That the two phases—the solid solute and the solution—are in dynamic equilibrium implies that we can use equilibrium concepts to discuss the composition of the saturated solution. The properties of aqueous solutions of electrolytes are commonly treated in terms of equilibrium constants, and in this section we shall confine our attention to them. We shall also limit our attention to **sparingly soluble** compounds, which are compounds that dissolve only slightly in water. This restriction is applied because the effects of ion–ion interactions are a complicating feature of more concentrated solutions and more advanced techniques are then needed before the calculations are reliable. Once again, we shall concentrate on general trends and properties rather than expecting to obtain numerically precise results.

8.8 The solubility constant

The equilibrium between a sparingly soluble ionic compound, such as calcium hydroxide, $Ca(OH)_2$, and its ions in aqueous solution is

$$Ca(OH)_2(s) \rightleftharpoons Ca^{2+}(aq) + 2\,OH^-(aq) \qquad K_s = a_{Ca^{2+}}a_{OH^-}^2$$

The equilibrium constant for an ionic equilibrium such as this, bearing in mind that the solid does not appear in the equilibrium expression because its activity is 1, is called the **solubility constant**.[5] As

usual, for very dilute solutions, we can replace the activity a_J of a species J by the numerical value of its molar concentration. Experimental values for solubility constants are given in Table 8.5.

We can interpret the solubility constant in terms of the numerical value of the molar solubility, S, of a sparingly soluble substance. For instance, it follows from the stoichiometry of the equilibrium equation written above that the molar concentration of Ca^{2+} ions in solution is equal to that of the $Ca(OH)_2$ dissolved in solution, so $S = [Ca^{2+}]$. Likewise, because the concentration of OH^- ions is twice that of $Ca(OH)_2$ formula units, it follows that $S = \frac{1}{2}[OH^-]$. Therefore, provided it is permissible to replace activities by molar concentrations,

$$K_s \approx S \times (2S)^2 = 4S^3$$

from which it follows that

$$S \approx (\tfrac{1}{4}K_s)^{1/3} \tag{8.14}$$

This expression is only approximate because ion–ion interactions have been ignored. However, because the solid is sparingly soluble, the concentrations of the ions are low and the inaccuracy is moderately low. Thus, from Table 8.5, $K_s = 5.5 \times 10^{-6}$, so $S \approx 1 \times 10^{-2}$ and the molar solubility is 1×10^{-2} mol L^{-1}. Solubility constants (which are determined by electrochemical measurements of the kind described in Chapter 9) provide a more accurate way of measuring solubilities of very sparingly soluble compounds than the direct measurement of the mass that dissolves.

8.9 The common-ion effect

The principle that an equilibrium constant remains unchanged whereas the individual concen-

---

[5] Other common names are the 'solubility product constant' or simply the 'solubility product'.

Table 8.5 *Solubility constants at 25°C*

Compound	Formula	K_s
Aluminium hydroxide	$Al(OH)_3$	1.0×10^{-33}
Antimony sulfide	Sb_2S_3	1.7×10^{-93}
Barium carbonate	$BaCO_3$	8.1×10^{-9}
fluoride	BaF_2	1.7×10^{-6}
sulfate	$BaSO_4$	1.1×10^{-10}
Bismuth sulfide	Bi_2S_3	1.0×10^{-97}
Calcium carbonate	$CaCO_3$	8.7×10^{-9}
fluoride	CaF_2	4.0×10^{-11}
hydroxide	$Ca(OH)_2$	5.5×10^{-6}
sulfate	$CaSO_4$	2.4×10^{-5}
Copper(I) bromide	$CuBr$	4.2×10^{-8}
chloride	$CuCl$	1.0×10^{-6}
iodide	CuI	5.1×10^{-12}
sulfide	Cu_2S	2.0×10^{-47}
Copper(II) iodate	$Cu(IO_3)_2$	1.4×10^{-7}
oxalate	CuC_2O_4	2.9×10^{-8}
sulfide	CuS	8.5×10^{-45}
Iron(II) hydroxide	$Fe(OH)_2$	1.6×10^{-14}
sulfide	FeS	6.3×10^{-18}
Iron(III) hydroxide	$Fe(OH)_3$	2.0×10^{-39}
Lead (II) bromide	$PbBr_2$	7.9×10^{-5}
chloride	$PbCl_2$	1.6×10^{-5}
fluoride	PbF_2	3.7×10^{-8}
iodate	$Pb(IO_3)_2$	2.6×10^{-13}
iodide	PbI_2	1.4×10^{-8}
sulfate	$PbSO_4$	1.6×10^{-8}
sulfide	PbS	3.4×10^{-28}
Magnesium		
ammonium phosphate	$MgNH_4PO_4$	2.5×10^{-13}
carbonate	$MgCO_3$	1.0×10^{-5}
fluoride	MgF_2	6.4×10^{-9}
hydroxide	$Mg(OH)_2$	1.1×10^{-11}
Mercury(I) chloride	Hg_2Cl_2	1.3×10^{-18}
iodide	Hg_2I_2	1.2×10^{-28}
Mercury(II) sulfide	HgS black:	1.6×10^{-52}
	red:	1.4×10^{-53}
Nickel(II) hydroxide	$Ni(OH)_2$	6.5×10^{-18}
Silver bromide	$AgBr$	7.7×10^{-13}
carbonate	Ag_2CO_3	6.2×10^{-12}
chloride	$AgCl$	1.6×10^{-10}
hydroxide	$AgOH$	1.5×10^{-8}
iodide	AgI	1.5×10^{-16}
sulfide	Ag_2S	6.3×10^{-51}
Zinc hydroxide	$Zn(OH)_2$	2.0×10^{-17}
sulfide	ZnS	1.6×10^{-24}

trations of species may change is applicable to solubility constants, and may be used to assess the effect of the addition of species to solutions. An example of particular importance is the effect on the solubility of a compound of the presence of another solute that provides an ion in common with the sparingly soluble compound already present. For example, we may consider the effect on the solubility of adding sodium chloride to a saturated solution of silver chloride, the common ion in this case being Cl^-.

The molar solubility of silver chloride in pure water is related to its solubility constant by $S \approx K_s^{1/2}$. To assess the effect of the common ion, we suppose that Cl^- ions are added to a concentration C mol L^{-1}, which greatly exceeds the concentration of the same ion that stems from the presence of the silver chloride. Therefore, we can write

$$K_s = a_{Ag^+} a_{Cl^-} \approx [Ag^+]C$$

It is very dangerous to neglect deviations from ideal behaviour in ionic solutions, so from now on the calculation will only be indicative of the kinds of changes that occur when a common ion is added to a solution of a sparingly soluble salt: the qualitative trends are reproduced, but the quantitative calculations are unreliable. With these remarks in mind, it follows that the solubility S' of silver chloride in the presence of added chloride ions is

$$S' \approx \frac{K_s}{C}$$

The solubility is greatly reduced by the presence of the common ion. For example, whereas the solubility of silver chloride in water is 1.3×10^{-5} mol L^{-1}, in the presence of 0.10 M NaCl(aq) it is only 2×10^{-9} mol L^{-1}, which is nearly ten thousand times less. The reduction of the solubility of a sparingly soluble salt by the presence of a common ion is called the **common-ion effect**.

Self-test 8.14

Estimate the molar solubility of calcium fluoride, CaF_2, in (a) water, (b) 0.010 M NaF(aq).

[*Answer:* (a) 2×10^{-4} mol L^{-1}; (b) 4×10^{-7} mol L^{-1}]

Exercises

8.1 Write the proton transfer equilibria for the following acids in aqueous solution and identify the conjugate acid–base pairs in each one: (a) H_2SO_4, (b) HF (hydrofluoric acid), (c) $C_6H_5NH_3^+$ (anilinium ion), (d) $H_2PO_4^-$ (dihydrogenphosphate ion), (e) HCOOH (formic acid), (f) $NH_2NH_3^+$ (hydrazinium ion).

8.2 Numerous acidic species are found in living systems. Write the proton transfer equilibria for the following biochemically important acids in aqueous solution: (a) lactic acid ($CH_3CHOHCOOH$), (b) glutamic acid (**5**), (c) glycine (NH_2CH_2COOH), (d) oxalic acid (HOOCCOOH).

5 Glutamic acid

8.3 For biological and medical applications we often need to consider proton transfer equilibria at body temperature (37°C). The value of K_w for water at body temperature is 2.5×10^{-14}. (a) What is the value of $[H_3O^+]$ and the pH of neutral water at 37°C? (b) What is the molar concentration of OH^- ions and the pOH of neutral water at 37°C?

8.4 Suppose that something had gone wrong in the Big Bang, and instead of ordinary hydrogen there was an abundance of deuterium in the universe. There would be many subtle changes in equilibria, particularly the deuteron transfer equilibria of heavy atoms and bases. The K_w for heavy water at 25°C is 1.35×10^{-15}. (a) Write the chemical equation for the autoprotolysis of D_2O. (b) Evaluate pK_w for D_2O at 25°C. (c) Calculate the molar concentrations of D_3O^+ and OD^- in neutral heavy water at 25°C. (d) Evaluate the pD and pOD of neutral heavy water at 25°C. (e) Find the relation between pD, pOD, and $pK_w(D_2O)$.

8.5 The molar concentration of H_3O^+ ions in the following solutions was measured at 25°C. Calculate the pH and pOH of the solution: (a) 1.5×10^{-5} mol L^{-1} (a sample of rain water), (c) 1.5 mmol L^{-1}, (d) 5.1×10^{-14} mol L^{-1}, (e) 5.01×10^{-5} mol L^{-1}.

8.6 Calculate the molar concentration of H_3O^+ ions and the pH of the following solutions: (a) 25.0 mL of 0.144 M HCl(aq) was added to 25.0 mL of 0.125 M NaOH(aq), (b) 25.0 mL of 0.15 M HCl(aq) was added to 35.0 mL of 0.15 M KOH(aq), (c) 21.2 mL of 0.22 M HNO_3(aq) was added to 10.0 mL of 0.30 M NaOH(aq).

8.7 Determine whether aqueous solutions of the following salts have a pH equal to, greater than, or less than 7; if pH > 7 or pH < 7, write a chemical equation to justify your answer. (a) NH_4Br, (b) Na_2CO_3, (c) KF, (d) KBr, (e) $AlCl_3$, (f) $Co(NO_3)_2$.

8.8 (a) A sample of potassium acetate, KCH_3CO_2, of mass 8.4 g is used to prepare 250 mL of solution. What is the pH of the solution? (b) What is the pH of a solution when 3.75 g of ammonium bromide, NH_4Br, is used to make 100 mL of solution? (c) An aqueous solution of volume 1.0 L contains 10.0 g of potassium bromide. What is the percentage of Br^- ions that are protonated?

8.9 There are many organic acids and bases in our cells, and their presence modifies the pH of the fluids inside them. It is useful to be able to assess the pH of solutions of acids and bases and to make inferences from measured values of the pH. A solution of equal concentrations of lactic acid and sodium lactate was found to have pH = 3.08. (a) What are the values of pK_a and K_a of lactic acid? (b) What would the pH be if the acid had twice the concentration of the salt?

8.10 Sketch reasonably accurately the pH curve for the titration of 25.0 mL of 0.15 M $Ba(OH)_2$(aq) with 0.22 M HCl(aq). Mark on the curve (a) the initial pH, (b) the pH at the stoichiometric point.

8.11 Determine the fraction of solute deprotonated or protonated in (a) 0.25 M C_6H_5COOH(aq), (b) 0.150 M NH_2NH_2(aq) (hydrazine), (c) 0.112 M $(CH_3)_3N$(aq) (trimethylamine).

8.12 Calculate the pH, pOH, and fraction of solute protonated or deprotonated in the following aqueous solutions: (a) 0.120 M $CH_3CH(OH)COOH$(aq) (lactic acid), (b) 1.4×10^{-4} M $CH_3CH(OH)COOH$(aq), (c) 0.10 M $C_6H_5SO_3H$(aq) (benzenesulfonic acid).

8.13 Show how the composition of an aqueous solution that contains 0.010 mol L^{-1} glycine varies with pH.

8.14 Show how the composition of an aqueous solution that contains 0.010 mol L^{-1} tyrosine varies with pH.

8.15 Calculate the pH of a solution of sodium hydrogen oxalate.

8.16 Calculate the pH of the following acid solutions at 25°C; ignore second deprotonations only when that approximation is justified. (a) 1.0×10^{-4} M H_3BO_3(aq) (boric acid acts as a monoprotic acid), (b) 0.015 M H_3PO_4(aq), (c) 0.10 M H_2SO_3(aq).

8.17 The weak base colloquially known as Tris, and more precisely as tris(hydroxymethyl)aminomethane, has pK_a = 8.3 at 20°C and is commonly used to produce a buffer for biochemical applications. At what pH would you expect Tris to act as a buffer in a solution that has equal molar concentrations of Tris and its conjugate acid?

8.18 The amino acid tyrosine has pK_a = 2.20 for deprotonation of its carboxylic acid group. What are the relative concentrations of tyrosine and its conjugate base at a pH of (a) 7, (b) 2.2, (c) 1.5?

8.19 (a) Calculate the molar concentrations of $(COOH)_2$, $HOOCCO_2^-$, $(CO_2)_2^{2-}$, H_3O^+, and OH^- in 0.15 M $(COOH)_2$(aq). (b) Calculate the molar concentrations of H_2S, HS^-, S^{2-}, H_3O^+, and OH^- in 0.065 M H_2S(aq).

8.20 A sample of 0.10 M CH_3COOH(aq) of volume 25.0 mL is titrated with 0.10 M NaOH(aq). The K_a for CH_3COOH is 1.8×10^{-5}. (a) What is the pH of 0.10 M CH_3COOH(aq)? (b) What is the pH after the addition of 10.0 mL of 0.10 M NaOH(aq)? (c) What volume of 0.10 M NaOH(aq) is required to reach halfway to the stoichiometric point? (d) Calculate the pH at that halfway point. (e) What volume of 0.10 M NaOH(aq) is required to reach the stoichiometric point? (f) Calculate the pH at the stoichiometric point.

8.21 A buffer solution of volume 100 mL consists of 0.10 M CH_3COOH(aq) and 0.10 M Na(CH_3CO_2)(aq). (a) What is its pH? (b) What is the pH after the addition of 3.3 mmol NaOH to the buffer solution? (c) What is the pH after the addition of 6.0 mmol HNO_3 to the initial buffer solution?

8.22 Predict the pH region in which each of the following buffers will be effective, assuming equal molar concentrations of the acid and its conjugate base: (a) sodium lactate and lactic acid, (b) sodium benzoate and benzoic acid, (c) potassium hydrogenphosphate and potassium phosphate, (d) potassium hydrogenphosphate and potassium dihydrogenphosphate, (e) hydroxylamine and hydroxylammonium chloride.

8.23 At the halfway point in the titration of a weak acid with a strong base the pH was measured as 4.66. What are the acidity constant and the pK_a of the acid? What is the pH of the solution that is 0.015 M in the acid?

8.24 Calculate the pH of (a) 0.15 M NH_4Cl(aq), (b) 0.15 M NaCH$_3CO_2$(aq), (c) 0.150 M CH_3COOH(aq).

8.25 Calculate the pH at the stoichiometric point of the titration of 25.00 mL of 0.100 M lactic acid with 0.175 M NaOH(aq).

8.26 Sketch the pH curve of a solution containing 0.10 M NaCH$_3CO_2$(aq) and a variable amount of acetic acid.

8.27 From the information in Tables 8.1 and 8.2, select suitable buffers for (a) pH = 2.2 and (b) pH = 7.0.

8.28 Write the expression for the solubility constants of the following compounds: (a) AgI, (b) Hg_2S, (c) $Fe(OH)_3$, (d) Ag_2CrO_4.

8.29 Use the data in Table 8.5 to estimate the molar solubilities of (a) $BaSO_4$, (b) Ag_2CO_3, (c) $Fe(OH)_3$, (d) Hg_2Cl_2 in water.

8.30 Use the data in Table 8.5 to estimate the solubility in water of each sparingly soluble substance in its respective solution: (a) silver bromide in 1.4×10^{-3} M NaBr(aq), (b) magnesium carbonate in 1.1×10^{-5} M Na_2CO_3(aq), (c) lead(II) sulfate in a 0.10 M $CaSO_4$(aq), (d) nickel(II) hydroxide in 2.7×10^{-5} M $NiSO_4$(aq).

8.31 Thermodynamic data can be used to predict the solubilities of compounds that would be very difficult to measure directly. Calculate the solubility of mercury(II) chloride in water at 25°C from standard Gibbs energies of formation.

Electrochemistry

Contents

The migration of ions

Electrochemical cells

Applications of standard potentials

Such apparently unrelated processes as combustion, respiration, photosynthesis, and corrosion are actually all closely related, for in each of them an electron, sometimes accompanied by a group of atoms, is transferred from one species to another. Indeed, together with the proton transfer typical of acid–base reactions, processes in which electrons are transferred, the so-called **redox reactions**, account for many of the reactions encountered in chemistry. Redox reactions—the principal topic of this chapter—are of immense practical significance, not only because they underlie many biochemical and industrial processes, but also because they are the basis of the generation of electricity by chemical reactions and the investigation of reactions by making electrical measurements.[1]

Measurements like the ones we describe in this chapter lead to a collection of data that are very useful for discussing the characteristics of electrolyte solutions and of a wide range of different types of equilibria in solution. They are also used throughout inorganic chemistry to assess the thermodynamic feasibility of reactions and the stabilities of compounds. They are used in physiology to discuss the details of the propagation of signals in neurons.

---

[1] This chapter makes use of a number of concepts related to electricity: they are reviewed in *Further information 6*.

The migration of ions

A^{N} ion in solution is mobile and responds to the presence of an electric field. As an ion migrates through a solution, it carries charge from one location to another and therefore gives rise to an electric current. Our bodies are electric conductors and some of the thoughts you are currently having as you read this sentence can be traced to the migration of ions through membranes in the enormously complex electrical circuits of our brains.

9.1 Conductivity

We can study the migration of ions in solution by measuring the electrical resistance of a solution of known concentration in a cell like that in Fig 9.1. The resistance, R (in ohms, Ω), of the solution is related to the current, I (in amperes, A), that flows when a potential difference, V (in volts, V), is applied between the two electrodes. Certain technicalities must be dealt with in practice, such as using an alternating current to minimize the effects of electrolysis, but the essential point is the determination of R.

It is found empirically that the resistance of a sample of matter is proportional to its length, l, and

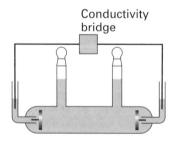

Conductivity bridge

Fig 9.1 A typical conductivity cell. The cell is made part of a 'bridge' and its resistance is measured. The conductivity is normally determined by comparison of its resistance to that of a solution of known conductivity. An alternating current is used to avoid the formation of decomposition products at the electrodes.

inversely proportional to its cross-sectional area, A. The constant of proportionality is called the **resistivity**, ρ (rho), and we write $R = \rho l/A$. The units of resistivity are ohm-metre (Ω m). The reciprocal of the resistivity is called the **conductivity**, κ (kappa), which is expressed in $\Omega^{-1}\,\text{m}^{-1}$. Reciprocal ohms appear so widely in electrochemistry that they are given their own name, siemens (S, $1\,\text{S} = 1\,\Omega^{-1}$); then conductivities are expressed in siemens per metre (S m^{-1}).

Once we have determined κ,[2] we find the **molar conductivity**, Λ_{m} (lambda), when the solute molar concentration is c by forming

$$\Lambda_{\text{m}} = \frac{\kappa}{c} \qquad (9.1)$$

As you might fear, at this point the units get even more involved. With molar concentration in moles per litre, molar conductivity is expressed in siemens per metre per (moles per litre), or S m^{-1} $(\text{mol L}^{-1})^{-1}$. These units can be simplified to siemens metre-squared per mole ($\text{S m}^2\,\text{mol}^{-1}$).[3]

The molar conductivity of a *strong* electrolyte varies with molar concentration in accord with the empirical law discovered by Friedrich Kohlrausch in 1876:

$$\Lambda_{\text{m}} = \Lambda_{\text{m}}^{\circ} - Kc^{1/2} \qquad (9.2)$$

The constant $\Lambda_{\text{m}}^{\circ}$, the **limiting molar conductivity**, is the molar conductivity in the limit of such low concentration that the ions no longer interact with one another. The constant K takes into account the effect of these interactions when the concentration is nonzero. We shall concentrate on the limiting conductivity.

When the ions are so far apart that their interactions can be ignored, we can suspect that the molar conductivity is due to the independent migration of cations in one direction and of anions in the opposite direction, and write

$$\Lambda_{\text{m}}^{\circ} = \lambda_{+} + \lambda_{-} \qquad (9.3)$$

---

[2] In practice, by calibrating the cell with a solution of known conductivity.

[3] The relation between units is $1\,\text{S m}^{-1}\,(\text{mol L}^{-1})^{-1} = 1\,\text{mS m}^2\,\text{mol}^{-1}$, where $1\,\text{mS} = 10^{-3}\,\text{S}$.

Table 9.1 *Ionic conductivities, $\lambda/(mS\ m^2\ mol^{-1})$* *

Cations		Anions	
$H^+(H_3O^+)$	34.96	OH^-	19.91
Li^+	3.87	F^-	5.54
Na^+	5.01	Cl^-	7.64
K^+	7.35	Br^-	7.81
Rb^+	7.78	I^-	7.68
Cs^+	7.72	CO_3^{2-}	13.86
Mg^{2+}	10.60	NO_3^-	7.15
Ca^{2+}	11.90	SO_4^{2-}	16.00
Sr^{2+}	11.89	$CH_3CO_2^-$	4.09
NH_4^+	7.35	HCO_2^-	5.46
$[N(CH_3)_4]^+$	4.49		
$[N(C_2H_5)_4]^+$	3.26		

*The same numerical values apply when we select the units $S\ m^{-1}\ (mol\ L^{-1})^{-1}$.

where λ_+ and λ_- are the **ionic conductivities** of the cations and anions (Table 9.1).

The molar conductivity of a *weak* electrolyte varies in a more complex way with concentration. This variation reflects the fact that the degree of ionization (or, in the case of weak acids and bases, the degree of protonation) varies with the concentration, with relatively more ions present at low concentrations than at high. Because we can use simple equilibrium-table techniques to relate the ion concentrations to the nominal (initial) concentration, we can use measurements of molar conductivity to determine acidity constants. The same kind of measurements can also be used to monitor the progress of reactions in solution, provided that they involve ions.

Example 9.1 *Determining the acidity constant from the conductivity of a weak acid*

The molar conductivity of 0.010 M $CH_3COOH(aq)$ is 1.65 mS m^2 mol^{-1}. What is the acidity constant of the acid?

Strategy Because acetic acid is a weak electrolyte, it is only partly deprotonated in aqueous solution.

Only the fraction of acid molecules present as ions contributes to the conduction, so we need to express Λ_m in terms of the fraction deprotonated. Set up an equilibrium table, find the molar concentration of H_3O^+ and $CH_3CO_2^-$ ions, and relate those concentrations to the observed molar conductivity.

Solution The equilibrium table is

Species:	CH_3COOH	H_3O^+	$CH_3CO_2^-$
Initial molar concentration/ (mol L^{-1})	0.10	0	0
Change/(mol L^{-1})	$-x$	$+x$	$+x$
Equilibrium concentration/ (mol L^{-1})	$0.10-x$	x	x

The value of x is found by substituting the entries in the last line into the expression for K_a:

$$K_a = \frac{[H_3O^+][CH_3CO_2^-]}{[CH_3COOH]} = \frac{x^2}{0.10-x}$$

On the assumption that x is small, the solution is $x = (0.010K_a)^{1/2}$. The fraction, α, of CH_3COOH molecules present as ions is therefore $x/0.010$, or $\alpha = (K_a/0.010)^{1/2}$. The molar conductivity of the solution is therefore this fraction multiplied by the molar conductivity of acetic acid calculated on the assumption that deprotonation is complete:

$$\Lambda_m = \alpha\Lambda_m^\circ = \alpha(\lambda_{H_3O^+} + \lambda_{CH_3CO_2^-})$$

Because

$$\lambda_{H_3O^+} + \lambda_{CH_3CO_2^-} = 34.96 + 4.09\ mS\ m^2\ mol^{-1}$$
$$= 39.05\ mS\ m^2\ mol^{-1}$$

it follows that $\alpha = (1.65\ mS\ m^2\ mol^{-1})/(39.05\ mS\ m^2\ mol^{-1}) = 0.0423$. Therefore,

$$K_a = 0.010\alpha^2 = 0.010 \times (0.0423)^2 = 1.8 \times 10^{-5}$$

This value corresponds to p$K_a = 4.75$.

Self-test 9.1

The molar conductivity of 0.0250 M $HCOOH(aq)$ is 4.61 mS m^2 mol^{-1}. What is the pK_a of formic acid?

[*Answer*: 3.49]

9.2 Ion mobility

The conductivities of ions depend on the ease with which they can pass through the solution: large ions in viscous liquids can be expected to have low conductivities. When an ion is subjected to an electric field, E, it accelerates. However, the faster it travels through the solution, the greater the retarding force it experiences from the viscosity of the medium. As a result, it settles down into a limiting velocity called its **drift velocity**, s. This drift velocity governs the rate at which charge is transported through the solution, and therefore determines the conductivity of the ions.

Derivation 9.1 *The mobilities of ions*

We put this argument on a quantitative footing by supposing that the drift velocity is proportional to the strength of the applied field, E, and writing

$$s = uE \qquad (9.4)$$

where u is called the **mobility** of the ion. The force acting on an ion of charge ze in the presence of a field E is zeE. However, the retarding force due to viscosity, η, on a spherical particle of radius a travelling at a speed s is given by **Stokes' law**:

$$F = 6\pi\eta as \qquad (9.5)$$

When the particle has reached its drift speed, the accelerating and viscous retarding forces are equal, so we can write

$$ezE = 6\pi\eta as$$

and solve this expression for s:

$$s = \frac{ezE}{6\pi\eta a}$$

At this point we can compare this expression for the drift speed with the expression defining the mobility, eqn 9.4, and find

$$u = \frac{ez}{6\pi\eta a} \qquad (9.6)$$

Equation 9.6 tells us that the mobility of an ion is high if it is highly charged, is small, and if it is in a solution with low viscosity. These features appear to contradict the trends in Table 9.2, which lists the mobilities of a number of ions. For instance, the mobilities of the Group 1 cations *increase* down the group despite their increasing radii. The explanation is that the radius to use in eqn 9.6 is the **hydrodynamic radius**, the *effective* radius for the migration of the ions taking into account the entire object that moves. When an ion migrates, it carries its hydrating water molecules with it and, as small ions are more extensively hydrated than large ions, ions of small radius actually have a large hydrodynamic radius. Thus, hydrodynamic radius *decreases* down Group 1 because the extent of hydration decreases with increasing ionic radius. One significant deviation from this trend is the very high mobility of the proton in water. It is believed that this high mobility reflects an entirely different mechanism for conduction, the **Grotthus mechanism**, in which the proton on one H_2O molecule migrates to its neighbour, the proton on that H_2O molecule migrates to its neighbour, and so on along a chain (Fig 9.2). The motion is therefore an *effective* motion of a proton, not the actual motion of a single proton.[4]

Table 9.2 *Ionic mobilities in water at 298 K, $u/(10^{-8}\,m^2\,s^{-1}\,V^{-1})$*

Cations		Anions	
H^+	36.23	OH^-	20.64
Li^+	4.01	F^-	5.74
Na^+	5.19	Cl^-	7.92
K^+	7.62	Br^-	8.09
Rb^+	8.06	I^-	7.96
Cs^+	8.00	CO_3^{2-}	7.18
Mg^{2+}	5.50	NO_3^-	7.41
Ca^{2+}	6.17	SO_4^{2-}	8.29
Sr^{2+}	6.16		
NH_4^+	7.62		
$[N(CH_3)_4]^+$	4.65		
$[N(C_2H_5)_4]^+$	3.38		

---

[4] For a detailed account of the modern version of this mechanism, see P.W. Atkins, *Physical chemistry*, 6th edn, Freeman/Oxford University Press (1998).

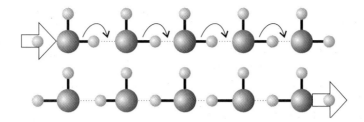

Fig 9.2 A simplified version of the 'Grotthus mechanism' of proton conduction through water. The proton leaving the chain on the right is not the same as the proton entering the chain on the left.

The mobility of an ion depends on its charge. If a protein can be contrived to have zero net charge, then it does not respond to an electric field. This 'isoelectric point' can be reached by varying the pH of the medium.

The **transport number**, $t_\pm$, of an ion is the fraction of the total current carried by that ion:

$$\text{Transport number,} \, t_\pm = \frac{\text{current carried by ion,} \, I_\pm}{\text{total current through solution,} \, I} \quad (9.7)$$

Because the total current is the sum of the currents carried by the cations and anions, it follows that

$$t_+ + t_- = 1 \quad (9.8)$$

where t_+ is the transport number of the cations and t_- that of the anions. The current carried by the ion is proportional to the mobility of the ion and, provided ion–ion interactions can be ignored

Example 9.2 *Determining the isoelectric point*

The speed with which bovine serum albumin (BSA) moves through water under the influence of an electric field was monitored at several values of pH, and the data are listed below. What is the isoelectric point of the protein?

pH	4.20	4.56	5.20	5.65	6.30	7.00
Velocity/(μm s^{-1})	0.50	0.18	−0.25	−0.60	−0.95	−1.20

Strategy The macromolecule is not influenced by the electric field when it is uncharged. Therefore, the isoelectric point is the pH at which it does not migrate in an electric field. We should therefore plot speed against pH and find by interpolation the pH at which the speed is zero.

Solution The data are plotted in Fig 9.3. The velocity passes through zero at pH = 4.8; hence pH = 4.8 is the isoelectric point.

Self-test 9.2

The following data were obtained for another protein:

pH	4.5	5.0	5.5	6.0
Velocity/(μm s^{-1})	−0.10	−0.20	−0.30	−0.35

Estimate the pH of the isoelectric point.

[*Answer*: 4.0]

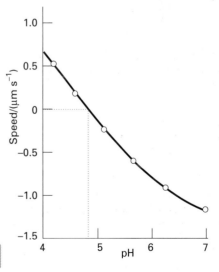

Fig 9.3 The plot of the speed of a moving macromolecule against pH allows the isoelectric point to be detected as the pH at which the speed is zero.

$$t_{\pm} = \frac{u_{\pm}}{u_+ + u_-} \qquad (9.9)$$

Transport numbers can be measured in a variety of ways,[5] so their determination is a way of measuring ionic mobilities.

Electrochemical cells

A^N electrochemical cell consists of two electronic conductors (metal or graphite, for instance) dipping into an electrolyte (an ionic conductor), which may be a solution, a liquid, or a solid. The electronic conductor and its surrounding electrolyte is an **electrode**. The physical structure containing them is called an **electrode compartment**. The two electrodes may share the same compartment (Fig 9.4). If the electrolytes are different, then the two compartments may be joined by a **salt bridge**, which is an electrolyte solution that completes the electrical circuit by permitting ions to move between the compartments (Fig 9.5). Alternatively, the two solutions may be in direct physical contact (for example, through a porous membrane) and form a **liquid junction**. However, a liquid junction introduces complications into the interpretation of measurements, and we shall not consider it further.

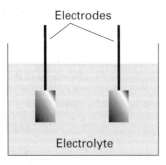

Fig 9.4 The arrangement for an electrochemical cell in which the two electrodes share a common electrolyte.

[5] For two of the techniques, see P.W. Atkins, *Physical chemistry*, 6th edn, W.H. Freeman/Oxford University Press (1998).

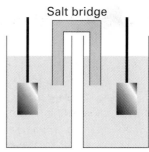

Fig 9.5 When the electrolytes in the electrode compartments of a cell are different, they need to be joined so that ions can travel from one compartment to another. One device for joining the two compartments is a salt bridge.

A **galvanic cell** is an electrochemical cell that produces electricity as a result of the spontaneous reaction occurring inside it.[6] An **electrolytic cell** is an electrochemical cell in which a nonspontaneous reaction is driven by an external source of direct current. The commercially available dry cells, mercury cells, nickel–cadmium cells, and lithium ion cells used to power electrical equipment are all galvanic cells and produce electricity as a result of the spontaneous chemical reaction between the substances built into them at manufacture. A **fuel cell** is a galvanic cell in which the reagents, such as hydrogen and oxygen or methane and oxygen, are supplied from outside. Fuel cells are used on manned spacecraft, and gas supply companies hope that one day they may be used as a convenient, compact source of electricity in homes. Electric eels and electric catfish are biological versions of fuel cells in which the fuel is food and the cells are adaptations of muscle cells. Electrolytic cells include the arrangement used to electrolyse water into hydrogen and oxygen and to obtain aluminium from its oxide in the Hall process. Electrolysis is the only commercially viable means for the production of fluorine. The electron transfer processes that occur in respiration and photosynthesis can be modelled by electrochemical cells in which electrons are transferred between cytochrome molecules.

[6] The term *voltaic cell* is also used.

9.3 Half-reactions and electrodes

A redox reaction is the outcome of the loss of electrons, and perhaps atoms, from one species and their gain by another species. It will be familiar from introductory chemistry that we identify the loss of electrons (oxidation) by noting whether an element has undergone an increase in oxidation number. We identify the gain of electrons (reduction) by noting whether an element has undergone a decrease in oxidation number. The requirement to break and form covalent bonds in some redox reactions, as in the conversion of PCl_3 to PCl_5 or of NO_2^- to NO_3^-, is one of the reasons why they often achieve equilibrium quite slowly, often much more slowly than acid–base proton transfer reactions.

Self-test 9.3

Identify the species that have undergone oxidation and reduction in the reaction $CuS(s) + O_2(g) \rightarrow Cu(s) + SO_2(g)$.

[*Answer:* Cu(II) reduced, S^{2-} oxidized to S(IV), O reduced]

Any redox reaction may be expressed as the difference of two reduction **half-reactions**. Two examples are

Reduction of Cu^{2+}: $Cu^{2+}(aq) + 2\ e^- \rightarrow Cu(s)$
Reduction of Zn^{2+}: $Zn^{2+}(aq) + 2\ e^- \rightarrow Zn(s)$
Difference: $Cu^{2+}(aq) + Zn(s) \rightarrow Cu(s) + Zn^{2+}(aq)$ (A)

A half-reaction in which atom transfer accompanies electron transfer is[7]

Reduction of MnO_4^-:
$MnO_4^-(aq) + 8\ H^+(aq) + 5\ e^- \rightarrow Mn^{2+}(aq) + 4\ H_2O(l)$ (B)

Half-reactions are *conceptual*. Redox reactions normally proceed by a much more complex mechanism in which the electron is never free. The electrons in these conceptual reactions are re-

garded as being 'in transit' and are not ascribed a state.

The oxidized and reduced species in a half-reaction form a **redox couple**, denoted Ox/Red. Thus, the redox couples mentioned so far are Cu^{2+}/Cu, Zn^{2+}/Zn, and $MnO_4^-,H^+/Mn^{2+}$. In general, we adopt the notation

Couple: Ox/Red   Half-reaction: Ox + $\nu e^- \rightarrow$ Red

A chemical reaction need not be a redox reaction for it to be expressed in terms of reduction half-reactions. For instance, the expansion of a gas

$H_2(g, p_i) \rightarrow H_2(g, p_f)$

can be expressed as the difference of two reductions:

$2\ H^+(aq) + 2\ e^- \rightarrow H_2(g, p_f)$
$2\ H^+(aq) + 2\ e^- \rightarrow H_2(g, p_i)$

The two couples are both H^+/H_2 with the gas is at a different pressure in each case. Similarly, the dissolution of the sparingly soluble salt silver chloride

$AgCl(s) \rightarrow Ag^+(aq) + Cl^-(aq)$

can be expressed as the difference of the following two reduction half-reactions:

$AgCl(s) + e^- \rightarrow Ag(s) + Cl^-(aq)$ $Ag^+(aq) + e^- \rightarrow Ag(s)$

Example 9.3 *Expressing a reaction in terms of half-reactions*

Express the oxidation of NADH (nicotinamide adenine dinucleotide, **1**), which participates in the chain of oxidations that constitutes respiration, to NAD^+ (**2**) by oxygen, when the latter is reduced to H_2O_2, in aqueous solution as the difference of two reduction half-reactions. The overall reaction is $NADH(aq) + O_2(g) + H^+(aq) \rightarrow NAD^+(aq) + H_2O_2(aq)$.

Strategy To decompose a reaction into reduction half-reactions, identify one reactant species that undergoes reduction, its corresponding reduction product, and write the half-reaction for this process. To find the second half-reaction, subtract the overall

---

[7] In the discussion of redox reactions, the hydrogen ion is commonly denoted simply $H^+(aq)$ rather than treated as a hydronium ion, $H_3O^+(aq)$, as proton transfer is less of an issue and the chemical equations are simplified.

1 Nicotinamide adenine dinucleotide, reduced form (NADH)

2 Nicotinamide adenine dinucleotide (NAD$^+$)

reaction from this half-reaction and rearrange the species so that all the stoichiometric coefficients are positive and the equation is written as a reduction.

Solution Oxygen undergoes reduction to H_2O_2, so one half-reaction is

$$O_2(g) + 2\,H^+(aq) + 2\,e^- \rightarrow H_2O_2(aq)$$

Subtraction of this half-reaction from the overall equation gives

$$NADH(aq) - H^+(aq) - 2\,e^- \rightarrow NAD^+(aq)$$

Addition of $H^+(aq) + 2\,e^-$ to both sides gives

$$NADH(aq) \rightarrow NAD^+(aq) + H^+(aq) + 2\,e^-$$

This is an oxidation half-reaction. We reverse it to find the corresponding reduction half-reaction:

$$NAD^+(aq) + H^+(aq) + 2\,e^- \rightarrow NADH(aq)$$

Self-test 9.4

Express the formation of H_2O from H_2 and O_2 in acidic solution as the difference of two reduction half-reactions.

[*Answer:* $4\,H^+(aq) + 4\,e^- \rightarrow 2\,H_2(g)$,
$O_2(g) + 4\,H^+(aq) + 4\,e^- \rightarrow 2\,H_2O(l)$]

We have already seen that, for thermodynamic considerations, a natural way to express the composition of a system is in terms of the reaction quotient Q (because Q occurs in a number of thermodynamic formulas, particularly the formula for the reaction Gibbs energy, eqn 7.6). The quotient for a half-reaction is defined like the quotient for the overall reaction, but with the electrons ignored. Thus, for the half-reaction of the NAD$^+$/NADH couple in Example 9.3 we would write

$$NAD^+(aq) + H^+(aq) + 2\,e^- \rightarrow NADH(aq)$$

$$Q = \frac{a_{NADH}}{a_{NAD^+}\,a_{H^+}} \approx \frac{[NADH]}{[NAD^+][H^+]}$$

where, in elementary work and provided the solution is very dilute, the activities are interpreted as the numerical values of the molar concentrations (see Table 6.2). The replacement of activities by molar concentrations is very hazardous for ionic solutions, so wherever possible we delay taking that final step. The replacement is not always necessary: as we shall see, electrochemical measurements are widely used to determine pH and equilibrium constants, both of which are defined in terms of activities.

9.4 Reactions at electrodes

In an electrochemical cell, oxidation takes place at the **anode** and reduction takes place at the **cathode**. As the reaction proceeds in a galvanic cell, the electrons released at the anode travel through the external circuit (Fig 9.6). They re-enter the cell at the cathode, where they bring about reduction. This flow of current in the external circuit, from anode to cathode, corresponds to the cathode having a higher potential than the anode.[8] In an electrolytic cell, the anode is also the location of oxidation (by definition). Now, though, electrons must be withdrawn from the species in the anode compartment, so the anode

---

[8] Negatively charged electrons tend to travel to regions of higher potential.

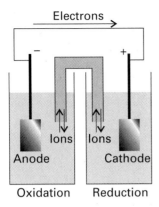

Fig 9.6 The flow of electrons in the external circuit is from the anode of a galvanic cell, where they have been lost in the oxidation reaction, to the cathode, where they are used in the reduction reaction. Electrical neutrality is preserved in the electrolytes by the flow of cations and anions in opposite directions through the salt bridge.

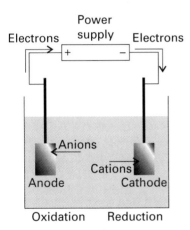

Fig 9.7 The flow of electrons and ions in an electrolytic cell. An external supply forces electrons into the cathode, where they are used to bring about a reduction, and withdraws them from the anode, which results in an oxidation reaction at that electrode. Cations migrate towards the negatively charged cathode and anions migrate towards the positively charged anode. An electrolytic cell usually consists of a single compartment, but a number of industrial versions have two compartments.

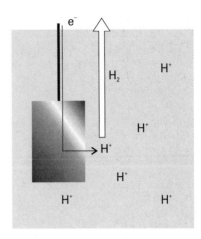

Fig 9.8 The schematic structure of a hydrogen electrode, which is like other gas electrodes. Hydrogen is bubbled over a black (that is, finely divided) platinum surface that is in contact with a solution containing hydrogen ions. The platinum, as well as acting as a source or sink for electrons, speeds the electrode reaction because hydrogen attaches to (adsorbs on) the surface as atoms.

must be connected to the positive terminal of an external supply. Similarly, electrons must pass from the cathode to the species undergoing reduction, so the cathode must be connected to the negative terminal of a supply (Fig 9.7).

In a **gas electrode** (Fig 9.8), a gas is in equilibrium with a solution of its ions in the presence of an inert metal. The inert metal, which is often platinum, acts as a source or sink of electrons but takes no other part in the reaction except perhaps acting as a catalyst. One important example is the *hydrogen electrode*, in which hydrogen is bubbled through an aqueous solution of hydrogen ions and the redox couple is H^+/H_2. This electrode is denoted $Pt(s)|H_2(g)|H^+(aq)$. The vertical lines denote junctions between phases. In this electrode, the junctions are between the platinum and the gas and between the gas and the liquid containing its ions.

Example 9.4 *Writing the half-reaction for a gas electrode*

Write the half-reaction and the reaction quotient for the reduction of oxygen to water in acidic solution.

Strategy Write the chemical equation for the half-reaction. Then express the reaction quotient in terms of the activities and the corresponding stoichiometric numbers, with products in the numerator and reactants in the denominator. Pure (and nearly pure) solids and liquids do not appear in Q; nor does the electron. The activity of a gas is set equal to the numerical value of its partial pressure.

Solution The equation for the reduction of O_2 in acidic solution is

$$O_2(g) + 4\,H^+(aq) + 4\,e^- \rightarrow 2\,H_2O(l)$$

The reaction quotient for the half-reaction is therefore

$$Q = \frac{1}{p_{O_2}\,a_{H^+}^4}$$

Note the very strong dependence of Q on the hydrogen ion activity.

Self-test 9.5

Write the half-reaction and the reaction quotient for a chlorine gas electrode.

[*Answer*: $Cl_2(g) + 2\,e^- \rightarrow 2\,Cl^-(aq)$, $Q = a_{Cl^-}^2/p_{Cl_2}$]

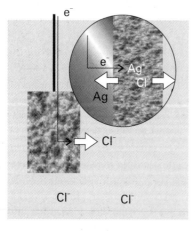

Fig 9.9 The schematic structure of a silver–silver chloride electrode (as an example of an insoluble-salt electrode). The electrode consists of metallic silver coated with a layer of silver chloride in contact with a solution containing Cl^- ions.

An **insoluble-salt electrode** consists of a metal M covered by a porous layer of insoluble salt MX, the whole being immersed in a solution containing X^- ions (Fig 9.9). The electrode is denoted $M|MX|X^-$. An example is the silver–silver-chloride electrode, $Ag(s)|AgCl(s)|Cl^-(aq)$, for which the reduction half-reaction is

$$AgCl(s) + e^- \rightarrow Ag(s) + Cl^-(aq) \qquad Q = a_{Cl^-} \approx [Cl^-]$$

The activities of both solids are 1. Note that the reaction quotient (and therefore, as we see later, the potential of the electrode) depends on the activity of chloride ions in the electrolyte solution.

Example 9.5 *Writing the half-reaction for an insoluble-salt electrode*

Write the half-reaction and the reaction quotient for the lead–lead-sulfate electrode of the lead–acid battery, in which Pb(II), as lead(II) sulfate, is reduced to metallic lead in the presence of hydrogensulfate ions in the electrolyte.

Strategy Begin by identifying the species that is reduced, and writing the half-reaction. Balance that half-reaction by using H_2O molecules if O atoms are required, hydrogen ions (because the solution is acidic) if H atoms are needed, and electrons for the charge. Then write the reaction quotient in terms of the stoichiometric coefficients and activities of the species present. Products appear in the numerator, reactants in the denominator.

Solution The electrode is

$$Pb(s)\,|\,PbSO_4(s)\,|\,HSO_4^-(aq)$$

in which Pb(II) is reduced to metallic lead. The equation for the reduction half-reaction is therefore

$$PbSO_4(s) + H^+(aq) + 2\,e^- \rightarrow Pb(s) + HSO_4^-(aq)$$

and the reaction quotient is

$$Q = \frac{a_{HSO_4^-}}{a_{H^+}}$$

The term **redox electrode** is normally reserved for an electrode in which the couple consists of two nonzero oxidation states of the same element (Fig 9.10). An example is an electrode in which the couple is Fe^{3+}/Fe^{2+}. In general, the equilibrium is

$$Ox + v\ e^- \rightarrow Red \qquad Q = \frac{a_{Red}}{a_{Ox}}$$

A redox electrode is denoted M|Red,Ox, where M is an inert metal (typically platinum) making electrical contact with the solution. The electrode corresponding to the Fe^{3+}/Fe^{2+} couple is therefore denoted Pt(s)|Fe^{2+}(aq),Fe^{3+}(aq) and the reduction half-reaction and reaction quotient are

$$Fe^{3+}(aq) + e^- \rightarrow Fe^{2+}(aq) \qquad Q = \frac{a_{Fe^{2+}}}{a_{Fe^{3+}}}$$

Another example of a similar kind is the electrode Pt(s)|NADH(aq),NAD$^+$(aq),H$^+$(aq) used to study the NAD$^+$/NADH couple.

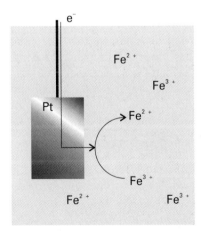

Fig 9.10 The schematic structure of a redox electrode. The platinum metal acts as a source or sink for electrons required for the interconversion of (in this case) Fe^{2+} and Fe^{3+} ions in the surrounding solution.

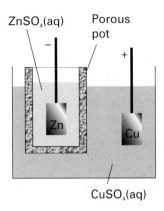

Fig 9.11 A Daniell cell consists of copper in contact with copper(II) sulfate solution and zinc in contact with zinc sulfate solution; the two compartments are in contact through the porous pot that contains the zinc sulfate solution. The copper electrode is the cathode and the zinc electrode is the anode.

9.5 **Varieties of cell**

The simplest type of galvanic cell has a single electrolyte common to both electrodes (as in Fig 9.4). In some cases it is necessary to immerse the electrodes in different electrolytes, as in the *Daniell cell* (Fig 9.11), in which the redox couple at one electrode is Cu^{2+}/Cu and at the other is Zn^{2+}/Zn. In an **electrolyte concentration cell,** which would be constructed like the cell in Fig 9.5, the electrode compartments are of identical composition except for the concentrations of the electrolytes. An electrolyte concentration cell is a model of a neuron, which consists of a cell membrane with different concentrations of Na$^+$ and K$^+$ ions on either side (Box 9.1). In an **electrode concentration cell** the electrodes themselves have different concentrations, either because they are gas electrodes operating at different pressures or because they are amalgams (solutions in mercury) with different concentrations.

In a cell with two different electrolyte solutions in contact, as in the Daniell cell or an electrolyte concentration cell, the **liquid junction potential**, E_j, the potential difference across the interface of the two electrolytes, contributes to the overall potential of the cell. The contribution of the liquid

Box 9.1 *Action potentials*

A concentration cell is a simple model of the electrochemical processes that occur across the membranes of neurons. For a cell reaction M(aq, R) → M(aq, L) in which ions at one concentration migrate into a region where their concentration is different, the Nernst equation for the potential across the membrane (see later) is

$$E = -\frac{RT}{F} \ln \frac{a_L}{a_R}$$

We set $E^{\ominus} = 0$ because the standard states of the two electrodes are identical. If R is the more concentrated solution $a_L/a_R < 1$, the logarithm is negative, and $E > 0$. Physically, a positive potential arises because positive ions tend to be reduced, so tending to withdraw electrons from the external circuit, and this process is dominant in the right-hand electrode compartment.

The cell wall of a neuron is more permeable to K^+ ions than to either Na^+ or Cl^- ions. The concentration of K^+ ions inside the cell is about 20 to 30 times that on the outside, and is maintained at that level by a specific pumping operation fuelled by ATP and governed by enzymes. If the system is approximately at equilibrium, the potential difference between the two sides is predicted to be

$$E \approx -(25.7 \text{ mV}) \times \ln \frac{1}{20} = 77 \text{ mV}$$

The potential difference across a membrane plays a particularly interesting role in the transmission of nerve impulses. When a neuron is inactive, there is a high concentration of K^+ ions inside the cell and a high concentration of Na^+ ions outside. The potential difference across the cell wall is about 77 mV. When the cell wall is subjected to a pulse of about 20 mV, the structure of the membrane adjusts and it becomes permeable to Na^+ ions. As a result, there is a decrease in membrane potential as the Na^+ ions flood into the interior of the cell. The change in potential difference triggers the adjacent part of the cell membrane, and the pulse of collapsing potential passes along the nerve. Behind the pulse the sodium and potassium pumps restore the concentration difference ready for the next pulse.

Exercise 1 Under certain stress conditions, such as viral infection or hypoxia, plants have been shown to have an intercellular pH increase of about 0.1 pH. Suppose this pH change also occurs in the mitochondrial intermembrane space. How much ATP can now be synthesized for the transport of 2 mol H^+ assuming no other changes occur?

Exercise 2 If the mitochondrial electric potential between matrix and the intermembrane space were 70 mV, as is common for other membranes, how much ATP could be synthesized from the transport of 4 mol H^+, assuming the pH difference remains the same?

junction to the potential can be decreased (to about 1 to 2 mV) by joining the electrolyte compartments through a salt bridge consisting of a saturated electrolyte solution (usually KCl) in agar jelly (as in Fig 9.5). The reason for the success of the salt bridge is that the liquid junction potentials at either end are largely independent of the concentrations of the two more dilute solutions in the electrode compartments, and so nearly cancel.

In the notation for cells, an interface between phases is denoted by a vertical bar. For example, a cell in which the left-hand electrode is a hydrogen electrode and the right-hand electrode is a silver–silver-chloride electrode is denoted

$$Pt(s)|H_2(g)|HCl(aq)|AgCl(s)|Ag(s)$$

A double vertical line || denotes an interface for which it is assumed that the junction potential has been eliminated. Thus a cell in which the left-hand electrode, in an arrangement like that in Fig 9.5, is zinc in contact with aqueous zinc sulfate and the right-hand electrode is copper in contact with aqueous copper(II) sulfate is denoted

$$Zn(s)|ZnSO_4(aq)||CuSO_4(aq)|Cu(s)$$

Self-test 9.7

Give the notation for a cell in which the oxidation of NADH by oxygen could be studied (recall Example 9.3).

[*Answer*: Pt(s) | NADH(aq), NAD$^+$(aq), H$^+$(aq) || H$_2$O$_2$(aq), H$^+$(aq) | O$_2$(g) | Pt(s)]

9.6 The cell reaction

The current produced by a galvanic cell arises from the spontaneous reaction taking place inside it. The **cell reaction** is the reaction in the cell written on the assumption that the right-hand electrode is the cathode, and hence that reduction is taking place in the right-hand compartment. Later we see how to predict if the right-hand electrode is in fact the cathode; if it is, then the cell reaction is spontaneous as written. If the left-hand electrode turns out to be the cathode, then the reverse of the cell reaction is spontaneous.

To write the cell reaction corresponding to the cell diagram, we first write the half-reactions at both electrodes as reductions, and then subtract the left-hand equation from the right-hand equation. Thus, in the cell used to study the reaction of NADH,

$$\text{Pt(s)|NADH(aq),NAD}^+\text{(aq),H}^+\text{(aq)||H}_2\text{O}_2\text{(aq),H}^+\text{(aq)|}$$
$$\text{O}_2\text{(g)|Pt(s)}$$

we saw in Example 9.3 that the two reduction half-reactions are

Right (R): $\text{O}_2\text{(g)} + 2\,\text{H}^+\text{(aq)} + 2\,\text{e}^- \rightarrow \text{H}_2\text{O}_2\text{(aq)}$
Left (L): $\text{NAD}^+\text{(aq)} + \text{H}^+\text{(aq)} + 2\,\text{e}^- \rightarrow \text{NADH(aq)}$

The equation for the cell reaction is the difference:

Overall (R − L): $\text{NADH(aq)} + \text{O}_2\text{(g)} + \text{H}^+\text{(aq)} \rightarrow$
$$\text{NAD}^+\text{(aq)} + \text{H}_2\text{O}_2\text{(aq)}$$

In other cases, it may be necessary to match the numbers of electrons in the two half-reactions by multiplying one of the equations through by a numerical factor: there should be no electrons showing in the overall equation.

Self-test 9.8

Write the chemical equation for the reaction in the cell Ag(s)|AgBr(s)|NaBr(aq)‖NaCl(aq)|Cl₂(g)|Pt(s).
[*Answer*: 2 Ag(s) + 2 Br⁻(aq) + Cl₂(g) → 2 AgBr(s) + 2 Cl⁻(aq)]

9.7 The cell potential

A galvanic cell does electrical work as the reaction drives electrons through an external circuit. The work done by a given transfer of electrons depends on the **cell potential**, the potential difference between the two electrodes. Cell potential is measured in volts (V, where $1\,\text{V} = 1\,\text{J C}^{-1}$). When the cell potential is large (for instance, 2 V), a given number of electrons travelling between the electrodes can do a large amount of electrical work. When the cell potential is small (such as 2 mV), the same number of electrons can do only a small amount of work. A cell in which the reaction is at equilibrium can do no work and its potential is zero.

According to the discussion in Section 4.9, we know that the maximum electrical work, w'_{max}, that a system (in this context, the cell) can do is given by the value of ΔG, and in particular that

At constant temperature and pressure: $w'_{max} = \Delta G$ (9.10)

Therefore, by measuring the electrical potential and converting that potential to the work done by the reaction, we have a means of determining a thermodynamic quantity. Conversely, if we know ΔG for a reaction, then we have a route to the prediction of the cell potential. However, we need to recall that maximum work is achieved only when a process occurs reversibly. In the present context, reversibility means that the cell should be connected to an external source of potential difference that opposes and exactly matches the potential generated by the cell. Then an infinitesimal change of the external potential will allow the reaction to proceed in its spontaneous direction and an opposite infinitesimal change will drive the reaction in its reverse direction.[9] The potential difference measured when a cell is balanced against an external source of potential is called the **zero-current cell potential**, E (Fig 9.12).[10] Our task is to relate this quantity to the Gibbs energy of the cell reaction.

---

[9] We saw in Section 2.3 that the criterion of thermodynamic reversibility is the reversal of a process by an infinitesimal change in the external conditions.

[10] An older name for this quantity (which is still widely used), is the *electromotive force*, or emf, of the cell.

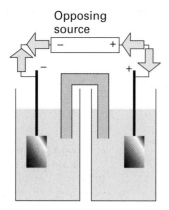

Fig 9.12 The zero-current cell potential is measured by balancing the cell against an external potential that opposes the reaction in the cell. When there is no current flow, the external potential difference is equal to the cell potential.

Derivation 9.2 *The cell potential*

Suppose that the cell reaction can be broken down into half-reactions of the form $A + v\,e^- \rightarrow B$. Then, when the reaction takes place, vN_A electrons are transferred from the reducing agent to the oxidizing agent per mole of reaction events, so the charge transferred between the electrodes is $vN_A \times (-e)$, or $-vF$. The constant F is the **Faraday constant**, and is the magnitude of electric charge per mole of electrons:

$$F = eN_A = 96.485 \text{ kC mol}^{-1}$$

The electrical work w' done when this charge travels from the anode to the cathode is equal to the product of the charge and the potential difference E:

$$w' = -vF \times E$$

Finally, provided the work is done reversibly at constant temperature and pressure, we equate this electrical work to the reaction Gibbs energy:

$$-vFE = \Delta_r G \qquad (9.11)$$

Recall that $\Delta_r G$ is the slope of a graph of G plotted against the composition of the reaction mixture (Section 7.1). When the reaction is spontaneous in the forward direction, corresponding to an equilib-

rium composition that is rich in reactants, $\Delta_r G < 0$, which implies that the cell potential is positive. When $\Delta_r G > 0$, corresponding to an equilibrium composition that is rich in products, the reverse reaction is spontaneous and $E < 0$. At equilibrium $\Delta_r G = 0$ and therefore $E = 0$ too.

Equation 9.11 provides an electrical method for measuring a reaction Gibbs energy at any composition of the reaction mixture: we simply measure the cell potential and convert the potential to $\Delta_r G$. Conversely, if we know the value of $\Delta_r G$ at a particular composition, we can predict the cell potential.

Illustration 9.1

Suppose $\Delta_r G \approx -1 \times 10^2 \text{ kJ mol}^{-1}$ and $v = 1$, then

$$E = -\frac{\Delta_r G}{v\,F} = -\frac{(-1 \times 10^5 \text{ J mol}^{-1})}{1 \times (9.6485 \times 10^4 \text{ C mol}^{-1})} = 1 \text{ V}$$

Most electrochemical cells bought commercially are indeed rated at between 1 and 2 V.

Our next step is to see how E varies with composition by combining eqn 9.11 with eqn 7.6 showing how the reaction Gibbs energy varies with the composition:

$$\Delta_r G = \Delta_r G^{\ominus} + RT \ln Q$$

In this expression, $\Delta_r G^{\ominus}$ is the standard reaction Gibbs energy and Q is the reaction quotient for the cell reaction. When we substitute this relation into eqn 9.11 in the form $E = -\Delta_r G/vF$ we obtain the **Nernst equation**:

$$E = E^{\ominus} - \frac{RT}{v\,F} \ln Q \qquad (9.12)$$

In this expression $E^{\ominus}$ is the **standard cell potential**:

$$E^{\ominus} = -\frac{\Delta_r G^{\ominus}}{v\,F} \qquad (9.13)$$

The standard cell potential is often interpreted as the cell potential when all the reactants and products are in their standard states (unit activity for all solutes, pure gases and solids, a pressure of 1 bar). However, because such a cell is not in general attainable, it is better to regard it simply as the standard Gibbs energy of the reaction expressed as a potential (Box 9.2).

Box 9.2 *Chemiosmotic theory*

The **chemiosmotic theory** of oxidative phosphorylation, the conversion of ADP to ATP, in mitochondria was proposed by Peter Mitchell. It earned him the Nobel Prize for chemistry in 1978. The structure of a mitochondrion is shown in the illustration. Phosphorylation takes place in the intermembrane space and is powered by the flux of protons through the inner membrane. These protons have been transported into the mitochondrion matrix by a series of reactions, such as the dehydrogenation of NADH to NAD^+. Dehydration is mediated by enzymes such as NADH dehydrogenase, that are orientated asymmetrically in the inner membrane so that the protons detached from NADH on one side of the membrane are deposited on the other side.

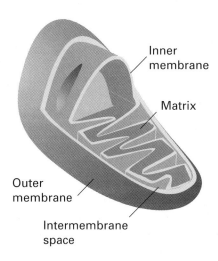

The general structure of a mitochondrion.

We can estimate the Gibbs energy available for phosphorylation by considering the proton concentration gradient across the inner membrane. Let's suppose that the concentration of protons (as hydronium ions) is $[H^+]_{in}$ in the mitochondrion matrix and $[H^+]_{out}$ outside the membrane in the intermembrane space. The difference in molar Gibbs energy due to this concentration gradient is

$$\Delta G_m = G_{m, in} - G_{m, out} = RT \ln \frac{[H^+]_{in}}{[H^+]_{out}}$$

(We have used eqn 9.12, replacing activities by molar concentrations.) Expressed in terms of the pH gradient defined as

$$\Delta pH = pH_{in} - pH_{out} = -\log [H^+]_{in} + \log [H^+]_{out}$$

the difference in molar Gibbs energy is

$$\Delta G_m = -(\ln 10) RT \Delta pH \approx -2.303 \ RT \Delta pH$$

(Here we have used $\ln x = (\ln 10) \log x$.) The pH is lower in the matrix than in the intermembrane space, with $\Delta pH \approx -1.4$, so at 25°C, $\Delta G_m \approx +8.0$ kJ mol^{-1}. At this stage, we conclude that the outward flow of stored protons could do about 8 kJ mol^{-1} of non-expansion work.

The expression we have derived in terms of the concentration difference would apply even if the stored particles were uncharged. Put another way, it is the ideal value of ΔG_m disregarding proton–proton electrical repulsions. If these repulsions are different on each side of the membrane, with weaker repulsions in the intermembrane space than in the matrix, there will be a potential difference $\Delta \phi = \phi_{in} - \phi_{out}$ across the membrane, with $\Delta \phi < 0$. Then even more work could be done as the protons escaped from their positively charged environment. The additional 'non-ideal' contribution to the work per mole of protons is $F \Delta \phi$, so the overall difference in molar Gibbs energy is

$$\Delta G_m \approx F \Delta \phi - 2.303 \ RT \Delta pH$$

The difference in electric potential between the matrix and the intermembrane space is about 0.14 V, corresponding to 13.5 kJ mol^{-1}, so overall $\Delta G_m \approx +21.5$ kJ mol^{-1}. Because 31 kJ mol^{-1} is needed for phosphorylation, we conclude that at least 2 mol of protons (and probably much more) must flow through the membrane into the intermembrane space for the phosphorylation of 2 mol ADP.

The coupling of the proton flow to the phosphorylation event is itself of considerable interest. Structural studies show that the channel through which the protons flow is linked in tandem to a unit composed of six protein molecules arranged in pairs to form three interlocked segments (see the second illustration). The conformations of the three pairs may be loose (L), tight (T), and open (O), with steric requirements ensuring that one of each type is present at each stage. At the start of a cycle, a T unit holds an ATP molecule. Then ADP and inorganic phosphorus migrate into the L site, and as it closes into T, the earlier T site opens into O and releases its ATP. The ADP and P$_i$ in the T site meanwhile condense into ATP, and the new L site is ready for the cycle to begin again. The proton flux

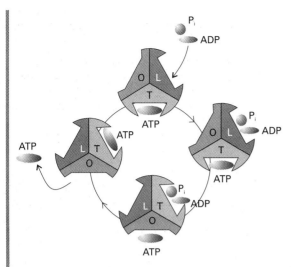

The sequence of steps involved in the phosphorylation of ADP to ATP. The channel involved consists of six proteins arranged in groups of two to give loose (L), tight (T), and open (O) structures.

drives the conformational changes of the proteins as well as providing the energy for the condensation reaction itself.

Exercise 1 A well known example of active transport is the Na^+/K^+ pump. Active transport is the transport of a substance from a lower to a higher chemical potential. The energy of ATP is used to move Na^+ out of the cell and K^+ ions into the cell. There are not only concentration differences, but also potential differences of about 77mV across the cell membrane (the inside is negative relative to the outside). The concentrations of Na^+ and K^+ inside the cell are 10 mmol L^{-1} and 100 mmol L^{-1}, respectively; the concentrations outside are $[Na^+]$ = 140 mmol L^{-1} and $[K^+]$ = 5 mmol L^{-1}. Calculate ΔG for the transport of 1 mol Na^+ ions out of the cell and the ΔG for the transport of 1 mol K^+ ions into the cell.

Exercise 2 The overall reaction for Na^+/K^+ transport is:

$$3\ Na^+(inside) + 2\ K^+(outside) + ATP \rightarrow$$
$$ADP + P_i + 3\ Na^+(outside) + 2\ K^+(inside)$$

At 310 K, $\Delta_r G^\ominus$ for the hydrolysis of ATP is -31.3 kJ mol^{-1}. Given that the ATP/ADP ratio is of the order of 100, is the hydrolysis of 1 mol ATP sufficient to provide the energy for the transport of Na^+ and K^+ according to the above equation?

Take $[P_i]$ = 1 mol L^{-1}.

At 25°C, RT/F = 25.7 mV. It follows from the Nernst equation that for a reaction in which $v = 1$, if Q is decreased by a factor of 10, then the cell potential becomes more positive by 59.2 mV. The reaction has a greater tendency to form products. If Q is increased by a factor of 10, then the cell potential falls by 59.2 mV and the reaction has a lower tendency to form products.

9.8 Cells at equilibrium

A special case of the Nernst equation has great importance in chemistry. Suppose the reaction has reached equilibrium; then $Q = K$, where K is the equilibrium constant of the cell reaction. However, because a chemical reaction at equilibrium cannot do work, it generates zero potential difference between the electrodes. Setting $Q = K$ and $E = 0$ in the Nernst equation gives

$$\ln K = \frac{vFE^\ominus}{RT} \qquad (9.14)$$

This very important equation lets us predict equilibrium constants from standard cell potentials.[11]

Illustration 9.2

Because the standard potential of the Daniell cell is +1.10 V, the equilibrium constant for the cell reaction (reaction A) is

$$\ln K = \frac{2 \times (9.6485 \times 10^4\ C\ mol^{-1}) \times (1.10\ V)}{(8.3145\ J K^{-1} mol^{-1}) \times (298.15\ K)}$$

$$= \frac{2 \times 9.6485 \times 1.10 \times 10^4}{8.3145 \times 298.15}$$

and therefore $K = 1.5 \times 10^{37}$. Hence, the displacement of copper by zinc goes virtually to completion in the sense that the ratio of concentrations of Zn^{2+} ions to Cu^{2+} ions at equilibrium is about 10^{37}.

Note that, if $E^\ominus > 0$, then $K > 1$ and at equilibrium the cell reaction lies in favour of products. The opposite is true if $E^\ominus < 0$, for then $K < 1$ and the reactants are favoured at equilibrium.

---

[11] Equation 9.14, of course, is simply eqn 7.8 expressed electrochemically.

9.9 **Standard potentials**

Each electrode in a galvanic cell makes a characteristic contribution to the overall cell potential. Although it is not possible to measure the contribution of a single electrode, one electrode can be assigned a value zero and the others assigned relative values on that basis. The specially selected electrode is the **standard hydrogen electrode** (SHE):

$$Pt(s)|H_2(g)|H^+(aq) \qquad E^{\ominus} = 0 \text{ at all temperatures}$$

The **standard potential**, $E^{\ominus}(Ox/Red)$, of a couple Ox/Red is then measured by constructing a cell in which the couple of interest forms the right-hand electrode and the standard hydrogen electrode is on the left. For example, the standard potential of the Ag^+/Ag couple is the standard potential of the cell

$$Pt(s)|H_2(g)|H^+(aq)||Ag^+(aq)|Ag(s)$$

and is +0.80 V. Table 9.3 lists a selection of standard potentials; a longer list will be found in Appendix 2.

To calculate the standard potential of a cell formed from any pair of electrodes we take the difference of their standard potentials:

$$E^{\ominus} = E_R^{\ominus} - E_L^{\ominus} \tag{9.15}$$

Here $E_R^{\ominus}$ is the standard potential of the right-hand electrode and $E_L^{\ominus}$ is that of the left. Because $\Delta_r G^{\ominus} = -\nu F E^{\ominus}$, it follows that, if $E^{\ominus} > 0$, then the corresponding cell reaction is spontaneous in the direction written (in the sense that $K > 1$). Once we have the value of $E^{\ominus}$ we can use it in eqn 9.14 to calculate the equilibrium constant of the cell reaction.

Example 9.6 *Identifying the spontaneous direction of a reaction*

One of the reactions important in corrosion in an acidic environment is

$$Fe(s) + 2H^+(aq) + \tfrac{1}{2}O_2(g) \rightarrow Fe^{2+}(aq) + H_2O(l)$$

Does the equilibrium constant favour the formation of $Fe^{2+}(aq)$?

Strategy We need to decide whether $E^{\ominus}$ for the reaction is positive or negative: $K > 1$ if $E^{\ominus} > 0$. We express the overall reaction as the difference of reduction half-reactions, look up the standard potentials of the two couples, and take their difference.

Solution The two reduction half-reactions are

Right: $2H^+(aq) + \tfrac{1}{2}O_2(g) + 2e^- \rightarrow H_2O(l)$
$E^{\ominus}(H^+,O_2,H_2O) = +1.23$ V
Left: $Fe^{2+}(aq) + 2e^- \rightarrow Fe(s)$ $E^{\ominus}(Fe^{2+},Fe) = -0.44$ V

The difference Right − Left is

$$Fe(s) + 2H^+(aq) + \tfrac{1}{2}O_2(g) \rightarrow Fe^{2+}(aq) + H_2O(l) \quad E^{\ominus} = +1.67 \text{ V}$$

Therefore, because $E^{\ominus} > 0$, it follows that $K > 1$, favouring products.

Self-test 9.9

Is the equilibrium constant for the displacement of lead from lead(II) solutions by copper greater or less than 1?

[*Answer*: $K < 1$]

Example 9.7 *Calculating an equilibrium constant*

Calculate the equilibrium constant for the disproportionation reaction $2Cu^+(aq) \rightleftharpoons Cu(s) + Cu^{2+}(aq)$ at 298 K.

Strategy The aim is to find the values of $E^{\ominus}$ and ν corresponding to the reaction, for then we can use eqn 9.14. To do so, we express the equation as the difference of two reduction half-reactions. The stoichiometric number of the electron in these matching half-reactions is the value of ν we require. We then look up the standard potentials for the couples corresponding to the half-reactions and calculate their difference to find $E^{\ominus}$. In calculations of this kind a useful value is $RT/F = 25.69$ mV.

Solution The two half-reactions are

Right: $Cu^+(aq) + e^- \rightarrow Cu(aq)$ $E^{\ominus}(Cu^+,Cu) = +0.52$ V
Left: $Cu^{2+}(aq) + e^- \rightarrow Cu^+(aq)$ $E^{\ominus}(Cu^{2+},Cu^+) = +0.15$ V

The standard potential for the overall reaction is

$$E^{\ominus} = (0.52 \text{ V}) - (0.15 \text{ V}) = +0.37 \text{ V}$$

Table 9.3 *Standard potentials at 25 °C*

Reduction half-reaction					$E^{\ominus}/V$
Oxidizing agent				**Reducing agent**	
Strongly oxidizing					
F_2	+	2 e⁻	→	2 F⁻	+2.87
$S_2O_8^{2-}$	+	2 e⁻	→	$2 SO_4^{2-}$	+2.05
Au^+	+	e⁻	→	Au	+1.69
Pb^{4+}	+	2 e⁻	→	Pb^{2+}	+1.67
Ce^{4+}	+	e⁻	→	Ce^{3+}	+1.61
$MnO_4^- + 8 H^+$	+	5 e⁻	→	$Mn^{2+} + 4 H_2O$	+1.51
Cl_2	+	2 e⁻	→	2 Cl⁻	+1.36
$Cr_2O_7^{2-} + 14 H^+$	+	6 e⁻	→	$2 Cr^{3+} + 7 H_2O$	+1.33
$O_2 + 4 H^+$	+	4 e⁻	→	$2 H_2O$	+1.23, +0.82 at pH = 7
Br_2	+	2 e⁻	→	2 Br⁻	+1.09
Ag^+	+	e⁻	→	Ag	+0.80
Hg_2^{2+}	+	2 e⁻	→	2 Hg	+0.79
Fe^{3+}	+	e⁻	→	Fe^{2+}	+0.77
I_2	+	2 e⁻	→	2 I⁻	+0.54
$O_2 + 2 H_2O$	+	4 e⁻	→	4 OH⁻	+0.40, +0.82 at pH = 7
Cu^{2+}	+	2 e⁻	→	Cu	+0.34
AgCl	+	e⁻	→	Ag + Cl⁻	+0.22
$2 H^+$	+	2 e⁻	→	H_2	0, by definition
Fe^{3+}	+	3 e⁻	→	Fe	−0.04
$O_2 + H_2O$	+	2 e⁻	→	$HO_2^- + OH^-$	−0.08
Pb^{2+}	+	2 e⁻	→	Pb	−0.13
Sn^{2+}	+	2 e⁻	→	Sn	−0.14
Fe^{2+}	+	2 e⁻	→	Fe	−0.44
Zn^{2+}	+	2 e⁻	→	Zn	−0.76
$2 H_2O$	+	2 e⁻	→	$H_2 + 2 OH^-$	−0.83, −0.42 at pH = 7
Al^{3+}	+	3 e⁻	→	Al	−1.66
Mg^{2+}	+	2 e⁻	→	Mg	−2.36
Na^+	+	e⁻	→	Na	−2.71
Ca^{2+}	+	2 e⁻	→	Ca	−2.87
K^+	+	e⁻	→	K	−2.93
Li^+	+	e⁻	→	Li	−3.05
				Strongly reducing	

For a more extensive table, see Appendix 2.

It then follows from eqn 9.5 with $\nu = 1$, that

$$\ln K = \frac{0.37\ V}{25.69 \times 10^{-3}\ V} = \frac{3.7 \times 10}{2.569}$$

Therefore, $K = 1.8 \times 10^6$. The equilibrium lies strongly towards the right of the reaction as written, so Cu⁺ disproportionates almost totally in aqueous solution.

Self-test 9.10

Calculate the equilibrium constant for the reaction $Sn^{2+}(aq) + Pb(s) \rightleftharpoons Sn(s) + Pb^{2+}(aq)$ at 298 K.

[*Answer*: 0.46]

9.10 The variation of potential with pH

The half-reactions of many redox couples involve hydrogen ions. For example, the reduction of fumaric acid, $HOOCCH=CHCOOH$, to succinic acid, $HOOCCH_2CH_2COOH$, which plays a role in the citric acid cycle, is

$$HOOCCH=CHCOOH(aq) + 2\,H^+(aq) + 2\,e^- \rightarrow$$
$$HOOCCH_2CH_2COOH(aq)$$

Half-reactions of this kind have potentials that depend on the pH of the medium. In this example, in which the hydrogen ions occur as reactants, an increase in pH, corresponding to a decrease in hydrogen ion activity, favours the formation of reactants, so the fumaric acid has a lower thermodynamic tendency to become reduced. We expect, therefore, that the potential of the fumaric/succinic acid couple should decrease as the pH is increased.

We can establish the quantitative variation of reduction potential with pH for a reaction by using the Nernst equation for the half-reaction and noting that

$$\ln a_{H^+} = (\ln 10) \times \log a_{H^+} = -\ln 10 \times pH$$

with $\ln 10 = 2.303\ldots$[12] If we suppose that fumaric acid and succinic acid have fixed concentrations, the potential of the fumaric/succinic redox couple is

$$E = E^\ominus - \frac{RT}{2F} \ln \frac{a_{suc}}{a_{fum}\,a_{H^+}^2} = \overbrace{E^\ominus - \frac{RT}{2F} \ln \frac{a_{suc}}{a_{fum}}}^{E'} + \frac{RT}{F} \ln a_{H^+}$$

which is easily rearranged into

$$E = E' - \frac{RT \ln 10}{F} \times pH \qquad (9.16a)$$

At 25°C,

$$E = E' - (59.2\ \text{mV}) \times pH \qquad (9.16b)$$

We see that each increase of 1 unit in pH decreases the potential by 59.2 mV, which is in agreement with the remark above, that the reduction of fumaric acid is discouraged by an increase in pH.

We use the same approach to convert standard potentials to **biological standard potentials**, $E^\oplus$, which correspond to neutral solution (pH = 7).[13] If the hydrogen ions appear as reactants in the reduction half-reaction, then the potential is decreased below its standard value (for the fumaric/succinic couple, by $7 \times 59.2\ \text{mV} = 414\ \text{mV}$, or about 0.4 V). If the hydrogen ions appear as products, then the biological standard potential is higher than the thermodynamic standard potential. The precise change depends on the number of electrons and protons participating in the half-reaction.

Example 9.8 *Converting a standard potential to a biological standard value*

Estimate the biological standard potential of the $NAD^+/NADH$ couple at 25°C (Example 9.3). The reduction half-reaction is

$$NAD^+(aq) + H^+(aq) + 2\,e^- \rightarrow NADH(aq) \quad E^\ominus = -0.11\ \text{V}$$

Strategy We write the Nernst equation for the potential, and express the reaction quotient in terms of the activities of the species. All species except H^+ are in their standard states, so their activities are all equal to 1. The remaining task is to express the hydrogen ion activity in terms of the pH, exactly as was done in the text, and set pH = 7.

Solution The Nernst equation for the half-reaction, with $\nu = 2$, is

$$E = E^\ominus - \frac{RT}{2F} \ln \frac{\overbrace{a_{NADH}}^{1}}{\underbrace{a_{H^+}\,a_{NAD^+}}_{1}} = E^\ominus + \frac{RT}{2F} \ln a_{H^+}$$

We rearrange this expression to

$$E = E^\ominus + \frac{RT}{2F} \ln a_{H^+} = E^\ominus - \frac{RT \ln 10}{2F} \times pH$$
$$= E^\ominus - (29.59\ \text{mV}) \times pH$$

---

[12] We have used the mathematical relation that $\ln x = \ln 10 \times \log x$, where log is the common logarithm (to base 10).

[13] Biological standard states were introduced in Section 7.5.

The biological standard potential (at pH = 7) is therefore

$$E^\oplus = (-0.11 \text{ V}) - (29.59 \times 10^{-3} \text{ V}) \times 7 = -0.32 \text{ V}$$

Self-test 9.11

Calculate the biological standard potential of the half-reaction $O_2(g) + 4 H^+(aq) + 4 e^- \rightarrow 2 H_2O(l)$ at 25°C given its value +1.23 V under thermodynamic standard conditions.

[*Answer*: +0.81 V]

Example 9.9 *Calculating a biological equilibrium constant*

The reduced and oxidized forms of riboflavin form a couple with $E^\oplus = -0.21$ V and the acetate/acetaldehyde couple has $E^\oplus = -0.60$ V under the same conditions. What is the equilibrium constant for the reduction of riboflavin (Rib) by acetaldehyde in neutral solution at 25°C? The reaction is

$$RibO(aq) + CH_3CHO(aq) \rightleftharpoons Rib(aq) + CH_3COOH(aq)$$

where RibO is the oxidized form of riboflavin and Rib is the reduced form.

Strategy The procedure is the same as before: we express the overall reaction as the difference of two half-reactions, and identify the value of v required to match them. The standard potential for the reaction is then the difference of the two standard potentials for the half-reactions.

Solution The two reduction half-reactions are

Right: $RibO(aq) + 2 H^+(aq) + 2 e^-$
$\rightarrow Rib(aq) + H_2O(l)$ $\quad E^\oplus = -0.21$ V
Left: $CH_3COOH(aq) + 2 H^+(aq) + 2 e^-$
$\rightarrow CH_3CHO(aq) + H_2O(l)$ $\quad E^\oplus = -0.60$ V

and their difference is the redox reaction required. Note that $v = 2$. The corresponding standard potential is

$$E^\oplus = (-0.21 \text{ V}) - (-0.60 \text{ V}) = +0.39 \text{ V}$$

It follows that $K = 1.5 \times 10^{13}$. We conclude that riboflavin can be reduced by acetaldehyde in neutral

solution. However, there may be mechanistic reasons—the energy required to break covalent bonds, for instance—that make the reduction too slow to be feasible in practice. Note that, because hydrogen ions do not appear in the chemical equation, the equilibrium constant is independent of pH.

Self-test 9.12

What is the equilibrium constant for the reduction of riboflavin with rubredoxin, a bacterial iron–sulfur protein, in the reaction

$$\text{Riboflavin(ox)} + \text{rubredoxin(red)} \rightleftharpoons$$
$$\text{riboflavin(red)} + \text{rubredoxin(ox)}$$

given the biological standard potential of the rubredoxin couple is −0.06 V?

[*Answer*: 8.5×10^{-6}; the reactants are favoured]

9.11 The determination of pH

The potential of a hydrogen electrode is directly proportional to the pH of the solution. For example, for the cell

$$Hg(l)|Hg_2Cl_2(s)|Cl^-(aq)||H^+(aq)|H_2(g)|Pt(s)$$

in which the cell reaction is

$$Hg_2Cl_2(s) + H_2(g) \rightarrow 2 Hg(l) + 2 H^+(aq) + 2 Cl^-(aq)$$

the Nernst equation gives

$$E = E^\ominus - \frac{RT}{2F} \ln Q \qquad Q = \frac{a_{H^+}^2 a_{Cl^-}^2}{p_{H_2}}$$

We shall suppose that the hydrogen pressure has its standard value of $p^\ominus$ (small variations in pressure make very little change to the cell potential). We also note that the activity of the Cl$^-$ ions is constant because it depends on the composition of the left-hand electrode, which is independent of that of the hydrogen electrode. Then

$$E = \overbrace{E^\ominus - \frac{RT}{F} \ln a_{Cl^-}}^{E'} - \frac{RT}{2F} \ln a_{H^+}^2 = E' - \frac{RT}{F} \ln a_{H^+}$$
$$= E' + \frac{RT \ln 10}{F} \times \text{pH} \qquad (9.17)$$

where E' is a constant. We see that the pH of a solution can be measured by determining the potential of a cell in which a hydrogen electrode is one component.

> **Self-test 9.13**
>
> What range should a voltmeter have (in volts) to display changes of pH from 1 to 14 at 25°C if it is arranged to give a reading of 0 when pH = 7?
>
> [*Answer*: from +0.41 V to –0.36 V, a range of 0.77 V]

In practice, indirect methods are much more convenient to use than one based on the standard hydrogen electrode, and the hydrogen electrode is replaced by a *glass electrode* (Fig 9.13). This electrode is sensitive to hydrogen ion activity and has a potential that depends linearly on the pH. It is filled with a phosphate buffer containing Cl^- ions, and conveniently has $E \approx 0$ when the external medium is at pH = 7. The glass electrode is much more convenient to handle than the gas electrode itself, and can be calibrated using solutions of known pH (for example, one of the buffer solutions described in Section 8.6).

Finally, it should be noted that we now have a method for measuring the pK_a of an acid electrically. As we saw in Section 8.5, the pH of a solution containing equal amounts of the acid and its conjugate base is pH = pK_a. We now know how to determine pH and hence can determine pK_a in the same way.

Applications of standard potentials

THE measurement of zero-current cell potential is a convenient source of data on the Gibbs energies, enthalpies, and entropies of reactions (Box 9.3). In practice the standard values (and the biological standard values) of these quantities are the ones normally determined.

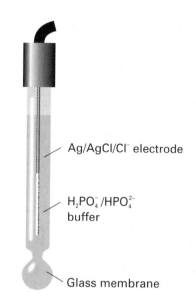

Fig 9.13 A glass electrode has a potential that varies with the hydrogen ion concentration in the medium in which it is immersed. It consists of a thin glass membrane containing an electrolyte and a silver chloride electrode. The electrode is used in conjunction with a calomel (Hg_2Cl_2) electrode that makes contact with the test solution through a salt bridge.

The labels on the figure read:
- Ag/AgCl/Cl^- electrode
- $H_2PO_4^-$/HPO_4^{2-} buffer
- Glass membrane

9.12 The electrochemical series

We have seen that a cell reaction is spontaneous as written if it has a positive cell potential, $E^{\ominus} > 0$. We have also seen that $E^{\ominus}$ may be written as the difference of the standard potentials of the redox couples in the right and left electrodes (eqn 9.15). A cell reaction is therefore spontaneous as written if $E_R^{\ominus} > E_L^{\ominus}$. Because the reduced species in the left-hand electrode compartment (the anode, the site of oxidation) reduces the oxidized species in the right-hand electrode compartment (the cathode, the site of reduction) in the cell reaction, we can conclude that

A species with a low standard potential has a thermodynamic tendency to reduce a species with a high standard potential.

More briefly: *low reduces high* and, equivalently, *high oxidizes low*. For example,

$$E^{\ominus}(Zn^{2+},Zn) = -0.76 \text{ V} < E^{\ominus}(Cu^{2+},Cu) = +0.34 \text{ V}$$

Box 9.3 *Cytochrome cascades*

One of the problems that animals have had to solve is how to achieve the oxidation of foods in a more controlled way than by combustion. Because oxidation by oxygen amounts to the reduction of oxygen to water, animals needed to discover a way to transfer electrons to O_2 molecules in a controlled manner, the overall reduction half-reaction being

$$O_2(g) + 4\,H^+(aq) + 4\,e^- \rightarrow 2\,H_2O(l) \qquad E^\ominus = +0.82\ V$$

The controlled reduction is achieved by an electron transport chain that makes use of a group of yellow–brown metalloproteins known as *cytochromes*. The cytochromes are classified as *a*, *b*, and *c* according to the wavelength of their absorption maxima, and into various subvarieties.

   Each cytochrome consists of a haemoglobin-like protein containing an iron atom in a porphyrin-like quadridentate ligand (as in haemoglobin and myoglobin) and, for cytochromes *a* and a_3, a nearby copper atom. The oxidation–reduction cycle that accepts and then passes on an electron involves the Fe^{3+}/Fe^{2+} and Cu^{2+}/Cu couples, as depicted in the first illustration. The redox couples represented by the cytochromes act like a series of lock gates on a canal and achieve reduction of oxygen in a series of steps rather than in a single step like a waterfall. The cascade of electron transfer is shown in the second illustration, together with the reaction Gibbs energy that accompanies each step. The chain is shown starting at NADH, which lies at the head of the electron-transfer chain. As we see, eight steps are used to lower the electrons cautiously down on to oxygen. The three steps with appreciable reaction Gibbs energy can be coupled, through an enzyme chain, to a biochemical process such as the phosphorylation of ADP to ATP. Each step, therefore, can be considered to be a small galvanic cell that powers a constructive process.

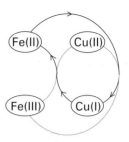

The coupling of the copper and iron redox processes, which shuttles the elements between two of their oxidation states.

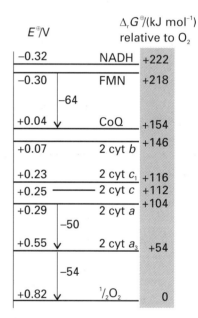

The cytochrome cascade showing the changes in the Gibbs energy as electrons are lowered down on to oxygen.

Exercise 1 Cytochrome *c* oxidase receives electrons from cyt c_{red} and transmits them to molecular oxygen. The net reaction is

$$2\ \text{cyt}\ c_{red} + \tfrac{1}{2}O_2 + 2\,H^+ \rightarrow 2\ \text{cyt}\ c_{ox} + H_2O \qquad E^\ominus = +0.562\ V$$

(a) Calculate the value of $\Delta_rG^\ominus$ for the reaction. (b) Calculate the equilibrium constant for the reaction at pH = 7 and 25°C.

Exercise 2 Photosynthesis utilizes the energy derived from the absorption of sunlight to drive an electron transport chain. This electron transport chain ends up with removal of electrons from water molecules to make molecular oxygen. The reaction can be written as $CO_2 + H_2O \rightarrow (CH_2O) + O_2$, where (CH_2O) represents carbon fixed as carbohydrate. How much energy is required to drive this reaction at constant temperature and pressure?

and Zn(s) has a thermodynamic tendency to reduce $Cu^{2+}(aq)$ under standard conditions. Hence, the reaction

$$Zn(s) + CuSO_4(aq) \rightleftharpoons ZnSO_4(aq) + Cu(s)$$

can be expected to have $K > 1$ (in fact, as we have seen, $K = 1.5 \times 10^{37}$ at 298 K).

Self-test 9.14

Does acidified dichromate ($Cr_2O_7^{2-}$) have a thermodynamic tendency to oxidize mercury metal to mercury(I)?

[*Answer*: yes]

9.13 The determination of thermodynamic functions

We have seen that the cell potential is related to the reaction Gibbs energy by eqn 9.13 ($\Delta_r G^\ominus = -vFE^\ominus$). Therefore, by measuring the standard potential of a cell driven by the reaction of interest we can obtain the standard reaction Gibbs energy. If we were interested in the biological standard state, we would use the same expression but with the standard potential at pH = 7 ($\Delta_r G^\oplus = -vFE^\oplus$).

The relation between the standard potential of a cell and the standard reaction Gibbs energy is a convenient route for the calculation of the standard potential of a couple from two others. We make use of the fact that G is a state function, and that the Gibbs energy of an overall reaction is the sum of the Gibbs energies of the reactions into which it can be divided. In general, we cannot combine the $E^\ominus$ values directly because they depend on the value of v, which may be different for the two couples.

Example 9.10 *Calculating a standard potential from two others*

Given the standard potentials $E^\ominus(Cu^{2+},Cu) = +0.340$ V and $E^\ominus(Cu^+,Cu) = +0.522$ V, calculate $E^\ominus(Cu^{2+},Cu^+)$.

Strategy We need to convert the $E^\ominus$ values to $\Delta_r G^\ominus$ values by using eqn 9.13, adding them appropriately,

and converting the overall $\Delta_r G^\ominus$ so obtained to the required $E^\ominus$ by using eqn 9.13 again. Because the Fs cancel at the end of the calculation, carry them through.

Solution The electrode reactions are as follows:

(a) $Cu^{2+}(aq) + 2\,e^- \rightarrow Cu(s)$ $E^\ominus = +0.340$ V
$$\Delta_r G^\ominus(a) = -2F \times (0.340\text{ V}) = (-0.680\text{ V}) \times F$$
(b) $Cu^+(aq) + e^- \rightarrow Cu(s)$ $E^\ominus = +0.522$ V
$$\Delta_r G^\ominus(b) = -F \times (0.522\text{ V}) = (-0.522\text{ V}) \times F$$

The required reaction is

(c) $Cu^{2+}(aq) + e^- \rightarrow Cu^+(aq)$ $\Delta_r G^\ominus(c) = -FE^\ominus$

Because (c) = (a) – (b), it follows that

$$\Delta_r G^\ominus(c) = \Delta_r G^\ominus(a) - \Delta_r G^\ominus(b)$$

Therefore, from eqn 9.13,

$$FE^\ominus(c) = -(-0.680\text{ V})F - (-0.522\text{ V})F$$

The Fs cancel, and we are left with $E^\ominus(c) = +0.158$ V.

Self-test 9.15

Given the standard potentials $E^\ominus(Fe^{3+},Fe) = -0.04$ V and $E^\ominus(Fe^{2+},Fe) = -0.44$ V, calculate $E^\ominus(Fe^{3+},Fe^{2+})$.

[*Answer*: +0.76 V]

Once we have measured $\Delta_r G^\ominus$ we can use thermodynamic relations to determine other properties. For instance, the entropy of the cell reaction can be obtained from the change in the cell potential with temperature.

Derivation 9.3 *The reaction entropy from the cell potential*

In Section 5.3 we used the fact that, at constant pressure, when the temperature changes by the infinitesimal amount dT, the Gibbs energy changes by $dG = -SdT$. Because this equation applies to the reactants and the products, it follows that

$$d(\Delta_r G^\ominus) = -\Delta_r S^\ominus \times dT$$

Substitution of $\Delta_r G^\ominus = -vFE^\ominus$ then gives

$$vF \times dE^\ominus = \Delta_r S^\ominus \times dT \qquad (9.18)$$

This equation is exact, but it applies only to infinitesimal changes in the temperature. If we suppose that the reaction entropy is independent of temperature over the range of interest, this expression becomes

$$\nu F \times (E^{\ominus} - E^{\ominus\prime}) = \Delta_r S^{\ominus} \times (T - T')$$

which is easily rearranged into

$$\Delta_r S^{\ominus} = \frac{\nu F(E^{\ominus} - E^{\ominus\prime})}{T - T'} \qquad (9.19)$$

We see from eqn 9.19 that the standard potential of a cell increases with temperature if the standard reaction entropy is positive, and that the slope of a plot of potential against temperature is proportional to the reaction entropy (Fig 9.14). An implication is that, if the cell reaction produces a lot of gas, then its potential will increase with temperature. The opposite is true for a reaction that consumes gas.

Finally, we can combine the results obtained so far by using $G = H - TS$ in the form $H = G + TS$ to obtain the standard reaction enthalpy:

$$\Delta_r H^{\ominus} = \Delta_r G^{\ominus} + T\Delta_r S^{\ominus} \qquad (9.20)$$

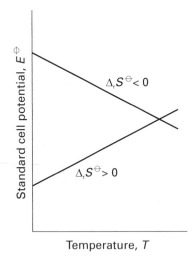

Fig 9.14 The variation of the standard potential of a cell with temperature depends on the standard entropy of the cell reaction.

with $\Delta_r G^{\ominus}$ determined from the cell potential and $\Delta_r S^{\ominus}$ from its temperature variation. Thus, we now have a noncalorimetric method of measuring a reaction enthalpy.

Example 9.11 *Using the temperature dependence of the cell potential*

The standard potential of the cell

$$Pt(s)\,|\,H_2(g)\,|\,HCl(aq)\,|\,Hg_2Cl_2(s)\,|\,Hg(l)$$

was found to be +0.2699 V at 293 K and +0.2669 V at 303 K. Evaluate the standard Gibbs energy, enthalpy, and entropy at 298 K of the reaction

$$Hg_2Cl_2(s) + H_2(g) \rightarrow 2\,Hg(l) + 2\,HCl(aq)$$

Strategy We find the standard reaction Gibbs energy from the standard potential by using eqn 9.13 and making a linear interpolation between the two temperatures (in this case, we take the mean $E^{\ominus}$ because 298 K lies midway between 293 K and 303 K). The standard reaction entropy is obtained by substituting the data into eqn 9.19. Then the standard reaction enthalpy is obtained by combining these two quantities by using eqn 9.20.

Solution Because the mean standard cell potential is +0.2684 V and $\nu = 2$ for the reaction,

$$\Delta_r G^{\ominus} = -\nu F E^{\ominus} = -2 \times (9.6485 \times 10^4\ C\ mol^{-1}) \times (0.2684\ V)$$
$$= -51.79\ kJ\ mol^{-1}$$

Then, from eqn 9.19, the standard reaction entropy is

$$\Delta_r S^{\ominus} = 2 \times (9.6485 \times 10^4\ C\ mol^{-1})$$
$$\times \left(\frac{0.2699\ V - 0.2669\ V}{293\ K - 303\ K} \right)$$
$$= -57.9\ J\ K^{-1}\ mol^{-1}$$

For the next stage of the calculation it is convenient to write the last value as $-5.79 \times 10^{-2}\ kJ\ K^{-1}\ mol^{-1}$. Then, from eqn 9.20 we find

$$\Delta_r H^{\ominus} = (-51.79\ kJ\ mol^{-1})$$
$$+ (298\ K) \times (-5.79 \times 10^{-2}\ kJ\ K^{-1}\ mol^{-1})$$
$$= -69.0\ kJ\ mol^{-1}$$

One difficulty with this procedure lies in the accurate measurement of small temperature variations of cell potential. Nevertheless, it is another example of the striking ability of thermodynamics to relate the apparently unrelated, in this case to relate electrical measurements to thermal properties.

Self-test 9.16

Predict the standard potential of the *Harned cell*

$$Pt(s)|H_2(g)|HCl(aq)|AgCl(s)|Ag(s)$$

at 303 K from tables of thermodynamic data for 298 K.

[*Answer*: +0.2191 V]

Exercises

9.1 Is the conversion of puruvate ion to lactate ion in the reaction $CH_3COCO_2^-(aq) + NADH(aq) + H^+(aq) \rightarrow CH_3CH(OH)CO_2^-(aq) + NAD^+(aq)$ a redox reaction?

9.2 Express the reaction in Exercise 9.1 as the difference of two half-reactions.

9.3 Express the reaction in which ethanol is converted to acetaldehyde (propanal) by NAD^+ in the presence of alcohol dehydrogenase as the difference of two half-reactions and write the corresponding reaction quotients for each half-reaction and the overall reaction.

9.4 Express the oxidation of cysteine (**3**) to cystine (**4**) as the difference of two half-reactions, one of which is $O_2(g) + 4 H^+(aq) + 4 e^- \rightarrow 2 H_2O(l)$.

9.5 One of the steps in photosynthesis is the reduction of $NADP^+$ by ferredoxin (fd) in the presence of ferredoxin–NADP reductase: $2 fd_{red}(aq) + NADP^+(aq) + 2 H^+(aq) \rightarrow 2 fd_{ox}(aq) + NADPH(aq)$. Express this reaction as the difference of two half-reactions. How many electrons are transferred in the reaction event?

9.6 From the biological standard half-cell potentials $E^\ominus(O_2,H^+,H_2O) = +0.82$ V and $E^\ominus(NAD^+,H^+,NADH) = -0.32$ V, calculate the standard potential arising from the reaction in which NADH is oxidized to NAD^+ and the corresponding biological standard reaction Gibbs energy.

9.7 Consider a hydrogen electrode in HBr(aq) at 25°C operating at 1.45 bar. Estimate the change in the elec-

trode potential when the solution is changed from 5.0 mmol L^{-1} to 25.0 mmol L^{-1}.

9.8 A hydrogen electrode can, in principle, be used to monitor changes in the molar concentrations of weak acids in biologically active solutions. Consider a hydrogen electrode in a solution of lactic acid as part of an overall galvanic cell at 25°C and 1 bar. Estimate the change in the electrode potential when the solution is changed from 5.0 mmol L^{-1} to 25.0 mmol L^{-1}.

9.9 Devise a cell in which the cell reaction is $Mn(s) + Cl_2(g) \rightarrow MnCl_2(aq)$. Give the half-reactions for the electrodes and from the standard cell potential of +2.54 V deduce the standard potential of the Mn^{2+}/Mn couple.

9.10 Write the cell reactions and electrode half-reactions for the following cells:

(a) $Ag(s)|AgNO_3(aq,b_L)||AgNO_3(aq,b_R)|Ag(s)$
(b) $Pt(s)|H_2(g,p_L)|HCl(aq)|H_2(g,p_R)|Pt(s)$
(c) $Pt(s)|K_3[Fe(CN)_6](aq),K_4[Fe(CN)_6](aq)||Mn^{2+}(aq),$
  $H^+(aq)|MnO_2(s)|Pt(s)$
(d) $Pt(s)|Cl_2(g)|HCl(aq)||HBr(aq)|Br_2(l)|Pt(s)$
(e) $Pt(s)|Fe^{3+}(aq),Fe^{2+}(aq)||Sn^{4+}(aq),Sn^{2+}(aq)|Pt(s)$
(f) $Fe(s)|Fe^{2+}(aq)||Mn^{2+}(aq),H^+(aq)|MnO_2(s)|Pt(s)$

9.11 Write the Nernst equations for the cells in the preceding exercise.

9.12 Devise cells in which the following are the reactions. In each case state the value for v to use in the Nernst equation.

(a) $Fe(s) + PbSO_4(s) \rightarrow FeSO_4(aq) + Pb(s)$
(b) $Hg_2Cl_2(s) + H_2(g) \rightarrow 2 HCl(aq) + 2 Hg(l)$
(c) $2 H_2(g) + O_2(g) \rightarrow 2 H_2O(l)$
(d) $H_2(g) + O_2(g) \rightarrow H_2O_2(aq)$
(e) $H_2(g) + I_2(g) \rightarrow 2 HI(aq)$
(f) $2 CuCl(aq) \rightarrow Cu(s) + CuCl_2(aq)$

3 Cysteine(Cys)   **4** Cystine

9.13 Devise cells to study the following biochemically important reactions. In each case state the value for ν to use in the Nernst equation.

(a) $CH_3CH_2OH(aq) + NAD^+(aq) \rightarrow$
$$CH_3CHO(aq) + NADH(aq) + H^+(aq)$$
(b) $ATP^{4-}(aq) + Mg^{2+}(aq) \rightarrow MgATP^{2-}(aq)$
(c) $2\ Cyt\text{-}c(red, aq) + CH_3COCO_2^-(aq) + 2\ H^+(aq) \rightarrow$
$$2\ Cyt\text{-}c(ox, aq) + CH_3CH(OH)CO_2^-(aq)$$

9.14 Use the standard potentials of the electrodes to calculate the standard potentials of the cells in Exercise 9.10.

9.15 Use the standard potentials of the electrodes to calculate the standard potentials of the cells devised in Exercise 9.12.

9.16 Express the standard potentials of the cells devised in Exercise 9.13 in terms of the $E^\ominus$ for the electrodes.

9.17 State what you would expect to happen to the cell potential when the following changes are made to the corresponding cells in Exercise 9.10. Confirm your prediction by using the Nernst equation in each case.

(a) The molar concentration of silver nitrate in the left-hand compartment is increased.
(b) The pressure of hydrogen in the left-hand compartment is increased.
(c) The pH of the right-hand compartment is decreased.
(d) The concentration of HCl is increased.
(e) Some iron(III) chloride is added to both compartments.
(f) Acid is added to both compartments.

9.18 State what you would expect to happen to the cell potential when the following changes are made to the corresponding cells devised in Exercise 9.12. Confirm your prediction by using the Nernst equation in each case.

(a) The molar concentration of $FeSO_4$ is increased.
(b) Some nitric acid is added to both cell compartments.
(c) The pressure of oxygen is increased.
(d) The pressure of hydrogen is increased.
(e) Some (i) hydrochloric acid, (ii) hydroiodic acid is added to both compartments.
(f) Hydrochloric acid is added to both compartments.

9.19 State what you would expect to happen to the cell potential when the following changes are made to the corresponding cells devised in Exercise 9.13. Confirm your prediction by using the Nernst equation in each case.

(a) The pH of the solution is raised.
(b) A solution of Epsom salts (magnesium sulfate) is added.
(c) Sodium lactate is added to the solution.

9.20 (a) Calculate the standard potential of the cell $Hg(l)|HgCl_2(aq)||TlNO_3(aq)|Tl(s)$ at 25°C. (b) Calculate the cell potential when the molar concentration of the Hg^{2+} ion is 0.150 mol L^{-1} and that of the Tl^+ ion is 0.93 mol L^{-1}.

9.21 Calculate the standard Gibbs energies at 25°C of the following reactions from the standard potential data in Appendix 2.

(a) $Ca(s) + 2\ H_2O(l) \rightarrow Ca(OH)_2(aq) + H_2(g)$
(b) $2\ Ca(s) + 4\ H_2O(l) \rightarrow 2\ Ca(OH)_2(aq) + 2\ H_2(g)$
(c) $Fe(s) + 2\ H_2O(l) \rightarrow Fe(OH)_2(aq) + H_2(g)$
(d) $Na_2S_2O_8(aq) + 2\ NaI(aq) \rightarrow I_2(s) + 2\ Na_2SO_4(aq)$
(e) $Na_2S_2O_8(aq) + 2\ KI(aq) \rightarrow I_2(s) + Na_2SO_4(aq) + K_2SO_4(aq)$
(f) $Pb(s) + Na_2CO_3(aq) \rightarrow PbCO_3(aq) + 2\ Na(s)$

9.22 Calculate the biological standard Gibbs energies of the following reactions and half-reactions:

(a) $2\ NADH(aq) + O_2(g) + 2\ H^+(aq) \rightarrow 2\ NAD^+(aq) +$
$$2\ H_2O(l) \quad E^\oplus = +1.14\ V$$
(b) $Malate(aq) + NAD^+(aq) \rightarrow oxaloacetate(aq) +$
$$NADH(aq) + H^+(aq) \quad E^\oplus = -0.154\ V$$
(c) $O_2(g) + 4\ H^+(aq) + 4\ e^- \rightarrow 2\ H_2O(l) \quad E^\oplus = +0.82\ V$

9.23 Tabulated thermodynamic data can be used to predict the standard potential of a cell even if it cannot be measured directly. The standard Gibbs energy of the reaction $K_2CrO_4(aq) + 2\ Ag(s) + 2\ FeCl_3(aq) \rightarrow Ag_2CrO_4(s) + 2\ FeCl_2(aq) + 2\ KCl(aq)$ is -62.5 kJ mol^{-1} at 298 K. (a) Calculate the standard potential of the corresponding galvanic cell and (b) the standard potential of the $Ag_2CrO_4/Ag,CrO_4^{2-}$ couple.

9.24 Estimate the potential of the cell

$$Ag(s)|AgCl(s)|KCl(aq,\ 0.025\ mol\ kg^{-1})||AgNO_3(aq,$$
$$0.010\ mol\ kg^{-1})|Ag(s)$$

at 25°C.

9.25 Use the information in Appendix 2 to calculate the standard potential of the cell $Ag(s)|AgNO_3\ (aq)||Cu(NO_3)_2(aq)|Cu(s)$ and the standard Gibbs energy and enthalpy of the cell reaction at 25°C. Estimate the value of $\Delta_rG^\ominus$ at 35°C.

9.26 Calculate the standard potential of the cell Pt(s)|cystine(aq),cysteine(aq)||H$^+$(aq)|O$_2$(g)|Pt(s) and the standard Gibbs energy of the cell reaction at 25°C. Use $E^\ominus = -0.34$ V for the cysteine/cystine couple.

9.27 Calculate the biological standard values of the potentials (the two electrode potentials and the cell potential) for the system in Exercise 9.26 at 310 K.

9.28 The biological standard potential of the redox couple pyruvic acid/lactic acid is -0.19 V and that of fumaric acid/succinic acid is $+0.03$ V at 25°C. What is the equilibrium constant for the reaction P + S $\rightleftharpoons$ L + F in pH = 7 (where P = pyruvic acid, S = succinic acid, L = lactic acid, and F = fumaric acid)?

9.29 The biological standard potential of the couple pyruvic acid/lactic acid is -0.19 V at 310 K. What is the thermodynamic standard potential of the couple? Pyruvic acid is CH$_3$COCOOH and lactic acid is CH$_3$CH(OH)COOH.

9.30 One ecologically important equilibrium is that between carbonate and hydrogencarbonate (bicarbonate) ions in natural water. (a) The standard Gibbs energies of formation of CO$_3^{2-}$(aq) and HCO$_3^-$(aq) are -527.81 kJ mol^{-1} and -586.77 kJ mol^{-1}, respectively. What is the standard potential of the HCO$_3^-$/CO$_3^{2-}$,H$_2$ couple? (b) Calculate the standard potential of a cell in which the cell reaction is Na$_2$CO$_3$(aq) + H$_2$O(l) → NaHCO$_3$(aq) + NaOH(aq). (c) Write the Nernst equation for the cell, and (d) predict and calculate the change in potential when the pH is changed to 7.0. (e) Calculate the value of pK_a for HCO$_3^-$(aq).

9.31 Calcium phosphate is the principal inorganic component of bone. Its solubility characteristics are important for the stabilities of skeletons, and they can be assessed electrochemcically. The solubility constant of Ca$_3$(PO$_4$)$_2$ is 1.3×10^{-37}. Calculate (a) the solubility of Ca$_3$(PO$_4$)$_2$, (b) the potential of the cell Pt(s)|H$_2$(g)| HCl(aq, pH = 0)|| Ca$_3$(PO$_4$)$_2$(aq, satd.)|Ca(s) at 25°C.

9.32 Calculate the equilibrium constants of the following reactions at 25°C from standard potential data:

(a) Sn(s) + Sn^{4+}(aq) $\rightleftharpoons$ 2 Sn^{2+}(aq)
(b) Sn(s) + 2 AgBr(s) $\rightleftharpoons$ SnBr$_2$(aq) + 2 Ag(s)
(c) Fe(s) + Hg(NO$_3$)$_2$(aq) $\rightleftharpoons$ Hg(l) + Fe(NO$_3$)$_2$(aq)
(d) Cd(s) + CuSO$_4$(aq) $\rightleftharpoons$ Cu(s) + CdSO$_4$(aq)

(e) Cu^{2+}(aq) + Cu(s) $\rightleftharpoons$ 2 Cu$^+$(aq)
(f) 3 Au^{2+}(aq) $\rightleftharpoons$ Au(s) + 2 Au^{3+}(aq)

9.33 The molar solubilities of AgCl and BaSO$_4$ in water are 1.34×10^{-5} mol L^{-1} and 9.51×10^{-4} mol L^{-1}, respectively, at 25°C. Calculate the $E^\ominus$ of the insoluble-salt electrode from the solubility data and $E^\ominus$ for the metal electrode.

9.34 The dichromate ion in acidic solution is a common oxidizing agent for organic compounds. Derive an expression for the potential of an electrode for which the half-reaction is the reduction of Cr$_2$O$_7^{2-}$ ions to Cr^{3+} ions in acidic solution.

9.35 The zero-current potential of the cell Pt(s)|H$_2$(g)|HCl(aq)|AgCl(s)|Ag(s) is 0.312 V at 25°C. What is the pH of the electrolyte solution?

9.36 The molar solubility of AgBr is 2.6 μmol L^{-1} at 25°C. What is the zero-current potential of the cell Ag(s)|AgBr(aq)|AgBr(s)|Ag(s) at that temperature?

9.37 The standard potential of the cell Ag(s)|AgI(s)|AgI(aq)|Ag(s) is $+0.9509$ V at 25°C. Calculate (a) the molar solubility of AgI and (b) its solubility constant.

9.38 Devise a cell in which the overall reaction is Bi(s) + Hg$_2$SO$_4$(s) → BiSO$_4$(s) + 2 Hg(l). What is its potential when the electrolyte is saturated with both salts at 25°C?

9.39 A fuel cell develops an electric potential from the chemical reaction between reagents supplied from an outside source. What is the zero-current potential of a cell fuelled by (a) hydrogen and oxygen, (b) the complete oxidation of benzene at 1.0 bar and 298 K?

9.40 A fuel cell is constructed in which both electrodes make use of the oxidation of methane. The left-hand electrode makes use of the complete oxidation of methane to carbon dioxide and water; the right-hand electrode makes use of the partial oxidation of methane to carbon monoxide and water. (a) Which electrode is the cathode? (b) What is the zero-current cell potential at 25°C when all gases are at 1 bar?

9.41 The permanganate ion is a common oxidizing agent. What is the standard potential of the MnO$_4^-$, H$^+$/ Mn^{2+} couple at (a) pH = 6.00, (b) general pH?

The rates of reactions

Contents

THE branch of physical chemistry called **chemical kinetics** is concerned with the rates of chemical reactions. Chemical kinetics deals with how rapidly reactants are consumed and products formed, how reaction rates respond to changes in the conditions or the presence of a catalyst, and the identification of the steps by which a reaction takes place.

One reason for studying the rates of reactions is the practical importance of being able to predict how quickly a reaction mixture approaches equilibrium. The rate might depend on variables under our control, such as the pressure, the temperature, and the presence of a catalyst, and we might be able to optimize it by the appropriate choice of conditions. Another reason is that the study of reaction rates leads to an understanding of the **mechanism** of a reaction, its analysis into a sequence of elementary steps. For example, we might discover that the reaction of hydrogen and bromine to form hydrogen bromide proceeds by the dissociation of a Br_2 molecule, the attack of a Br atom on an H_2 molecule, and several subsequent steps. By analysing the rate of a biochemical reaction we may discover how an enzyme acts. **Enzyme kinetics**, the study of the effect of enzymes on the rates of reactions, is also an important window on how these proteins work.

When studying biologically important processes, we need to cope with a wide variety of different rates and a process that appears to be slow may be the outcome of many faster steps. Photobiological processes like those responsible for photosynthesis and the slow growth of a plant may take place in about 1 ps. The binding of a neurotransmitter can have an effect after about 1 μs. Once a gene has been activated, a protein may emerge in about 100 s; but even that time-scale incorporates many others, including the wriggling of a newly formed polypeptide chain into its working conformation, each step of which may take about 1 ps. On a grander view, some of the equations of chemical kinetics are applicable to the behaviour of whole populations of organisms; such societies change on time-scales of 10^7–10^9 s.

Table 10.1 *Kinetic techniques for fast reactions*

Technique	Range of time-scales/s
Femtochemistry	$>10^{-15}$
Flash photolysis	$>10^{-12}$
Fluorescence decay	10^{-10}–10^{-6}
Ultrasonic absorption	10^{-10}–10^{-4}
EPR*	10^{-9}–10^{-4}
Electric field jump	10^{-7}–1
Temperature jump	10^{-6}–1
Phosphorescence	10^{-6}–10
NMR*	10^{-5}–1
Pressure jump	$>10^{-5}$
Stopped flow	$>10^{-3}$

*EPR is electron paramagnetic resonance (or electron spin resonance); NMR is nuclear magnetic resonance; see Chapter 19.

Empirical chemical kinetics

THE first stage in the investigation of the rate and mechanism of a reaction is the determination of the overall stoichiometry of the reaction and the identification of any side reactions. The next step is to determine how the concentrations of the reactants and products change with time after the reaction has been initiated. Because the rates of chemical reactions are sensitive to temperature, the temperature of the reaction mixture must be held constant throughout the course of the reaction, for otherwise the observed rate would be a meaningless average of the rates for different temperatures.

10.1 Experimental techniques

The method used to monitor the concentrations of reactants and products and their variation with time depends on the substances involved and the rapidity with which their concentrations change (Table 10.1).

Spectrophotometry, the measurement of the intensity of absorption in a particular spectral region, is widely used to monitor concentration. It is especially useful when one substance in the reaction mixture has a strong characteristic absorption in a conveniently accessible region of the spectrum. If a reaction changes the number or type of ions present in a solution, then concentrations may be followed by monitoring the conductivity of the solution. Reactions that change the concentration of hydrogen ions may be studied by monitoring the pH of the solution with a glass electrode. Other methods of monitoring the composition include the detection of fluorescence and phosphorescence, titration, mass spectrometry, gas chromatography, and magnetic resonance (both EPR and NMR, Chapter 19). Polarimetry, the observation of the optical activity of a reaction mixture, is occasionally applicable.

10.2 Application of the techniques

In a **real-time analysis**, the composition of a system is analysed while the reaction is in progress by direct spectroscopic observation of the reaction mixture. In the **quenching method**, the reaction is stopped after it has been allowed to proceed for a certain time and the composition is analysed at leisure. The quenching (of the entire mixture or of a sample drawn from it) can be achieved either by cooling suddenly, by adding the mixture to a large volume of solvent, or by rapid neutralization of an acid reagent. This method is suitable only for reactions that are slow enough for there to be little reaction during the time it takes to quench the mixture.

In the **flow method**, the reactants are mixed as they flow together in a chamber (Fig 10.1). The reaction continues as the thoroughly mixed solutions flow through a capillary outlet tube at about 10 m s^{-1}, and different points along the tube correspond to different times after the start of the reaction. Spectrophotometric determination of the composition at different positions along the tube is equivalent to the determination of the composition of the reaction mixture at different times after mixing. This technique was originally developed in connection with the study of the rate at which oxygen combined with haemoglobin. Its disadvantage is that a large volume of reactant solution is necessary, because the mixture must flow continuously through the apparatus. This disadvantage is particularly important for reactions that take place very rapidly, because the flow must be rapid if it is to spread the reaction over an appreciable length of tube.

The **stopped-flow technique** avoids this disadvantage (Fig 10.2). The two solutions are mixed very rapidly (in less than 1 ms) by injecting them into a mixing chamber designed to ensure that the flow is turbulent and that complete mixing occurs very quickly. Behind the reaction chamber there is an observation cell fitted with a plunger that moves back as the liquids flood in, but which comes up against a stop after a certain volume has been admitted. The filling of that chamber corresponds to

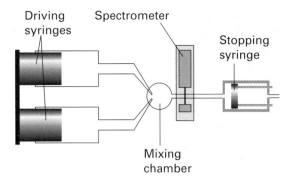

Fig 10.2 In the stopped-flow technique the reagents are driven quickly into the mixing chamber and then the time-dependence of the concentrations is monitored.

the sudden creation of an initial sample of the reaction mixture. The reaction then continues in the thoroughly mixed solution and is monitored spectrophotometrically. Because only a small, single charge of the reaction chamber is prepared, the technique is much more economical than the flow method. The suitability of the stopped-flow technique to the study of small samples means that it is appropriate for biochemical reactions, and it has been widely used to study the kinetics of enzyme action. Modern techniques of monitoring composition spectrophotometrically can span repetitively a wavelength range of about 200 nm at 1 ms intervals.

In **flash photolysis**, the gaseous or liquid sample is exposed to a brief photolytic or photoactivating flash of light or ultraviolet radiation, and then the contents of the reaction chamber are monitored spectrophotometrically. Lasers can be used to generate nanosecond flashes routinely, picosecond flashes quite readily, and flashes as brief as a few femtoseconds in special arrangements (Box 10.1). Either emission or absorption spectroscopy can be used to monitor the reaction, and the spectra are recorded electronically at a series of times following the flash.

In a **relaxation technique** the reaction mixture is initially at equilibrium but is then disturbed by a rapid change in conditions, such as a sudden increase in temperature in a **temperature-jump experiment** or pressure in a **pressure-jump experiment**. The equilibrium composition before the application of the perturbation becomes the initial state for the return of the system to its equilibrium

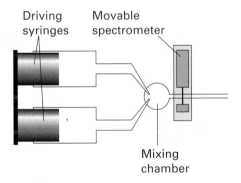

Fig 10.1 The arrangement used in the flow technique for studying reaction rates. The reactants are squirted into the mixing chamber at a steady rate from the syringes or by using peristaltic pumps (pumps that squeeze the fluid through flexible tubes, like in our intestines). The location of the spectrometer corresponds to different times after initiation.

Box 10.1 *Ultrafast reactions: femtochemistry*

Developments in laser technology have provided ultraviolet flashes on a femtosecond time-scale (1 fs = 10^{-15} s), and have opened up unsurpassed opportunities for studying chemical reactions in progress. In a typical pulse–probe experiment, a femtosecond pulse is used to excite a molecule to a dissociative state, and then a second femtosecond probe pulse is fired at an interval after the dissociating pulse. The frequency of the probe is set at an absorption of one of the free fragmentation products, so its absorption is a measure of the abundance of the dissociation product. For example, when ICN is dissociated by the first pulse, the emergence of CN from the photoexcited state can be monitored by watching the growth of the free CN absorption (or, more commonly, its laser-induced fluorescence). In this way it has been found that the CN signal remains zero until the fragments have separated by about 600 pm, which takes about 205 fs.

Some sense of the progress that has been made in the study of the intimate mechanism of chemical reactions can be obtained by considering the decay of the ion pair Na^+X^-, where X is a halogen. This reaction has been studied by exciting the ion pair with a femtosecond pulse to an excited state that corresponds to a covalently bonded NaX molecule. The probe pulse examines the system at an absorption frequency either of the free Na atom or at a frequency at which the atom absorbs when it is a part of the complex. The latter frequency depends on the Na–X distance, so an absorption (in practice, a laser-induced fluorescence) is obtained each time the vibration of the complex returns it to that separation.

The illustration shows a typical set of results for NaI. The bound Na absorption intensity shows up as a series of pulses that recur in about 1 ps, showing that the complex vibrates with about that period. The decline in intensity shows the rate at which the complex can dissociate as the two atoms swing away from each other. The complex does not dissociate on every outward-going swing because there is a chance that the I atom can be captured again, in which case it fails to make good its escape. The free Na absorption also grows in an oscillating manner, showing the periodicity of the vibration of the complex, each swing of which gives it a chance to dissociate. The precise period of the oscillation in NaI is 1.25 ps, and the complex survives for about ten oscillations. In contrast, although the oscillation frequency of NaBr is similar, it barely survives one oscillation.

Biological processes that are open to study include the energy-converting processes of photosynthesis and the photostimulated processes of vision, in which the pri-

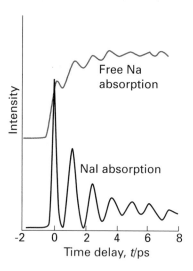

The absorption spectra of the species NaI and Na immediately after a femtosecond flash. The oscillations show how the species incipiently form and then reform their precursors before finally forming products. (Adapted from A.H. Zewail, *Science* **242**, 1645 (1988).)

mary energy and electron transfer reactions occur on the femtosecond and picosecond time-scales. In other experiments, the photoejection of carbon monoxide from myoglobin and the attachment of oxygen to the exposed site have been studied to obtain the rate constants for the two processes.

Exercise 1 Laser flash photolysis is often used to measure the binding rate of CO to haem proteins because CO photodissociates from the bound state relatively easily. The reaction is usually run under pseudofirst-order conditions (see Section 10.6). For a reaction in which $[Mb]_0 = 10$ mmol L^{-1}, $[CO] = 400$ mmol L^{-1}, and the rate constant is 5.8×10^5 mol^{-1} s^{-1}, plot a curve of [Mb] against time. The observed reaction is Mb + CO → MbCO.

Exercise 2 A protein dimerizes according to the reaction 2 A → A_2 (forward k, reverse k'). The usual expression for the relaxation time, τ, is

$$\tau = \frac{1}{k' + 4\,k[A]}$$

where [A] is the concentration of monomer at equilibrium. Derive an alternative form of the equation in terms of $[A]_{tot}$, the total concentration of protein.

composition at the new temperature or pressure, and the return to equilibrium—the 'relaxation' of the system—is monitored spectroscopically.

Reaction rates

THE raw data from experiments to measure reaction rates are the concentrations of reactants and products at a series of times after the reaction is initiated. Ideally, information on any intermediates should also be obtained, but often they cannot be studied because their existence is so fleeting or their concentration so low. More information about the reaction can be extracted if data are obtained at a series of different temperatures. The next few sections look at these observations in more detail.

10.3 The definition of rate

The rate of a reaction is defined in terms of the rate of change of the concentration of a designated species. However, because the rates at which reactants are consumed and products are formed change in the course of a reaction, it is necessary to consider the **instantaneous rate** of the reaction, its rate at a specific instant. The instantaneous rate of consumption of a reactant is the slope of a graph of its molar concentration plotted against the time, with the slope evaluated at the instant of interest (Fig 10.3).[1] The steeper the slope, the greater the rate of consumption of the reactant. Similarly, the rate of formation of a product is the slope of the graph of its concentration plotted against time. With the concentration measured in moles per litre and the time in seconds, the reaction rate is reported in moles per litre per second ($mol\ L^{-1}\ s^{-1}$).

In general, the various reactants in a given reaction are consumed at different rates, and the various products are also formed at different rates. For

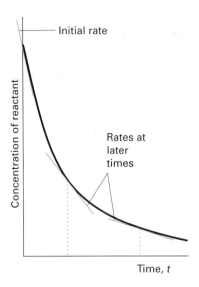

Fig 10.3 The rate of a chemical reaction is the slope of the tangent to the curve showing the variation of concentration of a species with time. This graph is a plot of the concentration of a reactant, which is consumed as the reaction progresses. The rate of consumption decreases in the course of the reaction as the concentration of reactant decreases.

example, in the decomposition of urea, $(NH_2)_2CO$, in acidic solution

$$(NH_2)_2CO(aq) + 2\ H_2O(l) \rightarrow 2\ NH_4^+(aq) + CO_3^{2-}(aq)$$

provided any intermediates are not present in significant quantities the rate of formation of NH_4^+ is twice the rate of disappearance of $(NH_2)_2CO$, because, for 1 mol $(NH_2)_2CO$ consumed, 2 mol NH_4^+ is formed. Once we know the rate of formation or consumption of one substance, we can use the reaction stoichiometry to deduce the rates of formation or consumption of the other participants in the reaction.

Self-test 10.1

The rate of formation of NH_3 in the reaction $N_2(g) + 3\ H_2(g) \rightarrow 2\ NH_3(g)$ was reported as $1.2\ mmol\ L^{-1}\ s^{-1}$ under a certain set of conditions. What is the rate of consumption of H_2?

[*Answer*: $1.8\ mmol\ L^{-1}\ s^{-1}$]

---

[1] All rates of reactions are reported as positive quantities.

10.4 Rate laws and rate constants

An empirical observation of the greatest importance is that *the rate of reaction is often found to be proportional to the molar concentrations of the reactants raised to a simple power*. For example, it may be found that the rate is directly proportional to the concentrations of the reactants A and B, so

Rate of reaction = $k[A][B]$

The coefficient k, which is characteristic of the reaction being studied, is called the **rate constant**. The rate constant is independent of the concentrations of the species taking part in the reaction but depends on the temperature. An *experimentally determined* equation of this kind is called the 'rate law' of the reaction. More formally, a **rate law** is an equation that expresses the rate of reaction in terms of the molar concentrations of the species in the overall reaction (including, possibly, the products).

The units of k are always such as to convert the product of concentrations into a rate expressed as a change in concentration divided by time. For example, if the rate law is the one shown above, with concentrations expressed in moles per litre (mol L^{-1}), then the units of k will be litres per mole per second (L mol^{-1} s^{-1}) because

$$\overbrace{\text{L mol}^{-1}\text{ s}^{-1}}^{k} \times \overbrace{\text{mol L}^{-1}}^{[A]} \times \overbrace{\text{mol L}^{-1}}^{[B]} = \overbrace{\text{mol L}^{-1}\text{ s}^{-1}}^{rate}$$

Once we know the rate law and the rate constant of the reaction, we can predict the rate of the reaction for any given composition of the reaction mixture. We shall also see that we can use a rate law to predict the concentrations of the reactants and products at any time after the start of the reaction. A rate law is also an important guide to the mechanism of the reaction, for any proposed mechanism must be consistent with the observed rate law.

10.5 Reaction order

A rate law provides a basis for the classification of reactions according to their kinetics. The advantage of having such a classification is that reactions belonging to the same class have similar kinetic behaviour—their rates and the concentrations of the reactants and products vary with composition in a similar way. The classification of reactions is based on their **order**, the power to which the concentration of a species is raised in the rate law. For example, a reaction with the rate law

Rate = $k[A][B]$ (10.1)

is *first-order* in A and first-order in B. A reaction with the rate law

Rate = $k[A]^2$ (10.2)

is *second-order* in A.

The **overall order** of a reaction is the sum of the orders of all the components. The two rate laws just quoted both correspond to reactions that are *second-order* overall. An example of the first type of reaction is the reformation of a DNA double helix after the double helix has been separated into two strands by raising the temperature or the pH:

Strand + complementary strand → double helix
Rate = k[strand][complementary strand]

This reaction is first-order in each strand and second-order overall. An example of the second type is the reduction of nitrogen dioxide by carbon monoxide,

$NO_2(g) + CO(g) \rightarrow NO(g) + CO_2(g)$ Rate = $k[NO_2]^2$

which is second-order in NO_2 and, because no other species occurs in the rate law, second-order overall. The rate of the latter reaction is independent of the concentration of CO provided that some CO is present. This independence of concentration is expressed by saying that the reaction is *zero-order* in CO, because a concentration raised to the power zero is 1 ($[CO]^0 = 1$, just as $x^0 = 1$ in algebra).

A reaction need not have an integral order, and many gas-phase reactions do not. For example, if a reaction is found to have the rate law

Rate = $k[A]^{1/2}[B]$ (10.3)

then it is *half-order* in A, first-order in B, and three-halves order overall. If a rate law is not of the form $[A]^x[B]^y[C]^z\ldots$ then the reaction does not have an overall order. Thus, the experimentally determined rate law for the gas-phase reaction $H_2(g) + Br_2(g) \rightarrow 2\, HBr(g)$ is

$$\text{Rate of formation of HBr} = \frac{k[H_2][Br]^{3/2}}{[Br_2] + k'[HBr]} \quad (10.4)$$

Although the reaction is first-order in H_2, it has an indefinite order with respect to both Br_2 and HBr and an indefinite order overall. Similarly, a typical rate law for the action of an enzyme E on a substrate S is

$$\text{Rate of formation of product} = \frac{k[E][S]}{[S] + K_M} \quad (10.5)$$

where K_M is a constant, the 'Michaelis constant' (see Section 11.9). This rate law is first-order in the enzyme but does not have a specific order with respect to the substrate.

Under certain circumstances a complicated rate law without an overall order may simplify into a law with a definite order. For example, if the substrate concentration in the enzyme-catalysed reaction is so low that $[S] \ll K_M$, then eqn 10.5 simplifies to

$$\text{Rate of formation of product} = \frac{k}{K_M}[E][S]$$

which is first-order in S, first-order in E, and second-order overall.

It is very important to note that *a rate law is established experimentally, and cannot in general be inferred from the chemical equation for the reaction.* The reaction of hydrogen and bromine, for example, has a very simple stoichiometry, but its rate law (eqn 10.4) is very complicated. In some cases, however, the rate law does happen to reflect the reaction stoichiometry. This is the case with the renaturation of DNA mentioned earlier and the oxidation of nitrogen oxide, NO, which under certain conditions is found to have a third-order rate law:

$$2\, NO(g) + O_2(g) \rightarrow 2\, NO_2(g)$$
$$\text{Rate of formation of } NO_2 = k[NO]^2[O_2]$$

10.6 The determination of the rate law

The determination of a rate law is simplified by the **isolation method**, in which all the reactants except one are present in large excess. We can find the dependence of the rate on each of the reactants by isolating each of them in turn—by having all the other substances present in large excess—and piecing together a picture of the overall rate law.

If a reactant B is in large excess, for example, it is a good approximation to take its concentration as constant throughout the reaction. Then, although the true rate law might be

$$\text{Rate} = k[A][B]^2$$

we can approximate $[B]$ by its initial value $[B]_0$ (which hardly changes in the course of the reaction) and write

$$\text{Rate} = k'[A], \text{ with } k' = k[B]_0^2$$

Because the true rate law has been forced into first-order form by assuming a constant B concentration, the effective rate law is classified as **pseudofirst-order** and k' is called the **effective rate constant** for a given, fixed concentration of B. If, instead, the concentration of A were in large excess, and hence effectively constant, the rate law would simplify to

$$\text{Rate} = k''[B]^2, \text{ with } k'' = k[A]_0$$

This **pseudosecond-order rate law** is also much easier to analyse and identify than the complete law.

In a similar manner, a reaction may even appear to be zeroth-order. For instance, the oxidation of ethanol to acetaldehyde by NAD^+ in the liver in the presence of the enzyme liver alcohol dehydrogenase

$$CH_3CH_2OH(aq) + NAD^+(aq) + H_2O(l)$$
$$\rightarrow CH_3CHO(aq) + NADH(aq) + H_3O^+(aq)$$

is zeroth-order overall as the ethanol is in excess and the concentration of the NAD^+ is maintained at a constant level by normal metabolic processes.

Many reactions in aqueous solution that are reported as first- or second-order are actually pseudo-first- or pseudosecond-order: the solvent water participates in the reaction but it is in such large excess that its concentration remains constant.

In the method of **initial rates**, which is often used in conjunction with the isolation method, the instantaneous rate is measured at the beginning of the reaction for several different initial concentrations of reactants. For example, suppose the rate law for a reaction with A isolated is

$$r = k'[A]^a$$

where r denotes rate. Then the initial rate of the reaction, r_0, is given by the initial concentration of A:

$$r_0 = k'[A]_0^a$$

Taking logarithms gives

$$\log r_0 = \log k' + a \log [A]_0 \qquad (10.6)$$

This equation has the form of the equation for a straight line:

$$y = \text{intercept} + \text{slope} \times x$$

with $y = \log r_0$ and $x = \log [A]_0$. It follows that, for a series of initial concentrations, a plot of the logarithms of the initial rates against the logarithms of the initial concentrations of A should be a straight line, and that the slope of the graph will be a, the order of the reaction with respect to the species A (Fig 10.4).

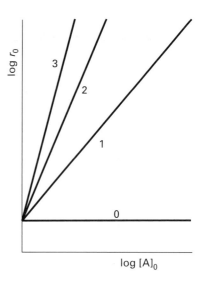

Fig 10.4 The plot of log r_0 against log $[A]_0$ gives straight lines with slopes equal to the order of the reaction.

$[I]_0/(10^{-5}\,\text{mol L}^{-1})$		1.0	2.0	4.0	6.0
$r_0/(\text{mol L}^{-1}\,\text{s}^{-1})$	(a)	1.070×10^{-3}	3.48×10^{-3}	1.39×10^{-2}	3.13×10^{-2}
	(b)	4.35×10^{-3}	1.74×10^{-2}	6.96×10^{-2}	1.57×10^{-1}
	(c)	1.069×10^{-2}	3.47×10^{-2}	1.38×10^{-1}	3.13×10^{-1}

The Ar concentrations are (a) 1.0×10^{-3} mol L^{-1}, (b) 5.0×10^{-3} mol L^{-1}, and (c) 1.0×10^{-2} mol L^{-1}. Find the orders of reaction with respect to I and Ar and the rate constant.

Strategy For constant $[Ar]_0$, the initial rate law has the form $r_0 = k'[I]_0^a$, with $k' = k[Ar]_0^b$, so

$$\log r_0 = \log k' + a \log [I]_0$$

We need to make a plot of log r_0 against log $[I]_0$ for a given $[Ar]_0$ and find the order a from the slope and the value of k' from the intercept at log $[I]_0 = 0$. Then, because

$$\log k' = \log k + b \log [Ar]_0$$

plot log k' against log $[Ar]_0$ to find log k from the intercept and b from the slope.

Example 10.1 *Using the method of initial rates*

The recombination of I atoms in the gas phase in the presence of argon (which removes the energy released by the formation of an I–I bond, and so prevents the immediate dissociation of a newly formed I_2 molecule) was investigated and the order of the reaction was determined by the method of initial rates. The initial rates of reaction of 2 I(g) + Ar(g) → I_2(g) + Ar(g) were as follows:

---

[2] Note that, when taking the logarithm of a number of the form $x.xx \times 10^n$, there are *four* significant figures in the answer: the figure before the decimal point is simply the power of 10. Strictly, the logarithms are of the quantity divided by its units.

Solution The data give the following points for the graph:[2]

log([I]$_0$/mol L^{-1})		−5.00	−4.70	−4.40	−4.22
log(r$_0$/mol L^{-1} s^{-1})	(a)	−2.971	−2.458	−1.857	−1.504
	(b)	−2.362	−1.760	−1.157	−0.804
	(c)	−1.971	−1.460	−0.860	−0.504

The graph of the data is shown in Fig 10.5. The slopes of the lines are 2 and the effective rate constants k' are as follows:

[Ar]$_0$/(mol L^{-1})	1.0 × 10^{-3}	5.0 × 10^{-3}	1.0 × 10^{-2}
log([Ar]$_0$/mol L^{-1})	−3.00	−2.30	−2.00
log(k'/mol^{-1} L s^{-1})	6.94	7.64	7.94

Figure 10.6 is the plot of log k' against log [Ar]$_0$. The slope is 1, so $b = 1$. The intercept at log [Ar]$_0$ = 0 is log k = 9.94, so $k = 8.7 \times 10^9$ mol^{-2} L^2 s^{-1}. The overall (initial) rate law is

$$r = k[I]_0^2[Ar]_0$$

Self-test 10.2

The initial rate of a certain reaction depended on concentration of a substance J as follows:

[J]$_0$/(10^{-3} mol L^{-1})	5.0	10.2	17	30
r$_0$/(10^{-7} mol L^{-1} s^{-1})	3.6	9.6	41	130

Find the order of the reaction with respect to J and the rate constant.

[*Answer*: 2, 1.4 × 10^{-2} mol^{-1} L s^{-1}]

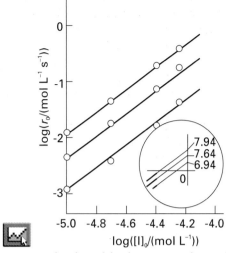

Fig 10.5 The plots of the data in Example 10.1 for finding the order with respect to I.

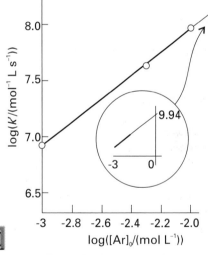

Fig 10.6 The plots of the data in Example 10.1 for finding the order with respect to Ar.

The method of initial rates might not reveal the entire rate law, for in a complex reaction the products themselves might affect the rate. That is the case for the synthesis of HBr, for eqn 10.4 shows that the rate law depends on the concentration of HBr, none of which is present initially.

10.7 Integrated rate laws

A rate law tells us the rate of the reaction at a given instant (when the reaction mixture has a particular composition). That is rather like being given the speed of a car at each point of its journey. For a car journey, we may want to know the distance that a car has travelled at a certain time given its varying speed. Similarly, for a chemical reaction, we may want to know the composition of the reaction mixture at a given time given the varying rate of the reaction. An **integrated rate law** is an expression that gives the concentration of a species as a function of the time.

Integrated rate laws have two principal uses. One is to predict the concentration of a species at any time after the start of the reaction. Another is to help find the rate constant and order of the reaction. Indeed, although we have introduced rate laws through a discussion of the determination of reaction rates, these rates are rarely measured directly because slopes are so difficult to determine accurately. Almost all experimental work in chemical kinetics deals with integrated rate laws, their great advantage being that they are expressed in terms of the experimental observables of concentration and time. Computers can be used to find numerical versions of the integrated form of even the most complex rate laws. However, in a number of simple cases analytical solutions are easily obtained, and prove to be very useful.

Derivation 10.1 *First-order integrated rate laws*

Our first step is to express the rate of consumption of a reactant A mathematically. In mathematics, the slope of a plot of [A] against t is the 'derivative' of [A] with respect to t and is denoted $d[A]/dt$. The rate of consumption of a reactant, a positive quantity, is defined as the negative of this slope, so we can interpret the rate as $-d[A]/dt$. It follows that a first-order rate equation has the form

$$-\frac{d[A]}{dt} = k[A]$$

This is an example of a 'differential equation'. To solve the equation, we rearrange it into

$$-\frac{d[A]}{[A]} = kdt$$

and then integrate both sides. Integration from $t = 0$, when the concentration of A is $[A]_0$, to the time of interest, t, when the molar concentration of A is [A], is written

$$\int_{[A]_0}^{[A]} \frac{d[A]}{[A]} = -k \int_0^t dt$$

Therefore,

$$\ln \frac{[A]_0}{[A]} = kt \qquad (10.7a)$$

To obtain this result, we have used the standard integral

$$\int \frac{dx}{x} = \ln x + \text{constant}$$

Two alternative forms of eqn 10.7a are

$$\ln [A] = \ln [A]_0 - kt \qquad (10.7b)$$
$$[A] = [A]_0 e^{-kt} \qquad (10.7c)$$

Equation 10.7c has the form of an **exponential decay** (Fig 10.7). A common feature of all first-order reactions, therefore, is that *the concentration of the reactant decays exponentially with time.*

Equation 10.7c lets us predict the concentration of A at any time after the start of the reaction. Equation 10.7b shows that, if we plot $\ln [A]$ against t, then we will get a straight line if the reaction is first-order. If the experimental data do not give a straight line when plotted in this way, the reaction is not first-order. If the line is straight, it follows from eqn 10.7b that its slope is $-k$, so we can also determine the rate constant from the graph. Some rate constants determined in this way are given in Table 10.2.

Example 10.2 *Analysing a first-order reaction*

The variation in the partial pressure p of azomethane with time was followed at 460 K, with the results given below. Confirm that the decomposition

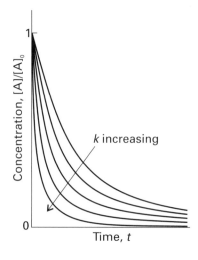

Fig 10.7 The exponential decay of the reactant in a first-order reaction. The greater the rate constant, the more rapid is the decay.

Table 10.2 *Kinetic data for first-order reactions*

Reaction	Phase	$\theta/°C$	k/s^{-1}	$t_{1/2}$
$2\,N_2O_5 \rightarrow 4\,NO_2 + \mathbf{O_2}$	g	25	3.38×10^{-5}	2.85 h
$2\,N_2O_5 \rightarrow 4\,NO_2 + \mathbf{O_2}$	$Br_2(l)$	25	4.27×10^{-5}	2.25 h
$\mathbf{C_2H_6} \rightarrow 2\,CH_3$	g	700	5.46×10^{-4}	21.2 min
Cyclopropane $\rightarrow$ propene	g	500	6.17×10^{-4}	17.2 min

The rate constant is for the rate of formation or consumption of the species in bold type. The rate laws for the other species may be obtained from the reaction stoichiometry.

$CH_3N_2CH_3(g) \rightarrow CH_3CH_3(g) + N_2(g)$ is first-order in $CH_3N_2CH_3$, and find the rate constant at this temperature.

t/s	0	1000	2000	3000	4000
$p/(10^{-2}\,\text{Torr})$	10.20	5.72	3.99	2.78	1.94

Strategy The easiest procedure is to plot $\ln(p/\text{Torr})$ against t/s and expect to obtain a straight line. If the graph is straight, then the slope is $-k$. Although graphs are good for judging the straightness of lines and assessing whether a point is unlikely to be valid because it is so far out of line, the analysis of the line—the determination of its slope and intercept—is best done using statistical software after using a graph to weed out rogue points.

Solution The graph of the data is shown in Fig 10.8. The plot is straight, confirming a first-order reaction. Its least-squares best-fit slope is -4.04×10^{-4}, so $k = 4.04 \times 10^{-4}\,s^{-1}$.

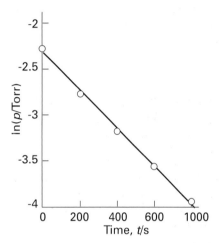

Fig 10.8 The determination of the rate constant of a first-order reaction. A straight line is obtained when $\ln[A]$ (or $\ln p$, where p is the partial pressure of the species of interest) is plotted against t; the slope gives the rate constant as $-k$. The data are from Example 10.2.

Now we need to see how the concentration varies with time for a second-order reaction with rate law

$$\text{Rate of consumption of A} = k[A]^2 \qquad (10.8)$$

As before, we suppose that the concentration of A at $t = 0$ is $[A]_0$.

Self-test 10.3

The concentration of N_2O_5 in liquid bromine varied with time as follows:

t/s	0	200	400	600	1000
$[N_2O_5]/(\text{mol L}^{-1})$	0.110	0.073	0.048	0.032	0.014

Confirm that the reaction is first-order in N_2O_5 and determine the rate constant.

[*Answer*: $2.1 \times 10^{-3}\,s^{-1}$]

Derivation 10.2 *Second-order integrated rate laws*

As before, the rate of consumption of the reactant A is $-d[A]/dt$, so the differential equation for the rate law is

$$-\frac{d[A]}{dt} = k[A]^2$$

To solve this equation, we rearrange it into

$$\frac{d[A]}{[A]^2} = -kdt$$

and integrate it between $t = 0$, when the concentration of A is $[A]_0$, and the time of interest t, when the concentration of A is $[A]$:

$$\int_{[A]_0}^{[A]} \frac{d[A]}{[A]^2} = -k \int_0^t dt$$

The term on the right is $-kt$. We evaluate the integral on the left by using the standard form

$$\int \frac{dx}{x^2} = -\frac{1}{x} + \text{constant}$$

The result is

$$\frac{1}{[A]_0} - \frac{1}{[A]} = kt \qquad (10.9a)$$

Two alternative forms of eqn 10.9a are

$$\frac{1}{[A]} = \frac{1}{[A]_0} + kt \qquad (10.9b)$$

$$[A] = \frac{[A]_0}{1 + kt[A]_0} \qquad (10.9c)$$

Equation 10.9b shows that to test for a second-order reaction we should plot $1/[A]$ against t and expect a straight line. If the line is straight, the reaction is second-order in A and the slope of the line is equal to the rate constant. Some rate constants

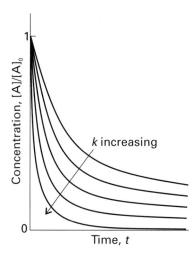

Fig 10.9 The variation with time of the concentration of a reactant in a second-order reaction.

determined in this way are given in Table 10.3. Equation 10.9c enables us to predict the concentration of A at any time after the start of the reaction (Fig 10.9).

A point to note about second-order reactions is that, when $[A]$ is plotted against t, the concentration of A approaches zero more slowly than in a first-order reaction with the same initial rate (Fig 10.10). That is, reactants that decay by a second-order process die away more slowly at low concentrations than would be expected if the decay were first-order. A point of interest in this connection is that

Table 10.3 *Kinetic data for second-order reactions*

Reaction	Phase	$\theta/°C$	$k/(\text{L mol}^{-1}\,\text{s}^{-1})$
$2\,NOBr \rightarrow 2\,NO + \mathbf{Br_2}$	g	10	0.80
$2\,NO_2 \rightarrow 2\,NO + \mathbf{O_2}$	g	300	0.54
$\mathbf{H_2} + I_2 \rightarrow 2\,HI$	g	400	2.42×10^{-2}
$\mathbf{D_2} + HCl \rightarrow DH + DCl$	g	600	0.141
$2\,I \rightarrow \mathbf{I_2}$	g	23	7×10^9
	hexane	50	1.8×10^{10}
$\mathbf{CH_3Cl} + CH_3O^-$	$CH_3OH(l)$	20	2.29×10^{-6}
$\mathbf{CH_3Br} + CH_3O^-$	$CH_3OH(l)$	20	9.23×10^{-6}
$\mathbf{H^+} + OH \rightarrow H_2O$	water	25	1.5×10^{11}

The rate constant is for the rate of formation or consumption of the species in bold type. The rate laws for the other species may be obtained from the reaction stoichiometry.

pollutants commonly disappear by second-order processes, so it takes a very long time for them to decline to acceptable levels.

Table 10.4 summarizes the integrated rate laws for a variety of simple reaction types.

10.8 Half-lives

A useful indication of the rate of a first-order chemical reaction is the **half-life**, $t_{1/2}$, of a reactant, which is the time it takes for the concentration of the species to fall to half its initial value. We can find the half-life of a species A that decays in a first-order reaction by substituting $[A] = \frac{1}{2}[A]_0$ and $t = t_{1/2}$ into eqn 10.7a:

$$kt_{1/2} = \ln \frac{[A]_0}{\frac{1}{2}[A]_0} = \ln 2$$

It follows that[3]

$$t_{1/2} = \frac{\ln 2}{k} \qquad (10.10)$$

For example, because the rate constant for the first-order reaction

$$2\ N_2O_5(g) \rightarrow 4\ NO_2(g) + O_2(g)$$
Rate of consumption of $N_2O_5 = k[N_2O_5]$

is equal to $6.76 \times 10^{-5}\ s^{-1}$ at 25°C, the half-life of N_2O_5 is 2.85 h. Hence, the concentration of N_2O_5 falls to half its initial value in 2.85 h, and then to half that

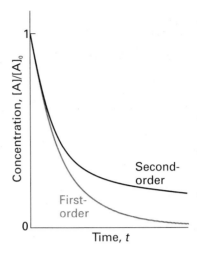

Fig 10.10 Although the initial decay of a second-order reaction may be rapid, later the concentration approaches zero more slowly than in a first-order reaction with the same initial rate (compare Fig 10.7).

concentration again in a further 2.85 h, and so on (Fig 10.11).

The main point to note about eqn 10.10 is that, *for a first-order reaction, the half-life of a reactant is independent of its initial concentration.* It follows that, if the concentration of A at some arbitrary stage of the reaction is $[A]$, then the concentration will fall to $\frac{1}{2}[A]$ after an interval of $0.693/k$ whatever the actual value of $[A]$ (Fig 10.12). Some half-lives are given in Table 10.2.

Table 10.4 *Integrated rate laws*

Order	Reaction type	Rate law	Integrated rate law
0	$A \rightarrow P$	$r = k$	$[P] = kt$ for $kt \leq [A]_0$
1	$A \rightarrow P$	$r = k[A]$	$[P] = [A]_0(1 - e^{-kt})$
2	$A \rightarrow P$	$r = k[A]^2$	$[P] = \dfrac{kt[A]_0^2}{1 + kt[A]_0}$
	$A + B \rightarrow P$	$r = k[A][B]$	$[P] = \dfrac{[A]_0[B]_0(1 - e^{([B]_0 - [A]_0)kt})}{[A]_0 - [B]_0 e^{([B]_0 - [A]_0)kt}}$

[3] $\ln 2 = 0.693\ldots$

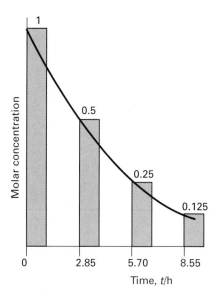

Fig 10.11 The molar concentration of N_2O_5 after a succession of half-lives.

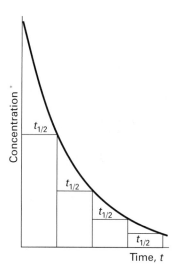

Fig 10.12 In each successive period of duration $t_{1/2}$, the concentration of a reactant in a first-order reaction decays to half its value at the start of that period. After n such periods, the concentration is $(\frac{1}{2})^n$ of its initial concentration.

Illustration 10.1

In acidic solution, the disaccharide sucrose (cane sugar) is converted to a mixture of the monosaccharides glucose and fructose in a pseudofirst-order reaction. Under certain conditions of pH, the half-life of sucrose is 28.4 min. To calculate how long it takes for the concentration of a sample to fall from 8.0 mmol L^{-1} to 1.0 mmol L^{-1} we note that

Molar concentration/(mmol L^{-1}):

$$8.0 \xrightarrow{\;28.4\ min\;} 4.0 \xrightarrow{\;28.4\ min\;} 2.0 \xrightarrow{\;28.4\ min\;} 1.0$$

The total time required is $3 \times 28.4\ min = 85.2\ min$.

Self-test 10.4

The half-life of a substrate in a certain enzyme-catalysed first-order reaction is 138 s. How long is required for the concentration of substrate to fall from 1.28 mmol L^{-1} to 0.040 mmol L^{-1}?

[*Answer:* 690 s]

In contrast to first-order reactions, the half-life of a second-order reaction does depend on the initial concentration of the reactant (see the answer to Self-test 10.5), and lengthens as the concentration of reactant falls. It is therefore not characteristic of the reaction itself, and for that reason is rarely used.

Self-test 10.5

Derive an expression for the half-life of a second-order reaction in terms of the rate constant k.

[*Answer:* $t_{1/2} = 1/k[A]_0$]

We can use the half-life of a substance to recognize first-order reactions. All we need do is inspect a set of data of composition against time. If we see that the initial concentration falls to half its value in a certain time, and that another concentration falls to half its value in the same time, then we can infer that the reaction is first-order. The first-order character can then be confirmed by plotting ln [A] against t and obtaining a straight line, as indicated earlier.

The temperature dependence of reaction rates

Tʜᴇ rates of most chemical reactions increase as the temperature is raised. Many organic reactions in solution fall somewhere in the range spanned by the hydrolysis of methyl ethanoate (for which the rate constant at 35°C is 1.8 times that at 25°C) and the hydrolysis of sucrose (for which the factor is 4.1). Reactions in the gas phase typically have rates that are only weakly sensitive to the temperature. Enzyme-catalysed reactions may show a more complex temperature dependence because raising the temperature may provoke conformational changes and even denaturation and degradation that lowers the effectiveness of the enzyme. Indeed, one of the reasons why we fight infection with a fever is to upset the balance of reaction rates in the infecting organism, and hence destroy it, by the increase in temperature. There is a fine line, though, between killing an invader and killing the invaded!

10.9 The Arrhenius parameters

As data on reaction rates were accumulated towards the end of the nineteenth century, the Swedish chemist Svante Arrhenius noted that almost all of them showed a similar dependence on the temperature. In particular, he noted that a graph of ln k, where k is the rate constant for the reaction, against $1/T$, where T is the (absolute) temperature at which k is measured, gives a straight line with a slope that is characteristic of the reaction (Fig 10.13). The mathematical expression of this conclusion is that the rate constant varies with temperature as

$$\ln\ k = \text{intercept} + \text{slope} \times \frac{1}{T}$$

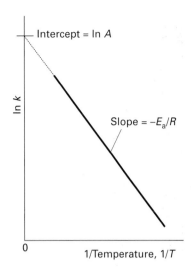

Fig 10.13 The general form of an Arrhenius plot of ln k against $1/T$. The slope is equal to $-E_a/R$ and the intercept at $1/T = 0$ is equal to ln A.

This expression is normally written as the **Arrhenius equation**

$$\ln\ k = \ln\ A - \frac{E_a}{RT} \tag{10.11}$$

or alternatively as

$$k = Ae^{-E_a/RT} \tag{10.12}$$

The parameter A (which has the same units as k) is called the **pre-exponential factor**, and E_a (which, like RT, is a molar energy and has the units of kilojoules per mole) is called the **activation energy**. Collectively, A and E_a are called the **Arrhenius parameters** of the reaction (Table 10.5).

A practical point to note from Fig 10.14 is that a high activation energy corresponds to a reaction rate that is very sensitive to temperature (the Arrhenius plot has a steep slope). Conversely, a small activation energy indicates a reaction rate that varies only slightly with temperature (the slope is shallow). A reaction with zero activation energy, such as for some radical recombination reactions in the gas phase, has a rate that is largely independent of temperature.

Example 10.3 *Determining the Arrhenius parameters*

The rate of the second-order decomposition of acetaldehyde (ethanal, CH_3CHO) was measured over the temperature range 700–1000 K, and the rate constants that were found are reported below. Find the activation energy and the pre-exponential factor.

T/K	700	730	760	790	810	840	910	1000
$k/(\text{L mol}^{-1}\text{s}^{-1})$	0.011	0.035	0.105	0.343	0.789	2.17	20.0	145

Strategy We plot $\ln k$ against $1/T$ and expect a straight line. The slope is $-E_a/R$ and the intercept of the extrapolation to $1/T = 0$ is $\ln A$. After sifting the data graphically, it is best to do a least-squares fit of the data to a straight line. Note that A has the same units as k.

Solution The Arrhenius plot is shown in Fig 10.15. The least-squares best fit of the line has slope -2.265×10^4 and intercept (which is well off the graph) 27.71. Therefore,

$$E_a = (2.265 \times 10^4 \text{ K}) \times (8.3145 \text{ J K}^{-1}\text{mol}^{-1}) = 188 \text{ kJ mol}^{-1}$$

and

$$A = e^{27.71} \text{ L mol}^{-1}\text{ s}^{-1} = 1.08 \times 10^{12} \text{ L mol}^{-1}\text{ s}^{-1}$$

Self-test 10.6

Determine A and E_a from the following data:

T/K	300	350	400	450	500
$k/(\text{L mol}^{-1}\text{ s}^{-1})$	7.9×10^6	3.0×10^7	7.9×10^7	1.7×10^8	3.2×10^8

[*Answer*: 8×10^{10} L mol^{-1} s^{-1}, 23 kJ mol^{-1}]

Once the activation energy of a reaction is known, it is a simple matter to predict the value of a rate constant k' at a temperature T' from its value k at another temperature T. To do so, we write

$$\ln k' = \ln A - \frac{E_a}{RT'}$$

and then subtract eqn 10.11, so obtaining

$$\ln k' - \ln k = -\frac{E_a}{RT'} + \frac{E_a}{RT}$$

We can rearrange this expression to

$$\ln \frac{k'}{k} = \frac{E_a}{R}\left(\frac{1}{T} - \frac{1}{T'}\right) \tag{10.13}$$

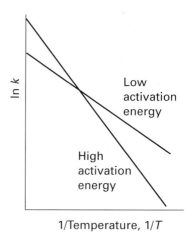

Fig 10.14 These two Arrhenius plots correspond to two different activation energies. Note that the plot corresponding to the higher activation energy is steeper, which indicates that the rate of that reaction is more sensitive to temperature.

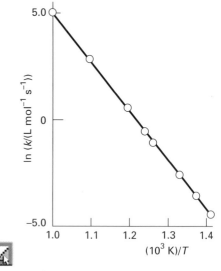

Fig 10.15 The Arrhenius plot for the decomposition of CH_3CHO, and the best (least squares) straight line fitted to the data points. The data are from Example 10.3.

Illustration 10.2

For a reaction with an activation energy of 50 kJ mol^{-1}, an increase in the temperature from 25 °C to 37 °C (body temperature) corresponds to

Table 10.5 *Arrhenius parameters*

Reaction	A/s^{-1}	$E_a/(kJ\ mol^{-1})$
First-order		
Cyclopropene → propane	1.58×10^{15}	272
$CH_3NC \rightarrow CH_3CN$	3.98×10^{13}	160
cis-CHD=CHD → *trans*-CHD=CHD	3.16×10^{12}	256
Cyclobutane → 2 C_2H_4	3.98×10^{15}	261
2 $N_2O_5 \rightarrow 4\ NO_2 + O_2$	4.94×10^{13}	103
$N_2O \rightarrow N_2 + O$	7.94×10^{11}	250
	$A/(L\ mol^{-1}\ s^{-1})$	$E_a/(kJ\ mol^{-1})$
Second-order, gas phase		
$O + N_2 \rightarrow NO + N$	1×10^{11}	315
$OH + H_2 \rightarrow H_2O + H$	8×10^{10}	42
$Cl + H_2 \rightarrow HCl + H$	8×10^{10}	23
$CH_3 + CH_3 \rightarrow C_2H_6$	2×10^{10}	0
$NO + Cl_2 \rightarrow NOCl + Cl$	4×10^{9}	85
Second-order, solution		
$NaC_2H_5O + CH_3I$ in ethanol	2.42×10^{11}	81.6
$C_2H_5Br + OH^-$ in water	4.30×10^{11}	89.5
$CH_3I + S_2O_3^{2-}$ in water	2.19×10^{12}	78.7
Sucrose + H_2O in acidic water	1.50×10^{15}	107.9

$$\ln \frac{k'}{k} = \frac{50 \times 10^3\ J\ mol^{-1}}{8.3145\ J\ K^{-1}\ mol^{-1}} \left(\frac{1}{298\ K} - \frac{1}{310\ K} \right)$$

$$= \frac{50 \times 10^3}{8.3145} \left(\frac{1}{298} - \frac{1}{310} \right)$$

(The right-hand side evaluates to 0.7812 . . ., but we take the next step before evaluating the answer.) By taking natural antilogarithms (that is, by forming e^x), $k' = 2.18k$. This result corresponds to slightly more than a doubling of the rate constant.

Self-test 10.7

The activation energy of one of the reactions in the Krebs citric acid cycle is 87 kJ mol^{-1}. What is the change in rate constant when the temperature falls from 37 °C to 15 °C?

[*Answer: k' = 0.076k*]

10.10 Collision theory

We can understand the origin of the Arrhenius parameters most simply by considering a class of gas-phase reactions in which reaction occurs when two molecules meet.[4] In this **collision theory** of reaction rates it is supposed that reaction occurs only if two molecules collide with a certain minimum kinetic energy along their line of approach (Fig 10.16). In collision theory, a reaction resembles the collision of two defective billiard balls: the balls bounce apart if they collide with only a small energy, but might smash each other into fragments (products) if they collide with more than a certain minimum kinetic energy. This model of a reaction is a reasonable first approximation to the types of process that take place in planetary atmospheres and govern their compositions and temperature profiles.

---

[4] In the terminology to be introduced in Section 11.3, we are considering bimolecular gas-phase reactions.

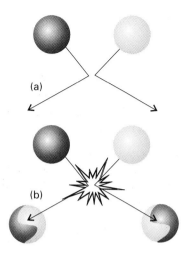

Fig 10.16 In the collision theory of gas-phase chemical reactions, reaction occurs when two molecules collide, but only if the collision is sufficiently vigorous. (a) An insufficiently vigorous collision: the reactant molecules collide but bounce apart unchanged. (b) A sufficiently vigorous collision results in a reaction.

A **reaction profile** in collision theory is a graph showing the variation in potential energy as one reactant molecule approaches another and the products then separate (Fig 10.17).[5] On the left, the horizontal line represents the potential energy of the two reactant molecules that are far apart from one another. The potential energy rises from this value only when the separation of the molecules is so small that they are in contact, when it rises as bonds bend and start to break. The potential energy reaches a peak when the two molecules are highly distorted. Then it starts to decrease as new bonds are formed. At separations to the right of the maximum, the potential energy rapidly falls to a low value as the product molecules separate. For the reaction to be successful, the reactant molecules must approach with sufficient kinetic energy along their line of approach to carry them over the **activation barrier**, the peak in the reaction profile. As we shall see, we can identify the height of the activation barrier with the activation energy of the reaction.

----

[5]   Remember that the potential energy of an object is the energy arising from its position (not speed), in this case the separation of the two reactant molecules as they approach, react, and then separate as products.

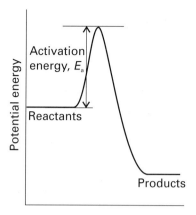

Fig 10.17 A reaction profile. The graph depicts schematically the changing potential energy of two species that approach, collide, and then go on to form products. The activation energy is the height of the barrier above the potential energy of the reactants.

With the reaction profile in mind, it is quite easy to establish that collision theory accounts for Arrhenius behaviour. Thus, the rate of collisions between species A and B is proportional to both their concentrations: if the concentration of B is doubled, then the rate at which A molecules collide with B molecules is doubled, and, if the concentration of A is doubled, then the rate at which B molecules collide with A molecules is also doubled. It follows that the rate of collision of A and B molecules is directly proportional to the concentrations of A and B, and we can write

Rate of collision $\propto$ [A][B]

Next, we need to multiply the collision rate by a factor f that represents the fraction of collisions that occur with at least a kinetic energy E_a along the line of approach (Fig 10.18), for only these collisions will lead to the formation of products. Molecules that approach with less than a kinetic energy E_a will behave like a ball that rolls toward the activation barrier, fails to surmount it, and rolls back. It follows from very general arguments concerning the probability that a molecule has a specified energy[6] that the fraction of collisions that occur with at least a kinetic energy E_a is

----

[6]   See the discussion of the Boltzmann distribution in Chapter 20.

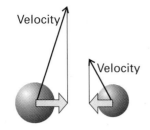

(a) Low kinetic energy of approach

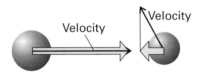

(b) High kinetic energy of approach

Fig 10.18 The criterion for a successful collision is that the two reactant species should collide with a kinetic energy along their line of approach that exceeds a certain minimum value E_a that is characteristic of the reaction. The two molecules might also have components of velocity (and an associated kinetic energy) in other directions (for example, the two molecules depicted here might be moving up the page as well as towards each other); but only the energy associated with their mutual approach can be used to overcome the activation energy.

$$f = e^{-E_a/RT} \tag{10.14}$$

Self-test 10.8

What is the fraction of collisions that have sufficient energy for reaction if the activation energy is 50 kJ mol^{-1} and the temperature is (a) 25°C, (b) 500°C?

[*Answer:* (a) 1.7×10^{-9}, (b) 4.2×10^{-4}]

At this stage we can conclude that the rate of reaction, which is proportional to the rate of collision multiplied by the fraction of successful collisions, is

$$\text{Reaction rate} \propto [\text{A}][\text{B}]e^{-E_a/RT}$$

If we compare this expression with a second-order rate law,

$$\text{Reaction rate} = k[\text{A}][\text{B}]$$

it follows that

$$k \propto e^{-E_a/RT}$$

This expression has exactly the Arrhenius form (eqn 10.12) if we identify the constant of proportionality with A. Collision theory therefore suggests the following interpretations:

The *pre-exponential factor*, A, is the constant of proportionality between the concentrations of the reactants and the rate at which the reactant molecules collide.

The *activation energy*, E_a, is the minimum kinetic energy required for a collision to result in reaction.

The value of A can be calculated from the kinetic theory of gases (Chapter 1):

$$A = \sigma \left(\frac{8 kT}{\pi \mu} \right)^{1/2} N_A^2 \qquad \mu = \frac{m_A m_B}{m_A + m_B} \tag{10.15}$$

where m_A and m_B are the masses of the molecules A and B and σ is the collision cross-section (Section 1.8). However, it is often found that the experimental value of A is smaller than that calculated from the kinetic theory. One possible explanation is that, not only must the molecules collide with sufficient kinetic energy, but they must also come together in a specific relative orientation (Fig 10.19). It follows that the reaction rate is proportional to the probability that the encounter occurs in the correct relative orientation. The pre-exponential factor A should therefore include a **steric factor**, P, which usually lies between 0 (no relative orientations lead to reaction) and 1 (all relative orientations lead to reaction). As an example, for the reactive collision

$$\text{NOCl} + \text{NOCl} \rightarrow \text{NO} + \text{NO} + \text{Cl}_2$$

in which two NOCl molecules collide and break apart into two NO molecules and a Cl$_2$ molecule, $P \approx 0.16$. For the hydrogen addition reaction

$$\text{H}_2 + \text{H}_2\text{C}=\text{CH}_2 \rightarrow \text{H}_3\text{C}-\text{CH}_3$$

in which a hydrogen molecule attaches directly to an ethene molecule to form an ethane molecule, P

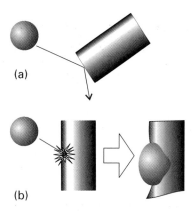

(a)

(b)

Fig 10.19 Energy is not the only criterion of a successful reactive encounter, for relative orientation may also play a role. (a) In this collision, the reactants approach in an inappropriate relative orientation, and no reaction occurs even though their energy is sufficient. (b) In this encounter, both the energy and the orientation are suitable for reaction.

is only 1.7×10^{-6}, which suggests that the reaction has very stringent orientational requirements.

Some reactions have $P > 1$. Such a value may seem absurd, because it appears to suggest that the reaction occurs more often than the molecules meet! An example of a reaction of this kind is

$$K + Br_2 \rightarrow KBr + Br$$

in which a K atom plucks a Br atom out of a Br_2 molecule; for this reaction the experimental value of P is 4.10. In this reaction, the distance of approach at which reaction can occur seems to be considerably larger than the distance needed for deflection of the path of the approaching molecules in a nonreactive collision! To explain this surprising conclusion, it has been proposed that the reaction proceeds by a 'harpoon mechanism'. This brilliant name is based on a model of the reaction that pictures the K atom as approaching the Br_2 molecules, and when the two are close enough an electron (the harpoon) flips across to the Br_2 molecule. In place of two neutral particles there are now two ions, and so there is a Coulombic attraction between them: this attraction is the line on the harpoon. Under its influence the ions move together (the line is wound in), the reaction takes place, and KBr and Br emerge. The harpoon extends the cross-section for the reactive

encounter and we would greatly underestimate the reaction rate if we used for the collision cross-section the value for simple mechanical contact between K and Br_2.

10.11 **Activated complex theory**

There is a more sophisticated theory of reaction rates that can be applied to reactions taking place in solution as well as in the gas phase. In the **activated complex theory** of reactions,[7] it is supposed that, as two reactants approach, their potential energy rises and reaches a maximum, as illustrated by the reaction profile in Fig 10.20. This maximum corresponds to the formation of an **activated complex**, a cluster of atoms that is poised to pass on to products or to collapse back into the reactants from which it was formed (Fig 10.21). An activated complex is not a reaction intermediate that can be isolated and studied like ordinary molecules. The concept of an activated complex is applicable to reactions in solutions as well as to the gas phase, because we can think of the activated complex as perhaps involving any solvent molecules that may be present.

Initially, only the reactants A and B are present. As the reaction event proceeds, A and B come into

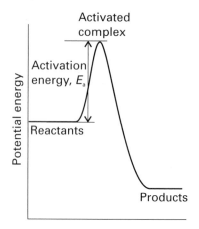

Fig 10.20 The same type of graph as in Fig 10.17 represents the reaction profile that is considered in activated complex theory. The activation energy is the potential energy of the activated complex relative to that of the reactants.

....

[7] The theory is also called *transition state theory*.

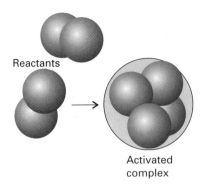

Reactants

Activated
complex

Fig 10.21 In the activated complex theory of chemical reactions, two reactants encounter each other (either in a gas-phase collision or as a result of diffusing together through a solvent) and, if they have sufficient energy, form an activated complex. The activated complex is depicted here by a relatively loose cluster of atoms that may undergo rearrangement into products. In an actual reaction, only some atoms—those at the actual reaction site—might be significantly loosened in the complex, the bonding of the others remaining almost unchanged. This would be the case for CH_3 groups attached to a carbon atom that was undergoing substitution.

contact, distort, and begin to exchange or discard atoms. The potential energy rises to a maximum, and the cluster of atoms that corresponds to the region close to the maximum is the activated complex. The potential energy falls as the atoms rearrange in the cluster, and reaches a value characteristic of the products. The climax of the reaction is at the peak of the potential energy. Here, two reactant molecules have come to such a degree of closeness and distortion that a small further distortion will send them in the direction of products. This crucial configuration is called the **transition state** of the reaction. Although some molecules entering the transition state might revert to reactants, if they pass through this configuration it is probable that products will emerge from the encounter.

The **reaction coordinate** is an indication of the stage reached in this process. On the left, we have undistorted, widely separated reactants. On the right are the products. Somewhere in the middle is the stage of the reaction corresponding to the formation of the activated complex. At the transition state, motion along the reaction coordinate corresponds to some complicated collective vibration-like motion of all the atoms in the complex (and the motion of the solvent molecules if they are involved too).

Derivation 10.3 *Activated complex theory*

In a simple form of activated complex theory, we suppose that the activated complex is in equilibrium with the reactants, and that we can express its abundance in the reaction mixture in terms of an equilibrium constant, $K^{\ddagger}$:

$$A + B \rightleftharpoons C^{\ddagger} \qquad K^{\ddagger} = \frac{[C^{\ddagger}]}{[A][B]}$$

Then, if we suppose that the rate at which products are formed is proportional to the concentration of the activated complex, we can write

Rate of formation of products $\propto [C^{\ddagger}] = K^{\ddagger}[A][B]$

The full activated complex theory gives an estimate of the constant of proportionality as kT/h, where k is Boltzmann's constant ($k = R/N_A$, Section 20.1) and h is Planck's constant (Section 12.1). Therefore

Rate of formation of products $= \dfrac{kT}{h} \times K^{\ddagger}[A][B]$

Now, by comparing this expression with the form of the rate law[8]

Rate of formation of products $= k_{rate}[A][B]$

we see that the rate constant k_{rate} is related to $K^{\ddagger}$ by

$$k_{rate} = \frac{kT}{h} \times K^{\ddagger} \qquad (10.16)$$

We saw in Section 7.3 that an equilibrium constant may be expressed in terms of the standard reaction Gibbs energy ($-RT \ln K = \Delta_r G^{\ominus}$). In this context, the Gibbs energy is called the **activation Gibbs energy**, and written $\Delta^{\ddagger}G$. It follows that

$$K^{\ddagger} = e^{-\Delta^{\ddagger}G/RT}$$

Therefore, by writing

$$\Delta^{\ddagger}G = \Delta^{\ddagger}H - T\Delta^{\ddagger}S \qquad (10.17)$$

we conclude that

$$k_{rate} = \frac{kT}{h} e^{-(\Delta^{\ddagger}H - T\Delta^{\ddagger}S)/RT} = \left(\frac{kT}{h} e^{\Delta^{\ddagger}S/R}\right) e^{-\Delta^{\ddagger}H/RT} (10.18)$$

This expression has the form of the Arrhenius expression, eqn 10.12, if we identify the **enthalpy of activation**, $\Delta^{\ddagger}H$, with the activation energy and the term in parentheses, which depends on the **entropy of activation**, $\Delta^{\ddagger}S$, with the pre-exponential factor.

The advantage of activated complex theory over collision theory is that it is applicable to reactions

---

[8] As is becoming apparent, we are running out of ks, so are using k_{rate} to denote the rate constant.

in solution as well as in the gas phase. It also gives some clue to the calculation of the steric factor P, for the orientation requirements are carried in the entropy of activation. Thus, if there are strict orientation requirements (for example, in the approach of a substrate molecule to an enzyme), then the entropy of activation will be strongly negative (representing a decrease in disorder when the activated complex forms), and the pre-exponential factor will be small. In practice, it is occasionally possible to estimate the sign and magnitude of the entropy of activation and hence to estimate the rate constant. The general importance of activated complex theory is that it shows that even a complex series of events—not only a collisional encounter in the gas phase—displays Arrhenius-like behaviour, and that the concept of activation energy is applicable.

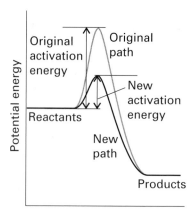

Fig 10.22 A catalyst provides an alternative reaction pathway with a lower activation energy than the uncatalysed reaction.

Self-test 10.9

In a particular reaction in water, it is proposed that two ions of opposite charge come together to form an electrically neutral activated complex. Is the contribution of the solvent to the entropy of activation likely to be positive or negative?

[*Answer:* positive, as H_2O is less organized around the neutral species]

10.12 Catalysis

Raising the temperature is one way of accelerating a reaction. Another way would be to find a means of lowering the activation energy, for then at the same temperature a higher proportion of molecules would be able to pass over the activation barrier. The height of the activation barrier for a given reaction path is, however, outside our control: it is determined by the electronic structures of the reactants and the arrangement of atoms in the activated complex. To change an activation barrier, we have to provide another route for the reaction—another reaction mechanism.

A **catalyst**, a species that increases the reaction rate but is not itself consumed in a reaction, acts by providing an alternative reaction path with a lower activation energy (Fig 10.22). A catalyst that is in the same phase as the reactants (for example, is dissolved in the same solvent) is called a **homoge-**

neous catalyst. A catalyst that is in a different phase—most commonly a solid introduced into a gas-phase reaction—is called a **heterogeneous catalyst**. Many industrial processes make use of heterogeneous catalysts, which include platinum, rhodium, zeolites,[9] and various metal oxides, but increasingly attention is turning to homogeneous catalysts, partly because they are easier to cool.

A strong acid can act as a homogenous catalyst for some reactions, and its action illustrates the general principle of catalysis, that a new reaction pathway is being provided. For instance, a strong acid can donate a proton to an organic species, and the resulting cation (the conjugate acid of the organic compound) may have a lower activation energy for reaction with another reactant (Fig 10.23). A metal acts as a heterogeneous catalyst for certain gas-phase reactions by providing a surface to which a reactant can attach by the process of **chemisorption**, the formation of chemical bonds to a surface. For example, hydrogen molecules may attach as atoms to a nickel surface, and these atoms react much more readily with another species (such as a hydrocarbon) than the original molecules (Fig 10.24). The chemisorption step therefore results in a reaction pathway with a lower activation energy than in the absence of the catalyst. We see how to formulate rate laws for such processes in the next chapter. Modern homogeneous catalysts in-

---

[9] Zeolites are aluminosilicates with an open, porous structure.

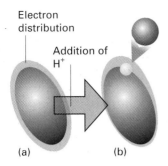

Fig 10.23 In acid catalysis, (a) the attachment of a proton to a species may so distort the electron distribution that (b) subsequent attack by another reactant may be facilitated.

clude complexes of rhodium or palladium, which can form bonds to organic molecules, and enable them to undergo rearrangements at temperatures far below those required in the absence of a catalyst

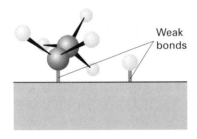

Fig 10.24 In a reaction catalysed by a surface, a reactant chemisorbed to the surface may react with a hydrogen atom formed by the chemisorption of hydrogen.

(Fig 10.25). These complexes can be regarded as the limit of a metal surface, for their active site is just a single metal atom.

Nature stumbled on catalysts long before chemists painstakingly fabricated them. Natural catalysts are the protein molecules called *enzymes*.

Despite the complexity of these large molecules, the central mode of action is the same as we have described. The enzyme provides a reaction pathway with a low activation energy and accelerates the reaction for which it has evolved. We examine the kinetics of enzyme reactions in the following chapter.

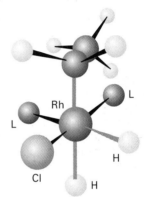

Fig 10.25 Certain metal complexes may also act as highly localized analogues of solid surfaces: an organic material may bind to the metal atom, and hence be prepared for a subsequent reaction step.

Exercises

10.1 The rate of formation of C in the reaction $2\,A + B \to 3\,C + 2\,D$ is $2.2\ \text{mol L}^{-1}\,\text{s}^{-1}$. State the rates of formation and consumption of A, B, and D.

10.2 The rate law for the reaction in Exercise 10.1 was reported as rate $= k[A][B][C]$. What are the units of k?

10.3 If the rate laws are expressed with (a) concentrations in numbers of molecules per metre cubed (molecules m^{-3}), (b) pressures in kilopascals, what are the units of the second-order and third-order rate constants?

10.4 The following initial-rate data were obtained on the rate of binding of glucose with the enzyme hexokinase (obtained from yeast) present at a concentration of 1.34 mmol L^{-1}. What is (a) the order of reaction with respect to glucose, (b) the rate constant?

| $[C_6H_{12}O_6]/(\text{mmol L}^{-1})$ | 1.00 | 1.54 | 3.12 | 4.02 |
| Initial rate/$(\text{mol L}^{-1}\,\text{s}^{-1})$ | 5.0 | 7.6 | 15.5 | 20.0 |

10.5 The following data were obtained on the initial rates of a reaction of a d-metal complex in aqueous solution. What is (a) the order of reaction with respect to the complex and the reactant Y, (b) the rate constant? For the experiments (a), $[Y] = 2.7$ mmol L^{-1} and for experiments (b) $[Y] = 6.1$ mmol L^{-1}.

[Complex]/(mmol L^{-1})		8.01	9.22	12.11
Rate/(mol L^{-1} s^{-1})	(a)	125	144	190
	(b)	640	730	960

10.6 The rate constant for the first-order decomposition of N_2O_5 in the reaction $2\ N_2O_5(g) \rightarrow 4\ NO_2(g) + O_2(g)$ is $k = 3.38 \times 10^{-5}\,s^{-1}$ at 25°C. What is the half-life of N_2O_5? What will be the total pressure, initially 500 Torr for the pure N_2O_5 vapour, (a) 10 s, (b) 10 min after initiation of the reaction?

10.7 In a study of the alcohol dehydrogenase catalysed oxidation of ethanol, the molar concentration of ethanol decreased in a first-order reaction from 220 mmol L^{-1} to 56.0 mmol L^{-1} in 1.22×10^4 s. What is the rate constant of the reaction?

10.8 The elimination of carbon dioxide from pyruvate ions by a decarboxylase enzyme was monitored by measuring the partial pressure of the gas as it was formed in a 250 mL flask at 20°C. In one experiment, the partial pressure increased from zero to 100 Pa in 522 s in a first-order reaction when the initial concentration of pyruvate ions in 100 mL of solution was 3.23 mmol L^{-1}. What is the rate constant of the reaction?

10.9 In the study of a second-order gas-phase reaction, it was found that the molar concentration of a reactant fell from 220 mmol L^{-1} to 56.0 mmol L^{-1} in 1.22×10^4 s. What is the rate constant of the reaction?

10.10 Carbonic anhydrase is a zinc-based enzyme that catalyses the conversion of carbon dioxide to carbonic acid. In an experiment to study its effect, it was found that the molar concentration of carbon dioxide in solution decreased from 220 to 56.0 mmol L^{-1} in 1.22×10^4 s. What is the rate constant of the first-order reaction?

10.11 The formation of NOCl from NO in the presence of a large excess of chlorine is pseudosecond-order in NO. In an experiment to study the reaction, the partial pressure of NOCl increased from zero to 100 Pa in 522 s. What is the rate constant of the reaction given that the initial partial pressure of NO is 300 Pa?

10.12 A number of reactions that take place on the surfaces of catalysts are zero-order in the reactant. One ex-

ample is the decomposition of ammonia on hot tungsten. In one experiment, the partial pressure of ammonia decreased from 21 kPa to 10 kPa in 770 s. (a) What is the rate constant for the zero-order reaction? (b) How long will it take all the ammonia to disappear?

10.13 The half-life of pyruvic acid in the presence of an aminotransferase enzyme (which converts it to alanine) was found to be 221 s. How long will it take for the concentration of pyruvic acid to fall to $\frac{1}{64}$ of its initial value in this first-order reaction?

10.14 The half-life for the (first-order) radioactive decay of ^{14}C is 5730 y (it emits β rays with an energy of 0.16 MeV). An archaeological sample contained wood that had only 69 per cent of the ^{14}C found in living trees. What is its age?

10.15 One of the hazards of nuclear explosions is the generation of ^{90}Sr and its subsequent incorporation in place of calcium in bones. This nuclide emits β rays of energy 0.55 MeV, and has a half-life of 28.1 y. Suppose 1.00 μg was absorbed by a newly born child. How much will remain after (a) 19 y, (b) 75 y if none is lost metabolically?

10.16 The second-order rate constant for the reaction $CH_3COOC_2H_5(aq) + OH^-(aq) \rightarrow CH_3CO_2^-(aq) + CH_3CH_2OH(aq)$ is 0.11 L mol^{-1} s^{-1}. What is the concentration of ester after (a) 15 s, (b) 15 min when ethyl acetate is added to sodium hydroxide so that the initial concentrations are $[NaOH] = 0.055$ mol L^{-1} and $[CH_3COOC_2H_5] = 0.150$ mol L^{-1}?

10.17 A reaction $2\ A \rightarrow P$ has a second-order rate law with $k = 1.24$ mL mol^{-1} s^{-1}. Calculate the time required for the concentration of A to change from 0.260 mol L^{-1} to 0.026 mol L^{-1}.

10.18 The composition of a liquid-phase reaction $2\ A \rightarrow B$ was followed spectrophotometrically with the following results:

t/min	0	10	20	30	40	∞
[B]/(mol L^{-1})	0	0.089	0.153	0.200	0.230	0.312

Determine the order of the reaction and its rate constant.

10.19 A rate constant is 1.78×10^{-4} L mol^{-1} s^{-1} at 19°C and 1.38×10^{-3} L mol^{-1} s^{-1} at 37 °C. Evaluate the Arrhenius parameters of the reaction.

10.20 The activation energy for the decomposition of benzene diazonium chloride is 99.1 kJ mol^{-1}. At what temperature will the rate be 10 per cent greater than its rate at 25°C?

10.21 Which reaction responds more strongly to changes of temperature, one with an activation energy of 52 kJ mol^{-1} or one with an activation energy of 25 kJ mol^{-1}?

10.22 The rate constant of a reaction increases by a factor of 1.23 when the temperature is increased from 20°C to 27°C. What is the activation energy of the reaction?

10.23 Food rots about 40 times more rapidly at 25°C than when it is stored at 4°C. Estimate the overall activation energy for the processes responsible for its decomposition.

10.24 Suppose that the rate constant of a reaction *de*creases by a factor of 1.23 when the temperature is increased from 20°C to 27°C. How should you report the activation energy of the reaction?

10.25 The enzyme urease catalyses the reaction in which urea is hydrolysed to ammonia and carbon dioxide. The half-life of urea in the pseudofirst-order reaction for a certain amount of urease doubles when the temperature is lowered from 20°C to 10°C and the Michaelis constant is largely unchanged. What is the activation energy of the reaction?

10.26 The activation energy of the first-order decomposition of dinitrogen oxide into N_2 and O is 251 kJ mol^{-1}. The half-life of the reactant is 6.5 Ms (1 Ms = 10^6 s) at 455°C. What will it be at 550°C?

10.27 Estimate the pre-exponential factor for the reaction between molecular hydrogen and ethene at 400°C.

10.28 Suppose an electronegative reactant needs to come to within 500 pm of a reactant with low ionization energy before an electron can flip across from one to the other (as in the harpoon mechanism). Estimate the reaction cross-section.

10.29 Estimate the activation Gibbs energy for the decomposition of urea in the reaction $CO(NH_2)(aq) + 2 H_2O(l) \rightarrow 2 NH_4^+(aq) + CO_3^{2-}(aq)$ for which the pseudofirst-order rate constant is 1.2×10^{-7} s^{-1} at 60°C and 4.6×10^{-7} s^{-1} at 70°C.

10.30 Calculate the entropy of activation of the reaction in Exercise 10.29 at the two temperatures.

10.31 Bovine rhodopsin (see the discussion of the processes of vision in Box 18.1 for the background to this question) undergoes a transition from one form (metarhodopsin-I) to another form (metarhodopsin-II) with a half-life of 600 μs at 37°C to 1 s at 0°C. On the other hand, studies of a frog retina show that the same transformation has a half-life that increases by only a factor of 6 over the same temperature range. Suggest an explanation and speculate on the survival advantages that this difference represents for the frog.

10.32 The activation Gibbs energy is composed of two terms: the activation enthalpy and the activation entropy. Differences in the latter can lead to the activation Gibbs energy for a process having the same value despite species inhabiting environments that differ widely in temperature. Show how the data shown in Fig 10.26 support this remark. The data relate to the enthalpy and entropy of activation of myofibrillar ATPase in different species of fish living in environments ranging from the Arctic to hot springs.

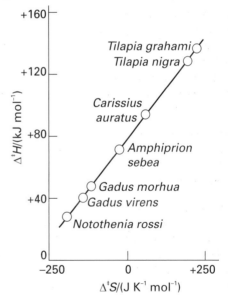

Fig 10.26 The correlation of the enthalpy and entropy of activation of the reaction catalysed by myosin ATPase in a variety of fish species. Data from I.A. Johnson and G. Goldspink, *Nature* **257**, 620 (1970), recalculated by H. Guttfreund, *Kinetics for the life sciences*, Cambridge University Press (1995).

Accounting for the rate laws

EVEN quite simple rate laws can give rise to complicated behaviour. That the heart maintains a steady pulse throughout a lifetime, but may break into fibrillation during a heart attack, is one sign of that complexity. On a less personal scale, reaction intermediates come and go, and all reactions approach equilibrium. However, the complexity of the behaviour of reaction rates means that the study of reaction rates can give deep insight into the way that reactions actually take place. As remarked in Chapter 10, rate laws are a window on to the **mechanism**, the sequence of elementary molecular events that lead from the reactants to the products, of the reactions they summarize.

Reaction schemes

So far, we have considered very simple rate laws, in which reactants are consumed or products formed. However, all reactions actually proceed towards a state of equilibrium in which the reverse reaction becomes increasingly important. Moreover, many reactions—particularly those in organisms—proceed to products through a series of intermediates. In organisms, as in industry, one of the intermediates may be of crucial importance and the ultimate products may represent waste.

11.1 The approach to equilibrium

All forward reactions are accompanied by their reverse reactions. At the start of a reaction, when little or no products are present, the rate of the reverse reaction is negligible. However, as the concentration of products increases, the rate at which they decompose into reactants becomes greater. At equilibrium, the reverse rate matches the forward rate and the reactants and products are present in abundances given by the equilibrium constant for the reaction.

We can analyse this behaviour by thinking of a very simple reaction of the form

Forward: A → B   rate of formation of B = $k[A]$
Reverse: B → A   rate of decomposition
                          of B = $k'[B]$

For instance, we could envisage this scheme as the interconversion of coiled (A) and uncoiled (B) DNA molecules. The *net* rate of formation of B, the difference of its rates of formation and decomposition, is

Net rate of formation of B = $k[A] - k'[B]$

If the initial concentrations of A and B are $[A]_0$ and $[B]_0$, respectively, then at any stage of the reaction

$$[A] + [B] = [A]_0 + [B]_0$$

and therefore

$$[A] = [A]_0 + [B]_0 - [B]$$

For simplicity we shall suppose that no B is present initially, so $[B]_0 = 0$. The integrated form of the rate law is then

$$[B] = \frac{k(1 - e^{-(k+k')t})[A]_0}{k+k'}$$

$$[A] = \frac{(k' + k e^{-(k+k')t})[A]_0}{k+k'} \qquad (11.1)$$

These solutions are quite complicated, but the graph of the concentrations (Fig 11.1) shows that they are plausible. The concentrations start from their initial values and move gradually towards

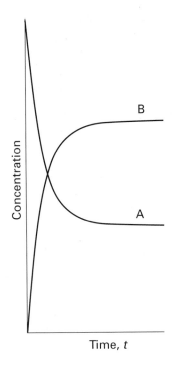

Fig 11.1 The approach to equilibrium of a reaction that is first-order in both directions. Here we have taken $k = 2k'$. Note how, at equilibrium, the ratio of concentrations is 2:1, corresponding to $K = 2$.

their final equilibrium values. We find the latter by setting t equal to infinity and using $e^{-x} = 0$ at $x = \infty$:

$$[B]_{eq} = \frac{k[A]_0}{k+k'} \qquad [A]_{eq} = \frac{k'[A]_0}{k+k'} \qquad (11.2)$$

The important conclusion we can draw from eqn 11.2 is that the equilibrium constant of the reaction is

$$K = \frac{[B]_{eq}}{[A]_{eq}} = \frac{k}{k'} \qquad (11.3)$$

This equation tells us that *the equilibrium constant for the reaction is the ratio of the forward and reverse rate constants*. Therefore, if the forward rate constant is much larger than the reverse rate constant, $K \gg 1$. If the opposite is true, then $K \ll 1$. This result is a crucial connection between the kinetics of a reaction and its equilibrium properties. It is also very useful in practice, for we may be able to measure the equilibrium constant and one of the rate constants, and can then calculate the missing rate constant from eqn 11.3. Alternatively, we can use the relation to calculate the equilibrium constant from kinetic measurements.

Illustration 11.1

The rate of constants of the forward and reverse reactions for the dimerization of proflavin (**1**), an antibacterial agent that acts by inhibiting the biosynthesis of DNA by intercalating between adjacent base pairs, were measured by a temperature jump method and found to be 8.1×10^8 L mol^{-1} s^{-1} and 2.0×10^6 s^{-1}, respectively. The equilibrium constant for the dimerization is therefore

$$K = \frac{8.1 \times 10^8}{2.0 \times 10^6} = 4.0 \times 10^2$$

Note how we discard the units, provided the rate constants are expressed in litres per mole per second. That ensures that the equilibrium constant is dimensionless and matches the conventions used in Chapters 7 and 8.

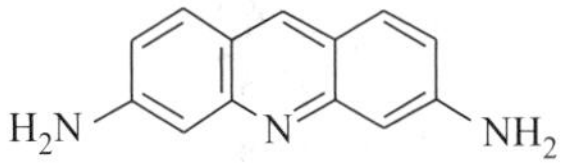

---

1 Proflavin

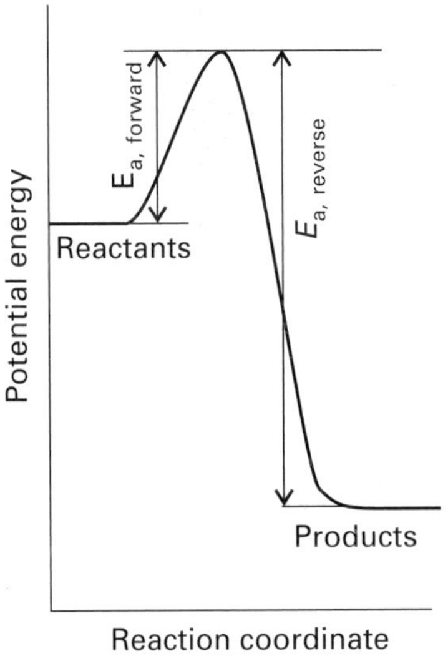

Fig 11.2 The reaction profile for an exothermic reaction. The activation energy is greater for the reverse reaction than for the forward reaction, so the rate of the forward reaction increases less sharply with temperature. As a result, the equilibrium constant shifts in favour of the reactants as the temperature is raised.

Equation 11.3 also gives us insight into the temperature dependence of equilibrium constants. First, we suppose that both the forward and reverse reactions show Arrhenius behaviour (Section 10.9). As we see from Fig 11.2, for an exothermic reaction the activation energy of the forward reaction is smaller than that of the reverse reaction. Therefore, the forward rate constant increases less sharply with temperature than the reverse reaction does. Consequently, when we increase the temperature of a system at equilibrium, k' increases more steeply than k does, and the ratio k/k', and therefore K, decreases. This is exactly the conclusion we drew from the van't Hoff equation (eqn 7.14), which was based on thermodynamic arguments.

11.2 Consecutive reactions

It is commonly the case that a reactant produces an intermediate, which subsequently decays into a product. Biochemical processes are often elaborate versions of this simple model. For instance, the restriction enzyme EcoRI catalyses the cleavage of DNA at a specific sequence of nucleotides (at GAATTC, making the cut between G and A on both strands). The reaction sequence it brings about is

Supercoiled DNA → open-circle DNA → linear DNA

In general in biology there is a whole cascade of intermediates preceding the formation of a final product. Industrial processes also often proceed by the formation of a series of intermediates. Indeed, in some cases, it is the 'intermediate' that is the desired substance. That intermediate must be extracted from the reaction mixture before a useless final 'product' has time to form.

To illustrate the kinds of considerations involved, let's suppose that the reaction takes place in two steps, in one of which the intermediate I (the open-circle DNA) is formed from the reactant A (the supercoiled DNA) in a first-order reaction, and then I decays in a first-order reaction to form the product P (the linear DNA):

A → I	rate of formation of I = $k_1[\text{A}]$
I → P	rate of formation of P = $k_2[\text{I}]$

For simplicity, we are ignoring the reverse reactions, which is valid if they are slow. The solution of these two equations can be found reasonably simply. For instance, the first rate law given above implies that A decays with a first-order rate law, and therefore that

$$[A] = [A]_0 e^{-k_1 t} \tag{11.4a}$$

The net rate of formation of I is the difference between its rate of formation and its rate of consumption, so we can write

Net rate of formation of I $= k_1[A] - k_2[I]$

with [A] given by eqn 11.4a. This equation is more difficult to solve, but it turns out that

$$[I] = \frac{k_1}{k_2 - k_1} (e^{-k_1 t} - e^{-k_2 t})[A]_0 \tag{11.4b}$$

Finally, we can insert this solution into the rate law for the formation of P, and solve the resulting equation to find

$$[P] = \left(1 + \frac{k_1 e^{-k_2 t} - k_2 e^{-k_1 t}}{k_2 - k_1}\right)[A]_0 \tag{11.4c}$$

These solutions are illustrated in Fig 11.3. We see that the intermediate grows in concentration initially, then decays as A is exhausted. Meanwhile, the concentration of P rises smoothly to its final value. The intermediate reaches its maximum concentration at[1]

$$t = \frac{1}{k_1 - k_2} \ln \frac{k_1}{k_2} \tag{11.5}$$

So, this is the optimum time for a manufacturer trying to make the intermediate in a batch process to extract it. For instance, if $k_1 = 0.120 \text{ h}^{-1}$ and $k_2 = 0.012 \text{ h}^{-1}$, then the intermediate is at a maximum at $t = 21$ h after the start of the process.

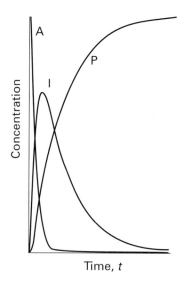

Fig 11.3 The concentrations of the substances involved in a consecutive reaction of the form A → I → P, where I is an intermediate and P a product. We have used $k_1 = 5k_2$. Note how, at each time, the sum of the three concentrations is a constant.

Reaction mechanisms

Wᴇ have seen how two simple types of reaction—approach to equilibrium and consecutive reactions—result in a characteristic dependence of the concentration on the time. We can suspect that other variations with time will act as the signatures of other reaction mechanisms.

11.3 Elementary reactions

Many reactions occur in a series of steps called **elementary reactions**, each of which involves only one or two molecules. We shall denote an elementary reaction by writing its chemical equation without displaying the physical state of the species, as in[2]

$$H + Br_2 \rightarrow HBr + Br$$

---

[1] We derive this result by differentiating the expression for [I] in eqn 11.4b and setting the derivative equal to 0.

[2] We have already used this convention without comment in some of the reactions discussed in Chapter 10.

This equation signifies that a specific H atom attacks a specific Br_2 molecule to produce a molecule of HBr and a Br atom. Ordinary chemical equations summarize the overall stoichiometry of the reaction and do not imply any specific mechanism.

The **molecularity** of an elementary reaction is the number of molecules coming together to react. In a **unimolecular reaction** a single molecule shakes itself apart or its atoms into a new arrangement (Fig 11.4). An example is the isomerization of cyclopropane to propene (**2** → **3**). The radioactive decay of nuclei (for example, the emission of a β particle from the nucleus of a tritium atom, which is used in mechanistic studies to follow the course of particular groups of atoms) is 'unimolecular' in the sense that a single nucleus shakes itself apart. In a **bimolecular reaction**, two molecules collide and exchange energy, atoms, or groups of atoms, or undergo some other kind of change, as in the reaction between H and F_2 or between H and Br_2 (Fig 11.5).

It is important to distinguish molecularity from order: the *order* of a reaction is an empirical quantity, and is obtained by inspection of the experimentally determined rate law; the *molecularity* of a reaction refers to an individual elementary reaction that has been postulated as a step in a proposed mechanism. Many substitution reactions in organic chemistry (for instance, S_N2 nucleophilic substitutions) are bimolecular and involve an activated complex that is formed from two reactant species. Enzyme-catalysed reactions can be regarded, to a good approximation, as bimolecular in the sense that they depend on the encounter of a substrate molecule and an enzyme molecule.

We can write down the rate law of an elementary reaction from its chemical equation. First, consider a unimolecular reaction. In a given interval, ten times as many A molecules decay when there are initially 1000 A molecules as when there are only

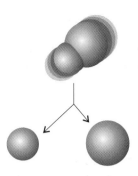

Fig 11.4 In a unimolecular elementary reaction, an energetically excited species decomposes into products: it simply shakes itself apart.

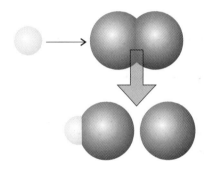

Fig 11.5 In a bimolecular elementary reaction, two species are involved in the process.

100 A molecules present. Therefore the rate of decomposition of A is proportional to its concentration and we can conclude that *a unimolecular reaction is first-order*:

$$A \rightarrow products \qquad rate = k[A] \qquad (11.6)$$

The rate of a bimolecular reaction is proportional to the rate at which the reactants meet, which in turn is proportional to both their concentrations. Therefore, the rate of the reaction is proportional to the product of the two concentrations and *an elementary bimolecular reaction is second-order overall*:

$$A + B \rightarrow products \qquad rate = k[A][B] \qquad (11.7)$$

We shall see below how to string simple steps together into a mechanism and how to arrive at the corresponding overall rate law. For the present we

H H
H—⟩⟨—H
H   H

2 Cyclopropane

CH_3 ⟍⟋ CH_2

3 Propene

emphasize that, if the reaction is an elementary bi-molecular process, then it has second-order kinetics; however, if the kinetics are second-order, then the reaction could be bimolecular but might be complex.

11.4 The formulation of rate laws

Suppose we propose that a particular reaction is the outcome of a sequence of elementary steps. How do we arrive at the rate law implied by the mechanism? We introduce the technique by considering the rate law for the gas-phase oxidation of nitrogen monoxide, NO:

$$2\,NO(g) + O_2(g) \rightarrow 2\,NO_2(g)$$

Nitrogen monoxide is a very important component of polluted atmospheres. It is formed in the hot exhausts of vehicles and the jet engines of aircraft and its oxidation is a step in the formation of acid rain. The compound is also a neurotransmitter involved in the physiological changes taking place during sexual arousal. Experimentally, the reaction is found to be third-order overall:

$$\text{Rate of formation of } NO_2 = k[NO]^2[O_2]$$

One explanation of the observed reaction order might be that the reaction is a single termolecular (three-molecule) elementary step involving the simultaneous collision of two NO molecules and one O_2 molecule. However, such collisions occur very infrequently. Therefore, although termolecular collisions may contribute, the rate of reaction by this mechanism is so slow that another mechanism usually dominates.

The following mechanism has been proposed:

Step 1 Two NO molecules combine to form a dimer:

$$NO + NO \rightarrow N_2O_2 \quad \text{rate of formation of } N_2O_2 = k_a[NO]^2$$

This step is plausible, because NO is an odd-electron species, a radical, and two radicals can form a covalent bond when they meet. That the N_2O_2 dimer is

also known in the solid makes the suggestion plausible: it is often a good strategy to decide whether a proposed intermediate is the analogue of a known compound.

Step 2 The N_2O_2 dimer decomposes into NO molecules:

$$N_2O_2 \rightarrow NO + NO \quad \text{rate of decomposition of } N_2O_2 = k_a'[N_2O_2]$$

This step, the reverse of step 1, is a unimolecular decay: the dimer shakes itself apart.[3]

Step 3 Alternatively, an O_2 molecule collides with the dimer and results in the formation of NO_2:

$$N_2O_2 + O_2 \rightarrow NO_2 + NO_2$$
$$\text{rate of consumption of } N_2O_2 = k_b[N_2O_2][O_2]$$

The rate at which NO_2 is formed in this step is

$$\text{Rate of formation of } NO_2 = 2k_b[N_2O_2][O_2]$$

The 2 appears in the rate law because two NO_2 molecules are formed in each reaction event, so the concentration of NO_2 increases at twice the rate that the concentration of N_2O_2 decays.

11.5 The steady-state approximation

Now we proceed to derive the rate law on the basis of this proposed mechanism. The rate of formation of product comes directly from step 3:

$$\text{Rate of formation of } NO_2 = 2k_b[N_2O_2][O_2]$$

However, this expression is not an acceptable overall rate law because it is expressed in terms of the concentration of the intermediate N_2O_2: an acceptable rate law for an overall reaction is expressed solely in terms of the species that appear in the overall reaction. Therefore, we need to find

---

[3] We adopt the convention in which the rate constant of a reverse reaction is marked with a prime (as in k_a for the forward reaction and k_a' for its reverse).

an expression for the concentration of N_2O_2. To do so, we consider the net rate of formation of the intermediate, the difference between its rates of formation and decay. Because N_2O_2 is formed by step 1 but decays by steps 2 and 3, its net rate of formation is

$$\text{Net rate of formation of } N_2O_2 = k_a[NO]^2 - k_a'[N_2O_2]$$
$$- k_b[N_2O_2][O_2]$$

Notice that formation terms occur with a positive sign and decay terms occur with a negative sign as they reduce the net rate of formation.

At this stage we introduce the **steady-state approximation**, in which we suppose that *the concentrations of all intermediates remain constant and small throughout the reaction* (except right at the beginning and right at the end). An **intermediate** is any species that does not appear in the overall reaction but which has been invoked in the mechanism. For our mechanism the intermediate is N_2O_2, so we write

$$k_a[NO]^2 - k_a'[N_2O_2] - k_b[N_2O_2][O_2] = 0$$

We can rearrange this equation into an equation for the concentration of N_2O_2:

$$[N_2O_2] = \frac{k_a[NO]^2}{k_a' + k_b[O_2]}$$

It follows that the rate of formation of NO_2 is

$$\text{Rate of formation of } NO_2 = 2k_b[N_2O_2][O_2]$$

$$= \frac{2k_ak_b[NO]^2[O_2]}{k_a' + k_b[O_2]} \quad (11.8a)$$

At this stage, the rate law is more complex than the observed law, but the numerator resembles it. The two expressions become identical if we suppose that the rate of decomposition of the dimer is much greater than its rate of reaction with oxygen, for then $k_a'[N_2O_2] \gg k_b[N_2O_2][O_2]$, or, after cancelling the $[N_2O_2]$, $k_a' \gg k_b[O_2]$. When this condition is satisfied, we can approximate the denominator in the overall rate law by k_a' alone and conclude that

$$\text{Rate of formation of } NO_2 = \left(\frac{2k_ak_b}{k_a'}\right)[NO]^2[O_2] \quad (11.8b)$$

This expression has the observed overall third-order form. Moreover, we can identify the observed rate constant as the following combination of rate constants for the elementary reactions:

$$k = \frac{2k_ak_b}{k_a'} \quad (11.9)$$

Self-test 11.1

An alternative mechanism that may apply when the concentration of O_2 is high and that of NO is low is one in which the first step is $NO + O_2 \rightarrow NO\cdots O_2$ and its reverse, followed by $NO\cdots O_2 + NO \rightarrow NO_2 + NO_2$. Confirm that this mechanism also leads to the observed rate law when the concentration of NO is low.

[*Answer*: rate = $2k_ak_b[NO]^2[O_2]/(k_a' + k_b[NO])$
$\approx (2k_ak_b/k_a')[NO]^2[O_2]$]

11.6 The rate-determining step

The oxidation of nitrogen monoxide introduces another important concept. Let's suppose that step 3 is very fast, so k_a' may be neglected relative to $k_b[O_2]$ in eqn 11.8a.[4] Then the rate law simplifies to

$$\text{Rate of formation of } NO_2 = \frac{2k_ak_b[NO]^2[O_2]}{k_b[O_2]} \quad (11.10)$$

$$= 2k_a[NO]^2$$

Now the reaction is second-order in NO and the concentration of O_2 does not appear in the rate law. The explanation is that the rate of reaction of N_2O_2 is so great on account of the high concentration of O_2 in the system that N_2O_2 reacts as soon as it is formed. Therefore, under these conditions, the rate of formation of NO_2 is determined by the rate at which N_2O_2 is formed. This step is an example of a **rate-determining step**, the slowest step in a reaction mechanism, which controls the rate of the overall reaction.

---

[4] One way to achieve this condition is to increase the concentration of O_2 in the reaction mixture.

The rate-determining step is not just the slowest step: it must be slow *and* be a crucial gateway for the formation of products. If a faster reaction can also lead to products, then the slowest step is irrelevant because the slow reaction can then be side-stepped (Fig 11.6). The rate-determining step is like a slow ferry crossing between two fast highways: the overall rate at which traffic can reach its destination is determined by the rate at which it can make the ferry crossing.

The rate law of a reaction that has a rate-determining step can often be written down almost by inspection. If the first step in a mechanism is rate-determining, then the rate of the overall reaction is equal to the rate of the first step because all subsequent steps are so fast that once the first intermediate is formed it results immediately in the formation of products. Figure 11.7 shows the reaction profile for a mechanism of this kind in which the slowest step is the one with the highest activation energy. Once over the initial barrier, the intermediates cascade into products. However, a rate-determining step may also stem from the low concentration of a crucial reactant or catalyst and

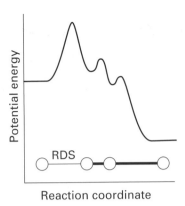

Fig 11.7 The reaction profile for a mechanism in which the first step (RDS) is rate-determining.

need not correspond to the step with highest activation barrier. A rate-determining step arising from the low activity of a crucial enzyme can sometimes be identified by determining whether or not the reactants and products for that step are in equilibrium: if the reaction is not at equilibrium it suggests that the step may be slow enough to be rate-determining.

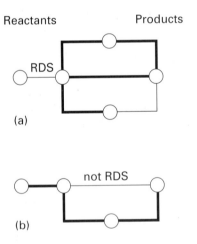

Fig 11.6 The rate-determining step (RDS) is the slowest step of a reaction *and* acts as a bottleneck. In this schematic diagram, fast reactions are represented by heavy lines (freeways) and slow reactions by thin lines (country lanes). Circles represent substances. (a) The first step is rate-determining. (b) Although the second step is the slowest, it is not rate-determining because it does not act as a bottleneck (there is a faster route that circumvents it).

Illustration 11.2

An analysis of the composition of heart tissue gave the following results:

	Concentration/ (mmol L^{-1})
Fructose-1,6-bis(phosphate), F16bP	0.019
Fructose-6-phosphate, F6P	0.089
ADP	1.30
ATP	11.4

The equilibrium constant for the reaction

$$F6P + ATP \xrightarrow{\text{phosphofructinase}} F16bP + ADP$$

is 1.2×10^3. However, the reaction quotient is

$$Q = \frac{[F16bP][ADP]}{[F6P][ATP]} = \frac{(1.9 \times 10^{-5}) \times (1.30 \times 10^{-3})}{(8.9 \times 10^{-5}) \times (1.14 \times 10^{-2})}$$
$$= 0.024$$

Because $Q \ll K$, we conclude that the reaction step is far from equilibrium. Therefore, the step is so slow that it may be rate-determining.

11.7 Reactions on surfaces

We saw in Section 10.12 that heterogeneous catalysis commonly depends on the chemisorption of a reactant on to the surface of a metal. We are now equipped to formulate the rate laws for this type of process. First, we need to know how the extent of chemisorption depends on the partial pressure of the reactants.

Suppose there are N possible sites for chemisorption and that, when the partial pressure of a gas A is p_A, the **fractional coverage**, the fraction of sites occupied, is θ (theta). We need the relation between θ and p_A. To obtain it, we suppose that the rate at which A adsorbs to the surface is proportional to the partial pressure (because the rate at which molecules strike the surface is proportional to the pressure), and to the number of sites that are not occupied at the time, which is $(1 - \theta)N$:

Rate of adsorption $= k_{ad}N(1 - \theta)p_A$

The rate at which the adsorbed molecules leave the surface is proportional to the number currently on the surface ($N\theta$):

Rate of desorption $= k_{des}N\theta$

At equilibrium, these two rates are equal, so we can write

$k_{ad}N(1 - \theta)p_A = k_{des}N\theta$

The Ns cancel, so

$$\theta = \frac{p_A}{p_A + K} \qquad K = \frac{k_{des}}{k_{ad}} \qquad (11.11)$$

This expression, which is plotted in Fig 11.8, is called the **Langmuir isotherm**. We see that, as the partial pressure of A increases, the fractional coverage increases towards 1. Half the surface is covered when $p_A = K$. Note that the model assumes that, as soon as a site is occupied, no further adsorption can take place there: that is, only direct *monolayer* adsorption to the metal surface is allowed.

First, suppose that the reactant decomposes when it is adsorbed. This is the case, for instance,

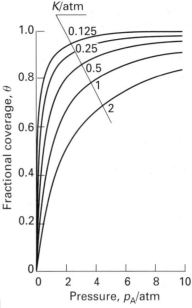

Fig 11.8 The Langmuir adsorption isotherm shows how the fraction of adsorption sites occupied depends on the pressure of the gas, provided no more than monolayer coverage is allowed.

with phosphine, PH_3, adsorbed on tungsten. Then the rate of decomposition is proportional to the fractional coverage, and we can write

$$\text{Rate of reaction} = k\theta = \frac{kp_A}{p_A + K} \qquad (11.12a)$$

An interesting aspect of this mechanism comes to light if the partial pressure of A is so high that $p_A \gg K$. Then K in the denominator can be ignored, the pressures cancel, and the rate of reaction is

$$\text{Rate of reaction} = k \qquad (11.12b)$$

That is, the reaction becomes zeroth-order in A. The rate-determining step is the decomposition: the rate of adsorption is so great at high pressures that the surface coverage is effectively constant and the reaction rate is determined by the rate at which adsorbed molecules decompose. At pressures so low that $p_A \ll K$, the rate law simplifies instead to

$$\text{Rate of reaction} = \frac{kp_A}{K} \qquad (11.12c)$$

Under these conditions, the reaction is first-order in the reactant gas. That order of reaction is common for many heterogeneously catalysed reactions. The rate-determining step is now adsorption of the molecules to the surface, for as soon as they adsorb they decompose.

Now suppose a gas-phase molecule of another reactant B reacts when it strikes an A molecule that is already bound to the surface. The rate of these collisions is proportional to the partial pressure of B and to the number of adsorbed A molecules, and therefore to the fractional coverage of A. We can write

$$\text{Rate of reaction} = k\theta p_B = \frac{k p_A p_B}{p_A + K} \qquad (11.13a)$$

This mechanism is called the **Eley–Rideal mechanism** of heterogeneous catalysis. We see that at high partial pressures of A, when $p_A \gg K$, the rate law simplifies to

$$\text{Rate of reaction} = k p_B \qquad (11.13b)$$

and the reaction is first-order in B. In this case, the rate-determining step is the impact of B with the surface. At high pressures, so much A is adsorbed that we can be confident that B will strike an adsorbed A molecule wherever it hits the surface.[5]

11.8 Unimolecular reactions

A number of gas-phase reactions follow first-order kinetics, as in the isomerization of cyclopropane mentioned earlier, in which the strained triangular molecule bursts open into an acyclic alkene:

$$cyclo\text{-}C_3H_6 \rightarrow CH_3CH{=}CH_2 \qquad \text{rate} = k[cyclo\text{-}C_3H_6]$$

The problem with reactions like this is that the reactant molecule presumably acquires the energy it needs to react by collisions with other molecules. Collisions, though, are simple *bimolecular* events, so how can they result in a first-order rate law? First-order gas-phase reactions are widely called 'unimolecular reactions' because (as we shall see) the

rate-determining step is an elementary unimolecular reaction in which the reactant molecule changes into the product. This term must be used with caution, however, because the composite mechanism has bimolecular as well as unimolecular steps.

The first successful explanation of unimolecular reactions is ascribed to Frederick Lindemann in 1921.[6] The **Lindemann mechanism** is as follows:

Step 1 A reactant molecule A becomes energetically excited (denoted A*) by collision with another A molecule:

$$A + A \rightarrow A^* + A \qquad \text{rate of formation of } A^* = k_a[A]^2$$

Step 2 The energized molecule might lose its excess energy by collision with another molecule:

$$A^* + A \rightarrow A + A \quad \text{rate of deactivation of } A^* = k_a'[A^*][A]$$

Step 3 Alternatively, the excited molecule might shake itself apart (as may happen with vibrationally excited cyclopropane) and form products P. That is, it might undergo the unimolecular decay

$$A^* \rightarrow P \quad \text{rate of formation of } P = k_b[A^*]$$
$$\text{rate of consumption of } A^* = k_b[A^*]$$

If the unimolecular step, Step 3, is slow enough to be the rate-determining step, the overall reaction will have first-order kinetics, as observed. We can demonstrate this explicitly by applying the steady-state approximation to the net rate of formation of the intermediate A*:

Net rate of formation of A *

$$= \underbrace{k_a[A]^2}_{\text{Formation}} - \underbrace{k_a'[A^*][A]}_{\text{Deactivation}} - \underbrace{k_b[A^*]}_{\text{Reaction}}$$

$$\underbrace{= 0}_{\text{Steady state}}$$

---

[5] Great advances in the study of surfaces have been made as a result of the invention of the scanning tunnelling microscope: see Box 12.1.

[6] K.J. Laidler, in his *Chemical kinetics* (Harper and Row, 1987), gives an interesting historical summary of the origin of the mechanism. Apparently, Lindemann sketched the mechanism at a meeting and published a brief note; almost simultaneously a young Danish doctoral student, J.A. Christiansen, published his Ph.D. thesis in which the same mechanism was proposed and developed in much greater detail. The 'Lindemann–Christiansen mechanism' would therefore appear to be a fairer name than the conventional 'Lindemann mechanism'.

The solution of this equation is

$$[A^*] = \frac{k_a [A]^2}{k_b + k_a' [A]}$$

It follows that the rate law for the formation of products is

$$\text{Rate of formation of } P = k_b [A^*] = \frac{k_a k_b [A]^2}{k_b + k_a' [A]} \quad (11.14a)$$

At this stage the rate law is not first-order in A. However, let's suppose that the rate of deactivation of A^* by (A^*, A) collisions is much greater than the rate of unimolecular decay of A^* to products. That is, we suppose that the unimolecular decay of A^* is the rate-determining step. Then $k_a'[A^*][A] \gg k_b[A^*]$, which corresponds to $k_a'[A] \gg k_b$. If that is the case, we can neglect k_b in the denominator of the rate law and obtain

$$\text{Rate of formation of } P = k[A], \text{ with } k = \frac{k_a k_b}{k_a'} \quad (11.14b)$$

Equation 11.14b is a first-order rate law, as we set out to show.

Self-test 11.2

Suppose that an inert gas M is present and dominates the excitation of A and de-excitation of A^*. Devise the rate law for the formation of products.

[*Answer*: Rate = $k_a k_b [A][M]/(k_b + k_a'[M])$]

Enzyme reactions

ONE major application of the kind of analysis we have been introducing is to the elucidation of the mechanisms of enzyme-catalysed reactions (Box 11.1). As we shall see, clever use of the steady-state approximation and graphical analyses of data can reveal detailed information about these reactions and how they are affected by inhibitors and coenzymes.

11.9 The mechanism of enzyme action

One of the earliest descriptions of the action of enzymes is the **Michaelis–Menten mechanism**. The proposed mechanism, with all species in an aqueous environment, is as follows.[7]

Step 1 The bimolecular formation of a combination, ES, of the enzyme and the substrate:

$$E + S \rightarrow ES \qquad \text{rate of formation of ES} = k_a[E][S]$$

Step 2 The unimolecular decomposition of the complex:

$$ES \rightarrow E + S \quad \text{rate of decomposition of ES} = k_a'[ES]$$

Step 3 The unimolecular formation of products and the release of the enzyme from its combination with the substrate:

$$ES \rightarrow P + E \qquad \text{rate of formation of P} = k_b[ES]$$
$$\text{rate of consumption of ES} = k_b[ES]$$

We seek the rate law for the rate of formation of product in terms of the concentrations of enzyme and substrate.

Derivation 11.1 *The Michaelis–Menten rate law*

The product is formed (irreversibly) in step 3, so we begin by writing

$$\text{Rate of formation of } P = k_b[ES]$$

To calculate the concentration [ES] we set up an expression for the net rate of formation of ES allowing for its formation in step 1 and its removal in steps 2 and 3. Then we set that net rate equal to zero:

$$\text{Net rate of formation of ES} = k_a[E][S] - k_a'[ES] - k_b[ES] = 0$$

---

[7] Michaelis and Menten derived their rate law in 1913 in a more restrictive way, by assuming a rapid equilibrium. The approach we take is a generalization using the steady-state approximation made by Briggs and Haldane in 1925.

It follows that

$$[ES] = \frac{k_a[E][S]}{k_a' + k_b}$$

However, [E] and [S] are the molar concentrations of the *free* enzyme and *free* substrate. If $[E]_0$ is the total concentration of enzyme, then

$$[E] + [ES] = [E]_0$$

Because only a little enzyme is added, the free substrate concentration is almost the same as the total substrate concentration and we can ignore the fact that [S] differs slightly from [S] + [ES]. Therefore,

$$[ES] = \frac{k_a([E]_0 - [ES])[S]}{k_a' + k_b}$$

This expression rearranges to

$$[ES] = \frac{k_a[E]_0[S]}{k_a' + k_b + k_a[S]}$$

It follows that the rate of formation of product is

Rate of formation of P = $k[E]_0$, with $k = \dfrac{k_b[S]}{[S] + K_M}$

(11.15)

where the **Michaelis constant**, K_M (which has the dimensions of a concentration), is

$$K_M = \frac{k_a' + k_b}{k_a} \qquad (11.16)$$

According to eqn 11.15, the rate of enzymolysis is first-order in the enzyme concentration, but the effective rate constant k depends on the concentration of substrate. When $[S] \ll K_M$, the effective rate constant is equal to $k_b[S]/K_M$. Therefore, the rate increases linearly with [S] at low concentrations. When $[S] \gg K_M$, the effective rate constant is equal to k_b, and the rate law in eqn 11.15 reduces to

Rate of formation of P = $k_b[E]_0$ \qquad (11.17)

The rate is independent of the concentration of S because there is so much substrate present that it remains at effectively the same concentration even though products are being formed. Under these conditions, the rate of formation of product is a maximum, and $k_b[E]_0$ is called the **maximum velocity**, v_{max}, of the enzymolysis:

$$v_{max} = k_b[E]_0 \qquad (11.18)$$

The constant k_b is called the **maximum turnover number**. The rate-determining step is step 3, because there is ample ES present (because S is so abundant), and the rate is determined by the rate at which ES reacts to form the product.

It follows from eqns 11.15 and 11.18 that the reaction rate v at a general substrate composition is related to the maximum velocity by

$$v = \frac{[S]v_{max}}{[S] + K_M} \qquad (11.19)$$

This relation is illustrated in Fig 11.9. Equation 11.19 is the basis of the analysis of enzyme kinetic data by using a **Lineweaver–Burk plot**, a graph of $1/v$ (the reciprocal of the reaction rate) against $1/[S]$ (the reciprocal of the substrate concentration). If we take the reciprocal of both sides of eqn 11.19 it becomes

$$\frac{1}{v} = \frac{[S] + K_M}{[S]v_{max}} = \frac{1}{v_{max}} + \left(\frac{K_M}{v_{max}}\right)\frac{1}{[S]} \qquad (11.20)$$

Because this expression has the form

$$y = \text{intercept} + \text{slope} \times x$$

with $y = 1/v$ and $x = 1/[S]$, we should obtain a straight line when we plot $1/v$ against $1/[S]$. The slope of the straight line is K_M/v_{max} and the extrapolated intercept at $1/[S] = 0$ is equal to $1/v_{max}$ (Fig 11.10). Therefore, the intercept can be used to find v_{max}, and then that value combined with the slope to find the value of K_M. Alternatively, note that the extrapolated intercept with the horizontal axis (where $1/v = 0$) occurs at $1/[S] = -1/K_M$.

Box 11.1 *Catalytic activity and catalytic antibodies*

One of the focuses of molecular biology has been the elucidation of the details at a molecular scale of the reaction sequence by which an enzyme carries out its function. The kind of information obtained has indicated how enzymes might be generated to carry out particular tasks.

One enzyme that has been studied in considerable detail is chymotrypsin, which is shown in the first illustration. The chymotrypsin functions by hydrolysing peptide bonds in polypeptides in the small intestine. The sequence of steps by which chymotrypsin carries out the first part of its task—to snip through the C—N bond of the peptide link—is shown in the next illustration. The crucial point to notice is the formation of a tetrahedral transition state in the course of the reaction. The second sequence of steps, by which the carboxylic acid group is eliminated from the polypeptide is shown in the third illustration. This step involves the attack by a water molecule on the carboxyl group and the subsequent cleavage of the original C—O bond. Once again, the crucial point is the formation of a tetrahedral transition state. In each case, the catalytic activity of the enzyme can be traced to the structure of the active site, and in particular to the presence of a serine and histidine group, which serve to stabilize the transition state.

Catalytic antibodies combine the insight that studies on molecules like chymotrypsin provide with an organism's natural defence systems to provide routes to

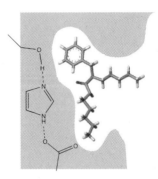

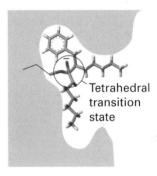

Tetrahedral transition state

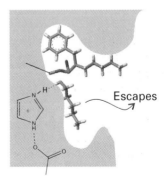

Escapes

The sequence of steps by which chymotrypsin cuts through the C—N bond of a peptide link and releases an amine.

A representation of the chymotrypsin molecule showing the regions of helix and sheet. The dots surrounding the structure are the locations of water molecules.

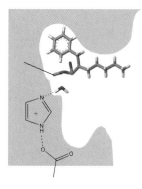

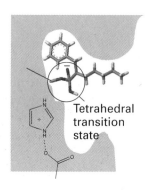

Tetrahedral transition state

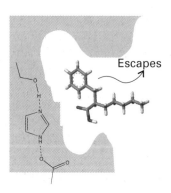

Escapes

The following sequence of steps by which chymotrypsin cuts through the C—O bond and releases a carboxylic acid.

alternative enzymes for carrying out particular reactions. The key idea that we need to incorporate is that an organism generates a flood of antibodies when an antigen—a foreign body—is introduced. The organism maintains a wide range of latent antibodies, but they proliferate in the presence of the antigen. It follows that, if we can introduce an antigen that emulates the tetrahedral transition state typical of an enzymolysis reaction, then an organism should produce a supply of antibodies that may be able to act as enzymes for that and related reactions.

This procedure has been applied to the search for enzymes for the hydrolysis of esters. The compound used to mimic the tetrahedral transition state is a tetrahedral phosphonate (**4**). When the monoclonal antibody stimulated to form by this antigen is used to catalyse the hydrolysis of an ester, pronounced activity is indeed found, with $K_M = 1.9$ μmol L^{-1} and an enhancement of rate over the uncatalysed reaction by a factor of 10^3. The hope is that catalytic antibodies can be formed that catalyse reactions currently untouched by enzymes, such as the proteins in viruses and tumours.

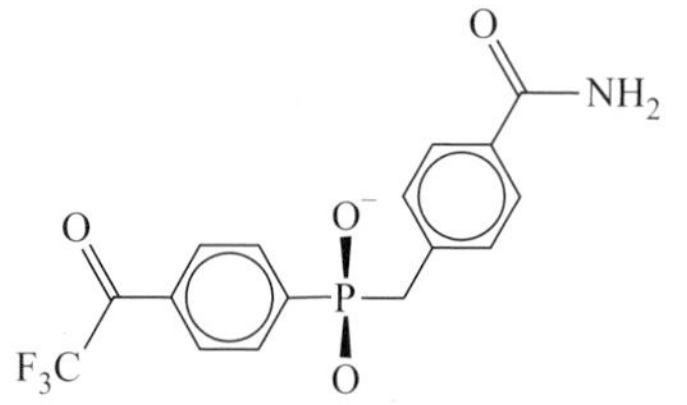

4 Phosphate transition state analogue

Exercise 1 The estimated half-life for P—O bonds is $t_{1/2} = 1.3 \times 10^5$ y. Approximately 10^9 such bonds are present in a strand of DNA. How long (in terms of its half-life) would a single strand of DNA survive with no cleavage in the absence of repair enzymes?

Exercise 2 The equilibrium for peptide assembly from individual amino acids is highly unfavourable ($K = 0.02$). It has been suggested that proteins might be formed by a spontaneous assembly of amino acids in the presence of enzymes. Is such a reaction feasible? What would be the equilibrium concentration of a tetrapeptide formed from a solution in which the concentration of each of the four amino acids is 1 mol L^{-1}?

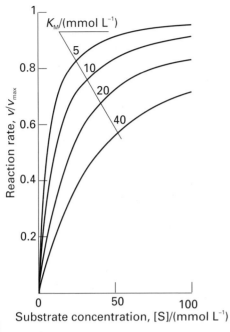

Fig 11.9 The variation of the rate of an enzyme-catalysed reaction with concentration of the substrate according to the Michaelis–Menten model. When $[S] \ll K_M$, the rate is proportional to $[S]$; when $[S] \gg K_M$, the rate is independent of $[S]$.

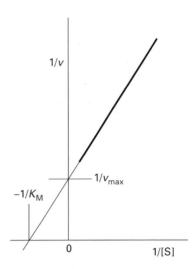

Fig 11.10 A Lineweaver–Burk plot is used to analyse kinetic data on enzyme-catalysed reactions. The reciprocal of the rate of formation of products $(1/v)$ is plotted against the reciprocal of the substrate concentration $(1/[S])$. All the data points (which typically lie in the black region of the line) correspond to the same overall enzyme concentration, $[E]_0$. The intercept of the extrapolated (grey) straight line with the horizontal axis is used to obtain the Michaelis constant, K_M. The intercept with the vertical axis, is used to determine $v_{max} = k_b[E]_0$, and hence k_b. The slope may also be used, for it is equal to K_M/v_{max}.

Example 11.1 *Analysing a Lineweaver-Burk plot*

Several solutions containing a substrate at different concentrations were prepared and the same small amount of enzyme was added to each one. The following data were obtained on the initial rates of the formation of product:

$[S]/(\mu mol\ L^{-1})$	10.0	20.0	40.0	80.0	120.0	180.0	300.0
$v/(\mu mol\ L^{-1}\ s^{-1})$	0.32	0.58	0.90	1.22	1.42	1.58	1.74

Determine the maximum velocity and the Michaelis constant for the reaction.

Strategy We construct a Lineweaver–Burk plot by drawing up a table of $1/[S]$ and $1/v$. The intercept at $1/[S] = 0$ is $1/v_{max}$ and the slope of the line through the points is K_M/v_{max}, so K_M is found from the slope divided by the intercept.

Solution We draw up the following table:

$10^3/([S]/(\mu mol\ L^{-1}))$	100	50.0	25.0	12.5	8.33	5.56	3.33
$1/(v/(\mu mol\ L^{-1}\ s^{-1}))$	3.1	1.7	1.1	0.820	0.704	0.633	0.575

The graph is plotted in Fig 11.11. A least squares analysis (or, more unreliably, inspection of the graph) gives an intercept at 0.476 and a slope of 26.17. It follows that

$$v_{max}/(\mu mol\ L^{-1}\ s^{-1}) = \frac{1}{intercept} = \frac{1}{0.476} = 2.10$$

and

$$K_M/(\mu mol\ L^{-1}) = \frac{slope}{intercept} = \frac{26.17}{0.476} = 55.0$$

Self-test 11.3

Repeat the analysis for the following data:

$[S]/(mmol\ L^{-1})$	10.0	20.0	30.0	40.0	80.0	160.0	320.0
$v/(mmol\ L^{-1}\ s^{-1})$	65.1	72.0	75.2	77.3	79.3	80.7	81.4

[*Answer*: 82.0 mmol L^{-1} s^{-1}, 2.62 mmol L^{-1}]

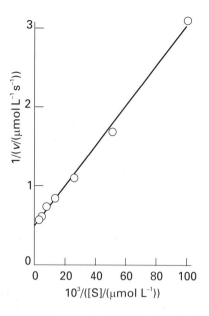

Fig 11.11 The Lineweaver–Burk plot based on the data in Example 11.1.

11.10 Enzyme inhibition

The action of an enzyme may be partially suppressed by the presence of a foreign substance, which is called an **inhibitor**. An inhibitor, which we denote I, may be a poison that has been administered to the organism, or it may be a substance that is naturally present in a cell and involved in its regulatory mechanism. In **competitive inhibition** the inhibitor competes for the active site and reduces the ability of the enzyme to bind the substrate (Fig 11.12). In **noncompetitive inhibition** the inhibitor attaches to another part of the enzyme molecule, thereby distorting it and reducing its ability to bind the substrate (Fig 11.13). We shall now see how to distinguish the two kinds of inhibition by making use of kinetic data.

First, consider competitive inhibition. We suppose that the inhibitor molecule, I, is in equilibrium with the complex EI it forms when it is bound to the active site:

$$EI \rightleftharpoons E + I \qquad K_I = \frac{[E][I]}{[EI]} \qquad (11.21)$$

The rate of formation of product turns out to be

$$v = \frac{[S]v_{max}}{[S] + \alpha K_M} \qquad \alpha = 1 + \frac{[I]}{K_I} \qquad (11.22)$$

This expression is illustrated in Fig 11.14. If we compare it with eqn 11.19 we see that the role of the inhibitor is to modify K_M to αK_M. Therefore, in a Lineweaver–Burk plot, the slope and the intercept with the horizontal axis change as [I] is changed, but the intercept with the vertical axis (the value of $v_{max} = k_b[E]_0$) remains unchanged (Fig 11.15).

Now consider noncompetitive inhibition. We suppose that the inhibitor is in equilibrium with a bound state IE, but that the site occupied by I is not the active site for the attachment of S (which is why we write it IE and not EI, to sug-

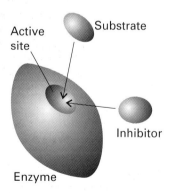

Fig 11.12 In competitive inhibition, both the substrate (the egg shape) and the inhibitor compete for the active site, and reaction ensues only if the substrate is successful in attaching there.

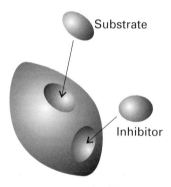

Fig 11.13 In one version of noncompetitive inhibition, the substrate and the inhibitor attach to distant sites of the enzyme molecule, and a complex in which they are both attached (IES) does not lead to the formation of product.

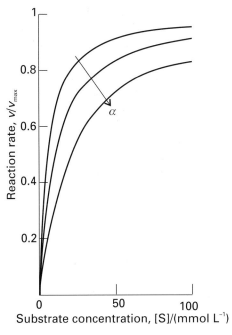

Fig 11.14 The variation of the rate of an enzyme-catalysed re-action with concentration of the substrate when a competitive inhibitor is present. The uppermost curve corresponds to [I] = 0 (no inhibitor present).

gest the use of a site that is distant from the one used to form ES). Moreover, because I and S are not in competition for the same site, I may also bind to the complex ES to give a complex IES:

$$IES \rightleftharpoons I + ES \qquad K_I = \frac{[I][ES]}{[IES]} \qquad (11.23)$$

We suppose that, although I and S may both bind to E, the enzyme can bring about a change in S only if I is not present.[8] Therefore, only ES can give rise to products; IES cannot. In this scenario, the rate of formation of product turns out to follow the rate law

$$v = \frac{[S]v_{max}}{\alpha([S] + K_M)} \qquad (11.24)$$

with α the same as in eqn 11.22 (Fig 11.16). In a Lineweaver–Burk plot with different values of [I], the straight lines now pass through a common intercept with the horizontal axis (because $1/v = 0$ at

---

[8] The presence of I in IE allows S to bind, but so affects the structure of the enzyme that the latter cannot carry out its function.

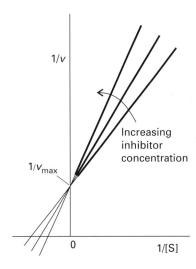

Fig 11.15 A Lineweaver–Burk plot may be used to distinguish competitive and noncompetitive inhibition kinetically. An indication that competitive inhibition is occurring is that the intercept with the vertical axis does not move as the concentration of inhibitor is increased.

$1/[S] = -1/K_M$ independent of the inhibitor properties), but the slope and the intercept with the vertical axis both increase as the concentration of inhibitor is increased (Fig 11.17).

Example 11.2 *Distinguishing between types of inhibition*

Five solutions of a substrate, S, were prepared with the concentrations given in the first column below and each one was divided into five equal volumes. The same concentration of enzyme was present in each one. An inhibitor, I, was then added in five different concentrations to the samples, and the initial rate of formation of product was determined with the results given below. Does the inhibitor act competitively or noncompetitively? Determine K_I and K_M.

[S]/ (mmol L^{-1})	[I]/(mmol L^{-1})				
	0	0.20	0.40	0.60	0.80
0.050	0.033	0.026	0.021	0.018	0.016
0.10	0.055	0.045	0.038	0.033	0.029
0.20	0.083	0.071	0.062	0.055	0.050
0.40	0.111	0.100	0.091	0.084	0.077
0.60	0.126	0.116	0.108	0.101	0.094

$v/(\mu mol\ L^{-1}\ s^{-1})$

Strategy We draw a series of Lineweaver–Burk plots for different inhibitor concentrations. If the plots resemble those in Fig 11.15, then the inhibition is competitive. On the other hand, if the plots resemble those in Fig 11.17, then the inhibition is noncompetitive. To find K_I, we need to determine the slope at each value of [I], which is equal to $\alpha K_M / v_{max}$, or $K_M / v_{max} + K_M [I] / K_I v_{max}$, then plot this slope against [I]: the intercept at [I] = 0 is the value of K_M / v_{max} and the slope is $K_M / K_I v_{max}$.

Solution First, we draw up a table of 1/[S] and $1/v$ for each value of [I]:

$1/(v/(\mu mol\ L^{-1}\ s^{-1}))$ for [I]/(mmol L^{-1}) =

$1/([S]/(mmol\ L^{-1}))$	0	0.20	0.40	0.60	0.80
20	30	38	48	56	62
10	18	22	26	30	34
5.0	12	14	16	18	20
2.5	9.01	10.0	11.0	11.9	13.0
1.7	7.94	8.62	9.26	9.90	10.6

The five plots (one for each [I]) are given in Fig 11.18. We see that they pass through the same intercept on the vertical axis, so the inhibition is competitive. The mean of the (least squares) intercepts is 5.88, so $v_{max} = 0.17\ \mu mol\ L^{-1}\ s^{-1}$ (note how it picks up the units for v in the data). The (least squares) slopes of the lines are as follows:

[I]/(mmol L^{-1})	0	0.20	0.40	0.60	0.80
Slope	1.219	1.627	2.090	2.489	2.832

These values are plotted in Fig 11.19. The intercept at [I] = 0 is 1.235, so $K_M = 0.21$ mmol L^{-1}. The (least squares) slope of the line is 1.992, so

$$K_I /(mmol\ L^{-1}) = \frac{K_M}{slope} = \frac{1.234}{2.044} = 0.60$$

Self-test 11.4

Repeat the question using the following data:

$v/(\mu mol\ L^{-1}\ s^{-1})$ for [I]/(mmol L^{-1}) =

[S]/(mmol L^{-1})	0	0.20	0.40	0.60	0.80
0.050	0.020	0.015	0.012	0.0098	0.00084
0.10	0.035	0.026	0.021	0.017	0.015
0.20	0.056	0.042	0.033	0.028	0.024
0.40	0.080	0.059	0.047	0.039	0.034
0.60	0.093	0.069	0.055	0.046	0.039

[*Answer*: non-competitive, $K_M = 0.30$ mmol L^{-1}, $K_I = 0.58$ mmol L^{-1}]

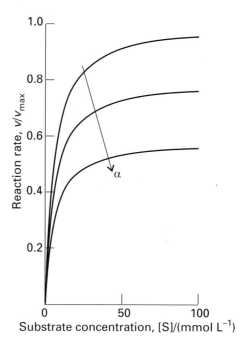

Fig 11.16 The variation of the rate of an enzyme-catalysed reaction with concentration of the substrate when a noncompetitive inhibitor is present. The uppermost curve corresponds to [I] = 0 (no inhibitor present).

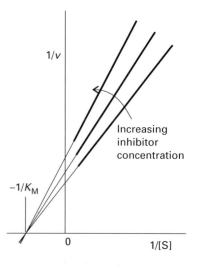

Fig 11.17 When there is noncompetitive inhibition, the intercept with the horizontal axis does not move as the concentration of inhibitor is increased.

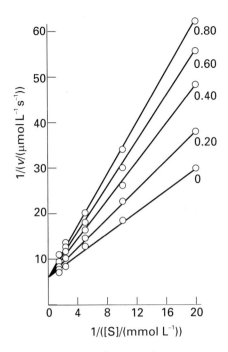

Fig 11.18 Lineweaver–Burk plots for the data in Example 11.2. Each line corresponds to a different concentration of inhibitor.

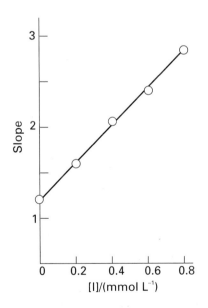

Fig 11.19 Plot of the slopes of the plots in Fig 11.18 against [I] based on the data in Example 11.2.

Chain reactions

MANY gas-phase reactions and liquid-phase polymerization reactions are **chain reactions**, reactions in which an intermediate produced in one step generates a reactive intermediate in a subsequent step, then that intermediate generates another reactive intermediate, and so on.

11.11 The structure of chain reactions

The intermediates responsible for the propagation of a chain reaction are called **chain carriers**. In a **radical chain reaction** the chain carriers are radicals. Ions may also propagate chains, and in nuclear fission the chain carriers are neutrons.

The first chain carriers are formed in the **initiation step** of the reaction. For example, Cl atoms are formed by the dissociation of Cl_2 molecules either as a result of vigorous intermolecular collisions in a thermolysis reaction or as a result of absorption of a photon in a photolysis reaction. The chain carriers produced in the initiation step attack other reactant molecules in the **propagation steps**, and each attack gives rise to a new chain carrier. An example is the attack of a methyl radical on ethane:

$$\cdot CH_3 + CH_3CH_3 \rightarrow CH_4 + \cdot CH_2CH_3$$

The dot signifies the unpaired electron and marks the radical. In some cases the attack results in the production of more than one chain carrier. An example of such a **branching step** is

$$\cdot O\cdot + H_2O \rightarrow HO\cdot + HO\cdot$$

where the attack of one O atom on an H_2O molecule forms two $\cdot OH$ radicals.[9]

---

[9] In the notation to be introduced in Chapter 13, an O atom has the configuration $[He]2s^2 2p^4$, with two unpaired electrons.

The chain carrier might attack a product molecule formed earlier in the reaction. Because this attack decreases the net rate of formation of product, it is called a **retardation step**. For example, in a photochemical reaction in which HBr is formed from H_2 and Br_2, an H atom might attack an HBr molecule, leading to H_2 and Br:

$$\cdot H + HBr \rightarrow H_2 + \cdot Br$$

Retardation does not end the chain, because one radical ($\cdot H$) gives rise to another ($\cdot Br$), but it does deplete the concentration of the product. Elementary reactions in which radicals combine and end the chain are called **termination steps**, as in

$$CH_3CH_2\cdot + \cdot CH_2CH_3 \rightarrow CH_3CH_2CH_2CH_3$$

In an **inhibition step**, radicals are removed other than by chain termination, such as by reaction with the walls of the vessel or with foreign radicals:

$$CH_3CH_2\cdot + \cdot R \rightarrow CH_3CH_2R$$

The NO molecule has an unpaired electron and is a very efficient chain inhibitor. The observation that a gas-phase reaction is quenched when NO is introduced is a good indication that a radical chain mechanism is in operation.

11.12 The rate laws of chain reactions

A chain reaction often leads to a complicated rate law (but not always). As a first example, consider the thermal reaction of H_2 with Br_2. The overall reaction and the observed rate law are

$$H_2(g) + Br_2(g) \rightarrow 2\,HBr(g)$$

$$\text{Rate of formation of } HBr = \frac{k[H_2][Br_2]^{3/2}}{[Br_2] + k'[HBr]} \quad (11.25)$$

The complexity of the rate law suggests that a complicated mechanism is involved. The following radical chain mechanism has been proposed:

Step 1 Initiation: $Br_2 \rightarrow Br\cdot + Br\cdot$
  Rate of consumption of $Br_2 = k_a[Br_2]$

At low pressures this elementary reaction is bimolecular and second-order in Br_2.

Step 2 Propagation: $Br\cdot + H_2 \rightarrow HBr + H\cdot$
  rate $= k_b[Br][H_2]$

  $H\cdot + Br_2 \rightarrow HBr + Br\cdot$
  rate $= k_b'[H][Br_2]$

In this and the following steps, 'rate' means either the rate of formation of one of the products or the rate of consumption of one of the reactants. We shall specify the species only if the rates differ.

Step 3 Retardation: $H\cdot + HBr \rightarrow H_2 + Br\cdot$
  rate $= k_c[H][HBr]$

Step 4 Termination: $Br\cdot + \cdot Br + M \rightarrow Br_2 + M$
  rate of formation of $Br_2 = k_d[Br]^2$

The 'third body', M, a molecule of an inert gas, removes the energy of recombination; the constant concentration of M has been absorbed into the rate constant k_d. Other possible termination steps include the recombination of H atoms to form H_2 and the combination of H and Br atoms; however, it turns out that only Br atom recombination is important.

Now we establish the rate law for the reaction. The experimental rate law is expressed in terms of the rate of formation of product, HBr, so we start by writing an expression for its net rate of formation. Because HBr is formed in step 2 (by both reactions) and consumed in step 3,

$$\text{Net rate of formation of } HBr = k_b[Br][H_2] + k_b'[H][Br_2] - k_c[H][HBr] \quad (11.26)$$

To make progress, we need the concentrations of the intermediates Br and H. Therefore, we set up the expressions for their net rate of formation and apply the steady-state assumption to both:

$$\text{Net rate of formation of } H = k_b[Br][H_2] - k_b'[H][Br_2] - k_c[H][HBr] = 0$$

Net rate of formation of Br =

$2k_a[Br_2] - k_b[Br][H_2] + k_b'[H][Br_2] + k_c[H][HBr] - 2k_d[Br]^2$

$$= 0$$

The steady-state concentrations of the intermediates are found by solving these two equations and are

$$[Br] = \left(\frac{k_a[Br_2]}{k_d}\right)^{1/2} \qquad [H] = \frac{k_b(k_a/k_d)^{1/2}[H_2][Br_2]^{1/2}}{k_b'[Br_2] + (k_c/k_b)[HBr]}$$

When we substitute these concentrations into eqn 11.26 we obtain

$$\text{Rate of formation of HBr} = \frac{2k_b(k_a/k_d)^{1/2}[H_2][Br_2]^{3/2}}{[Br_2] + (k_c/k_b')[HBr]}$$

$$(11.27)$$

This equation has the same form as the empirical rate law, and we can identify the two empirical rate coefficients as

$$k = 2k_b\left(\frac{k_a}{k_d}\right)^{1/2} \qquad k' = \frac{k_c}{k_b'}$$

$$(11.28)$$

We can conclude that the proposed mechanism is at least consistent with the observed rate law. Additional support for the mechanism would come from the detection of the proposed intermediates (by spectroscopy), and the measurement of individual rate constants for the elementary steps and confirming that they correctly reproduce the observed composite rate constants.

11.13 Explosions

A **thermal explosion** is due to the rapid increase of reaction rate with temperature. If the energy released in an exothermic reaction cannot escape, the temperature of the reaction system rises, and the reaction goes faster. The acceleration of the rate results in a faster rise of temperature, and so the reaction goes even faster—catastrophically fast. A **chain-branching explosion** may occur when there are chain-branching steps in a reaction, for then the number of chain carriers grows

exponentially and the rate of reaction may cascade into an explosion.

An example of both types of explosion is provided by the reaction between hydrogen and oxygen:

$$2\,H_2(g) + O_2(g) \rightarrow 2\,H_2O(g)$$

Although the net reaction is very simple, the mechanism is very complex and has not yet been fully elucidated. It is known that a chain reaction is involved, and that the chain carriers include $\cdot H$, $\cdot O\cdot$, $\cdot OH$, and $\cdot O_2H$. Some steps are:

Initiation: $\qquad H_2 + \cdot(O_2)\cdot \rightarrow \cdot OH + \cdot OH$

Propagation: $\quad H_2 + \cdot OH \rightarrow \cdot H + H_2O$

$\qquad\qquad \cdot(O_2)\cdot + \cdot H \rightarrow \cdot O\cdot + \cdot OH$ (branching)

$\qquad\qquad \cdot O\cdot + H_2 \rightarrow \cdot OH + \cdot H$ (branching)

$\qquad\qquad \cdot H + \cdot(O_2)\cdot + M \rightarrow \cdot HO_2 + M^*$

The two branching steps can lead to a chain-branching explosion.

The occurrence of an explosion depends on the temperature and pressure of the system, and the **explosion regions** for the reaction are shown in Fig 11.20. At very low pressures, the system is outside the explosion region and the mixture reacts smoothly. At these pressures the chain carriers produced in the branching steps can reach the walls of the container where they combine (with an efficiency that depends on the composition of the walls). Increasing the pressure of the mixture along the broken line in the illustration takes the system through the **lower explosion limit** (provided that the temperature is greater than about 730 K). The mixture then explodes because the chain carriers react before reaching the walls and the branching reactions are explosively efficient. The reaction is smooth when the pressure is above the **upper explosion limit**. The concentration of molecules in the gas is then so great that the radicals produced in the branching reaction combine in the body of the gas, and gas-phase reactions such as $O_2 + \cdot H \rightarrow \cdot O_2H$ can occur. Recombination reactions like this are facilitated by three-body collisions, because the third body

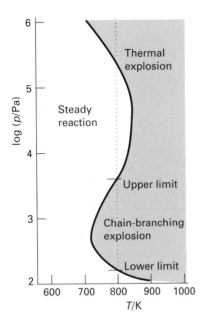

Fig 11.20 The explosion limits of the reaction between hydrogen and oxygen. In the explosive regions, the reaction proceeds explosively when heated homogeneously.

(M) can remove the excess energy and allow the formation of a bond:

$$O_2 + \cdot H + M \rightarrow \cdot O_2H + M^*$$

The radical $\cdot OH_2$ is relatively unreactive and can reach the walls, where it is removed. At low pressures three-particle collisions are unimportant and recombination is much slower. At higher pressures, when three-particle collisions are important, the explosive propagation of the chain by the radicals produced in the branching step is partially quenched because $\cdot O_2H$ is formed in place of $\cdot O \cdot$ and $\cdot OH$. If the pressure is increased to above the third explosion limit the reaction rate increases so much that a thermal explosion occurs.

Photochemical processes

MANY reactions can be initiated by the absorption of light. The most important of all are the photochemical processes that capture the Sun's radiant energy. Some of these reactions lead to the heating of the atmosphere during the daytime by absorption in the ultraviolet region as a result of reactions like those depicted in Fig 11.21. Others include the absorption of red and blue light by chlorophyll and the subsequent use of the energy to bring about the synthesis of carbohydrates from carbon dioxide and water. Without photochemical processes the world would be simply a warm, sterile, rock (Box 11.2).

11.14 Quantum yield

A molecule acquires enough energy to react by absorbing a photon. As we shall see in Section 12.1, a photon is a packet of electromagnetic radiation; if the frequency of the radiation is v (nu), then the energy of each photon is hv, where h is Planck's constant, $h = 6.626\ 08 \times 10^{-34}$ J s. Therefore, when a molecule absorbs radiation of frequency v, it acquires the energy brought to it by the photon and can use that energy to undergo further reaction. However, not every excited molecule may form a specific primary product (atoms, radicals, or ions, for instance) because there are many ways in which the excitation may be lost other than by dissociation or ionization.[10] We therefore speak of the **primary quantum yield**, ϕ (phi), which is the number of reactant molecules producing specified primary products (atoms or ions, for instance) for each photon absorbed. If each molecule that absorbs a photon undergoes dissociation (for instance), then $\phi = 1$. If none does, because the excitation energy is lost before the molecule has time to dissociate, then $\phi = 0$.

---

[10] One route to the loss of energy is *fluorescence*. That topic, and its kinetic analysis, are treated in Section 18.6.

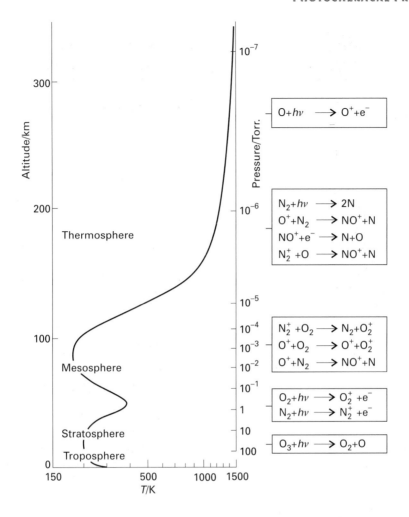

Fig 11.21 The temperature profile through the atmosphere and some of the reactions that occur. The temperature peak at about 50 km is due to the absorption of solar radiation by the O_2 and N_2 reactions.

One successfully excited molecule might initiate the consumption of more than one reactant molecule. We therefore need to introduce the **overall quantum yield**, Φ (uppercase phi), which is the number of reactant molecules that react for each photon absorbed. In the photolysis of HI, for example, the processes are

$$HI + h\nu \rightarrow H + I$$
$$H + HI \rightarrow H_2 + I$$
$$I + I + M \rightarrow I_2$$

The overall quantum yield is 2 because the absorption of one photon leads to the destruction of two HI molecules. In a photochemically initiated chain reaction, Φ may be very large, and values of about 10^4 are common. In such cases the chain reaction acts as a chemical amplifier of the initial absorption step.

Box 11.2 *Photobiology*

The absorption of electromagnetic radiation is the starting point for a variety of important processes in biology, among them photosynthesis, mutation, and vision.[11] Photobiology also includes processes that repair damage that would otherwise result in mutation or disease.

Damage to DNA, for instance, can be caused by ultraviolet (UV) radiation, most effectively near 260 nm. In one type of process the irradiation of pyrimidine bases, particularly but not only thymines, in adjacent nucleotides form covalent bonds (**5**) and distort the chain, so hindering replication. The damage can be repaired either by chemical excision and replacement of the dimerized region, or by *enzymatic photoreactivation*. In the latter case, an enzyme, *DNA photolyase*, binds to the damaged region of the chain. The resulting complex absorbs visible light (most effectively at 370 nm) and the absorbed energy cleaves the bonds within the pyrimidine dimer.

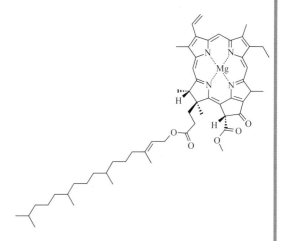

6 Chlorophyll

5 Nucleotide dimer

The photobiological process of enormous importance is *oxygenic photosynthesis*, the formation of carbohydrates from carbon dioxide and water in the deceptively simple reaction

$$CO_2(g) + H_2O(l) \rightarrow \text{'}CH_2O(aq)\text{'} + O_2(g)$$

The simplicity of this reaction conceals a network of processes of considerable complexity. Broadly speaking, the process has two components, a *light reaction* and a *dark reaction*. The light reaction depends on the absorption of visible light by two organized regions of a cell, one is called *photosystem I* (PS I) and the other *photosystem II* (PS II). Each photosystem consists of a collection of 'antenna' chlorophyll molecules (**6**). When a photon is absorbed, the excitation it produces hops between chlorophyll molecules until it ends up in a reaction centre

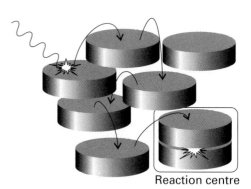

Reaction centre

The energy brought by a photon is absorbed by a chlorophyll molecule (represented by the disk) and then hops around on to other chlorophyll molecules. The hopping terminates when the energy excites a special pair of chlorophyll molecules.

that contains a special pair called P700 in PS I and P680 in PS II (see illustration); the numbers indicate the wavelengths (in nanometres) up to which the reaction centre absorbs radiation. The energy of the excited reaction centre is slightly less than that of an unassociated chlorophyll molecule, so the excitation is trapped in the centre.

The electron excited out of P680 in PS II is replenished by the oxidation of water in the reaction $2\,H_2O(l) \rightarrow 4\,H^+(aq) + O_2(g) + 4\,e^-$. The excited electron passes down a chain of molecules to replace an electron that has been excited out of P700 in PS I. This electron transfer releases energy to produce ATP from ADP (see the discussion of the chemiosmotic theory in Box 9.2). The electron excited out of P700 by the absorption of another photon is

---

[11] See Box 18.1 for a discussion of vision.

passed down another chain of molecules and reduces NADP$^+$ (**7**) to NADPH. At this point the sequence of dark reactions can begin, in which NADPH and ATPase are used to generate carbohydrate in a series of biochemical processes mediated by enzymes.

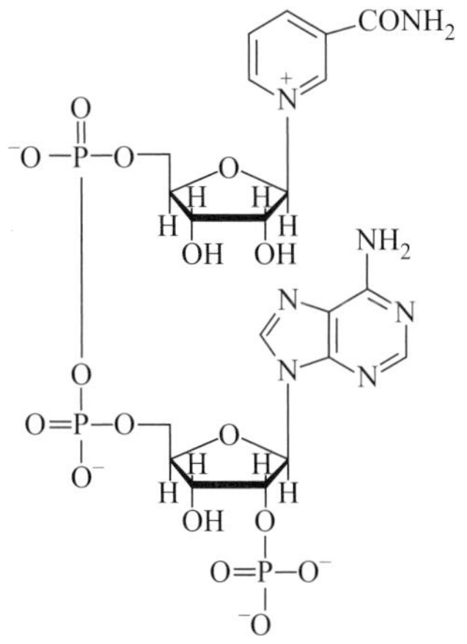

7 Nicotinamide adenine dinucleotide phosphate, NADP$^+$

Exercise 1 Electromagnetic radiation of wavelength 260 nm is absorbed by DNA. If a photon of radiation at this wavelength can cause a transition resulting in bond cleavage, what is the energy of this transition in kilocalories per mole (kcal mol^{-1})?

Exercise 2 A DNA sample has an absorbance at 260 nm of 2.0 in a cell of pathlength 1.0 cm. What percentage of light is being transmitted? If a sample of 2.0 mg mL^{-1} of DNA has an absorbance of 0.1 at 260 nm, what is the concentration of the sample with an absorbance of 2.0? (*Hint.* See Chapter 18.)

Example 11.3 *Using the quantum yield*

The overall quantum yield for the formation of ethene from 4-heptanone with 313 nm light is 0.21. How many molecules of 4-heptanone per second, and what chemical amount per second, are destroyed when the sample is irradiated with a 50 W, 313 nm source under conditions of total absorption?

Strategy We need to calculate the number of photons emitted by the lamp per second; all are absorbed (by assertion); the number of molecules destroyed per second is the number of photons absorbed multiplied by the overall quantum yield, Φ. The number of photons emitted by the source per second is the power (joules per second) divided by the energy of a single photon ($E = h\nu$, with $\nu = c/\lambda$).

Solution The energy of a photon of wavelength λ is $E = hc/\lambda$, so a source of power P generates photons at a rate P divided by E:

Rate of photon production

$$= \frac{P}{(hc/\lambda)} = \frac{P\lambda}{hc}$$

$$= \frac{(50 \text{ J s}^{-1}) \times (313 \times 10^{-9} \text{ m})}{(6.62608 \times 10^{-34} \text{ J s}) \times (2.99792 \times 10^8 \text{ m s}^{-1})}$$

$$= 7.9 \times 10^{19} \text{ s}^{-1}$$

The number of 4-heptanone molecules destroyed per second is therefore 0.21 times this quantity, or $1.7 \times 10^{19} \text{ s}^{-1}$, corresponding (after division by N_A) to $2.8 \times 10^{-5} \text{ mol s}^{-1}$.

Self-test 11.5

The overall quantum yield for another reaction at 290 nm is 0.30. For what length of time must irradiation with a 100 W source continue in order to destroy 1.0 mol of molecules?

[*Answer:* 3.8 h]

11.15 **Photochemical rate laws**

As an example of how to incorporate the photo-chemical activation step into a mechanism, con-sider the photochemical activation of the reaction

$$H_2(g) + Br_2(g) \rightarrow 2\,HBr(g)$$

In place of the first step in the thermal reaction we have

$$Br_2 + h\nu \rightarrow Br + Br \quad \text{rate of consumption of } Br_2 = I_{abs}$$

where I_{abs} is the number of photons of the appropri-ate frequency absorbed divided by the time interval and the volume. Because the thermal reaction mechanism had $k_a[Br_2]$ in place of I_{abs} for the equiva-lent of this step, it follows that I_{abs} should take the place of $k_a[Br_2]$ in the rate law we derived for the thermal reaction scheme. Therefore, from eqn 11.27 we can write

$$\text{Rate of formation of HBr} = \frac{2\,k_b\,(1/k_d)^{1/2}\,[H_2][Br_2]I_{abs}^{1/2}}{[Br_2]+(k_c/k_b')[HBr]} \quad (11.29)$$

Although the details of this expression are compli-cated, the essential prediction is clear: the reaction rate should depend on the square root of the ab-sorbed light intensity. This prediction is confirmed experimentally.

Exercises

11.1 The equilibrium constant for the attachment of a substrate to the active site of an enzyme was measured as 235. In a separate experiment, the rate constant for the second-order attachment was found to be 7.4×10^7 L mol^{-1} s^{-1}. What is the rate constant for the loss of the unreacted substrate from the active site?

11.2 Confirm (by differentiation) that the expressions in eqn 11.1 are the correct solutions of the rate laws for approach to equilibrium.

11.3 Find the solutions of the same rate laws that led to eqn 11.1, but for some B present initially. Go on to con-firm that the solutions you find reduce to those in eqn 11.1 when $[B]_0 = 0$. (*Hint.* This exercise requires calculus.)

11.4 Sketch, without carrying out the calculation, the variation of concentration with time for the approach to equilibrium when both forward and reverse reactions are second order. How does your graph differ from that in Fig 11.1?

11.5 Confirm (by differentiation) that the three expres-sions in eqn 11.4 are correct solutions of the rate laws for consecutive first-order reactions.

11.6 Confirm that the time at which the intermediate reaches its maximum concentration is given by eqn 11.5. (*Hint.* See footnote 1.)

11.7 Two radioactive nuclides decay by successive first-order processes:

$$X \xrightarrow{22.5\,d} Y \xrightarrow{33.0\,d} Z$$

(The times are half-lives in days.) Suppose that Y is an iso-tope that is required for medical applications. At what stage after X is first formed will Y be most abundant?

11.8 The reaction $2\,H_2O_2(aq) \rightarrow 2\,H_2O(l) + O_2(g)$ is catal-ysed by Br$^-$ ions. If the mechanism is:

$$H_2O_2 + Br^- \rightarrow H_2O + BrO^- \text{ (slow)}$$
$$BrO^- + H_2O_2 \rightarrow H_2O + O_2 + Br^- \text{ (fast)}$$

give the predicted order of the reaction with respect to the various participants.

11.9 The reaction mechanism

$$A_2 \rightleftharpoons A + A \text{ (fast)}$$
$$A + B \rightarrow P \text{ (slow)}$$

involves an intermediate A. Deduce the rate law for the formation of P.

11.10 Consider the following mechanism for renaturation of a double helix from its strands A and B:

$$A + B \rightleftharpoons \text{unstable helix (fast)}$$

unstable helix → stable double helix (slow)

Derive the rate equation for the formation of the double helix and express the rate constant of the renaturation reaction in terms of the rate constants of the individual steps.

11.11 The following mechanism has been proposed for the decomposition of ozone in the atmosphere:

(1) $O_3 \rightarrow O_2 + O$ and its reverse (k_1, k_1')

(2) $O + O_3 \rightarrow O_2 + O_2$ $(k_2$; the reverse reaction is negligibly slow)

Use the steady-state approximation, with O treated as the intermediate, to find an expression for the rate of decomposition of O_3. Show that, if step 2 is slow, then the rate is second-order in O_3 and −1 order in O_2.

11.12 The condensation reaction of acetone, $(CH_3)_2CO$ (propanone), in aqueous solution is catalysed by bases, B, which react reversibly with acetone to form the carbanion $C_3H_5O^-$. The carbanion then reacts with a molecule of acetone to give the product. A simplified version of the mechanism is

(1) $AH + B \rightarrow BH^+ + A^-$

(2) $A^- + BH^+ \rightarrow AH + B$

(3) $A^- + HA \rightarrow$ product

where AH stands for acetone and A⁻ its carbanion. Use the steady-state approximation to find the concentration of the carbanion and derive the rate equation for the formation of the product.

11.13 Consider the acid-catalysed reaction

$HA + H^+ \rightleftharpoons HAH^+$ (fast)

$HAH^+ + B \rightarrow BH^+ + AH$ (slow)

Deduce the rate law and show that it can be made independent of the specific term $[H^+]$.

11.14 Show that the Langmuir isotherm (eqn 11.11) implies that the constant K can be determined experimentally by plotting $1/\theta$ against $1/p_A$.

11.15 The data below are for the adsorption of CO on charcoal at 273 K. Confirm that they fit the Langmuir isotherm and determine (a) the constant K and (b) the

volume corresponding to complete coverage. (All volumes have been corrected to 1.00 atm.) (*Hint*: Write $\theta = V/V_{mon}$, where V_{mon} is the volume corresponding to complete (monolayer) coverage.)

p_{CO}/Torr	100	200	300	400	500	600	700
V/cm³	10.2	18.6	25.5	31.5	36.9	41.6	46.1

11.16 The values of K for the adsorption of CO on charcoal are 1.0×10^{-3} Torr⁻¹ at 273 K and 2.7×10^{-3} Torr⁻¹ at 250 K. Estimate the enthalpy of adsorption. (*Hint.* Recall the van't Hoff equation, eqn 7.14.)

11.17 If two reactants A and B compete for the same sites on a surface, their fractional coverages have the form

$$\theta_J = \frac{K_J p_J}{1 + K_A p_A + K_B p_B}$$

According to the Langmuir–Hinshelwood mechanism of surface-catalysed reactions, the rate of reaction between A and B depends on the rate at which the adsorbed species meet. (a) Write the rate law for the reaction according to this mechanism. (b) Find the limiting form of the rate law when the partial pressures of the reactants are low. (c) Could this mechanism ever account for zero-order kinetics?

11.18 Consider the Lindemann mechanism of unimolecular reactions, but modify it by supposing that glass wool has been inserted into the reaction vessel. Derive the rate law on the basis that A* can be deactivated by collisions with the surface of the glass wool, of area S. Show what happens to the rate law when the surface area is increased so much that this mode of quenching dominates.

11.19 The enzyme-catalysed conversion of a substrate at 25°C has a Michaelis constant of 0.045 mol L⁻¹. The rate of the reaction is 1.15 mmol L⁻¹ s⁻¹ when the substrate concentration is 0.110 mol L⁻¹. What is the maximum velocity of this enzymolysis?

11.20 Find the condition for which the reaction rate of an enzymolysis that follows Michaelis–Menten kinetics is half its maximum value.

11.21 Show that a plot of the rate of enzymolysis, v, plotted against $v/[S]$ is an alternative route to the determination of the value of K_M.

11.22 Calculate the ratio of rates of enzyme catalysed to non-catalysed reactions at 37°C given that the Gibbs energy of activation for a particular reaction is reduced from 100 kJ mol^{-1} to 10 kJ mol^{-1}.

11.23 The rate, v, of an enzyme-catalysed reaction was measured when various amounts of substrate S were present and the concentration of enzyme was 12.5 μmol L^{-1}. The following results were obtained:

[S]/(mmol L^{-1})	1.0	2.0	3.0	4.0	5.0
v/(μmol L^{-1} s^{-1})	1.1	1.8	2.3	2.6	2.9

Determine the Michaelis–Menten constant, the maximum velocity of the reaction, and the maximum turnover number of the enzyme.

11.24 The following results were obtained for the action of an ATPase on ATP at 20°C, when the concentration of the ATPase is 20 nmol L^{-1}:

[ATP]/(μmol L^{-1})	0.60	0.80	1.4	2.0	3.0
v/(μmol L^{-1} s^{-1})	0.81	0.97	1.30	1.47	1.69

Determine the Michaelis–Menten constant, the maximum velocity of the reaction, and the maximum turnover number of the enzyme.

11.25 The following results were obtained when the rate of an enzymolysis was monitored (a) without inhibitor, (b) with inhibitor at a concentration of 15 μmol L^{-1}:

[S]/(10^{-4} mol L^{-1})	1.0	3.0	7.0	12.0	18.0
v/(μmol L^{-1} s^{-1}) (a)	0.49	0.95	1.3	1.5	1.6
v/(μmol L^{-1} s^{-1}) (b)	0.27	0.52	0.71	0.81	0.86

Is the inhibition competitive or non-competitive?

11.26 Consider the following chain mechanism:

(1) $AH \rightarrow A\cdot + H\cdot$

(2) $A\cdot \rightarrow B\cdot + C$

(3) $AH + B\cdot \rightarrow A\cdot + D$

(4) $A\cdot + B\cdot \rightarrow P$

Identify the initiation, propagation, and termination steps, and use the steady-state approximation to deduce that the decomposition of AH is first-order in AH.

11.27 Consider the following mechanism for the thermal decomposition of R_2:

(1) $R_2 \rightarrow R + R$

(2) $R + R_2 \rightarrow P_B + R'$

(3) $R' \rightarrow P_A + R$

(4) $R + R \rightarrow P_A + P_B$

where R_2, P_A, and P_B are stable hydrocarbons and R and R' are radicals. Find the dependence of the rate of decomposition of R_2 on the concentration of R_2.

11.28 Refer to Fig. 11.20 and determine the pressure range for branching-chain explosion in the hydrogen–oxygen reaction at (a) 700 K, (b) 800 K, and (c) 900 K.

11.29 In a photochemical reaction A → 2 B + C, the overall quantum yield with 500 nm light is 2.1×10^2 mol einstein^{-1}, where 1 einstein = 1 mol photons. After exposure of 300 mmol A to the light, 2.15 mmol B is formed. How many photons were absorbed by A?

11.30 In an experiment to measure the quantum yield of a photochemical reaction, the absorbing substance was exposed to 490 nm light from a 100 W source for 45 minutes. The intensity of the transmitted light was 35 per cent of the intensity of the incident light. As a result of irradiation, 0.297 mol of the absorbing substance decomposed. Find the quantum yield.

Quantum theory

Contents

THE phenomena of chemistry cannot be understood thoroughly without a firm understanding of the principal concepts of quantum mechanics. The same is true of virtually all the spectroscopic techniques that are now so central to investigations of composition and structure. Present day techniques for studying chemical reactions have progressed to the point where the information is so detailed that quantum mechanics has to be used in its interpretation. And, of course, the very currency of chemistry—the electronic structures of atoms and molecules—cannot be discussed without making use of quantum mechanical concepts.

The role—indeed, the existence—of quantum mechanics was appreciated only during the twentieth century. Until then it was thought that the motion of atomic and subatomic particles could be expressed in terms of the laws of classical mechanics introduced in the seventeenth century by Isaac Newton (see *Further information 3*), for these laws were very successful at explaining the motion of planets and everyday objects such as pendulums and projectiles. However, towards the end of the nineteenth century, experimental evidence accumulated showing that classical mechanics failed when it was applied to very small particles, such as individual atoms, nuclei, and electrons, and when the transfers of energy were very small. It took until 1926 to identify the appropriate concepts and equations for describing them.

The failures of classical physics

IN this section, we see how it came to be realized that classical mechanics has severe shortcomings, particularly when applied to systems in which only small energies are transferred. To appreciate these shortcomings, we need to know that classical physics is based on three 'obvious' assumptions:

1 A particle travels in a **trajectory**, a path with a precise position and momentum[1] at each instant.

2 Any type of motion can be excited to a state of arbitrary energy.

3 Waves and particles are distinct concepts.

These assumptions agree with everyday experience. For example, a pendulum swings with a precise oscillating motion and can be made to oscillate with any energy simply by pulling it back to an arbitrary angle and then letting it swing freely. Classical mechanics lets us predict the angle of the pendulum and the speed at which it is swinging at any instant. Everyday experience, however, does not extend to the behaviour of individual atoms and subatomic particles. Careful experiments of the type described below have shown that the laws of classical mechanics, and particularly the three basic assumptions, fail to account for the observed behaviour of very small particles. Classical mechanics is in fact only an *approximate* description of the motion of particles, and the approximation is invalid when it is applied to molecules, atoms, and electrons.

12.1 Black-body radiation

Electromagnetic radiation is a wave of electric and magnetic fields travelling through space.[2] Such waves are generated by the acceleration of electric charge, as in the oscillating motion of electrons in the antenna of a radio transmitter. A hot object emits electromagnetic radiation because its atoms and electrons are ceaselessly being accelerated: the atoms vibrate around their mean positions and their electrons are moved from location to location. At high temperatures, an appreciable proportion of the radiation is in the visible region of the spectrum. A higher proportion of short-wavelength blue light is generated as the temperature is raised. We observe this behaviour when an iron bar glowing red hot becomes white hot when heated further, because then more blue light mixes into the red light and changes the perceived colour towards white. The precise dependence is illustrated in Fig 12.1, which shows how the energy output varies with wavelength at a series of temperatures. The curves are those of an ideal emitter called a **black body**, which is a body capable of emitting and absorbing all frequencies of radiation. A good approximation to a black body is a pinhole in a container, because the radiation leaking out of the hole has been absorbed and re-emitted inside the container so

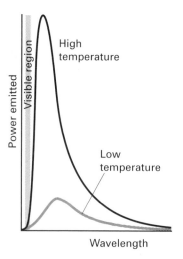

Fig 12.1 The power emitted by a black body at two temperatures. Note how the power increases in the visible region (the tinted range of wavelengths) as the temperature is raised, and how the peak maximum moves to shorter wavelengths. The total power (the area under the curve) increases as the temperature is increased (as T^4).

---

[1] Momentum, p, is the product of mass, m, and velocity, v: $p = mv$.

[2] We do not need to know many of the properties of electromagnetic radiation, but those we do need are summarized in *Further information 7*.

many times that it has come to thermal equilibrium with the walls.[3]

We shall consider the **energy density** of the radiation in a cavity, the total energy in the cavity divided by the volume of the cavity,[4] and in particular the contribution to the total energy density from radiation of different wavelengths. Figure 12.1 shows two main features. The first is that shorter wavelengths contribute more to the energy density as the temperature is raised. As a result, the perceived colour shifts towards the blue, as already mentioned. An analysis of the data led Wilhelm Wien (in 1893) to summarize this shift by a statement now known as **Wien's displacement law**:

$$T\lambda_{max} = \text{constant} \tag{12.1}$$

where the value of the constant is 2.9 mm K. In this expression, λ_{max} is the wavelength of the radiation that makes the greatest contribution to the energy density when the temperature is T. Wien's law implies that, as T increases, λ_{max} decreases enough to preserve the value of $T\lambda_{max}$.

Illustration 12.1

One application of Wien's law is to the estimation of the temperatures of stars, and other inaccessible hot objects. For example, the maximum emission of the Sun occurs at $\lambda_{max} \approx 490$ nm, so its surface temperature must be close to

$$T = \frac{\text{constant}}{\lambda_{max}} = \frac{2.9 \times 10^{-3} \text{ m K}}{4.9 \times 10^{-7} \text{ m}} = 5.9 \times 10^3 \text{ K}$$

or about 6000 K.

Self-test 12.1

Estimate the wavelength at the maximum energy output of an incandescent lamp if the filament is at 3000°C.

[*Answer*: 890 nm]

---

[3] Thermal equilibriun. in the sense that the temperature of the radiation-filled space inside the container is the same as that of the walls.

[4] Just as we get the mass of an object by multiplying its volume by its mass density, so we get the total energy in a container by multiplying the energy density by the volume of the container.

The second feature of black-body radiation had been noticed in 1879 by Josef Stefan, who considered the sharp rise in the **emittance**, M, the total power emitted (the rate of energy output) by a black body divided by the surface area of the body, as the temperature is raised. He established what is now called the **Stefan–Boltzmann law**:

$$M = aT^4 \tag{12.2}$$

with $a = 56.7$ nW m^{-2} K^{-4} (where 1 nW = 10^{-9} W). This law implies that each square centimetre of the surface of a black body at 1000 K radiates about 5.7 W when all wavelengths are taken into account. It radiates $3^4 = 81$ times that power (460 W) when the temperature is increased by a factor of 3, to 3000 K. The law is the basis of seeking as high a temperature as possible for an incandescent lamp, for then the emission is as strong as possible.

Self-test 12.2

Suppose technological advances made it possible to produce a ceramic material that could be used as a filament at 3800°C instead of 3000°C. By what factor would the power output of a lamp that used the new material increase?

[*Answer*: 2.4]

The physicist Lord Rayleigh studied black-body radiation from a classical viewpoint. In his day (at the end of the nineteenth century), electromagnetic radiation was regarded as waves in a ubiquitous 'ether'. The idea of the time was that, if the ether could oscillate at a certain frequency v (nu), then radiation of that frequency would be present. Rayleigh took the view that the ether could oscillate with any frequency, so waves could exist in it of any wavelength. He calculated the contribution to the energy density in a narrow wavelength range of width $\Delta\lambda$ at any wavelength λ (lambda). With minor help from James Jeans, Rayleigh arrived at the **Rayleigh–Jeans law**:

Contribution to the energy density from radiation in the range λ to $\lambda + \Delta\lambda = \rho\Delta\lambda$, with

$$\rho = \frac{8\pi kT}{\lambda^4} \tag{12.3}$$

The power emitted in the wavelength range $\Delta\lambda$ is proportional to the energy density in that range.

Unfortunately (for Rayleigh, Jeans, and classical physics), although the Rayleigh–Jeans law is quite successful at long wavelengths (low frequencies), it fails badly at short wavelengths (high frequencies). Thus, as λ decreases, ρ goes on increasing (Fig 12.2). The law therefore predicts that oscillations of very short wavelength (high frequency, corresponding to ultraviolet radiation, X-rays, and even γ-rays) are strongly excited even at room temperature. So, according to classical physics, every time you strike a match, you blast the surroundings with γ-rays! This absurd result is called the **ultraviolet catastrophe**.

In 1900, the German physicist Max Planck found that he could account for the characteristics of black-body radiation by proposing that *the energy of each electromagnetic oscillator is limited to discrete values and cannot be varied arbitrarily.* Thus, the oscillation of the electromagnetic field that corresponds to yellow light, for instance, can be stimulated only if a certain energy is provided. This limitation of the energy to discrete values is called the **quantization of energy**. Specifically, Planck proposed that the energy of an oscillator of frequency v is restricted to an integral multiple of the quantity hv, where h is a fundamental constant now known as **Planck's constant**:

$$E = nhv \quad \text{where} \quad h = 6.626 \times 10^{-34} \text{ J s} \quad (12.4)$$

with $n = 0, 1, 2, \ldots$ (Fig 12.3).[5]

Self-test 12.3

What is the minimum energy that can be used to excite an oscillator corresponding to yellow light of frequency 5.2×10^{14} Hz?

[*Answer*: 3.4×10^{-19} J]

When Planck calculated ρ using his quantization postulate, he obtained the **Planck distribution**:

$$\rho = \frac{8\pi hc}{\lambda^5}\left(\frac{1}{e^{hc/\lambda kT} - 1}\right) \quad (12.5)$$

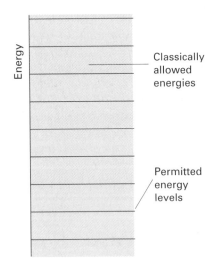

Fig 12.3 According to classical physics, an oscillator (including the oscillators that correspond to vibrations of the electromagnetic field and correspond to radiation of a particular frequency) can have any energy (as depicted by the tinted range of energies). Planck's proposal implied that an oscillator could be excited only in discrete steps, for it can possess only certain energies (those depicted by the horizontal lines in the illustration).

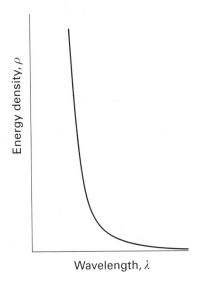

Fig 12.2 The Rayleigh–Jeans law leads to an infinite energy density at short wavelengths and gives rise to the ultraviolet catastrophe.

---

[5] To use this expression, we need to note that 1 Hz = 1 s^{-1}, so 1 Hz $\times$ 1 s = 1.

This function is another example of a slightly daunting expression that is better to understand than remember:

1 Figure 12.4 shows a plot of the function: it compares very favourably with the experimental curve shown in Fig 12.1. Note that the curve passes through a maximum.

2 As λ approaches zero, $hc/\lambda kT$ approaches infinity and $e^{hc/\lambda kT}$ approaches infinity too. Therefore, the denominator in the Planck distribution becomes infinite and ρ approaches zero. This behaviour eliminates the ultraviolet catastrophe.

3 When the wavelength is very long, $hc/\lambda kT$ is very small. We can then replace $e^{hc/\lambda kT}$ by $1 + hc/\lambda kT + \cdots$ and keep just the first two terms.[6] Then the denominator becomes simply $hc/\lambda kT$ and the Planck distribution becomes the same as the Rayleigh–Jeans distribution.

Self-test 12.4

Calculate the ratio of the energy densities predicted by the Planck and Rayleigh–Jeans formulae for yellow (580 nm) light at 1000 K.

[*Answer:* $\rho_{Rayleigh}/\rho_{Planck} = 2.4 \times 10^9$]

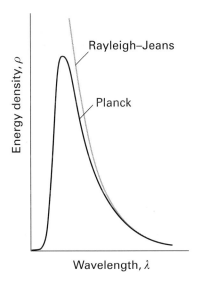

Fig 12.4 The Planck distribution results in a very low energy density for high-frequency, short-wavelength oscillators and is in excellent agreement with experiment.

What is the physical reason for the success of Planck's quantization hypothesis? The atoms in the walls of the black body undergo thermal motion, and this motion excites the oscillators of the electromagnetic field. According to classical mechanics, all the electromagnetic oscillators are excited, even those of very high frequency, and the corresponding wavelengths of radiation are emitted, including radiation of very short wavelength. According to quantum mechanics, however, the oscillators are excited only if they can acquire an energy of at least $h\nu$. This minimum energy is too large for the walls to supply in the case of the high-frequency oscillators, so the latter remain unexcited. The effect of quantization is to quench the contribution from the high-frequency oscillators, for they cannot be excited with the energy available, and hence the very short wavelength radiation is not emitted.

The Planck distribution accounts quantitatively for the Stefan–Boltzmann and Wien laws. Thus, when the area under the graph in Fig 12.4 is calculated (to obtain the total energy density over the entire wavelength range), the resulting expression is proportional to T^4, in accord with the Stefan–Boltzmann law. Similarly, when we calculate the wavelength λ_{max} corresponding to the maximum point of the curve, we find that its value is inversely proportional to the temperature, which is in agreement with Wien's law.

12.2 Heat capacities

In Section 2.4 we saw that heat capacity, C, is the constant of proportionality between the rise in temperature, ΔT, of a sample and the heat, q, supplied:

$$q = C\Delta T$$

We can suspect that there are similarities between black-body radiation and heat capacities. The former involves examining how energy is taken up by the oscillations of the electromagnetic field. The latter involves examining how energy is taken up by

---

[6] A very important expansion is $e^x = 1 + x + \frac{1}{2}x^2 + \cdots$.

the oscillation of atoms about their mean positions in the solid.

On the basis of some somewhat slender experimental evidence, the French scientists Pierre-Louis Dulong and Alexis-Thérèse Petit had proposed in 1819 that the molar heat capacity of all monatomic solids—such as metals—was equal (in modern units) to about 25 J K^{-1} mol^{-1}. Classical physics was able to account for this value quite readily for, if we assume that the atomic oscillators can be excited to any energy, then the predicted molar heat capacity is $3R$, where R is the gas constant, and $3R = 25$ J K^{-1} mol^{-1}.[7]

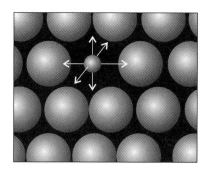

Fig 12.5 An atom in a solid can oscillate about its position in three perpendicular directions and, when a solid is heated, the vigour of the motion increases. This illustration depicts a view of one layer of atoms in a solid and shows the oscillations of one of the atoms (which is drawn smaller for clarity).

Derivation 12.1 *The heat capacity of a monatomic solid*

The N atoms of a solid can oscillate in any of three perpendicular directions (Fig 12.5). The solid is therefore equivalent to a collection of $3N$ oscillators. The second step makes use of a conclusion from classical physics known as the *equipartition theorem*, which implies that, at a temperature T, the average energy of an oscillator is kT, where k is Boltzmann's constant. It follows that the total energy of the N vibrating atoms is $3N \times kT$. The energy per mole of atoms is therefore $3N_A kT$, or $3RT$, because the number of atoms per mole is N_A. Finally, we note that, because the molar energy is $3RT$, when the temperature of the sample increases by ΔT, the molar energy increases by $3R\Delta T$. This increase in energy must be supplied as heat from the surroundings, so the heat required to raise the temperature by ΔT is $3R\Delta T$. It follows by comparison with the definition of molar heat capacity ($q = C_m\Delta T$) that $C_m = 3R$.

The apparent success of classical mechanics in accounting for observed heat capacities was short-lived for, when technological advances made it possible to measure heat capacities at low temperatures, all substances were found to have values significantly lower than 25 J K^{-1} mol^{-1}. At very low temperatures, the heat capacity was found to approach zero (Fig 12.6). Even some quite common

substances at room temperature were found to have molar heat capacities well below the expected value: the value for diamond, for instance, is only 6.1 J K^{-1} mol^{-1} at 25°C.

In 1905 Einstein set out to explain these observations. He took the view that each atom oscillates about its mean position with a single frequency v. He then borrowed Planck's hypothesis and asserted

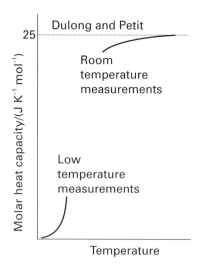

Fig 12.6 The Dulong and Petit law implies that the heat capacity should be the same at all temperatures. However, all solids have a lower heat capacity as the temperature is lowered and, as T approaches zero, the heat capacity approaches zero too.

---

[7] We have already remarked that the gas constant crops up in expressions for all manner of 'non-gas' properties. It is actually a more fundamental constant (Boltzmann's constant, k) in disguise ($R = N_A k$).

that the permitted energy of any oscillating atom is an integral multiple of $h\nu$ (exactly as for electromagnetic oscillators). On the basis of this model, Einstein deduced that

$$C_{V,\,m} = 3Rf^2 \quad \text{where} \quad f = \frac{h\nu}{kT}\left(\frac{e^{h\nu/2kT}}{e^{h\nu/kT} - 1}\right) \quad (12.6)$$

This expression is plotted in Fig 12.7. We see that it does indeed predict a decrease in heat capacity as the temperature is lowered, and that C_m approaches zero as T approaches zero.

The physical reason for the success of Einstein's model is that an atom can start to oscillate only if it can acquire a certain minimum energy ($h\nu$). At low temperatures there is only enough energy available for a few atoms to be able to oscillate. Because so few atoms can be involved in taking up energy, the solid cannot absorb heat readily and consequently its heat capacity is low. At higher temperatures there is enough energy available for all the oscillators to become active: all $3N$ oscillators contribute, and the molar heat capacity approaches its classical value of $3R$.

The Dutch physicist Peter Debye carried out a more refined approach to the calculation. He allowed for the atoms to oscillate with a range of frequencies rather than the single frequency supposed by Einstein. The graph of his more complicated expression is similar to Einstein's, but the numerical agreement with the experimental data is better (see Fig 12.7). An important practical conclusion from Debye's calculation is that, at low temperatures, the heat capacity of a solid is expected to be proportional to T^3. This dependence, which is called the **Debye T^3 law**, is used to extrapolate measurements of heat capacities to $T = 0$ in the experimental determination of entropies (Section 4.5).

12.3 The photoelectric effect

So far, we have seen that two observations—on the electromagnetic field and the heat capacities of solids—have led to the overthrow of the classical view that oscillators can have any energy. We shall now see how three other experimental observations upset another central concept of classical physics, the distinction between waves and particles.

Planck's discovery that an electromagnetic oscillator of frequency ν can possess only the energies 0, $h\nu$, $2h\nu$, . . . inspired a new view of the nature of electromagnetic radiation. Instead of thinking of radiation of a given frequency as the excitation of the electromagnetic field to one of its permitted states of oscillation at that frequency, we can think of it as a stream of 0, 1, 2, . . . particles travelling at the speed c, each particle having an energy $h\nu$. When there is only one such particle present, the energy of the radiation is $h\nu$, when there are two particles of that frequency, their total energy is $2h\nu$, and so on. These particles of electromagnetic radiation are now called **photons**. According to the photon picture of radiation, a ray of light of frequency ν consists of a stream of photons, each one having an energy $h\nu$ and speed c. As the intensity of the ray is increased, the number of photons increases, but each one continues to have the energy $h\nu$. An intense beam of monochromatic (single-frequency) radiation consists of a dense stream of photons; a weak beam of radiation of the same frequency consists of a relatively small number of the same type of photons.

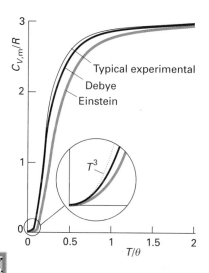

Fig 12.7 The Einstein formula predicts the general temperature dependence quite well, but is everywhere too low. Debye's modification of Einstein's calculation gives very good agreement with experiment. The inset shows a magnified view of the graphs close to $T = 0$. The Debye curve is proportional to T^3 in this region.

Example 12.1 *Calculating the number of photons*

Calculate the number of photons emitted by a 100 W yellow lamp in 10.0 s. Take the wavelength of yellow light as 560 nm and assume 100 per cent efficiency.

Strategy The total energy emitted by a lamp in a given interval is its power multiplied by the time interval of interest (1 J = 1 W s). The number of photons emitted in that time is therefore the total energy divided by the energy of one photon. We calculate the energy of a single photon from the formula $E = hv$. For this calculation we need to know that the wavelength and frequency are related by $v = c/\lambda$.

Solution The electromagnetic energy emitted by the lamp (if all the energy it consumes is converted into radiation of a single frequency) is

Total energy = $(100\ W) \times (10.0\ s) = 100 \times 10.0\ J$

Each photon has an energy

$$E = hv = \frac{hc}{\lambda} = \frac{(6.626 \times 10^{-34}\ J\ s) \times (2.998 \times 10^{8}\ m\ s^{-1})}{5.60 \times 10^{-7}\ m}$$

$$= \frac{6.626 \times 2.998 \times 10^{-19}}{5.60}\ J$$

The number of photons required to carry away the total energy is therefore

$$N = \frac{\text{total energy}}{\text{energy of one photon}}$$

$$= \frac{100 \times 10.0\ J}{(6.626 \times 2.998 \times 10^{-19})/5.60\ J} = 2.82 \times 10^{21}$$

Self-test 12.5

How many 1000 nm photons does a 1 mW infrared rangefinder emit in 0.1 s?

[*Answer:* 5×10^{14}]

Evidence that confirmed the view that radiation can be interpreted as a stream of particles comes from the **photoelectric effect**, the ejection of electrons from metals when they are exposed to ultraviolet radiation (Fig 12.8). The characteristics of the photoelectric effect are as follows:

1 No electrons are ejected, regardless of the intensity of the radiation, unless the frequency exceeds a threshold value characteristic of the metal.

2 The kinetic energy of the ejected electrons varies linearly with the frequency of the incident radiation but is independent of its intensity.

3 Even at low light intensities, electrons are ejected immediately if the frequency is above the threshold value.

These observations strongly suggest an interpretation of the photoelectric effect in which an electron is ejected in a collision with a particle-like projectile, provided the projectile carries enough energy to expel the electron from the metal. If we suppose that the projectile is a photon of energy hv, where v is the frequency of the radiation, then the conservation of energy requires that the kinetic energy of the electron (which is equal to $\frac{1}{2}m_{e}v^{2}$, when the speed of the electron is v) should be equal to the energy supplied by the photon less the energy Φ (phi) required to remove the electron from the metal (Fig 12.9):

$$\frac{1}{2}m_{e}v^{2} = hv - \Phi \tag{12.7}$$

The quantity Φ is called the **work function** of the metal.

When $hv < \Phi$, photoejection (the ejection of electrons by light) cannot occur because the photon supplies insufficient energy to expel the electron: this conclusion is consistent with observation 1. Equation 12.7 predicts that the kinetic energy of an ejected electron should increase linearly with the frequency, in agreement with observation 2. When

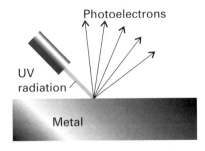

Fig 12.8 The experimental arrangement to demonstrate the photoelectric effect. A beam of ultraviolet radiation is used to irradiate a patch of the surface of a metal, and electrons are ejected from the surface if the frequency of the radiation is above a threshold value that depends on the metal.

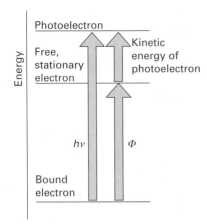

Fig 12.9 In the photoelectric effect, an incoming photon brings a definite quantity of energy, $h\nu$. It collides with an electron close to the surface of the metal target, and transfers its energy to it. The difference between the work function, Φ, and the energy $h\nu$ appears as the kinetic energy of the ejected electron.

Self-test 12.6

The work function of rubidium is 2.09 eV (1 eV = 1.60 $\times 10^{-19}$ J). Can blue (470 nm) light eject electrons from the metal?

[*Answer*: yes]

a photon collides with an electron, it gives up all its energy, so we should expect electrons to appear as soon as the collisions begin, provided the photons carry sufficient energy: this conclusion agrees with observation 3. Thus, the photoelectric effect is strong evidence for the existence of photons.

12.4 The diffraction of electrons

The photoelectric effect shows that light has certain properties of particles. Although contrary to the long-established wave theory of light, a similar view had been held before, but discarded. No significant scientist, however, had taken the view that matter is wave-like. Nevertheless, experiments carried out in 1925 forced people to even that conclusion. The crucial experiment was performed by the American physicists Clinton Davisson and Lester Germer, who observed the diffraction of electrons by a crystal (Fig

12.10). **Diffraction** is the interference between waves caused by an object in their path, and results in a series of bright and dark fringes where the waves are detected. It is a typical characteristic of waves.

The Davisson–Germer experiment, which has since been repeated with other particles (including molecular hydrogen), shows clearly that 'particles' have wave-like properties. We have also seen that 'waves' have particle-like properties. Thus we are brought to the heart of modern physics. When examined on an atomic scale, the concepts of particle and wave melt together, particles taking on the characteristics of waves, and waves the characteristics of particles. This joint wave–particle character of matter and radiation is called **wave–particle duality**.

As these concepts emerged there was an understandable confusion about how to combine both aspects of matter into a single description. Some progress was made by Louis de Broglie when, in 1924, he suggested that any particle travelling with a linear momentum, p, should have (in some sense) a wavelength λ given by the **de Broglie relation**:

$$\lambda = \frac{h}{p} \tag{12.8}$$

The de Broglie relation implies that the wavelength of a particle should decrease as its speed increases (Fig 12.11). The relation also implies that, for a given speed, heavy particles should have shorter wavelengths than lighter particles. Equation 12.8 was

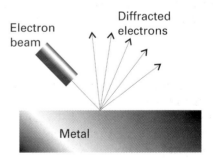

Fig 12.10 In the Davisson–Germer experiment, a beam of electrons was directed on a single crystal of nickel, and the scattered electrons showed a variation in intensity with angle that corresponded to the pattern that would be expected if the electrons had a wave character and were diffracted by the layers of atoms in the solid.

confirmed by the Davisson–Germer experiment, for the wavelength it predicts for the electrons they used in their experiment agrees with the details of the diffraction pattern they observed. We shall build on the relation, and understand it more, in the next section.

12.5 **Atomic and molecular spectra**

The most directly compelling evidence for the quantization of energy comes from the frequencies of radiation absorbed or emitted by atoms and molecules. We shall only mention this point here, and leave it for a much more complete treatment later (Chapters 17 and 18). Figure 12.12 shows a typical atomic emission spectrum and Fig 12.13 shows a typical molecular absorption spectrum. The obvious feature of both is that *radiation is emitted and absorbed at a series of discrete frequencies*. This observation can be understood if the energy of the atoms or molecules is also confined to discrete values, for then energy can be discarded or accepted only in packets (Fig 12.14). For example, if the energy of an atom decreases by ΔE, then the energy is carried

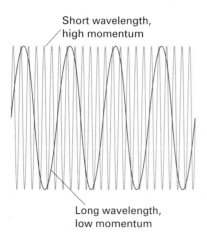

Short wavelength, high momentum

Long wavelength, low momentum

Fig 12.11 According to the de Broglie relation, a particle with low momentum has a long wavelength whereas a particle with high momentum has a short wavelength. A high momentum can result either from a high mass or from a high velocity (because $p = mv$). Macroscopic objects have such large masses that, even if they are travelling very slowly, their wavelengths are undetectably short.

Example 12.2 *Estimating the de Broglie wavelength*

Estimate the wavelength of electrons that have been accelerated from rest through a potential difference of 1.00 kV.

Strategy To use the de Broglie relation, we need to know the linear momentum, $p = m_e v$, of the accelerated electron. We would know that if we knew its kinetic energy $E_K = \frac{1}{2} m_e v^2$, because we can combine these two expressions and rearrange them into $p = (2 m_e E_K)^{1/2}$. The kinetic energy acquired by an electron that is accelerated from rest by falling through a potential difference V is eV, where e is the magnitude of its charge (see *Further information 3*), so we can write $E_K = eV$ and obtain $p = (2 m_e eV)^{1/2}$. It follows that the de Broglie wavelength is

$$\lambda = \frac{h}{(2 m_e eV)^{1/2}}$$

At this stage, all we need do is to substitute the data and use the relations 1 C V = 1 J and 1 J = 1 kg m² s⁻².

Solution Substituting the data and the fundamental constants (from inside the front cover) gives

$$\lambda = \frac{6.626 \times 10^{-34} \text{ J s}}{\{2 \times (9.110 \times 10^{-31} \text{ kg}) \times (1.602 \times 10^{-19} \text{ C}) \times (1.00 \times 10^3 \text{ V})\}^{1/2}}$$

$$= 3.88 \times 10^{-11} \text{ m}$$

The wavelength of 38.8 pm is comparable to typical bond lengths in molecules (about 100 pm). Electrons accelerated in this way are used in the technique of *electron diffraction* for the determination of molecular structure.

Self-test 12.7

Calculate the wavelength of an electron in a 10 MeV particle accelerator (1 MeV = 10^6 eV).

[*Answer*: 0.39 pm]

away as a photon of that energy, and therefore of frequency $v = \Delta E/h$. As a result, radiation of frequency v, a so-called **spectroscopic line**, appears in the spectrum.

Classical mechanics utterly failed in its attempts to account for the existence of discrete spectroscopic lines, just as it failed to account for the other

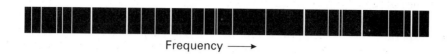

Fig 12.12 The spectrum of radiation emitted by energetically excited mercury atoms consists of radiation at a series of discrete frequencies, called spectral lines. The lines are images of the slit that the radiation passes through before being separated into its components by a prism or diffraction grating.

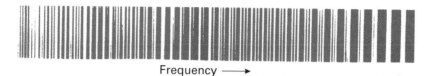

Fig 12.13 When a molecule changes its state it does so by absorbing radiation at definite frequencies. This suggests that it can possess only discrete energies, not an arbitrary energy. The illustration shows a part of the ultraviolet absorption spectrum of ScF: the lines stem from excitations of the electrons of the molecule and its vibrational and rotational state.

experiments described above. Such total failure showed that the basic concepts of classical mechanics were false. A new mechanics, quantum mechanics, had to be devised to take its place.

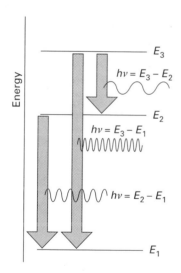

Fig 12.14 Spectral lines can be accounted for if we assume that a molecule emits a photon as it changes between discrete energy levels. High-frequency radiation is emitted when the two states involved in the transition are widely separated in energy; low-frequency radiation is emitted when the two states are close in energy.

The dynamics of microscopic systems

WE shall take the de Broglie relation as our starting point, and abandon the classical concept of particles moving along trajectories. From now on, we adopt the quantum mechanical view that *a particle is spread through space like a wave*. As for a wave in water where the water accumulates in some places but is low in others, there are regions where the particle is more likely to be found than others. To describe this distribution, we introduce the concept of **wavefunction**, ψ (psi), in place of the trajectory, and then set up a scheme for calculating and interpreting ψ. To a very crude first approximation, we can visualize a wavefunction as a blurred version of a trajectory (Fig 12.15); however, we refine this picture in the following sections.

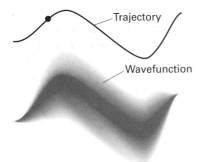

Fig 12.15 According to classical mechanics, a particle may have a well defined trajectory, with a precisely specified position and momentum at each instant (as represented by the precise path in the diagram). According to quantum mechanics, a particle cannot have a precise trajectory; instead, there is only a probability that it may be found at a specific location at any instant. The wavefunction that determines its probability distribution is a kind of blurred version of the trajectory. Here, the wavefunction is represented by areas of shading: the darker the area, the greater the probability of finding the particle there.

12.6 The Schrödinger equation

In 1926, the Austrian physicist Erwin Schrödinger proposed an equation for finding the wavefunction of any collection of particles. The **Schrödinger equation** for a single particle of mass m moving in one dimension with energy E is

$$-\frac{\hbar^2}{2m}\frac{d^2\psi}{dx^2} + V\psi = E\psi \qquad (12.9)$$

In this expression, V, which may depend on the position x of the particle, is the potential energy; $\hbar$ (which is read h-bar) is a convenient modification of Planck's constant:

$$\hbar = \frac{h}{2\pi} = 1.054\ 59 \times 10^{-34}\ \text{J s}$$

The fact that the Schrödinger equation is a 'differential equation', an equation in terms of the derivatives of a function, should not cause too much consternation. We shall not need to solve it explicitly. The rare cases where we need to see the explicit forms of its solution will involve mathematical functions no more complicated than $\sin x$ and e^{-x}.

Derivation 12.2 *The Schrödinger equation*

We can justify the form of the Schrödinger equation to a certain extent by the following remarks. Consider a region where the potential energy is zero (like a bead on a horizontal wire). Then the equation simplifies to

$$-\frac{\hbar^2}{2m}\frac{d^2\psi}{dx^2} = E\psi$$

and a solution is

$$\psi = \sin kx \qquad k = \frac{(2mE)^{1/2}}{\hbar}$$

as may be verified by substitution of the solution into both sides of the equation and using

$$\frac{d}{dx}\sin kx = k\cos kx \qquad \frac{d}{dx}\cos kx = -k\sin kx$$

The function $\sin kx$ is a wave of wavelength $\lambda = 2\pi/k$, as we can see by comparing $\sin kx$ with the standard form of a harmonic wave of wavelength λ, which is $\sin(2\pi x/\lambda)$; see Fig 12.16. Next, we note that the energy of the particle is entirely kinetic (because $V = 0$ everywhere), so the total energy of the particle is just its kinetic energy:

$$E = \tfrac{1}{2}mv^2 = \frac{(mv)^2}{2m} = \frac{p^2}{2m}$$

Because E is related to k by

$$E = \frac{k^2\hbar^2}{2m}$$

it follows from a comparison of the two equations that $p = k\hbar$. Therefore, the linear momentum is related to the wavelength of the wavefunction by

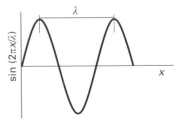

Fig 12.16 The wavelength of a harmonic wave of the form $\sin(2\pi x/\lambda)$. The amplitude of the wave is the maximum height above the centre line.

$$p = \frac{2\pi}{\lambda} \times \frac{h}{2\pi} = \frac{h}{\lambda}$$

which is de Broglie's relation. We see, in the case of freely moving particles, that the Schrödinger equation has led to an experimentally verified conclusion.

One feature of the solution of the Schrödinger equation, which is common to all differential equations, is that an infinite number of possible solutions are allowed mathematically. For instance, if $\sin x$ is a solution of the equation, then so too is $a \sin bx$, where a and b are arbitrary constants.[8] However, it turns out that only some of these solutions are acceptable physically. To be acceptable, a solution must satisfy certain constraints called **boundary conditions** that we describe shortly (Fig 12.17). Suddenly, we are at the heart of quantum mechanics: *the fact that only some solutions are acceptable, together with the fact that each solution corresponds to a characteristic value of E, implies that only certain values of the energy are acceptable.* That is, *when the Schrödinger equation is solved subject to the boundary conditions that the solutions must satisfy, we find that the energy of the system is quantized.* Planck and his

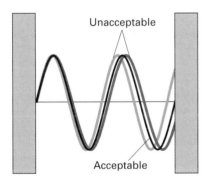

Unacceptable

Acceptable

Fig 12.17 Although an infinite number of solutions of the Schrödinger equation exist, not all of them are physically acceptable. Acceptable wavefunctions have to satisfy certain boundary conditions, which vary from system to system. In the example shown here, where the particle is confined between two impenetrable walls, the only acceptable wavefunctions are those that fit between the walls (like the vibrations of a stretched string). Because each wavefunction corresponds to a characteristic energy, and the boundary conditions rule out many solutions, only certain energies are permissible.

immediate successors had to postulate the quantization of energy for each system they considered: now we see that quantization is an automatic feature of a single equation, the Schrödinger equation, which is applicable to all systems. Later in this chapter and the next we shall see exactly which energies are allowed in a variety of systems, the most important of which (for chemistry) are atoms.

12.7 The Born interpretation

Before we demonstrate the nature and role of boundary conditions, it will be helpful to understand the physical significance of a wavefunction. The interpretation of ψ that is widely used is based on a suggestion made by the German physicist Max Born. He made use of an analogy with the wave theory of light, in which the square of the amplitude of an electromagnetic wave is interpreted as its intensity and therefore (in quantum terms) as the number of photons present. The **Born interpretation** asserts that:

The probability of finding a particle in a small (strictly, infinitesimal) region of space of volume δV is proportional to $\psi^2 \delta V$, where ψ is the value of the wavefunction in the region.

In other words, ψ^2 is a **probability density**.[9] This interpretation implies that, wherever ψ^2 is large, there is a high probability of finding the particle. Wherever ψ^2 is small, there is only a small chance of finding the particle. The density of shading in Fig 12.18 represents this **probabilistic interpretation**, an interpretation that accepts that we can make predictions only about the probability of finding a particle somewhere. This interpretation is in contrast to classical physics, which claims to be able to predict precisely that a particle will be at a given point on its path at a given instant.

---

[8] This conclusion is easily verified by substituting $\psi = a \sin bx$ into the equation in *Derivation 12.2* and noting that it is a solution for all values of a and b.

[9] As before, for other kinds of density, we get the probability itself by multiplying the density by the volume of the region of interest.

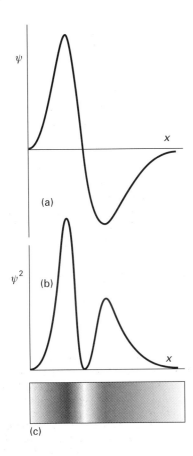

Fig 12.18 (a) A wavefunction does not have a direct physical interpretation. However, (b) its square (its square modulus if it is complex) tells us the probability of finding a particle at each point. The probability density implied by the wavefunction shown here is depicted by the density of shading in (c).

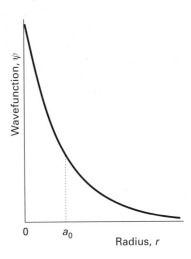

Fig 12.19 The wavefunction for an electron in the ground state of a hydrogen atom is an exponentially decaying function of the form e^{-r/a_0}, where a_0 is the Bohr radius.

$$\text{Probability} \propto \psi^2 \delta V$$

with ψ evaluated at the point in question.

Solution (a) At the nucleus, $r = 0$, so there $\psi^2 \propto 1.0\, \delta V$. (b) At a distance $r = a_0$ in an arbitrary direction, $\psi^2 \propto e^{-2} \times \delta V = 0.14 \delta V$. Therefore, the ratio of probabilities is $1.0/0.14 = 7.1$. It is more probable (by a factor of 7.1) that the electron will be found at the nucleus than in the same volume element located at a distance a_0 from the nucleus.

Self-test 12.8

The wavefunction for the lowest energy state in the ion He^+ is proportional to e^{-2r/a_0}. Repeat the calculation for this ion. Any comment?

[*Answer*: 55; a more compact wavefunction]

12.8 The uncertainty principle

We have seen that, according to the de Broglie relation, a wave of constant wavelength, the wavefunction $\sin(2\pi x/\lambda)$, corresponds to a particle with a definite linear momentum $p = h/\lambda$. However, a wave does not have a definite location at a single point in space, so we cannot speak of the precise position of the particle if it has a definite momentum. Indeed, because a sine wave spreads throughout the whole

Example 12.3 *Interpreting a wavefunction*

The wavefunction of an electron in the lowest energy state of a hydrogen atom is proportional to e^{-r/a_0}, with $a_0 = 52.9$ pm and r the distance from the nucleus (Fig 12.19). Calculate the relative probabilities of finding the electron inside a small volume located at (a) the nucleus, (b) a distance a_0 from the nucleus.

Strategy The probability is proportional to $\psi^2 \delta V$ evaluated at the specified location. The volume of interest is so small (even on the scale of the atom) that we can ignore the variation of ψ within it and write

of space we cannot say anything about the location of the particle: because the wave spreads everywhere, the particle may be found anywhere in the whole of space. This statement is one half of the **uncertainty principle** proposed by Werner Heisenberg in 1927, in one of the most celebrated results of quantum mechanics:

It is impossible to specify simultaneously, with arbitrary precision, both the momentum and the position of a particle.

Before discussing the principle further, we must establish the other half: that, if we know the position of a particle exactly, then we can say nothing about its momentum. If the particle is at a definite location, then its wavefunction must be nonzero there and zero everywhere else (Fig 12.20). We can simulate such a wavefunction by forming a **superposition** of many wavefunctions; that is, by adding together the amplitudes of a large number of sine functions (Fig 12.21). This procedure is successful because the amplitudes of the waves add together at one location to give a nonzero total amplitude, but cancel everywhere else. In other words, we can create a sharply localized wavefunction by adding together wavefunctions corresponding to many different wavelengths, and therefore, by the de Broglie relation, of many different linear momenta.

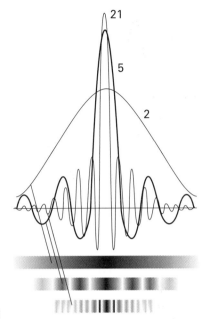

Fig 12.21 The wavefunction for a particle with an ill-defined location can be regarded as the sum (superposition) of several wavefunctions of different wavelength that interfere constructively in one place but destructively elsewhere. As more waves are used in the superposition, the location becomes more precise at the expense of uncertainty in the particle's momentum. An infinite number of waves is needed to construct the wavefunction of a perfectly localized particle. The number against each curve is the number of sine waves used in the superposition.

The superposition of a few sine functions gives a broad, ill-defined wavefunction. As the number of functions increases, the wavefunction becomes sharper because of the more complete interference between the positive and negative regions of the components. When an infinite number of components are used, the wavefunction is a sharp, infinitely narrow spike like that in Fig 12.20, which corresponds to perfect localization of the particle. Now the particle is perfectly localized, but at the expense of discarding all information about its momentum.

The quantitative version of the position–momentum uncertainty relation is

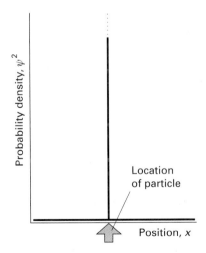

Fig 12.20 The wavefunction for a particle with a well defined position is a sharply spiked function that has zero amplitude everywhere except at the particle's position.

$$\Delta p \Delta x \geq \tfrac{1}{2}\hbar \qquad\qquad (12.10)$$

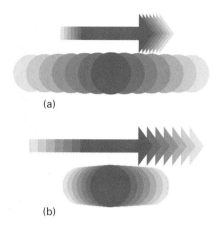

(a)

(b)

Fig 12.22 A representation of the content of the uncertainty principle. The range of locations of a particle is shown by the circles, and the range of momenta by the arrows. In (a), the position is quite uncertain, and the range of momenta is small. In (b), the location is much better defined, and now the momentum of the particle is quite uncertain.

The quantity Δp is the 'uncertainty' in the linear momentum and Δx is the uncertainty in position (which is proportional to the width of the peak in Fig 12.21).[10] Equation 12.10 expresses quantitatively the fact that, the more closely the location of a particle is specified (the smaller the value of Δx), then the greater the uncertainty in its momentum (the larger the value of Δp) parallel to that coordinate, and vice versa (Fig 12.22).

The uncertainty principle applies to location and momentum *along the same axis*. It is silent on location on one axis and momentum along a perpendicular axis. The restrictions it implies are summarized in Table 12.1.

Example 12.4 *Using the uncertainty principle*

The speed of a certain projectile of mass 1.0 g is known to within 1.0×10^{-6} m s^{-1}. What is the minimum uncertainty in its position along its line of flight?

---
[10] Strictly, the uncertainty in momentum is the root mean square (r.m.s.) deviation of the momentum from its mean value, $\Delta p = (\langle p^2 \rangle - \langle p \rangle^2)^{1/2}$, where the angle brackets denote mean values. Likewise, the uncertainty in position is the r.m.s. deviation in the mean value of position, $\Delta x = (\langle x^2 \rangle - \langle x \rangle^2)^{1/2}$.

Strategy We can estimate Δp from $m\Delta v$, where Δv is the uncertainty in the speed; then we use eqn 12.10 to estimate the minimum uncertainty in position, Δx, where x is the direction in which the projectile is travelling.

Solution The uncertainty in position is

$$\Delta x \geq \frac{\hbar}{2\Delta p} = \frac{1.054 \times 10^{-34} \text{ J s}}{2 \times (1.0 \times 10^{-3} \text{ kg}) \times (1.0 \times 10^{-6} \text{ m s}^{-1})}$$

$$= 5.3 \times 10^{-26} \text{ m}$$

This degree of uncertainty is completely negligible for all practical purposes. However, when the mass is that of an electron, the same uncertainty in speed implies an uncertainty in position far larger than the diameter of an atom, so the concept of a trajectory—the simultaneous possession of a precise position and momentum—is untenable.

Self-test 12.9

Estimate the minimum uncertainty in the speed of an electron in a hydrogen atom (taking its diameter as 100 pm).

[*Answer*: 580 km s^{-1}]

The uncertainty principle epitomizes the difference between classical and quantum mechanics. Classical mechanics supposed, falsely as we now know, that the position and momentum of a particle can be specified simultaneously with arbitrary precision. However, quantum mechanics shows

Table 12.1 *Constraints of the uncertainty principle*

Variable 1:	x	y	z	p_x	p_y	p_z
Variable 2						
x	✓	✓	✓	✗	✓	✓
y	✓	✓	✓	✓	✗	✓
z	✓	✓	✓	✓	✓	✗
p_x	✗	✓	✓	✓	✓	✓
p_y	✓	✗	✓	✓	✓	✓
p_z	✓	✓	✗	✓	✓	✓

Observables that can be determined simultaneously with arbitrary precision are marked with a ✓.

that position and momentum are **complementary**, that is, not simultaneously specifiable. Quantum mechanics requires us to make a choice: we can specify position at the expense of momentum, or momentum at the expense of position.

Applications of quantum mechanics

Wᴇ shall now illustrate some of the concepts that have been introduced. Many more will be introduced in the following chapters, for quantum mechanics pervades the whole of chemistry. Here we describe three basic types of motion: translation (motion in a straight line), rotation, and vibration.

12.9 Translation: a particle in a box

First, we consider the translational motion of a 'particle in a box', a particle of mass m that can travel in a straight line in one dimension (along the x-axis) but is confined between two walls separated by a distance L. The potential energy of the particle is zero inside the box but rises abruptly to infinity at the walls (Fig 12.23). The particle might be a bead free to slide along a horizontal wire between two stops.

The boundary conditions for this system are the requirement that each acceptable wavefunction of the particle must fit inside the box exactly, like the vibrations of a violin string (as in Fig 12.24).[11] It follows that the wavelength, λ, of the permitted wavefunctions must be one of the values

$$\lambda = 2L, L, \tfrac{2}{3}L, \ldots \text{ or } \lambda = \frac{2L}{n}, \text{ with } n = 1, 2, 3, \ldots$$

Each wavefunction is a sine wave with one of these wavelengths; therefore, because a sine wave of

----

[11] More precisely, the boundary conditions stem from the requirement that the wavefunction is continuous everywhere: because the wavefunction is zero outside the box, it must therefore be zero at $x = 0$ and at $x = L$.

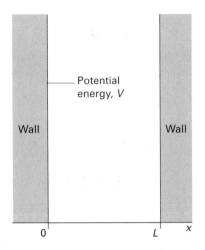

Fig 12.23 A particle in a one-dimensional region with impenetrable walls at either end. Its potential energy is zero between $x = 0$ and $x = L$ and rises abruptly to infinity as soon as the particle touches either wall.

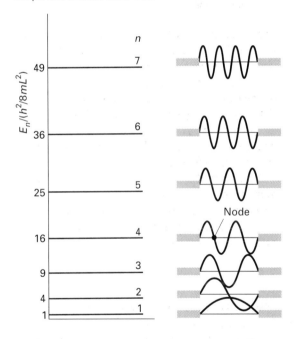

Fig 12.24 The allowed energy levels and the corresponding (sine wave) functions for a particle in a box. Note that the energy levels increase as n^2, and so their spacing increases as n increases. Each wavefunction is a standing wave, and successive functions possess one more half wave and a corresponding shorter wavelength.

wavelength λ has the form $\sin(2\pi x/\lambda)$, the permitted wavefunctions are

$$\psi_n = N \, \sin\frac{n\pi x}{L} \quad n = 1, 2, \ldots \qquad (12.11)$$

The constant N is called the **normalization constant**. It is chosen so that the total probability of finding the particle inside the box is 1.

Derivation 12.3 *The normalization constant*

According to the formal statement of the Born interpretation, the probability of finding a particle in the infinitesimal region of length dx at the point x, given that its normalized wavefunction has the value ψ at that point, is equal to $\psi^2 dx$. Therefore, the total probability of finding the particle between $x = 0$ and $x = L$ is the sum (integral) of all the probabilities of its being in each infinitesimal region. That total probability is 1 (the particle is certainly in the range somewhere), so we know that

$$\int_0^L \psi^2 \, dx = 1$$

Substitution of the form of the wavefunction turns this expression into

$$N^2 \int_0^L \sin^2 \frac{n\pi x}{L} \, dx = 1$$

Our task is to solve this equation for N. Because

$$\int \sin^2 ax \; dx = \tfrac{1}{2}x - \frac{\sin 2ax}{4a} + \text{constant}$$

it follows that, because the sine term is zero at $x = 0$ and $x = L$,

$$\int_0^L \sin^2 \frac{n\pi x}{L} \, dx = \tfrac{1}{2}L$$

Therefore,

$$N^2 \times \tfrac{1}{2}L = 1$$

and hence $N = (2/L)^{1/2}$. Note that, in this case but not in general, the same normalization factor applies to all the wavefunctions regardless of the value of n.

It is a simple matter to find the permitted energy levels because the only contribution to the energy is the kinetic energy of the particle: the potential energy is zero everywhere inside the box, and the particle is never outside the box. First, we note that it follows from the de Broglie relation, eqn 12.8, that

the only values of the linear momentum that are acceptable are

$$p = \frac{h}{\lambda} = \frac{nh}{2L} \quad n = 1, 2, \ldots$$

Then, because the kinetic energy of a particle of momentum p and mass m is $E = p^2/2m$, it follows that the permitted energies of the particle are

$$E_n = \frac{n^2 h^2}{8mL^2} \quad n = 1, 2, \ldots \qquad (12.12)$$

As we see in eqns 12.11 and 12.12, the energies and wavefunctions of a particle in a box are labelled with the number n. A **quantum number**, of which n is an example, is an integer (in certain cases, as we shall see, a half-integer) that labels the state of the system. As well as acting as a label, a quantum number specifies certain physical properties of the system: in the present example, n specifies the energy of the particle through eqn 12.12.

The permitted energies of the particle are shown in Fig 12.24 together with the shapes of the wavefunctions for $n = 1$ to 5. All the wavefunctions except the one of lowest energy ($n = 1$) possess points called **nodes** where the function passes through zero.[12] The number of nodes in the wavefunctions shown in Fig 12.24 increases from 0 (for $n = 1$) to 6 (for $n = 7$), and is $n - 1$ for a particle in a box in general. It is a general feature of quantum mechanics that the state of lowest energy has no nodes and, as the number of nodes in a wavefunction increases, the energy increases too.

The solutions of a particle in a box introduce another important general feature of quantum mechanics. Because the quantum number n cannot be zero (for this system), the lowest energy that the particle may possess is not zero, as would be allowed by classical mechanics, but $h^2/8mL^2$ (the energy when $n = 1$). This lowest, irremovable energy is called the **zero-point energy**. The existence of a zero-point energy is consistent with the uncertainty principle. If a particle is confined to a finite region, its location is not completely indefinite;

---

[12] Passing *through* zero is an essential part of the definition: just becoming zero is not sufficient. The points at the edges of the box where $\psi = 0$ are not nodes, because the wavefunction does not pass through zero there.

consequently, its momentum cannot be specified precisely as zero, and therefore its kinetic energy cannot be precisely zero either. The zero-point energy is not a special, mysterious kind of energy. It is simply the last remnant of energy that a particle cannot give up. For a particle in a box it can be interpreted as the energy arising from a ceaseless fluctuating motion of the particle between the two confining walls of the box.

The energy difference between adjacent levels is

$$\Delta E = E_{n+1} - E_n = (2n+1)\frac{h^2}{8mL^2} \qquad (12.13)$$

This expression shows that the difference decreases as the length L of the box increases, and that it becomes zero when the walls are infinitely far apart (Fig 12.25). Atoms and molecules free to move in laboratory-sized vessels may therefore be treated as though their translational energy is not quantized, because L is so large. The expression also shows that the separation decreases as the mass of the particle increases. Particles of macroscopic mass (like balls and planets, and even minute specks of dust) behave as though their translational motion is unquantized. Both these conclusions are true in general:

1   The greater the size of the system, the less important are the effects of quantization.

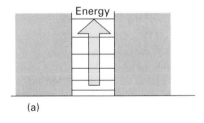

(a)

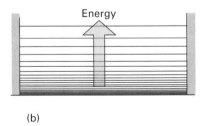

(b)

Fig 12.25 (a) A narrow box has widely spaced energy levels; (b) a wide box has closely spaced energy levels. (In each case, the separations depend on the mass of the particle too.)

2   The greater the mass of the particle, the less important are the effects of quantization.

> **Self-test 12.10**
>
> Consider an electron that is a part of a conjugated polyene (such as a carotene molecule) of length 2.0 nm. Treat the long molecule as a one-dimensional box and the electron as a particle confined to the box. What is the energy in electronvolts (1 eV = 1.602 × 10^{-19} J) to excite it from the level with $n = 5$ to the next higher level?
>
> [*Answer*: 1.0 eV]

12.10 Rotation: a particle on a ring

The discussion of translational motion focused on linear momentum, p. When we turn to rotational motion we have to focus on the analogous **angular momentum**, J. The magnitude of the angular momentum of a particle that is travelling on a circular path of radius r is defined as

$$J = pr \qquad (12.14)$$

where p is its linear momentum ($p = mv$) at any instant. A particle that is travelling at high speed in a circle has a higher angular momentum than a particle of the same mass travelling more slowly. An object with a high angular momentum (like a flywheel) requires a strong braking force (more precisely, a strong torque) to bring it to a standstill.

To see what quantum mechanics tells us about rotational motion, we consider a particle of mass m moving in a horizontal circular path of radius r in the xy plane. The energy of the particle is entirely kinetic because the potential energy is constant and can be set equal to zero everywhere. We can therefore write $E = p^2/2m$. By using eqn 12.14, we can express this energy in terms of the angular momentum as

$$E = \frac{J_z^2}{2mr^2}$$

where J_z is the angular momentum around the z-axis (the axis perpendicular to the plane). The

quantity mr^2 is the **moment of inertia** of the particle about the z-axis, and denoted I: a heavy particle
in a path of large radius has a large moment of
inertia (Fig 12.26). It follows that the energy of the
particle is

$$E = \frac{J_z^2}{2I} \tag{12.15}$$

Now we use the de Broglie relation to see that the
energy of rotation is quantized. To do so, we express
the angular momentum in terms of the wavelength
of the particle:

$$J_z = pr = \frac{hr}{\lambda}$$

Suppose for the moment that λ can take an
arbitrary value. In that case, the amplitude of the
wavefunction depends on the angle as shown in
Fig 12.27. When the angle increases beyond 2π (that
is, 360°), the wavefunction continues to change. For
an arbitrary wavelength it gives rise to a different
amplitude at each point and the interference between the waves on successive circuits cancels the
amplitude of the wave on its previous circuit. Thus,
this particular arbitrary wave cannot survive in the
system. An acceptable solution is obtained only if
the wavefunction reproduces itself on successive
circuits: we say that the wavefunction must satisfy
cyclic boundary conditions. Specifically, acceptable wavefunctions match after each circuit, and
therefore have wavelengths that are given by the
expression[13]

$$\lambda = \frac{2\pi r}{n} \qquad n = 0, 1, \ldots$$

It follows that the permitted energies are

$$E_n = \frac{(hr/\lambda)^2}{2I} = \frac{(nh/2\pi)^2}{2I} = \frac{n^2\hbar^2}{2I}$$

with $n = 0, \pm1, \pm2, \ldots$.
We need to make two points about the expression
for the energy before we use it. One is that a particle
can travel either clockwise or counterclockwise
around a ring. We represent these different direc-

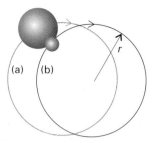

Fig 12.26 A particle travelling on a circular path has a
moment of inertia I that is given by mr^2. (a) This heavy particle
has a large moment of inertia about the central point; (b) this
light particle is travelling on a path of the same radius, but it
has a smaller moment of inertia. The moment of inertia plays
a role in circular motion that is the analogue of the mass for
linear motion: a particle with a high moment of inertia is
difficult to accelerate into a given state of rotation, and
requires a strong braking force to stop its rotation.

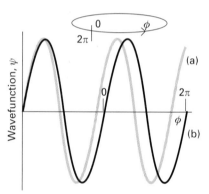

Fig 12.27 Two solutions of the Schrödinger equation for a
particle on a ring. The circumference has been opened out
into a straight line; the points at $\phi = 0$ and 2π are identical.
The solution in (a) is unacceptable because it has different
values after each circuit, and so interferes destructively with
itself. The solution in (b) is acceptable because it reproduces
itself on successive circuits.

tions by positive and negative values of n, with positive values representing clockwise rotation seen
from below (like a right-handed screw), and negative values representing counterclockwise rotation.
The energy depends on n^2, so the difference in
sign—the direction of rotation—has no effect on
the energy. Second, in the discussion of rotational
motion it is conventional to denote the quantum
number by m_l in place of n.[14] Therefore, the final expression for the energy levels is

---

[13] The value $n = 0$, which gives an infinite wavelength, corresponds to a uniform amplitude.

[14] This convention is elaborated in Chapter 13.

$$E_{m_l} = \frac{m_l^2 \hbar^2}{2I} \qquad m_l = 0, \pm 1, \dots \qquad (12.16)$$

These energy levels are drawn in Fig 12.28.

As we have remarked, the occurrence of m_l^2 in the expression for the energy means that two states of motion, such as those with $m_l = +1$ and $m_l = -1$, both correspond to the same energy. Such a condition, in which more than one state has the same energy, is called **degeneracy**. All the states with $|m_l| > 0$ are doubly degenerate because two states correspond to the same energy for each value of $|m_l|$. The state with $m_l = 0$, the lowest energy state of the particle, is **nondegenerate**, meaning that only one state has a particular energy (in this case, zero).

An important additional conclusion is that *the angular momentum of the particle is quantized*. We can use the relation between angular momentum and linear momentum (angular momentum $= pr$), and between linear momentum and the allowed wavelengths of the particle ($\lambda = 2\pi r/m_l$), to conclude that the angular momentum of a particle around the z-axis is confined to the values

$$J_z = pr = \frac{hr}{\lambda} = \frac{hr}{2\pi r/m_l} = m_l \times \frac{h}{2\pi}$$

That is, the angular momentum of the particle around the axis is confined to the values

$$J_z = m_l \hbar \qquad (12.17)$$

with $m_l = 0, \pm 1, \pm 2, \dots$. Positive values of m_l correspond to clockwise rotation (as seen from below) and negative values correspond to counterclockwise rotation (Fig 12.29). The quantized motion can be thought of in terms of the rotation of a bicycle wheel that can rotate only with a discrete series of angular momenta so that, as the wheel is accelerated, the angular momentum jerks from the value 0 (when the wheel is stationary) to $\hbar$, $2\hbar$, ... but can have no intermediate value.

A final point concerning the rotational motion of a particle is that it does not have a zero-point energy: m_l may take the value 0, so E may be zero. This conclusion is also consistent with the uncertainty principle. Although the particle is certainly be-

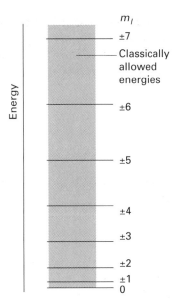

Fig 12.28 The energy levels of a particle that can move on a circular path. Classical physics allowed the particle to travel with any energy (as represented by the continuous grey band); quantum mechanics, however, allows only discrete energies. Each energy level, other than the one with $m_l = 0$, is doubly degenerate, because the particle may rotate either clockwise or counterclockwise with the same energy.

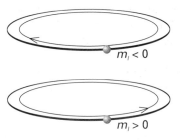

Fig 12.29 The significance of the sign of m_l. When $m_l < 0$, the particle travels in a counterclockwise direction as viewed from below; when $m_l > 0$, the motion is clockwise.

tween the angles 0 and 360° on the ring, that range is equivalent to not knowing anything about where it is on the ring. Consequently, the angular momentum may be specified exactly, and a value of zero is possible. When the angular momentum is zero precisely, the energy of the particle is also zero precisely.

12.11 Vibration: the harmonic oscillator

In the type of vibrational motion known as **harmonic oscillation**, a particle vibrates backwards and forwards, restrained by a spring that obeys **Hooke's law** of force. Hooke's law states that the restoring force is proportional to the displacement, x:

$$\text{Restoring force} = -kx \qquad (12.18a)$$

The constant of proportionality k is called the **force constant**: a stiff spring has a high force constant. The potential energy of a particle subjected to this force increases as the square of the displacement, and specifically

$$V = \tfrac{1}{2}kx^2 \qquad (12.18b)$$

The variation of V with x is shown in Fig 12.30: it has the shape of a parabola (a curve of the form $y = ax^2$), and we say that a particle undergoing harmonic motion has a 'parabolic potential energy'.

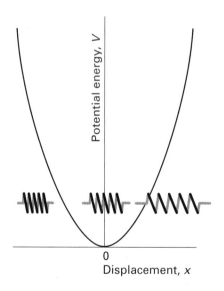

Fig 12.30 The parabolic potential energy characteristic of a harmonic oscillator. Positive displacements correspond to extension of the spring; negative displacements correspond to compression of the spring.

The solutions of the Schrödinger equation are quite hard to find, but once found they turn out to be very simple. The only allowed energies are

$$E_v = (v + \tfrac{1}{2})h\nu \quad v = 0,1,2,\ldots \quad \nu = \frac{1}{2\pi}\left(\frac{k}{m}\right)^{1/2} \quad (12.19)$$

where m is the mass of the particle and v is the vibrational **quantum number**.[15] These energies form a uniform ladder of values separated by $h\nu$ (Fig 12.31). The quantity ν is a frequency (in cycles per second, or hertz, Hz), and is in fact the frequency that a classical oscillator of mass m and force constant k would be calculated to have. In quantum mechanics, though, ν tells us (through $h\nu$) the separation of any pair of adjacent energy levels. The separation is large for stiff springs and small masses.

Illustration 12.2

The force constant for the H–Cl bond is 516 N m^{-1}, where the newton (N) is the SI unit of force (1 N = 1 kg m s^{-2}). If we suppose that, because the chlorine atom is relatively very heavy, only the hydrogen atom moves, we take m as the mass of the H atom (1.67 × 10^{-27} kg for ^{1}H). We find

$$\nu = \frac{1}{2\pi}\left(\frac{k}{m}\right)^{1/2} = \frac{1}{2\pi}\left(\frac{516 \text{ N m}^{-1}}{1.67 \times 10^{-27} \text{ kg}}\right)^{1/2}$$

$$= 8.85 \times 10^{13} \text{ Hz}$$

The separation between adjacent levels is h times this frequency, or 5.86×10^{-20} J.

Figure 12.32 shows the shapes of the first few wavefunctions of a harmonic oscillator. The ground-state wavefunction (corresponding to $v = 0$ and having the zero-point energy $\tfrac{1}{2}h\nu$) is a bell-shaped curve with no nodes. This shape shows that the particle is most likely to be found at $x = 0$ (zero displacement), but may be found at greater displacements with decreasing probability. The first excited wavefunction has a node at $x = 0$ and peaks on either side. Therefore, in this state, the particle will be found most probably with the 'spring' stretched or compressed to the same amount.

---

[15] Be very careful to distinguish the quantum number v (italic vee) from the frequency ν (Greek nu).

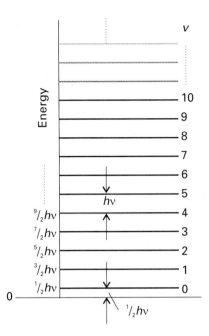

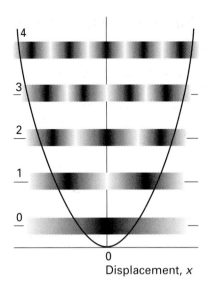

Fig 12.33 A schematic illustration of the probability density for finding a harmonic oscillator at a given displacement. Classically, the oscillator cannot be found at displacements at which its total energy is less than its potential energy (because the kinetic energy cannot be negative). A quantum oscillator, though, may tunnel into regions that are classically forbidden.

Fig 12.31 The array of energy levels of an harmonic oscillator. The separation depends on the mass and the force constant. Note the zero-point energy.

One intriguing aspect of a harmonic oscillator is that the wavefunctions extend beyond the limits of motion of a classical oscillator (Fig 12.33). The pene-tration of a wavefunction into classically disallowed regions—not just for vibrational motion—is called **tunnelling**. Tunnelling is most important for light

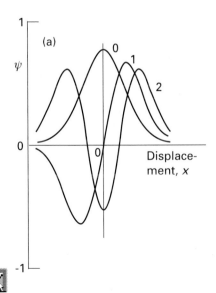

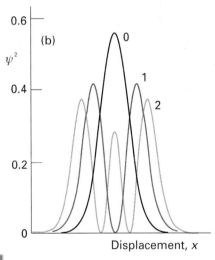

Fig 12.32 (a) The wavefunctions and (b) the probability densities of the first three states of an harmonic oscillator. Note how the probability of finding the oscillator at large displacements increases as the state of excitation increases.

particles, so it is most likely to be noticed for electrons and protons, the two lightest particles commonly encountered in chemistry. It has profoundly important consequences. For instance, the rapid equilibration of acids and bases in water, and by implication some of the functions of enzymes, is due in part to the ease with which protons can tunnel out of the bonds attaching them to other atoms. One of the most significant recent developments in the study of surfaces and of molecules adsorbed on them makes use of the ability of electrons to tunnel through regions (Box 12.1).

The relevance of the harmonic oscillator itself to chemistry is that atoms vibrate relative to one another in molecules with the bond acting like a spring. Therefore, eqn 12.19 describes the allowed vibrational energy levels of molecules. The equation is enormously important for the interpretation of vibrational (infrared) spectroscopy, as we see in Chapter 17.

Box 12.1 *Scanning tunnelling microscopy*

An extraordinary revitalization of the study of surfaces took place about a decade ago with the introduction of a technique based on the quantum mechanical ability of electrons to tunnel into and through regions where they were forbidden to go classically.

The central component in *scanning tunnelling microscopy* (STM) is a platinum–rhodium or tungsten needle, which is scanned in successive rows across the surface of a conducting solid. When the tip of the needle is brought very close to the surface, electrons can tunnel across the narrow gap (see the first illustration). In the constant-current mode of operation, the stylus moves up and down according to the rise and fall of the surface, and the

it. In the constant-z mode, the vertical height of the tip of the stylus is held constant and the current through it is monitored. Because the tunnelling probability is very sensitive to the size of the gap, the microscope can detect tiny, atomic-scale variations in the height of the surface. An example of the kind of image obtained with a clean surface is shown in the second illustration: the cliff is one atom high. There is hope that STM can be used to 'feel' along a strand of DNA and identify each nucleotide base as

An STM image of caesium atoms on a gallium arsenide surface.

A scanning tunnelling microscope makes use of the current of electrons that tunnel between the surface and the tip. That current is very sensitive to the distance of the tip above the surface.

topography of the surface, including the presence of any molecules adsorbed on the surface, can be mapped on an atomic scale. The vertical motion of the stylus is achieved by fixing it to a piezoelectric cylinder, which contracts or expands according to the potential difference applied to

it comes to it, or sequence (determine the sequence of amino acids in) a protein molecule that has been denatured into a long polypeptide strand.

In *atomic force microscopy* (AFM), a sharpened stylus attached to a cantilevered arm is scanned in successive rows across a surface (see the third illustration). The force exerted by the surface and any adsorbed molecule pushes or pulls the stylus and deflects the arm. The deflection is monitored either by interferometry or by using a laser beam. Because no current is needed between the sample and the probe (so no tunnelling is involved in this variety of microscopy), the technique can be applied to nonconducting surfaces too and to surfaces immersed in water. In a further variation, the tip may be used to nudge single atoms around on the surface and hence—in principle—to achieve atomic-level control in chemical synthesis.

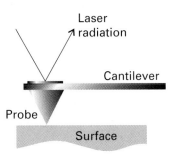

In atomic force microscopy, a laser beam is used to monitor the tiny changes in position of a probe as it is attracted to or repelled from atoms on a surface.

Exercise 1 Electron microscopes can obtain images with several hundred fold higher resolution than light microscopes. Their higher resolution is due to the short wavelength obtainable from a beam of electrons. The reativistic expression for the de Broglie wavelength of electrons accelerated from rest by a potential, V, is

$$\lambda = \frac{h}{\left\{ 2m_e \, eV \left(1 + \dfrac{eV}{2m_e c^2} \right) \right\}^{1/2}}$$

where the quantities have their normal meaning. (a) Calculate the de Broglie wavelengths of electrons accelerated through 50 kV. (b) Is the relativistic correction important?

Exercise 2 The rate, ν, at which electrons tunnel through a potential barrier of height 2eV, like that in a scanning tunnelling microscope, and thickness d can be expressed $\nu = Ae^{-d/l}$, with $A = 5 \times 10^{14}\,\mathrm{s}^{-1}$ and $l = 70$ pm. (a) Calculate the rate at which electrons tunnel across a barrier of width 750 pm. (b) By what factor is the current reduced when the probe is moved away by a further 100 pm?

Exercises

12.1 In early 1999 it was thought that the first planet outside the solar system had been photographed; by midsummer of that year, the 'planet' had been reclassified as a cool star with surface temperature 2500°C. What is the wavelength of its maximum emission?

12.2 A photodetector produces 0.68 μW when exposed to radiation of wavelength 245 nm. How many photons does it detect per second?

12.3 A glow-worm emits red light of wavelength 650 nm. If that were radiation from a hot source, what would be the temperature of the source? What is your conclusion?

12.4 Incandescent lamps are a common feature of everyday life. Calculate the total power radiated by a 5.0 cm × 2.0 cm section of the surface of a hot body at 3000 K.

12.5 Show that the Planck distribution, eqn 12.5, becomes identical to the Rayleigh–Jeans distribution, eqn 12.3, at long wavelengths.

12.6 The Planck distribution, eqn 12.5, gives the energy of electromagnetic radiation in the wavelength range $\delta\lambda$ at the wavelength λ. Calculate the energy density in the range 650–655 nm inside an oven when its temperature is (a) 25°C, (b) 3000°C.

12.7 Find an expression for the location of the maximum in the Planck distribution. (*Hint*. Differentiate eqn 12.5 and set the result equal to 0. You will need to make the approximation that the wavelength is short ($\lambda \ll hc/kT$).)

12.8 The wavelength of the emission maximum from a small pinhole in an electrically heated container was determined at a series of temperatures, and the results are given below. Deduce a value for Planck's constant.

$\theta/°C$	1000	1500	2000	2500	3000	3500
λ_{max}/nm	2181	1600	1240	1035	878	763

12.9 Calculate the size of the quantum involved in the excitation of (a) an electronic motion of frequency $1.0 \times$

10^{15} Hz, (b) a molecular vibration of period 2.0×10^{-14} s, (c) a pendulum of period 0.50 s. Express the results in joules and in kilojoules per mole.

12.10 The 'Einstein frequency' of the atoms, the frequency to use in eqn 12.6 when calculating the heat capacity of a solid, is higher for lithium than for sodium. Which solid will have the higher molar heat capacity at 0°C? Why?

12.11 A certain lamp emits blue light of wavelength 350 nm. How many photons does it emit each second if its power is (a) 1.00 W, (b) 100 W?

12.12 An FM radio transmitter broadcasts at 98.4 MHz with a power of 45 kW. How many photons does it generate per second?

12.13 The work function for metallic caesium is 2.14 eV. Calculate the kinetic energy and the speed of the electrons ejected by light of wavelength (a) 750 nm, (b) 250 nm.

12.14 A diffraction experiment requires the use of electrons of wavelength 550 pm. Calculate the velocity of the electrons.

12.15 Calculate the de Broglie wavelength of (a) a mass of 1.0 g travelling at 1.0 m s^{-1}, (b) the same, travelling at 1.00×10^5 km s^{-1}, (c) a He atom travelling at 1000 m s^{-1} (a typical speed at room temperature).

12.16 Calculate the de Broglie wavelength of an electron accelerated from rest through a potential difference, V, of (a) 1.00 V, (b) 1.00 kV, (c) 100 kV. (*Hint.* The electron is accelerated to a kinetic energy equal to eV.)

12.17 Calculate the linear momentum of photons of wavelength (a) 725 nm, (b) 75 pm, (c) 20 m.

12.18 Calculate the energy per photon and the energy per mole of photons for radiation of wavelength (a) 600 nm (red), (b) 550 nm (yellow), (c) 400 nm (violet), (d) 200 nm (ultraviolet), (e) 150 pm (X-ray), (f) 1.0 cm (microwave).

12.19 How fast would a particle of mass 1.0 g need to travel to have the same linear momentum as a photon of radiation of wavelength 300 nm?

12.20 Recall (Section 0.3) that pressure is force divided by area, and that force is rate of change of momentum.

Suppose that you designed a spacecraft to work by photon pressure. The sail was a completely absorbing fabric of area 1.0 km^2 and you directed a 1.0 kW red laser beam of wavelength 650 nm on to it from a base on the Moon. What is (a) the force, (b) the pressure exerted by the radiation on the sail? (c) Suppose the mass of the spacecraft was 1.0 kg. Given that, after a period of acceleration from standstill, speed = (force/mass) × time, how long would it take for the craft to accelerate to a speed of 1.0 m s^{-1}?

12.21 The energy required for the ionization of a certain atom is 3.44 aJ (1 aJ = 10^{-18} J; a denotes 'atto'). The absorption of a photon of unknown wavelength ionizes the atom and ejects an electron with velocity 1.03×10^6 m s^{-1}. Calculate the wavelength of the incident radiation.

12.22 In an X-ray photoelectron experiment, a photon of wavelength 150 pm ejects an electron from the inner shell of an atom and it emerges with a speed of 2.24×10^7 m s^{-1}. Calculate the binding energy of the electron.

12.23 Calculate the probability that an electron will be found (a) between $x = 0.1$ and 0.2 nm, (b) between 4.9 and 5.2 nm in a box of length $L = 10$ nm when its wavefunction is $\psi = (2/L)^{1/2} \sin(2\pi x/L)$.

12.24 The speed of a certain proton is 350 km s^{-1}. If the uncertainty in its momentum is 0.0100 per cent, what uncertainty in its location must be tolerated?

12.25 Calculate the minimum uncertainty in the speed of a ball of mass 500 g that is known to be within 5.0 μm of a certain point on a bat.

12.26 What is the minimum uncertainty in the position of a bullet of mass 5.0 g that is known to have a speed somewhere between 350.000 001 m s^{-1} and 350.000 000 m s^{-1}?

12.27 An electron is confined to a linear region with a length of the same order as the diameter of an atom (ca. 100 pm). Calculate the minimum uncertainties in its position and speed.

12.28 A hydrogen atom, treated as a point mass, is confined to an infinite one-dimensional square well of width 1.0 nm. How much energy does it have to give up to fall from the level with $n = 2$ to the lowest energy level?

12.29 The pores in zeolite catalysts are so small that quantum mechanical effects on the distribution of atoms and molecules within them can be significant. Calculate the location in a box of length L at which the probability of a particle being found is 50 per cent of its maximum probability when $n = 1$.

12.30 The blue solution formed when an alkali metal dissolves in liquid ammonia consists of the metal cations and electrons trapped in a cavity formed by ammonia molecules. (a) Calculate the spacing between the levels with $n = 4$ and $n = 5$ of an electron in a one-dimensional box of length 5.0 nm. (b) What is the wavelength of the radiation emitted when the electron makes a transition between the two levels?

12.31 A certain wavefunction is zero everywhere except between $x = 0$ and $x = L$, where it has the constant value A. Normalize the wavefunction.

12.32 As indicated in Self-test 12.10, the particle in a box is a crude model of the distribution and energy of electrons in conjugated polyenes, such as carotene and related molecules. Carotene itself is a molecule in which 21 bonds, 10 single and 11 double, alternate along a chain of carbon atoms. Take each CC bond length to be about 140 pm and suppose that the first possible upward transition (for reasons related to the Pauli principle, Section 13.9) is from $n = 11$ to $n = 12$. Estimate the wavelength of this transition.

12.33 Treat a rotating HI molecule as a stationary I atom around which an H atom circulates in a plane at a distance of 161 pm. Calculate (a) the moment of inertia of the molecule, (b) the greatest wavelength of the radiation that can excite the molecule into rotation.

12.34 A bee of mass 1 g lands on the end of a horizontal twig, which starts to oscillate up and down with a period of 1 s. Treat the twig as a massless spring, and estimate its force constant.

12.35 Treat a vibrating HI molecule as a stationary I atom with the H atom oscillating towards and away from the I atom. Given that the force constant of the HI bond is 314 N m^{-1}, calculate (a) the vibrational frequency of the molecule, (b) the wavelength required to excite the molecule into vibration.

Atomic structure

Contents

CHAPTER 12 provided enough quantum mechanical background for us now to be able to move on to the discussion of the atomic structure. Atomic structure—the description of the arrangement of electrons in atoms—is an essential part of chemistry because it is the basis of understanding molecular and solid structures and all the physical and chemical properties of elements and their compounds.

A **hydrogenic atom** is a one-electron atom or ion of general atomic number Z. Hydrogenic atoms include H, He^+, Li^{2+}, C^{5+}, and even U^{91+}. Such very highly ionized atoms may be found in the outer regions of stars. A **many-electron atom** is an atom or ion that has more than one electron. Many-electron atoms include all neutral atoms other than H. For instance, helium, with its two electrons, is a many-electron atom in this sense. Hydrogenic atoms, and H in particular, are important because the Schrödinger equation can be solved for them and their structures can be discussed exactly. They provide a set of concepts that are used to describe the structures of many-electron atoms and (as we shall see in the next chapter) the structures of molecules too.

Hydrogenic atoms

ENERGETICALLY excited atoms are produced when an electric discharge is passed through a gas or vapour or when an element is exposed to a hot flame. These atoms emit electromagnetic radiation of discrete frequencies as they discard energy and return to the **ground state**, their state of lowest energy (Fig 13.1). The record of frequencies, v, wavenumbers ($\tilde{v} = v/c$), or wavelengths ($\lambda = c/v$), of the radiation emitted is called the **emission spectrum** of the atom.[1] In its earliest form, the radiation was detected photographically as a series of lines (the focused image of the slit that the light was sampled through), and the components of radiation present in a spectrum are still widely referred to as spectroscopic 'lines'.

13.1 The spectra of hydrogenic atoms

The first important contribution to understanding the spectrum of atomic hydrogen, which is observed when an electric discharge is passed through hydrogen gas, was made by the Swiss schoolteacher Johann Balmer. In 1885 he pointed out that (in modern terms) the wavenumbers of the light in the visible region of the electromagnetic spectrum fit the expression

$$\tilde{v} = R_H \left(\frac{1}{2^2} - \frac{1}{n^2} \right)$$

with $n = 3, 4, \ldots$ and $R_H = 109\ 677\ \text{cm}^{-1}$. The lines described by this formula are now called the **Balmer series** of the spectrum. Later, another set of lines was discovered in the ultraviolet region of the spectrum, and is called the **Lyman series**. Yet another set was discovered in the infrared region when detectors became available for that region, and is called the **Paschen series**. With this additional information available, the Swedish spectroscopist

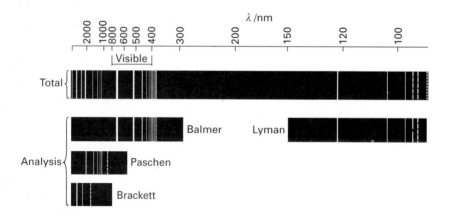

Fig 13.1 The spectrum of atomic hydrogen. The spectrum is shown at the top, and is analysed into overlapping series below. The Balmer series lies largely in the visible region.

---

[1] The essential properties of electromagnetic radiation are summarized in *Further information* 7.

Johannes Rydberg noted (in 1890) that all the lines are described by the expression

$$\tilde{v} = R_H \left(\frac{1}{n_1^2} - \frac{1}{n_2^2} \right) \tag{13.1}$$

with $n_1 = 1, 2, \ldots$ and $n_2 = n_1 + 1, n_1 + 2, \ldots$. The constant R_H is now called the **Rydberg constant** for hydrogen. The first five series of lines then correspond to n_1 taking the values 1 (Lyman), 2 (Balmer), 3 (Paschen), 4 (Brackett), and 5 (Pfund).

The existence of discrete spectroscopic lines strongly suggests that the energy of the electron in the hydrogen atom is quantized. The total energy is conserved when a **transition**, a change of state, occurs from one energy level to another. Therefore, when an atom changes its energy by ΔE, this difference must be carried away as a photon of frequency v (as was illustrated in Fig 12.14), where

$$\Delta E = hv \tag{13.2}$$

This relation is called the **Bohr frequency condition**. It follows that we can expect to observe discrete lines if an electron in an atom can exist only in certain energy states.

13.2 **The permitted energies of hydrogenic atoms**

The quantum mechanical description of the structure of a hydrogenic atom is based on Rutherford's **nuclear model**, in which the atom is pictured as consisting of an electron outside a central nucleus of charge Ze. To derive the details of the structure of this type of atom, we have to set up and solve the Schrödinger equation in which the potential energy, V, is the **Coulomb potential energy** for the interaction between the nucleus of charge $+Ze$ and the electron of charge $-e$:

$$V = -\frac{Ze^2}{4\pi\varepsilon_0 r} \tag{13.3}$$

Here ε_0 is the **vacuum permittivity** (a fundamental constant, see inside front cover) and e is the elementary electric charge. We also need to identify the appropriate boundary conditions. For the hydrogen atom, these conditions are that the wave-

function must not become infinite anywhere and that it must repeat itself as we circle the nucleus either over the poles or round the equator.

With a lot of work, the Schrödinger equation with this potential energy and these boundary conditions can be solved, and we shall summarize the results. The need to satisfy boundary conditions leads to the conclusion that only certain energies can occur, which is qualitatively in accord with the spectroscopic evidence. Schrödinger himself found that, for a hydrogenic atom of atomic number Z with a nucleus of mass m_N, the allowed energy levels are given by the expression

$$E_n = -\frac{hcRZ^2}{n^2} \quad hcR = \frac{\mu e^4}{32\pi^2\varepsilon_0^2\hbar^2} \quad \mu = \frac{m_e m_N}{m_e + m_N} \tag{13.4}$$

with $n = 1, 2, \ldots$. The quantity μ is the **reduced mass**. The mass of the nucleus is so much bigger than the mass of the electron for all except the most precise considerations that the latter may be neglected in the denominator of μ, and then $\mu \approx m_e$. The constant R is numerically identical to the experimental Rydberg constant R_H when m_N is not ignored but set equal to the mass of the proton. Schrödinger must have been thrilled to find that, when he calculated R_H, the value he obtained was in exact agreement with the experimental value.[2]

The quantum number n is called the **principal quantum number**. We use it to calculate the energy of the electron in the atom by substituting its value into eqn 13.4. The resulting energy levels are depicted in Fig 13.2. Note how they are widely separated at low values of n, but then converge as n increases. At low values of n the electron is confined close to the nucleus by the pull between opposite charges and the energy levels are widely spaced like those of a particle in a narrow box. At high values of n, when the electron has such a high energy that it can travel out to large distances, the energy levels are close together, like those of a particle in a large box.

---

[2] Neils Bohr had already derived the same expression for R_H, but his model of the atom—with an electron orbiting the nucleus—is now known to be erroneous.

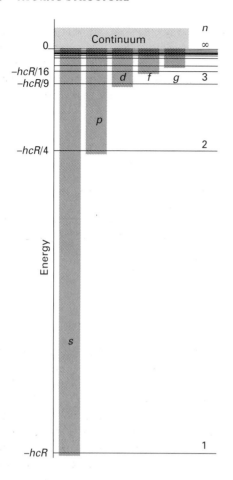

Fig 13.2 The energy levels of the hydrogen atom. The energies are relative to a proton and an infinitely distant, stationary electron. The degeneracy of the orbitals with the same value of n but different values of l is a special characteristic of hydrogenic atoms.

Self-test 13.1

The shortest wavelength transition in the Paschen series in hydrogen occurs at 821 nm; at what wavelength does it occur in Li^{2+}? (*Hint.* Think about the variation of energies with atomic number Z.)

[*Answer:* $\frac{1}{9} \times 821$ nm = 91.2 nm]

All the energies given by eqn 13.4 are negative, which signifies that an electron in an atom has a lower energy than when it is free. The zero of energy (which occurs at $n = \infty$) corresponds to the infinitely widely separated and stationary electron and nucleus. The state of lowest energy, the ground

state of the atom, is the one with $n = 1$ (the lowest permitted value of n and hence the largest negative value of the energy). The energy of this state is

$$E_1 = -hcRZ^2$$

The negative sign means that the ground state lies $hcRZ^2$ *below* the energy of the infinitely separated electron and nucleus. The first excited state of the atom, the state with $n = 2$, lies at

$$E_2 = -\tfrac{1}{4}hcRZ^2$$

This energy level is $\tfrac{3}{4}hcRZ^2$ above the ground state.

We can now explain the empirical expression for the spectroscopic lines observed in the spectrum of hydrogen. In a transition, an electron jumps from an energy level with one quantum number (n_2) to a level with a lower energy (with quantum number n_1). As a result, the magnitude of the energy change is

$$\Delta E = \frac{hcR}{n_1^2} - \frac{hcR}{n_2^2}$$

This energy is carried away by a photon of energy $hc\tilde{\nu}$. By equating this energy to ΔE, we immediately obtain eqn 13.1.

The energy needed to remove an electron completely from an atom is called the **ionization energy**, I. For a hydrogen atom, the ionization energy is the energy required to raise the electron from the ground state (with $n = 1$ and energy $E_1 = -hcR_H$) to the state corresponding to complete removal of the electron (the state with $n = \infty$ and zero energy). Therefore, the energy that must be supplied is

$$I = hcR_H = 2.179 \times 10^{-18} \text{ J}$$

which corresponds to 1312 kJ mol^{-1} or 13.59 eV.

Self-test 13.2

Predict the ionization energy of He^+ given that the ionization energy of H is 13.59 eV. (*Hint.* Decide how the energy of the ground state varies with Z.)

[*Answer:* 54.36 eV]

13.3 Quantum numbers

The wavefunction of the electron in a hydrogenic atom is called an **atomic orbital**.[3] An electron that is described by a particular wavefunction is said to 'occupy' that orbital. Each atomic orbital is specified by three quantum numbers that act as a kind of 'address' of the electron in the atom. One quantum number is the principal quantum number n, which we have already met. Another is the **orbital angular momentum quantum number**, l.[4] This quantum number is restricted to the values

$$l = 0, 1, 2, \ldots, n - 1$$

For a given value of n, there are n allowed values of l: all the values are positive. The third is the **magnetic quantum number**, m_l. This quantum number is confined to the values

$$m_l = l, l - 1, l - 2, \ldots, -l$$

For a given value of l, there are $2l + 1$ values of m_l (for example, when $l = 3$, m_l may have any of the seven values +3, +2, +1, 0, −1, −2, −3).

Illustration 13.1

It follows from these restrictions that there is only one orbital with $n = 1$, because when $n = 1$ the only value that l can have is 0, and that in turn implies that m_l can have only the value 0. Likewise, there are four orbitals with $n = 2$, because l can take the values 0 and 1, and in the latter case m_l can have the three values +1, 0, and −1. In general, there are n^2 orbitals with a given value of n.

Although we need all three quantum numbers to specify a given orbital, eqn 13.4 reveals that for hydrogenic atoms—and, as we shall see, *only* in hydrogenic atoms—the energy depends only on the principal quantum number, n. Therefore, in hydrogenic atoms, *all orbitals of the same value of n but different values of l and m_l have the same energy.* This degeneracy is one reason why all orbitals with the same value of n are said to belong to the same **shell** of the atom.[5] It is common to refer to successive shells by letters:

n	1	2	3	4 . . .
	K	L	M	N . . .

Thus, all four orbitals of the shell with $n = 2$ form the L shell of the atom.

Orbitals with the same value of n but different values of l belong to different **subshells** of a given shell. These subshells are denoted by the letters s, p, . . . using the following correspondence:

l	0	1	2	3 . . .
	s	p	d	f . . .

Only these four types of subshell are important in practice. For the shell with $n = 1$, there is only one subshell, the one with $l = 0$. For the shell with $n = 2$, there are two subshells, namely the 2s subshell (with $l = 0$) and the 2p subshell (with $l = 1$). The general pattern of the first three shells and their subshells are listed in Fig 13.3. In a hydrogenic atom, all the subshells of a given shell correspond to the same energy (because, as we have seen, the energy depends on n and not on l).

Each subshell contains $2l + 1$ individual orbitals (corresponding to the $2l + 1$ values of m_l for each value of l). Thus, in any given subshell, the number of orbitals is

s	p	d	f . . .
1	3	5	7 . . .

An orbital with $l = 0$ (and necessarily $m_l = 0$) is called an **s orbital**. A p subshell ($l = 1$) consists of three **p orbitals** (corresponding to $m_l = +1, 0, -1$). An electron that occupies an s orbital is called an **s electron**. Similarly, we can speak of p, d, . . . electrons according to the orbitals they occupy.

---

[3] The name is intended to express something less definite than the 'orbit' of classical mechanics.

[4] This quantum number is also called by its older name, the *azimuthal quantum number*.

[5] The other reason is that an electron is most likely to be found at about the same distance from the nucleus whichever orbital it occupies in a given shell.

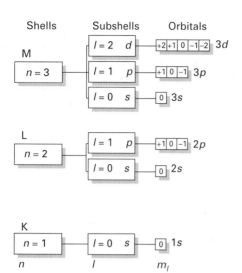

Fig 13.3 The structures of atoms are described in terms of shells of electrons that are labelled by the principal quantum number *n*, and a series of *n* subshells of these shells, with each subshell of a shell being labelled by the quantum number *l*. Each subshell consists of 2*l* + 1 orbitals.

Self-test 13.3

How many orbitals are there in a shell with *n* = 5?

[*Answer*: 25]

13.4 The wavefunctions: *s* orbitals

The mathematical form of a 1*s* orbital (the wavefunction with *n* = 1, *l* = 0, and m_l = 0) for a hydrogen atom is

$$\psi = \frac{1}{(\pi a_0^3)^{1/2}} e^{-r/a_0} \qquad a_0 = \frac{4\pi\varepsilon_0\hbar^2}{m_e e^2} \qquad (13.5)$$

The constant a_0 is called the **Bohr radius** and has the value 52.9177 pm. This wavefunction is normalized to 1 (Section 12.9), so the probability of finding the electron in a small volume of magnitude δV at a given point is *equal* to $\psi^2 \delta V$, with ψ evaluated at the point of interest.

Illustration 13.2

We can calculate the probability of finding the electron in a volume of 1.0 pm³ centred on the nucleus in a hydrogen atom by setting *r* = 0 in the expression for ψ and taking δV = 1.0 pm³:

$$\text{Probability} = \frac{1}{\pi a_0^3} \times \delta V = \frac{1}{\pi \times (52.9 \text{ pm})^3} \times (1.0 \text{ pm}^3)$$

$$= \frac{(1.0)^3}{\pi \times (52.9)^3} = 2.2 \times 10^{-6}$$

This result means that the electron will be found in the volume on one observation in 455 000.

Self-test 13.4

Repeat the calculation for finding the electron in the same volume located at the Bohr radius.

[*Answer*: 2.9×10^{-7}, 1 in 3 400 000 observations]

The wavefunction in eqn 13.5 depends only on the radius, *r*, of the point of interest and is independent of angle (the latitude and longitude of the point). Therefore, the orbital has the same amplitude at all points at the same distance from the nucleus regardless of direction. Because the probability of finding an electron is proportional to the square of the wavefunction, we now know that the electron will be found with the same probability in any direction (for a given distance from the nucleus). We summarize this angular independence by saying that a 1*s* orbital is **spherically symmetrical**.

The wavefunction decays exponentially towards zero from a maximum value of $1/(\pi a_0^3)^{1/2}$ at the nucleus (at *r* = 0). It follows that *the most probable point at which the electron will be found is at the nucleus itself*. A method of depicting the probability of finding the electron at each point in space is to represent ψ^2 by the density of shading in a diagram (Fig 13.4). A simpler procedure is to show only the **boundary surface**, the shape that captures about 90 per cent of the electron probability. For the 1*s* orbital, the boundary surface is a sphere (Fig 13.5).

We often need to know the probability that an electron will be found at a given distance from a nucleus regardless of its angular position (Fig 13.6). We can find this probability by combining the wavefunction in eqn 13.5 with the Born interpretation.

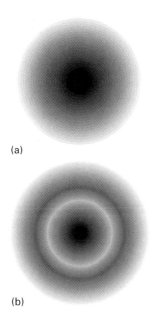

(a)

(b)

Fig 13.4 Representations of the first two s hydrogenic atomic orbitals, (a) 1s, (b) 2s, in terms of the electron densities (as represented by the density of shading).

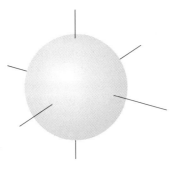

Fig 13.5 The boundary surface of an s orbital within which there is a high probability of finding the electron.

Derivation 13.1 *The radial distribution function*

Consider two spherical shells centred on the nucleus, one of radius r and the other of radius $r + \delta r$. The probability of finding the electron at a radius r regardless of its direction is equal to the probability of finding it between these two spherical surfaces. The volume of the region of space between the surfaces is equal to the surface area of the inner shell, $4\pi r^2$, multiplied by

the thickness, δr, of the region, and is therefore $4\pi r^2 \delta r$. According to the Born interpretation, the probability of finding an electron inside a small volume of magnitude δV is given, for a normalized wavefunction, by the value of $\psi^2 \delta V$. Therefore, interpreting δV as the volume of the shell, we obtain

$$\text{Probability} = 4\pi r^2 \psi^2 \delta r$$

If we write

$$P(r) = 4\pi r^2 \psi^2 \tag{13.6a}$$

where P is the **radial distribution function**, then the probability of finding the electron between the two shells is

$$\text{Probability} = P(r)\delta r \tag{13.6b}$$

The radial distribution function tells us the probability that an electron is a distance r from the nucleus regardless of its direction. Because r^2 increases from 0 as r increases but ψ^2 decreases towards 0 exponentially, P starts at 0, goes through a maximum, and declines to 0 again. The location of

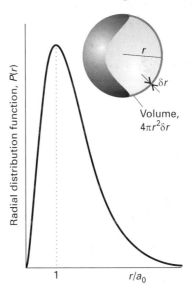

Fig 13.6 The radial distribution function gives the probability that the electron will be found anywhere in a shell of radius r and thickness δr regardless of angle. The graph shows the output from an imaginary shell-like detector of variable radius and fixed thickness δr.

the maximum marks the most probable *radius* (not point) at which the electron will be found. For a 1*s* orbital of hydrogen, the maximum occurs at a_0, the Bohr radius. An analogy that might help to fix the significance of the radial distribution function for an electron is the corresponding distribution for the population of the Earth regarded as a perfect sphere. The radial distribution function is zero at the centre of the Earth and for the next 6400 km (to the surface of the planet), when it peaks sharply and then rapidly decays again to zero. It remains virtually zero for all radii more than about 10 km above the surface. Almost all the population will be found very close to $r = 6400$ km, and it is not relevant that people are dispersed nonuniformly over a very wide range of latitudes and longitudes. The small probabilities of finding people above and below 6400 km anywhere in the world corresponds to the population that happens to be down mines or living in places as high as Denver or Tibet at the time.

Illustration 13.3

To find the probability that the electron will be found between a shell of radius a_0 and a shell of radius 1.0 pm greater, we take $\delta r = 1.0$ pm and set $r = a_0$ in eqn 13.6b:

$$\text{Probability} = 4\pi r^2 \overbrace{\psi^2}^{P} \times \delta r = 4\pi r^2 \times \overbrace{\frac{1}{\pi a_0^3} e^{-2r/a_0}}^{\psi^2} \times \delta r$$

$$= 4\pi a_0^2 \times \frac{1}{\pi a_0^3} e^{-2} \times \delta r = \frac{4e^{-2} \times \delta r}{a_0}$$

$$= \frac{4e^{-2} \times (1.0 \text{ pm})}{52.9 \text{ pm}} = 0.010$$

or about 1 inspection in 100.

A 2*s* orbital (an orbital with $n = 2$, $l = 0$, and $m_l = 0$) is also spherical, so its boundary surface is a sphere. Because a 2*s* orbital spreads further out from the nucleus than a 1*s* orbital—because the electron it describes has more energy to climb away from the nucleus—its boundary surface is a sphere of larger radius. The orbital also differs from a 1*s* orbital in its

radial dependence (Fig 13.7) for, although the wavefunction has a nonzero value at the nucleus (like all *s* orbitals), it passes through zero before commencing its exponential decay towards zero at large distances. We summarize the fact that the wavefunction passes through zero everywhere at a certain radius by saying that the orbital has a **radial node**. A 3*s* orbital has two radial nodes, a 4*s* orbital has three radial nodes.

13.5 **The wavefunctions: *p* and *d* orbitals**

All *p* orbitals (orbitals with $l = 1$) have a double-lobed appearance like that shown in Fig 13.8. The two lobes are separated by a **nodal plane** that cuts through the nucleus. There is zero probability density for an electron on this plane, and therefore (because the nucleus lies in the plane) zero probability of finding the electron at the nucleus.

The exclusion of the electron from the nucleus is a common feature of all atomic orbitals except *s* orbitals. To understand its origin, we need to know that the value of the quantum number *l* tells us the magnitude of the angular momentum of the electron around the nucleus (in classical terms, how rapidly it is circulating around the nucleus) through the expression

$$\text{Magnitude of angular momentum} = \{l(l + 1)\}^{1/2}\hbar \qquad (13.7)$$

For an *s* orbital, the orbital angular momentum is zero (because $l = 0$), and in classical terms the electron does not circulate around the nucleus. Because $l = 1$ for a *p* orbital, the magnitude of the angular momentum of a *p* electron is $2^{1/2}\hbar$. As a result, a *p* electron is flung away from the nucleus by the centrifugal force arising from its motion, but an *s* electron is not. The same centrifugal effect appears in all orbitals with angular momentum (those for which $l > 0$), such as *d* orbitals and *f* orbitals, and all such orbitals have nodal planes that cut through the nucleus.

Each *p* subshell consists of three individual orbitals ($m_l = +1, 0, -1$). The three orbitals are normally represented by their boundary surfaces, as depicted

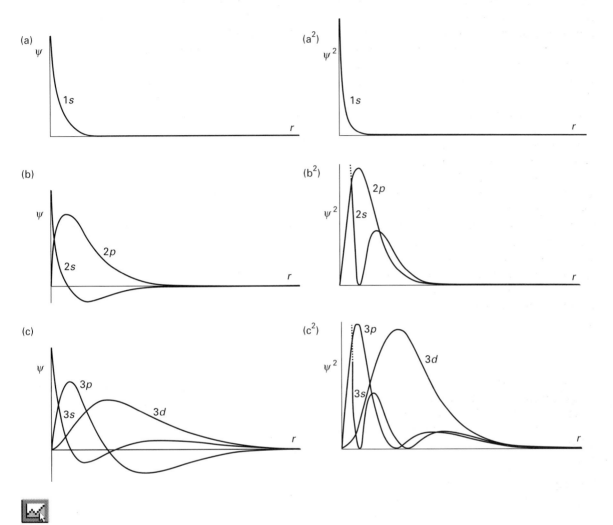

Fig 13.7 (a,b,c) The radial wavefunctions of the first few states of the hydrogen atom and (a^2,b^2,c^2) the probability densities. Note that the s orbitals have a nonzero and finite value at the nucleus. The vertical scales are different in each case.

in Fig 13.8. The p_x orbital has a symmetrical double-lobed shape directed along the x-axis, and similarly the p_y and p_z orbitals are directed along the y- and z-axes, respectively.[6] As n increases, the p orbitals become bigger (for the same reason as s orbitals), and acquire a more complex radial nodal structure. However, their boundary surfaces retain the double-lobed shape shown in the illustration.

----

[6] There is no *direct* correspondence between the values of m_l and the labels x, y, and z for the p orbitals.

There are five d orbitals in each shell (for $n \geq 3$), each one corresponding to one of the values $m_l = +2$, $+1$, 0, -1, and -2. Figure 13.9 shows the five boundary surfaces of the orbitals. Similar boundary surfaces are obtained for all values of n.

We can now explain the physical significance of the quantum number m_l. It indicates the component of the electron's orbital angular momentum around an arbitrary axis passing through the nucleus. Positive values of m_l correspond to clockwise

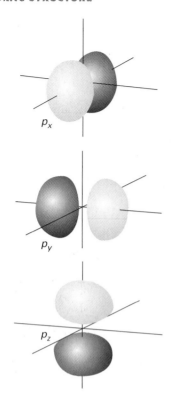

Fig 13.8 The boundary surfaces of p orbitals. A nodal plane passes through the nucleus and separates the two lobes of each orbital. The light and dark areas denote regions of opposite sign of the wavefunction.

motion seen from below and negative values correspond to counterclockwise motion. The larger the value of $|m_l|$, the higher the angular momentum around the arbitrary axis. Specifically:

$$\text{Component of angular momentum} = m_l \hbar \quad (13.8)$$

An s electron has $m_l = 0$, and has no angular momentum about any axis. A p electron can circulate clockwise about an axis as seen from below ($m_l = +1$). Of its total angular momentum of $2^{1/2}\hbar = 1.414\hbar$, an amount $\hbar$ is due to motion around the selected axis (the rest is due to motion around the other two axes). A p electron can also circulate counterclockwise as seen from below ($m_l = -1$), or not at all ($m_l = 0$) about that selected axis. An electron in the d subshell can circulate with five different amounts of angular momentum about an arbitrary axis ($+2\hbar$, $+\hbar$, 0, $-\hbar$, $-2\hbar$).

13.6 Electron spin

To complete the description of the state of a hydrogenic atom, we need to introduce one more concept, that of electron spin. The **spin** of an electron is an *intrinsic* angular momentum that every electron possesses and that cannot be changed or eliminated (just like its mass or its charge). The name 'spin' is evocative of a ball spinning on its axis, and (so long as it is treated with caution) this classical interpretation can be used to help to visualize the motion. However, in fact, spin is a purely quantum mechanical phenomenon and has no classical counterpart, so the analogy must be used with care.

We shall make use of two properties of electron spin:

1 Electron spin is described by a **spin quantum number**, s (the analogue of l for orbital angular momentum), with s fixed at the single (positive) value of $\frac{1}{2}$ for all electrons at all times.

2 The spin can be clockwise or counterclockwise; these two states are distinguished by the **spin magnetic quantum number**, m_s, which can take the values $+\frac{1}{2}$ or $-\frac{1}{2}$ but no other values.

An electron with $m_s = +\frac{1}{2}$ is called an **α electron** and commonly denoted α or $\uparrow$; an electron with $m_s = -\frac{1}{2}$ is called a **β electron** and denoted β or $\downarrow$.

The existence of electron spin was confirmed by an experiment performed by Otto Stern and Walther Gerlach in 1921, who shot a beam of silver atoms through an inhomogeneous magnetic field (Fig 13.10). A silver atom has 47 electrons, and (for reasons that will become clear later) 23 of the spins are $\uparrow$ and 23 spins are $\downarrow$; the one remaining spin may be either $\uparrow$ or $\downarrow$.[7] Because the spin angular momenta of the $\uparrow$ and $\downarrow$ electrons cancel each other, the atom behaves as if it had the spin of a single electron. The idea behind the Stern–Gerlach experiment was that a rotating, charged body—in this case an electron—behaves like a magnet and interacts with the applied field. Because the magnetic field pushes or pulls the electron according to the orientation of the electron's spin, the initial beam

---

[7] As will probably be familiar from introductory chemistry, 46 of the electrons are paired and occupy orbitals in accord with the building-up principle.

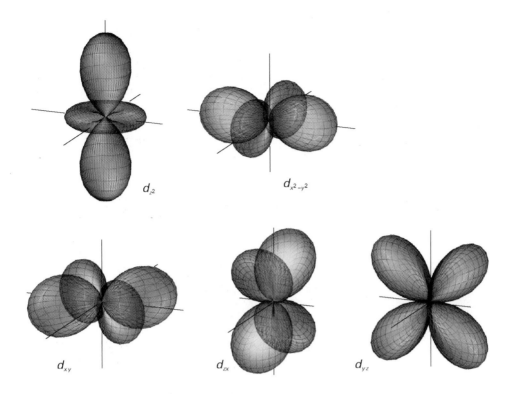

d_{z^2}

$d_{x^2-y^2}$

d_{xy}

d_{zx}

d_{yz}

Fig 13.9 The boundary surfaces of *d* orbitals. Two nodal planes in each orbital intersect at the nucleus and separate the four lobes of each orbital.

of atoms should split into two beams, one corresponding to atoms with ↑ spin and the other to atoms with ↓ spin. This result was observed.

Other fundamental particles also have characteristic spins. For example, protons and neutrons are spin-$\frac{1}{2}$ particles (that is, for them $s = \frac{1}{2}$) so invariably spin with a single, irremovable angular momentum. Because the masses of a proton and a neutron are so much greater than the mass of an electron, yet they all have the same spin angular momentum, the classical picture of proton and neutron spin would be of particles spinning much more slowly than an electron. Some elementary particles have $s = 1$ and therefore have a higher intrinsic angular momentum than an electron. For our purposes the most important spin-1 particle is the photon. It is a very deep feature of nature, that the fundamental particles from which matter is built have half-integral spin (such as electrons and quarks, all of which have $s = \frac{1}{2}$). The particles that transmit forces between these particles, so binding them together

into entities like nuclei, atoms, and planets, all have integral spin (such as $s = 1$ for the photon, which transmits the electromagnetic interaction between charged particles). Fundamental particles with half-integral spin are called **fermions**; those with integral spin are called **bosons**. Matter therefore consists of fermions bound together by bosons.

13.7 **Spectral transitions and selection rules**

We have already seen that, when an electron makes a transition from an orbital in a shell with a principal quantum number n_2 into an orbital of a shell with principal quantum number n_1, the excess energy is emitted as a photon and contributes to one of the spectral lines given by the Rydberg formula, eqn 13.1. We can think of the sudden change in the distribution of the electron as it changes its spatial distribution from one orbital to another orbital as

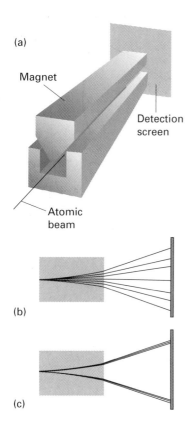

Magnet

Detection
screen

Atomic
beam

(b)

(c)

Fig 13.10 (a) The experimental arrangement for the Stern–Gerlach experiment: the magnet is the source of an inhomogeneous field. (b) The classically expected result, when the orientations of the electron spins can take all angles. (c) The observed outcome using silver atoms, when the electron spins can adopt only two orientations (↑ and ↓).

jolting the electromagnetic field into oscillation, and that oscillation corresponds to the generation of a photon of light.

It turns out, however, that not all transitions between all available orbitals are possible. For example, it is not possible for an electron in a $3d$ orbital to make a transition to a $1s$ orbital. Transitions are classified as either **allowed**, if they can contribute to the spectrum, or **forbidden**, if they cannot. The forbidden or allowed character of a transition can be traced to the role of the photon spin, which we mentioned above. When a photon, with its one unit of angular momentum, is generated in a transition, the angular momentum of the electron must change by one unit to compensate for the angular

momentum carried away by the photon. That is, the angular momentum must be conserved—neither created nor destroyed—just as linear momentum is conserved in collisions. Thus, an electron in a d orbital (with $l = 2$) cannot make a transition into an s orbital (with $l = 0$) because the photon cannot carry away enough angular momentum. Similarly, an s electron cannot make a transition to another s orbital, because then there is no change in the electron's angular momentum to make up for the angular momentum carried away by the photon.

A **selection rule** is a statement about which spectroscopic transitions are allowed. They are derived (for atoms) by identifying the transitions that conserve angular momentum when a photon is emitted or absorbed. The selection rules for hydrogenic atoms are

$$\Delta l = \pm 1 \qquad \Delta m_l = 0, \pm 1$$

The principal quantum number n can change by any amount consistent with the Δl for the transition because it does not relate directly to the angular momentum.

Derivation 13.2 *The transition dipole moment*

The derivation of selection rules is based on an analysis of the **transition dipole moment**, μ_{fi}, which is defined as[8]

$$\mu_{fi} = \int \psi_f \mu \psi_i \, d\tau$$

where ψ_i is the wavefunction for the initial state and ψ_f is that for the final state. The μ in the integrand stands for $-er$ (in a hydrogenic atom). The integration is over all space. If it turns out that this integral is zero, then the transition is forbidden and is not observed in the spectrum. If the integral is nonzero, then the transition is allowed and its intensity is proportional to the square of μ_{fi}. The physical significance of μ_{fi} is that it is a measure of the strength of the coupling between the electronic transition and the surrounding electromagnetic field. In other words, it is a measure of the magnitude of the 'kick' that an electronic transition delivers to the surrounding field when it occurs.

---

[8] If the wavefunctions are complex (in the sense of having real and imaginary components), ψ_f is replaced by its complex conjugate, ψ_f^*.

Example 13.1 *Using the selection rules*

To what orbitals may a 4*d* electron make spectroscopic transitions?

Strategy We apply the selection rules, principally the rule concerning *l*, and identify the orbital to which the transition may occur.

Solution Because *l* = 2, the final orbital must have *l* = 1 or 3. Thus, an electron may make a transition from a 4*d* orbital to any *np* orbital (subject to $\Delta m_l = 0, \pm 1$) and to any *nf* orbital (subject to the same rule). However, it cannot undergo a transition to any other orbital, so a transition to any *ns* orbital or another *nd* orbital is forbidden.

Self-test 13.5 To what orbitals may a 4*s* electron make spectroscopic transitions?

[*Answer: np* orbitals only]

Selection rules enable us to construct a **Grotrian diagram** (Fig 13.11), which is a diagram that summarizes the energies of the states and the allowed transitions between them. The thicknesses of the transition lines in the diagram indicate in a general way their relative intensities in the spectrum.

The structures of many-electron atoms

THE Schrödinger equation for a many-electron atom is highly complicated because all the electrons interact with one another. Even for a He atom,

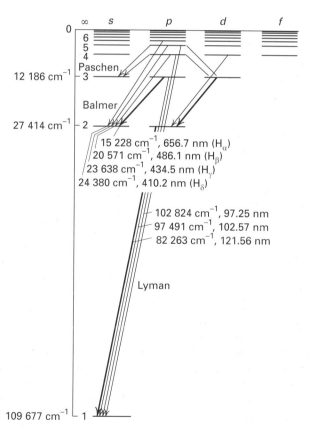

Fig 13.11 A Grotrian diagram that summarizes the appearance and analysis of the spectrum of atomic hydrogen. The thicker the line, the more intense the transition.

with its two electrons, no mathematical expression for the orbitals and energies can be given, and we are forced to make approximations. Modern computational techniques, though, are able to refine the approximations we are about to make, and permit highly accurate numerical calculations of energies and wavefunctions.

13.8 The orbital approximation

In the **orbital approximation** we suppose that a reasonable first approximation to the exact wavefunction is obtained by letting each electron occupy its 'own' orbital, and writing

$$\psi = \psi(1)\psi(2) \cdots \qquad (13.9)$$

where $\psi(1)$ is the wavefunction of electron 1, $\psi(2)$ that of electron 2, and so on. We can think of the individual orbitals as resembling the hydrogenic orbitals, but with nuclear charges that are modified by the presence of all the other electrons in the atom. This description is only approximate, but it is a useful model for discussing the properties of atoms, and is the starting point for more sophisticated descriptions of atomic structure.

Illustration 13.4

The wavefunction for a 1s electron in helium is $\psi = (8/\pi a_0^3)^{1/2}e^{-2r/a_0}$. If electron 1 is at a radius r_1 and electron 2 is at a radius r_2, the overall wavefunction for the two-electron atom is

$$\psi = \psi(1)\,\psi(2) = \left(\frac{8}{\pi a_0^3}\right)^{1/2} e^{-2r_1/a_0} \times \left(\frac{8}{\pi a_0^3}\right)^{1/2} e^{-2r_2/a_0}$$

$$= \left(\frac{8}{\pi a_0^3}\right) e^{-2(r_1+r_2)/a_0}$$

The orbital approximation allows us to express the electronic structure of an atom by reporting its configuration, the list of occupied orbitals (usually, but not necessarily, in its ground state). For example, because the ground state of a hydrogen atom consists of a single electron in a 1s orbital, we report its configuration as $1s^1$ (read 'one-s-one'). A helium atom has two electrons. We can imagine forming the atom by adding the electrons in succession to the orbitals of the bare nucleus (of charge $2e$). The first electron occupies a hydrogenic 1s orbital but, because $Z = 2$, the orbital is more compact than in H itself. The second electron joins the first in the same 1s orbital, and so the electron configuration of the ground state of He is $1s^2$ (read 'one-s-two').

13.9 The Pauli principle

Lithium, with $Z = 3$, has three electrons. Two of its electrons occupy a 1s orbital drawn even more closely than in He around the more highly charged nucleus. The third electron, however, does not join the first two in the 1s orbital because a $1s^3$ configuration is forbidden by a fundamental feature of nature summarized by the **Pauli exclusion principle**:

No more than two electrons may occupy any given orbital, and if two electrons do occupy one orbital, then their spins must be paired.

Electrons with **paired spins**, denoted ↑↓, have zero net spin angular momentum because the spin angular momentum of one electron is cancelled by the spin of the other. The exclusion principle is the key to understanding the structures of complex atoms, to chemical periodicity, and to molecular structure. It was proposed by the Austrian Wolfgang Pauli in 1924 when he was trying to account for the absence of some lines in the spectrum of helium.

Lithium's third electron cannot enter the 1s orbital because that orbital is already full: we say the K shell is **complete** and that the two electrons form a **closed shell**. Because a similar closed shell occurs in the He atom, we denote it [He]. The third electron is excluded from the K shell ($n = 1$) and must occupy the next available orbital, which is one with $n = 2$ and hence belonging to the L shell. However, we now have to decide whether the next available orbital is the 2s orbital or a 2p orbital, and therefore whether the lowest energy configuration of the atom is $[He]2s^1$ or $[He]2p^1$.

13.10 **Penetration and shielding**

Unlike in hydrogenic atoms, in many-electron atoms the $2s$ and $2p$ orbitals (and, in general, all orbitals of a given shell) are not degenerate. For reasons we shall now explain, s electrons generally lie lower in energy than p electrons of a given shell, and p electrons lie lower than d electrons.

An electron in a many-electron atom experiences a Coulombic repulsion from all the other electrons present. When the electron is at a distance r from the nucleus, the repulsion it experiences from the other electrons can be modelled by a point negative charge located on the nucleus and having a magnitude equal to the charge of the electrons within a sphere of radius r (Fig 13.12). The effect of the point negative charge is to lower the full charge of the nucleus from Ze to $Z_{eff}e$, the **effective nuclear charge**. To express the fact that an electron experiences a nuclear charge that has been modified by the other electrons present, we say that the electron experiences a **shielded nuclear charge**. The electrons do not actually 'block' the full Coulombic attraction of the nucleus: the effective charge is simply a way of expressing the net outcome of the nuclear attraction and the electronic repulsions in terms of a single equivalent charge at the centre of the atom.

The effective nuclear charges experienced by s and p electrons are different because the electrons have different wavefunctions and therefore different distributions around the nucleus (Fig 13.13). An s electron has a greater **penetration** through inner shells than a p electron of the same shell in the sense that an s electron is more likely to be found close to the nucleus than a p electron of the same shell (the p orbital, remember, has an angular node passing through the nucleus). As a result of this greater penetration, an s electron experiences less shielding than a p electron of the same shell and therefore experiences a larger Z_{eff}. Consequently, by the combined effects of penetration and shielding, an s electron is more tightly bound than a p electron of the same shell. Similarly, a d electron penetrates less than a p electron of the same shell, and it therefore experiences more shielding and an even smaller Z_{eff}.

The consequence of penetration and shielding is that, in general, the energies of orbitals in the same shell of a many-electron atom lie in the order

$$s < p < d < f$$

The individual orbitals of a given subshell (such as the three p orbitals of the p subshell) remain degenerate because they all have the same radial charac-

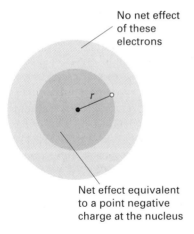

No net effect of these electrons

r

Net effect equivalent to a point negative charge at the nucleus

Fig 13.12 An electron at a distance r from the nucleus experiences a Coulombic repulsion from all the electrons within a sphere of radius r and which is equivalent to a point negative charge located on the nucleus. The effect of the point charge is to reduce the apparent nuclear charge of the nucleus from Ze to $Z_{eff}e$.

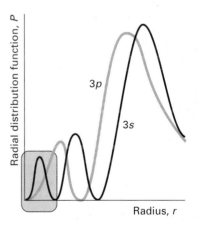

Radial distribution function, P

$3p$

$3s$

Radius, r

Fig 13.13 An electron in an s orbital (here a 3s orbital) is more likely to be found close to the nucleus than an electron in a p orbital of the same shell. Hence it experiences less shielding and is more tightly bound.

teristics and so experience the same effective nuclear charge.

We can now complete the Li story. Because the shell with $n = 2$ has two nondegenerate subshells, with the $2s$ orbital lower in energy than the three $2p$ orbitals, the third electron occupies the $2s$ orbital. This arrangement results in the ground-state configuration $1s^22s^1$, or $[\text{He}]2s^1$. It follows that we can think of the structure of the atom as consisting of a central nucleus surrounded by a complete helium-like shell of two $1s$ electrons, and around that a more diffuse $2s$ electron. The electrons in the outermost shell of an atom in its ground state are called the **valence electrons** because they are largely responsible for the chemical bonds that the atom forms (and, as we shall see, the extent to which an atom can form bonds is called its 'valence'). Thus, the valence electron in Li is a $2s$ electron, and lithium's other two electrons belong to its core, where they take little part in bond formation.

13.11 The building-up principle

The extension of the procedure used for H, He, and Li to other atoms is called the **building-up principle**.[9] The building-up principle specifies an order of occupation of atomic orbitals that reproduces the experimentally determined ground-state configurations of neutral atoms.

We imagine the bare nucleus of atomic number Z, and then feed into the available orbitals Z electrons one after the other. The first two rules of the building-up principle are:

1 The order of occupation of orbitals is[10]

   $1s\ 2s\ 2p\ 3s\ 3p\ 4s\ 3d\ 4p\ 5s\ 4d\ 5p\ 6s\ 5d\ 4f\ 6p\ \ldots$

2 According to the Pauli exclusion principle, each orbital may accommodate up to two electrons.

The order of occupation is approximately the order of energies of the individual orbitals, because, in general, the lower the energy of the orbital, the

---

[9] The building-up principle is still widely the *Aufbau principle*, from the German word for building up.

[10] This order is best remembered by noting that it follows the layout of the periodic table.

lower the total energy of the atom as a whole when that orbital is occupied. An s subshell is complete as soon as two electrons are present in it. Each of the three p orbitals of a shell can accommodate two electrons, so a p subshell is complete as soon as six electrons are present in it. A d subshell, which consists of five orbitals, can accommodate up to ten electrons.

As an example, consider a carbon atom. Because $Z = 6$ for carbon, there are six electrons to accommodate. Two enter and fill the $1s$ orbital, two enter and fill the $2s$ orbital, leaving two electrons to occupy the orbitals of the $2p$ subshell. Hence its ground configuration is $1s^22s^22p^2$, or more succinctly $[\text{He}]2s^22p^2$, with $[\text{He}]$ the helium-like $1s^2$ core. However, it is possible to be more specific. On electrostatic grounds, we can expect the last two electrons to occupy different $2p$ orbitals, for they will then be further apart on average and repel each other less than if they were in the same orbital. Thus, one electron can be thought of as occupying the $2p_x$ orbital and the other the $2p_y$ orbital, and the lowest energy configuration of the atom is $[\text{He}]2s^22p_x^12p_y^1$. The same rule applies whenever degenerate orbitals of a subshell are available for occupation. Therefore, another rule of the building-up principle is:

3 Electrons occupy different orbitals of a given subshell before doubly occupying any one of them.

It follows that a nitrogen atom ($Z = 7$) has the configuration $[\text{He}]2s^22p_x^12p_y^12p_z^1$. Only when we get to oxygen ($Z = 8$) is a $2p$ orbital doubly occupied, giving the configuration $[\text{He}]2s^22p_x^22p_y^12p_z^1$.

An additional point arises when electrons occupy degenerate orbitals (such as the three $2p$ orbitals) singly, as they do in C, N, and O, for there is then no requirement that their spins should be paired. We need to know whether the lowest energy is achieved when the electron spins are the same (both ↑, for instance, denoted ↑↑, if there are two electrons in question, as in C) or when they are paired (↑↓). This problem is resolved by **Hund's rule**:

4 In its ground state, an atom adopts a configuration with the greatest number of unpaired electrons.

The explanation of Hund's rule is complicated, but it reflects the quantum mechanical property of **spin correlation**, that electrons in different orbitals with parallel spins have a tendency to stay well apart and hence repel each other less.[11] We can now conclude that in the ground state of a C atom, the two $2p$ electrons have the same spin, that all three $2p$ electrons in an N atom have the same spin, and that the two electrons that singly occupy different $2p$ orbitals in an O atom have the same spin (the two in the $2p_x$ orbital are necessarily paired).

Neon, with $Z = 10$, has the configuration $[He]2s^2 2p^6$, which completes the L shell. This closed-shell configuration is denoted $[Ne]$, and acts as a core for subsequent elements. The next electron must enter the $3s$ orbital and begin a new shell, and so a Na atom, with $Z = 11$, has the configuration $[Ne]3s^1$. Like lithium with the configuration $[He]2s^1$, sodium has a single s electron outside a complete core.

Self-test 13.6

Predict the ground-state electron configuration of sulfur.

[*Answer*: $[Ne]3s^2 3p_x^2 3p_y^1 3p_z^1$]

This analysis has brought us to the origin of chemical periodicity. The L shell is completed by eight electrons, and so the element with $Z = 3$ (Li) should have similar properties to the element with $Z = 11$ (Na). Likewise, Be ($Z = 4$) should be similar to Mg ($Z = 12$), and so on up to the noble gases He ($Z = 2$), Ne ($Z = 10$), and Ar ($Z = 18$).

13.12 The occupation of *d* orbitals

Argon has complete $3s$ and $3p$ subshells and, as the $3d$ orbitals are high in energy, the atom effectively has a closed-shell configuration. Indeed, the $4s$ orbitals are so lowered in energy by their ability to penetrate close to the nucleus that the next elec-

tron (for potassium) occupies a $4s$ orbital rather than a $3d$ orbital and the K atom resembles a Na atom. The same is true of a Ca atom, which has the configuration $[Ar]4s^2$, resembling that of its congener Mg, which is $[Ne]3s^2$. However, at this point, the $3d$ orbitals become comparable in energy to the $4s$ orbitals (Fig 13.14), and they commence to be filled.

Ten electrons can be accommodated in the five $3d$ orbitals, which accounts for the electron configurations of scandium to zinc. However, the building-up principle has less clear-cut predictions about the ground-state configurations of these elements because electron–electron repulsions are comparable to the energy difference between the $4s$ and $3d$ orbitals, and a simple analysis no longer works. At gallium, the energy of the $3d$ orbitals has fallen so far below those of the $4s$ and $4p$ orbitals that they (the full $3d$ orbitals) can be largely ignored, and the building-up principle can be used in the same way as in preceding periods. Now the $4s$ and $4p$ subshells constitute the valence shell, and the period terminates with krypton. Because 18 electrons have intervened since argon, this period is the first long period of the periodic table. The existence of the *d*-block elements (the 'transition metals') reflects the stepwise occupation of the $3d$ orbitals, and the subtle shades of energy differences along this series give rise to the rich complexity of inorganic (and bioinorganic) *d*-metal chemistry. A similar intrusion of the *f* orbitals in Periods 6 and 7 accounts for the existence of the *f* block of the periodic table (the lanthanides and actinides).

13.13 The configurations of cations and anions

The configurations of cations of elements in the s, p, and d blocks of the periodic table are derived by removing electrons from the ground-state configuration of the neutral atom in a specific order. First, we remove valence p electrons (if any are present), then valence s electrons, and then as many d electrons as are necessary to achieve the stated charge. For instance, because the configuration of Fe is $[Ar]3d^6 4s^2$, an Fe^{3+} cation has the configuration $[Ar]3d^5$.

---

[11] The effect of spin correlation is to allow the atom to shrink slightly, so the electron–nucleus interaction is improved when the spins are parallel.

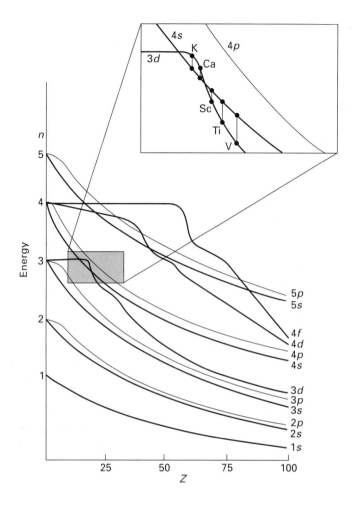

Fig 13.14 Energy levels of many-electron atoms in the periodic table. The inset shows a magnified view close to $Z = 20$.

The configurations of anions are derived by continuing the building-up procedure and adding electrons to the neutral atom until the configuration of the next noble gas has been reached. Thus, the configuration of an O^{2-} ion is achieved by adding two electrons to $[He]2s^2 2p^4$, giving $[He]2s^2 2p^6$, the configuration of Ne.

Self-test 13.7

Predict the electron configurations of (a) a Cu^{2+} ion and (b) an S^{2-} ion.

[*Answer*: (a) $[Ar]3d^9$, (b) $[Ne]3s^2 3p^6$]

Periodic trends in atomic properties

THE periodic recurrence of analogous ground-state electron configurations as the atomic number increases accounts for the periodic variation in the properties of atoms. Here we concentrate on two aspects of atomic periodicity: atomic radius and ionization energy.

13.14 Atomic radius

The **atomic radius** of an element is half the distance between the centres of neighbouring atoms in a solid (such as Cu) or, for nonmetals, in a homonuclear molecule (such as H_2 or S_8). It is of great significance in chemistry, for the size of an atom is one of the most important controls on the number of chemical bonds the atom can form. Moreover, the size and shape of a molecule depend on the sizes of the atoms of which it is composed, and molecular shape and size are crucial aspects of a molecule's biological function. Atomic radius also has an important technological aspect, because the similarity of the atomic radii of the *d*-block elements is the main reason why they can be blended together to form so many different alloys, particularly varieties of steel. If there is one single attribute of an element that determines its chemical properties (either directly, or indirectly through the variation of other properties), then it is atomic radius (Box 13.1).

In general, atomic radii decrease from left to right across a period and increase down each group (Table 13.1 and Fig 13.15). The decrease across a period can be traced to the increase in nuclear charge, which draws the electrons in closer to the nucleus.

Table 13.1 *Atomic radii of main-group elements, r/pm*

Li	Be	B	C	N	O	F
157	112	88	77	74	66	64
Na	**Mg**	**Al**	**Si**	**P**	**S**	**Cl**
191	160	143	118	110	104	99
K	**Ca**	**Ga**	**Ge**	**As**	**Se**	**Br**
235	197	153	122	121	117	114
Rb	**Sr**	**In**	**Sn**	**Sb**	**Te**	**I**
250	215	167	158	141	137	133
Cs	**Ba**	**Tl**	**Pb**	**Bi**	**Po**	
272	224	171	175	182	167	

The increase in nuclear charge is partly cancelled by the increase in the number of electrons but, because electrons are spread over a region of space, one electron does not fully shield one nuclear charge, so the increase in nuclear charge dominates. The increase in atomic radius down a group (despite the increase in nuclear charge) is explained by the fact that the valence shells of successive periods correspond to higher principal quantum

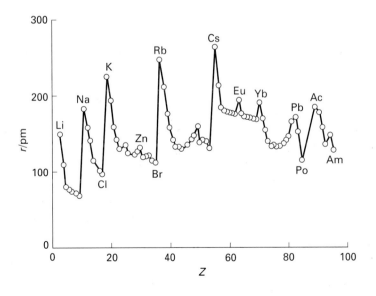

Fig 13.15 The variation of radius through the periodic table. Note the contraction of radii following the lanthanides in Period 6 (following Yb, ytterbium).

Box 13.1 *Atomic radius and respiration*

Nature makes unconscious use of subtle differences in atomic and ionic radius to pump and store oxygen through our bodies. Here we concentrate on haemoglobin (hemoglobin, Hb), the protein used to transport oxygen through our bodies, and myoglobin (Mb), the protein used to store oxygen in muscle tissue and to release it on demand. Haemoglobin is a tetramer of four myoglobin-like molecules, and each unit, as in myoglobin, has a single iron atom at its centre (see illustration). The oxygenated form of haemoglobin is called the *relaxed state* (R state) and the deoxygenated form is called the *tense state* (T state).

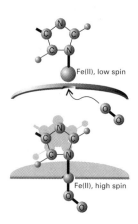

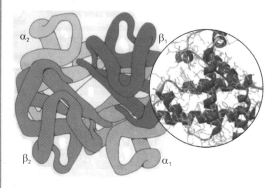

The structure of haemoglobin showing the four polypeptide chains that contribute to the structure. The inset shows a polypeptide chain in more detail. There is an iron atom at the centre of each of the four chains.

The action takes place at the *haem group* (heme group) an almost flat ring-like structure at the centre of which there is an iron atom present as iron(II) (see the illustration). When the iron–O_2 complex forms, there is a favourable interaction between the filled d_{xy} and d_{yz} orbitals on Fe(II) with the half-full π^* orbitals of O_2.[12] There is also an interaction between the empty Fe$3d_{z^2}$ orbital and the full σ orbital of O_2. The O_2 adopts a bent orientation with respect to the iron atom, partly because that orientation maximizes overlap but also because it is consistent with the arrangement of polypeptide residues in the pocket of the protein containing the haem group.

One important change that occurs when the iron atom is oxygenated is the transition from a high-spin d^6

The change in molecular geometry that takes place when an oxygen molecule attaches to an iron atom in a haemoglobin molecule. Note how the histidine residue is pulled into a different location by the motion of the iron atom.

$(d_{xy}^2 d_{yz}^1 d_{zx}^1 d_{x^2-y^2}^1 d_{z^2}^1)$ paramagnetic configuration in which the maximum number of electrons have parallel spins, to a low-spin d^6 $(d_{xy}^2 d_{yz}^2 d_{zx}^2)$ diamagnetic configuration, the formation of which accompanies the change in coordination number of the iron atom from five to six. In the T state the fifth location is taken up by the N atom of a histidine residue (His); in the R state, that link remains but the O_2 molecule occupies a coordination site on the other side of the ring. The change from high spin to low spin results in a slightly smaller atom. As a result, instead of lying 60 pm above the plane of the haem ring, the iron can fall back almost into the plane of the ring, and in the R state it lies only 20 pm above the plane. As it falls back, it pulls the histidine residue with it. That has further consequences in haemoglobin, for one pair of myoglobin-like molecules rotate through 15° relative to the other pair and become offset by 80 pm. This realignment of two of the molecules relative to the other two disrupts an ionic $His^+ \cdots Asp^-$ ionic interaction which helps to stabilize the deoxygenated form, and as a result the oxygenated molecule is more capable of taking up the next O_2 molecule than the originally fully deoxygenated form was. In other words, in haemoglobin there is a *cooperative* uptake of O_2 molecules, and a similar cooperative unloading of O_2 when conditions demand it. The result of this cooperativity is that Hb can release its O_2 under conditions where Mb cannot, which is an ideal arrangement for a transport protein rather than a storage protein.

---

[12] This notation is introduced in Section 14.10, but will be familiar from introductory chemistry.

Exercise 1 In the laboratory, the Fe atom in the haem group of haemoglobin can be removed and replaced by a Zn atom. (a) Write the electron configuration for Zn^{2+}. (b) Discuss whether this modified protein is likely to bind oxygen efficiently.

Exercise 2 Ferritin is an iron storage protein, one function of which is to bind insoluble Fe(III) and increase its solubility so that it does not precipitate in the body. The solubility of Fe(III) at pH = 7 is 1.0×10^{-18} mol L^{-1}. Concentrations as high as 3×10^{-4} mol L^{-1} have been achieved with solutions of ferritin. What volume of water would a person have to drink in one day in the presence of ferritin to solubilize the iron released from old red blood cells (25 mg d^{-1})? What volume would it be necessary to drink in the absence of ferritin?

numbers. That is, successive periods correspond to the start and then completion of successive (and more distant) shells of the atom that surround each other like the successive layers of an onion. The need to occupy a more distant shell leads to a larger atom despite the increased nuclear charge.

A modification of the increase down a group is encountered in Period 6, for the radii of the atoms late in the d block and in the following regions of the p block are not as large as would be expected by simple extrapolation down the group. The reason can be traced to the fact that in Period 6 the f orbitals are in the process of being occupied. An f electron is a very inefficient shielder of nuclear charge (for reasons connected with its radial extension) and, as the atomic number increases from La to Yb, there is a considerable contraction in radius. By the time the d block resumes (at lutetium, Lu), the poorly shielded but considerably increased nuclear charge has drawn in the surrounding electrons, and the atoms are compact. They are so compact, that the

metals in this region of the periodic table (iridium to lead) are very dense. The reduction in radius below that expected by extrapolation from preceding periods is called the **lanthanide contraction**.

13.15 Ionization energy and electron affinity

The minimum energy necessary to remove an electron from a many-electron atom is its **first ionization energy**, I_1. The **second ionization energy**, I_2, is the minimum energy needed to remove a second electron (from the singly charged cation). The variation of the first ionization energy through the periodic table is shown in Fig 13.16 and some numerical values are given in Table 13.2. The ionization energy of an element plays a central role in determining the ability of its atoms to participate in bond formation (for bond formation, as we shall see in Chapter 14, is a consequence of the relocation of

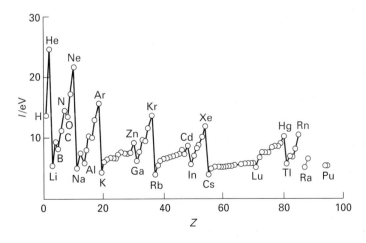

Fig 13.16 The periodic variation of the first ionization energies of the elements.

Table 13.2 *First ionization energies of main-group elements, I/eV**

H							He
13.60							24.59
Li	Be	B	C	N	O	F	Ne
5.32	9.32	8.30	11.26	14.53	13.62	17.42	21.56
Na	Mg	Al	Si	P	S	Cl	Ar
5.14	7.65	5.98	8.15	10.49	10.36	12.97	15.76
K	Ca	Ga	Ge	As	Se	Br	Kr
4.34	6.11	6.00	7.90	9.81	9.75	11.81	14.00
Rb	Sr	In	Sn	Sb	Te	I	Xe
4.18	5.70	5.79	7.34	8.64	9.01	10.45	12.13
Cs	Ba	Ti	Pb	Bi	Po	At	Rn
3.89	5.21	6.11	7.42	7.29	8.42	9.64	10.78

*1 eV = 96.485 kJ mol^{-1}. See also Table 3.2.

electrons from one atom to another). After atomic radius, it is the most important property for determining an element's chemical characteristics. Lithium has a low first ionization energy: its outermost electron is well-shielded from the nucleus by the core ($Z_{eff} = 1.3$ compared with $Z = 3$) and it is easily removed. Beryllium has a higher nuclear charge than lithium, and its outermost electron (one of the two 2s electrons) is more difficult to remove: its ionization energy is larger. The ionization energy decreases between beryllium and boron because in the latter the outermost electron occupies a 2p orbital and is less strongly bound than if it had been a 2s electron. The ionization energy increases between boron and carbon because the latter's outermost electron is also 2p and the nuclear charge has increased. Nitrogen has a still higher ionization energy because of the further increase in nuclear charge.

There is now a kink in the curve because the ionization energy of oxygen is lower than would be expected by simple extrapolation. At oxygen a 2p orbital must become doubly occupied, and the electron–electron repulsions are increased above what would be expected by simple extrapolation along the row. (The kink is less pronounced in the next

row, between phosphorus and sulfur, because their orbitals are more diffuse.) The values for oxygen, fluorine, and neon fall roughly on the same line, the increase of their ionization energies reflecting the increasing attraction of the nucleus for the outermost electrons.

The outermost electron in sodium is 3s. It is far from the nucleus, and the latter's charge is shielded by the compact, complete neon-like core. As a result, the ionization energy of sodium is substantially lower than that of neon. The periodic cycle starts again along this row, and the variation of the ionization energy can be traced to similar reasons.

The **electron affinity**, E_{ea}, is the difference in energy between a neutral atom and its anion. It is the energy *released* in the process

$$E(g) + e^-(g) \rightarrow E^-(g)$$

The electron affinity is positive if the anion has a lower energy than the neutral atom. Care should be taken to distinguish the electron affinity from the electron-gain enthalpy (Section 3.2): they have very similar numerical values but differ in sign.

Table 13.3 *Electron affinities of main-group elements, E_{ea}/eV^**

H							He†[†]
+0.75							< 0
Li	**Be**	**B**	**C**	**N**	**O**	**F**	**Ne**
+0.62	−0.19	+0.28	+1.26	−0.07	+1.46	+3.40	−0.30†
Na	**Mg**	**Al**	**Si**	**P**	**S**	**Cl**	**Ar**
+0.55	−0.22	+0.46	+1.38	+0.46	+2.08	+3.62	−0.36†
K	**Ca**	**Ga**	**Ge**	**As**	**Se**	**Br**	**Kr**
+0.50	−1.99	+0.3	+1.20	+0.81	+2.02	+3.37	−0.40†
Rb	**Sr**	**In**	**Sn**	**Sb**	**Te**	**I**	**Xe**
+0.49	+1.51	+0.3	+1.20	+1.05	+1.97	+3.06	−0.42†
Cs	**Ba**	**Tl**	**Pb**	**Bi**	**Po**	**At**	**Rn**
+0.47	−0.48	+0.2	+0.36	+0.95	+1.90	+2.80	−0.42†

*1 eV = 96.485 kJ mol^{-1}. See also Table 3.3.
†Calculated.

Electron affinities (Table 13.3) vary much less systematically through the periodic table than ionization energies. Broadly speaking, the highest electron affinities are found close to fluorine. In the halogens, the incoming electron enters the valence shell and experiences a strong attraction from the nucleus. The electron affinities of the noble gases are negative—which means that the anion has a higher energy than the neutral atom—because the incoming electron occupies an orbital outside the closed valence shell. It is then far from the nucleus and repelled by the electrons of the closed shells. The first electron affinity of oxygen is positive for the same reason as for the halogens. However, the second electron affinity (for the formation of O^{2-} from O^-) is strongly negative because, although the incoming electron enters the valence shell, it experiences a strong repulsion from the net negative charge of the O^- ion.

The spectra of complex atoms

THE spectra of many-electron atoms can be very complicated, yet that complexity contains a great deal of detailed information about the interactions between electrons. Here we consider the notation used to specify the states of atoms. Chemists need to know how to designate the states of atoms when they are describing photochemical events in the atmosphere. We also describe an interaction that has important consequences in molecular spectroscopy and magnetism.

13.16 Term symbols

For historical reasons, the energy level of an atom is called a **term** and the notation used to specify the term is called a **term symbol**. A term symbol looks like $^3D_2$, with each component (the 3, the D, and the 2) telling us something about the angular momentum of the electrons in the atom.

The letter (D, for instance) tells us the total orbital angular momentum of the electrons in the atom. To find it, we work out the **total orbital angular momentum quantum number**, L, in the manner described below, and then we use the following code:

$$L \quad 0 \quad 1 \quad 2 \quad 3\ldots$$
$$\quad\quad S \quad P \quad D \quad F\ldots$$

Note that the code is the same as for orbitals, but we use upper-case Roman letters. To find L we identify the orbital angular momentum quantum numbers (l_1 and l_2 for instance) of the electrons in the valence shell of the atom, and then form the following series:[13]

$$L = l_1 + l_2, l_1 + l_2 - 1, \ldots, |l_1 - l_2|$$

The modulus signs ($|\ldots|$) simply mean that the series terminates at a positive value. The highest total orbital angular momentum occurs when the two electrons are orbiting in the same direction (in classical terms, like the planets round the Sun); the lowest occurs when they are orbiting in opposite directions.

Illustration 13.5

Suppose we are considering the excited state configuration of carbon [He]$2s^2 2p^1 3p^1$ in which a $2p$ electron has been promoted to a $3p$ orbital. We concentrate on the p electrons because the s electrons have no orbital angular momentum. For each electron $l = 1$ (that is, $l_1 = 1$ and $l_2 = 1$ for the two electrons we are considering). It follows that

$$L = 1 + 1, 1 + 1 - 1, \ldots, |1-1| = 2, 1, 0$$

So the configuration gives rise to D, P, and S terms, corresponding to the three states of total orbital angular momentum.

Next, we consider the **total spin angular momentum quantum number**, S. This quantum number is obtained in the same way as L, by adding

together the individual spin angular momentum quantum numbers:

$$S = s_1 + s_2, s_1 + s_2 - 1, \ldots, |s_1 - s_2|$$

For electrons, $s = \tfrac{1}{2}$, so for two electrons

$$S = \tfrac{1}{2} + \tfrac{1}{2}, \tfrac{1}{2} + \tfrac{1}{2} - 1, \ldots, |\tfrac{1}{2} - \tfrac{1}{2}| = 1, 0$$

Then the value of S is represented in the term symbol by writing the **multiplicity** of the term, the value of $2S + 1$, as a left superscript. The higher the multiplicity of a term, the more electrons there are in the atom that are spinning in the same direction.

Illustration 13.6

For the excited configuration of carbon, [He]$2s^2 2p^1 3p^1$, that we are considering, the two p electrons each have $s = \tfrac{1}{2}$, so $S = 1,0$. The corresponding multiplicities are $2 \times 1 + 1 = 3$ (a 'triplet' term) and $2 \times 0 + 1 = 1$ (a 'singlet' term). The corresponding term symbols are

Triplet terms: ^{3}D, ^{3}P, ^{3}S   singlet terms: ^{1}D, ^{1}P, ^{1}S

Finally, we come to the right subscript. This label is the **total angular momentum quantum number**, J, the total angular momentum being the sum of the orbital angular momentum and the spin angular momentum. We find J (a positive number) by forming the series

$$J = L + S, L + S - 1, \ldots, |L - S|$$

If there are many electrons having spins in the same direction as their orbital motion, then J is large. If the spins are aligned against the orbital motion, then J is small. Each value of J corresponds to a particular **level** of a term.

Illustration 13.7

The levels that occur in the ^{3}D term are found by setting $L = 2$ and $S = 1$; then

$$J = 2 + 1, 2 + 1 - 1, \ldots, |2 - 1| = 3, 2, 1$$

[13] This and the analogous series introduced later are called *Clebsch–Gordan series*.

That is, the levels of the ^{3}D term are ^{3}D$_3$, ^{3}D$_2$, and ^{3}D$_1$ (note that there are three levels for this triplet term). In the ^{3}D$_3$ level, not only are the two p electrons orbiting in the same sense, the two spins are spinning in the same direction as each other, and the total spin is in the same direction as the orbital angular momentum. In ^{3}D$_1$ the total spin is aligned oppositely to the total orbital momentum and the overall total angular momentum is relatively low.

Self-test 13.8

What terms and levels can arise from the configuration . . . $4p^13d^1$?

[*Answer:* ^{1}F$_3$, ^{1}D$_2$, ^{1}P$_1$, ^{3}F$_{4,3,2}$, ^{3}D$_{3,2,1}$, ^{3}P$_{2,1,0}$]

The terms of a configuration in general have different energies because they correspond to the occupation of different orbitals and to different numbers of electrons with parallel spins. Typically, Hund's rule enables us to identify the term with the greatest number of parallel spins as having the lowest energy (Section 13.11), for parallel spins tend to stay apart. In other words, *the term with the greatest multiplicity lies lowest in energy.* In our example, the triplet terms lie lower than the singlet terms, so one of the terms ^{3}D, ^{3}P, ^{3}S lies lowest. It is also commonly found that, having sorted the terms by multiplicity, *the term with the greatest orbital angular momentum lies lowest.* Classically, we can think of the term with the greatest orbital angular momentum as having electrons circulating in the same direction, like cars on a traffic circle, and therefore being able to stay far apart. We therefore predict that the ^{3}D term will lie lowest of all the terms arising from the [He]$2s^22p^13p^1$ configuration. To find which of the three *levels* of this term lies lowest, we need another concept.

13.17 **Spin–orbit coupling**

An electron is a charged particle, so its orbital angular momentum gives rise to a magnetic field, just as an electric current in a loop gives rise to a magnetic field in an electromagnet. That is, an electron with orbital angular momentum acts like a tiny bar magnet. An electron also has a spin angular mo-

mentum, and this intrinsic 'spinning motion' means that it also acts as a tiny bar magnet. The interaction of the two magnets is called **spin–orbit coupling**.

The two magnets have a higher energy when they are parallel than when they are antiparallel (Fig 13.17). However, because the relative orientation of the magnets reflects the relative orientation of the orbital and spin angular momenta, the energy of the atom depends on the total angular momentum quantum number J (because its value also reflects the relative orientation of the two kinds of momentum). A high energy is obtained when the angular momenta, and therefore the bar magnets, are parallel to each other. That arrangement of angular momenta corresponds to a high value of J. Therefore, we can predict that *the level with lowest J lies lowest in energy.* In our current example, we predict that the lowest level of the ^{3}D term is ^{3}D$_1$.

The strength of the spin–orbit coupling increases sharply with atomic number. In Period 2 atoms it gives rise to splittings between levels of the order of

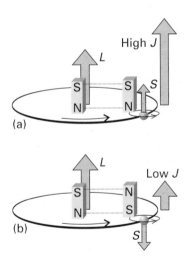

Fig 13.17 The magnetic interaction responsible for spin–orbit coupling. (a) A high total angular momentum corresponds to a parallel arrangement of magnetic moments (represented by the bar magnets), and hence a high energy. (b) A low total angular momentum corresponds to an antiparallel arrangement of magnetic moments, and hence a low energy. Note that the difference in energy is not due *directly* to the differences in total angular momentum: the total simply tells us the relative orientations of the two magnetic moments.

10^2 cm^{-1}, but in Period 3 the difference approaches 10^3 cm^{-1}. We can understand this increase by thinking about the source of the orbital magnetic field. To do so, imagine that we are riding on the electron as it orbits the nucleus. From our viewpoint, the nucleus appears to orbit around us (rather as the pre-Copernicans thought the Sun revolved around the Earth). If the nucleus has a high atomic number it will have a high charge, we shall be at the centre of a strong electric current, and we shall experience a strong magnetic field. If the nucleus has a low atomic number, we experience a feeble magnetic field arising from the low current that encircles us.

Spin–orbit coupling has important consequences in photochemistry and in particular for the existence of the property of 'phosphorescence', which we discuss in Chapter 18.

Exercises

13.1 Calculate the wavelength of the line with $n = 5$ in the Balmer series of the spectrum of atomic hydrogen.

13.2 The frequency of one of the lines in the Paschen series of the spectrum of atomic hydrogen is 2.7415×10^{14} Hz. Identify the principal quantum number of the upper state in the transition.

13.3 One of the terms of the H atom is at 27 414 cm^{-1}. What is (a) the wavenumber, (b) the energy of the term with which it combines to produce light of wavelength 486.1 nm?

13.4 The Rydberg constant, eqn 13.4, depends on the mass of the nucleus. What is the difference in wavenumbers of the $2p \rightarrow 1s$ transition in hydrogen and deuterium?

13.5 What transition in He$^+$ has the same frequency (disregarding mass differences) as the $2p \rightarrow 1s$ transition in H?

13.6 Predict the ionization energy of Li^{2+} given that the ionization energy of He$^+$ is 54.36 eV.

13.7 How many orbitals are present in the N shell of an atom?

13.8 The 'Humphreys series' is another group of lines in the spectrum of atomic hydrogen. It begins at 12 368 nm and has been traced to 3281.4 nm. (a) What are the transitions involved? (b) What are the wavelengths of the intermediate transitions?

13.9 At what wavelength would you expect the longest wavelength transition of the Humphreys series to occur

in He$^+$? (*Hint.* The energy levels of hydrogenic atoms and ions are proportional to Z^2.)

13.10 A series of lines in the spectrum of atomic hydrogen lies at 656.46, 486.27, 434.17, and 410.29 nm. (a) What is the wavelength of the next line in the series? (b) What is the ionization energy of the atom when it is in the lower state of the transitions?

13.11 The Li^{2+} ion is hydrogenic and has a Lyman series at 740 747, 877 924, 925 933 cm^{-1}, and beyond. Show that the energy levels are of the form $-hcR_{\mathrm{Li}}/n^2$ and find the value of R_{Li} for this ion. Go on to predict the wavenumbers of the two longest-wavelength transitions of the Balmer series of the ion and find the ionization energy of the ion.

13.12 At what radius does the probability of finding an electron in a small volume located at a point in the ground state of an H atom fall to 25 per cent of its maximum value?

13.13 At what radius in the H atom does the radial distribution function of the ground state have (a) 25 per cent, (b) 10 per cent of its maximum value?

13.14 What is the probability of finding an electron anywhere in one lobe of a p orbital given that it occupies the orbital?

13.15 What is the probability of finding the electron in a volume of 5.0 pm^3 centred on the nucleus in a hydrogen atom?

13.16 The (normalized) wavefunction for a 2s orbital in hydrogen is

$$\psi = \left(\frac{1}{32\pi a_0^3}\right)^{1/2}\left(2 - \frac{r}{a_0}\right)e^{-r/2a_0}$$

Calculate the probability of finding an electron that is described by this wavefunction in a volume of $1.0\ pm^3$ (a) centred on the nucleus, (b) at the Bohr radius, (c) at twice the Bohr radius.

13.17 Construct an expression for the radial distribution function of a hydrogenic 2s electron (see Exercise 13.16 for the form of the orbital), and plot the function against r. What is the most probable radius at which the electron will be found? (*Hint.* For a more accurate answer, differentiate the radial distribution function to find where it is a maximum.)

13.18 Locate the radial nodes in (a) the 3s orbital, (b) the 4s orbital of an H atom, which is proportional to $24 - 36\rho + 12\rho^2 - \rho^3$, with $\rho = r/2a_0$

13.19 The wavefunction of one of the d orbitals is proportional to $\sin\theta\cos\theta$. At what angles does it have nodal planes?

13.20 What is the orbital angular momentum (as multiples of $\hbar$) of an electron in the orbitals (a) 1s, (b) 3s, (c) 3d, (d) 2p, (e) 3p? Give the numbers of angular and radial nodes in each case.

13.21 State the orbital degeneracy of the levels in the hydrogen atom that have energy (a) $-hcR_H$, (b) $-\frac{1}{9}hcR_H$, and (c) $-\frac{1}{49}hcR_H$.

13.22 Which of the following transitions are allowed in the normal electronic emission spectrum of an atom: (a) $2s \rightarrow 1s$, (b) $2p \rightarrow 1s$, (c) $3d \rightarrow 2p$, (d) $5d \rightarrow 2s$, (e) $5p \rightarrow 3s$?

13.23 To what orbitals may a 5f electron make spectroscopic transitions?

13.24 How many electrons can occupy subshells with the following values of l: (a) 0, (b) 3, (c) 5?

13.25 Give the electron configurations of the ground states of the first 18 elements in the periodic table.

13.26 When ultraviolet radiation of wavelength 58.4 nm from a helium lamp is directed on to a sample of krypton, electrons are ejected with a speed of $1.59 \times 10^6\ m\ s^{-1}$. Calculate the ionization energy of krypton.

13.27 If we lived in a four-dimensional world, there would be one s orbital, four p orbitals, and nine d orbitals in their respective subshells. (a) Suggest what form the periodic table might take for the first 24 elements. (b) Which elements (using their current names) would be noble gases?

13.28 What terms (expressed as S, D, etc.) can arise from the $[He]2s^2 2p^1 3d^1$ excited configuration of carbon?

13.29 What are the total spin angular momenta (reported as the value of S) that can arise from four electrons? (*Hint.* Use the Clebsch–Gordan series successively.)

13.30 What levels can the following terms possess: (a) ^1S, (b) ^3F, (c) ^5S, (d) ^5P?

13.31 The ground configuration of a Ti^{2+} ion is $[Ar]3d^2$. (a) What is the term of lowest energy and which level of that term lies lowest? (b) How many states belong to that lowest level?

The chemical bond

Contents

THE **chemical bond**, a link between atoms, is central to all aspects of chemistry. Reactions make them and break them, and the structures of solids and individual molecules depend on them. The physical properties of individual molecules and of bulk samples of matter also stem in large part from the shifts in electron density that take place when atoms form bonds to one another. The theory of the origin of the numbers, strengths, and three-dimensional arrangements of chemical bonds between atoms is called **valence theory**.

Valence theory is an attempt to explain the properties of molecules ranging from the smallest to the largest. For instance, it explains why N_2 is so inert that it acts as a diluent for the aggressive oxidizing power of atmospheric oxygen. At the other end of the scale, valence theory deals with the structural origins of the function of protein molecules and the molecular biology of DNA. The description of chemical bonding has become highly developed through the use of computers, and it is now possible to compute details of the electron distribution in molecules of almost any complexity. However, much can also be achieved in terms of a simple qualitative understanding of bond formation, and that is the focus of this chapter.

There are two major approaches to the calculation of molecular structure, **valence bond theory** (VB theory) and **molecular orbital theory** (MO theory). Almost all modern computational work makes use of MO theory, and we shall concentrate on it in this chapter. Valence bond theory, though, has left its imprint on the language of chemistry, and it is important to know the significance of terms that chemists use every day.

Introductory concepts

CERTAIN ideas of valence theory should be well known from introductory chemistry. This section reviews this background.

14.1 The classification of bonds

We distinguish between two types of bond:

An **ionic bond** is formed by the transfer of electrons from one atom to another and the consequent attraction between the ions so formed.

A **covalent bond** is formed when two atoms share a pair of electrons.

The character of a covalent bond, which we concentrate on in this chapter, was identified by G.N. Lewis (in 1916, before quantum mechanics was fully developed). We shall assume that Lewis's ideas are familiar, but for convenience they are reviewed in *Further information 9*. In this chapter we develop the modern theory of chemical bond formation in terms of the quantum mechanical properties of electrons and set Lewis's ideas in a modern context. We shall see that ionic and covalent bonds are two extremes of a common type of bond. However, because there are certain aspects of ionic solids that require special attention, we treat them separately in Chapter 15.

Lewis's original theory was unable to account for the shapes adopted by molecules. The most elementary (but qualitatively quite successful) explanation of the shapes adopted by molecules is the **valence-shell electron pair repulsion model** (VSEPR model) in which we suppose that the shape of a molecule is determined by the repulsions between electron pairs in the valence shell. A brief review of the VSEPR model is given in *Further information 10*. Once again, the purpose of this chap-

ter is to extend these elementary arguments and indicate some of the contributions that quantum theory has made to understanding why a molecule adopts its characteristic shape.

14.2 Potential energy curves

All theories of molecular structure adopt the **Born–Oppenheimer approximation**. In this approximation, we suppose that the nuclei, being so much heavier than an electron, move relatively slowly and may be treated as stationary while the electrons move around them. We can therefore think of the nuclei as being fixed at arbitrary locations, and then solve the Schrödinger equation for the wavefunction of the electrons alone. The approximation is quite good for molecules in their electronic ground states, for calculations suggest that (in classical terms) the nuclei in H_2 move through only about 1 pm while the electron speeds through 1000 pm.

By invoking the Born–Oppenheimer approximation, we can select an internuclear separation in a diatomic molecule and solve the Schrödinger equation for the electrons for that nuclear separation. Then we can choose a different separation and repeat the calculation, and so on. In this way we can explore how the energy of the molecule varies with bond length and bond angles too and obtain a **molecular potential energy curve**. A typical example of such a curve is illustrated in Fig 14.1. It is called a *potential* energy curve because the nuclei are stationary and contribute no kinetic energy. Once the curve has been calculated, we can identify the **equilibrium bond length**, R_e, the internuclear separation at the minimum of the curve, and D_e, the depth of the minimum below the energy of the infinitely widely separated atoms. In Chapter 17 we shall also see that the narrowness of the potential well is an indication of the stiffness of the bond.

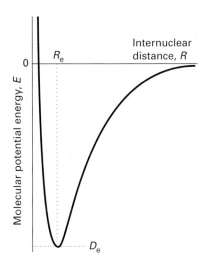

Fig 14.1 A molecular potential energy curve. The equilibrium bond length R_e corresponds to the energy minimum.

Valence bond theory

I N valence bond theory, a bond is regarded as forming when an electron in one atomic orbital pairs its spin with that of an electron supplied by another atomic orbital (Fig 14.2). To understand this model, we have to examine the wavefunction for the two electrons that form the bond.

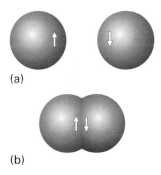

Fig 14.2 In the valence bond theory a σ bond is formed when two electrons in orbitals on neighbouring atoms pair and the orbitals merge to form a cylindrical electron cloud.

14.3 **Diatomic molecules**

We begin by considering the simplest possible chemical bond, the one in molecular hydrogen. When the two ground-state atoms are far apart, we can be confident that electron 1 is in the 1s orbital of atom A and electron 2 is in the 1s orbital of atom B. The two 1s orbitals are denoted ψ_A and ψ_B, respectively. The overall wavefunction for the two electrons is

$$\psi = \psi_A(1)\psi_B(2)$$

When the two atoms are at their bonding distance, it may still be true that electron 1 is on A and electron 2 is on B. However, an equally likely arrangement is for electron 1 to escape from A and be found on B and for electron 2 to be on A. In this case the wavefunction is

$$\psi = \psi_A(2)\psi_B(1)$$

Whenever two outcomes are equally likely, the rules of quantum mechanics tell us to add together the two corresponding wavefunctions. Therefore, the (unnormalized) wavefunction for the two electrons in a hydrogen molecule is

$$\psi(\text{H–H}) = \psi_A(1)\psi_B(2) + \psi_A(2)\psi_B(1) \qquad (14.1)$$

This wavefunction is the valence bond wavefunction for the bond in molecular hydrogen. It expresses the idea that we cannot keep track of either electron, and that their distributions blend together.

For technical reasons stemming from the Pauli exclusion principle, the wavefunction in eqn 14.1 can exist only if the two electrons it describes have opposite spins (that is, are paired). It follows that the merging of orbitals that gives rise to a bond is accompanied by the pairing of the two electrons that contribute to it. Bonds do not form *because* electrons tend to pair: bonds are *allowed* to form by the electrons pairing their spins.

Because ψ is built from the merging of H1s orbitals, we can expect the overall distribution of the

328   THE CHEMICAL BOND

electrons in the molecule to be sausage-shaped (as in Fig 14.2). A VB wavefunction with cylindrical symmetry around the internuclear axis is called a σ bond.[1] All VB wavefunctions are constructed in a similar way, by using the atomic orbitals available on the participating atoms. In general, therefore, the (unnormalized) VB wavefunction for an A–B bond is

$$\psi(\text{A–B}) = \psi_A(1)\psi_B(2) + \psi_A(2)\psi_B(1) \qquad (14.2)$$

To calculate the energy of a molecule with a given internuclear separation R, we substitute the VB wavefunction, eqn 14.2, into the Schrödinger equation and carry out the necessary mathematical manipulations. When the energy is plotted against R, we get a curve like the one shown in Fig 14.1. As R decreases from infinity, the energy falls below that of two separated H atoms as each electron becomes free to migrate to the other atom. However, this decrease in energy is counteracted by an increase in energy from the Coulombic repulsion between the two positively charged nuclei, which has the form

$$V_{\text{nuc,nuc}} = \frac{Z_A Z_B e^2}{4\pi\varepsilon_0 R} \qquad (14.3)$$

(For H_2, $Z_A = Z_B = 1$.) This positive contribution to the energy becomes large as R becomes small. As a result, the total energy curve passes through a minimum and then climbs to a strongly positive value as the two nuclei are pressed together.

We can use a similar description for molecules built from atoms that contribute more than one electron to the bonding. For example, to construct the valence bond description of N_2, we consider the valence electron configuration of each atom, which is $2s^2 2p_x^1 2p_y^1 2p_z^1$. It is conventional to take the z-axis to be the internuclear axis, so we can imagine each atom as having a $2p_z$ orbital pointing towards a $2p_z$ orbital on the other atom, with the $2p_x$ and $2p_y$ orbitals perpendicular to the axis (Fig 14.3). Each of these p orbitals is occupied by one electron, so we can think of the molecule as forming by the merging of matching orbitals on neighbouring

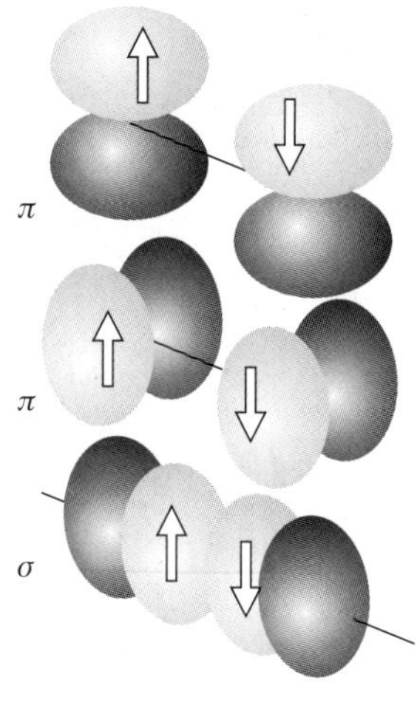

Fig. 14.3 The bonds in N_2 are built by allowing the electrons in the $N2p$ orbitals to pair. However, only one orbital on each atom can form a σ bond: the orbitals perpendicular to the axis form π bonds.

atoms and the pairing of the electrons that occupy them. We get a cylindrically symmetric σ bond from the merging of the two $2p_z$ orbitals and the pairing of the electrons they contain. However, the remaining p orbitals cannot merge to give σ bonds because they do not have cylindrical symmetry around the internuclear axis. Instead, the $2p_x$ orbitals merge and the two electrons pair to form a π bond.[2] Similarly, the $2p_y$ orbitals merge and their electrons pair to form another π bond. In general, a π bond arises from the merging of two p orbitals that approach side-by-side and the pairing of the electrons that they contain. It follows that the overall bonding pattern in N_2 is a σ bond plus two π bonds (Fig 14.4), which is consistent with the Lewis structure :N≡N: in which the atoms are linked by a triple bond.

Self-test 14.1

Describe the valence-bond ground state of a Cl_2 molecule.

[*Answer*: one $\sigma(Cl3p_z,Cl3p_z)$ bond]

---

[1] It is so called because, when viewed along the bond, it resembles a pair of electrons in an s orbital (and σ, sigma, is the Greek equivalent of s).

[2] A π bond is so called because, viewed along the internuclear axis, it resembles a pair of electrons in a p orbital (and π is the Greek equivalent of p).

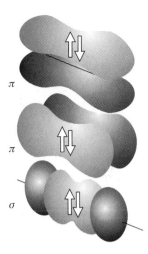

Fig 14.4 The electrons in the 2*p* orbitals of two neighbouring N atoms merge to form σ and π bonds. The electrons in the N2p_z orbitals pair to form a bond of cylindrical symmetry. Electrons in the N2*p* orbitals that lie perpendicular to the axis also pair to form two π bonds.

14.4 Polyatomic molecules

We can readily extend the basic VB concept of the merging of singly occupied orbitals to polyatomic species. Each σ bond in a polyatomic molecule is formed by the merging of orbitals with cylindrical symmetry about the internuclear axis and the pairing of the spins of the electrons they contain. Likewise, π bonds are formed by pairing electrons that occupy atomic orbitals of the appropriate symmetry. A simple description of the electronic structure of H_2O should make this clear.

> ### Illustration 14.1
>
> The valence electron configuration of an O atom is $2s^2 2p_x^2 2p_y^1 2p_z^1$. The two unpaired electrons in the O2*p* orbitals can each pair with an electron in a H1*s* orbital, and each combination results in the formation of a σ bond (each bond has cylindrical symmetry about the respective O–H internuclear distance). Because the 2p_y and 2p_z orbitals lie at 90° to each other, the two σ bonds they form also lie at 90° to each other (Fig 14.5). We predict, therefore, that H_2O should be an angular molecule, which it is. However, the model predicts a bond angle of 90°, whereas the actual bond angle is 104°.

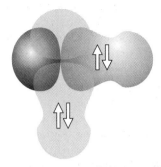

Fig 14.5 The bonding in an H_2O molecule can be pictured in terms of the pairing of an electron belonging to one H atom with an electron in an O2*p* orbital; the other bond is formed likewise, but using a perpendicular O2*p* orbital. The predicted bond angle is 90°, which is in poor agreement with the experimental bond angle (104°).

> ### Self-test 14.2
>
> Give a valence bond description of NH_3, and predict the bond angle of the molecule on the basis of this description. The experimental bond angle is 107°.
>
> [*Answer*: three σ(N2*p*,H1*s*) bonds; 90°]

While broadly correct, valence bond theory seems to have two deficiencies. One is the poor estimate it provides for the bond angle in H_2O (and other molecules, such as NH_3). Indeed, the theory appears to make worse predictions than the qualitative VSEPR model, which predicts HOH and HNH bond angles of slightly less than 109° in H_2O and NH_3, respectively. The second major deficiency is the apparent inability of VB theory to account for the number of bonds that atoms can form, and in particular the tetravalence of carbon. To appreciate the latter problem, we note that the ground-state valence configuration of a carbon atom is $2s^2 2p_x^1 2p_y^1$, which suggests that it should be capable of forming only two bonds, not four.

14.5 Promotion and hybridization

Two modifications solve all these problems. First, we allow for a valence electron to be **promoted** from one atomic orbital to another as a bond is formed. Promotion is worthwhile if the energy it requires can be more than recovered in the greater strength or number of bonds that can be formed. In carbon, for example, the promotion of a 2s electron to a 2p orbital leads to the configuration $2s^1 2p_x^1 2p_y^1 2p_z^1$, with four unpaired electrons in separate orbitals. These electrons may pair with four electrons in orbitals provided by four other atoms (such as four H1s orbitals if the molecule is CH_4), and as a result the atom can form four σ bonds. Although energy is required to promote the 2s electron, it is more than recovered by the atom's ability to form four bonds in place of the two bonds of the unpromoted atom.

We can now see why tetravalent carbon is so common. The promotion energy of carbon is small because the promoted electron leaves a doubly occupied 2s orbital and enters a vacant 2p orbital, hence significantly relieving the electron–electron repulsion it experiences in the former.

Promotion, however, appears to imply the presence of three σ bonds of one type (in CH_4, formed from the merging of H1s and C2p orbitals) and a fourth σ bond of a distinctly different type (formed from the merging of H1s and C2s). It is well known, however, that all four bonds in methane are exactly equivalent both in terms of their chemical properties and their physical properties (their lengths, strengths, and stiffnesses).

This problem is overcome by another technical feature of quantum mechanics that allows the same electron distribution to be described in different ways. In this case, we can describe the electron distribution in the promoted atom either as arising from four electrons in one s and three p orbitals, or as arising from four electrons in four different *mixtures* of these orbitals. Mixtures (more formally, linear combinations) of atomic orbitals on the same atom are called **hybrid orbitals**. We can picture them by thinking of the four original atomic orbitals, which are waves centred on a nucleus, as being like ripples spreading from a single point on the surface of a lake. These waves interfere destructively or constructively in different regions and give rise to four new shapes. The specific linear combinations that give rise to four equivalent hybrid orbitals are

$$h_1 = s + p_x + p_y + p_z \qquad h_2 = s - p_x - p_y + p_z$$
$$h_3 = s - p_x + p_y - p_z \qquad h_4 = s + p_x - p_y - p_z$$

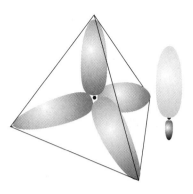

Fig 14.6 The 2s and three 2p orbitals of a carbon atom hybridize, and the resulting hybrid orbitals point towards the corners of a regular tetrahedon.

As a result of the constructive and destructive interference between the positive and negative regions of the component orbitals, each hybrid orbital has a large lobe pointing towards one corner of a regular tetrahedron (Fig 14.6). Because each hybrid is built from one s orbital and three p orbitals, it is called an ***sp*³ hybrid orbital**.

It is now easy to see how the valence bond description of the methane molecule leads to a tetrahedral molecule containing four equivalent C–H bonds. It is energetically favourable (in the end, after bonding has been taken into account) for the carbon atom to undergo promotion. The promoted configuration has a distribution of electrons that is equivalent to one electron occupying each of four tetrahedral hybrid orbitals. Each hybrid orbital of the promoted atom contains a single unpaired electron; a hydrogen 1s electron can pair with each one, giving rise to a σ bond pointing in a tetrahedral direction. Because each sp^3 hybrid orbital has the same composition, all four σ bonds are identical apart from their orientation in space (Fig 14.7).

Hybridization is also used in the VB description of alkenes. An ethene molecule is planar, with HCH and HCC bond angles close to 120°. To reproduce this σ-bonding structure, we think of each C atom as being promoted to a $2s^1 2p^3$ configuration. However, instead of using all four orbitals to form hybrids, we form ***sp*² hybrid orbitals** by allowing the s orbital and two of the p orbitals to interfere. As shown in Fig 14.8, the three hybrid orbitals

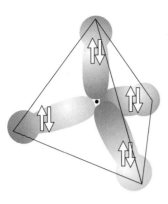

Fig 14.7 The valence bond description of the structure of CH_4. Each σ bond is formed by the pairing of an electron in an H1s orbital with an electron in one of the hybrid orbitals shown in Fig 14.6. The resulting molecule is regular tetrahedral.

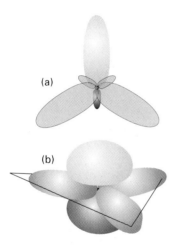

(a)

(b)

Fig 14.8 (a) Trigonal planar hybridization is obtained when an s and two p orbitals are hybridized. The three lobes lie in a plane and make an angle of 120° to each other. (b) The remaining p orbital in the valence shell of an sp^2-hybridized atom lies perpendicular to the plane of the three hybrids.

$$h_1 = s + 2^{1/2}p_x$$
$$h_2 = s + \left(\tfrac{3}{2}\right)^{1/2}p_x - \left(\tfrac{1}{2}\right)^{1/2}p_y$$
$$h_3 = s - \left(\tfrac{3}{2}\right)^{1/2}p_x - \left(\tfrac{1}{2}\right)^{1/2}p_y$$

lie in a plane and point towards the corners of an equilateral triangle. The third $2p$ orbital ($2p_z$) is not included in the hybridization, and its axis is perpendicular to the plane in which the hybrids lie. The coefficients $2^{1/2}$, etc. in the hybrids have been chosen to give the correct directional properties of the hybrids. The *squares* of the coefficients give the proportion of each atomic orbital in the hybrid. All three hybrids have s and p orbitals in the ratio 1:2, as indicated by the label 'sp^2'.

The sp^2-hybridized C atoms each form three σ bonds with either the h_1 hybrid of the other C atom or with the H1s orbitals. The σ framework therefore consists of bonds at 120° to each other. Moreover, provided the two CH_2 groups lie in the same plane, the two electrons in the unhybridized $C2p_z$ orbitals can pair and form a π bond (Fig 14.9). The formation of this π bond locks the framework into the planar arrangement, for any rotation of one CH_2 group relative to the other leads to a weakening of the π bond (and consequently an increase in energy of the molecule).

A similar description applies to a linear ethyne (acetylene) molecule, H–C≡C–H. Now the carbon atoms are **sp hybridized**, and the σ bonds are built from hybrid atomic orbitals of the form

$$h_1 = s + p_z \qquad h_2 = s - p_z$$

Note that the s and p orbitals contribute in equal proportions. The two hybrids lie along the z-axis. The electrons in them pair either with an electron in the corresponding hybrid orbital on the other C atom or with an electron in the H1s orbitals. Electrons in the two remaining p orbitals on each atom, which are perpendicular to the molecular axis, pair to form two perpendicular π bonds (as in Fig 14.10).

Other hybridization schemes, particularly those involving d orbitals, are often invoked to account

Fig 14.9 The valence bond description of the structure of a carbon–carbon double bond, as in ethene. The electrons in the two sp^2 hybrids that point towards each other pair and form a σ bond. Electrons in the two p orbitals that are perpendicular to the plane of the hybrids pair and form a π bond. The electrons in the remaining hybrid orbitals are used to form bonds to other atoms (in ethene itself, to H atoms).

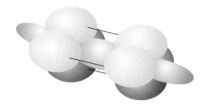

Fig 14.10 The electronic structure of ethyne (acetylene). The electrons in the two *sp* hybrids on each atom pair to form σ bonds either with the other C atom or with an H atom. The remaining two unhybridized *2p* orbitals on each atom are perpendicular to the axis: the electrons in corresponding orbitals on each atom pair to form two π bonds. The overall electron distribution is cylindrical.

for (or at least be consistent with) other molecular geometries (Table 14.1). An important point to note is that *the hybridization of N atomic orbitals always results in the formation of N hybrid orbitals*. For example, sp^3d^2 hybridization results in six equivalent hybrid orbitals pointing towards the corners of a regular octahedron. This octahedral hybridization scheme is sometimes invoked to account for the structure of octahedral molecules, such as SF_6.

Self-test 14.3

Describe the bonding in a PCl_5 molecule in VB terms.
[*Answer:* five σ bonds formed from sp^3d hybrids]

The 'pure' schemes in Table 14.1 are not the only possibilities: it is possible to form hybrid orbitals with intermediate proportions of atomic orbitals. For example, as more *p*-orbital character is included in an *sp*-hybridization scheme, the hybridization changes towards sp^2 and the angle between the

Table 14.1 *Hybrid orbitals*

Number	Shape	Hybridization*
2	Linear	sp
3	Trigonal planar	sp^2
4	Tetrahedral	sp^3
5	Trigonal bipyramidal	sp^3d
6	Octahedral	sp^3d^2

* Other combinations are possible.

hybrids changes continuously from 180° for pure *sp* hybridization to 120° for pure sp^2 hybridization. If the proportion of *p* character continues to be increased (by reducing the proportion of *s* orbital), then the hybrids eventually become pure *p* orbitals at an angle of 90° to each other. Now we can account for the structure of H_2O, with its bond angle of 104°. Each O–H σ bond is formed from an O atom hybrid orbital with a composition that lies between pure *p* (which would lead to a bond angle of 90°) and pure sp^2 (which would lead to a bond angle of 120°). The actual bond angle and hybridization adopted are found by calculating the energy of the molecule as the bond angle is varied, and looking for the angle at which the energy is a minimum.

14.6 **Resonance**

Another term introduced by VB theory into chemistry is **resonance**, the superposition of the wavefunctions representing different electron distributions in the same nuclear framework. To understand what this means, consider the VB description of a purely covalently bonded HCl molecule, which could be written

$$\psi_{cov} = \psi_H(1)\psi_{Cl}(2) + \psi_H(2)\psi_{Cl}(1)$$

We have supposed that the bond is formed by the spin pairing of electrons in the hydrogen 1*s* orbital, ψ_H, and the chlorine $2p_z$ orbital, ψ_{Cl}. However, there is something wrong with this description: it allows electron 1 to be on the H atom when electron 2 is on the Cl atom, and vice versa, but it does not allow both electrons to be on the Cl atom simultaneously. On physical grounds, we might expect the covalent character of HCl to be only a partial description of the molecule: because the Cl atom is so electronegative, we ought to expect a polar molecule with H^+Cl^- playing a role in its description. The wavefunction for this ionic structure, in which both electrons are in the $Cl2p_z$ orbital, is

$$\psi_{ion} = \psi_{Cl}(1)\psi_{Cl}(2)$$

However, this wavefunction alone is unrealistic, because HCl is not an ionic species. A better description of the wavefunction for the molecule is as a superposition of the covalent and ionic descriptions, and we write

$$\psi = \psi_{cov} + \lambda\psi_{ion} \qquad (14.4)$$

The square of λ gives the relative proportions of the covalent and ionic contributions. If λ is very small, the covalent description is dominant. If λ is very large, the ionic description is dominant.

We find the numerical value of λ, and hence judge whether covalent or ionic bonding is a good description, by using the **variation theorem**. First, we write down a plausible wavefunction, a **trial wavefunction**, for the molecule, such as the wavefunction in eqn 14.4 where λ is a variable parameter. The variation theorem then states that:

The energy of a trial wavefunction is never less than the true energy.

The theorem implies that, if we vary λ until we achieve the lowest energy, then that wavefunction is the best available. For instance, we might find that the lowest energy is reached when $\lambda = 0.1$, so the best description of the bond in the molecule is $\psi = \psi_{cov} + 0.1\psi_{ion}$.

The approach summarized by eqn 14.4, in which we express a wavefunction as the sum of wavefunctions corresponding to a variety of structures, is called **resonance**. In this case, where one structure is pure covalent and the other pure ionic, it is called **ionic–covalent resonance**. The interpretation of the wavefunction, which is called a **resonance hybrid**, is that, if we were to inspect the molecule, then the proportion of the time that it would be found with an ionic structure is proportional to λ^2. Resonance is not a flickering between the contributing states: it is a blending of their characteristics, much as a mule is a blend of a horse and a donkey.

One of the most famous examples of resonance is in the VB description of benzene, where the wavefunction of the molecule is written as a superposition of the wavefunctions of the two Kekulé structures (**1**) and (**2**):

$$\psi = \psi_{Kek1} + \psi_{Kek2} \qquad (14.5)$$

The two contributing structures have identical energies, so they contribute equally to the superposition. The effect of resonance in this case is to distribute double-bond character around the ring and to make all carbon–carbon bonds equivalent. Because the wavefunction is improved by allowing resonance, it follows from the variation theorem that the energy of the molecule is lowered relative to either Kekulé structure alone. This lowering is called the **resonance stabilization** of the molecule and is largely responsible for the unusual stability of aromatic rings. Resonance always lowers the energy (because it improves the wavefunction), and the lowering is greatest when the contributing structures have similar energies.

1 Kekulé structure1   **2** Kekulé structure 2

Molecular orbitals

IN molecular orbital theory, electrons are treated as spreading throughout the entire molecule: every electron contributes to the strength of every bond. This theory has been more fully developed than valence bond theory and provides the language that is widely used in modern discussions of bonding in small inorganic molecules, d-metal complexes, and solids. To introduce it, we follow the same strategy as in Chapter 13, where the one-electron hydrogen atom was taken as the fundamental species for discussing atomic structure, and then developed into a description of complex atoms. In this section we use the simplest molecule of all, the one-electron hydrogen molecule-ion, H_2^+, to introduce the essential features of bonding, and then use it as a guide to the structures of more complex systems.

14.7 Linear combinations of atomic orbitals

A **molecular orbital** is a one-electron wavefunction for an electron that spreads throughout the molecule. The mathematical forms of such orbitals are highly complicated, even for such a simple species as H_2^+, and they are unknown in general. All modern work builds approximations to the true molecular orbital by modelling them by combining together the atomic orbitals on the atoms making up the molecule.

First, we recall the general principle of quantum mechanics that, if there are several possible outcomes, then we add together the wavefunctions that represent those outcomes.[3] In H_2^+, there are two possible outcomes: an electron may be found either in an atomic orbital centred on A or it may be found in an orbital centred on B. Therefore, we write

$$\psi = c_A \psi_A + c_B \psi_B \qquad (14.6)$$

This wavefunction is called a **linear combination of atomic orbitals** (LCAO). An approximate molecular orbital formed from a linear combination of atomic orbitals is called an **LCAO-MO**.[4] The squares of the coefficients c in the LCAO-MO tell us the relative proportions of the atomic orbitals contributing to the molecular orbital. In a homonuclear diatomic molecule, an electron can be found with equal probability in orbital A or orbital B, so the *squares* of the coefficients must be equal. Therefore, the two possible wavefunctions are[5]

$$\psi = \psi_A \pm \psi_B$$

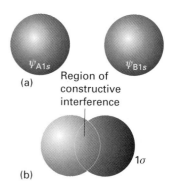

Fig 14.11 The formation of a bonding molecular orbital (a σ orbital). (a) Two H1s orbitals come together and, where they overlap, interfere constructively and give rise to an enhanced amplitude in the internuclear region. (b) The resulting orbital has cylindrical symmetry about the internuclear axis. When it is occupied by two paired electrons, to give the configuration σ^2, we have a σ bond.

First, we consider the LCAO-MO with the plus sign,

$$\psi = \psi_A + \psi_B \qquad (14.7)$$

as this molecular orbital will turn out to have the lower energy of the two. The form of this orbital is shown in Fig 14.11. It is called a σ **orbital** because it resembles an s orbital when viewed along the axis.[6] Because (as we shall see) it is the σ orbital of lowest energy, it is labelled 1σ. An electron that occupies a σ orbital is called a σ **electron**. In the ground state of the H_2^+ ion, there is a single 1σ electron, so we report the ground-state configuration of H_2^+ as $1\sigma^1$.

We can see by examining the LCAO-MO the origin of the lowering of energy that is responsible for the formation of the bond. The two atomic orbitals are like waves centred on adjacent nuclei. In the internuclear region, the amplitudes interfere constructively and the wavefunction has an enhanced amplitude there (Fig 14.12). Because the amplitude is increased, there is an increased probability of finding the electron between the two nuclei, where

---

[3] We used this principle to construct valence bond wavefunctions.

[4] It may be helpful to have a note of the approximations made so far. They are (1) the separation of electron and nuclear motions (the Born–Oppenheimer approximation), (2) the expression of a many-electron wavefunction as a product of one-electron wavefunctions (molecular orbitals), and (3) the construction of an MO as an LCAO.

[5] For simplicity, we are ignoring the overall normalization factor.

---

[6] More precisely, it is so called because an electron that occupies a σ orbital has zero orbital angular momentum around the internuclear axis, just as an s electron has zero orbital angular momentum around a nucleus.

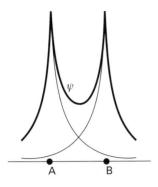

Fig 14.12 The wavefunction along the internuclear axis. Note that there is an enhancement of amplitude between the nuclei, so there is an increased probability of finding the bonding electrons in that region.

it is in a good position to interact strongly with both of them. Hence the energy of the molecule is lower than that of the separate atoms, where each electron can interact strongly with only one nucleus. In elementary MO theory, the bonding effect of an electron that occupies a molecular orbital is ascribed to its accumulation in the internuclear region as a result of the constructive interference of the contributing atomic orbitals.

14.8 **Bonding orbitals**

A 1σ orbital is an example of a **bonding orbital**, a molecular orbital which, if occupied, contributes to the strength of a bond between two atoms. As in VB theory, we can substitute the wavefunction in eqn 14.7 into the Schrödinger equation for the molecule-ion with the nuclei at a fixed separation R and solve the equation for the energy. The molecular potential energy curve obtained by plotting the energy against R is very similar to the one drawn in Fig 14.1. The energy of the molecule decreases as R is decreased from large values because the electron is increasingly likely to be found in the internuclear region as the two atomic orbitals interfere more effectively. However, at small separations, there is too little space between the nuclei for significant accumulation of electron density there. In addition, the nucleus–nucleus repulsion $V_{nuc,nuc}$ (eqn 14.3) becomes large. As a result, after an initial decrease, at small internuclear separations the potential energy

curve passes through a minimum and then rises sharply to high values. Calculations on H_2^+ give the equilibrium bond length as 130 pm and the bond dissociation energy as 171 kJ mol^{-1}; the experimental values are 106 pm and 250 kJ mol^{-1}, so this simple LCAO-MO description of the molecule, while inaccurate, is not absurdly wrong.

14.9 **Antibonding orbitals**

Now consider the alternative LCAO, the one with a minus sign:

$$\psi = \psi_A - \psi_B \tag{14.8}$$

Because this wavefunction is also cylindrically symmetrical around the internuclear axis it is also a σ orbital, and is denoted 2σ (Fig 14.13). When substituted into the Schrödinger equation, we find that it has a higher energy than the 1σ orbital and, indeed, it has a higher energy than either of the two atomic orbitals.

> **Self-test 14.4**
>
> Show that the molecular orbital written above is zero on a plane cutting through the internuclear axis at its midpoint. Take each atomic orbital to be of the form e^{-r/a_0}, with r_A measured from nucleus A and r_B measured from nucleus B.
>
> [*Answer*: the atomic orbitals cancel for values equidistant from the two nuclei]

We can trace the origin of the high energy of 2σ to the existence of a **nodal plane**, a plane on which the wavefunction passes through zero. This plane lies halfway between the nuclei and cuts through the internuclear axis. The two atomic orbitals cancel on this plane as a result of their destructive interference, because they have opposite signs. In drawings like those in Figs 14.11 and 14.13, we represent overlap of orbitals with the same sign (as in the formation of 1σ) by shading of the same tint; the overlap of orbitals of opposite sign (as in the formation of 2σ) is represented by one orbital of a light tint (or white) and another orbital of a dark tint.

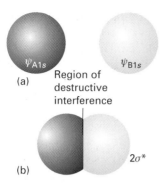

Fig 14.13 The formation of an antibonding molecular orbital (a σ* orbital). (a) Two H1s orbitals come together and, where they overlap with opposite signs (as depicted by different shades of grey), interfere destructively and give rise to a decreased amplitude in the internuclear region. (b) There is a nodal plane exactly halfway between the nuclei on which any electrons that occupy the orbital will not be found.

The 2σ orbital is an example of an **antibonding orbital**, an orbital that, if occupied, decreases the strength of a bond between two atoms. Antibonding orbitals are sometimes marked with an asterisk, as in $2\sigma^*$. The antibonding character of the 2σ orbital is partly a result of the exclusion of the electron from the internuclear region and its relocation outside the bonding region where it helps to pull the nuclei apart rather than pulling them together. An antibonding orbital is often slightly more strongly antibonding than the corresponding bonding orbital is bonding. This is partly because, although the 'gluing' effect of a bonding electron and the 'anti-gluing' effect of an antibonding electron are similar, the nuclei repel each other in both cases, and this repulsion pushes both levels up in energy.

14.10 The structures of diatomic molecules

In Chapter 13 we used the hydrogenic atomic orbitals and the building-up principle to deduce the ground electronic configurations of many-electron atoms. We use the same procedure for many-electron diatomic molecules (such as H_2 with two electrons and even Br_2 with 70), but using the H_2^+ molecular orbitals as a basis. The general procedure is as follows:

1 Construct molecular orbitals by forming linear combinations of all suitable valence atomic orbitals supplied by the atoms (the meaning of 'suitable' will be explained shortly); N atomic orbitals result in N molecular orbitals.

2 Accommodate the valence electrons supplied by the atoms so as to achieve the lowest overall energy subject to the constraint of the Pauli exclusion principle, that no more than two electrons may occupy a single orbital (and then must be paired).

3 If more than one molecular orbital of the same energy is available, add the electrons to each individual orbital before doubly occupying any one orbital (because that minimizes electron–electron repulsions).

4 Take note of Hund's rule (Section 13.11), that, if electrons occupy different degenerate orbitals, then they do so with parallel spins.

The following sections show how these rules are used in practice.

Self-test 14.5

How many molecular orbitals can be built from the valence shell orbitals in O_2?

[*Answer*: 8]

14.11 Hydrogen and helium molecules

The first step in the discussion of H_2, the simplest many-electron diatomic molecule, is to build the molecular orbitals. Because each H atom of H_2 contributes a 1s orbital (as in H_2^+), we can form the 1σ and $2\sigma^*$ bonding and antibonding orbitals from them, as we have seen already. At the equilibrium internuclear separation these orbitals will have the energies represented by the horizontal lines in Fig 14.14.

There are two electrons to accommodate (one from each atom). Both can enter the 1σ orbital by pairing their spins (Fig 14.15). The ground-state configuration is therefore $1\sigma^2$, and the atoms are joined by a bond consisting of an electron pair in a bonding σ orbital. These two electrons bind the two

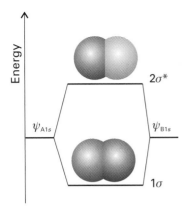

Fig 14.14 A molecular orbital energy level diagram for orbitals constructed from (1s,1s)-overlap, the separation of the levels corresponding to the equilibrium bond length.

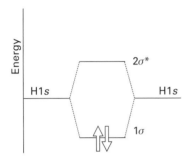

Fig 14.15 The ground electronic configuration of H_2 is obtained by accommodating the two electrons in the lowest available orbital (the bonding orbital).

nuclei together more strongly and closely than the single electron in H_2^+, and the bond length is reduced from 106 pm to 74 pm. A pair of electrons in a σ orbital is called a σ **bond**, and is very similar to the σ bond of valence bond theory.[7] We can conclude that *the importance of an electron pair in bonding stems from the fact that two is the maximum number of electrons that can enter each bonding molecular orbital.* Electrons do not 'want' to pair: they pair because in that way they are able to occupy a low-energy orbital.

---

[7] The two differ in certain details of the electron distribution between the two atoms joined by the bond, but both have an accumulation of density between the nuclei.

A similar argument shows why helium is a monatomic gas. Consider a hypothetical He_2 molecule. Each He atom contributes a $1s$ orbital to the linear combination used to form the molecular orbitals, and so we can construct 1σ and $2\sigma^*$ molecular orbitals. They differ in detail from those in H_2 because the $He1s$ orbitals are more compact, but the general shape is the same, and for qualitative discussions we can use the same molecular orbital energy level diagram as for H_2. Because each atom provides two electrons, there are four electrons to accommodate. Two can enter the 1σ orbital, but then it is full (by the Pauli exclusion principle). The next two electrons must enter the antibonding $2\sigma^*$ orbital (Fig 14.16). The ground electronic configuration of He_2 is therefore $1\sigma^2 2\sigma^{*2}$. Because an antibonding orbital is slightly more antibonding than a bonding orbital is bonding, the He_2 molecule has a higher energy than the separated atoms and is unstable. Hence, two ground-state He atoms do not form bonds to each other, and helium is a monatomic gas.

Example 14.1 *Judging the stability of diatomic molecules*

Decide whether Li_2 is likely to exist on the assumption that only the valence s orbitals contribute to its molecular orbitals.

Strategy Decide what molecular orbitals can be formed from the available valence orbitals, rank them in order of energy, then feed in the electrons supplied by the valence orbitals of the atoms. Judge whether there is a net bonding or net antibonding effect between the atoms.

Solution Each molecular orbital is built from $2s$ atomic orbitals, which give one bonding and one antibonding combination (1σ and $2\sigma^*$, respectively). Each Li atom supplies one valence electron; the two electrons fill the 1σ orbital, to give the configuration $1\sigma^2$, which is bonding.

Self-test 14.6

Is LiH likely to exist if the Li atom uses only its $2s$ orbital for bonding?

[*Answer:* yes, $(Li2s,H1s)\sigma^2$]

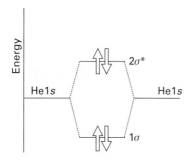

Fig 14.16 The ground electronic configuration of the four-electron molecule He$_2$ has two bonding electrons and two antibonding electrons. It has a higher energy than the separated atoms, and so He$_2$ is unstable relative to two He atoms.

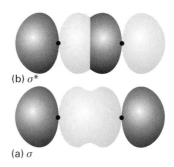

(b) σ^*

(a) σ

Fig 14.17 (a) The interference leading to the formation of a σ bonding orbital and (b) the corresponding antibonding orbital when two p orbitals overlap along an internuclear axis.

14.12 **Period 2 diatomic molecules**

We shall now see how the concepts we have introduced apply to other **homonuclear diatomic molecules**, which are diatomic molecules formed from identical atoms, such as N$_2$ and Cl$_2$, and diatomic ions such as O$_2^{2-}$. In line with the building-up procedure, we first consider the molecular orbitals that may be formed from the valence orbitals and do not (at this stage) trouble about how many electrons are available.

In Period 2, the valence orbitals are $2s$ and $2p$. Suppose first that we consider these two types of orbital separately. Then the $2s$ orbitals on each atom overlap to form bonding and antibonding combinations that we shall denote 1σ and $2\sigma^*$, respectively. Likewise, the two $2p_z$ orbitals (by convention, the internuclear axis is denoted the z-axis) have cylindrical symmetry around the internuclear axis. They may therefore participate in σ-orbital formation to give the bonding and antibonding combinations 3σ and $4\sigma^*$, respectively (Fig 14.17). The resulting energy levels of the σ orbitals are shown in the MO energy level diagram in Fig 14.18. Strictly, we should not consider the s and p_z orbitals separately, because both of them can contribute to the formation of σ orbitals. Therefore, in a more advanced treatment, we should combine all four orbitals together to form four σ molecular orbitals, each one of the form

$$\psi = c_1\psi_{A2s} + c_2\psi_{B2s} + c_3\psi_{A2p_z} + c_4\psi_{B2p_z}$$

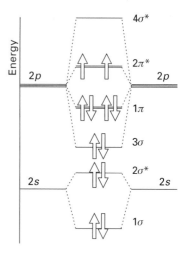

Fig 14.18 A typical molecular orbital energy level diagram for Period 2 homonuclear diatomic molecules. The valence atomic orbitals are drawn in the columns on the left and the right; the molecular orbitals are shown in the middle. Note that the π orbitals form doubly degenerate pairs. The sloping lines joining the molecular orbitals to the atomic orbitals show the principal composition of the molecular orbitals. This diagram is suitable for O$_2$ and F$_2$.

We find the four coefficients, c_i, which represent the different contributions that each atomic orbital makes to the overall molecular orbital, by using the variation theorem. However, in practice, the two lowest energy combinations of this kind are very similar to the combination 1σ and $2\sigma^*$ of $2s$ orbitals that we have described, and the two highest energy

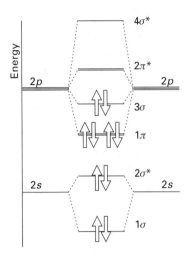

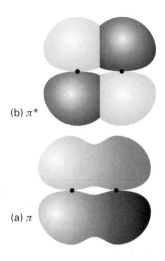

Fig 14.19 A typical molecular orbital energy level diagram for Period 2 homonuclear diatomic molecules up to and including N₂.

Fig 14.20 (a) The interference leading to the formation of a π bonding orbital and (b) the corresponding antibonding orbital.

combinations are very similar to the 3σ and $4\sigma^*$ combinations of $2p_z$ orbitals. In each case there will be small differences: the $2\sigma^*$ orbital, for instance, will be contaminated by some $2p_z$ character and the 3σ orbital will be contaminated by some $2s$ character, and their energies will be slightly shifted from where they would be if we considered only the 'pure' combinations. Nevertheless, the changes are not great, and we can continue to think of 1σ and $2\sigma^*$ as being one bonding and antibonding pair, and of 3σ and $4\sigma^*$ as being another pair. The four orbitals are shown in the centre column of Fig 14.18. There is no guarantee that $2\sigma^*$ and 3σ will be in the exact location shown in the illustration and the locations shown in Fig 14.19 are found in some molecules (see below).

Now consider the $2p_x$ and $2p_y$ orbitals of each atom, which are perpendicular to the internuclear axis and may overlap side-by-side. This overlap may be constructive or destructive and results in a bonding and an antibonding **π orbital**, which we label

1π and $2\pi^*$, respectively. The notation π is the analogue of p in atoms for, when viewed along the axis of the molecule, a π orbital looks like a p orbital (Fig 14.20).[8] The two $2p_x$ orbitals overlap to give a bonding and an antibonding π orbital, as do the two $2p_y$ orbitals too. The two bonding combinations have the same energy; likewise, the two antibonding combinations have the same energy. Hence, each π energy level is doubly degenerate and consists of two distinct orbitals. Two electrons in a π orbital constitute a **π bond**: such a bond resembles a π bond of valence bond theory, but the details of the electron distribution are slightly different.

The relative order of the σ and π orbitals in a molecule cannot be predicted readily (it varies with the energy separation between the $2s$ and $2p$ orbitals of the atoms), and in some molecules the order shown in Fig 14.18 applies, whereas others have the order shown in Fig 14.19. The change in order can be seen in Fig 14.21, which shows the calculated energy levels for the Period 2 homonuclear diatomic molecules. A useful rule is that, for neutral molecules, the order shown in Fig 14.18 is valid for O_2 and F_2, whereas the order shown in Fig 14.19 is valid for the preceding elements of the period.

[8] More precisely, an electron in a π orbital has one unit of orbital angular momentum about the internuclear axis.

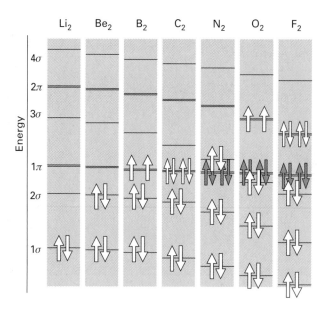

Fig 14.21 The variation of the orbital energies of Period 2 homonuclear diatomic molecules.

14.13 **Symmetry and overlap**

One central feature of molecular orbital theory can now be addressed. We have seen that s and p_z orbitals may contribute to the formation of σ orbitals, and that p_x and p_y orbitals may contribute to π orbitals. However, we never have to consider orbitals formed by the overlap of s and p_x orbitals (or p_y orbitals). When building molecular orbitals, *we need consider linear combinations only of atomic orbitals of the same symmetry with respect to the internuclear axis*. Because an s orbital has cylindrical symmetry around the internuclear axis, but a p_x orbital does not, the two atomic orbitals cannot contribute to the same molecular orbital. The reason for this distinction based on symmetry can be understood by considering the interference between an s orbital and a p_x orbital (Fig 14.22): although there is constructive interference between the two orbitals on one side of the axis, there is an exactly compensating amount of destructive interference on the other side of the axis, and the net bonding or antibonding effect is zero.

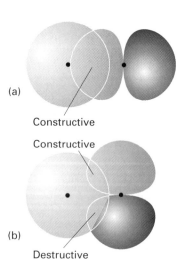

(a)

Constructive

Constructive

(b)

Destructive

Fig 14.22 Overlapping s and p orbitals. (a) End-on overlap leads to nonzero overlap and to the formation of an axially symmetric σ orbital. (b) Broadside overlap leads to no net accumulation or reduction of electron density and does not contribute to bonding.

Derivation 14.1 *Overlap integrals*

The extent to which two orbitals overlap is measured by the **overlap integral**, S:

$$S = \int \psi_A \psi_B \, d\tau \qquad (14.9)$$

where the integration is over all space.[9] If the atomic orbital ψ_A on A is small wherever the orbital ψ_B on B is large, or vice versa, then the product of their amplitudes is everywhere small and the integral—the sum of these products—is small (Fig 14.23a). If ψ_A and ψ_B are simultaneously large in some region of space, then S may be large (Fig 14.23b). If the two atomic orbitals are identical (for example, 1s orbitals on the same nucleus), $S = 1$. Typical values for orbitals with $n = 2$ are in the range 0.2 to 0.3.

Now consider the arrangement in Fig 14.23c in which an s orbital overlaps a p_x orbital of a different atom. At some point the product $\psi_A \psi_B$ may be large. However, there is a point where $\psi_A \psi_B$ has exactly the same magnitude but an opposite sign. When the integral is evaluated, these two contributions are added together and cancel. For every point in the upper half of the diagram, there is a point in the lower half that cancels it, so $S = 0$. Therefore, there is no net overlap between the s and p orbitals in this arrangement.

We now have the criteria for selecting atomic orbitals from which molecular orbitals are to be built:

1 Use all available valence orbitals from both atoms (in polyatomic molecules, from all the atoms).

2 Classify the atomic orbitals as having σ and π symmetry with respect to the internuclear axis, and build σ and π orbitals from all atomic orbitals of a given symmetry.

3 From N_σ atomic orbitals of σ symmetry, N_σ σ orbitals can be built with progressively higher energy from strongly bonding to strongly antibonding.

4 From N_π atomic orbitals of π symmetry, N_π π orbitals can be built with progressively higher energy from strongly bonding to strongly antibonding. The π orbitals occur in doubly degenerate pairs.

---

[9] In quantum mechanics, it is conventional to use $d\tau$ (where τ is tau) to represent an infinitesimal volume. In cartesian coordinates, $d\tau = dxdydz$. In spherical coordinates, $d\tau = r^2 dr \sin\theta \, d\theta \, d\phi$.

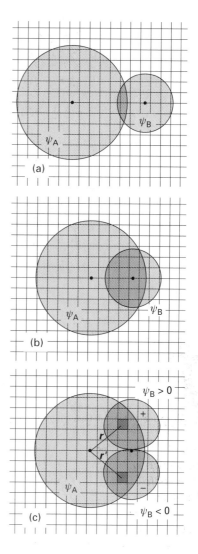

Fig 14.23 A schematic representation of the contributions to the overlap integral. (a) $S \approx 0$ because the orbitals are far apart and their product is always small. (b) S is large (but less than 1) because the product $\psi_A \psi_B$ is large over a substantial region. (c) $S = 0$ because the positive region of overlap is exactly cancelled by the negative region.

As a general rule, the energy of each type of orbital (σ or π) increases with the number of internuclear nodes. The lowest energy orbital of a given species has no internuclear nodes and the highest energy orbital has a nodal plane between each pair of adjacent atoms (Fig 14.24).

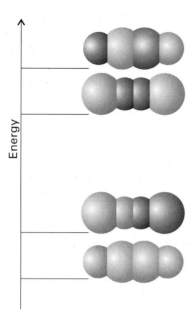

Fig 14.24 A schematic representation of the four molecular orbitals that can be formed from four s orbitals in a chain of four atoms. The lowest energy combination (the bottom diagram) is formed from atomic orbitals with the same sign, and there are no internuclear nodes. The next higher orbital has one node (at the centre of the molecule). The next higher orbital has two internuclear nodes, and the uppermost, highest energy orbital has three internuclear nodes, one between each neighbouring pair of atoms, and is fully antibonding. The sizes of the spheres reflect the contributions of each atom to the molecular orbital: the shading represents different signs.

Example 14.2 *Assessing the contribution of d orbitals*

Can d orbitals contribute to σ and π orbitals in diatomic molecules?

Strategy We need to assess the symmetry of d orbitals with respect to the internuclear z-axis: orbitals of the same symmetry can contribute to a given molecular orbital.

Solution A d_{z^2} orbital has cylindrical symmetry around z and so can contribute to σ orbitals. The d_{zx} and d_{yz} orbitals have π symmetry with respect to the axis (Fig 14.25), so they can contribute to π orbitals.

Self-test 14.7
Sketch the 'δ orbitals' (orbitals that resemble four-lobed d orbitals when viewed along the internuclear axis) that may be formed by the remaining two d orbitals (and which contribute to bonding in some d-metal cluster compounds).

[*Answer*: see Fig 14.25]

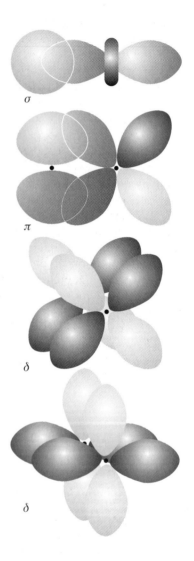

Fig 14.25 The types of molecular orbital to which d orbitals can contribute. The σ and π combinations can be formed with s, p, and d orbitals of the appropriate symmetry, but the δ orbitals can be formed only by the d orbitals of the two atoms.

14.14 The electronic structures of homonuclear diatomic molecules

Figures 14.18 and 14.19 show the general layout of the valence-shell atomic orbitals of Period 2 atoms on the left and right. The lines in the middle are an indication of the energies of the molecular orbitals that can be formed by overlap of atomic orbitals. From the eight valence shell orbitals (four from each atom), we can form eight molecular orbitals: four are σ orbitals and four, in two pairs, are doubly degenerate π orbitals. With the orbitals established, we derive the ground-state electron configurations of the molecules by adding the appropriate number of electrons to the orbitals and following the building-up rules. Charged species (such as the peroxide ion, O_2^{2-}, and C_2^+) need either more or fewer electrons (for anions and cations, respectively) than the neutral molecules.

We shall illustrate the procedure with N_2, which has ten valence electrons; for this molecule we use Fig 14.19. The first two electrons pair, enter, and fill the 1σ orbital. The next two electrons enter and fill the $2\sigma^*$ orbital. Six electrons remain. There are two 1π orbitals, so four electrons can be accommodated in them. The two remaining electrons enter the 3σ orbital. The ground-state configuration of N_2 is therefore $1\sigma^2 2\sigma^{*2} 1\pi^4 3\sigma^2$. This configuration is also depicted in Fig 14.19.

The strength of a bond in a molecule is the net outcome of the bonding and antibonding effects of the electrons in the orbitals. The **bond order**, b, in a diatomic molecule is defined as

$$b = \tfrac{1}{2}(n - n^*) \tag{14.10}$$

where n is the number of electrons in bonding orbitals and n^* is the number of electrons in antibonding orbitals. Each electron pair in a bonding orbital increases the bond order by 1 and each pair in an antibonding orbital decreases it by 1. For H_2, $b = 1$, corresponding to a single bond between the two atoms: this bond order is consistent with the Lewis structure H–H for the molecule. In He_2, which has equal numbers of bonding and antibonding electrons (with $n = 2$ and $n^* = 2$), the bond order is $b = 0$, and there is no bond. In N_2, 1σ, 3σ, and 1π are bonding orbitals, and $n = 2 + 2 + 4 = 8$; however, $2\sigma^*$ (the anti-

bonding partner of 1σ) is antibonding, so $n^* = 2$ and the bond order of N_2 is $b = \tfrac{1}{2}(8 - 2) = 3$. This value is consistent with the Lewis structure :N≡N:, in which there is a triple bond between the two atoms.

The bond order is a useful parameter for discussing the characteristics of bonds, because it correlates with bond length and, the greater the bond order between atoms of a given pair of atoms, the shorter the bond. The bond order also correlates with bond strength and, the greater the bond order, the greater the strength. The high bond order of N_2 is consistent with its high dissociation energy (942 kJ mol^{-1}).

Example 14.3 *Writing the electron configuration of a diatomic molecule*

Write the ground-state electron configuration of O_2 and calculate the bond order.

Strategy Decide which MO energy level diagram to use (Fig 14.18 or Fig 14.19). Count the valence electrons and accommodate them by using the building-up principle.

Solution Figure 14.18 is appropriate for oxygen. There are 12 valence electrons to accommodate. The first 10 electrons recreate the N_2 configuration (with a reversal of the order of the 3σ and 1π orbitals); the remaining two electrons must occupy the $2\pi^*$ orbitals. The configuration is therefore $1\sigma^2 2\sigma^{*2} 3\sigma^2 1\pi^4 2\pi^{*2}$. This configuration is also depicted in Fig 14.18. Because 1σ, 3σ, and 1π are bonding and $2\sigma^*$ and $2\pi^*$ are antibonding, the bond order is $b = \tfrac{1}{2}(8 - 4) = 2$. This bond order accords with the classical view that oxygen has a double bond.

Self-test 14.8

Write the electron configuration of F_2 and deduce its bond order.

[*Answer:* $1\sigma^2 2\sigma^{*2} 3\sigma^2 1\pi^4 2\pi^{*4}$, $b = 1$]

We see from Example 14.3 that the electron configuration of O_2 is $1\sigma^2 2\sigma^{*2} 3\sigma^2 1\pi^4 2\pi^{*2}$. According to the building-up principle, the two $2\pi^*$ electrons in O_2 will occupy different orbitals. One enters the $2\pi^*$ orbital formed by overlap of $2p_x$. The other enters its degenerate partner, the $2\pi^*$ orbital formed from overlap of the $2p_y$ orbitals. Because the two electrons occupy different orbitals, by Hund's rule they will have parallel spins (↑↑). Therefore, we can predict that an O_2 molecule will be magnetic because the magnetic fields generated by the two unpaired spins do not cancel. Specifically, oxygen is predicted to be a **paramagnetic** substance, a substance that is drawn into a magnetic field. Most substances (those with paired electron spins) are **diamagnetic**, and are pushed out of a magnetic field. That O_2 is in fact a paramagnetic gas is a striking confirmation of the superiority of the molecular orbital description of the molecule over the Lewis and valence bond descriptions (which require all the electrons to be paired). The property of paramagnetism is utilized to monitor the oxygen content of incubators by measuring the magnetism of the gases they contain.

An F_2 molecule has two more electrons than an O_2 molecule, so its configuration is $1\sigma^2 2\sigma^{*2} 3\sigma^2 1\pi^4 2\pi^{*4}$ and its bond order is 1. We conclude that F_2 is a singly bonded molecule, in agreement with its Lewis structure $:\ddot{F}-\ddot{F}:$. The low bond order is consistent with the low dissociation energy of F_2 (154 kJ mol^{-1}). A hypothetical Ne_2 molecule would have two further electrons: its configuration would be $1\sigma^2 2\sigma^{*2} 3\sigma^2 1\pi^4 2\pi^{*4} 4\sigma^{*2}$ and its bond order 0. The bond order of zero—which implies that two neon

atoms do not bond together—is consistent with the monatomic character of neon.

14.15 Parity

We have seen a little of the importance of the symmetry of atomic orbitals in the construction of molecular orbitals. Symmetry plays a role in many discussions of molecular orbitals themselves, particularly in the interpretation of the electronic transitions that give rise to molecular spectra (Chapter 18).

The classification of molecular orbitals as σ and π is one aspect of molecular symmetry, for σ and π orbitals differ in their 'rotational' symmetry about the internuclear axis. The orbitals of *homonuclear* diatomic molecules may also be classified according to their **parity**, their behaviour under the process called **inversion** (Fig 14.26). To decide on the parity of an orbital, we consider an arbitrary point in a homonuclear diatomic molecule, and note the sign of the orbital. Then we imagine travelling through the centre of the molecule and out to the corresponding point on the other side. If the orbital has the same sign (the same shading) at that point, then it has 'even parity' and is denoted g (from *gerade*, the German word for even). If the orbital has opposite sign, then it has 'odd parity' and is denoted u (from *ungerade*, uneven). Heteronuclear diatomic molecules (such as HCl) do not have inversion symmetry of the kind that we have described: such molecules do not have a 'centre'.

The diagram in Fig 14.26 shows that a bonding σ orbital has even parity; so it is written σ_g; an antibonding σ orbital has odd parity and is written σ_u^*. A bonding π orbital has odd parity and is denoted π_u and an antibonding π orbital has even parity, denoted π_g^*. It follows that the full-dress version of the electron configuration of the ground state of N_2 is[10]

$$N_2 \quad 1\sigma_g^2 2\sigma_u^{*2} 1\pi_u^4 3\sigma_g^2$$

This detailed specification of a configuration is needed (in the main) only when discussing electronic transitions and molecular selection rules.

Example 14.4 *Judging the relative bond strengths of molecules and ions*

The superoxide ion, O_2^-, plays an important role in the ageing processes that take place in organisms. Judge whether O_2^- is likely to have a larger or smaller dissociation energy than O_2.

Strategy Because a species with the larger bond order is likely to have the larger dissociation energy, we should compare their electronic configurations, and assess their bond orders.

Solution From Fig 14.18,

O_2	$1\sigma^2 2\sigma^{*2} 3\sigma^2 1\pi^4 2\pi^{*2}$	$b = 2$
O_2^-	$1\sigma^2 2\sigma^{*2} 3\sigma^1 1\pi^4 2\pi^{*3}$	$b = 1.5$

Because the anion has the smaller bond order, we expect it to have the smaller dissociation energy.

Self-test 14.9

Which can be expected to have the higher dissociation energy, F_2 or F_2^+?

[*Answer*: F_2^+]

---

[10] In an alternative numbering scheme, the g and u orbitals are treated as separate sets and numbered independently. In this scheme, the configuration of N_2 would be reported as $1\sigma_g^2 1\sigma_u^{*2} 1\pi_u^4 2\sigma_g^2$.

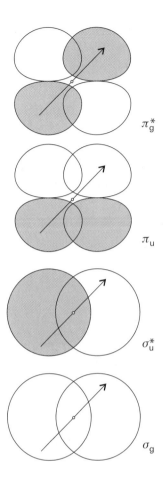

Fig 14.26 The parity of an orbital is even (g) if its amplitude is unchanged under inversion in the centre of symmetry of the molecule, but odd (u) if the amplitude changes sign. Heteronuclear diatomic molecules do not have a centre of inversion, and so the g,u classification is irrelevant.

Self-test 14.10

Write the full electron configuration of the ground state of the F_2 molecule.

[*Answer*: $1\sigma_g^2 2\sigma_u^{*2} 3\sigma_g^2 1\pi_u^4 2\pi_g^{*4}$]

14.16 Heteronuclear diatomic molecules

A **heteronuclear diatomic molecule** is a diatomic molecule formed from atoms of two different elements, such as CO and HCl. The electron distribution in the covalent bond between the atoms is not symmetrical between the atoms because it is ener-

getically favourable for a bonding electron pair to be found closer to one atom rather than the other. This imbalance results in a **polar bond**, which is a covalent bond in which the electron pair is shared unequally by the two atoms. The **electronegativity**, χ (chi), of an element is the power of its atoms to draw electrons to itself when it is part of a compound, so we can expect the polarity of a bond to depend on the relative electronegativities of the elements.

Linus Pauling formulated a numerical scale of electronegativity based on considerations of bond dissociation energies, $E(X–Y)$:

$$|\chi_A - \chi_B| = 0.102 \times \{\Delta E/(\text{kJ mol}^{-1})\}^{1/2} \qquad (14.11a)$$

with

$$\Delta E = E(A–B) - \tfrac{1}{2}\{E(A–A) + E(B–B)\} \qquad (14.11b)$$

Table 14.2 lists values for the main-group elements. Robert Mulliken proposed an alternative definition in terms of the ionization energy, I, and the electron affinity, E_{ea}, of the element expressed in electronvolts:

$$\chi = \tfrac{1}{2}(I + E_{ea}) \qquad (14.12)$$

Table 14.2 *Electronegativities of the main-group elements*

H						
2.1						
Li	Be	B	C	N	O	F
1.0	1.5	2.0	2.5	3.0	3.5	4.0
Na	Mg	Al	Si	P	S	Cl
0.9	1.2	1.5	1.8	2.1	2.5	3.0
K	Ca	Ga	Ge	As	Se	Br
0.8	1.0	1.6	1.8	2.0	2.4	2.8
Rb	Sr	In	Sn	Sb	Te	I
0.8	1.0	1.7	1.8	1.9	2.1	2.5
Cs	Ba	Tl	Pb	Bi	Po	
0.7	0.9	1.8	1.8	1.9	2.0	

This relation is plausible, because an atom that has a high electronegativity is likely to be one that has a high ionization energy (so that it is unlikely to lose electrons to another atom in the molecule) and a high electron affinity (so that it is energetically favourable for an electron to move towards it). The Mulliken electronegativities are broadly in line with the Pauling electronegativities. Electronegativities show a periodicity, and the elements with the highest electronegativities are those close to fluorine in the periodic table.

The location of the bonding electron pair close to one atom in a heteronuclear molecule results in that atom having a net negative charge, which is called a **partial negative charge** and denoted $\delta-$. There is a compensating **partial positive charge**, $\delta+$, on the other atom. In a typical heteronuclear diatomic molecule, the more electronegative element has the partial negative charge and the more electropositive element has the partial positive charge.

Self-test 14.11

Predict the (weak) polarity of a C–H bond.

[*Answer:* $^{\delta-}$C—H$^{\delta+}$]

14.17 Polar covalent bonds

Molecular orbital theory takes polar bonds into its stride. A polar bond consists of two electrons in an orbital of the form

$$\psi = c_A\psi_A + c_B\psi_B \tag{14.13}$$

with unequal coefficients. The proportion of the atomic orbital ψ_A in the bond is c_A^2 and that of ψ_B is c_B^2. These proportions can also be thought of as the relative lengths of time that an electron spends in each atomic orbital. A **nonpolar bond**, a covalent bond in which the electron pair is shared equally between the two atoms and there are zero partial charges on each atom, has $c_A^2 = c_B^2$. A pure ionic bond, in which one atom has obtained virtually sole possession of the electron pair (as in Cs^+F^-, to a first approximation), has one coefficient zero (so that A^+B^- would have $c_A^2 = 0$ and $c_B^2 = 1$).

A general feature of molecular orbitals between dissimilar atoms is that the atomic orbital with the

lower energy (that belonging to the more electronegative atom) makes the larger contribution to the lowest energy molecular orbital. The opposite is true of the highest (most antibonding) orbital, for which the principal contribution comes from the atomic orbital with higher energy (the less electronegative atom). Figure 14.27 shows a schematic representation of this point.

These features of polar bonds can be illustrated by considering HF. The general form of the molecular orbitals of HF is

$$\psi = c_H\psi_H + c_F\psi_F \tag{14.14}$$

where ψ_H is an H1s orbital and ψ_F is an F2p_z orbital. Because the ionization energy of a hydrogen atom is 13.6 eV, we know that the energy of the H1s orbital is –13.6 eV. As usual, the zero of energy is the infinitely separated electron and proton (Fig 14.28). Similarly, from the ionization energy of fluorine, which is 18.6 eV, we know that the energy of the F2p_z orbital is –18.6 eV, about 5 eV lower than the

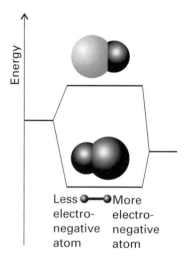

Fig 14.27 A schematic representation of the relative contributions of atoms of different electronegativities to bonding and antibonding molecular orbitals. In the bonding orbital, the more electronegative atom makes the greater contribution (represented by the larger sphere), and the electrons of the bond are more likely to be found on that atom. The opposite is true of an antibonding orbital. A part of the reason why an antibonding orbital is of high energy is that the electrons that occupy it are likely to be found on the more electropositive atom.

H1s orbital. It follows that the bonding σ orbital in HF is mainly $F2p_z$ and the antibonding σ orbital is mainly H1s orbital in character. The two electrons in the bonding orbital are most likely to be found in the $F2p_z$ orbital, so there is a partial negative charge on the F atom and a partial positive charge on the H atom.

A systematic way of finding the coefficients in the linear combinations is to use the variation theorem and to look for the values of the coefficients that result in the lowest energy (Section 14.6). For example, when the variation principle is applied to an H_2 molecule, the calculated energy is lowest when the two H1s orbitals contribute equally to a bonding orbital. However, when we apply the principle to HF, the lowest energy is obtained for the orbital

$$\psi = 0.33\psi_H + 0.94\psi_F$$

We see that indeed the $F2p_z$ orbital does make the greater contribution to the bonding σ orbital.

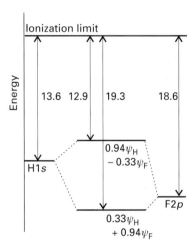

Fig 14.28 The atomic orbital energy levels of H and F atoms and the molecular orbitals they form. The bonding orbital has predominantly F atom character and the antibonding orbital has predominantly H atom character.

Fig 14.29 The computed electron density in hydrogen fluoride. There is an accumulation of electron density on the F atom.

Self-test 14.12

What percentage of its time does a σ electron in HF spend in a $F2p_z$ orbital?

[*Answer:* 88 per cent (= $(0.94)^2 \times 100$ per cent)]

Modern molecular graphics software has transformed the discussion of electron densities in molecules (Box 14.1). Commercial software is now widely available for plotting the individual molecular orbitals in molecules (as we see later) and for mapping the total electron density. These images are more striking in colour, for different colours can be used to represent regions of low and high electron density (corresponding to regions of net positive and negative charge, respectively). Figure 14.29 shows a less convincing grey-scale version of such a calculation on HF, and we see more examples later.[11]

Figure 14.30 shows the bonding scheme in CO and illustrates a number of points we have made. The ground configuration is $1\sigma^2 2\sigma^2 1\pi^4 3\sigma^2$. (Note that the parity designation is inapplicable because the molecule is heteronuclear.) The lowest energy orbitals are predominantly of O character as that is the more electronegative element. The **highest occupied molecular orbital** (HOMO) is 3σ, which is a

---

[11] Full colour versions of these illustrations are available on the web site for this book.

Box 14.1 *Computational chemistry*

Commercial software is now widely available for calculating the electronic structures of molecules and displaying the results graphically. All such calculations work within the Born–Oppenheimer approximation and express the molecular orbitals as linear combinations of atomic orbitals. There are then two principal approaches to solving the Schrödinger equation. In the *semi-empirical methods*, certain integrals that occur in the solution are set equal to parameters that have been chosen to lead to the best fit to experimental quantities, such as enthalpies of formation. Semi-empirical methods are applicable to a wide range of molecules with a virtually limitless number of atoms, and are widely popular. In the more fundamental *ab initio methods*, an attempt is made to calculate structures from first principles, using only the atomic numbers of the atoms present. Such an approach is intrinsically more reliable than a semi-empirical procedure, but is limited to molecules with no more than a dozen or so atoms. Both types of approach typically adopt a *self- consistent field* (SCF) procedure, in which an initial guess about the composition of LCAO-MOs is successively refined until the solution remains unchanged in a cycle of calculation.

The semi-empirical methods have grown in sophistication. When solving the Schrödinger equation, the calculation gives rise to a large number of integrals over the atomic orbitals, ψ, used to construct LCAO-MOs, and they have the form

$$\int \psi_A(r_1)\,\psi_B(r_2)\,\frac{1}{r_{12}}\,\psi_C(r_1)\,\psi_D(r_2)\,d\tau_1 d\tau_2$$

where r_{12} is the distance between the two electrons at distances r_1 and r_2 from the nuclei of their respective atoms. The evaluation of these integrals takes an enormous amount of computer time, and their approximation greatly shortens the time needed for the calculation and makes molecular structure calculations possible on desktop computers. Each semi-empirical procedure uses different criteria for neglecting or approximating these integrals rather than evaluating them explicitly. The earliest procedure of this kind, the *Hückel procedure*, simplified the calculation by neglecting all overlap integrals between atoms and all electron–electron repulsion terms. Some of the severe restrictions of this method were removed with the introduction of the *complete neglect of differential overlap* (CNDO) method. The phrase 'differential overlap' simply means the product of atomic orbitals $\psi_A\psi_B$: in the CNDO method the two-electron integral is set equal to zero unless A = B. The surviving integrals are then ascribed values that lead to a good fit with certain experimentally determined properties. The introduction of

CNDO opened the door to an avalanche of similar but improved methods and their accompanying acronyms, such as *intermediate neglect of differential overlap* (INDO), *modified neglect of differential overlap* (MNDO), and the *Austin Model 1* (AM1, version 2 of MINDO).

The *ab initio* methods also simplify the computation of four-electron integrals, but they do so by setting up the problem in a different manner. One popular approach is to replace the hydrogenic atomic orbitals used to form the LCAO-MOs by a *gaussian-type orbital* (GTO) in which the exponential function e^{-r} characteristic of actual orbitals is replaced by a sum of gaussian functions of the form e^{-r^2}. The advantage of this procedure is that the product of two GTOs is itself *one* GTO, so the four-centre integral is turned into a sum of two-centre integrals, which is much easier to evaluate.

Whatever the computational procedure, the aim is to calculate molecular properties such as enthalpies of formation, bond lengths, partial charge distribution, electron density, and dipole moment. Some of the illustrations in this book have been obtained by using software to compute an *elpot diagram* (an electric potential diagram) in which the net charge at each point is mapped as a colour (here, as density of shading) on the surface showing the distribution of electron density.

Computational chemistry is now a standard part of chemical research. One major application is in pharmaceutical chemistry, where the likely pharmacological activity of a molecule can be assessed computationally from its shape and electron density distribution before expensive clinical trials are started.

Exercise 1 Coulombic charge–charge interactions are important in computational chemistry for determining molecular structure. The potential energy of this interaction for two charges q_1 and q_2 separated by a distance r in a medium of permittivity ε is

$$V = \frac{q_1 q_2}{4\pi\varepsilon r}$$

The *relative* permittivity, $\varepsilon_r = \varepsilon/\varepsilon_0$, is 1 for a vacuum, about 2.0 for liquid hydrocarbons, and close to 80 for water. Calculate the interaction energy, in kilojoules per mole, in a vacuum between an electron and proton separated by the Bohr radius. The N–C distance of the hydrogen bonded groups in proteins, such as occur in an α helix (–NH···O=C–), is 0.29 nm. How much energy is required to break the hydrogen bond (a) in a membrane (essentially a liquid hydrocarbon) and (b) in water?

> **Exercise 2** In a study of the electronic structure of tyrosine, a computational chemist used a gaussian-tye orbital centred on the atoms. Each orbital was of the form $N_i e^{-a_i r^2}$ where i identifies the atom. Show that the four-centre integral shown above becomes a sum of two-centre integrals when expressed in terms of these GTOs.

largely nonbonding orbital centred on C, so the two electrons that occupy it can be regarded as a lone pair on the C atom. The **lowest unoccupied molecular orbital** (LUMO) is 2π, which is largely a doubly degenerate orbital of $2p$ character on carbon. This combination of a lone pair orbital on C and a pair of empty π orbitals also largely on C is at the root of the importance of carbon monoxide in *d*-block chemistry, because it enables it to form an extensive series of carbonyl complexes by a combination of electron donation from the 3σ orbital and electron acceptance into the 2π orbitals.

14.18 The structures of polyatomic molecules

The bonds in polyatomic molecules are built in the same way as in diatomic molecules, the only differ-

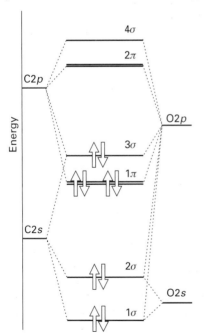

Fig 14.30 The molecular orbital energy level diagram for CO.

ence being that we use more atomic orbitals to construct the molecular orbitals, and these molecular orbitals spread over the entire molecule, not just the adjacent atoms of the bond. In general, a molecular orbital is a linear combination of all the atomic orbitals of all the atoms in the molecule. In H_2O, for instance, the atomic orbitals are the two H1s orbitals, the O2s orbital, and the three O2p orbitals (if we consider only the valence shell). From these six atomic orbitals we can construct six molecular orbitals that spread over all three atoms. The molecular orbitals differ in energy. The lowest energy, most strongly bonding orbital has the least number of nodes between adjacent atoms. The highest energy, most strongly antibonding orbital has the greatest numbers of nodes between neighbouring atoms.

According to MO theory, the bonding influence of a single electron pair is distributed over all the atoms, and each electron pair (the maximum number of electrons that can occupy any single molecular orbital) helps to bind all the atoms together. In the LCAO-MO approximation, each molecular orbital is modelled as a linear combination of atomic orbitals, with atomic orbitals contributed by all the atoms in the molecule. Thus, a typical molecular orbital in H_2O constructed from H1s orbitals (denoted ψ_A and ψ_B) and O2s and O2p orbitals (denoted ψ_{Os} and ψ_{Op}) will have the composition

$$\psi = c_1\psi_A + c_2\psi_{Os} + c_3\psi_{Op} + c_4\psi_B \qquad (14.15)$$

Because four atomic orbitals are being used to form the molecular orbital, there will be four possible molecular orbitals: the lowest energy (most bonding) orbital will have no internuclear nodes and the highest energy (most antibonding) orbital will have a node between each pair of neighbouring nuclei. Figure 14.31 shows the computed electron density distribution for H_2O, NH_3, and a selection of amino acids.

An important example of the application of molecular orbital theory is to the orbitals that may be

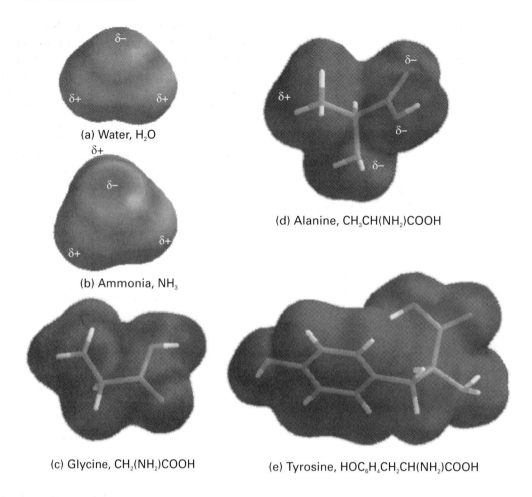

Fig 14.31 Electron densities in a selection of molecules. The regions of partial positive and negative charge are shown by different colours in the originals but here are indicated by δ+ and δ−. (No attempt has been made to show how the partial charges sum to zero overall.)

formed from the p orbitals perpendicular to the molecular plane of benzene, C_6H_6. Because there are six such atomic orbitals, it is possible to form six molecular orbitals of the form

$$\psi = c_1\psi_1 + c_2\psi_2 + c_3\psi_3 + c_4\psi_4 + c_5\psi_5 + c_6\psi_6 \quad (14.16)$$

The lowest energy, most strongly bonding orbital has no internuclear nodes, and has the form[12]

$$\psi = \psi_1 + \psi_2 + \psi_3 + \psi_4 + \psi_5 + \psi_6$$

and is illustrated in Fig 14.32. It is strongly bonding because the constructive interference between neighbouring p orbitals results in a good accumulation of electron density between the nuclei (but slightly off the internuclear axis, as in the π bonds of diatomic molecules). The most antibonding orbital has the form

$$\psi = \psi_1 - \psi_2 + \psi_3 - \psi_4 + \psi_5 - \psi_6$$

The alternation of signs in the linear combination results in destructive interference between neighbours, and the molecular orbital has a nodal plane between each pair of neighbours, as shown in the illustration. The remaining four molecular orbitals

[12] We are ignoring normalization factors, for clarity. In this and the following case it would be $(\tfrac{1}{6})^{1/2}$ if we ignore overlap.

and they occupy the lowest three orbitals (Fig 14.33). The resulting electron distribution is like a double doughnut (Fig 14.34). It is an important feature of the configuration that the only molecular orbitals occupied have a net bonding character, for this is one contribution to the stability (in the sense of low energy) of the benzene molecule. It may be helpful to note the similarity between the molecular orbital energy level diagram for benzene and that for N_2 (see Fig 14.19): the strong

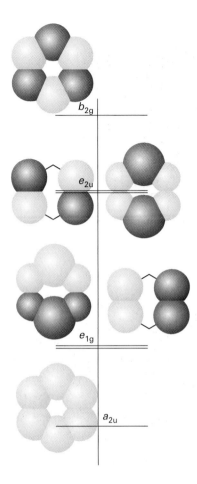

Fig 14.32 The π orbitals of benzene. The lowest energy orbital is fully bonding between neighbouring atoms but the uppermost orbital is fully antibonding. The two pairs of doubly degenerate molecular orbitals have an intermediate number of internuclear nodes. As usual, light and dark shading represents different signs of the wavefunction. The orbitals have opposite signs below the plane of the ring. The labels are a generalization of the σ and π labels used for linear molecules.

are more difficult to establish by qualitative arguments, but they have the form shown in Fig 14.32, and lie in energy between the most bonding and most antibonding orbitals. Note that the four intermediate orbitals form two doubly degenerate pairs, one net bonding and the other net antibonding.

We find the energies of the six π molecular orbitals in benzene by solving the Schrödinger equation; they are also shown in the molecular orbital energy level diagram. There are six electrons to be accommodated (one is supplied by each C atom),

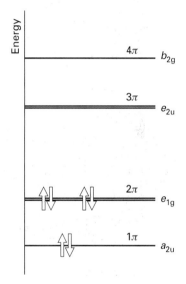

Fig 14.33 The π molecular orbital energy level diagram for benzene, and the configuration in its ground state.

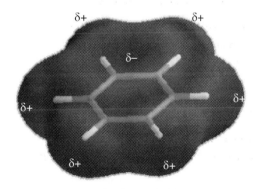

Fig 14.34 When six electrons occupy the three lowest orbitals, the resulting overall electron density is the same at each atom: this uniformity is represented by the electron density distribution that spreads round the ring.

bonding in benzene is echoed in the strong bonding in nitrogen.

A feature of the molecular orbital description of benzene is that each molecular orbital spreads either all round or partially round the C_6 ring. That is, π bonding is **delocalized**, and each electron pair helps to bind together several or all of the C atoms. The delocalization of bonding influence is a primary feature of molecular orbital theory and we shall encounter it in its extreme form when we come to consider the electronic structures of solids.

Exercises

14.1 Give the valence bond description of a P_2 molecule. Why is P_4 the stable form of molecular phosphorus?

14.2 Write down the valence bond wavefunction for a nitrogen molecule.

14.3 Calculate the molar energy of repulsion between two hydrogen nuclei at the separation in H_2 (74.1 pm). This is the energy that must be overcome by the attraction from the electrons that form the bond.

14.4 Give the valence bond description of SO_2 and SO_3 molecules.

14.5 Which of the following species are expected to be linear: (a) CO_2, (b) NO_2, (c) NO_2^+, (d) NO_2^-, (e) SO_2, (f) H_2O, (g) H_2O_2? Give reasons in each case.

14.6 Use the VSEPR model to predict the shapes of (a) H_2S, (b) SF_6, (c) XeF_4, (d) SF_4.

14.7 The structure of the visual pigment retinal is shown in (**3**). Label each atom with its state of hybridization and specify the composition of each of the different types of bond.

3 11-*cis*-retinal

14.8 Show that the orbitals $h_1 = s + p_x + p_y + p_z$ and $h_2 = s - p_x - p_y + p_z$ are mutually orthogonal, that is, have zero overlap. (*Hint*. The atomic orbitals are all mutually orthogonal and each one is individually normalized to 1.)

14.9 Show that the sp^2 hybrid orbital $(s + 2^{1/2}p)/3^{1/2}$ is normalized to 1 if the s and p orbitals are normalized to 1.

14.10 Find another sp^2 hybrid orbital that is orthogonal to (has zero overlap with) the hybrid orbital in the preceding problem.

14.11 Normalize the wavefunction $\psi = \psi_{cov} + \lambda\psi_{ion}$ in terms of the parameter λ and the overlap integral S between the covalent and ionic wavefunctions.

14.12 A normalized valence bond wavefunction turned out to have the form $\psi = 0.989\psi_{cov} + 0.150\psi_{ion}$. What is the chance that, in 1000 inspections of the molecule, both electrons of the bond will be found on one atom?

14.13 Benzene is commonly regarded as a resonance hybrid of the two Kekulé structures, but other possible structures can also contribute. Draw three other structures in which there are only covalent π bonds (allowing for bonding between some non-adjacent C atoms) and two structures in which there is one ionic bond. Why may these structures be ignored in simple descriptions of the molecule?

14.14 Show, if overlap is ignored, (a) that any molecular orbital expressed as a linear combination of two atomic orbitals may be written in the form $\psi = \psi_A \cos\theta + \psi_B \sin\theta$, where θ is a parameter that varies between 0 and $\frac{1}{2}\pi$, and (b) that if ψ_A and ψ_B are orthogonal and normalized to 1, then ψ is also normalized to 1. (c) To what values of θ do the bonding and antibonding orbitals in a homonuclear diatomic molecule correspond?

14.15 Draw diagrams to show the various orientations in which a p orbital and a d orbital on adjacent atoms may form bonding and antibonding molecular orbitals.

14.16 Give the ground-state electron configurations of (a) Li_2, (b) Be_2, and (c) C_2.

14.17 Give the ground-state electron configurations of (a) H_2^-, (b) N_2, and (c) O_2.

14.18 Three diatomic species, which are biologically important either because they promote or inhibit life, are (a) CO, (b) NO, and (c) CN⁻. The first binds to haemoglobin, the second is a neurotransmitter, and the third interrupts the electron-transfer chain. Their biochemical action is a reflection of their orbital structure. Deduce their ground-state electron configurations. For heteronuclear diatomic molecules, a good first approximation is that the energy level diagram is much the same as for homonuclear diatomic molecules.

14.19 From the ground-state electron configurations of B_2 and C_2, predict which molecule should have the greater dissociation energy.

14.20 Some chemical reactions proceed by the initial loss or transfer of an electron to a diatomic species. Which of the molecules N_2, NO, O_2, C_2, F_2, and CN would you expect to be stabilized by (a) the addition of an electron to form AB⁻, (b) the removal of an electron to form AB⁺?

14.21 The existence of compounds of the noble gases was once a great surprise and stimulated a great deal of theoretical work. Sketch the molecular orbital energy level diagram for XeF and deduce its ground-state electron configurations. Is XeF likely to have a shorter bond length than XeF⁺?

14.22 Where it is appropriate, give the parity of (a) $2\pi^*$ in F_2, (b) 3σ in NO, (c) 1δ in Tl_2, (d) $2\delta^*$ in Fe_2.

14.23 Give the parities of the first four levels of a particle-in-a-box wavefunction.

14.24 (a) Give the parities of the wavefunctions for the first four levels of a harmonic oscillator. (b) How may the parity be expressed in terms of the quantum number v?

14.25 State the parities of the six π orbitals of benzene (see Fig 14.32).

14.26 Two diatomic molecules that are important for the welfare of humanity are NO and N_2: the former is both a pollutant and a neurotransmitter, and the latter is the ultimate source of the nitrogen of proteins and other biomolecules. Use the electron configurations of NO and N_2 to predict which is likely to have the shorter bond length.

14.27 Put the following species in order of increasing bond length: F_2^-, F_2, F_2^+.

14.28 Construct the molecular orbital energy level diagrams of (a) ethene (ethylene) and (b) ethyne (acetylene) on the basis that the molecules are formed from the appropriately hybridized CH_2 or CH fragments.

14.29 Predict the electronic configurations of (a) the benzene anion, (b) the benzene cation.

14.30 Many of the colours of vegetation are due to electronic transitions in conjugated π-electron systems. In the *free-electron molecular orbital* (FEMO) theory, the electrons in a conjugated molecule are treated as independent particles in a box of length L. Sketch the form of the two occupied orbitals in butadiene predicted by this model and predict the minimum excitation energy of the molecule. The tetraene $CH_2=CHCH=CHCH=CHCH=CH_2$ can be treated as a box of length $8R$, where $R = 140$ pm (as in this case, an extra half bond-length is often added at each end of the box). Calculate the minimum excitation energy of the molecule and sketch the HOMO and LUMO.

Metallic and ionic solids

Contents

MODERN chemistry is closely concerned with the properties of solids. Apart from their intrinsic usefulness for construction, modern solids have made possible the semiconductor revolution and recent advances in ceramics have given rise to the hope that we may now be on the verge of a super-conductor revolution. Solids are used widely in the chemical industry as catalysts, where the details of their action often depend on the details of their electronic structure, particularly at the surfaces where the reaction takes place. Advances in our understanding of electron mobility in solids are also useful in biology, where electron transport processes are responsible for many biochemical processes, particularly photosynthesis and respiration.

The principal technique for investigating the arrangements of atoms in condensed phases, primarily crystalline solids, is X-ray diffraction. Information from X-ray diffraction is the basis of much of molecular biology, so the material presented here is the foundation for our discussion of biomolecular structures in Chapter 16. In each case, the observed crystal structure is Nature's solution to the problem of condensing objects of various shapes into an aggregate of minimum energy and, for temperatures above zero, of minimum Gibbs energy.

Bonding in solids

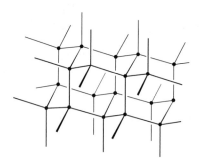

THE bonding within a solid may be of various kinds. Simplest of all (in principle) is the bonding in a **metallic solid**, in which electrons are delocalized over arrays of identical cations and bind the whole together into a rigid but malleable structure. Because the delocalized electrons can accommodate bonding patterns with very little directional character, the crystal structures of metals are determined largely by the geometrical problem of packing spherical atoms into a dense, orderly array. In an **ionic solid**, the ions (in general, of different radii, and not always spherical) are held together by their Coulombic interaction, and pack together to give an electrically neutral structure. In a **covalent solid** (or *network solid*), covalent bonds in a definite spatial orientation link the atoms in a network extending through the crystal. The stereochemical demands of valence now override the geometrical problem of packing spheres together, and elaborate and extensive structures may be formed. A famous example of a covalent solid is diamond (Fig 15.1), in which each sp^3-hybridized carbon is bonded tetrahedrally by σ bonds to its four neighbours. Covalent solids are often hard and unreactive. **Molecular solids**, which are the subject of the overwhelming majority of modern structural determinations, consist of discrete molecules attracted to one another by the interactions described in Chapter 16.

Some solids—notably the metals—conduct electricity because they have mobile electrons. These **electronic conductors** are classified on the basis of the variation of their electrical conductivity with temperature (Fig 15.2). A **metallic conductor** is an electronic conductor with a conductivity that *decreases* as the temperature is raised. Metallic conductors include the metallic elements, their alloys, and graphite. Some organic solids are metallic conductors. A **semiconductor** is an electronic conductor with a conductivity that *increases* as the temperature is raised. Semiconductors include silicon, diamond, and gallium arsenide. A semiconductor generally

Fig 15.1 A fragment of the structure of diamond. Each C atom is tetrahedrally bonded to four neighbours. This framework-like structure results in a rigid crystal with a high thermal conductivity.

has a lower conductivity than that typical of metals, but the magnitude of the conductivity is not relevant to the distinction. It is conventional to classify substances with very low electrical conductivities, such as most ionic solids, as **insulators**. We shall use this term, but it is one of convenience rather than one of fundamental significance. We shall not consider **superconductors**, which are substances that conduct electricity with zero resistance.

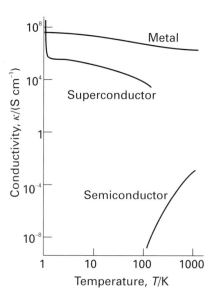

Fig 15.2 The typical variation with temperature of the electrical conductivities of different classes of electronic conductor.

15.1 The band theory of solids

Metallic and ionic solids can both be treated by molecular orbital theory. The advantage of that approach is that we can then see them as two extremes of a single kind of solid. In each case, the electrons responsible for the bonding are delocalized throughout the solid (like in benzene, but on a much bigger scale). In a metal, the electrons can be found on all the atoms with equal probability, which matches the elementary picture of a metal as consisting of cations embedded in a nearly uniform electron 'sea'. In an ionic solid the wavefunctions occupied by the delocalized electrons are almost entirely concentrated on the anions.

We shall consider initially a single, infinitely long line of identical atoms, each one having one s orbital available for forming molecular orbitals (as in sodium). One atom of the solid contributes one s orbital at a certain energy (Fig 15.3a). When a second atom is brought up it forms a bonding and antibonding orbital (Fig 15.3b). The orbital of the third atom overlaps its nearest neighbour (and only slightly the next-nearest), and three molecular orbitals are formed from these three atomic orbitals (Fig 15.3c). The fourth atom leads to the formation of a fourth molecular orbital (Fig 15.3d). At this stage we can begin to see that the general effect of bringing up successive atoms is to spread the range of energies covered by the molecular orbitals, and also to fill in the range of energies with more and more orbitals (one more for each additional atom). When N atoms have been added to the line, there are N molecular orbitals covering a band of finite width. When N is infinitely large, the difference between neighbouring energy levels in the band is infinitely small, but the band still has finite width. This band consists of N different molecular orbitals, the lowest-energy orbital being fully bonding, and the highest-energy orbital being fully antibonding between adjacent atoms (Fig 15.4).

A band formed from overlap of s orbitals is called an **s band**. If the atoms have p orbitals available, then the same procedure leads to a **p band** (as in the

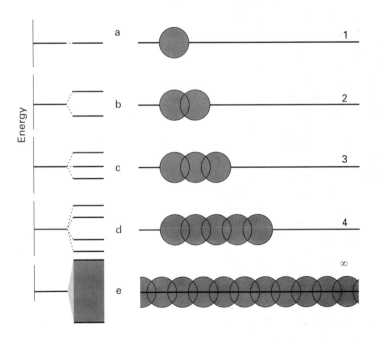

Fig 15.3 The formation of a band of N molecular orbitals by successive addition of N atoms to a line. Note that the band remains of finite width and, although it looks continuous when N is large, it consists of N different orbitals.

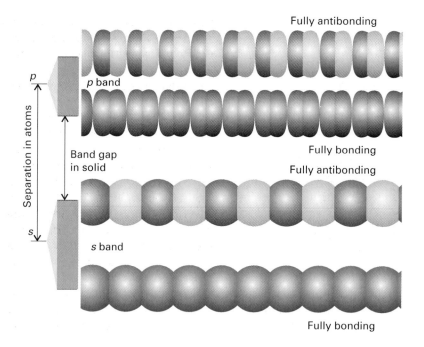

Fig 15.4 The overlap of *s* orbitals gives rise to an *s* band, and the overlap of *p* orbitals gives rise to a *p* band. In this case the *s* and *p* orbitals of the atoms are so widely spaced that there is a band gap. In many cases the separation is less, and the bands overlap.

upper half of Fig 15.4). If the atomic *p* orbitals lie higher in energy than the s orbitals, then the *p* band lies higher than the *s* band, and there may be a **band gap**, a range of energies for which no molecular orbitals exist.

15.2 The occupation of bands

Now consider the electronic structure of a solid formed from atoms each of which is able to contribute one electron (for example, the alkali metals). There are N atomic orbitals and therefore N molecular orbitals squashed into an apparently continuous band. There are N electrons to accommodate, and they enter the lowest $\frac{1}{2}N$ molecular orbitals (Fig 15.5). The highest occupied molecular orbital is called the **Fermi level**. However, unlike in the discrete molecules we considered in Chapter 14, there are empty orbitals just above and very close in energy to the Fermi level, so it requires hardly any energy to excite the uppermost electrons. Some of

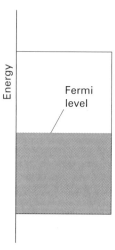

Fig 15.5 When N electrons occupy a band of N orbitals, it is only half full and the electrons near the Fermi level (the top of the filled levels) are mobile.

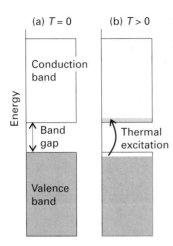

Fig 15.6 (a) When 2*N* electrons are present, the band is full and the material is an insulator at *T* = 0. (b) At temperatures above *T* = 0, electrons populate the levels of the conduction band at the expense of the valence band, and the solid is a semiconductor.

the electrons are therefore very mobile, and give rise to electrical conductivity.

An increase in temperature causes more vigorous thermal motion of the atoms, with the result that there are more collisions between the moving electrons and the atoms. That is, at high temperatures the electrons are scattered out of their paths through the solid and are less efficient at transporting charge.

When each atom provides two electrons, the 2*N* electrons fill the *N* orbitals of the *s* band. The Fermi level now lies at the top of the band and there is a gap before the next band begins (Fig 15.6a). As the temperature is increased, electrons can populate the empty orbitals of the upper band (Fig 15.6b). They are now mobile, and the solid has become an electronic conductor. In fact, it is a *semiconductor*, because the electrical conductivity depends on the number of electrons that are promoted across the gap and that number increases as the temperature is raised.

If the gap is large, very few electrons will be excited across it at ordinary temperatures, and the conductivity will remain close to zero, giving an insulator. Thus, the conventional distinction between

an insulator and a semiconductor is related to the size of the band gap and is not absolute like the distinction between a metal (incomplete bands at *T* = 0) and a semiconductor (full bands at *T* = 0).

Another method of increasing the number of charge carriers and enhancing the semiconductivity of a solid is to implant foreign atoms into an otherwise pure material. If these **dopants** can trap electrons (as indium or gallium atoms can in silicon), then they withdraw electrons from the filled band, leaving holes which allow the remaining electrons to move (Fig 15.7a). This doping procedure gives rise to **p-type semiconductivity**, the p indicating that the holes are positive relative to the electrons in the band. Alternatively, a dopant might carry excess electrons (for example, phosphorus atoms introduced into germanium), and these additional electrons occupy otherwise empty bands, giving **n-type semiconductivity** (Fig 15.7b), where n denotes the negative charge of the carriers.[1]

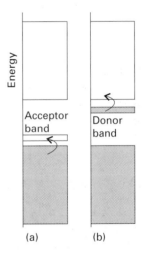

Fig 15.7 (a) A dopant with fewer electrons than its host can form a narrow band that accepts electrons from the valence band. The holes in the valence band are mobile, and the substance is a *p-type semiconductor*. (b) A dopant with more electrons than its host forms a narrow band that can supply electrons to the conduction band. The electrons it supplies are mobile, and the substance is an *n-type semiconductor*.

---

[1] The preparation of doped but otherwise ultrapure materials was described in Section 6.11.

15.3 The ionic model of bonding

Suppose we have a line of atoms with different electronegativities, such as a line of sodium and chlorine atoms, rather than the identical atoms treated so far. For simplicity, we suppose that each atom contributes an *s* orbital and one electron. That is the case with each sodium atom in sodium chloride. Although each chlorine atom in sodium chloride contributes a *p* orbital and its one electron, treating the orbital as an *s* orbital simplifies the discussion without undermining it in principle.

We use the *s* orbitals to build molecular orbitals spreading throughout the solid. Now, though, there is a crucial difference. The *s* orbitals on the two types of atom have markedly different energies, so (just as in the construction of molecular orbitals for diatomic molecules, Section 14.12) we consider them separately. The Cl3*s* orbitals form one band and the higher energy Na3*s* orbitals form another band. However, because the sodium atoms have very little overlap with one another (they are separated by a chlorine atom), the Na3*s* band is

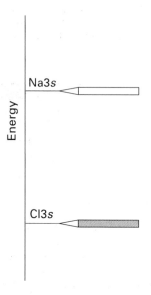

very narrow; so is the Cl3*s* band, for a similar reason. As a result, there is a big gap between two narrow bands (Fig 15.8).

Now consider the occupation of the bands. If there are *N* sodium atoms and *N* chlorine atoms, there will be 2*N* electrons to accommodate. These electrons occupy and fill the lower Cl3*s* band. As a result of the big band gap before the Na3*s* band becomes available, the substance is an insulator. Moreover, because only the Cl3*s* band is occupied, the electron density is almost entirely on the chlorine atoms. In other words, we can treat the solid as composed of Na^+ cations and Cl^- anions, just as in an elementary picture of ionic bonding.

Now that we know where the electron density is largely located, we can adopt a much simpler model of the solid. Instead of expressing the structure in terms of molecular orbitals, we treat it as a collection of cations and anions. This simplification is the basis of the **ionic model** of bonding.

15.4 Lattice enthalpy

The strength of a covalent bond is measured by its dissociation energy, the energy needed to separate the two atoms joined by the bond. For thermodynamic applications we express this energy in terms of the bond enthalpy. The strength of an ionic bond is measured similarly, but now we have to take into account the energy required to separate *all* the ions of a solid sample from one another and, for thermodynamic applications, express this energy as an enthalpy change. The **lattice enthalpy**, $\Delta H_L^\ominus$, is the standard enthalpy change accompanying the separation of the species that compose the solid (such as ions if the solid is ionic, and molecules if the solid is molecular). For example, the lattice enthalpy of an ionic solid such as sodium chloride is the standard molar enthalpy change accompanying the process[2,3]

$$NaCl(s) \rightarrow Na^+(g) + Cl^-(g) \quad \Delta H_L^\ominus = 786 \text{ kJ mol}^{-1}$$

----

[2] Because the lattice enthalpy is invariably a positive quantity, it is normally reported without its + sign.

[3] The lattice enthalpy of a molecular solid, such as ice, is the standard molar enthalpy of sublimation; the lattice enthalpy of a metal is its enthalpy of atomization.

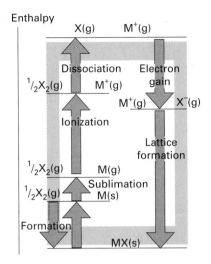

Fig 15.9 The Born–Haber cycle for the determination of one of the unknown enthalpies, most commonly the lattice enthalpy. Upward pointing arrows denote positive changes in enthalpy; downward pointing arrows denote negative enthalpy changes. All the steps in the cycle correspond to the same temperature.

Lattice enthalpies of solids are determined from other experimental data by using a **Born–Haber cycle**, which is a cycle (a closed path) of steps that includes lattice formation as one stage. The value of the lattice enthalpy—the only unknown in a well-chosen cycle—is found from the requirement that the sum of the enthalpy changes round a complete cycle is zero (because enthalpy is a state property).[4] A typical cycle for an ionic compound has the form shown in Fig 15.9. The following example illustrates how the cycle is used and Table 15.1 gives characteristic values.

> **Example 15.1** *Using a Born–Haber cycle to determine a lattice enthalpy*
>
> Calculate the lattice enthalpy of KCl(s) using a Born–Haber cycle and the information given on the right, which is for 25°C.

----

[4] All the data must be for a single temperature.

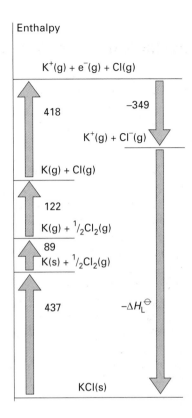

Fig 15.10 The Born–Haber cycle for the calculation of the lattice enthalpy of potassium chloride. The sum of the enthalpy changes around the cycle is zero. The numerical values are in kilojoules per mole.

Step	$\Delta H^{\ominus}/(\text{kJ mol}^{-1})$
Sublimation of K(s)	+89
Ionization of K(g)	+418
Dissociation of Cl_2(g)	+244
Electron attachment to Cl(g)	−349
Formation of KCl(s)	−437

Strategy First, draw the cycle, showing the atomization of the elements, their ionization, and the formation of the solid lattice; then complete the cycle (for the step *solid compound → original elements*) by using the enthalpy of formation. The sum of enthalpy changes round the cycle is zero, so include the numerical data and set the sum of all the terms equal to zero;

then solve the equation for the one unknown (the lattice enthalpy).

Solution Figure 15.10 shows the cycle required. The first step is the sublimation (atomization) of solid potassium:

$$\Delta H^{\ominus}/(\text{kJ mol}^{-1})$$

$K(s) \rightarrow K(g)$ +89

(the enthalpy of sublimation or atomization of potassium)

Chlorine atoms are formed by dissociation of Cl_2:

$\frac{1}{2}Cl_2(g) \rightarrow Cl(g)$ +122

(half the dissociation enthalpy of Cl_2)

Now, potassium ions are formed by ionization of the gas-phase atoms:

$K(g) \rightarrow K^+(g) + e^-(g)$ +418

(the ionization enthalpy of potassium)

and chloride ions are formed from the chlorine atoms:

$Cl(g) + e^-(g) \rightarrow Cl^-(g)$ −349

(the electron-gain enthalpy of chlorine)

The solid is now formed:

$K^+(g) + Cl^-(g) \rightarrow KCl(s)$ $-\Delta H_L^{\ominus}$

(the enthalpy change when the lattice *forms* is the negative of the lattice enthalpy)

and the cycle is completed by decomposing KCl(s) into its elements:

$KCl(s) \rightarrow K(s) + \frac{1}{2}Cl_2(g)$ +437

(the negative of the enthalpy of formation of KCl)

The sum of the enthalpy changes is $-\Delta H_L^{\ominus} + 717$ kJ mol^{-1}; however, the sum must be equal to zero, so $\Delta H_L^{\ominus} = 717$ kJ mol^{-1}.

Self-test 15.1

Calculate the lattice enthalpy of magnesium bromide from the data below and the information in Appendix 1.

Step	$\Delta H^{\ominus}/(\text{kJ mol}^{-1})$
Sublimation of Mg(s)	+148
Ionization of Mg(g) to Mg^{2+}(g)	+2187
Dissociation of Br$_2$(g)	+193
Electron attachment to Br(g)	−325

[*Answer:* 2433 kJ mol^{-1}]

15.5 Coulombic contributions to lattice enthalpies

Our next task is to account for the values of lattice enthalpies. The dominant interaction in an ionic lattice is the Coulombic interaction between ions, which is far stronger than any other attractive interaction, so we concentrate on that.

The starting point is the Coulombic potential energy for the interaction of two ions of charge numbers z_1 and z_2 (with cations having positive charge numbers and anions negative charge numbers) separated by a distance r_{12}:

$$V_{12} = \frac{(z_1 e) \times (z_2 e)}{4 \pi \varepsilon_0 r_{12}} \tag{15.1}$$

where ε_0 is the vacuum permittivity (see inside front cover). To calculate the total potential energy

Table 15.1 *Lattice enthalpies, $\Delta H_L^{\ominus}/(\text{kJ mol}^{-1})$*

LiF	1037	LiCl	852	LiBr	815	LiI	761
NaF	926	NaCl	786	NaBr	752	NaI	705
KF	821	KCl	717	KBr	689	KI	649
MgO	3850	CaO	3461	SrO	3283	BaO	3114
MgS	3406	CaS	3119	SrS	2974	BaS	2832
Al$_2$O$_3$	15.9×10^3						

of all the ions in a crystal, we have to sum this expression over all the pairs of ions in the solid. Nearest neighbours (which have opposite signs) attract and contribute a large negative term, second-nearest neighbours (which have the same sign) repel and contribute a slightly weaker positive term, and so on (Fig 15.11). The overall result, however, is that there is a net attraction between the cations and anions and a favourable (negative) contribution to the energy of the solid. It is instructive to return to our earlier one-dimensional model and do the calculation for a uniformly spaced line of alternating cations and anions for which $z_1 = +z$ and $z_2 = -z$, with d the distance between adjacent ions.

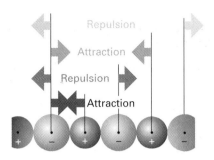

Fig 15.11 There are alternating positive and negative contributions to the potential energy of a crystal lattice on account of the repulsions between ions of like charge and attractions of ions of opposite charge. The total potential energy is negative, but the sum might converge quite slowly.

Derivation 15.1 *The lattice energy of a one-dimensional crystal*

Consider a line of alternating cations and anions extending in an infinite direction to the left and right of the ion of interest. The Coulombic potential energy of interaction with the ions on the right is the following sum of terms, where the negative terms represent attractions between ions of charge opposite to that of the ion of interest and the positive terms represent repulsions between ions of the same charge:

$$V = \frac{1}{4\pi\varepsilon_0} \times \left(-\frac{z^2 e^2}{d} + \frac{z^2 e^2}{2d} - \frac{z^2 e^2}{3d} + \frac{z^2 e^2}{4d} - \cdots \right)$$

$$= -\frac{z^2 e^2}{4\pi\varepsilon_0 d} \times \left(1 - \tfrac{1}{2} + \tfrac{1}{3} - \tfrac{1}{4} + \cdots \right) = -\frac{z^2 e^2}{4\pi\varepsilon_0 d} \times \ln 2$$

In the last step we have used

$$1 - \tfrac{1}{2} + \tfrac{1}{3} - \tfrac{1}{4} + \cdots = \ln 2$$

The interaction of the ion of interest with the ions to its left is the same, so the total potential energy of interaction is twice this expression

$$V = -\frac{z^2 e^2}{4\pi\varepsilon_0 d} \times 2\ln 2 \tag{15.2}$$

The negative sign in eqn 15.2 tells us that the potential energy of the ion is lower in the crystal than in a gas of widely separated ions. Although this calculation is based on an unrealistic model of a crystal, it is already showing the features we are seeking to explain: the energy of the ion is greatly lowered

if the charge numbers (z) of the ions are high and their diameters are small (so that d is small).

When the calculation is repeated for more realistic, three-dimensional arrays of ions it is also found that the potential energy depends on the charge numbers of the ions and the value of a single parameter d, which may be taken as the distance between the centres of nearest neighbours:

$$V = \frac{e^2}{4\pi\varepsilon_0} \times \frac{z_1 z_2}{d} \times A \tag{15.3}$$

where A is a number called the **Madelung constant**. Because the charge number of cations is positive and that of anions is negative, the product $z_1 z_2$ is negative. Therefore, V is also negative, which corresponds to a lowering in potential energy relative to the gas of widely separated ions. The value of the Madelung constant for a single line of ions is $2 \ln 2 = 1.386 \ldots$ as we have already seen. Table 15.2 gives the computed values of the Madelung constant for a variety of lattices with structures that we describe later in the chapter.

Table 15.2 *Madelung constants*

Structural type	A
Caesium chloride	1.763
Fluorite	2.519
Rock salt	1.748
Rutile	2.408

So far, we have considered only the Coulombic interaction between ions. However, regardless of their signs, the ions repel each other when they are pressed together and their wavefunctions overlap. These additional repulsions work against the net Coulombic attraction between ions, so they raise the energy of the solid. When their effect is taken into account, it turns out that the lattice enthalpy is given by the **Born–Mayer equation**:

$$\Delta H_L^{\ominus} = |z_1 z_2| \times \frac{N_A e^2}{4\pi\varepsilon_0 d} \times \left(1 - \frac{d^*}{d}\right) \times A \quad (15.4)$$

where d^* is an empirical parameter that is often taken as 34.5 pm (simply because that value is found to give reasonable agreement with experiment). The modulus signs ($|\ldots|$) mean that we should remove any minus sign from the product of z_1 and z_2, which results in a positive value for the lattice enthalpy. The important feature of this expression is that it shows that $\Delta H_L^{\ominus} \propto |z_1 z_2|/d$, which implies that *the lattice enthalpy increases with increasing charge number of the ions and with decreasing ionic radius*. The second conclusion follows from the fact that the smaller the ionic radii, the smaller the value of d. This feature is in accord with the variation in the experimental values in Table 15.1.

Self-test 15.2

Which can be expected to have the greater lattice enthalpy, magnesium oxide or strontium oxide?

[*Answer*: MgO]

Crystal structure

Now we turn to the structures adopted by atoms and ions when they stack together to give a crystalline solid. The structures of crystals are of considerable practical importance, for they have implications for geology, materials, technologically advanced materials such as semiconductors and high-temperature superconductors, and biology. The first, and often very demanding, step in an X-ray structural analysis of biological macromolecules is

to form crystals in which the large molecules lie in orderly ranks. On the other hand, the crystallization of a virus particle would take it out of circulation, and one of the strategies adopted by viruses for avoiding this kind of entombment makes unconscious use of the geometry of crystal packing.

15.6 Unit cells

The pattern that atoms, ions, or molecules adopt in a crystal is expressed in terms of an array of points making up the **lattice** that identify the locations of the individual species (Fig 15.12). A **unit cell** of a

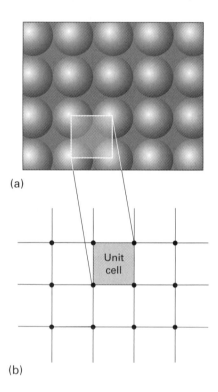

(a)

(b)

Fig 15.12 (a) A crystal consists of a uniform array of atoms, molecules, or ions, as represented by these spheres. In many cases, the components of the crystal are far from spherical, but this diagram illustrates the general idea. (b) The location of each atom, molecule, or ion can be represented by a single point; here (for convenience only), the locations are denoted by a point at the centre of the sphere. The unit cell, which is shown shaded, is the smallest block from which the entire array of points can be constructed without rotating or otherwise modifying the block.

crystal is the small three-dimensional figure obtained by joining typically eight of these points, and which may be used to construct the entire crystal lattice by purely translational displacements, much as a wall may be constructed from bricks (Fig 15.13). An infinite number of different unit cells can describe the same structure, but it is conventional to choose the cell with sides that have the shortest lengths and are most nearly perpendicular to one another.

Unit cells are classified into one of seven **crystal systems** according to the symmetry they possess under rotations about different axes. The *cubic system*, for example, has four threefold axes (Fig 15.14). A threefold axis is an axis of a rotation that restores the unit cell to the same appearance three times during a complete revolution, after rotations through 120°, 240°, and 360°. The four axes make the tetrahedral angle to each other. The *monoclinic system* has one twofold axis (Fig 15.15). A twofold axis is an axis of a rotation that leaves the cell apparently unchanged twice during a complete revolution, after rotations through 180° and 360°. The **essential symmetries**, the properties that must be present for the unit cell to belong to a particular system, are listed in Table 15.3.

A unit cell may have lattice points other than at its corners, so each crystal system can occur in a number of different varieties. For example, in some cases points may occur on the faces and in the body of the cell without destroying the cell's essential symmetry. These various possibilities give rise to fourteen distinct types of unit cell, which define the **Bravais lattices** (Fig 15.16).

15.7 **The identification of crystal planes**

The identification of the type of unit cell specifies the internal symmetry of the crystal. To specify a unit cell fully, we also need to know its size, such as the lengths of its sides. There is a useful relation between the spacing of the planes passing through the lattice points, which (as we shall see) we can measure, and the lengths we need to know.

Because two-dimensional arrays of points are easier to visualize than three-dimensional arrays, we

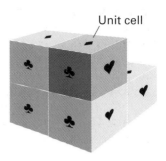

Fig 15.13 A unit cell, here shown in three dimensions, is like a brick used to construct a wall. Once again, only pure translations are allowed in the construction of the crystal. (Some bonding patterns for actual walls use rotations of bricks, so for these patterns a single brick is not a unit cell.)

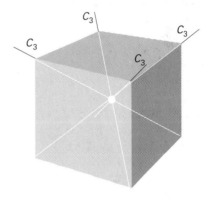

Fig 15.14 A unit cell belonging to the cubic system has four threefold axes (denoted C_3) arranged tetrahedrally.

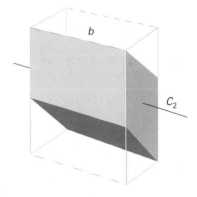

Fig 15.15 A unit cell belonging to the monoclinic system has one twofold (denoted C_2) axis (along b).

Table 15.3 *The essential symmetries of the seven crystal systems*

The systems	Essential symmetries
Triclinic	None
Monoclinic	One twofold axis
Orthorhombic	Three perpendicular twofold axes
Rhombohedral	One threefold axis
Tetragonal	One fourfold axis
Hexagonal	One sixfold axis
Cubic	Four threefold axes in a tetrahedral arrangement

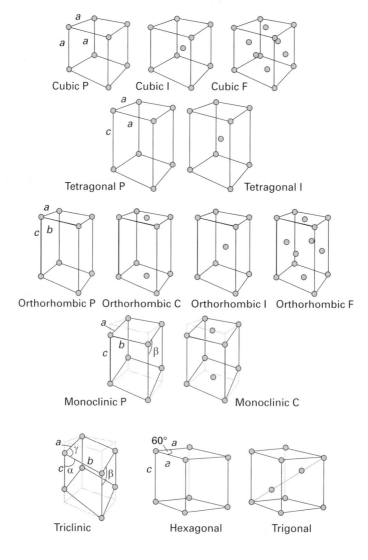

Fig 15.16 The fourteen Bravais lattices. The letter P denotes a primitive unit cell, I a body-centred unit cell, F a face-centred unit cell, and C (or A or B) a cell with lattice points on two opposite faces.

shall introduce the concepts we need by referring to two dimensions initially, and then extend the conclusions to three dimensions. Consider the two-dimensional rectangular lattice formed from a rectangular unit cell of sides a and b (Fig 15.17). We can distinguish the four sets of planes shown in the illustration by the distances at which they intersect the axes. One way of labelling the planes would therefore be to denote each set by the smallest intersection distances. For example, we could denote the four sets in the illustration as $(1a,1b)$, $(3a,2b)$, $(-1a,1b)$, and $(\infty a,1b)$. If, however, we agreed always to quote distances along the axes as multiples of the lengths of the unit cell, then we could omit the a and b and label the planes more simply as $(1,1)$, $(3,2)$, $(-1,1)$, and $(\infty,1)$.

Now let's suppose that the array in Fig 15.17 is the top view of a three-dimensional rectangular lattice in which the unit cell has a length c in the z direction. All four sets of planes intersect the z-axis at infinity, so the full labels of the sets of planes of lattice points are $(1,1,\infty)$, $(3,2,\infty)$, $(-1,1,\infty)$, and $(\infty,1,\infty)$.

The presence of infinity in the labels is inconvenient. We can eliminate it by taking the reciprocals of the numbers in the labels; this step also turns out to have further advantages, as we shall see. The resulting **Miller indices**, (hkl), are the reciprocals of the numbers in the parentheses with fractions cleared. For example, the $(1,1,\infty)$ planes in Fig 15.17 are the (110) planes in the Miller notation. Similarly, the $(3,2,\infty)$ planes become first $(\frac{1}{3},\frac{1}{2},0)$ when reciprocals are formed, and then $(2,3,0)$ when fractions are cleared by multiplication through by 6, so they are referred to as the (230) planes. We write negative indices with a bar over the number: Fig 15.17c shows the $(\bar{1}10)$ planes. Figure 15.18 shows some planes in three dimensions, including an example of a lattice with axes that are not mutually perpendicular.

Self-test 15.3

A representative member of a set of planes in a crystal intersects the axes at $3a$, $2b$, and $2c$; what are the Miller indices of the planes?

[*Answer:* (233)]

It is helpful to keep in mind the fact, as illustrated in Fig 15.17, that the smaller the value of h in the Miller index (hkl), the more nearly parallel the plane is to the a-axis. The same is true of k and the b-axis and l and the c-axis. When $h = 0$, the planes intersect the a-axis at infinity, so the $(0kl)$ planes are parallel to the a-axis. Similarly, the $(h0l)$ planes are parallel to b and the $(hk0)$ planes are parallel to c.

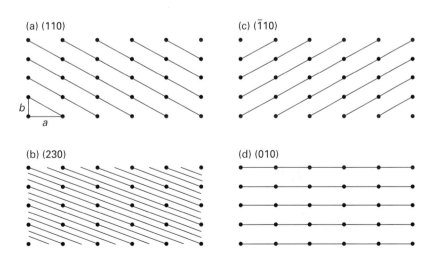

Fig 15.17 Some of the planes that can be drawn through the points of the space lattice and their corresponding Miller indices (hkl).

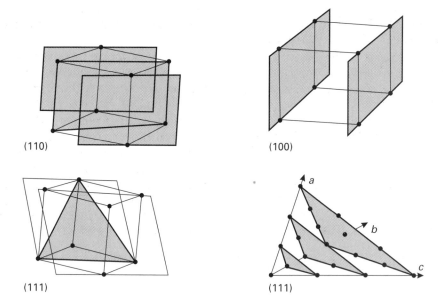

Fig 15.18 Some representative planes in three dimensions and their Miller indices. Note that a 0 indicates that a plane is parallel to the corresponding axis, and that the indexing may also be used for unit cells with nonorthogonal axes.

The Miller indices are very useful for calculating the separation of planes. For instance, they can be used to derive a very simple expression for the separation, d, of the (hkl) planes.

In three dimensions, this expression becomes

$$\frac{1}{d^2} = \frac{h^2}{a^2} + \frac{k^2}{b^2} + \frac{l^2}{c^2} \tag{15.5}$$

Derivation 15.2 *The separation of lattice planes*

Consider the ($hk0$) planes of a rectangular lattice built from an orthorhombic unit cell of sides of lengths a and b (Fig 15.19). We can write the following trigonometric expressions for the angle ϕ shown in the illustration:

$$\sin\phi = \frac{d}{(a/h)} = \frac{hd}{a} \quad \cos\phi = \frac{d}{(b/k)} = \frac{kd}{b}$$

Then, because $\sin^2\phi + \cos^2\phi = 1$, we obtain

$$\frac{h^2 d^2}{a^2} + \frac{k^2 d^2}{b^2} = 1$$

which we can rearrange into

$$\frac{1}{d^2} = \frac{h^2}{a^2} + \frac{k^2}{b^2}$$

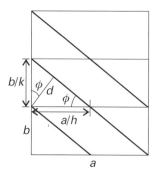

Fig 15.19 The geometrical construction used to relate the separation of planes to the dimensions of the unit cell.

Example 15.2 *Using the Miller indices*

Calculate the separation of (a) the (123) planes and (b) the (246) planes of an orthorhombic cell with $a = 0.82$ nm, $b = 0.94$ nm, and $c = 0.75$ nm.

Strategy For the first part, we simply substitute the information into eqn 15.5. For the second part, instead of repeating the calculation, we should examine how d in eqn 15.5 changes when all three Miller indices are multiplied by 2 (or by a more general factor, n).

Solution Substituting the data into eqn 15.5 gives

$$\frac{1}{d^2} = \frac{1^2}{(0.82\,\text{nm})^2} + \frac{2^2}{(0.94\,\text{nm})^2} + \frac{3^2}{(0.75\,\text{nm})^2} = \frac{22}{\text{nm}^2}$$

It follows that $d = 0.21$ nm. When the indices are all increased by a factor of 2, the separation becomes

$$\frac{1}{d^2} = \frac{(2 \times 1)^2}{(0.82\,\text{nm})^2} + \frac{(2 \times 2)^2}{(0.94\,\text{nm})^2} + \frac{(2 \times 3)^2}{(0.75\,\text{nm})^2} = 4 \times \frac{22}{\text{nm}^2}$$

So, for these planes $d = 0.11$ nm. In general, increasing the indices uniformly by a factor n decreases the separation of the planes by n.

Self-test 15.4

Calculate the separation of the (133) and (399) planes in the same lattice.

[*Answer*: 0.19 nm, 0.063 nm]

15.8 The determination of structure

One of the most important techniques for the determination of the structures of crystals is **X-ray diffraction**. In its simplest form, the technique is used to identify the lattice type and the separation of the planes of lattice points (and hence the distance between the centres of atoms and ions). In its most sophisticated version, X-ray diffraction provides detailed information about the location of all the atoms in molecules as complicated as proteins. The current considerable success of modern molecular biology has stemmed from X-ray diffraction techniques that have grown in sensitivity and scope as computing techniques have become more powerful. Here we concentrate on the

principles of the technique and illustrate how it may be used to determine the spacing of atoms in a crystal.

A characteristic property of waves is that they **interfere** with one another, which means that they give a greater amplitude where their displacements add and a smaller amplitude where their displacements subtract (Fig 15.20). Because the intensity of electromagnetic radiation is proportional to the square of the amplitude of the waves, the regions of constructive and destructive interference show up as regions of enhanced and diminished intensities. The phenomenon of **diffraction** is the interference caused by an object in the path of waves, and the pattern of varying intensity that results is called the **diffraction pattern**

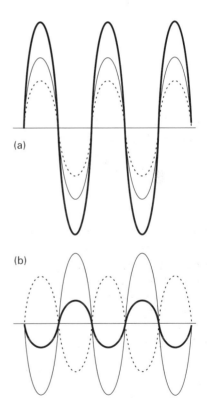

(a)

(b)

Fig 15.20 When two waves (drawn as thin lines) are in the same region of space they interfere. Depending on their relative phase, they may interfere (a) constructively, to give an enhanced amplitude, or (b) destructively, to give a smaller amplitude.

(Fig 15.21).[5] Diffraction occurs when the dimensions of the diffracting object are comparable to the wavelength of the radiation. Sound waves, with wavelengths of the order of 1 m, are diffracted by macroscopic objects. Light waves, with wavelengths of the order of 500 nm, are diffracted by narrow slits.

X-rays have wavelengths comparable to bond lengths in molecules and the spacing of atoms in crystals (about 100 pm), so they are diffracted by them. By analysing the diffraction pattern, it is possible to draw up a detailed picture of the location of atoms. Electrons moving at about 2×10^4 km s^{-1} (after acceleration through about 4 kV) have wavelengths of about 20 pm (recall Example 12.2), and may also be diffracted by molecules. Neutrons generated in a nuclear reactor, and then slowed to thermal velocities, have similar wavelengths and may also be used for diffraction studies.

The short-wavelength electromagnetic radiation we call X-rays is produced by bombarding a metal with high-energy electrons. The electrons decelerate as they plunge into the metal and generate radiation with a continuous range of wavelengths. This radiation is called **bremsstrahlung**.[6] Superimposed on the continuum are a few high-intensity, sharp peaks. These peaks arise from the interaction of the incoming electrons with the electrons in the inner shells of the atoms. A collision expels an electron (Fig 15.22), and an electron of higher energy drops into the vacancy, emitting the excess energy as an X-ray photon. An example of the process is the expulsion of an electron from the K shell (the shell with $n = 1$) of a copper atom, followed by the transition of an outer electron into the vacancy. The energy so released gives rise to copper's K_α radiation of wavelength 154 pm.

In 1923, the German physicist Max von Laue suggested that X-rays might be diffracted when passed through a crystal, for the wavelengths of X-rays are comparable to the separation of atoms. Laue's suggestion was confirmed almost immediately by Walter Friedrich and Paul Knipping, and then developed by William and Lawrence Bragg, who later jointly received the Nobel Prize. It has grown since then into a technique of extraordinary power.

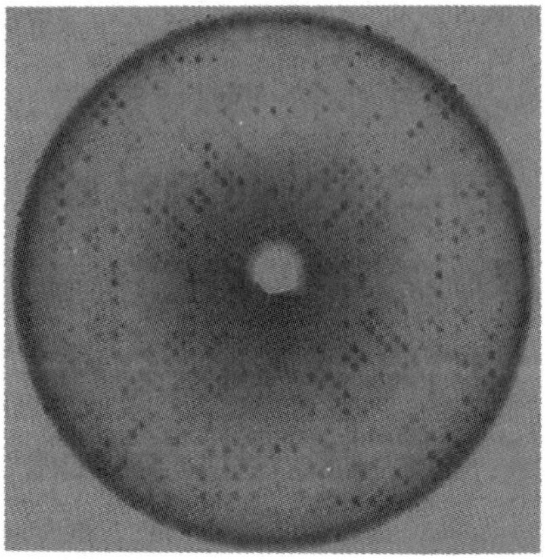

Fig 15.21 A typical diffraction pattern obtained in a version of the X-ray diffraction technique. The black dots are the reflections, the points of maximum constructive interference, that are used to determine the structure of the crystal.

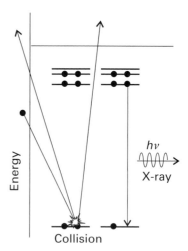

Fig 15.22 The formation of X-rays. When a metal is subjected to a high-energy electron beam, an electron in an inner shell of an atom is ejected. When an electron falls into the vacated orbital from an orbital of much higher energy, the excess energy is released as an X-ray photon.

---

[5] We first encountered diffraction in Section 12.4 in connection with the wave properties of electrons.

[6] *Bremse* is German for brake, *Strahlung* for ray.

15.9 The Bragg law

The earliest approach to the analysis of X-ray diffraction patterns treated a plane of atoms as a semi-transparent mirror and modelled the crystal as stacks of reflecting planes of separation d (Fig 15.23). The model makes it easy to calculate the angle the crystal must make to the incoming beam of X-rays for constructive interference to occur. It has also given rise to the name **reflection** to denote an intense spot arising from constructive interference.

The path-length difference of the two rays shown in the illustration is

$$AB + BC = 2d \sin \theta$$

where θ is the **glancing angle**. When the path-length difference is equal to one wavelength ($AB + BC = \lambda$), the reflected waves interfere constructively. It follows that a reflection should be observed when the glancing angle satisfies the **Bragg law**:

$$\lambda = 2d \sin \theta \qquad (15.6)$$

The primary use of the Bragg law is to determine the spacing between the layers of atoms for, once the angle θ corresponding to a reflection has been determined, d may readily be calculated.

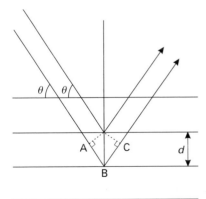

Fig 15.23 The derivation of the Bragg law treats each lattice plane as reflecting the incident radiation. The path lengths differ by AB + BC, which depends on the glancing angle θ. Constructive interference (a 'reflection') occurs when AB + BC is equal to an integral number of wavelengths.

Example 15.3 *Using the Bragg law*

A reflection from the (111) planes of a cubic crystal was observed at a glancing angle of 11.2° when Cu K_α X-rays of wavelength 154 pm were used. What is the length of the side of the unit cell?

Strategy We can find the separation, d, of the lattice planes from eqn 15.6 and the data. Then we find the length of the side of the unit cell by using eqn 15.5. Because the unit cell is cubic, $a = b = c$, so

$$\frac{1}{d^2} = \frac{h^2 + k^2 + l^2}{a^2}$$

which rearranges to

$$a = d \times (h^2 + k^2 + l^2)^{1/2}$$

Solution According to the Bragg law, the separation of the (111) planes responsible for the diffraction is

$$d = \frac{\lambda}{2 \sin \theta} = \frac{154 \text{ pm}}{2 \sin 11.2°}$$

It then follows that with $h = 1, k = 1, l = 1$,

$$a = \frac{154 \text{ pm}}{2 \sin 11.2°} \times 3^{1/2} = 687 \text{ pm}$$

Self-test 15.5

Calculate the angle at which the same lattice will give a reflection from the (123) planes.

[*Answer*: 24.8°]

15.10 Experimental techniques

Laue's original method consisted of passing a beam of X-rays of a wide range of wavelengths into a single crystal, and recording the diffraction pattern photographically. The idea behind the approach was that a crystal might not be suitably orientated to act as a diffraction grating for a single wavelength but, whatever its orientation, the Bragg law would be satisfied for at least one of the wavelengths when a range of wavelengths is present in the beam.

An alternative technique was developed by Peter Debye and Paul Scherrer and independently by Albert Hull. They used monochromatic (single-frequency radiation and a powdered sample. When the sample is a powder, we can be sure that some of the randomly distributed crystallites will be orientated so as to satisfy the Bragg law. For example, some of them will be orientated so that their (111) planes, of spacing d, give rise to a reflection at a particular angle and others will be orientated so that their (230) planes give rise to a reflection at a different angle. Each set of (*hkl*) planes gives rise to reflections at a different angle. In the modern version of the technique, which uses a **powder diffractometer**, the sample is spread on a flat plate and the diffraction pattern is monitored electronically. The major application is for qualitative analysis because the diffraction pattern is a kind of fingerprint and may be recognizable (Fig 15.24). The technique is also used for the initial determination of the dimensions and symmetries of unit cells.

Modern X-ray diffraction, which utilizes an **X-ray diffractometer** (Fig 15.25), is now a highly sophisticated technique. By far the most detailed information comes from developments of the techniques pioneered by the Braggs, in which a single crystal is employed as the diffracting grating and a monochromatic beam of X-rays is used to generate the dif-

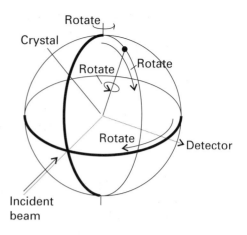

Fig 15.25 A four-circle diffractometer. The settings of the orientations of the components are controlled by computer; each reflection is monitored in turn, and their intensities are recorded.

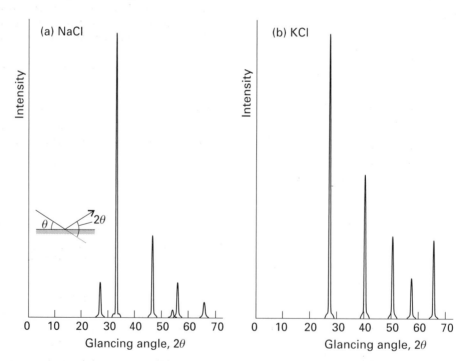

Fig 15.24 Typical X-ray powder diffraction patterns that can be used to identify the material and determine the size of its unit cell. (a) Sodium chloride, (b) potassium chloride.

fraction pattern. The single crystal (which may be only a fraction of a millimetre in length) is rotated relative to the beam, and the diffraction pattern is monitored and recorded electronically for each crystal orientation. A computer then analyses the diffraction pattern and the results are presented as a detailed structural map of the unit cell of the crystal, showing the relative locations of all the atoms it contains.

Derivation 15.3 *The determination of structure*

The primary data in X-ray diffraction is a set of intensities arising from the Miller planes (*hkl*). Each set of planes gives a reflection of intensity I_{hkl}. For our purposes, we focus on the (*h*00) planes and write the intensities I_h. First, we need the *amplitude* of the wave responsible for the signal. Because the intensity of electromagnetic radiation is given by the square of the amplitude, we need to form the **structure factors** $F_h = (I_h)^{1/2}$. Here is the first difficulty: we do not know the sign to take. For instance, if $I_h = 4$, then F_h can be either +2 or −2. This ambiguity is the **phase problem** of X-ray diffraction. However, once we have the structure factors we can calculate the electron density, $\rho(x)$, by forming the following sum:

$$\rho(x) = \frac{1}{V}\left\{F_0 + 2\sum_{h=1}^{\infty} F_h \cos(2h\pi x)\right\} \quad (15.7)$$

where *V* is the volume of the unit cell. This expression is called a **Fourier synthesis** of the electron density; we illustrate it in the following *Illustration*. Low values of *h* give the major features of the structure (they correspond to long-wavelength cosine terms) whereas the high values give the fine detail (short-wavelength cosine terms). Clearly, if we do not know the sign of F_h, we do not know whether the corresponding term in the sum is positive or negative and we get different electron densities, and hence crystal structures, for different choices of sign.

The phase problem can be overcome to some extent by the method of **isomorphous replacement**, in which heavy atoms are introduced into the crystal. These atoms dominate the diffraction pattern and greatly simplify its interpretation. The phase problem can also be resolved by judging whether the calculated structure is chemically plausible, whether the electron density is positive throughout, and by using more refined mathematical techniques.

Illustration 15.1

The following intensities were obtained in an experiment on an organic solid:

h	0	1	2	3	4	5	6	7	8	9
I_h	256	100	5	1	50	100	8	0	5	10

h	10	11	12	13	14	15
I_h	40	25	9	4	4	9

To find the structure factors, we take square-roots of the intensities:

h	0	1	2	3	4	5	6	7
F_h	±16	±10	±2.2	±1	±7.1	±10	±2.8	±3.2

h	8	9	10	11	12	13	14	15
F_h	±2.2	±3.2	±6.3	±5	±3	±2	±2	±3

Suppose the signs alternate + − + − . . .; then the electron density is

$$V\rho(x) = 16 - 20\cos(2\pi x) + 4.4\cos(4\pi x) - \cdots - 6\cos(30\pi x)$$

This function is shown in Fig 15.26a, and the locations of several types of atom are easy to identify as peaks in the electron density. If we use + signs up to h = 5 and − signs thereafter; the electron density is

$$V\rho(x) = 16 + 20\cos(2\pi x) + 4.4\cos(4\pi x) + \cdots - 9\cos(30\pi x)$$

This density is shown in Fig 15.26b. This structure has more regions of negative electron density, so it is less plausible than the structure obtained from the first choice of phases.

Typical structures

HUGE numbers of crystal structures have been determined. In this section we review how some of them can be rationalized. For metals and monatomic ions we can model the atoms and ions as hard spheres, and consider how such spheres can be stacked together in a regular, electrically neutral array.

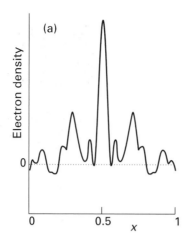
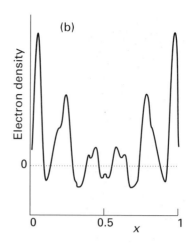

Fig 15.26 The Fourier synthesis of the electron density of a one-dimensional crystal using the data in *Illustration 15.1*. (a) Using alternating signs for the structure factors; (b) using positive signs for *h* up to 5, then negative signs.

15.11 **Metal crystals**

Most metallic elements crystallize in one of three simple forms, two of which can be explained in terms of stacking spheres to give the closest possible packing. In such **close-packed structures** the spheres representing the atoms are packed together with least waste of space and each sphere has the greatest possible number of nearest neighbours.

We can form a close-packed layer of identical spheres, one with maximum utilization of space, as shown in Fig 15.27a. Then we can form a second close-packed layer by placing spheres in the depressions of the first layer (Fig 15.27b). The third layer may be added in either of two ways, both of which result in the same degree of close packing. In one, the spheres are placed so that they reproduce the first layer (Fig 15.27c), to give an ABA pattern of layers. Alternatively, the spheres may be placed over the gaps in the first layer (Fig 15.27d), so giving an ABC pattern.

Two types of structures are formed if the two stacking patterns are repeated. The spheres are **hexagonally close-packed** (hcp) if the ABA pattern is repeated to give the sequence of layers ABABAB . . . . The name reflects the symmetry of the unit cell (Fig 15.28). Metals with hcp structures include beryllium, cadmium, cobalt, manganese, tita-

nium, and zinc. Solid helium (which forms only under pressure) also adopts this arrangement of atoms. Alternatively, the spheres are **cubic close-packed** (ccp) if the ABC pattern is repeated to give the sequence of layers ABCABC . . . . Here too, the name reflects the symmetry of the unit cell (Fig 15.29). Metals with this structure include silver, aluminium, gold, calcium, copper, nickel, lead, and platinum. The noble gases other than helium also adopt a ccp structure.

The compactness of the ccp and hcp structures is indicated by their **coordination number**, the number of atoms immediately surrounding any selected atom, which is 12 in both cases. Another measure of their compactness is the **packing fraction**, the fraction of space occupied by the spheres, which is 0.740. That is, in a close-packed solid of identical hard spheres, 74.0 per cent of the available space is occupied and only 26.0 per cent of the total volume is empty space.

The fact that many metals are close-packed accounts for one of their common characteristics, their high density. However, there is a difference between ccp and hcp metals. In cubic close packing, the faces of the cubes extend throughout the solid, and give rise to a **slip plane**. When the metal is under stress, the layers of atoms may slip past one another. As a result, a ccp metal is more malleable

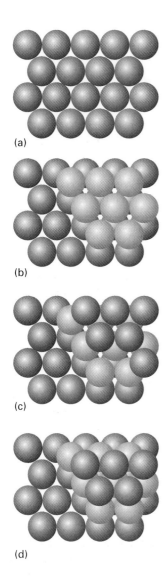

(a)

(b)

(c)

(d)

Fig 15.27 The close-packing of identical spheres. (a) The first layer of close-packed spheres. (b) The second layer of close-packed spheres occupies the dips of the first layer. The two layers are the AB component of the structure. (c) The third layer of close-packed spheres might occupy the dips lying directly above the spheres in the first layer, resulting in an ABA structure. (d) Alternatively, the third layer might lie in the dips that are not above the spheres in the first layer, resulting in an ABC structure.

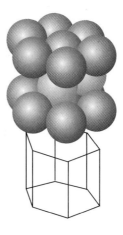

Fig 15.28 A hexagonal close-packed structure. The tinting of the spheres denoting the three layers of atoms is the same as in Fig 15.27.

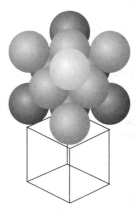

Fig 15.29 A cubic close-packed structure. The tinting of the spheres is the same as in Fig 15.27.

than an hcp metal, which does not have these slip planes. Thus, copper, which is highly malleable, is ccp, but zinc, which is hcp, is more brittle.

A number of common metals adopt structures that are not close-packed, which suggests that directional covalent bonding between neighbouring atoms is beginning to influence the structure and impose a specific geometrical arrangement. One such arrangement results in a **body-centred cubic** (bcc) lattice, with one sphere at the centre of a cube formed by eight others (Fig 15.30). The bcc structure is adopted by a number of common metals, including barium, caesium, chromium, iron, potassium, and tungsten. The coordination number of a bcc lattice is 8 and its packing fraction is only 0.68, showing that only about two-thirds of the available space is occupied.

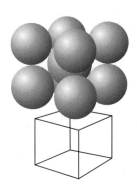

Self-test 15.6

What is the coordination number and the packing fraction of a primitive cubic lattice in which there is a lattice point at each corner of a cube?

[*Answer*: 6, 0.52]

Fig 15.30 A body-centred cubic unit cell. Note that the spheres on the corners touch the central sphere but the packing pattern leaves more empty space than in the two close-packed structures.

15.12 **Ionic crystals**

When we model the structures of ionic crystals by stacks of spheres, we must allow for the fact that the two or more types of ion present in the compound have different radii (generally with the cations smaller than the anions) and different charges.

The **coordination number** of an ion in an ionic crystal is the number of nearest neighbours of opposite charge. Even if, by chance, the ions have the same size, the problem of ensuring that the unit cells are electrically neutral makes it impossible to achieve 12-coordinate close-packed structures (which is a reason why ionic solids are generally less dense than metals). The closest packing that can be achieved is the 8-coordination of the **caesium-chloride structure** in which each cation is surrounded by eight anions and each anion is surrounded by eight cations (Fig 15.31). In the caesium-chloride structure, an ion of one charge occupies the centre of a cubic unit cell with eight ions of opposite charge at its corners. This structure is adopted by caesium chloride itself and by calcium sulfide, caesium cyanide (with some distortion), and one type of brass (CuZn).

When the radii of the ions differ by more than in caesium chloride, even 8-coordinate packing cannot be achieved. One common structure adopted is

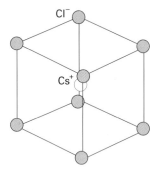

Fig 15.31 The caesium-chloride structure consists of two interpenetrating simple cubic lattices, one of cations and the other of anions, so that each cube of ions of one kind has a counterion at its centre. This illustration shows a single unit cell with a Cs^+ ion at the centre. By imagining eight of these unit cells stacked together to form a bigger cube, it should be possible to imagine an alternative form of the unit cell with Cs^+ at the corners and a Cl^- ion at the centre.

the 6-coordinated **rock-salt structure** typified by sodium chloride (rock salt is a mineral form of sodium chloride) in which each cation is surrounded by six anions and each anion is surrounded by six cations (Fig 15.32). The rock-salt structure is the structure of sodium chloride itself and of several other compounds of formula MX, including potassium bromide, silver chloride, and magnesium oxide.

The switch from the caesium-chloride structure to the rock-salt structure (in a number of examples) can be correlated with the **radius ratio**

$$\gamma = \frac{r_{smaller}}{r_{larger}} \qquad (15.8)$$

The two radii are those of the smaller and larger ions in the crystal. The **radius-ratio rule**, which is derived by analysing the geometrical problem of stacking together spheres of different radii, states that the caesium-chloride structure should be expected when

$$\gamma > 3^{1/2} - 1 = 0.732$$

and that the rock-salt structure should be expected when

$$2^{1/2} - 1 = 0.414 < \gamma < 0.732$$

For $\gamma < 0.414$, when the two types of ions have markedly different radii (like oranges and grape-fruit), the most efficient packing leads to four-coordination of the type exhibited by the sphalerite (or zinc-blende) form of zinc sulfide, ZnS (Fig 15.33). The radius-ratio rule is moderately well supported by observation. The deviation of a structure from the prediction is often taken to be an indication of a shift from ionic towards covalent bonding.

Self-test 15.7

Is sodium iodide likely to have a rock-salt or caesium-chloride structure?

[*Answer*: rock salt]

The **ionic radii** used to calculate γ, and wherever else it is important to know the sizes of ions, are derived from the distance between the centres of adjacent ions in a crystal. However, in a diffraction experiment we measure the distance between the centres of ions. It is necessary to apportion that total distance by defining the radius of one ion and reporting all others on that basis. One scale that is widely used is based on the value 140 pm for the radius of the O^{2-} ion (Table 15.4). Other scales are also available (such as one based on F^- for discussing halides), and it is essential not to mix values from different scales. Because ionic radii are so arbitrary, predictions based on them (such as those made by using the radius-ratio rule) must be viewed cautiously.

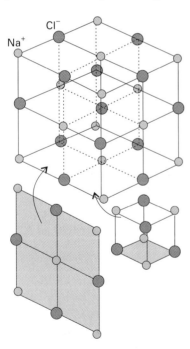

Fig 15.32 The rock-salt (NaCl) structure consists of two mutually interpenetrating slightly expanded face-centred cubic lattices. The additional diagrams in this illustration show various details of the structure.

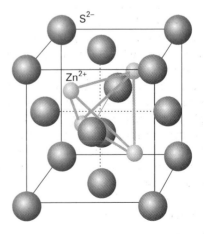

Fig 15.33 The sphalerite (zinc-blende, ZnS) structure. This structure is typical of ions that have markedly different radii and equal but opposite charges.

Table 15.4 *Ionic radii, r/pm*

Li$^+$	Be^{2+}	B^{3+}	N^{3-}	O^{2-}	F$^-$
59	27	12	171	140	133
Na$^+$	Mg^{2+}	Al^{3+}	P^{3-}	S^{2-}	Cl$^-$
102	72	53	212	184	181
K$^+$	Ca^{2+}	Ga^{3+}	As^{3-}	Se^{2-}	Br$^-$
138	100	62	222	198	196
Rb$^+$	Sr^{2+}				I$^-$
149	116				220
Cs$^+$	Ba^{2+}				
170	136				

Exercises

15.1 Classify as n-type or p-type a semiconductor formed by doping (a) germanium with phosphorus, (b) germanium with indium.

15.2 The electrical resistance of a sample increased from 100 Ω to 120 Ω when the temperature was changed from 0°C to 100°C. Is the substance a metallic conductor or a semiconductor?

15.3 Describe the bonding in magnesium oxide, MgO, in terms of bands composed of Mg and O atomic orbitals. How does this model justify the ionic model of this compound?

15.4 Estimate the lattice enthalpy of magnesium oxide from the data in Appendix 1 and the ionization and electron gain enthalpies in Chapter 3.

15.5 Estimate the lattice enthalpy of calcium chloride, $CaCl_2$, from thermodynamic data.

15.6 Calculate the potential energy of an ion at the centre of a diffuse 'spherical crystal' in which concentric spheres of ions of opposite charge surround the ion and the numbers of ions on the spherical surfaces fall away rapidly with distance. Let successive spheres lie at radii d, $2d$, . . . and the number of ions (all of the same charge) on each successive sphere is inversely proportional to the radius of the sphere. You will need the following sum:

$$1 - \frac{1}{2^2} + \frac{1}{3^2} - \frac{1}{4^2} + \cdots = \frac{\pi^2}{12}$$

15.7 Estimate the ratio of the lattice enthalpies of SrO and CaO from the Born–Mayer equation by using the ionic radii in Table 15.4.

15.8 Draw a set of points as a rectangular array based on unit cells of side a and b, and mark the planes with Miller indices (10), (01), (11), (12), (23), (41), (4$\bar{1}$).

15.9 Repeat Exercise 15.8 for an array of points in which the a and b axes make 60° to each other.

15.10 In a certain unit cell, planes cut through the crystal axes at $(2a, 3b, c)$, (a, b, c), $(6a, 3b, 3c)$, $(2a, -3b, -3c)$. Identify the Miller indices of the planes.

15.11 Draw an orthorhombic unit cell and mark on it the (100), (010), (001), (011), (101), and (10$\bar{1}$) planes.

15.12 Draw a triclinic unit cell and mark on it the (100), (010), (001), (011), (101), and (10$\bar{1}$) planes.

15.13 Calculate the separations of the planes (111), (211), and (100) in a crystal in which the cubic unit cell has sides of length 532 pm.

15.14 Calculate the separations of the planes (123) and (236) in an orthorhombic crystal in which the unit cell has sides of lengths 0.754, 0.623, and 0.433 nm.

15.15 The glancing angle of a Bragg reflection from a set of crystal planes separated by 97.3 pm is 19.85°. Calculate the wavelength of the X-rays.

15.16 Copper K_α radiation consists of two components of wavelengths 154.433 pm and 154.051 pm. Calculate the separation of the diffraction lines arising from the two components in a powder diffraction pattern recorded on a cylindrical film of radius 5.74 cm from planes of separation 73.2 pm.

15.17 The separation of (100) planes of lithium metal is 350 pm and its density is 0.53 g cm^{-3}. Is the structure of lithium fcc or bcc?

15.18 Copper crystallizes in an fcc structure with unit cells of side 361 pm. Predict the appearance of the powder diffraction pattern using 154 pm radiation. What is the density of copper?

15.19 Construct the electron density along the x-axis of a crystal given the following structure factors:

h	0	1	2	3	4	5	6	7	8	9
F_h	+30.0	+8.2	+6.5	+4.1	+5.5	−2.4	+5.4	+3.2	−1.0	+1.1

h	10	11	12	13	14	15
F_h	+6.5	+5.2	−4.3	−1.2	+0.1	+2.1

15.20 Calculate the packing fraction of (a) a stack of cylinders and (b) a primitive cubic lattice.

15.21 Suppose a virus can be regarded as a sphere and that it stacks together in a hexagonal close-packed arrangement. If the density of the virus is the same as that of water (1.00 g cm^{-3}), what is the density of the solid?

15.22 How many (a) nearest neighbours, (b) next-nearest neighbours are there in a body-centred cubic structure. What are their distances if the side of the cube is 500 nm?

15.23 How many (a) nearest neighbours, (b) next-nearest neighbours are there in a cubic close-packed structure. What are their distances if the side of the cube is 500 nm?

15.24 The thermal and mechanical processing of materials is an important step in ensuring that they have the appropriate physical properties for their intended application. Suppose a metallic element underwent a phase transition in which its crystal structure changed from cubic close-packed to body-centred cubic. (a) Would it become more or less dense? (b) By what factor would its density change?

15.25 The compound Rb_3TlF_6 has a simple tetragonal unit cell with dimensions $a = 651$ pm and $c = 934$ pm. Calculate the volume of the unit cell and the density of the solid.

15.26 The orthorhombic unit cell of $NiSO_4$ has the dimensions $a = 634$ pm, $b = 784$ pm, and $c = 516$ pm, and the density of the solid is estimated as 3.9 g cm^{-3}. Determine the number of formula units per unit cell and calculate a more precise value of the density.

15.27 The unit cells of $SbCl_3$ are orthorhombic with dimensions $a = 812$ pm, $b = 947$ pm, and $c = 637$ pm. Calculate the spacing of (a) the (321) planes, (b) the (642) planes.

15.28 Use the radius-ratio rule to predict the kind of crystal structure expected for magnesium oxide.

Molecular substances

Contents

Aᴛᴏᴍs and molecules with complete valence shells are still able to interact with one another. They attract one another over the range of several atomic diameters and they repel one another when pressed together. These residual forces are highly important. They account, for instance, for the condensation of gases to liquids and the structures of molecular solids. All organic liquids and solids, ranging from small molecules like benzene to virtually infinite cellulose and the polymers from which fabrics are made, are bound together by the forces of cohesion we explore in this chapter. These forces are also responsible for the structural organization of biological macromolecules, for they twist the long polypeptide chains of proteins into characteristic shapes and then pin them together in the arrangement essential to their function.

The origins of cohesion

A van der Waals force is an interaction between closed-shell molecules. The attractive contributions to these forces include the interactions between the partial electric charges of polar molecules and of polar functional groups in macromolecules. Van der Waals forces also include the repulsive interactions that prevent the complete collapse of matter to densities as high as those characteristic of atomic nuclei. The repulsive interactions arise from the exclusion of electrons from regions of space where the orbitals of closed-shell species overlap.

16.1 Interactions between partial charges

Atoms in molecules in general have partial charges. Table 16.1 gives the partial charges typically found in peptides. If these charges were separated by a vacuum, they would attract or repel each other in accord with Coulomb's law, and we would write[1]

$$V = \frac{q_1 q_2}{4 \pi \varepsilon_0 r} \qquad (16.1a)$$

where q_1 and q_2 are the partial charges and r is their separation. However, it is more accurate to take into account the possibility that other parts of the molecule, or other molecules, lie between the charges, and decrease the strength of the interaction (Fig 16.1). We therefore write

$$V = \frac{q_1 q_2}{4 \pi \varepsilon r} \qquad (16.1b)$$

where ε is the **permittivity** of the medium lying between the charges. The permittivity is usually expressed as a multiple of the vacuum permittivity by writing $\varepsilon = \varepsilon_r \varepsilon_0$, where ε_r is the **relative permittivity**.[2] The effect of the medium can be very large: for water, $\varepsilon_r = 78$, so the potential energy of two charges

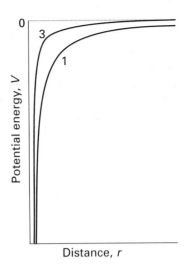

Fig 16.1 The Coulomb potential for two charges and its dependence on their separation. The two curves correspond to different relative permittivities (1 for a vacuum, 3 for a fluid).

separated by bulk water is reduced by nearly two orders of magnitude compared to the value it would have if the charges were separated by a vacuum. The problem is made worse in calculations on polypeptides (and nucleic acids) by the fact that two partial charges may have water and polypeptide chain lying between them. Various models have been proposed to take this awkward effect into account, the simplest being to set $\varepsilon_r = 3.5$ and to hope for the best.

16.2 Electric dipole moments

When the molecules or groups that we are considering are widely separated, it turns out to be simpler to express the principal features of their interaction in terms of the dipole moments associated with the charge distributions rather than with each partial charge. At its simplest, an **electric dipole** consists of two charges q and $-q$ separated by a distance l. The product ql is called the **electric di-**

Table 16.1 *Partial charges in polypeptides*

Atom	Partial charge/e
N	−0.36
H(−N)	+0.18
C(−CO)	+0.06
H(−C)	+0.02
C(=O)	+0.45
O	−0.38
H(−O)	+0.42

---

[1] Equation 16.1a is for the potential energy of the interaction. The force between the two charges is inversely proportional to the square of their separation, $F \propto 1/r^2$.

[2] The relative permittivity is still widely called the *dielectric constant*.

pole moment, μ. We represent dipole moments by an arrow with a length proportional to μ and pointing from the negative charge to the positive charge (**1**).[3] Because a dipole moment is the product of a charge (in coulombs, C) and a length (in metres, m), the SI unit of dipole moment is the coulomb-metre (C m). However, it is often much more convenient to report a dipole moment in **debye**, D, where

$$1 \text{ D} = 3.335\ 64 \times 10^{-30} \text{ C m}$$

because the experimental values for molecules are then close to 1 D (Table 16.2).[4] The dipole moment of charges e and $-e$ separated by 100 pm is 1.6×10^{-29} C m, corresponding to 4.8 D. Dipole moments of small molecules are typically smaller than that, at about 1 D.

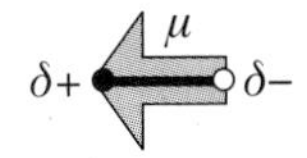

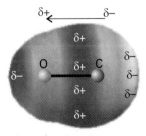

1

A **polar molecule** is a molecule with a permanent electric dipole moment arising from the partial charges on its atoms (Section 14.17). A **nonpolar molecule** is a molecule that has no permanent electric dipole moment. All heteronuclear diatomic molecules are polar because the difference in electronegativities of their two atoms results in nonzero partial charges. Typical dipole moments are 1.08 D for HCl and 0.42 D for HI (Table 16.2). A very approximate relation between the dipole moment and the difference in Pauling electronegativities (Table 14.2) of the two atoms, $\Delta\chi$, is

$$\mu/\text{D} \approx \Delta\chi \qquad (16.2)$$

Illustration 16.1

The electronegativities of hydrogen and bromine are 2.1 and 2.8 respectively. The difference is 0.7, so we predict an electric dipole moment of about 0.7 D for HBr. The experimental value is 0.80 D.

---

[3] Be careful with this convention: for historical reasons the opposite convention is still widely adopted.
[4] The unit is named after Peter Debye, the Dutch pioneer of the study of dipole moments of molecules.

Table 16.2 *Dipole moments (μ) and polarizability volumes (α')*

	μ/D	$\alpha'/(10^{-30}\text{ m}^3)$
Ar	0	1.66
CCl_4	0	10.5
C_6H_6	0	10.4
H_2	0	0.819
H_2O	1.85	1.48
NH_3	1.47	2.22
HCl	1.08	2.63
HBr	0.80	3.61
HI	0.42	5.45

The polarizability volumes are mean values over all orientations of the molecule.

Because it attracts the electrons more strongly, the more electronegative atom is usually the negative end of the dipole. However, there are exceptions, particularly when antibonding orbitals are occupied. Thus, the dipole moment of CO is very small (0.12 D) but the negative end of the dipole is on the C atom even though the O atom is more electronegative. This apparent paradox is resolved as soon as we realize that antibonding orbitals are occupied in CO (see Fig 14.30), and, because electrons in antibonding orbitals tend to be found closer to the less electronegative atom, they contribute a negative partial charge to that atom. If this contribution is larger than the opposite contribution from the electrons in bonding orbitals, the net effect will be a small negative partial charge on the *less* electronegative atom. Figure 16.2 shows a

Fig 16.2 The computed charge distribution in a CO molecule. Note that, although oxygen is more electronegative than carbon, the negative end of the dipole is on carbon.

computed electron density distribution in CO, and the small charge imbalance can be seen.

Molecular symmetry is of the greatest importance in deciding whether a polyatomic molecule is polar or not. Indeed, molecular symmetry is more important than the question of whether or not the atoms in the molecule belong to the same element. Homonuclear polyatomic molecules may be polar if they have low symmetry and the atoms are in inequivalent positions. For instance, the angular molecule ozone, O_3 (**2**), is homonuclear; however, it is polar because the central O atom is different from the outer two (it is bonded to two atoms, they are bonded only to one); moreover, the dipole moments associated with each bond make an angle to each other and do not cancel. Heteronuclear polyatomic molecules may be nonpolar if they have high symmetry, because individual bond dipoles may then cancel. The heteronuclear linear triatomic molecule CO_2, for example, is nonpolar because, although there are partial charges on all three atoms, the dipole moment associated with the OC bond points in the opposite direction to the dipole moment associated with the CO bond, and the two cancel (**3**).

> **Self-test 16.1**
>
> Use the VSEPR model to judge whether ClF_3 is polar or nonpolar. (*Hint.* Predict the structure first.)
>
> [*Answer:* polar]

To a first approximation, it is possible to resolve the dipole moment of a polyatomic molecule into contributions from various groups of atoms in the molecule and the directions in which these individual contributions lie (Fig 16.3). Thus, *p*-dichlorobenzene is nonpolar by symmetry on account of the cancellation of two equal but opposing C–Cl moments (exactly as in carbon dioxide). *o*-Dichlorobenzene has a dipole moment which is approximately the resultant of two chlorobenzene dipole moments arranged at 60° to each other. This technique of 'vector addition' can be applied with fair success

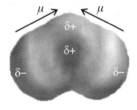

2 Ozone

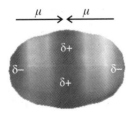

3 Carbon dioxide

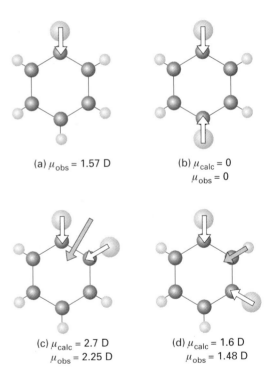

(a) μ_{obs} = 1.57 D

(b) μ_{calc} = 0
μ_{obs} = 0

(c) μ_{calc} = 2.7 D
μ_{obs} = 2.25 D

(d) μ_{calc} = 1.6 D
μ_{obs} = 1.48 D

Fig 16.3 The dipole moments of the dichlorobenzene isomers can be obtained approximately by vectorial addition of two chlorobenzene dipole moments (1.57 D).

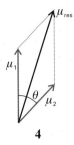

4

to other series of related molecules, and the resultant μ_{res} of two dipole moments μ_1 and μ_2 that make an angle θ to each other (4) is approximately

$$\mu_{res} \approx (\mu_1^2 + \mu_2^2 + 2\mu_1\mu_2 \cos \theta)^{1/2} \qquad (16.3)$$

Self-test 16.2

Estimate the ratio of the electric dipole moments of *ortho*- and *meta*-disubstituted benzenes.

[*Answer:* $\mu(ortho)/\mu(meta)$ = 1.7]

A better approach to the calculation of dipole moments is to take into account the locations and magnitudes of the partial charges on all the atoms. These partial charges are included in the output of many molecular structure software packages. Indeed, the programs calculate the dipole moments of the molecules in the manner we now describe.

Derivation 16.1 *The dipole moment of a molecule*

An electric dipole moment is a vector, $\boldsymbol{\mu}$, with three components, μ_x, μ_y, and μ_z. The direction of $\boldsymbol{\mu}$ shows the orientation of the dipole in the molecule and the length of the vector is the magnitude, μ, of the dipole moment. In common with all vectors, the magnitude is related to the components by

$$\mu = (\mu_x^2 + \mu_y^2 + \mu_z^2)^{1/2} \qquad (16.4a)$$

Each component is related to the magnitude of the partial charges and their positions relative to a point in the molecule:

μ_x = Sum of (partial charge
$\qquad \qquad \times$ distance along x-axis) (16.4b)

---

with similar expressions for the *y*- and *z*-components. For an electrically neutral molecule, the origin of the coordinates is arbitrary, so it is best chosen to simplify the measurements. For a planar molecule, there are no charges above or below the *xy*-plane, so the *z*-component (where *z* is perpendicular to the plane) is zero. The procedure is illustrated in the following example.

Example 16.1 *Calculating a molecular dipole moment*

Calculate the electric dipole moment of the peptide group using the partial charges (as multiples of *e*) in Table 16.1 and the locations of the atoms shown in Fig 16.4.

Strategy We use eqn 16.4b to calculate each of the components of the dipole moment and then eqn 16.4a to assemble the three components into the magnitude of the dipole moment. Note that the partial charges are multiples of the fundamental charge, $e = 1.609 \times 10^{-19}$ C (see inside front cover).

Solution The first few terms of the expression for μ_x are

$$\mu_x = (-0.36e) \times (-0.8 \text{ pm}) + (0.45e) \times (2.1 \text{ pm}) + \cdots$$
$$= -3.8 \times 10^{-18} \text{ C pm} = -3.8 \times 10^{-30} \text{ C m}$$

corresponding to -1.1 D. A similar calculation gives $\mu_y = -0.91$ D and $\mu_z = 0$. Therefore,

$$\mu = [(-1.1 \text{ D})^2 + (-0.96 \text{ D})^2]^{1/2} = 1.5 \text{ D}$$

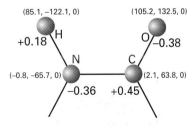

Fig 16.4 The locations of the atoms (in picometres relative to the centre of the molecule) and the partial charges (as multiples of *e*) used to calculate the dipole moment of a peptide group.

We can find the orientation of the dipole moment by arranging an arrow of length 1.5 units of length to have x-, y-, and z-components of −1.1, −0.96, and 0 units (**5**).
Self-test 16.3

Calculate the electric dipole moment of formaldehyde, using the information in (**6**).

[*Answer*: −3.2 D]

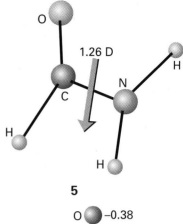

5

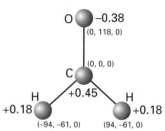

6 Formaldehyde, H₂CO

16.3 **Interactions between dipoles**

We calculate the potential energy of a dipole μ_1 in the presence of a charge q_2 by taking into account the interaction of the charge with the two partial charges of the dipole.

Derivation 16.2 *The interaction of a charge with a dipole*

For simplicity, suppose that the charge and dipole are collinear (**7**). Then the potential energy is

$$V = \frac{q_1 q_2}{4\pi\varepsilon_0 (r + \frac{1}{2}l)} - \frac{q_1 q_2}{4\pi\varepsilon_0 (r - \frac{1}{2}l)}$$

$$= \frac{q_1 q_2}{4\pi\varepsilon_0 r \left(1 + \frac{l}{2r}\right)} - \frac{q_1 q_2}{4\pi\varepsilon_0 r \left(1 - \frac{l}{2r}\right)}$$

Now we suppose that the separation of charges in the dipole is much smaller than the distance of the charge q_2 in the sense that $l/2r \ll 1$. Then we can use the expansion

$$\frac{1}{1+x} = 1 - x + x^2 - \cdots$$

to write

$$V = \frac{q_1 q_2}{4\pi\varepsilon_0 r} \left\{ \left(1 - \frac{l}{2r} + \cdots\right) - \left(1 + \frac{l}{2r} + \cdots\right) \right\}$$

$$= -\frac{q_1 q_2 l}{4\pi\varepsilon_0 r^2} + \cdots$$

Now we recognize that $q_1 l = \mu_1$, the dipole moment of molecule 1, and ignore all but the leading term (that is, we neglect the unwritten terms in this expression as they are so small). We find

$$V = -\frac{\mu_1 q_2}{4\pi\varepsilon_0 r^2} \tag{16.5a}$$

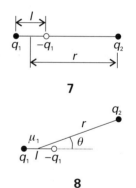

7

8

A similar calculation for the more general orientation shown in (**8**) gives

$$V = -\frac{\mu_1 q_2 \cos\theta}{4\pi\varepsilon_0 r^2} \tag{16.5b}$$

If q_2 is positive, the energy is lowest when $\theta = 180°$ (and $\cos\theta = -1$), because then the partial negative charge of the dipole lies closer than the partial positive charge to the point charge and the attraction outweighs the repulsion. This interaction energy decreases more rapidly with distance than that between two point charges (as r^2 rather than r) because, from the viewpoint of the single charge, the partial charges of the point dipole seem to merge and cancel as the distance r increases.

We can calculate the interaction energy between two dipoles μ_1 and μ_2 in the orientation shown in (9) in a similar way, taking into account all four charges of the two dipoles. The outcome is

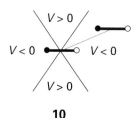

9

$$V = \frac{\mu_1 \mu_2 (1 - 3\cos^2 \theta)}{4\pi \varepsilon_0 r^3} \qquad (16.6)$$

This potential energy decreases even more rapidly than in eqn 16.5 because the charges of both dipoles seem to merge as the separation of the dipoles increases. The angular factor takes into account how the like or opposite charges come closer to one another as the relative orientation of the dipoles is changed. The energy is lowest when $\theta = 0$ or $180°$ (when $1 - 3\cos^2 \theta = -2$), because opposite partial charges then lie closer together than like partial charges. The potential energy is negative (attractive) in some orientations when $\theta < 54.7°$ (the angle at which $1 - 3\cos^2 \theta = 0$) because opposite charges are closer than like charges. It is positive (repulsive) when $\theta > 54.7°$ because then like charges are closer than unlike charges. The potential energy is zero on the line at $54.7°$ (and at $180 - 54.7 = 125.3°$) because at that angle the two attractions and the two repulsions cancel (10).

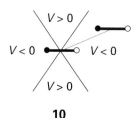

10

Illustration 16.2

To calculate the molar potential energy of the dipolar interaction between two peptide links separated by 3.0 nm in different regions of a polypeptide chain with $\theta = 180°$, we take $\mu_1 = \mu_2 = 1.4$ D, corresponding to 4.7×10^{-30} C m, and find

$$V = -\frac{(4.7 \times 10^{-30}\,\mathrm{C\,m})^2 \times (-2)}{4\pi \times (8.854 \times 10^{-12}\,\mathrm{J}^{-1}\mathrm{C}^2\mathrm{m}^{-1}) \times (3.0 \times 10^{-9}\,\mathrm{m})^3}$$

$$= -1.5 \times 10^{-23}\,\mathrm{J}$$

which corresponds to $-8.9\,\mathrm{J\,mol}^{-1}$.

The average potential energy of interaction between polar molecules that are freely rotating in a fluid (a gas or liquid) is zero because the attractions and repulsions cancel. However, because the potential energy of a dipole near another dipole depends on their relative orientations, the molecules exert forces on each other and therefore do not in fact rotate completely freely, even in a gas. As a result, the lower energy orientations are marginally favoured, so there is a nonzero interaction between polar molecules (Fig 16.5). The detailed calculation of the average interaction energy is quite complicated, but the final answer is very simple:

$$V = -\frac{2\mu_1^2 \mu_2^2}{3(4\pi \varepsilon_0)^2 kTr^6} \qquad (16.7)$$

The important features of this expression are the dependence of the average interaction energy on the inverse sixth power of the separation and its inverse dependence on the temperature. The temperature-dependence reflects the way that the greater thermal motion overcomes the mutual orientating effects of the dipoles at higher temperatures.

Ilustration 16.3

At 25°C the average interaction energy for pairs of molecules with $\mu = 1$ D is about -1.4 kJ mol^{-1} when the separation is 0.3 nm. This energy should be compared with the average molar kinetic energy of $\frac{3}{2}RT = 3.7$ kJ mol^{-1} at the same temperature: the two are not very dissimilar, but they are both much less than the energies involved in the making and breaking of chemical bonds.

Fig 16.5 A dipole–dipole interaction. When a pair of molecules can adopt all relative orientations with equal probability, the favourable orientations (a) and the unfavourable ones (b) cancel, and the average interaction is zero. In an actual fluid, the interactions in (a) slightly predominate.

16.4 Induced dipole moments

A nonpolar molecule may acquire a temporary **induced dipole moment**, μ^*, as a result of the influence of an electric field generated by a nearby ion or polar molecule. The field distorts the electron distribution of the polarizable molecule, and gives rise to an electric dipole in it. The magnitude of the induced dipole moment is proportional to the strength of the field, $\mathcal{E}$, and we write

$$\mu^* = \alpha \mathcal{E} \tag{16.8}$$

The proportionality constant α is the **polarizability** of the molecule. The larger the polarizability of the molecule, the greater the distortion that is caused by a given electric field. If the molecule has few electrons, they are tightly controlled by the nuclear charges and the polarizability of the molecule is low. If the molecule contains large atoms with electrons some distance from the nucleus, the nuclear control is less and the polarizability of the molecule is greater. The polarizability depends on the orientation of the molecule with respect to the field unless the molecule is tetrahedral (such as CCl_4), octahedral (such as SF_6), or icosahedral (C_{60}, buckminsterfullerene). Atoms, tetrahedral, octahedral, and icosahedral molecules have isotropic (orientation-independent) polarizabilities; all other molecules have anisotropic (orientation-dependent) polarizabilities.

The polarizabilities reported in Table 16.2 are given as **polarizability volumes**, α':

$$\alpha' = \frac{\alpha}{4\pi\varepsilon_0} \tag{16.9}$$

The polarizability volume has the dimensions of volume (hence its name) and is comparable in magnitude to the volume of a molecule.[5]

Self-test 16.4

What strength of electric field is required to induce an electric dipole moment of 1.0 μD in a molecule of polarizability volume 1.1×10^{-31} m³ (like CCl_4)?

[*Answer*: 2.7 kV cm⁻¹]

A polar molecule with dipole moment μ_1 can induce a dipole moment in a polarizable molecule (which may itself be either polar or nonpolar) because the partial charges of the polar molecule give rise to an electric field that distorts the second molecule. That induced dipole interacts with the permanent dipole of the first molecule, and the two are attracted together (Fig 16.6). The formula for the **dipole–induced-dipole interaction energy** is

$$V = -\frac{\mu_1^2 \alpha_2}{\pi\varepsilon_0 r^6} \tag{16.10}$$

where α_2 is the polarizability of molecule 2. The negative sign shows that the interaction is attractive. For a molecule with $\mu = 1$ D (such as HCl) near a molecule of polarizability volume $\alpha' = 1.0 \times 10^{-31}$ m³ (such as benzene, Table 16.2) the average interaction energy is about −33 J mol⁻¹ when the separation is 0.3 nm.

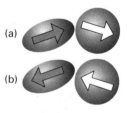

Fig 16.6 A dipole–induced-dipole interaction. The induced dipole (light arrows) follows the changing orientation of the permanent dipole (dark arrow).

---

[5] When using older compilations of data, it is useful to note that polarizability volumes expressed in cubic centimetres (cm³) have the same numerical values as the 'polarizabilities' reported using c.g.s. electrical units; so the tabulated values previously called 'polarizabilities' can be used directly as polarizability volumes.

16.5 Dispersion interactions

Finally, we consider the interactions between species that have neither a net charge nor a permanent electric dipole moment (such as two Xe atoms in a gas or two nonpolar groups on the peptide residues of a protein). We know that uncharged, nonpolar species can interact because they form condensed phases, such as benzene, liquid hydrogen, and liquid xenon.

The **dispersion interaction**, or **London force**, between nonpolar species arises from the transient dipoles that they possess as a result of fluctuations in the instantaneous positions of their electrons (Fig 16.7). Suppose, for instance, that the electrons in one molecule flicker into an arrangement that results in partial positive and negative charges and thus gives it an instantaneous dipole moment μ_1. While it exists, this dipole can polarize the other molecule and induce in it an instantaneous dipole moment μ_2. The two dipoles attract each other and the potential energy of the pair is lowered. Although the first molecule will go on to change the size and direction of its dipole (perhaps within 10^{-16} s), the second will follow it; that is, the two dipoles are *correlated* in direction like two meshing gears, with a positive partial charge on one molecule appearing close to a negative partial charge on the other molecule and vice versa. Because of this correlation of the relative positions of the partial charges, and their resulting attractive interaction, the attraction between the two instantaneous dipoles does not average to zero. Instead, it gives rise to a net attractive interaction. Polar molecules interact by a dispersion interaction as well as by dipole–dipole interactions.

The strength of the dispersion interaction depends on the polarizability of the first molecule because the magnitude of the instantaneous dipole moment μ_1 depends on the looseness of the control that the nuclear charge has over the outer electrons. If the control is loose, the electron distribution can undergo relatively large fluctuations. Moreover, if the control is loose, then the electron distribution can also respond strongly to applied electric fields and hence have a high polarizability. It follows that a high polarizability is a sign of large fluctuations in local charge density. The strength also depends on the polarizability of the second molecule, for that polarizability determines how readily a dipole can be induced in molecule 2 by molecule 1. We therefore expect $V \propto \alpha_1 \alpha_2$. The actual calculation of the dispersion interaction is quite involved, but a reasonable approximation to the interaction energy is the **London formula**:

$$V = -\frac{2}{3} \times \frac{a_1' a_2'}{r^6} \times \frac{I_1 I_2}{I_1 + I_2} \tag{16.11}$$

where I_1 and I_2 are the ionization energies of the two molecules. Once again, the interaction turns out to be proportional to the inverse sixth power of the separation. For two CH_4 molecules, $V \approx -5$ kJ mol^{-1} when $r = 0.3$ nm.

16.6 Hydrogen bonding

The strongest intermolecular interaction arises from the formation of a **hydrogen bond**, in which a hydrogen atom lies between two strongly electronegative atoms and binds them together. The bond is normally denoted X—H$\cdots$Y, with X and Y being nitrogen, oxygen, or fluorine. Unlike the other interactions we have considered, hydrogen bonding is not universal, but is restricted to molecules that contain these atoms.

The most elementary description of the formation of a hydrogen bond is that it is the result of a Coulombic interaction between the partly exposed positive charge of a proton bound to an electron-withdrawing X atom (in the fragment X–H) and the negative charge of a lone pair on the second atom Y, as in $^{\delta-}X—H^{\delta+}\cdots:Y^{\delta-}$.

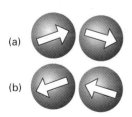

(a)

(b)

Fig 16.7 In the dispersion interaction, an instantaneous dipole on one molecule induces a dipole on another molecule, and the two dipoles then interact to lower the energy. The directions of the two instantaneous dipoles are correlated, and, although they occur in different orientations at different instants, the interaction does not average to zero.

Example 16.2 *Assessing a hydrogen bond energy*

Given that the partial charges on H and O are 0.45e and −0.83e, respectively, calculate the hydrogen bond energy for an O–H group next to an O atom in the arrangement shown in (**11**), and plot the molar energy as a function of the OOH angle. Take $R = 200$ pm and $r = 95.7$ pm.

Strategy The potential energy of two charges $z_1 e$ and $z_2 e$ separated by a distance r is given by the Coulomb law. We need to evaluate all the distances between pairs of atoms on different molecules and plot the result as a function of the two angles.

Solution The O–O distance is R, so the contribution to the molar potential energy is

$$V_{OO} = \frac{N_A z_O^2 e^2}{4 \pi \varepsilon_0 R}$$

By trigonometry, the O–H distance, d, is given by

$$d^2 = R^2 + r^2 - 2Rr \cos \theta$$

Therefore, the contribution of the O–H interaction to the molar energy is

$$V_{OH} = \frac{N_A z_O z_H e^2}{4 \pi \varepsilon_0 d} = \frac{N_A z_O z_H e^2}{4 \pi \varepsilon_0 (R^2 + r^2 - 2Rr \cos \theta)^{1/2}}$$

The total molar potential energy is the sum:

$$V = V_{OO} + V_{OH}$$

To plot V as a function of angle, we use

$$\frac{N_A e^2}{4 \pi \varepsilon_0} = \frac{(6.02214 \times 10^{23} \, \text{mol}^{-1}) \times (1.6022 \times 10^{-19} \, \text{C})^2}{4 \pi \times (8.85419 \times 10^{-12} \, \text{J}^{-1} \text{C}^2 \text{m}^{-1})}$$

$$= 1.389 \times 10^{-4} \, \text{J m}$$

The graph of V is shown in Fig 16.8. We see that at $\theta = 0$, a line of atoms, $V = -19 \, \text{kJ mol}^{-1}$. Note how sharply the energy depends on angle: it is negative only within ±12° of linearity.

Self-test 16.5

Calculate the energy of interaction between two acetic acid molecules in a dimer, as depicted in (**12**). Consider only the Coulombic interactions indicated by the dotted lines and their symmetry-related equivalents. At what distance R does the interaction become attractive?

[*Answer*: 271 pm]

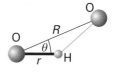

11

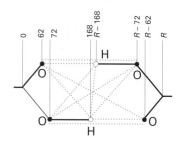

12 Acetic acid dimer

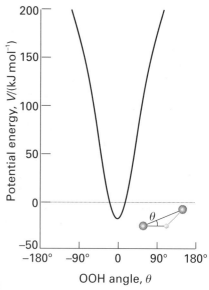

Fig 16.8 The variation of the energy of interaction (on the electrostatic model) of a hydrogen bond as the angle between the O–H and :O groups is changed.

A slightly more sophisticated version of the electrostatic description is to regard hydrogen bond formation as the formation of a Lewis acid–base complex in which the partly exposed proton of the

X–H group is the Lewis acid and :Y, with its lone pair, is the Lewis base:

$$X–H + :Y \rightarrow X–H\cdots Y$$

Molecular orbital theory provides an alternative description that is more in line with the concept of delocalized bonding and the ability of an electron pair to bind more than one pair of atoms (Section 14.18). Thus, if the X–H bond is regarded as formed from the overlap of an orbital on X, ψ_X, and a hydrogen 1s orbital, ψ_H, and the lone pair on Y occupies an orbital on Y, ψ_Y, then when the two molecules are close together, we can build three molecular orbitals from the three basis orbitals:

$$\psi = c_1 \psi_X + c_2 \psi_H + c_3 \psi_Y$$

One of the molecular orbitals is bonding, one almost nonbonding, and the third antibonding (Fig 16.9). These three orbitals need to accommodate four electrons (two from the original X–H bond and two from the lone pair of Y), and they may do so if two enter the bonding orbital and two enter the nonbonding orbital. Because the antibonding orbital remains empty, the net effect is a lowering of energy.

Hydrogen bond formation, which has a typical strength of the order of 20 kJ mol^{-1},[6] dominates all other van der Waals interactions when it can occur. It accounts for the rigidity of molecular solids such as sucrose and ice, the secondary structure of proteins (the formation of helices and sheets of polypeptide chains), the low vapour pressure of liquids such as water, and their high viscosities and surface tensions. Hydrogen bonding also contributes to the solubility in water of species such as ammonia and compounds containing hydroxyl groups and to the hydration of anions. In this last case, even ions such as Cl$^-$ and HS$^-$ can participate in hydrogen bond formation with water, for their charge enables them to interact with the hydroxylic protons of H$_2$O.

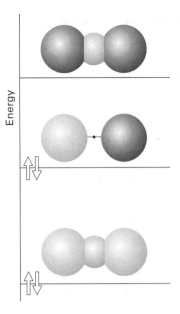

Fig 16.9 A schematic portrayal of the molecular orbitals that can be formed from an X, H, and Y orbital and which gives rise to an X–H$\cdots$Y hydrogen bond. The lowest energy combination is fully bonding, the next nonbonding, and the uppermost is antibonding. The antibonding orbital is not occupied by the electrons provided by the X–H bond and the :Y lone pair, so the configuration shown may result in a net lowering of energy in certain cases (namely when the X and Y atoms are N, O, or F).

Table 16.3 summarizes the strengths and distance dependences of the attractive forces that we have considered so far.

16.7 The total interaction

The total attractive interaction energy between rotating molecules that cannot participate in hydrogen bonding is the sum of the contributions from the dipole–dipole, dipole–induced-dipole, and dispersion interactions. Only the dispersion interaction contributes if both molecules are nonpolar. All three interactions vary as the inverse sixth power of the separation, so we may write

$$V = -\frac{C}{r^6} \tag{16.12}$$

where C is a coefficient that depends on the identity of the molecules.

---

[6] This figure is approximately half the enthalpy of vaporization of water: in water, there is an average of two hydrogen bonds per H$_2$O molecule.

Table 16.3 *Interaction potential energies*

Interaction type	Distance dependence of potential energy	Typical energy/ (kJ mol^{-1})	Comment
Ion–ion	$1/r$	250	Only between ions
Ion–dipole	$1/r^2$	15	
Dipole–dipole	$1/r^3$	2	Between stationary polar molecules
	$1/r^6$	0.3	Between rotating polar molecules
London (dispersion)	$1/r^6$	2	Between all types of molecules

The energy of a hydrogen bond X—H$\cdots$Y is typically 20 kJ mol^{-1} and occurs on contact for X, Y = N, O, or F.

Repulsive terms become important and begin to dominate the attractive forces when molecules are squeezed together (Fig 16.10), for instance, during the impact of a collision, under the force exerted by a weight pressing on a substance, or simply as a result of the attractive forces drawing the molecules together. These repulsive interactions arise in large measure from the Pauli exclusion principle, which forbids pairs of electrons being in the same region of space. The repulsions increase steeply with decreasing separation in a way that can be deduced

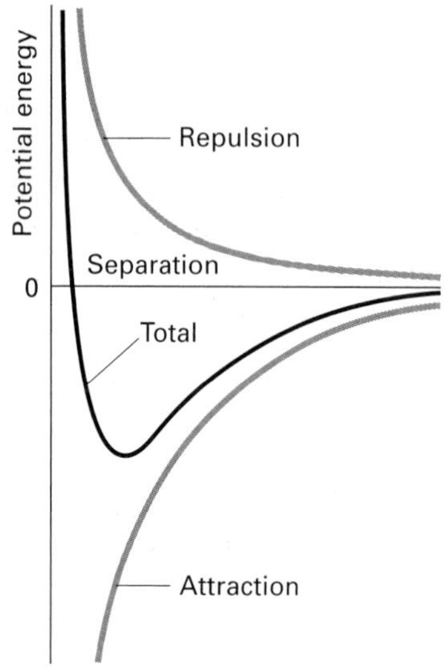

Fig 16.10 The general form of an intermolecular potential energy curve (the graph of the potential energy of two closed-shell species as the distance between them is changed). The attractive (negative) contribution has a long range, but the repulsive (positive) interaction increases more sharply once the molecules come into contact. The overall potential energy is shown by the heavy line.

only by very extensive, complicated molecular structure calculations. In many cases, however, progress can be made by using a greatly simplified representation of the potential energy, where the details are ignored and the general features expressed by a few adjustable parameters.

One such approximation is the **hard-sphere potential**, in which it is assumed that the potential energy rises abruptly to infinity as soon as the particles come within some separation σ (Fig 16.11):

$$V = \begin{cases} \infty & \text{for } r \leq \sigma \\ 0 & \text{for } r > \sigma \end{cases} \tag{16.13}$$

This very simple potential is surprisingly useful for assessing a number of properties.

Another widely used approximation is to express the short-range repulsive potential energy as inversely proportional to a high power of r:

$$V = +\frac{C^*}{r^n} \tag{16.14}$$

where C^* is another constant (the star signifies repulsion). Typically, n is set equal to 12, in which case the repulsion dominates the $1/r^6$ attractions strongly at short separations because then $C^*/r^{12} \gg C/r^6$. The sum of the repulsive interaction with $n = 12$ and the attractive interaction given by eqn 16.12 is called the **Lennard-Jones (12,6) potential**. It is normally written in the form

$$V = 4\varepsilon \left\{ \left(\frac{\sigma}{r}\right)^{12} - \left(\frac{\sigma}{r}\right)^6 \right\} \tag{16.15}$$

and is drawn in Fig 16.12. The two parameters are now ε, the depth of the well, and σ, the separation

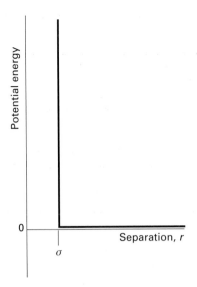

Fig 16.11 The true intermolecular potential can be modelled in a variety of ways. One of the simplest is this *hard-sphere potential*, in which there is no potential energy of interaction until the two molecules are separated by a distance σ when the potential energy rises abruptly to infinity as the impenetrable hard spheres repel each other.

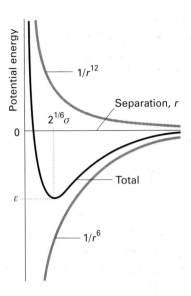

Fig 16.12 The Lennard-Jones potential is another approximation to the true intermolecular potential energy curves. It models the attractive component by a contribution that is proportional to $1/r^6$, and the repulsive component by a contribution that is proportional to $1/r^{12}$. Specifically, these choices result in the Lennard-Jones (12,6)-potential. Although there are good theoretical reasons for the former, there is plenty of evidence to show that $1/r^{12}$ is only a very poor approximation to the repulsive part of the curve.

at which $V = 0$. Some typical values are listed in Table 16.4. Although the (12,6)-potential has been used in many calculations, there is plenty of evidence to show that $1/r^{12}$ is a very poor representation of the repulsive potential, and that the exponential form $e^{-r/\sigma}$ is superior. An exponential function is more faithful to the exponential decay of atomic wavefunctions at large distances, and

hence to the distance-dependence of the overlap that is responsible for repulsion. However, a disadvantage of the exponential form is that it is slower to compute, which is important when considering the interactions between the large numbers of atoms in polypeptides and nucleic acids.[7]

Table 16.4 *Lennard–Jones parameters for the (12,6) potential*

	$\varepsilon/(kJ\ mol^{-1})$	σ/pm
Ar	128	342
Br$_2$	536	427
C$_6$H$_6$	454	527
Cl$_2$	368	412
H$_2$	34	297
He	11	258
Xe	236	406

Self-test 16.6

At what separation does the minimum of the potential energy curve occur for a Lennard-Jones potential?

[*Answer:* $r = 2^{1/6}\sigma$]

---

[7] A further computational advantage of the (12,6)-potential is that, once r^6 has been calculated, r^{12} is obtained by taking the square.

Biopolymers

BIOPOLYMERS like the polypeptides and nucleic acids provide an interesting and important illustration of how valence forces and van der Waals forces jointly determine the shape of a molecule. The overall shape of a polypeptide, for instance, is sustained by a variety of intermolecular forces of the kind we have encountered in this and earlier chapters, including hydrogen bonding, the hydrophobic effect, and interactions between partial charges.

16.8 Polypeptide structures

The **primary structure** of a biopolymer is the sequence of its monomer units: this sequence is determined by valence forces in the sense that the monomers are linked by covalent bonds. For polypeptides, which we consider here, the primary structure is an ordered list of the amino acid residues. The **secondary structure** of a polypeptide is the spatial arrangement of the polypeptide chain—its twisting into a specific shape—under the influence of interactions between the various peptide residues (the amino acid groups).

We can rationalize the secondary structures of proteins in large part in terms of the hydrogen bonds between the –NH– and –CO– groups of the peptide links (Fig 16.13). These bonds lead to two principal structures. One, which is stabilized by hydrogen bonding between peptide links of the same chain, is the **α helix**. The other, which is stabilized by hydrogen bonding links to different chains or more distant parts of the same chain, is the **β-sheet**.[8]

The α helix is illustrated in Fig 16.14. Each turn of the helix contains 3.6 amino acid residues, so there are 18 residues in 5 turns of the helix. The pitch of a

---

[8] The β-sheet is still widely known by its former name, the *β-pleated sheet.*

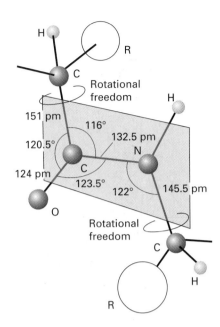

Fig 16.13 The dimensions of a typical peptide link. The C–CO–NH–C atoms define a plane (the C–N bond has partial double-bond character), but there is rotational freedom around the C–CO and N–C bonds.

single turn is 544 pm. The N–H···O bonds lie parallel to the axis and link every fifth group (so residue i is linked to residues $i - 4$ and $i + 4$). There is freedom for the helix to be arranged as either a right- or a left-handed screw, but the overwhelming majority of natural polypeptides are right-handed on account of the preponderance of the L-configuration of the naturally occurring amino acids. It turns out, in agreement with experience, that a right-handed α helix of L-amino acids has a marginally lower energy than a left-handed helix of the same acids.

Helical polypeptide chains are folded into a **tertiary structure** if there are other bonding influences between the residues of the chain that are strong enough to overcome the interactions responsible for the secondary structure. The folding influences include –S—S– **disulfide links**, ionic interactions (which depend on the pH), and strong hydrogen bonds (such as O—H···O–).

Proteins with $M > 50$ kg mol^{-1} (50 kDa) are often found to be aggregates of two or more polypeptide chains. The possibility of such **quaternary structure** often confuses the determination of their molar masses, because different techniques might

In contemporary physical chemistry and molecular biophysics, a great deal of work is being done on the rationalization and prediction of the structures of biomolecules such as polypeptides and nucleic acids using the interactions we have described in this section (Box 16.1).

16.9 Denaturation

Protein **denaturation**, or loss of structure, can be caused by several means, and different aspects of structure may be affected. The 'permanent waving' of hair, for example, is reorganization at the quaternary level. Hair is a form of the protein keratin, and its quaternary structure is thought to be a multiple helix, with the α helices bound together by disulfide links and hydrogen bonds. The process of permanent waving consists of disrupting these links, unravelling the keratin quaternary structure, and then reforming it into a more fashionable disposition. The 'permanence' is only temporary, however, because the structure of the newly formed hair is genetically controlled. Incidentally, normal hair grows at a rate that requires at least 10 twists of the keratin helix to be produced each second, so very close inspection of the human scalp would show it to be literally writhing with activity.

Denaturation at the secondary level is brought about by agents that destroy hydrogen bonds. Thermal motion may be sufficient, in which case denaturation is a kind of intramolecular melting. When eggs are cooked the albumin is denatured irreversibly, and the protein collapses into a structure resembling a random coil. The **helix–coil transition** is sharp, like ordinary melting, because it is a cooperative process in the sense that when one hydrogen bond has been broken it is easier to break its neighbours, and then even easier to break theirs, and so on.[9] The disruption cascades down the helix, and the transition occurs sharply. Denaturation may also be brought about chemically. For instance, a solvent that forms stronger hydrogen bonds than those within the helix will compete successfully for the NH and CO groups. Acids and bases can cause

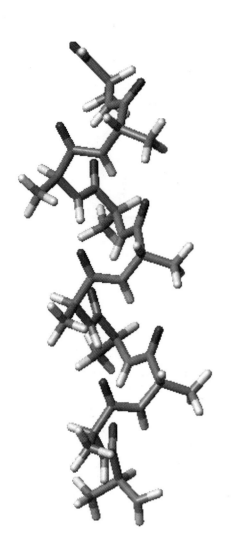

Fig 16.14 A tube representation of an α helix (polyalanine). There are 3.6 residues per turn, and a translation along the helix of 150 pm per residue, giving a pitch of 540 pm. The diameter (ignoring side chains) is about 600 pm.

give values differing by factors of 2 or more. Haemoglobin, which consists of four myoglobin-like chains (Fig 16.15), is an example of a quaternary structure. Myoglobin is an oxygen-storage protein. The subtle differences that arise when four such molecules coalesce to form haemoglobin result in the latter being an oxygen transport protein, able to load O_2 cooperatively and to unload it cooperatively too.

---

[9] This cooperative process is described more fully in Box 20.1.

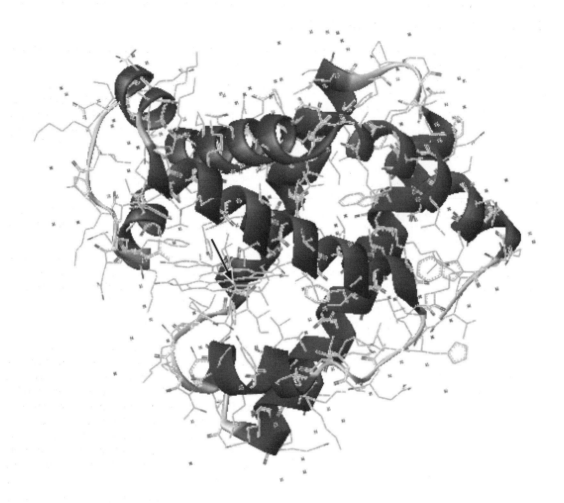

Fig 16.15 The molecular structure of myoglobin. A haemoglobin molecule consists of four units like this one. The O_2 molecule attaches to the iron atom indicated.

denaturation by protonation or deprotonation of various groups.

Liquids

THE starting point for the discussion of gases is the totally chaotic distribution of the molecules of a perfect gas. The starting point for the discussion of solids is the well ordered structure of perfect crystals. The liquid state is between these two extremes: there is some structure and some disorder. The particles of a liquid are held together by intermolecular forces, but their kinetic energies are comparable to their potential energies. As a result, although the molecules are not free to escape completely from the bulk, the whole structure is very mobile. The flow of molecules is like a crowd of spectators leaving a stadium.

Box 16.1 *The prediction of protein structure*

A polypeptide chain adopts a conformation corresponding to a minimum Gibbs energy. That simple perception, though, conceals a great deal of complexity. A part of the difficulty is that the Gibbs energy depends on the entropy as well as the energy of interaction between different parts of the chain. Secondly, the protein molecule is not isolated: its outer surface is covered by a mobile sheath of water molecules, and its interior contains pockets of water molecules. These water molecules play an important role in determining the conformation that the chain adopts through the hydrogen bonding in which it can participate and the hydrophobic effect (Box 4.1).

Even if we disregard these complications and concentrate on the energy alone, the problems are still severe. A simple view would be that the polypeptide adopts a conformation that corresponds to minimum energy. However, there is no guarantee that as the protein is formed in a cell it does not get trapped in a *local* minimum and, once fully formed and in the aqueous environment of a cell, it is unable to wriggle into a *global* minimum (see illustration). In other words, the conformation of a polypeptide may represent a kinetically trapped metastable structure rather than a true thermodynamically stable state.

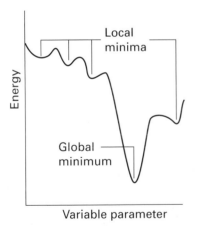

The variation of potential energy with the conformation of a molecule showing various local minima and a single global minimum.

To calculate the energy of a conformation we need to make use of many of the interactions described earlier in the chapter. The calculation is an elaboration of the one illustrated in Example 16.2, with several additional interactions taken into account:

1 *Bond stretch.* Bonds are not rigid, and it may be advantageous for some bonds to stretch and others to be compressed slightly as parts of the chain press against one another. If we liken a bond to a spring, then the potential energy depends on the displacement from equilibrium as

$$V_{stretch} = \tfrac{1}{2}k_{stretch}(R - R_e)^2$$

where R_e is the equilibrium bond length and $k_{stretch}$ is the force constant (a measure of the stiffness of the bond in question).

2 *Bond bend.* An O–C–H bond angle (or some other angle) may open out or close up slightly to enable the molecule as a whole to fit together better. If the equilibrium bond angle is θ_e, we write

$$V_{bend} = \tfrac{1}{2}k_{bend}(\theta - \theta_e)^2$$

3 *Interaction between partial charges.* The calculation is based on eqn 16.1. In some models of structure, the interaction between partial charges is judged to take into account the effect of hydrogen bonding; in other models, hydrogen bonding is added as another very short range interaction. The interaction between partial charges does away with the need to take dipole–dipole interactions into account, for they are taken care of by dealing with each partial charge explicitly.

4 *Dispersion interactions.* The London forces between atoms are taken into account by using the London formula, eqn 16.11, and its inverse-sixth power dependence on the separation.

5 *Repulsive interactions.* Other than the repulsion between like partial charges, which is taken into account by the charge–charge contribution, we need to take into account the effect of overlap of wavefunctions when two closed-shell atoms come into contact. That can be done by using a $1/r^{12}$ term, as in the Lennard-Jones (6,12)-potential, eqn 16.15.

These interactions—with a number of enhancements and modifications—are built into commercially available software packages. These packages also include schemes for modifying the locations of the atoms in a systematic way and searching for local and global minima.

Exercise 1 Theoretical studies have estimated that the lumiflavin isoalloazine ring system has an energy minimum at the bending angle of 15°, but that it requires only 8.5 kJ mol^{-1} to increase the angle to 30°. If there are no other compensating forces, what is the force constant for lumiflavin bending?

Exercise 2 The equilibrium bond length of a carbon–carbon single bond is 152 pm. Given a C–C force constant of 400 N m^{-1}, how much energy, in kilojoules per mole, would it take to stretch the bond to 165 pm?

16.10 The relative positions of molecules

We describe the average locations of the particles in the liquid in terms of the **pair distribution function**, g. This function is defined so that $g\delta r$ is the probability that a molecule will be found at a distance between r and $r + \delta r$ from another molecule.[10] It follows that, if g passes through a maximum at a radius of, for instance, 0.5 nm, then the most probable distance (regardless of direction) at which a second molecule will be found will be at 0.5 nm from the first molecule.

In a crystal, g is an array of sharp spikes, representing the certainty (in the absence of defects and thermal motion) that particles lie at definite locations. This regularity continues out to large distances (to the edge of the crystal, billions of molecules away), so we say that crystals have **long-range order**. When the crystal melts, the long-range order is lost and wherever we look at long distances from a given particle there is equal probability of finding a second particle. Close to the first particle, though, there may be a remnant of order (Fig 16.16). Its nearest neighbours might still adopt approximately their original positions, and even if they are displaced by newcomers the new particles might adopt their vacated positions. It may still be possible to detect, on average, a sphere of nearest neighbours at a distance r_1, and perhaps beyond them a sphere of next-nearest neighbours at r_2. The existence of this **short-range order** means that g can be expected to have a broad but pronounced peak at r_1, a smaller and broader peak at r_2, and perhaps some more structure beyond that.

---

[10] Recall the analogous quantity used to describe the distance of an electron from an atom, Section 13.4.

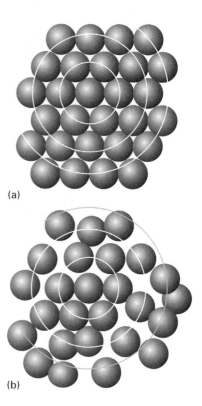

(a)

(b)

Fig 16.16 (a) In a perfect crystal at $T = 0$, the distribution of molecules (or ions) is highly regular, and the pair distribution function shows a series of sharp peaks showing the regular organization of rings of neighbours around any selected central molecule or ion. (b) In a liquid, there remain some elements of structure close to each molecule, but the further the distance, the less the correlation. The pair distribution function now shows a pronounced (but broadened) peak corresponding to the nearest neighbours of the molecule of interest (which are only slightly more disordered than in the solid), and a suggestion of a peak for the next ring of molecules, but little structure at greater distances.

The pair distribution function can be determined experimentally by X-ray diffraction, for g can be extracted from the diffuse diffraction pattern characteristic of liquid samples in much the same way as a crystal structure is obtained from X-ray diffraction of crystals (see Section 15.8). The shells of local structure shown in the example in Fig 16.17 (for water) are unmistakable. Closer analysis shows that any given H_2O molecule is surrounded by other molecules at the corners of a tetrahedron, similar to the arrangement in ice (Fig 16.18). The form of g at 100°C shows that the intermolecular forces (in this case, largely hydrogen bonds) are strong enough to affect the local structure right up to the boiling point.

16.11 Molecular motion in liquids

A molecule in a liquid is surrounded by other molecules, and it can move only a fraction of a diameter, perhaps because its neighbours move aside momentarily, before colliding. Molecular motion in liquids is a series of short steps, with incessantly changing directions, like people in an aimless, milling crowd.

The process of migration by means of a random jostling motion though a liquid is called **diffusion**. We can think of the motion of the molecule

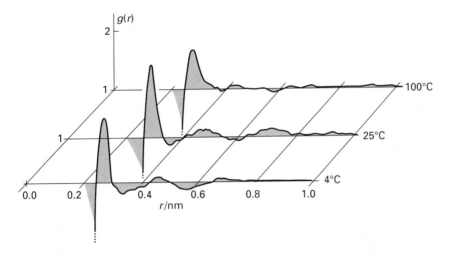

Fig 16.17 The experimentally determined radial distribution function of the oxygen atoms in liquid water at three temperatures. Note the expansion as the temperature is raised.

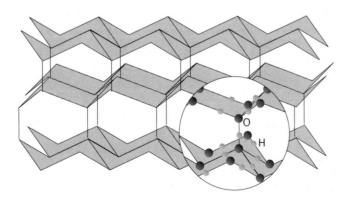

Fig 16.18 A fragment of the crystal structure of ice. Each O atom is at the centre of a tetrahedron of four O atoms at a distance of 276 pm. The central O atom is attached by two short O–H bonds to two H atoms and by two relatively long O⋯H bonds to two neighbouring H_2O molecules. Overall, the structure consists of planes of hexagonal puckered rings of H_2O molecules (like the chair form of cyclohexane).

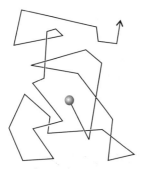

Fig 16.19 One possible path of a random walk in three dimensions. In this general case, the step length is also a random variable.

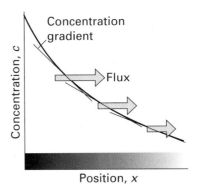

Fig 16.20 The flux of solute particles is proportional to the concentration gradient. Here we see a solution in which the concentration falls from left to right (as depicted by the shaded band and the curve). The gradient is negative (down from left to right) and the flux is positive (towards the right). The greatest flux is found where the gradient is steepest (at the left).

as a series of short jumps in random directions, a so-called **random walk** (Fig 16.19). If there is an initial concentration gradient in the liquid (for instance, a solution may have a high concentration of solute in one region), then the rate at which the molecules spread out is proportional to the concentration gradient, and we write

Rate of diffusion = D × concentration gradient

The coefficient D is called the **diffusion coefficient**: if it is large, molecules diffuse rapidly. Some values are given in Table 16.5. To express this relation mathematically, we introduce the **flux**, J, which is the number of particles passing through an imaginary window in a given time interval, divided by the area of the window and the duration of the interval:

Flux, J =

$$\frac{\text{number of particles passing through window}}{\text{area of window} \times \text{time interval}}$$

Then,

$$J = -D \times \text{concentration gradient} \qquad (16.16)$$

The negative sign simply means that, if the concentration gradient is negative (down from left to right, Fig 16.20), then the flux is positive (flowing from left to right). To get the number passing through a given window in a given time, we multiply the flux by the area of the window and the time interval.

Diffusion coefficients are of the greatest importance for discussing the spread of pollutants in lakes and through the atmosphere. In both cases, the spread of pollutant may be assisted by bulk motion of the fluid as a whole (as when a wind blows in the atmosphere). This motion is called **convection**. Because diffusion is often a slow process, we speed up the spread of solute molecules by inducing convection by stirring a fluid or turning on an extractor fan.

One of the most important equations in the physical chemistry of fluids is the *diffusion equation*, which enables us to predict the rate at which the concentration of a solute changes in a nonuniform solution. In essence, the diffusion equation expresses the fact that wrinkles in the concentration tend to disperse.

Table 16.5 *Diffusion coefficients at 25°C, D/$(10^{-9}\,m^2\,s^{-1})$*

H_2O in water	2.26
Ar in tetrachloromethane	3.63
CH_3OH in water	1.58
$C_{12}H_{22}O_{11}$ (sucrose) in water	0.522
NH_2CH_2COOH in water	0.673
O_2 in tetrachloromethane	3.82

Derivation 16.3 *The diffusion equation*

To express this idea quantitatively, we consider the arrangement in Fig 16.21. The number of solute particles passing through the window of area A located at x in an interval dt is $J(x)A\,dt$, where $J(x)$ is the flux at the location x. The number of particles passing out of the region through a window of area A at $x + dx$ is $J(x + dx)A\,dt$, where $J(x + dx)$ is the flux at the location of this window. The flux in and the flux out will be different if the concentration gradients are different at the two windows. The net change in the number of solute particles in the region between the two windows is

$$\text{Net change in number} = J(x)A\,dt - J(x + dx)A\,dt$$
$$= \{J(x) - J(x + dx)\}A\,dt$$

Now we express the two fluxes in terms of the concentration gradients at the locations of the two windows by using eqn 16.16, which gives

$$\text{Net change in number} =$$
$$\left\{-D\left(\frac{dc}{dx}\right)_{\text{at } x} + D\left(\frac{dc}{dx}\right)_{\text{at } x + dx}\right\}A\,dt$$

Now we come to the only tricky part of the calculation. The concentration gradients are slightly different at x and $x + dx$, so we write

$$\left(\frac{dc}{dx}\right)_{\text{at } x + dx} = \left(\frac{dc}{dx}\right)_{\text{at } x} + \left(\frac{d^2c}{dx^2}\right)_{\text{at } x}dx$$

where d^2c/dx^2 is the 'second derivative' of the concentration; in effect, the second derivative is just the curvature of the graph of the concentration. When we substitute this expression into the preceding one, the concentration gradients at x cancel and we are left with

$$\text{Net change in number} = D\left(\frac{d^2c}{dx^2}\right)A\,dx\,dt$$

The change in concentration inside the region between the two windows is the net change in number divided by the volume of the region (which is Adx):

$$\text{Change in concentration} = D\left(\frac{d^2c}{dx^2}\right)dt$$

The rate of change in concentration, dc/dt, is therefore[11]

$$\frac{dc}{dt} = D\frac{d^2c}{dx^2} \tag{16.17}$$

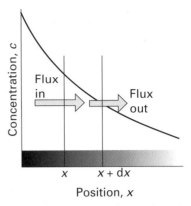

Fig 16.21 To calculate the change in concentration in the region between the two walls, we need to consider the net effect of the influx of particles from the left and their efflux towards the right. Only if the slope of the concentration is different at the two walls will there be a net change.

The **diffusion equation**, eqn 16.17, tells us that the rate of change of concentration in a small region is proportional to the curvature of the concentration there. A uniform concentration and a concentration with unvarying slope through the region results in no net change in concentration because the rate of influx through one wall is equal to the rate of efflux through the opposite wall. Only if the slope of the concentration varies through the region—only if the concentration is wrinkled—is there a change in concentration. Where the curvature is positive (a dip, Fig 16.22) the change in concentration is positive: the dip tends to fill. Where the curvature is negative (a heap), the change in concentration is negative: the heap tends to spread.

We can understand the nature of diffusion more deeply by considering it as the outcome of a random walk. Although a molecule undergoing a random walk may take many steps in a given time, it has only a small probability of being found far from its starting point because some of the steps lead it away from the starting point but others lead it back (Fig 16.23). The net distance travelled in a time t from the starting point is measured by the **root mean square distance**, d, with

---

[11] Because the concentration is a function of both time and location, the derivatives are in fact *partial* derivatives; but we are not using that notation in this book.

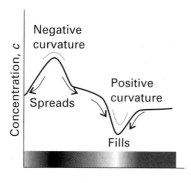

Fig 16.22 Nature abhors a wrinkle. The diffusion equation tells us that peaks in a distribution (regions of negative curvature) spread and troughs (regions of positive curvature) fill in.

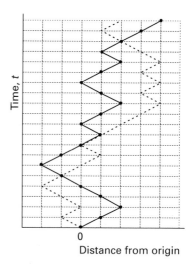

Fig 16.23 The directions of a typical sequence of steps in a one-dimensional random walk. Note how the walker might still be close to the origin after many steps. Two walks are shown here. The equations in the text refer to an average outcome of many such walks.

$$d = (2Dt)^{1/2} \tag{16.18}$$

Thus, the net distance increases only as the square root of the time, so for a particle to be found twice as far (on average) from its starting point, we must wait four times as long.

Self-test 16.7

The diffusion coefficient of H_2O in water is 2.26×10^{-9} $m^2\ s^{-1}$ at 25°C. How long does it take for an H_2O molecule to travel (a) 1.0 cm, (b) 2.0 cm from its starting point in a sample of unstirred water?

[*Answer*: (a) 6.1 h, (b) 25 h]

The relation between the diffusion coefficient and the rate at which the molecule takes its steps and the distance of each step is called the **Einstein–Smoluchowski equation**:

$$D = \frac{\lambda^2}{2\tau} \tag{16.19}$$

where λ is the length of each step and τ (tau) is the time each step takes. This equation tells us that a molecule that takes rapid, long steps has a high diffusion coefficient. We can interpret τ as the average lifetime of a molecule near another molecule before it makes a sudden jump to its next position.

Self-test 16.8

Suppose an H_2O molecule moves through one molecular diameter (about 200 pm) each time it takes a step in a random walk. What is the time for each step at 25°C?

[*Answer*: 9 ps]

The diffusion coefficient increases with temperature because an increase in temperature enables a molecule to escape more easily from the attractive forces exerted by its neighbours. If we suppose that the rate $(1/\tau)$ of the random walk follows an Arrhenius temperature dependence with an activation energy E_a, then the diffusion coefficient will follow the relation

$$D = D_0\ e^{-E_a/RT} \tag{16.20}$$

The rate at which particles diffuse through a liquid is related to the viscosity, and we should expect a high diffusion coefficient to be found for fluids that have a low viscosity. That is, we can suspect that $\eta \propto 1/D$, where η (eta) is the **coefficient of viscosity**. In fact, the **Einstein relation** states that

$$D = \frac{kT}{6\pi\eta a} \tag{16.21}$$

where a is the radius of the molecule. It follows that[12]

$$\eta = \eta_0\, e^{E_a/RT} \qquad\qquad (16.22)$$

This temperature dependence is observed, at least over reasonably small temperature ranges (Fig 16.24). The intermolecular potentials govern the magnitude of E_a, but the problem of calculating it is immensely difficult and still largely unsolved.

Self-test 16.9

Estimate the activation energy for the viscosity of water from the graph in Fig 16.24, by using the viscosities at 40°C and 80°C. (*Hint.* Use an equation like eqn 16.22 to formulate an expression for the logarithm of the ratio of the two viscosities.)

[*Answer*: 14 kJ mol^{-1}]

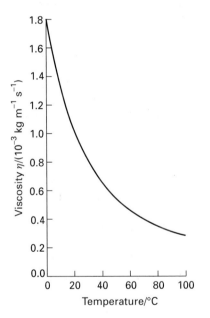

Fig 16.24 The experimental temperature dependence of the viscosity of water. As the temperature is increased, more molecules are able to escape from the potential wells provided by their neighbours, so the liquid becomes more fluid.

----

[12] Note the change in sign of the exponent: viscosity decreases as the temperature is raised. We are supposing that the strong temperature dependence of the exponential term dominates the weak linear dependence on T in the numerator of eqn 16.21.

Mesophases and disperse systems

A mesophase is a bulk phase that is intermediate in character between a solid and a liquid. The most important type of mesophase is a **liquid crystal**, which is a substance having liquid-like imperfect short-range order in some directions but some aspects of crystal-like long-range order in other directions. Liquid crystals can be used as models of biological cell walls and studied to gain insight into the process of transport through membranes. They are also of considerable technological importance for their use in liquid-crystal displays on electronic equipment. A **disperse system** is a dispersion of small particles of one material in another. The small particles are commonly called **colloids**. In this context, 'small' means something less than about 500 nm in diameter (about the wavelength of light). In general, they are aggregates of numerous atoms or molecules, but are too small to be seen with an ordinary optical microscope. They pass through most filter papers, but can be detected by light-scattering, sedimentation, and osmosis.

16.12 Liquid crystals

There are three important types of liquid crystal; they differ in the type of long-range order that they retain. One type of retained long-range order gives rise to a **smectic phase** (from the Greek word for soapy), in which the molecules align themselves in layers (Fig 16.25). Other materials, and some smectic liquid crystals at higher temperatures, lack the layered structure but retain a parallel alignment (Fig 16.26): this mesophase is the **nematic phase** (from the Greek for thread). The strongly anisotropic optical properties of nematic liquid crystals, and their response to electric fields, is the basis of their use as data displays. In the **cholesteric phase**, which is so-called because some derivatives of cholesterol form them, the molecules lie

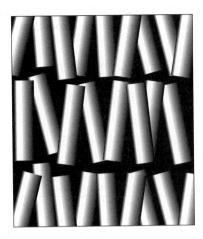

Fig 16.25 The arrangement of molecules in the smectic phase of a liquid crystal.

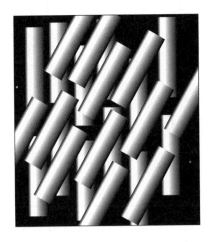

Fig 16.27 The arrangement of molecules in the cholesteric phase of a liquid crystal. Two layers are shown; the relative orientation of these layers is repeated in successive layers, to give a helical array of molecules.

16.13 **Classification of disperse systems**

The name given to the system depends on the nature of the substances involved. A **sol** is a dispersion of a solid in a liquid (such as clusters of gold atoms in water) or of a solid in a solid (such as ruby glass, which is a gold-in-glass sol, and achieves its colour by scattering). An **aerosol** is a dispersion of a liquid in a gas (like fog and many sprays) and of a solid in a gas (such as smoke): the particles are often large enough to be seen with a microscope. An **emulsion** is a dispersion of a liquid in a liquid (such as milk and some paints).

A further classification of colloids is as **lyophilic** (solvent attracting) and **lyophobic** (solvent repelling); in the case of water as solvent, the terms **hydrophilic** and **hydrophobic** are used instead. Lyophobic colloids include the metal sols. Lyophilic colloids generally have some chemical similarity to the solvent, such as OH groups able to form hydrogen bonds. A **gel** is a semirigid mass of a lyophilic sol in which all the dispersion medium has been absorbed by the sol particles.

The preparation of aerosols can be as simple as sneezing (which produces an aerosol). Laboratory and commercial methods make use of several tech-

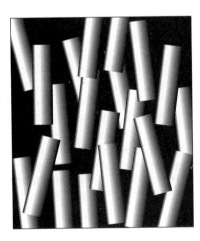

Fig 16.26 The arrangement of molecules in the nematic phase of a liquid crystal.

in sheets at angles that change slightly between neighbouring sheets (Fig 16.27), so forming helical structures. The pitch of the helix varies with temperature and, as a result, cholesteric liquid crystals diffract light and appear to have colours that depend on the temperature. They are used for detecting temperature distributions in living material, including human patients, and have even been incorporated into fabrics.

niques. Material (for example, quartz) may be ground in the presence of the dispersion medium. Passing a heavy electric current through a cell may lead to the crumbling of an electrode into colloidal particles; arcing between electrodes immersed in the support medium also produces a colloid. Chemical precipitation sometimes results in a colloid. A precipitate (for example, silver iodide) already formed may be converted to a colloid by the addition of **peptizing agent**, a substance that disperses a colloid. An example of a peptizing agent is potassium iodide, which provides ions that adhere to the colloidal particles and cause them to repel one another. Clays may be peptized by alkalis, the OH^- ion being the active agent.

Emulsions are normally prepared by shaking the two components together, although some kind of **emulsifying agent** has to be used in order to stabilize the product. This emulsifier may be a soap (a long-chain fatty acid), a surfactant, or a lyophilic sol that forms a protective film around the dispersed phase. In milk, which is an emulsion of fats in water, the emulsifying agent is casein, a protein containing phosphate groups. That casein is not completely successful in stabilizing milk is apparent from the formation of cream: the dispersed fats coalesce into oily droplets, which float to the surface. This separation may be prevented by ensuring that the emulsion is dispersed very finely initially: violent agitation with ultrasonics or extrusion through a very fine mesh brings this about, the product being 'homogenized' milk.

Aerosols are formed when a spray of liquid is torn apart by a jet of gas. The dispersal is aided if a charge is applied to the liquid, for then the electrostatic repulsions blast the jet apart into droplets. This procedure may also be used to produce emulsions, for the charged liquid phase may be squirted into another liquid.

Disperse systems are often purified by dialysis (recall Box 6.2). The aim is to remove much (but not all, for reasons explained later) of the ionic material that may have accompanied their formation. A membrane (for example, cellulose) is selected that is permeable to solvent and ions, but not to the bigger colloid particles. Dialysis is very slow, and is normally accelerated by applying an electric field and making use of the charge carried by many colloids; the technique is then called **electrodialysis**.

16.14 Surface, structure, and stability

The principal feature of colloids is the very great surface area of the dispersed phase in comparison with the same amount of ordinary material. For example, a cube of side 1 cm has a surface area of 6 cm^2. When it is dispersed as 10^{18} little 10 nm cubes the total surface area is $6 \times 10^6 \text{ cm}^2$ (about the size of a tennis court). This dramatic increase in area means that surface effects are of dominating importance in the chemistry of disperse systems.

As a result of their great surface area, colloids are thermodynamically unstable with respect to the bulk: that is, colloids have a thermodynamic tendency to reduce their surface area (like a liquid). Their apparent stability must therefore be a consequence of the kinetics of collapse: disperse systems are kinetically nonlabile, not thermodynamically stable. At first sight, though, even the kinetic argument seems to fail: colloidal particles attract one another over large distances by the dispersion interaction, so there is a long-range force tending to collapse them down into a single blob.

Several factors oppose the long-range dispersion attraction. There may be a protective film at the surface of the colloid particles that stabilizes the interface and cannot be penetrated when two particles touch. For example, the surface atoms of a platinum sol in water react chemically and are turned into $-Pt(OH)_3H_3$, and this layer encases the particle like a shell. A fat can be emulsified by a soap because the long hydrocarbon tails penetrate the oil droplet but the $-CO_2^-$ head groups (or other hydrophilic groups in detergents) surround the surface, form hydrogen bonds with water, and give rise to a shell of negative charge that repels a possible approach from another similarly charged particle.

By a **surfactant** we mean a species that accumulates at the interface of two phases or substances (one of which may be air) and modifies the properties of the surface. A typical surfactant consists of a long hydrocarbon tail that dissolves in hydrocarbon and other nonpolar materials, and a hydrophilic

head group, such as a carboxylate group, $-CO_2^-$, that dissolves in a polar solvent (typically water). In other words, a surfactant is an **amphipathic** substance,[13] meaning that it has both hydrophobic and hydrophilic regions. Soaps, for example, consist of the alkali metal salts of long-chain carboxylic acids, and the surfactant in detergents is typically a long-chain benzenesulfonic acid ($R-C_6H_4SO_3H$). The mode of action of a surfactant in a detergent, and of soap, is to dissolve in both the aqueous phase and the hydrocarbon phase where their surfaces are in contact, and hence to solubilize the hydrocarbon phase so that it can be washed away (Fig 16.28).

Surfactant molecules can group together as **micelles**, colloid-sized clusters of molecules, even in the absence of grease droplets, for their hydrophobic tails tend to congregate, and their hydrophilic heads provide protection (Fig 16.29). Micelles form only above the **critical micelle concentration** (CMC) and above the **Krafft temperature**. Nonionic surfactant molecules may cluster together in swarms of 1000 or more, but ionic species tend to

be disrupted by the Coulombic repulsions between head groups and are normally limited to groups of between 10 and 100 molecules. The shapes of the individual micelles vary with concentration. Although spherical micelles do occur, they are more commonly flattened spheres close to the CMC, and rod-like at higher concentrations. The interior of a micelle is like a droplet of oil, and magnetic resonance shows that the hydrocarbon tails are mobile, but slightly more restricted than in the bulk.

Micelles are important in industry and biology on account of their solubilizing function: matter can be transported by water after it has been dissolved in their hydrocarbon interiors. For this reason, micellar systems are used as detergents and drug carriers, and for organic synthesis, froth flotation, and petroleum recovery. They can be perceived as a part of a family of similar structures formed when amphipathic substances are present in water (Fig 16.30). A **monolayer** forms at the air–water interface, with the hydrophilic head groups facing the water. Micelles are like monolayers that enclose a region. A **bilayer vesicle** is like a double-micelle, with an inward pointing inner surface of molecules surrounded by and outward pointing outer layer. The 'flat' version of a bilayer vesicle is the analogue of a cell membrane (Box 16.2).

The thermodynamics of micelle formation shows that the enthalpy of formation in aqueous systems is probably positive (that is, that they are endothermic) with $\Delta H \approx 1$–2 kJ per mole of surfactant. That

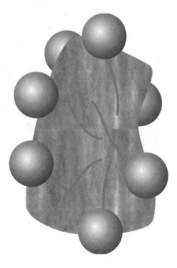

Fig 16.28 A surfactant molecule in a detergent or soap acts by sinking its hydrophobic hydrocarbon tail into the grease, so leaving its hydrophilic head groups on the surface of the grease where they can interact attractively with the surrounding water.

[13] The *amphi-* part of the name is from the Greek word for 'both', and the *-pathic* part is from the same root as sympathetic (meaning 'feeling').

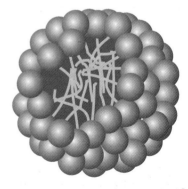

Fig 16.29 A representation of a spherical micelle. The hydrophilic groups are represented by spheres, and the hydrophobic hydrocarbon chains are represented by the stalks: the latter are mobile.

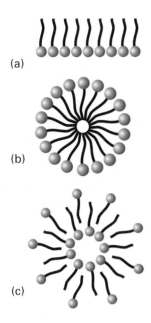

(a)

(b)

(c)

Fig 16.30 Amphipathic molecules form a variety of related structures in water: (a) a monolayer; (b) a spherical micelle; (c) a bilayer vesicle.

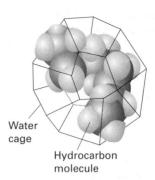

Water cage

Hydrocarbon molecule

Fig 16.31 When a hydrocarbon molecule is surrounded by water, the water molecules form a clathrate cage. As a result of this acquisition of structure, the entropy of the water decreases, so the dispersal of the hydrocarbon into water is entropy-opposed; the coalescence of many hydrocarbon molecules into a single large blob is entropy-favoured.

they do form above the CMC indicates that the entropy change accompanying their formation must then be positive (in order for the Gibbs energy accompanying the formation process to be negative), and measurements suggest a value of about +140 $J K^{-1} mol^{-1}$ at room temperature. That the entropy change is positive even though the molecules are clustering together shows that there must be a contribution to the entropy from the solvent and that its molecules must be more free to move once the solute molecules have herded into small clusters. This interpretation is plausible, because each individual solute molecule is held in an organized solvent cage (Fig 16.31), but once the micelle has formed the solvent molecules need form only a single (admittedly larger) cage. The increase in entropy when hydrophobic groups cluster together and reduce their structural demands on the solvent is the origin of the **hydrophobic interaction** that tends to stabilize groupings of hydrophobic groups in biological macromolecules (see Box 4.1). The hydrophobic interaction is an example of an ordering process that is stabilized by a tendency toward greater disorder of the solvent.

16.15 **The electric double layer**

Apart from the physical stabilization of disperse systems, a major source of kinetic stability is the existence of an electric charge on the surfaces of the colloidal particles. On account of this charge, ions of opposite charge tend to cluster nearby.

Two regions of charge must be distinguished. First, there is a fairly immobile layer of ions that stick tightly to the surface of the colloidal particle, and which may include water molecules (if that is the support medium). The radius of the sphere that captures this rigid layer is called the **radius of shear**, and is the major factor determining the mobility of the particles (Fig 16.32). The electric potential at the radius of shear relative to its value in the distant, bulk medium is called the **zeta potential**, ζ (zeta), or the **electrokinetic potential**. The charged unit attracts an oppositely charged ionic atmosphere. The inner shell of charge and the outer atmosphere jointly constitute the **electric double layer**.

At high concentrations of ions of high charge number, the atmosphere is dense and the potential falls to its bulk value within a short distance. In this case there is little electrostatic repulsion to hinder the close approach of two colloid particles. As a result, **flocculation**, the coalescence of the colloidal particles, occurs as a consequence of the van der

Box 16.2 *Cell membranes*

Although micelles, bilayers, and vesicles are convenient starting points for the discussion of the nature of cell membranes, actual membranes are highly sophisticated, complex structures. The basic structural element of a membrane is a phospholipid (**13**), which has the formula $CH_2(OCOR)CH(OCOR)CH_2OPO(OH)OX$, where R is a long hydrocarbon chain (typically in the range C_{14}–C_{24}) and X is one of a variety of polar groups, such as $-CH_2CH_2NH_2$ or $-CH_2CH(NH_2)COOH$. These twin-tailed amphipathic molecules stack together to form an extensive bilayer; the head groups are different on the outside of the cell from those on the inside, so the wall is asymmetric. The membrane is about 5 nm across, and each layer of the bilayer is called a *leaflet*. The lipid molecules form layers rather than micelles because the hydrocarbon chains are too bulky to allow packing into nearly spherical clusters. Interspersed among the phospholipids are sterols, such as cholesterol (**14**), which are also amphipathic: the –OH group is hydrophilic and the rest of the molecule hydrophobic. The presence of the sterols—which are present in different proportions in different types of cell—prevents the phospholipid tails freezing into a solid array and, by disrupting the packing of these tails, spreads the melting point of the membrane over a range of temperatures.

13 A phospholipid

The bilayer is a highly mobile structure. Not only are the hydrocarbon tails ceaselessly twisting and turning in the region between the two leaflets, but the phospholipid and cholesterol molecules are migrating over the

14 Cholesterol

surface. It is better to think of the membrane as a viscous fluid rather than a permanent structure, with a viscosity about 100 times that of water, rather like olive oil. The average distance, l, a phospholipid molecule diffuses in two dimensions in a time t is given by the formula $l = (4Dt)^{1/2}$, where D is the diffusion constant (which is of the order of 10^{-12} m^2 s^{-2}). This formula implies that a molecule migrates through about 1 μm (the diameter of a cell) in about 1 min. There is very little likelihood that a phospholipid molecule will undergo the process called *flip-flop* and migrate from one leaflet to another: estimates of the rate of this process suggest that one flip-flop requires several hours. Certain specialized enzymes, however, can accelerate this process.

The mobile but viscous lipid bilayer is like a sea in which there are immersed *integral proteins* that may span one or both leaflets of the bilayer and on which there float *peripheral proteins* that remain largely outside the outer leaflet. The structure of an integral protein is typically a sequence of hydrophilic residues, which remain outside the leaflet, followed by a helix of about 20 hydrophobic residues that sit comfortably within the hydrocarbon region of the bilayer, and then another sequence of hydrophilic residues reaching into the interior of the cell. Several of these units may form a quaternary structure; others are monomers with a more complex folded structure with a preponderance of hydrophobic regions in the interleaflet region of the bilayer (see the structure of rhodopsin in Box 18.1). The *fluid mosaic model* shown in the illustration, suggests how the proteins and lipids are organized. Once again, the proteins are mobile, but their diffusion constants are much smaller than those of the lipids.

The mobility of the bilayer enables it to flow round a molecule close to the outer surface, to engulf it, and incorporate it into the cell by the process called *endocytosis* (like an amoeba feeding). Alternatively, material from the cell interior wrapped in cell membrane may coalesce with the cell wall itself, which then withdraws and ejects the material in the process called *exocytosis*. The function of the proteins embedded in the bilayer, though, is

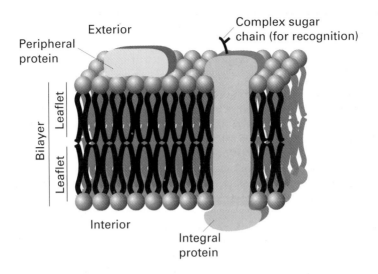

The mosaic model of a bilayer showing the various types of associated proteins.

to act as devices for transporting matter into and out of the cell in a more subtle manner. By providing hydrophilic channels through an otherwise alien hydrophobic environment, these proteins act as *ion channels* and *ion pumps* (see Box 9.2 on chemiosmotic theory).

Exercise 1 Lipid diffusion in a cell plasma membrane has a diffusion constant of 1.0×10^{-8} cm^2 s^{-1} and the same lipid in a lipid bilayer has a diffusion constant of 1.0×10^{-7} cm^2 s^{-1}. How long will it take the lipid to diffuse 10 nm in a plasma membrane and a lipid bilayer?

Exercise 2 Diffusion constants of proteins are often used as a measure of molar mass. For a spherical protein $D \propto M^{-1/2}$. Considering only two-dimensional diffusion, compare the length of time it would take ribonuclease ($M = 13.683$ kDa) to diffuse 10 nm to the length of time it would take the enzyme catalase ($M = 250$ kDa) to diffuse the same distance.

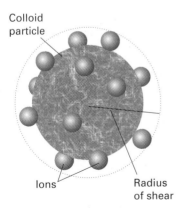

Fig 16.32 The definition of the radius of shear for a colloidal particle. The spheres are ions attached to the surface of the particle.

Waals forces. Flocculation is often reversible, and should be distinguished from **coagulation**, which is the irreversible collapse of the colloid into a bulk phase. When river water containing colloidal clay flows into the sea, the brine induces coagulation and is a major cause of silting in estuaries.

Metal oxide sols tend to be positively charged whereas sulfur and the noble metals tend to be negatively charged. Naturally occurring macromolecules also acquire a charge when dispersed in water, and an important feature of proteins and other natural macromolecules is that their overall charge depends on the pH of the medium. For instance, in acid environments protons attach to basic groups, and the net charge of the macromolecule is positive; in basic media the net charge is negative as a result of proton loss. At the **isoelectric point**, the pH is such that there is no net charge on the macro-molecule.

The primary role of the electric double layer is to render the colloid kinetically nonlabile. Colliding colloidal particles break through the double layer and coalesce only if the collision is sufficiently energetic to disrupt the layers of ions and solvating molecules, or if thermal motion has stirred away the surface accumulation of charge. This kind of disruption of the double layer may occur at high temperatures, which is one reason why sols precipitate when they are heated. The protective role of the double layer is the reason why it is important not to remove all the ions when a colloid is being purified by dialysis, and why proteins coagulate most readily at their isoelectric point.

The presence of charge on colloidal particles and natural macromolecules also permits us to control their motion, as in dialysis and electrophoresis. Apart from its application to the determination of molar mass, electrophoresis has several analytical and technological applications. One analytical application is to the separation of different macromolecules, and a typical apparatus is illustrated in Fig 16.33. Technical applications include silent ink-jet printers, the painting of objects by airborne charged paint droplets, and electrophoretic rubber forming by deposition of charged rubber molecules on anodes formed into the shape of the desired product (for example, surgical gloves).

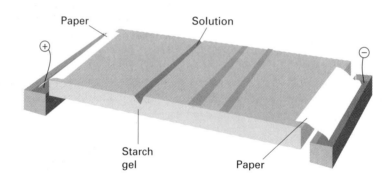

Fig 16.33 The layout of a simple electrophoresis apparatus. The sample is introduced into the trough in the gel, and the different components form separated bands under the influence of the potential difference.

Exercises

16.1 Estimate the dipole moment of an HCl molecule from the electronegativities of the elements and express the answer in debye and coulomb-metres.

16.2 Use the VSEPR model to judge whether SF_4 is polar.

16.3 The electric dipole moment of toluene (methylbenzene) is 0.4 D. Estimate the dipole moments of the three xylenes (dimethylbenzenes). Which value can you be sure about?

16.4 From the information in the preceding problem, estimate the dipole moments of (a) 1,2,3-trimethylbenzene, (b) 1,2,4-trimethylbenzene, and (c) 1,3,5-trimethylbenzene. Which value can you be sure about?

16.5 Calculate the resultant of two dipoles of magnitude 1.5 D and 0.80 D that make an angle 109.5° to each other.

16.6 At low temperatures a substituted 1,2-dichloroethane molecule can adopt the three conformations (**15**), (**16**), and (**17**) with different probabilities. Suppose that the dipole moment of each bond is 1.5 D.

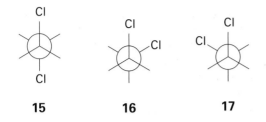

15   **16**   **17**

Calculate the mean dipole moment of the molecule when (a) all three conformations are equally likely, (b) only conformation (**16**) occurs, (c) the three conformations occur with probabilities in the ratio 2:1:1 and (d) 1:2:2.

16.7 Calculate the electric dipole moment of a glycine molecule using the partial charges in Table 16.1 and the locations of the atoms shown in (**18**).

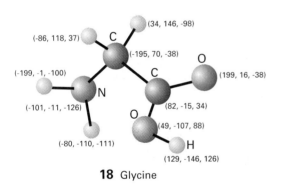

18 Glycine

16.8 (a) Plot the magnitude of the electric dipole moment of hydrogen peroxide as the H–OO–H (azimuthal) angle ϕ changes. Use the dimensions shown in (**19**). (b) Devise a way for depicting how the angle as well as the magnitude changes.

19 Hydrogen peroxide

16.9 Calculate the molar energy required to reverse the direction of a water molecule located (a) 100 pm, (b) 300 pm from a Li^+ ion. Take the dipole moment of water as 1.85 D.

16.10 Show, by following the procedure in Derivation 16.2, that eqn 16.6 describes the potential energy of two electric dipole moments in the orientation shown in structure (**9**).

16.11 What is the contribution to the total molar energy of (a) the kinetic energy, (b) the potential energy of interaction between hydrogen chloride molecules in a gas at 298 K when 1.00 mol of molecules is confined to 10 L? Is the kinetic theory of gases justifiable in this case?

16.12 The electric field at a distance r from a point charge q is equal to $q/4\pi\varepsilon_0 r^2$. How close to a water molecule (of polarizability volume 1.48×10^{-30} m^3) must a proton approach before the dipole moment it induces is equal to the permanent dipole moment of the molecule (1.85 D)?

16.13 Phenylanine (Phe, **20**) is a naturally occurring amino acid with a benzene ring. What is the maximum energy of interaction between its benzene ring and the electric dipole moment of a neighbouring peptide group? Take the distance between the groups as 4.0 nm and treat the benzene ring as benzene itself. The dipole moment of the peptide group is 1.26 D.

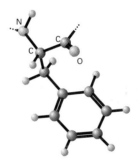

20 Phe

16.14 Now consider the London interaction between the benzene rings of two Phe residues (see Exercise 16.13). Estimate the potential energy of attraction between two such rings (treated as benzene molecules) separated by 4.0 nm. For the ionization energy, use $I = 5.0$ eV.

16.15 In a region of the oxygen-storage protein myoglobin, the OH group of a tyrosine residue is hydrogen bonded to the N atom of a histidine residue in the geometry shown in (**21**). Use the partial charges in Table 16.1 to estimate the potential energy of this interaction.

His

97.5pm

104.3pm

Tyr

21

16.16 Given that force is the negative slope of the potential, calculate the distance-dependence of the force acting between two nonbonded groups of atoms in a polypeptide chain that have a London dispersion interaction with each other. What is the separation at which the force is zero? (*Hint*. Calculate the slope by considering the potential energy at R and $R + \delta R$, with $\delta R \ll R$, and evaluating $[V(R + \delta R) - V(R)]/\delta R$. You should use the expansion in Derivation 16.2 together with

$$(1 \pm x + \cdots)^6 = 1 \pm 6x + \cdots$$
$$(1 \pm x + \cdots)^{12} = 1 \pm 12x + \cdots$$

At the end of the calculation, let δR become vanishingly small.)

16.17 Acetic acid vapour contains a proportion of planar, hydrogen-bonded dimers (**22**). The apparent dipole moment of molecules in pure gaseous acetic acid increases with increasing temperature. Suggest an interpretation of the latter observation.

22

16.18 The potential energy of a CH_3 group in ethane as it is rotated around the C–C bond can be written $V = \frac{1}{2}V_0(1 + \cos 3\phi)$, where ϕ is the azimuthal angle (**23**) and $V_0 = 11.6$ kJ mol^{-1}. (a) What is the change in potential energy be-

tween the *trans* and fully eclipsed conformations? (b) Show that, for small variations in angle, the torsional (twisting) motion around the C–C bond can be expected to be that of a harmonic oscillator. (c) Estimate the vibrational frequency of this torsional oscillation.

23

16.19 Suppose you distrusted the Lennard-Jones (12,6)-potential for assessing a particular polypeptide conformation, and replaced the repulsive term by an exponential function of the form $e^{-r/\sigma}$. Sketch the form of the potential energy and locate the distance at which it is a minimum.

16.20 A diffuse 'spherical crystal' was described in Exercise 15.6. Sketch the pair distribution function for such a substance.

16.21 Many liquids preserve some local features of the structure of the solid from which they form or to which they freeze. Sketch the form of the radial distribution function for a liquid that locally resembles (a) a cubic close-packed structure and (b) a body-centred cubic structure. In each case, show only the first two spheres of neighbours (the nearest and the next nearest).

16.22 What is (a) the flux of nutrient molecules down a concentration gradient of 0.10 mol L^{-1} m^{-1}, (b) the amount of molecules (in moles) passing through an area of 5.0 mm^2 in 1.0 min? Take for the diffusion coefficient the value for sucrose in water (5.22×10^{-10} m^2 s^{-1}).

16.23 How long does it take a sucrose molecule in water at 25°C to diffuse (a) 1 mm, (b) 1 cm, (c) 1 m from its starting point?

16.24 The mobility of species through fluids is of the greatest importance for nutritional processes. (a) Estimate the diffusion coefficient for a molecule that leaps 150 pm each 1.8 ps. (b) What would be the diffusion coefficient if the molecule travelled only half as far on each step?

16.25 How long will it take a small molecule of radius 200 pm to diffuse across a phospholipid bilayer of thickness 0.50 nm at 37°C if the viscosity within the bilayer is 0.010 kg m^{-1} s^{-1}?

16.26 Pollutants spread through the environment by convection (winds and currents) and by diffusion. How many steps must a molecule take to be 1000 step lengths away from its origin if it undergoes a one-dimensional random walk?

16.27 Suppose a toxin is injected into one end of a horizontal tube full of water, so initially there is a uniform concentration of the toxin for the first 5 mm of the tube. Sketch a sequence of illustrations to show how the concentration profile of the toxin changes with time. Ignore convection.

16.28 (a) Combine eqns 16.18 and 16.19 to obtain an expression for the total path length of a molecule that diffuses to a distance d in a series of steps of length λ. (b) Use the expression you derive to estimate the total journey length of a lysozyme molecule ($D = 1.2 \times 10^{-6}$ cm^2 s^{-1}) in water given that each step takes 10 ps and in the interval in question the molecule diffuses through a total path length of 1.0 cm.

16.29 In the technique called *isoelectric focusing* (IEF), buffers are used to establish a pH gradient in a gel between two electrodes. High pH is established at the negative electrode and low pH at the positive electrode. The sample of protein is introduced into the gel at the negative electrode. What will be observed?

Molecular rotations and vibrations

Contents

General features of spectroscopy

Rotational spectroscopy

Vibrational spectroscopy

SPECTROSCOPY is the analysis of the electromagnetic radiation emitted, absorbed, or scattered by molecules. We saw in Chapter 13 that photons act as messengers from inside atoms and that we can use atomic spectra to obtain detailed information about electronic structure. Photons of radiation ranging from long radio waves to the ultraviolet also bring information to us about molecules. The difference between molecular and atomic spectroscopy, however, is that the energy of a molecule can change not only as a result of electronic transitions but also because it can make transitions between its rotational and vibrational states. Molecular spectra are more complicated but they contain more information. As well as its role in structural investigation—the determination of electronic energy levels, bond lengths, bond angles, bond strengths, and other features—molecular spectroscopy is used to monitor changing concentrations in kinetic studies (Section 10.1).

This and the next two chapters survey several principal types of spectroscopy, using radiation that ranges over eight orders of magnitude, from radiofrequencies (10^8 Hz) up to the ultraviolet (10^{16} Hz).

General features of spectroscopy

IN **emission spectroscopy**, a molecule undergoes a transition from a state of high energy, E_2, to a state of lower energy, E_1, and emits the excess energy as a photon (Fig 17.1). In **absorption spectroscopy**, the absorption radiation is monitored as the frequency of the radiation is swept over a range. In **Raman spectroscopy**, an intense, monochromatic (single-frequency) incident beam is passed through the sample and we record the frequencies present in the radiation scattered by the sample (Fig 17.2).

The energy of a photon emitted or absorbed, and therefore the frequency, v, of the radiation emitted or absorbed, is given by the Bohr frequency condition (Section 13.1):

$$h v = |E_1 - E_2| \qquad (17.1)$$

Here E_1 and E_2 are the energies of the two states between which the transition occurs and h is Planck's constant.[1] This relation is often expressed in terms of the wavelength, λ, of the radiation by using the relation

$$\lambda = \frac{c}{v} \qquad (17.2a)$$

where c is the speed of light, or the wavenumber, $\tilde{v}$:

$$\tilde{v} = \frac{1}{\lambda} = \frac{v}{c} \qquad (17.2b)$$

The units of wavenumber are almost always chosen as reciprocal centimetres (cm^{-1}). The chart in Fig 17.3 summarizes the frequencies, wavelengths, and wavenumbers of the various regions of the electromagnetic spectrum.

---

[1] Raman scattering is a special case, and we deal with it later.

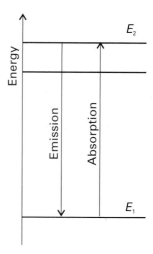

Fig 17.1 In emission spectroscopy, a molecule returns to a lower state (typically the ground state) from an excited state, and emits the excess energy as a photon. The same transition may be observed in absorption, when the incident radiation supplies a photon that can excite the molecule from its ground state to an excited state.

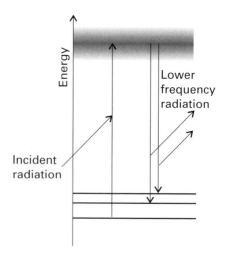

Fig 17.2 In Raman spectroscopy, an incident photon is scattered from a molecule with either an increase in frequency (if the radiation collects energy from the molecule) or—as shown here—with a lower frequency if it loses energy to the molecule. The process can be regarded as taking place by an excitation of the molecule to a wide range of states (represented by the shaded band), and the subsequent return of the molecule to a lower state; the net energy change is then carried away by the photon.

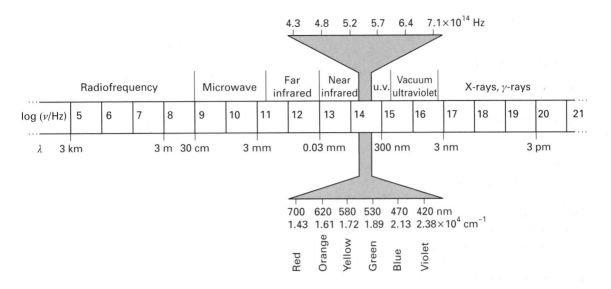

Fig 17.3 The electromagnetic spectrum and the classification of the spectral regions.

17.1 **Experimental techniques**

Emission, absorption, and Raman spectroscopy give the same information about energy level separations, but practical considerations generally determine which technique is employed. Emission spectroscopy, if it is used at all, is used mainly for the study of energetically excited reaction products. Absorption spectroscopy is much more widely used, and we shall concentrate on it.

The source in a spectrometer typically produces radiation spanning a range of frequencies, but in a few cases (including lasers) it generates nearly monochromatic radiation. For the far infrared, the source is a mercury arc inside a quartz envelope, most of the radiation being generated by the hot quartz. A **Nernst filament** is used to generate radiation in the near infrared: it consists of a heated ceramic filament containing rare-earth oxides, and emits radiation closely resembling that of a hot black body (of the kind considered in Section 12.1). For the visible region of the spectrum, a tungsten–iodine lamp is used, which gives out intense white light. A discharge through deuterium gas or xenon in quartz is still widely used for the near ultraviolet. A **klystron** (which is also used in radar installations and microwave ovens) or, more commonly now, a semiconductor device known as a **Gunn diode**, is used to generate microwaves. Radiofrequency radiation is generated by causing an electric current to oscillate in a coil of wire.

In all but specialized techniques using monochromatic microwaves and certain types of laser radiation, a spectrometer includes a component for separating the frequencies of the radiation so that the variation of the absorption with frequency can be monitored. In conventional spectrometers, this component is a **dispersing element** which separates different frequencies into rays that travel in different directions. The simplest dispersing element is a glass or quartz prism, but a diffraction grating is more widely used. A **diffraction grating** consists of a glass or ceramic plate into which fine grooves have been cut about 1000 nm apart (a spacing comparable to the wavelength of visible light) and covered with a reflective aluminium coating. The grating causes interference between waves reflected from its surface, and constructive interference occurs at specific angles that depend on the frequency of the radiation being used. Thus, each wavelength of light is directed into a specific direction.

The third common component of spectrometers is the **detector**, the device that converts incident radiation into an electric current for the appropriate signal processing or plotting. Radiation-sensitive semiconductor devices dominate this role. The radiation is chopped by a shutter that rotates in the beam so that an alternating signal is obtained from the detector (an oscillating signal is easier to amplify than a steady signal).

The highest resolution is obtained when the sample is gaseous and of such low pressure that collisions between the molecules are infrequent. Gaseous samples are essential for rotational (microwave) spectroscopy, for only in that phase can molecules rotate freely. To achieve sufficient absorption, the path lengths of gaseous samples must be very long, of the order of metres. Long path lengths are achieved by multiple passage of the beam between two parallel mirrors at each end of the sample cavity.

For infrared spectroscopy, the sample is typically a liquid held between windows of sodium chloride (which is transparent down to 700 cm^{-1}) or potassium bromide (down to 400 cm^{-1}). Other ways of preparing the sample include grinding it into a paste with 'Nujol', a hydrocarbon oil, or pressing it into a solid disk, perhaps with powdered potassium bromide.

17.2 **Intensities and linewidths**

The intensity of a spectral line depends on the number of molecules that are in the initial state and the strength with which individual molecules are able to interact with the electromagnetic field and generate or absorb photons. If we confine our attention to vibrational and electronic spectroscopy, then the situation is very simple: *almost all vibrational absorptions and all electronic absorptions occur from the ground state of a molecule*, because that is the only state populated at room temperature. However, molecules can be prepared in short-lived excited states as a result of chemical reaction, electric discharge, or photolysis. In these cases the populations may be quite different from those at thermal equilibrium, and absorption and emission spectra—if they can be recorded quickly enough—then arise from transitions from all the populated levels.

Spectroscopic lines are not infinitely narrow. In condensed media an electronic transition may spread over several thousand reciprocal centime-

tres: its width stems from the simultaneous excitation of molecular vibrations, with the individual spectral lines blending together to give a broad band. An important broadening process in gaseous samples is the **Doppler effect**, in which radiation is shifted in frequency when the source is moving towards or away from the observer. When a source emitting radiation of frequency v recedes with a speed s, the observer detects radiation of frequency

$$v' = \left(\frac{1 - s/c}{1 + s/c} \right)^{1/2} v \tag{17.3a}$$

where c is the speed of the radiation (the speed of light for electromagnetic radiation, the speed of sound for sound waves). A source approaching the observer appears to be emitting radiation of frequency

$$v' = \left(\frac{1 + s/c}{1 - s/c} \right)^{1/2} v \tag{17.3b}$$

Similar expressions apply to the wavenumber of the radiation.

Self-test 17.1

A laser line occurs at 628.443 cm^{-1}. What wavenumber will an observer detect when approaching the laser at (a) 1 m s^{-1}, (b) 1000 m s^{-1}?

[*Answer*: (a) 628.443 cm^{-1}, (b) 628.445 cm^{-1}]

Molecules reach high speeds in all directions in a gas, and a stationary observer detects the corresponding Doppler-shifted range of frequencies. Some molecules approach the observer, some move away; some move quickly, others slowly. The detected spectroscopic 'line' is the absorption or emission profile arising from all the resulting Doppler shifts. The profile reflects the Maxwell distribution of molecular speeds (Section 1.6) towards or away from the observer, and the outcome is that we observe a bell-shaped Gaussian curve (a curve of the form e$^{-x^2}$, Fig 17.4). When the temperature is T and the molar mass of the molecule is M, the width of the line at half its maximum height is

$$\delta\lambda = \frac{2\lambda}{c} \left(\frac{2RT \ln 2}{M} \right)^{1/2} \tag{17.4}$$

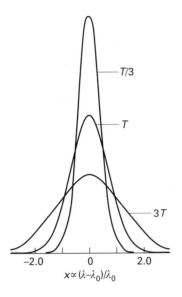

Fig 17.4 The shape of a Doppler-broadened spectral line reflects the Maxwell distribution of speeds in the sample at the temperature of the experiment. Notice that the line broadens as the temperature is increased. The width at half height is given by eqn 17.4.

The Doppler width increases with temperature because the molecules acquire a wider range of speeds. Therefore, to obtain spectra of maximum sharpness, it is best to work with cold gaseous samples.

Self-test 17.2

The Sun emits a spectral line at 677.4 nm which has been identified as arising from a transition in highly ionized ^{57}Fe. Its width at half-height is 5.3 pm. What is the temperature of the Sun's surface?

[*Answer*: 6.7×10^3 K]

Another source of line broadening is the finite lifetime of the states involved in the transition. When the Schrödinger equation is solved for a system that is changing with time, it is found that the states of the system do not have precisely defined energies. If the time-constant for the decay of a state

is τ (tau), which is called the **lifetime** of the state,[2] then its energy levels are blurred by δE, where

$$\delta E \approx \frac{\hbar}{\tau} \qquad (17.5a)$$

We see that, the shorter the lifetime of a state, the less well defined is its energy. The energy spread inherent to the states of systems that have finite lifetimes is called **lifetime broadening**.[3] When we express the energy spread as a wavenumber by writing $\delta E = hc\delta\tilde{v}$ and use the values of the fundamental constants, the practical form of this relation becomes

$$\delta\tilde{v} \approx \frac{5.3 \text{ cm}^{-1}}{\tau/\text{ps}} \qquad (17.5b)$$

Only if τ is infinite can the energy of a state be specified exactly (with $\delta E = 0$). However, no excited state has an infinite lifetime; therefore, all states are subject to some lifetime broadening and, the shorter the lifetimes of the states involved in a transition, the broader the spectral lines.

Self-test 17.3

What is the width (expressed as a wavenumber) of a transition from a state with a lifetime of 5.0 ps?

[*Answer*: 1.1 cm^{-1}]

Two processes are principally responsible for the finite lifetimes of excited states, and hence for the widths of transitions to or from them. The dominant one is **collisional deactivation**, which arises from collisions between molecules or with the walls of the container. If the collisional lifetime is τ_{col}, then the resulting collisional linewidth is $\delta E_{col} \approx \hbar/\tau_{col}$. In gases, the collisional lifetime can be lengthened, and the broadening minimized, by working at low pressures. The second contribution is **spontaneous emission**, the emission of radiation when an excited state collapses into a lower state. The rate of spontaneous emission depends on details of the wavefunctions of the excited and lower states. Because the rate of spontaneous emission cannot be changed (without changing the molecule), it is a

---

[2] The decay of the state is assumed to be exponential and proportional to $e^{-t/\tau}$.

[3] Lifetime broadening is also called *uncertainty broadening*.

natural limit to the lifetime of an excited state. The resulting lifetime broadening is the **natural linewidth** of the transition.

The natural linewidth of a transition cannot be changed by modifying the temperature or pressure. Natural linewidths depend strongly on the transition frequency v (they increase as v^3), so low-frequency transitions (such as the microwave transitions of rotational spectroscopy) have very small natural linewidths; for such transitions, collisional and Doppler line-broadening processes are dominant. The natural lifetimes of electronic transitions are very much shorter than for vibrational transitions, so the natural linewidths of electronic transitions are much greater than those of vibrational and rotational transitions. For example, a typical electronic excited state natural lifetime is about 10^{-8} s (10^4 ps), corresponding to a natural width of about 5×10^{-4} cm^{-1} (equivalent to 15 MHz).

Rotational spectroscopy

THE energy levels of molecules free to rotate are quantized, and transitions between these levels give rise to the **rotational spectrum** of a molecule. Very little energy is needed to change the state of rotation of a molecule, and the electromagnetic radiation emitted or absorbed lies in the microwave region, with wavelengths of the order of 1 cm. The rotational spectroscopy of gas-phase samples is therefore also known as **microwave spectroscopy**.

17.3 The rotational energy levels of molecules

To a first approximation, the rotational states of molecules are based on a model system called a **rigid rotor**, a body that is not distorted by rotation. The simplest type of rigid rotor is called a **linear**

rotor, and corresponds to a linear molecule, such as HCl, CO_2, or HC≡CH, that is supposed not to be able to bend or stretch under the stress of rotation. When the Schrödinger equation is solved for a linear rotor, the energy levels turn out to be

$$E_J = hBJ(J + 1) \qquad J = 0,1,2, \ldots \qquad (17.6)$$

where J is the rotational quantum number. The constant B (a frequency, with the units hertz, Hz) is called the **rotational constant** of the molecule, and is defined as

$$B = \frac{\hbar}{4\pi I} \qquad (17.7)$$

where I is the **moment of inertia** of the molecule. Note that, because B is inversely proportional to I, the larger the moment of inertia (corresponding to a long bond and heavy atoms), the smaller the rotational constant. The moment of inertia plays a role in rotation analogous to mass in translation. A body with a high moment of inertia (like that of a flywheel or a heavy molecule) undergoes only a small rotational acceleration when a twisting force (a torque) is applied, but a body with a small moment of inertia undergoes a large acceleration when subjected to the same torque. The moment of inertia of a molecule depends on the masses of the atoms and their distances from the centre of mass of the molecule (the point about which rotation occurs). For a diatomic molecule of bond length R and atomic masses m_A and m_B the moment of inertia is

$$I = \mu R^2 \qquad n = \frac{m_A m_B}{m_A + m_B} \qquad (17.8)$$

Figure 17.5 shows the energy levels predicted by eqn 17.6. Note that the separation increases with J. Note also that, because J may be 0, the lowest possible energy is 0: there is no zero-point rotational energy for molecules. The rotational quantum number also specifies the angular momentum of the molecule: a molecule with $J = 0$ has zero angular momentum and, as J increases, so does the molecule's angular momentum. In general, the rotational angular momentum is $\{J(J + 1)\}^{1/2}\hbar$, the same relation as that between the orbital angular momentum of an electron and the quantum number l.

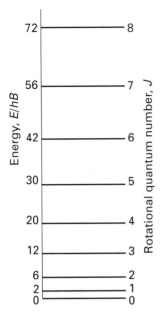

Fig 17.5 The energy levels of a linear rigid rotor as multiples of hB.

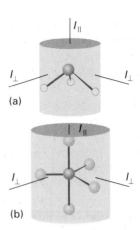

Fig 17.6 The two different moments of inertia of (a) a trigonal pyramidal molecule and (b) a trigonal bipyramidal molecule.

A number of nonlinear molecules can be modelled as a **symmetric rotor**, a rigid rotor in which the moments of inertia about two axes are the same but different from a third (and all three are nonzero).[4] An example is ammonia, NH_3, and another is phosphorus pentachloride, PCl_5 (Fig 17.6). The energy levels of a symmetric rotor are determined by two quantum numbers, J and K, and are

Fig 17.7 The two rotational constants of a symmetric rotor, which are inversely proportional to the moments of inertia parallel and perpendicular to the axis of the molecule.

$$E_{J,K} = hBJ(J + 1) + h(A - B)K^2$$
$$J = 0, 1, 2, \ldots$$
$$K = J, J - 1, \ldots, -J \qquad (17.9)$$

The rotational constants A and B are inversely proportional to the moments of inertia parallel and perpendicular to the axis of the molecule (Fig 17.7):

$$A = \frac{\hbar}{4\pi I_{\parallel}} \qquad B = \frac{\hbar}{4\pi I_{\perp}} \qquad (17.10)$$

The quantum number K tells us the extent to which the molecule is rotating about its axis (Fig 17.8). When $K = 0$, the molecule is rotating end-over-end. When $K = \pm J$ (the greatest values in its range), the

---

[4] The formal criterion of a molecule being a symmetric rotor is that it has an axis of threefold or higher symmetry.

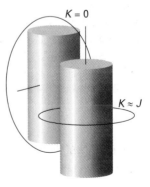

Fig 17.8 When $K = 0$ for a symmetric rotor, the entire motion of the molecule is around an axis perpendicular to the symmetry axis of the rotor. When the value of $|K|$ is close to J, almost all the motion is around the symmetry axis.

molecule is rotating mainly about its symmetry axis. Intermediate values of K correspond to a combination of the two modes of rotation.

A special case of a symmetric rotor is a **spherical rotor**, a rigid body with three equal moments of inertia (like a sphere). Tetrahedral, octahedral, and icosahedral molecules (CH_4, SF_6, and C_{60}, for instance) are spherical rotors. Their energy levels are very simple: when $I_\parallel = I_\perp$, the rotational constants A and B are equal and eqn 17.9 simplifies to eqn 17.6.

17.4 **Rotational transitions: microwave spectroscopy**

We met the concept of a *selection rule* in Section 13.7 as a statement about which spectroscopic transitions are forbidden and which are allowed. Selection rules also apply to molecular spectra, and the form they take depends on the type of transition. The idea to keep in mind is that, for the molecule to interact with the electromagnetic field and absorb or create a photon of frequency v, it must possess, at least transiently, an electric dipole oscillating at that frequency. The transient dipole moment associated with the transition is called a **transition moment**. A large transition moment gives a strong jolt to the electromagnetic field and results in intense emission or absorption.

A **gross selection rule** specifies the general features that a molecule must possess if it is to have a spectrum of a given kind. For a rotation of a molecule to give rise to an absorption or emission spectrum, the gross selection rule is that *the molecule must be polar*. The classical basis of this rule is that a stationary observer watching a rotating polar molecule sees its partial charges moving backwards and forwards and their motion shakes the electromagnetic field into oscillation (Fig 17.9). Because the molecule must be polar, it follows that tetrahedral (CH_4, for instance), octahedral (SF_6), symmetric linear (CO_2), and homonuclear diatomic (H_2) molecules do not have rotational spectra. On the other hand, heteronuclear diatomic (HCl) and less symmetrical polar polyatomic molecules (NH_3) are polar and do have rotational spectra. We say that polar molecules are **rotationally active** whereas nonpolar molecules are **rotationally inactive**.

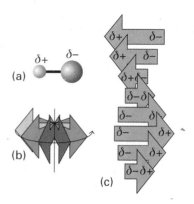

Fig 17.9 To an external observer, (a) a rotating polar molecule has (b) an electric dipole (the arrow) that (c) appears to oscillate. This oscillating dipole can interact with the electromagnetic field.

A **specific selection rule** identifies the details of the transitions, commonly by stating the values by which quantum numbers may change. For rotational transitions, the specific selection rules are

$$\Delta J = \pm 1 \qquad \Delta K = 0$$

The first of these selection rules can be traced, like the rule $\Delta l = \pm 1$ for atoms (Section 13.7), to the conservation of angular momentum when a photon is absorbed or created. A photon is a spin-1 particle, and when one is absorbed or created the angular momentum of the molecule must change by a compensating amount. Because J is a measure of the angular momentum of the molecule, J can change only by ± 1. The second selection rule ($\Delta K = 0$; that is, the quantum number K may not change) can be traced to the fact that the dipole moment of a polar molecule does not move when a molecule rotates around its symmetry axis (think of NH_3 rotating around its threefold axis). As a result, there can be no acceleration or deceleration of the rotation of the molecule about that axis by the absorption or emission of electromagnetic radiation.

When a molecule changes its rotational quantum number from J to $J + 1$ in an absorption, the change in rotational energy of the molecule is

$$\Delta E = E_{J+1} - E_J = hB(J + 1)(J + 2) - hBJ(J + 1)$$
$$= 2hB(J + 1) \qquad (17.11)$$

The energies of these transitions are $2hB$, $4hB$, $6hB$, ..., and the frequencies of the radiation absorbed are therefore $2B$, $4B$, $6B$, ...[5] A rotational spectrum of a polar linear molecule (HCl) and of a polar symmetric rotor (NH_3), therefore consists of a series of lines at frequencies separated by $2B$ (Fig 17.10).

Example 17.1 *Estimating the frequency of a rotational transition*

Estimate the frequency of the $J = 0 \rightarrow 1$ transition of the $^1H^{35}Cl$ molecule. The masses of the two atoms are 1.673×10^{-27} kg and 5.807×10^{-26} kg, respectively, and the equilibrium bond length is 127.4 pm.

Strategy The calculation depends on the value of B, which we obtain by substituting the data into eqn 17.7. The frequency of the transition is $2B$.

Solution The moment of inertia of the molecule is

$$I = \mu R^2$$

$$= \frac{(1.673 \times 10^{-27}\,\text{kg}) \times (5.807 \times 10^{-26}\,\text{kg})}{(1.673 \times 10^{-27}\,\text{kg}) + (5.807 \times 10^{-26}\,\text{kg})}$$

$$\times (1.274 \times 10^{-10}\,\text{m})^2$$

$$= 2.639 \times 10^{-47}\,\text{kg m}^2$$

Therefore, the rotational constant is

$$B = \frac{\hbar}{4\pi I}$$

$$= \frac{1.054\,57 \times 10^{-34}\,\text{J s}}{4\pi(2.639 \times 10^{-47}\,\text{kg m}^2)} = 3.180 \times 10^{11}\,\text{s}^{-1}$$

or 318.0 GHz (1 GHz = 10^9 Hz). It follows that the frequency of the transition is

$$v = 2B = 636.0\,\text{GHz}$$

This frequency corresponds to the wavelength 0.4712 mm.

Self-test 17.4

What is the frequency and wavelength of the same transition in the $^2H^{35}Cl$ molecule? The mass of ^2H is 3.344×10^{-27} kg. Before commencing the calculation, decide whether the frequency should be higher or lower than for $^1H^{35}Cl$.

[*Answer*: 327.0 GHz, 0.9168 mm]

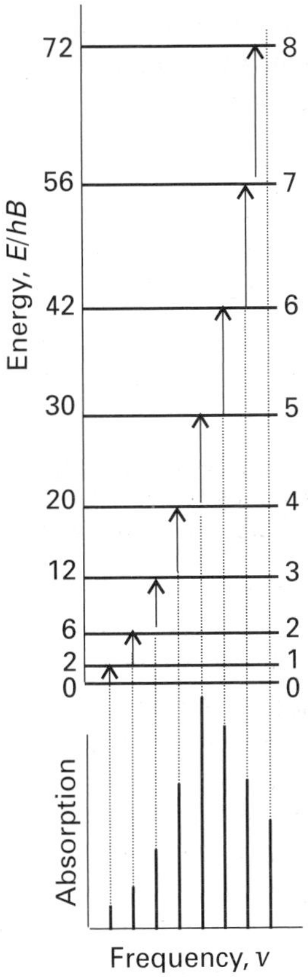

Fig 17.10 The allowed rotational transitions (shown as absorptions) for a linear molecule.

Once we have measured the separation between adjacent lines in a rotational spectrum of a molecule and converted it to B, we can use the value of B to obtain a value for the moment of inertia $I_\perp$. For a diatomic molecule, we can convert that value to a value of the bond length, R, by using eqn 17.8. Highly accurate bond lengths can be obtained in this way. More complicated procedures can be used to obtain the bond lengths in simple polyatomic molecules and to measure the electric dipole moments of polar molecules from the modification of the appearance of their rotational spectra caused by applied electric fields.

---

[5] To obtain the latter, we have identified ΔE with the energy of a photon, $h v$, and then cancelled the h.

17.5 Rotational Raman spectra

In a Raman spectrum, the incident radiation—which is typically monochromatic blue light or ultraviolet radiation from a laser—is scattered by the molecules in the sample and monitored by a detector placed at 90° to the incident beam (Fig 17.11). It is found that the scattered radiation has components with many different frequencies. Lines shifted to lower frequency than the incident radiation are called **Stokes lines** and lines shifted to higher frequency are called **anti-Stokes lines**. The Stokes lines arise when the collision of a photon with a molecule results in the excitation of rotation. When that happens, the photon responsible for the excitation loses some energy and hence travels to the detector with a lower energy and therefore a lower frequency. The anti-Stokes radiation arises from photon–molecule collisions in which the incoming photon acquires energy from the rotation of the molecule. As a result, the scattered photon travels to the detector with higher energy and hence a higher frequency.

The gross selection rule for rotational Raman spectra is that the polarizability of *the molecule must be anisotropic.* The *polarizability* of a molecule is a measure of the extent to which an applied electric field can induce an electric dipole moment (recall Section 16.4). The *anisotropy* of this polarizability is its variation with the orientation of the molecule. Tetrahedral (CH_4), octahedral (SF_6), and icosohedral (C_{60}) molecules, like all spherical rotors, have the same polarizability regardless of their orientations, so these molecules are **rotationally Raman inactive**: they do not have rotational Raman spectra. All other molecules, including homonuclear diatomic molecules such as H_2, are **rotationally Raman active**.

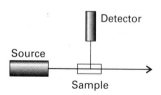

Fig 17.11 The experimental arrangement for the observation of Raman spectra. The source is a laser, and the scattered radiation is detected electronically.

The specific selection rules for the rotational Raman transitions of linear molecules (the only ones we consider) are

$\Delta J = +2$ (Stokes lines)
$\Delta J = -2$ (anti-Stokes lines)

It follows that the change in energy when a molecule makes the transition $J \rightarrow J + 2$ is

$$\Delta E = hB(J + 2)(J + 3) - hBJ(J + 1)$$
$$= 2hB(2J + 3) \tag{17.12}$$

Therefore, when a photon scatters from molecules in the rotational states $J = 0, 1, 2, 3, \ldots$, and transfers some of its energy to the molecule, the energy of the photon is decreased by $6hB, 10hB, 14hB, \ldots$ and its frequency is reduced by $6B, 10B, 14B, \ldots$ from the frequency of the incident radiation. If the photon acquires energy during the collision, then a similar argument shows that the anti-Stokes lines occur with frequencies $6B, 10B, 14B, \ldots$ higher than the incident radiation (Fig 17.12). It follows that, from a measurement of the separation of the Raman lines, we can determine the value of B and hence calculate the bond length. Because homonuclear diatomic species are rotationally Raman active, this technique can be applied to them as well as to heteronuclear species.

Vibrational spectroscopy

ALL molecules are capable of vibrating, and complicated molecules may do so in a large number of different modes. Even benzene, with 12 atoms, can vibrate in 30 different modes, some of which involve the periodic swelling and shrinking of the ring and others its buckling into various distorted shapes. A molecule as big as a protein can vibrate in tens of thousands of different ways, twisting, stretching, and buckling in different regions and in different manners. Vibrations can be excited by the

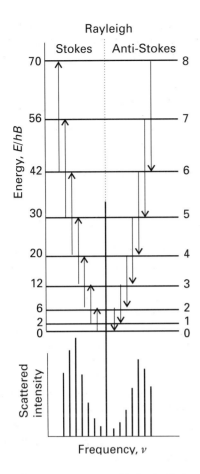

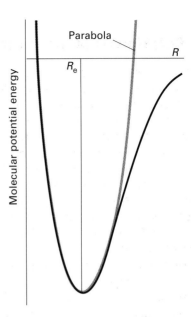

Fig 17.13 The molecular potential energy curve can be approximated by a parabola near the bottom of the well. A parabolic potential results in harmonic oscillation. At high vibrational excitation energies the parabolic approximation is poor.

approximate the potential energy by a parabola (a curve of the form $y = x^2$), and write

$$V = \tfrac{1}{2} k (R - R_e)^2 \tag{17.13}$$

where k is the *force constant* of the bond, as in the discussion of vibrations in Section 12.11. The steeper the walls of the potential (the stiffer the bond), the greater the force constant (Fig 17.14).

The potential energy in eqn 17.13 has the same form as that for the harmonic oscillator (Section 12.11), so we can use the solutions of the Schrödinger equation given there. The only complication is that both atoms in the bond move, so the 'mass' of the oscillator has to be interpreted carefully. Detailed calculation shows that, for two atoms of masses m_A and m_B joined by a bond of force constant k, the energy levels are[6]

$$E_\nu = (\nu + \tfrac{1}{2}) h \nu \qquad \nu = 0, 1, 2, \ldots \tag{17.14a}$$

---

[6] We have previously warned about the importance of distinguishing between the quantum number ν (vee) and the frequency ν (nu).

Fig 17.12 The transitions responsible for the Stokes and anti-Stokes lines of a rotational Raman spectrum of a linear molecule. The 'Rayleigh line' is scattered radiation at the incident frequency.

absorption of electromagnetic radiation. Observing the frequencies at which this absorption occurs gives very valuable information about the identity of the molecule and provides quantitative information about the flexibility of its bonds.

17.6 The vibrations of molecules

We base our discussion on Fig 17.13, which shows a typical potential energy curve (it is a reproduction of Fig 14.1) of a diatomic molecule as its bond is lengthened by pulling one atom away from the other. In regions close to the equilibrium bond length R_e (at the minimum of the curve) we can

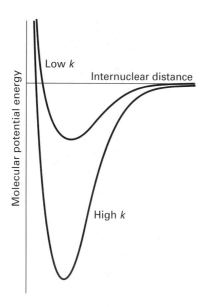

Fig 17.14 A low value of k indicates a loose bond; a high value indicates a stiff bond. Although the value of k is not directly related to the strength of the bond, this illustration indicates that it is likely that a strong bond (one with a deep minimum) has a large force constant.

where

$$v = \frac{1}{2\pi}\left(\frac{k}{\mu}\right)^{1/2} \qquad \mu = \frac{m_A m_B}{m_A + m_B} \qquad (17.14b)$$

These energy levels are illustrated in Fig 17.15 (which repeats Fig 12.31): we see that they form a uniformly spaced ladder with separation hv.

The quantity μ is called the **effective mass** of the vibrational mode of the molecule.[7] At first sight it might be puzzling that the effective mass appears rather than the total mass of the two atoms. However, the presence of μ is physically plausible. If atom A were as heavy as a brick wall, it would not move at all during the vibration and the vibrational frequency would be determined by the lighter, mobile atom. Indeed, if A were a brick wall, we could neglect m_B compared with m_A in the denominator of μ and find $\mu \approx m_B$, the mass of the lighter atom. This is approximately the case in HI, for example, where the I atom barely moves and $\mu \approx m_H$. In the case of a

---

[7] Because, in the special case of a diatomic molecule, the effective mass has the same form as a similar quantity in eqn 17.8, the effective mass is also sometimes called the *reduced mass* of the mode.

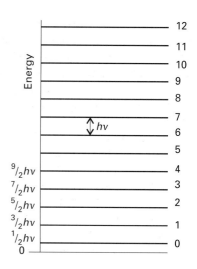

Fig 17.15 The energy levels of an harmonic oscillator. The quantum number v ranges from 0 to infinity, and the permitted energy levels form a uniform ladder with spacing hv.

homonuclear diatomic molecule, for which $m_A = m_B = m$, the effective mass is half the mass of one atom: $\mu = \tfrac{1}{2}m$.

Self-test 17.5

An $^1H^{35}Cl$ molecule has a force constant of $516\,N\,m^{-1}$, a reasonably typical value. Calculate the vibrational frequency, v, of the molecule and the energy separation between any two neighbouring vibrational energy levels.

[*Answer*: 8.96×10^{13} Hz (89.6 THz); 5.94×10^{-20} J]

17.7 Vibrational transitions

Because a typical vibrational excitation energy is of the order of 10^{-20}–10^{-19} J, the frequency of the radiation should be of the order of 10^{13}–10^{14} Hz (from $\Delta E = hv$). This frequency corresponds to infrared radiation, so vibrational transitions are observed by **infrared spectroscopy**. In infrared spectroscopy, transitions are normally expressed in terms of their wavenumbers and lie typically in the range 300–3000 cm^{-1}.

The gross selection rule for vibrational spectra is that *the electric dipole moment of the molecule must change during the vibration*. The basis of this rule is that the molecule can shake the electromagnetic field into oscillation only if it has an electric dipole moment that oscillates as the molecule vibrates (Fig 17.16). The molecule need not have a permanent dipole: the rule requires only a *change* in dipole moment, possibly from zero. The stretching motion of a homonuclear diatomic molecule does not change its electric dipole moment from zero, so the vibrations of such molecules neither absorb nor generate radiation. We say that homonuclear diatomic molecules are **infrared inactive**, because their dipole moments remain zero however long the bond. Heteronuclear diatomic molecules, which have a dipole moment that changes as the bond lengthens and contracts, are **infrared active**.

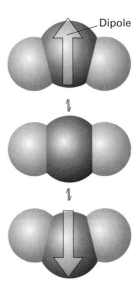

Fig 17.16 The oscillation of a molecule, even if it is nonpolar, may result in an oscillating dipole that can interact with the electromagnetic field. Here we see a representation of a bending mode of CO_2.

Example 17.2 *Using the gross selection rule*

State which of the following molecules are infrared active: N_2, CO_2, OCS, H_2O, $CH_2=CH_2$, C_6H_6.

Strategy Molecules that are infrared active (that is, have vibrational spectra) have dipole moments that change during the course of a vibration. Therefore, judge whether a distortion of the molecule can change its dipole moment (including changing it from zero).

Solution All the molecules except N_2 possess at least one vibrational mode that results in a change of dipole moment, so all except N_2 are infrared active. It should be noted that not all the modes of complicated molecules are infrared active. For example, a vibration of CO_2 in which the O–C–O bonds stretch and contract symmetrically is inactive because it leaves the dipole moment unchanged (at zero). A bending motion of the molecule, however, is active and can absorb radiation.

Self-test 17.6

Repeat the question for H_2, NO, N_2O, CH_4.

[*Answer*: NO, N_2O, CH_4]

The specific selection rule for vibrational transitions is

$$\Delta v = \pm 1$$

The change in energy for the transition from a state with quantum number v to one with quantum number $v + 1$ is

$$\Delta E = (v + \tfrac{3}{2})hv - (v + \tfrac{1}{2})hv = hv \qquad (17.15)$$

It follows that absorption occurs when the incident radiation provides photons with this energy, and therefore when the incident radiation has a frequency v given by eqn 17.14b (and wavenumber $\tilde{v} = v/c$). Molecules with stiff bonds (large k) joining atoms with low masses (small μ) have high vibrational frequencies. Bending modes are usually less stiff than stretching modes, so bends tend to occur at lower frequencies in the spectrum than stretches.

At room temperature, almost all the molecules will be in their vibrational ground states initially (the state with $v = 0$). Therefore, the most important spectral transition is from $v = 0$ to $v = 1$.

Illustration 17.1

It follows from the calculation of ν for HCl (in Self-test 17.5), that ν = 89.6 THz, so the infrared spectrum of the molecule will be an absorption at that frequency. The corresponding wavenumber and wavelength are 2990 cm^{-1} and 3.35 µm, respectively.

Self-test 17.7

The force constant of the bond in the CO group of a peptide link is approximately 1.2 kN m^{-1}. At what wavenumber would you expect it to absorb? (*Hint*. For the effective mass, treat the group as a CO molecule.)

[*Answer*: at approximately 1700 cm^{-1}]

The vibrational spectra of gas-phase molecules are more complicated than this discussion implies, because the excitation of a vibration also results in the excitation of rotation. The effect is rather like that when ice skaters throw out or draw in their arms: they rotate more slowly or more rapidly. The effect on the spectrum is to break the single line resulting from a vibrational transition into a multitude of lines with separations between neighbours that depend on the rotational constant of the molecule. We shall not go into this complication here, except to say that analysis of the separation of the lines gives information like that obtained from pure rotational spectroscopy (principally bond lengths).

17.8 Vibrational Raman spectra of diatomic molecules

In **vibrational Raman spectroscopy** the incident photon leaves some of its energy in the vibrational modes of the molecule it strikes, or collects additional energy from a vibration that has already been excited.

The gross selection rule for vibrational Raman transitions is that *the molecular polarizability must change as the molecule vibrates*. The polarizability plays a role in vibrational Raman spectroscopy because the molecule must be squeezed and stretched by the incident radiation in order that a vibrational excitation may occur during the photon–molecule collision. Both homonuclear and heteronuclear diatomic molecules swell and contract during a vibration, and the control of the nuclei over the electrons, and hence the molecular polarizability, changes too. Both types of diatomic molecule are therefore vibrationally Raman active.

The specific selection rule for vibrational Raman transitions is the same as for infrared transitions ($\Delta\nu$ = ±1). The photons that are scattered with a lower wavenumber than that of the incident light, the Stokes lines, are those for which $\Delta\nu$ = +1. The Stokes lines are more intense than the anti-Stokes lines (for which $\Delta\nu$ = −1), because very few molecules are in an excited vibrational state initially.

The information available from vibrational Raman spectra adds to that from infrared spectroscopy because homonuclear diatomic molecules can also be studied. The spectra can be interpreted in terms of the force constants, dissociation energies, and bond lengths, and some of the information obtained is included in Table 17.1.

17.9 The vibrations of polyatomic molecules

How many modes of vibration are there in a polyatomic molecule? We can answer this question by thinking about how each atom may change its location.

Table 17.1 *Properties of diatomic molecules*

	$\tilde{\nu}$/cm^{-1}	R_e/pm	k/(N m^{-1})	D/(kJ mol^{-1})
$^1H_2^+$	2333	106	160	256
$^1H_2$	4400	74	575	432
$^2H_2$	3118	74	577	440
$^1H^{19}F$	4138	92	955	564
$^1H^{35}Cl$	2991	127	516	428
$^1H^{81}Br$	2649	141	412	363
$^1H^{127}I$	2308	161	314	295
$^{14}N_2$	2358	110	2294	942
$^{16}O_2$	1580	121	1177	494
$^{19}F_2$	892	142	445	154
$^{35}Cl_2$	560	199	323	239

Derivation 17.1 *The number of normal modes*

Each atom may move along any of three perpendicular axes. Therefore, the total number of such displacements in a molecule consisting of N atoms is $3N$. Three of these displacements correspond to movement of the centre of mass of the molecule, so these three displacements correspond to the translational motion of the molecule as a whole. The remaining $3N - 3$ displacements are 'internal' modes of the molecule that leave its centre of mass unchanged. Three angles are needed to specify the orientation of a nonlinear molecule in space (Fig 17.17). Therefore three of the $3N - 3$ internal displacements leave all bond angles and bond lengths unchanged but change the orientation of the molecule as a whole. These three displacements are therefore rotations. That leaves $3N - 6$ displacements which change neither the centre of mass of the molecule nor the orientation of the molecule in space. These $3N - 6$ displacements are the vibrational modes. A similar calculation for a linear molecule, which requires only two angles to specify its orientation in space, gives $3N - 5$ as the number of vibrational modes. We summarize this conclusion as follows:

Nonlinear molecules:
Number of vibrational modes = $3N - 6$

Linear molecules:
Number of vibrational modes = $3N - 5$

Illustration 17.2

A water molecule, H_2O, is triatomic and nonlinear, and has three modes of vibration. Naphthalene, $C_{10}H_8$, has 48 distinct modes of vibration. Any diatomic molecule ($N = 2$) has one mode; carbon dioxide ($N = 3$) has four modes.

The description of the vibrational motion of a polyatomic molecule is much simpler if we consider combinations of the stretching and bending motions of individual bonds. For example, although we could describe two of the four vibrations of a CO_2 molecule as individual carbon–oxygen bond stretches, ν_L and ν_R in Fig 17.18, the description of the motion is much simpler if we use two combinations of these vibrations. One combination is ν_1 in Fig 17.19: this combination is the **symmetric stretch**. The other combination is ν_3, the **antisymmetric stretch**, in which the two O atoms always move in the same directions and opposite to the C atom. The

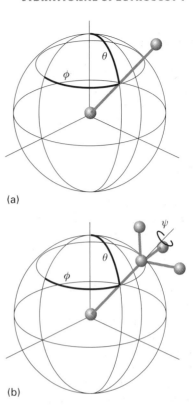

(a)

(b)

Fig 17.17 (a) The orientation of a linear molecule requires the specification of two angles (the latitude and longitude of its axis). (b) The orientation of a nonlinear molecule requires the specification of three angles (the latitude and longitude of its axis and the angle of twist around that axis).

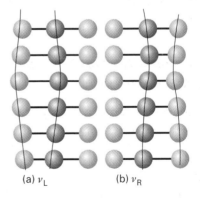

(a) ν_L (b) ν_R

Fig 17.18 The stretching vibrations of a CO_2 molecule can be represented in a number of ways. In this representation, (a) one O=C bond vibrates and the remaining O atom is stationary, and (b) the C=O bond vibrates while the other O atom is stationary. Because the stationary atom is linked to the C atom, it does not remain stationary for long. That is, if one vibration begins, it rapidly stimulates the other to occur.

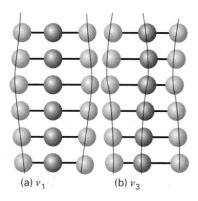

(a) ν_1 (b) ν_3

Fig 17.19 Alternatively, linear combinations of the two modes can be taken to give these two normal modes of the molecule. The mode in (a) is the symmetric stretch and that in (b) is the antisymmetric stretch. The two modes are independent and, if either of them is stimulated, the other remains unexcited. Normal modes greatly simplify the description of the vibrations of the molecule.

two modes are independent in the sense that, if one is excited, then its motion does not excite the other. They are two of the four 'normal modes' of the molecule, its independent, collective vibrational displacements. The two other normal modes are the **bending modes**, ν_2. In general, a **normal mode** is an independent, synchronous motion of atoms or groups of atoms that may be excited without leading to the excitation of any other normal mode.

Self-test 17.8

How many normal modes of vibration are there in (a) ethyne (HC≡CH) and (b) a protein molecule of 4000 atoms?

[*Answer*: (a) 7, (b) 11 994]

The four normal modes of CO_2, and the $3N - 6$ (or $3N - 5$) normal modes of polyatomic molecules in general, are the key to the description of molecular vibrations. Each normal mode behaves like an independent harmonic oscillator and the energies of the vibrational levels are given by the same expression as in eqn 17.14, but with an effective mass that depends on the extent to which each of the atoms contributes to the vibration. Atoms that do not move, such as the C atom in the symmetric stretch of CO_2, do not contribute to the effective mass. The force constant also depends in a complicated way on the

extent to which bonds bend and stretch during a vibration. Typically, a normal mode that is largely a bending motion has a lower force constant (and hence a lower frequency) than a normal mode that is largely a stretching motion.

The gross selection rule for the infrared activity of a normal mode is that *the motion corresponding to a normal mode must give rise to a changing dipole moment*. Deciding whether this is so can sometimes be done by inspection. For example, the symmetric stretch of CO_2 leaves the dipole moment unchanged (at zero), so this mode is infrared inactive and makes no contribution to the molecule's infrared spectrum. The antisymmetric stretch, however, changes the dipole moment because the molecule becomes unsymmetrical as it vibrates, so this mode is infrared active. The fact that the mode does absorb infrared radiation enables carbon dioxide to act as a 'greenhouse gas' by absorbing infrared radiation emitted from the surface of the Earth. Because the dipole moment change is parallel to the molecular axis in the antisymmetric stretching mode, the transitions arising from this mode are classified as **parallel bands** in the spectrum. Both bending modes are also infrared active:

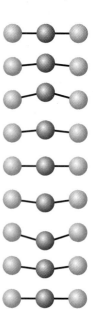

Fig 17.20 The bending mode of CO_2 results in the formation of an electric dipole, which oscillates as the molecule vibrates.

they are accompanied by a changing dipole perpendicular to the molecular axis (Fig 17.20), so transitions involving them lead to a **perpendicular band** in the spectrum.

Self-test 17.9

State the ways in which the infrared spectrum of dinitrogen oxide (nitrous oxide, N_2O) will differ from that of carbon dioxide.

[*Answer*: different frequencies on account of different atomic masses and force constants; all four modes infrared active]

Some of the normal modes of organic molecules can be regarded as motions of individual functional groups. Others cannot be regarded as localized in this way and are better regarded as collective motions of the molecule as a whole. The latter are generally of relatively low frequency, and occur below about 1500 cm^{-1} in the spectrum. The resulting whole-molecule region of the absorption spectrum is called the **fingerprint region** of the spectrum, for it is characteristic of the molecule. The matching of the fingerprint region with a spectrum of a known compound in a library of infrared spectra is a very powerful way of confirming the presence of a particular substance.

The characteristic vibrations of functional groups that occur outside the fingerprint region are very useful for the identification of an unknown compound. Most of these vibrations can be regarded as stretching modes, for the lower frequency bending

modes usually occur in the fingerprint region and so are less readily identified. The characteristic wavenumbers of some functional groups are listed in Table 17.2.

Example 17.3 *Interpreting an infrared spectrum*

The infrared spectrum of an organic compound is shown in Fig 17.21. Suggest an identification.

Strategy Some of the features at wavenumbers above 1500 cm^{-1} can be identified by comparison with the data in Table 17.2.

Solution (a) C–H stretch of a benzene ring, indicating a substituted benzene; (b) carboxylic acid O–H stretch, indicating a carboxylic acid; (c) the strong absorption of a conjugated C≡C group, indicating a substituted alkyne; (d) this strong absorption is also characteristic of a carboxylic acid that is conjugated to a carbon–carbon multiple bond; (e) a characteristic vibration of a benzene ring, confirming the deduction drawn from (a); (f) a characteristic absorption of a nitro group (–NO_2) connected to a multiply bonded carbon–carbon system, suggesting a nitro-substituted benzene. The molecule contains as components a benzene ring, an aromatic carbon–carbon bond, a –COOH group, and a –NO_2 group. The molecule is in fact O_2N–C_6H_4–C≡C–COOH. A more detailed analysis and comparison of the fingerprint region shows it to be the 1,4-isomer.

Self-test 17.10

Suggest an identification of the organic compound responsible for the spectrum shown in Fig 17.22. (*Hint*. The molecular formula of the compound is C_3H_5ClO.)

[*Answer*: CH_2=CClCH_2OH]

Table 17.2 *Typical vibrational wavenumbers*

Vibration type	$\tilde{v}/cm^{-1}$
C—H stretch	2850–2960
C—H bend	1340–1465
C—C stretch, bend	700–1250
C=C stretch	1620–1680
C≡C stretch	2100–2260
O—H stretch	3590–3650
C=O stretch	1640–1780
C≡N stretch	2215–2275
N—H stretch	3200–3500
Hydrogen bonds	3200–3570

17.10 **Vibrational Raman spectra of polyatomic molecules**

The gross selection rule for the vibrational Raman spectrum of a polyatomic molecule is that *the normal mode of vibration is accompanied by a changing polarizability*. However, it is often quite difficult to judge by inspection when this is so. The symmetric stretch of CO_2, for example, alternately swells and contracts the molecule: this motion changes its polarizability, so the mode is Raman active. The other modes of CO_2 leave the polarizability unchanged

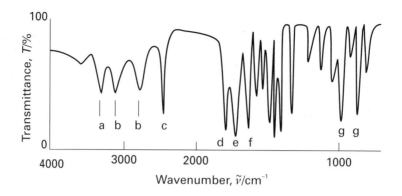

Fig 17.21 A typical infrared absorption spectrum taken by forming a sample into a disk with potassium bromide. As explained in the example, the substance can be identified as $O_2N-C_6H_4-C\equiv C-COOH$.

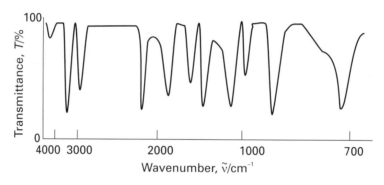

Fig 17.22 The spectrum considered in Self-test 17.10.

(although that is hard to justify pictorially), so they are Raman inactive.

In some cases it is possible to make use of a very general rule about the infrared and Raman activity of vibrational modes. The **exclusion rule** states:

If the molecule has a centre of inversion, then no modes can be both infrared and Raman active.

(A mode may be inactive in both.) A molecule has a centre of inversion if it looks unchanged when each atom is projected through a single point and out an equal distance on the other side (Fig 17.23). Because we can often judge intuitively when a mode changes the molecular dipole moment, we can use this rule to identify modes that are not Raman active. The rule applies to CO_2 but to neither H_2O nor CH_4 because they have no centre of symmetry.

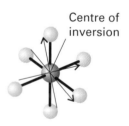

Fig 17.23 In an inversion operation, we consider every point in a molecule, and project them all through the centre of the molecule out to an equal distance on the other side.

Self-test 17.11

One vibrational mode of benzene is a 'breathing mode' in which the ring alternately expands and contracts. May it be vibrationally Raman active?

[*Answer:* yes]

One application of vibrational Raman spectroscopy is to the determination of the bond lengths of non-polar molecules such as XeF_4 and SF_6. Another application makes use of the fact that the intensity characteristics of Raman transitions, which depend on molecular polarizabilities, are more readily transferred from molecule to molecule than the intensities of infrared spectra, which depend on dipole moments and are more sensitive to the other groups present in a molecule and to the solvent. Hence, Raman spectra are useful in the identification of organic and inorganic species in solution. An example of the technique is shown in Fig 17.24, which shows the vibrational Raman spectrum of an aqueous solution of lysozyme and, for comparison, a superposition of the Raman spectra of the constituent amino acids. The differences are indications of the effects of conformation, environment, and specific interactions (such as disulfide links, S–S) in the enzyme molecule.

A modification called **resonance Raman spectroscopy** uses incident radiation that coincides with the frequency of an electronic transition of the sample (Fig 17.25; compare Fig 17.2, where the inci-

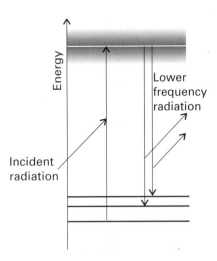

Fig 17.25 In the *resonance Raman effect*, the incident radiation has a frequency corresponding to an actual electronic excitation of the molecule. A photon is emitted when the excited state returns to a state close to the ground state.

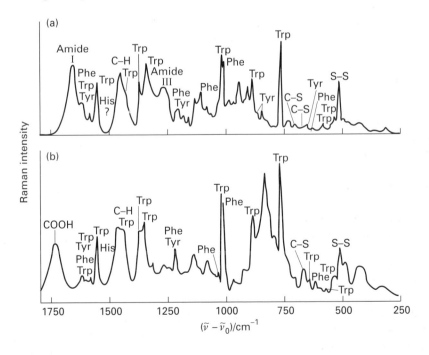

Fig 17.24 (a) The vibrational Raman spectrum of lysozyme in water and (b) the superposition of the Raman spectra of the constituent amino acids. (From *Raman spectroscopy*, D. A. Long. Copyright 1977, McGraw-Hill Inc. Used with the permission of the McGraw-Hill Book Company.)

dent radiation did not coincide with an actual electronic transition of the sample). The technique is characterized by a much greater intensity in the scattered radiation and, because only a few vibrational modes contribute to the scattering, to a greatly simplified Raman spectrum. The high intensity of the resonance Raman transitions is employed to examine the metal ions in biological macromolecules (such as the iron in haemoglobin and cytochromes or the cobalt in vitamin B_{12}), which are present in such low abundances that conventional Raman spectroscopy cannot detect them.

Exercises

For these exercises, use $m(^1H) = 1.0078\,u$, $m(^2H) = 2.0140\,u$, $m(^{12}C) = 12.0000\,u$, $m(^{13}C) = 13.0034\,u$, $m(^{16}O) = 15.9949\,u$, $m(^{19}F) = 18.9984\,u$, $m(^{127}I) = 126.9045\,u$.

17.1 Express a wavelength of 670 nm as (a) a frequency, (b) a wavenumber.

17.2 What is (a) the wavenumber, (b) the wavelength of the radiation used by an FM radio transmitter broadcasting at 92.0 MHz?

17.3 What is the Doppler-shifted wavelength of a red (660 nm) traffic light approached at 55 m.p.h.? At what speed would it appear green (520 nm)?

17.4 A spectral line of $^{48}Ti^{8+}$ in a distant star was found to be shifted from 654.2 nm to 706.5 nm and to be broadened to 61.8 pm. What is the speed of recession and the surface temperature of the star?

17.5 Estimate the lifetime of a state that gives rise to a line of width (a) $0.1\,cm^{-1}$, (b) $1\,cm^{-1}$, (c) 1.0 GHz.

17.6 A molecule in a liquid undergoes about 1×10^{13} collisions in each second. Suppose that (a) every collision is effective in deactivating the molecule vibrationally and (b) that one collision in 200 is effective. Calculate the width (in cm^{-1}) of vibrational transitions in the molecule.

17.7 The kinetic energy of a bicycle wheel rotating once per second is about 0.2 J. To what rotational quantum number does that correspond? For the moment of inertia, let the mass of the wheel (which is concentrated in its rim) be 0.75 kg and its radius be 70 cm.

17.8 Calculate the moment of inertia of (a) $^1H_2$, (b) $^2H_2$, (c) $^{12}C^{16}O_2$, (d) $^{13}CO_2$. Use $R_e(CO) = 112$ pm.

17.9 Calculate the rotational constants of the molecules in Exercise 17.8; express your answer in hertz (Hz).

17.10 (a) Express the moment of inertia of an octahedral AB_6 molecule in terms of its bond lengths and the masses of the B atoms. (b) Calculate the rotational constant of $^{32}S^{19}F_6$ for which the S–F bond length is 158 pm.

17.11 (a) Derive expressions for the two moments of inertia of a square-planar AB_4 molecule in terms of its bond lengths and the masses of the B atoms.

17.12 Suppose you were seeking the presence of (planar) SO_3 molecules in the microwave spectra of interstellar gas clouds. (a) You would need to know the rotational constants A and B. Calculate these parameters for $^{32}S^{16}O_3$, for which the S–O bond length is 143 pm. (b) Could you use microwave spectroscopy to distinguish the relative abundances of $^{32}S^{16}O_3$ and $^{33}S^{16}O_3$?

17.13 Which of the following molecules can have a pure rotational spectrum? (a) HCl, (b) N_2O, (c) O_3, (d) SF_4, (e) XeF_4.

17.14 Which of the molecules in Exercise 17.13 can have a rotational Raman spectrum?

17.15 A rotating methane molecule is described by the quantum numbers J, M_J, and K. How many rotational states have an energy equal to $hBJ(J+1)$ with $J = 10$?

17.16 Suppose the methane molecule in Exercise 17.15 is replaced by chloromethane. How many rotational states now have an energy equal to $hBJ(J+1)$ with $J = 10$?

17.17 The rotational constant of $^1H^{35}Cl$ is 318.0 GHz. What is the separation of the lines in its pure rotational spectrum (a) in gigahertz, (b) in reciprocal centimetres?

17.18 Suppose that hydrogen is replaced by deuterium in $^1H^{35}Cl$. Would you expect the $J = 1 \rightarrow 0$ transition to move to higher or lower wavenumber?

17.19 The rotational constant of $^{12}C^{16}O_2$ (from Raman spectroscopy) is 11.70 GHz. What is the CO bond length in the molecule?

17.20 The microwave spectrum of $^1H^{127}I$ consists of a series of lines separated by 384 GHz. Compute its bond length. What would be the separation of the lines in $^2H^{127}I$?

17.21 Suppose the C=O group in a peptide bond can be regarded as isolated from the rest of the molecule. Given that the force constant of the bond in a carbonyl group is 908 N m^{-1}, calculate the vibrational frequency of (a) $^{12}C=^{16}O$, (b) $^{13}C=^{16}O$.

17.22 The wavenumber of the fundamental vibrational transition of Cl_2 is 565 cm^{-1}. Calculate the force constant of the bond.

17.23 The hydrogen halides have the following fundamental vibrational wavenumbers:

	HF	HCl	HBr	HI
$\tilde{v}/cm^{-1}$	4141.3	2988.9	2649.7	2309.5

Calculate the force constants of the hydrogen–halogen bonds.

17.24 From the data in Exercise 17.23, predict the fundamental vibrational wavenumbers of the deuterium halides.

17.25 Which of the following molecules may show infrared absorption spectra: (a) H_2, (b) HCl, (c) CO_2, (d) H_2O, (e) CH_3CH_3, (f) CH_4, (g) CH_3Cl, (h) N_2?

17.26 How many normal modes of vibration are there for (a) NO_2, (b) N_2O, (c) cyclohexane, (d) hexane?

17.27 Suggest an interpretation of the infrared spectrum in Fig 17.26.

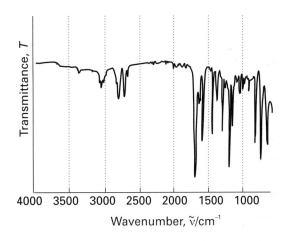

Fig 17.26

Electronic transitions

Contents

THE energy needed to change the distribution of an electron in a molecule is of the order of several electronvolts.[1] Consequently, the photons emitted or absorbed when such changes occur lie in the visible and ultraviolet regions of the spectrum, which spread from about $14\,000$ cm^{-1} for red light to $21\,000$ cm^{-1} for blue, and on to $50\,000$ cm^{-1} for ultraviolet radiation (Table 18.1). Indeed, many of the colours of the objects in the world around us, including the green of vegetation, the colours of flowers and of synthetic dyes, and the colours of pigments and minerals, stem from transitions in which an electron makes a transition from one orbital of a molecule or ion into another. The change in location of an electron that takes place when chlorophyll absorbs red and blue light (leaving green to be reflected) is the primary energy-harvesting step by which our planet captures energy from the Sun and uses it to drive the non-spontaneous reactions of photosynthesis (see Box 11.2). In some cases the relocation of an electron may be so extensive that it results in the breaking of a bond and the dissociation of the molecule: such processes give rise to the numerous reactions of photochemistry, including the reactions that sustain or damage the atmosphere.

---

[1] 1 eV, the energy acquired by an electron when it is moved through a potential difference of 1 V, corresponds to 8065.5 cm^{-1} and 96.485 kJ mol^{-1}.

Table 18.1 *Colour, frequency, and energy of light*

Colour	λ/nm	$\nu/(10^{14}\ Hz)$	$\tilde{\nu}/(10^4\ cm^{-1})$	E/eV	$E/(kJ\ mol^{-1})$
Infrared	1000	3.00	1.00	1.24	120
Red	700	4.28	1.43	1.77	171
Orange	620	4.84	1.61	2.00	193
Yellow	580	5.17	1.72	2.14	206
Green	530	5.66	1.89	2.34	226
Blue	470	6.38	2.13	2.64	254
Violet	420	7.14	2.38	2.95	285
Near ultraviolet	300	10.0	3.33	4.15	400
Far ultraviolet	200	15.0	5.00	6.20	598

Ultraviolet and visible spectra

WHITE light is a mixture of light of all different colours. The removal, by absorption, of any one of these colours from white light results in the complementary colour being observed. For instance, the absorption of red light from white light by an object results in that object appearing green, the complementary colour of red. Conversely, the absorption of green results in the object appearing red. The pairs of complementary colours are neatly summarized by the artist's colour wheel shown in Fig 18.1, where complementary colours lie opposite one another along a diameter.

It should be stressed, however, that the perception of colour is a very subtle phenomenon. Although an object may appear green because it absorbs red light, it may also appear green because it absorbs all colours from the incident light *except* green. This is the origin of the colour of vegetation, because chlorophyll absorbs in two regions of the spectrum, leaving green to be reflected (Fig 18.2). Moreover, an absorption band

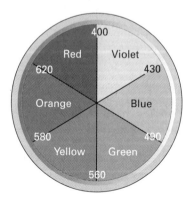

Fig 18.1 An artist's colour wheel: complementary colours are opposite one another on a diameter.

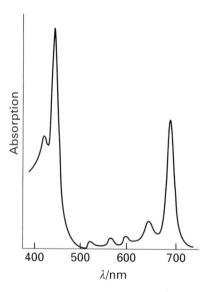

Fig 18.2 The absorption spectrum of chlorophyll in the visible region. Note that it absorbs in the red and blue regions, and that green light is not absorbed significantly.

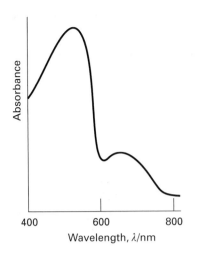

Fig 18.3 An electronic absorption of a species in solution is typically very broad and consists of several broad bands.

may be very broad and, although it may be a maximum at one particular wavelength, it may have a long tail that spreads into other regions (Fig 18.3). In such cases, it is very difficult to predict the perceived colour from the location of the absorption maximum.

18.1 **The Franck–Condon principle**

Whenever an electronic transition takes place, it is accompanied by the excitation of vibrations of the molecule. In the electronic ground state of a molecule, the nuclei take up locations in response to the Coulombic forces acting on them. These forces arise from the electrons and the other nuclei. After an electronic transition, when an electron has migrated to a different part of the molecule, the nuclei are subjected to different Coulombic forces from the surrounding electrons. The molecule may respond to the change in forces by bursting into vibration. As a result, some of the energy used to redistribute an electron is in fact used to stimulate the vibrations of the absorbing molecules. Therefore, instead of a single, sharp, and purely electronic absorption line being observed, the absorption spectrum consists of many lines. This **vibrational structure** of an electronic transition can

be resolved if the sample is gaseous, but in a liquid or solid the lines usually merge together and result in a broad, almost featureless band (Fig 18.4).

The vibrational structure of a band is explained by the **Franck–Condon principle**:

Because nuclei are so much more massive than electrons, an electronic transition takes place faster than the nuclei can respond.

In an electronic transition, electron density is lost rapidly from some regions of the molecule and is built up rapidly in others. As a result, the initially stationary nuclei suddenly experience a new force field. They respond by beginning to vibrate, and (in classical terms) swing backwards and forwards from their original separation, which was maintained during the rapid electronic excitation. The equilibrium separation of the nuclei in the initial electronic state therefore becomes a **turning point,** one of the end points of a nuclear swing, in the final electronic state (Fig 18.5). We can predict the most likely final vibrational state by drawing a vertical line from the minimum of the lower curve (the starting point for the transition) up to the point at which the line intersects the curve repre-

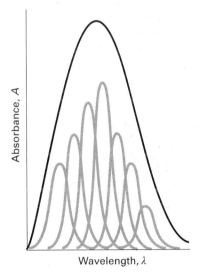

Fig 18.4 An electronic absorption band consists of many superimposed bands that merge together to give a single broad band with unresolved vibrational structure.

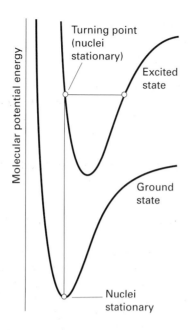

Fig 18.5 According to the Franck–Condon principle, the most intense electronic transition is from the ground vibrational state to the vibrational state that lies vertically above it in the upper electronic state. Transitions to other vibrational levels also occur, but with lower intensity.

senting the upper electronic state (the turning point of the newly stimulated vibration). This procedure gives rise to the term **vertical transition** for a transition in accord with the Franck–Condon principle. In practice, the electronically excited molecule may be formed in one of several excited vibrational states, so the absorption occurs at several different frequencies. As remarked above, in a condensed medium, the individual transitions merge together to give a broad, largely featureless band of absorption.

18.2 **Measures of intensity**

The theoretical basis of the intensity of an electronic transition is the transition dipole moment introduced in Section 13.7. We saw there that the transition dipole moment is a measure of the strength of the coupling between the electronic transition and the electromagnetic field. If the transition dipole moment is zero for a specified pair of states, the transition between them is forbidden. If the transition dipole moment is nonzero, the transition is allowed and its intensity is proportional to the square of the transition dipole moment. The transition dipole moment may be large if there is a substantial relocation of an electron between the initial and final states.

The intensity with which radiation is absorbed depends on the identity of the absorbing species (which we denote J), the frequency v of the radiation, the molar concentration [J] of the species in the sample, and the path length, l, of the radiation through the sample. It is found *experimentally* that the **transmittance**, T, the ratio of the emerging intensity, I, to the incident intensity, I_0, is given by the **Beer–Lambert law**:

$$\log T = -\varepsilon[\mathrm{J}]l \qquad T = \frac{I}{I_0} \tag{18.1}$$

The coefficient ε is the **molar absorption coefficient** (formerly the 'extinction coefficient') of the species. This coefficient depends on the frequency of the incident radiation and is proportional to the square of the transition dipole moment. Its units are those of 1/(concentration × length) and it is commonly reported in litres per mole per centimetre (L mol^{-1} cm^{-1}).[2] The Beer–Lambert law is empirical, but it can be understood on the basis that the intensity of the intensity of absorption is proportional to the intensity of the radiation inside the sample.

Derivation 18.1 *The Beer–Lambert law*

We think of the sample as consisting of a stack of infinitesimal slices, like sliced bread (Fig 18.6). The thickness of each layer is dx. The change in intensity, dI, that occurs when electromagnetic radiation passes through one particular slice is proportional to the thickness of the layer, the concentration of the absorber, and the intensity of the incident radiation at that slice of the sample, so d$I \propto$ [J]Idx. Because dI is negative (the intensity is reduced by absorption), we can write

$$\mathrm{d}I = -\kappa[\mathrm{J}]I\mathrm{d}x$$

---

[2] The alternative units of square centimetres per mole (cm^2 mol^{-1}) bring out the point that ε is a molar cross-section for absorption and, the greater the cross-section of the molecule for absorption, the greater the reduction in the intensity of the beam for a given path length, concentration, and frequency.

where κ (kappa) is the proportionality coefficient. Division by I gives

$$\frac{dI}{I} = -\kappa[\,J\,]dx$$

This expression applies to each successive slice. To obtain the intensity, I, that emerges from a sample of thickness I when the incident intensity is I_0, we sum (that is, integrate) all the successive changes:

$$\int_{I_0}^{I} \frac{dI}{I} = -\kappa \int_{0}^{l} [\,J\,]dx$$

When the concentration of the absorber is uniform, $[J]$ is independent of x, and the expression integrates to

$$\ln \frac{I}{I_0} = -\kappa[\,J\,]l$$

Because the relation between natural and common logarithms is $\ln x = \ln 10 \times \log x$, we can write $\varepsilon = \kappa/\ln 10$ and obtain

$$\log \frac{I}{I_0} = -\varepsilon[\,J\,]l \qquad (18.2)$$

which, on substituting $T = I/I_0$, is the Beer–Lambert law.

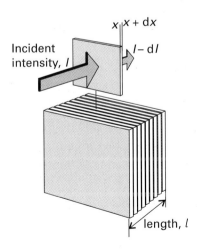

Fig 18.6 To establish the Beer–Lambert law, the sample is supposed to be sliced into a large number of planes. The reduction in intensity caused by one plane is proportional to the intensity incident on it (after passing through the preceding planes), the thickness of the plane, and the concentration of absorbing species.

The Beer–Lambert law is sometimes written in the form

$$I = I_0 \times 10^{-\varepsilon[J]l} \qquad (18.3)$$

This expression shows that the transmitted intensity decreases exponentially with the length of the sample (Fig 18.7). The dimensionless product

$$A = \varepsilon[\,J\,]l \qquad (18.4)$$

is called the **absorbance** (formerly the 'optical density') of the sample.

Example 18.1 *Calculating the molar absorption coefficient*

Radiation of wavelength 256 nm passed through 1.0 mm of a solution that contained benzene at a concentration of 0.050 mol L^{-1}. The light intensity is reduced to 16 per cent of its initial value (so $T = 0.16$). Calculate the absorbance and the molar absorption coefficient of the sample. What would be the transmittance through a cell of thickness 2.0 mm?

Strategy For ε we use eqn 18.1 rearranged into

$$\varepsilon = -\frac{\log T}{[\,J\,]l}$$

and then combine this expression with the definition of A in eqn 18.4:

$$A = \varepsilon[J]l = -\log T$$

For the transmittance through the thicker cell, we use eqn 18.1 and the value of ε calculated here.

Solution The molar absorption coefficient is

$$\varepsilon = -\frac{\log 0.16}{(0.050 \, \text{mol L}^{-1}) \times (1.0 \, \text{mm})} = 16 \, \text{L mol}^{-1} \text{mm}^{-1}$$

These units are convenient for the rest of the calculation (but the outcome could be reported as 1.6×10^2 L mol^{-1} cm^{-1} if desired). The absorbance is

$$A = -\log 0.16 = 0.80$$

The absorbance of a sample of length 2.0 mm is

$$A = (16 \, \text{L mol}^{-1} \text{mm}^{-1}) \times (0.050 \, \text{mol L}^{-1}) \\ \times (2.0 \, \text{mm}) \\ = 1.6$$

It follows that the transmittance is

$$T = 10^{-A} = 10^{-1.6} = 0.025$$

That is, the emergent light is reduced to 2.5 per cent of its incident intensity.

Self-test 18.1

The transmittance of an aqueous solution that contained Cu^{2+} ions at a molar concentration of 0.10 mol L^{-1} was measured as 0.30 at 600 nm in a cell of length 5.0 mm. Calculate the molar absorption coefficient of $Cu^{2+}(aq)$ at that wavelength, and the absorbance of the solution. What would be the transmittance through a cell of length 1.0 mm?

[*Answer*: 10 L mol^{-1} cm^{-1}, $A = 0.52$, $T = 0.79$]

The **maximum molar absorption coefficient,** ε_{max}, is an indication of the intensity of a transition. Typical values for strong transitions are of the order of 10^4–10^5 L mol^{-1} cm^{-1}, indicating that in a solution of molar concentration 0.01 mol L^{-1} the intensity of light (of frequency corresponding to the maximum absorption) falls to 10 per cent of its initial value after passing through about 0.1 mm of solution.

Beer's law (as it is normally called) is used to determine the concentrations of species of known molar absorption coefficients. To do so, we measure the absorbance of a sample and rearrange eqn 18.4 into

$$[J] = \frac{A}{\varepsilon l} \tag{18.5}$$

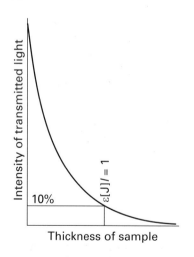

Fig 18.7 The intensity of light transmitted by an absorbing sample decreases exponentially with the path length through the sample.

More importantly, particularly in biological applications, we can make measurements at two wavelengths and use them to find the individual concentrations of two components A and B in a mixture. For this analysis, we write the total absorbance at a given wavelength as

$$A = A_A + A_B = \varepsilon_A[A]l + \varepsilon_B[B]l = (\varepsilon_A[A] + \varepsilon_B[B])l$$

Then, for two measurements of the total absorbance at wavelengths λ_1 and λ_2 at which the molar absorption coefficients are ε_1 and ε_2 (Fig 18.8), we have

$$A_1 = (\varepsilon_{A1}[A] + \varepsilon_{B1}[B])l \quad A_2 = (\varepsilon_{A2}[A] + \varepsilon_{B2}[B])l$$

We can solve these two simultaneous equations for the two unknowns (the molar concentrations of A and B), and we find

$$[A] = \frac{\varepsilon_{B2} A_1 - \varepsilon_{B1} A_2}{(\varepsilon_{A1} \varepsilon_{B2} - \varepsilon_{A2} \varepsilon_{B1})l} \quad [B] = \frac{\varepsilon_{A1} A_2 - \varepsilon_{A2} A_1}{(\varepsilon_{A1} \varepsilon_{B2} - \varepsilon_{A2} \varepsilon_{B1})l}$$

$$\tag{18.6}$$

There may be a wavelength, $\lambda°$, called the **iso-sbestic wavelength,** at which the molar extinction coefficients of the two species are equal; we write

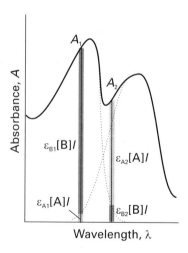

Fig 18.8 The concentrations of two absorbing species in a mixture can be determined from their molar absorption coefficients and the measurement of their absorbances at two different wavelengths lying within their joint absorption region.

this common value as $\varepsilon°$. The total absorbance of the mixture at the isosbestic wavelength is

$$A° = \varepsilon°([A] + [B])l \qquad (18.7)$$

Even if A and B are interconverted in a reaction of the form A → B or its reverse, because their total concentration remains constant, so does $A°$. As a result, one or more **isosbestic points,** which are invariant points in the absorption spectrum, may be observed (Fig 18.9). It is very unlikely that three or more species would have the same molar extinction coefficients at a single wavelength. Therefore, the observation of an isosbestic point, or at least not more than one such point, is compelling evidence that a solution consists of only two solutes in equilibrium with each other with no intermediates.

18.3 **Circular dichroism**

When plane-polarized radiation passes through samples of certain kinds of matter, the plane of polarization is rotated. This rotation is the phenomenon of **optical activity**. Optical activity is observed when the molecules in the sample are **chiral**, which means distinguishable from their mirror image (Fig 18.10). In many cases, organic chiral compounds are easy to identify, because they contain a carbon atom to which are bonded four different groups. The amino acid alanine,

$NH_2CH(CH_3)COOH$, is an example. Mirror image pairs of molecules, which are called **enantiomers,** rotate light of a given frequency through exactly the same angle but in opposite directions.

Chiral molecules have a second characteristic: they absorb left and right circularly polarized light to different extents. In a circularly polarized ray of light, the electric field describes a helical path as the wave travels through space (Fig 18.11),

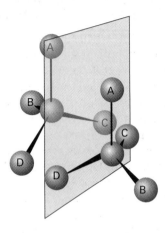

Fig 18.10 A chiral molecule is one that is not superimposable to its mirror image. A carbon atom attached to four different groups is an example of a chiral centre in a molecule. Such molecules are optically active.

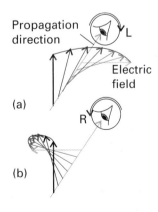

Fig 18.11 In circularly polarized light, the electric field at different points along the direction of propagation rotates. The arrays of arrows in these illustrations show the view of the electric field when looking toward the oncoming ray: (a) left-circularly polarized, (b) right-circularly polarized light.

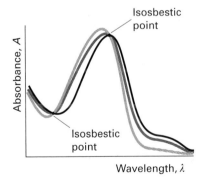

Fig 18.9 One or more isosbestic points are formed when there are two interrelated absorbing species in solution. The three curves correspond to three different stages of the reaction A → B.

and the rotation may be either clockwise or counterclockwise. The differential absorption of left- and right-circularly polarized light is called **circular dichroism**. In terms of the absorbances for the two components, A_L and A_R, the circular dichroism of a sample of molar concentration $[J]$ is reported as

$$\Delta\varepsilon = \varepsilon_L - \varepsilon_R = \frac{A_L - A_R}{[J]l} \qquad (18.8)$$

where l is the path length of the sample.

Circular dichroism is a useful adjunct to visible and UV spectroscopy. For example, the CD spectra of chiral d-metal complexes are distinctly different, whereas there is little difference between their absorption spectra (Fig 18.12). Moreover, CD spectra can be used to assign the absolute configuration of complexes by comparing the observed spectrum with the CD spectrum of a similar complex of known handedness. The CD spectra of polypeptides and nucleic acids give similar structural information. In these cases the spectrum of the polymer chain arises from the chirality of individual monomer units and, in addition, a contribution from the helical structure of the polymer itself. By subtracting the CD spectra of a mixture of monomers, the remaining structure is due largely to the secondary structure of the polymer, and in this way its conformation may be investigated. Circular dichroism can also be used to identify and follow changes in the conformation of biological macromolecules (Fig 18.13).

18.4 **Specific types of transitions**

The absorption of a photon can often be traced to the excitation of an electron that is localized on a small group of atoms. For example, an absorption at about 290 nm is normally observed when a carbonyl group is present. Groups with characteristic optical absorptions are called **chromophores** (from the Greek for 'colour bringer'), and their presence often accounts for the colours of many substances.

A d-metal complex may absorb light as a result of the transfer of an electron from the ligands into the d orbitals of the central atom, or vice versa

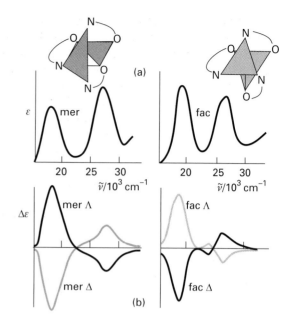

Fig 18.12 (a) The absorption spectra of two isomers of $[Co(ala)_3]$, where ala is the conjugate base of alanine, and (b) the corresponding CD spectra. The left- and right-handed forms of these isomers give identical absorption spectra. However, the CD spectra are distinctly different, and the absolute configurations have been assigned by comparison with the CD spectra of a complex of known absolute configuration.

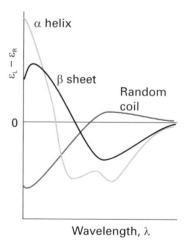

Fig 18.13 CD spectra typical of a polypeptide in three different conformations.

(Fig 18.14). In such **charge-transfer transitions** the electron moves through a considerable distance, which means that the redistribution of charge as measured by the transition dipole moment may be large and the absorption correspondingly intense. This mode of chromophore activity is shown by the permanganate ion, MnO_4^-: the charge redistribution that accompanies the migration of an electron from the O atoms to the central Mn atom accounts for its intense purple colour (resulting from absorption in the range 420 to 700 nm).

The transition responsible for absorption in carbonyl compounds can be traced to the lone pairs of electrons on the O atom. One of these electrons may be excited into an empty π^* orbital of the carbonyl group (Fig 18.15), which gives rise to an **n-to-π^* transition**, where n denotes a nonbonding orbital (one that is neither bonding nor antibonding, such as that occupied by a lone pair). Typical absorption energies are about 4 eV.

Self-test 18.2

Estimate the wavelength of maximum absorption for a transition of energy 4.3 eV.

[*Answer*: 290 nm]

A C=C double bond acts as a chromophore because the absorption of a photon excites a π electron into an antibonding π^* orbital (Fig 18.16). The chromophore activity is therefore due to a **π-to-π^* transition**. Its energy is around 7 eV for an unconjugated double bond, which corresponds to an absorption at 180 nm (in the ultraviolet). When the double bond

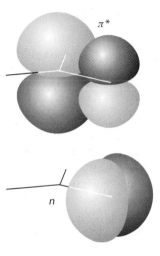

Fig 18.15 A carbonyl group acts as a chromophore primarily on account of the excitation of a nonbonding O lone-pair electron to an antibonding CO π^* orbital.

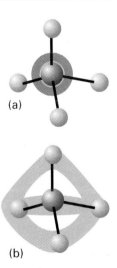

(a)

(b)

Fig 18.14 A schematic representation of a charge-transfer transition in which an electron initially on the outer atoms, represented by the grey tint in (a), migrates to the central ion in (b).

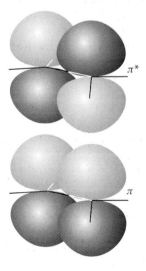

Fig 18.16 A carbon–carbon double bond acts as a chromophore. One of its important transitions is the π-to-π^* transition illustrated here, in which an electron is promoted from a π orbital to the corresponding antibonding orbital.

is part of a conjugated chain, the energies of the molecular orbitals lie closer together and the transition shifts into the visible region of the spectrum. Many of the reds and yellows of vegetation are due to transitions of this kind. For example, the carotenes that are present in green leaves (but are concealed by the intense absorption of the chlorophyll until the latter decays in the autumn) collect some of the solar radiation incident on the leaf by a π-to-π^* transition in their long conjugated hydrocarbon chains. A similar type of absorption is responsible for the primary process of vision (Box 18.1).

Box 18.1 *The photochemistry of vision*

The eye is an exquisite photochemical organ that acts as a transducer, converting radiant energy into electrical signals that travel along neurons. Here we concentrate on the events taking place in the human eye, but similar processes occur in all animals. Indeed, a single type of molecule, rhodopsin, is the primary receptor for light throughout the animal kingdom, which indicates that vision emerged very early in evolutionary history, no doubt because of its enormous value for survival.

Photons enter the eye through the cornea, pass through the ocular fluid that fills the eye, and fall on the retina. The ocular fluid is principally water, and passage of light through this medium is largely responsible for the *chromatic aberration* of the eye, the blurring of the image as a result of different frequencies being brought to slightly different focuses. Water, like any other transparent medium, disperses light because its refractive index varies with frequency. This effect can be traced to the connection between polarizability and refractive index: the greater the polarizability, the greater the refractive index and the greater the bending of a ray of light. Water is more polarizable in response to blue (high-frequency) light than red (low-frequency) light because each photon of blue light has more energy and can more readily interact with the water molecules. The chromatic aberration is reduced to some extent by the tinted region called the *macular pigment* that covers part of the retina. The pigments in this region are the carotene-like xanthophylls (**1**), which remove some of the blue light and hence help to sharpen the image. They also protect the photoreceptor molecules from too great a flux of potentially dangerous high-energy photons. The xanthophylls have de-localized electrons that spread along the chain of conjugated double bonds, and the π-to-π^* transition lies in the visible. Much the same type of protection is found in leaves, where carotene molecules help to protect the delicate chlorophyll molecules (see Box 11.2).

1 A xanthophyll

About 57 per cent of the photons that enter the eye reach the retina; the rest are scattered or absorbed by the ocular fluid. Here the primary act of vision takes place, in which the chromophore of a rhodopsin molecule absorbs a photon in another π-to-π^* transition. A rhodopsin molecule consists of an opsin protein molecule to which is attached an 11-*cis*-retinal molecule (**2**). The latter resembles half a carotene molecule, showing Nature's economy in its use of available materials. The attachment is by the formation of a Schiff's base, utilizing the –CHO group of the chromophore. The free 11-*cis*-retinal molecule absorbs in the ultraviolet, but attachment to the opsin protein molecule shifts the absorption into the visible region. The rhodopsin molecules are situated in the membranes of the rods and cones that cover the retina. The opsin molecule is anchored into the cell membrane by two hydrophobic groups and largely surrounds the chromophore (see illustration).

2 11-*cis*-retinal

Immediately after the absorption of a photon, the 11-*cis*-retinal molecule undergoes photoisomerization into all-*trans*-retinal (**3**). Photoisomerization takes about 200 fs and occurs with a primary quantum yield of about 0.67. The process is able to occur because the π-to-π^* excitation of an electron loosens one of the π bonds (the one indicated by the arrow in **3**), its torsional rigidity is lost, and one part of the molecule swings round into its new position. At that point, the electron returns to its ground state, and the molecule is trapped in its new conformation. The straightened tail of all-*trans*-retinal results in the molecule taking up more space than 11-*cis*-retinal did, so the molecule presses against the coils of the opsin molecule that surrounds it. Thus, in about 0.25–0.50 ms

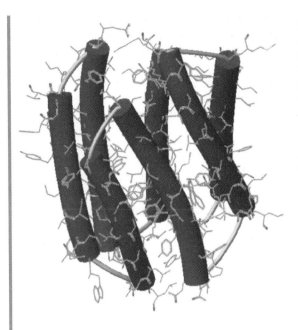

The structure of the rhodopsin molecule, consisting of an opsin protein to which is attached an 11-*cis*-retinal molecule embedded in the region surrounded by the helical regions.

3 All-*trans*-retinal

from the initial absorption even, the rhodopsin molecule is activated.

Now a sequence of biochemical events—the *biochemical cascade*—converts the altered configuration of the rhodopsin molecule into a pulse of electric potential that travels through the optical nerve into the optical cortex, where it is interpreted as a signal and incorporated into the web of events we call 'vision'. At the same time, the resting state of the rhodopsin molecule is restored by a series of nonradiative chemical events powered by ATP. The process involves the escape of all-*trans*-retinal as all-*trans*-retinol (in which –CHO has been reduced to –CH_2OH) from the opsin molecule by a process catalysed by the enzyme rhodopsin kinase and the attachment of another protein molecule, arrestin. The free all-*trans*-retinol molecule now undergoes enzyme-catalysed isomerization into 11-*cis*-retinol followed by dehydrogenation to form 11-*cis*-retinal, which is then delivered back into an opsin molecule. At this point, the cycle of excitation, photoisomerization, and regeneration is ready to begin again.

Exercise 1   The flux of visible photons reaching Earth from the North Star is about 4×10^3 mm^{-2} s^{-1}. Of these photons, 30 per cent are absorbed or scattered by the atmosphere and 25 per cent of the surviving photons are scattered by the surface of the cornea of the eye. A further 9 per cent are absorbed inside the cornea. The area of the pupil at night is about 40 mm^2 and the response time of the eye is about 0.1 s. Of the photons passing through the pupil, about 43 per cent are absorbed in the ocular medium. How many photons from the North Star are focused on to the retina in 0.1 s? For a continuation of this story, see R. W. Rodieck, *The first steps in seeing*, Sinauer (1998).

Exercise 2   In the free-electron molecular orbital theory of electronic structure, the π electrons in a conjugated molecule are treated as noninteracting particles in a box with a length equal to the length of the conjugated system. On the basis of this model, at what wavelength would you expect all-*trans*-retinal to absorb? Take the mean carbon–carbon bond length to be 140 pm.

Radiative decay

In most cases, the excitation energy of a molecule that has absorbed a photon is degraded into the disordered thermal motion of its surroundings. However, one process by which an electronically excited molecule can discard its excess energy is by **radiative decay**, in which an electron relaxes back into a lower energy orbital and in the process generates a photon. As a result, an observer sees the sample glowing (if the emitted radiation is in the visible region of the spectrum).

There are two principal modes of radiative decay, fluorescence and phosphorescence. In **fluorescence**, the spontaneously emitted radiation ceases immediately after the exciting radiation is extinguished. In **phosphorescence**, the spontaneous emission may persist for long periods (even hours, but characteristically seconds or fractions of seconds). The difference suggests that fluorescence is an immediate conversion of absorbed light into re-emitted radiant energy and that phosphorescence involves the storage of energy in a reservoir from which it slowly leaks.

Other than thermal degradation, a nonradiative fate for an electronically excited molecule is **dissociation,** or fragmentation (Fig 18.17). The onset of dissociation can be detected in an absorption spectrum by seeing that the vibrational structure of a band terminates at a certain energy. Absorption occurs in a continuous band above this **dissociation limit,** the highest frequency before the onset of continuous absorption, because the final state is unquantized translational motion of the fragments. Locating the dissociation limit is a valuable way of determining the bond dissociation energy.

18.5 **Fluorescence**

Figure 18.18 is a simple example of a **Jablonski diagram,** a schematic portrayal of molecular electronic and vibrational energy levels, which shows the sequence of steps involved in fluorescence. The initial absorption takes the molecule to an excited electronic state and, if the absorption spectrum were monitored, it would look like the one shown in Fig 18.19a. The excited molecule is subjected to collisions with the surrounding molecules, and as it gives up energy it steps down the ladder of vibrational levels. The surrounding molecules, however, might be unable to accept the larger energy needed to lower the molecule to the ground electronic state. The excited state might therefore survive long enough to generate a photon and emit the remaining excess energy as radiation. The down-

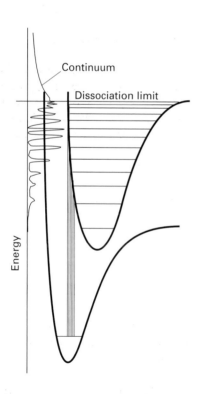

Fig 18.17 When absorption occurs to unbound states of the upper electronic state, the molecule dissociates and the absorption is a continuum. Below the dissociation limit the electronic spectrum has a normal vibrational structure.

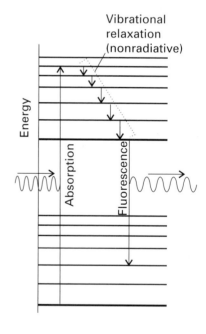

Fig 18.18 A Jablonski diagram showing the sequence of steps leading to fluorescence. After the initial absorption the upper vibrational states undergo radiationless decay—the process of vibrational relaxation—by giving up energy to the surroundings. A radiative transition then occurs from the ground state of the upper electronic state. In practice, the separation of the ground states of the electronic states (the heavy horizontal lines) is 10 to 100 times greater than the separation of the vibrational levels.

ward electronic transition is **vertical**, which means in accord with the Franck–Condon principle, and the fluorescence spectrum has a vibrational structure characteristic of the lower electronic state (Fig 18.19b).

Fluorescence occurs at a lower frequency than that of the incident radiation because the fluorescence radiation is emitted after some vibrational energy has been discarded into the surroundings. The vivid oranges and greens of fluorescent dyes are an everyday manifestation of this effect: they absorb in the ultraviolet and blue, and fluoresce in the visible. The mechanism also suggests that the intensity of the fluorescence ought to depend on the ability of the solvent molecules to accept the electronic and vibrational quanta. It is indeed found that a solvent composed of molecules with widely spaced vibrational levels (such as water) may be able to accept the large quantum of electronic energy and so quench the fluorescence.

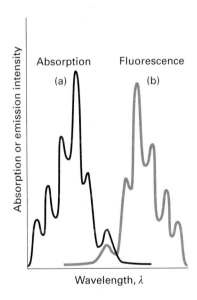

Fig 18.19 The absorption spectrum (a) shows a vibrational structure characteristic of the upper state. The fluorescence spectrum (b) shows a structure characteristic of the lower state; it is also displaced to lower frequencies (longer wavelengths) and resembles a mirror image of the absorption.

18.6 Fluorescence quenching

The excitation energy may be removed by another molecule, which therefore acts to **quench** the fluorescence. The study of the effect of quencher concentration on the intensity of fluorescence is a valuable source of information about the rates of photochemical processes in solution.

The intensity of fluorescence from an excited state, A*, is proportional to the concentration of that excited state, so we can write

$$\text{Fluorescence intensity} = k_f[A^*]$$

The rate of formation of the excited state is

$$\text{Rate of formation of } A^* = I_{abs}$$

The excited state is lost by two processes: radiative decay and transfer of energy to the quencher:

$$\text{Rate of loss of } A^* = k_f[A^*] + k_q[A^*][Q] \qquad (18.9)$$

Here we are supposing that the quenching is bimolecular: the quenching molecule needs to approach the excited molecule closely if it is to remove its excitation energy. In a steady state, the rate of loss of A* is equal to its rate of formation, so we can write

$$I_{abs} = k_f[A^*] + k_q[A^*][Q]$$

The solution of this equation is

$$[A^*] = \frac{I_{abs}}{k_f + k_q[Q]}$$

and the fluorescence intensity, I_f, is

$$I_f = k_f[A^*] = \frac{k_f I_{abs}}{k_f + k_q[Q]}$$

To determine the constants that appear in this expression, we invert both sides, and obtain the **Stern–Volmer equation**:

$$\frac{1}{I_f} = \frac{1}{I_{abs}} + \left(\frac{k_q}{I_{abs}k_f}\right)[Q] \qquad (18.10)$$

This expression shows that, if we plot $1/I_f$ against the quencher concentration, [Q], which is called a

Stern–Volmer plot, then we should get a straight line with intercept $1/I_{abs}$ and slope $k_q/I_{abs}k_f$ (Fig 18.20).

At this stage, by combining the intercept and the slope, we can find only the ratio k_q/k_f, not the individual rate constants. To separate these two observables, we need to make another type of measurement. Suppose that, instead of sustaining a steady state, we extinguish the exciting radiation; then the concentration of A^*, and therefore the fluorescence intensity, decays towards zero in accord with eqn 18.9, which has the solution

$$I_f = I_{f,0}e^{-t/\tau} \qquad \frac{1}{\tau} = k_f + k_q[Q] \qquad (18.11)$$

That is, the fluorescence decays exponentially with a time constant τ, the **fluorescence lifetime,** which depends on the quencher concentration. If we measure τ for a series of quencher concentrations and plot $1/\tau$ against [Q], the slope gives k_q and the intercept gives k_f (Fig 18.21).

Example 18.2 *Analysing the kinetics of fluorescence*

The fluorescence of a dye when it was irradiated by 350 nm light was studied in the presence of a quenching agent. The following intensities were observed:

[Q]/(mmol L^{-1})	1.0	2.0	3.0	4.0	5.0
I_{abs}/I_f	5.2	9.4	13.7	18.0	22.2

In a second series of experiments, the incident light was extinguished, and the decay of the fluorescence was observed:

[Q]/(mmol L^{-1})	1.0	2.0	3.0	4.0	5.0
τ/ns	91	51	34	25	22

Determine the half-life of the fluorescence and the quenching rate constant.

Strategy From the first set of data and eqn 18.10 expressed in the form

$$\frac{I_{abs}}{I_f} = 1 + \left(\frac{k_q}{k_f}\right)[Q]$$

we can find k_q/k_f by plotting the intensity ratio against [Q] and determining the slope of the line. From the second set of data and eqn 18.11 we plot $1/\tau$ against [Q] and determine k_f from the intercept and k_q from

the slope. Because the fluorescent decay is a unimolecular, first-order process, we can use eqn 10.10 ($t_{1/2} = \ln 2/k$, with $k = k_f$) to find the half-life.

Solution Figure 18.22 shows I_{abs}/I_f plotted against [Q]. The slope of the line is 4.2×10^3, so $k_q/k_f = 4.2 \times 10^3$ L mol^{-1}. For the second plot we draw up the following table:

[Q]/(mmol L^{-1})	1.0	2.0	3.0	4.0	5.0
$10^7/(\tau/s)$	1.1	2.0	2.9	4.0	4.5

These points are plotted in Fig 18.23. The intercept at [Q] = 0 lies at 2.1×10^6 s^{-1} and the slope is 8.9×10^9, so $k_f = 2.1 \times 10^6$ s^{-1} and $k_q = 8.9 \times 10^9$ L mol^{-1} s^{-1}, in accord with the ratio found from the first plot. It follows that the half-life of fluorescent decay is

$$t_{1/2} = \frac{\ln 2}{2.1 \times 10^6 \text{ s}^{-1}} = 3.3 \times 10^{-7} \text{ s}$$

Self-test 18.3

Repeat the calculations using the following data:

[Q]/(mmol L^{-1})	1.0	3.0	5.0	7.0	9.0
I_{abs}/I_f	3.9	9.7	15.5	21.3	27.1
τ/ns	10.0	3.9	2.4	1.8	1.4

[*Answer:* $k_q = 7.7 \times 10^{10}$ L mol^{-1} s^{-1}, $t_{1/2} = 0.06$ µs]

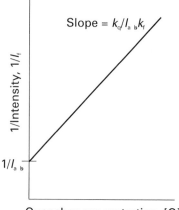

Slope = $k_q/I_{a\;b}k_f$

1/Intensity, $1/I_f$

$1/I_{a\;b}$

Quencher concentration, [Q]

Fig 18.20 A Stern–Volmer plot for the determination of the kinetics of fluorescent decay and quenching.

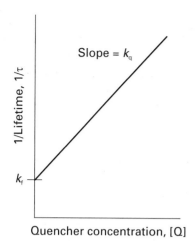

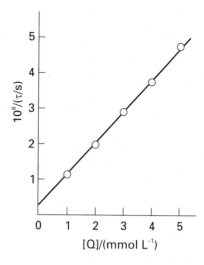

Fig 18.21 From measurements of the lifetime of fluorescent decay in the presence of a quencher, we can determine the individual rate constants from the intercept and slope of this graph.

Fig 18.23 The determination of the rate constants for fluorescence and quenching from the information in Example 18.2.

Quenching may also be brought about by energy transfer from one chromophore in a molecule to another chromophore. One mechanism, called the **Förster mechanism,** arises from a dipole–dipole coupling between the *transition* dipoles on the two groups. It turns out that the quenching lifetime on the basis of this mechanism is proportional to the sixth power of the separation of the groups:

$$\frac{\tau_T}{\tau_0} = \left(\frac{r_F}{r}\right)^6 \tag{18.12}$$

In this expression, τ_0 is the lifetime in the absence of the acceptor chromophore, τ_T is the lifetime arising from transfer to the second chromophore at a distance r; the constant r_F, the **Förster length**, is a collection of quantities that include the intensities of transitions on the two chromophores. The strong dependence on separation of the chromophores means that a study of quenching rates by this mechanism is an excellent probe of conformational changes in macromolecules, such as RNA and oligopeptides to which chromophores have been attached in known locations.

18.7 **Phosphorescence**

Figure 18.24 is a Jablonski diagram showing the events leading to phosphorescence. The first steps are the same as in fluorescence, but the presence of a triplet state plays a decisive role. A **triplet state** is a state in which two electrons in different orbitals have parallel spins: the ground state of O_2, which

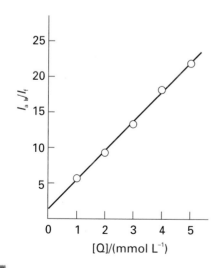

Fig 18.22 The Stern–Volmer plot for the data in Example 18.2.

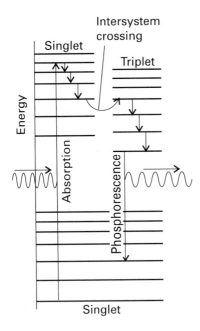

Fig 18.24 The sequence of steps leading to phosphorescence. The important step is the intersystem crossing from an excited singlet to an excited triplet state. The triplet state acts as a slowly radiating reservoir because the return to the ground state is very slow.

was discussed in Section 9.5, is an example. The name 'triplet' reflects the (quantum mechanical) fact that the total spin of two parallel electron spins ($\uparrow\uparrow$) can adopt only three orientations with respect to an axis. An ordinary spin-paired state ($\uparrow\downarrow$) is called a **singlet state** because there is only one orientation in space for such a pair of spins.[3]

The ground state of a typical phosphorescent molecule is a singlet because its electrons are all paired; the excited state to which the absorption excites the molecule is also a singlet. The peculiar feature of a phosphorescent molecule, however, is that it possesses an excited triplet state of an energy similar to that of the excited singlet state and into which the excited singlet state may convert. Hence,

if there is a mechanism for unpairing two electron spins (and so converting $\uparrow\downarrow$ into $\uparrow\uparrow$), then the molecule may undergo **intersystem crossing** and become a triplet state. The unpairing of electron spins is possible if the molecule contains a heavy atom, such as an atom of sulfur, with strong spin–orbit coupling (Section 13.17). Then the angular momentum needed to convert a singlet state into a triplet state may be acquired from the orbital motion of the electrons.

After an excited singlet molecule crosses into a triplet state, it continues to discard energy into the surroundings and to step down the ladder of vibrational states. However, it is now stepping down the triplet's ladder, and at the lowest vibrational energy level it is trapped. The solvent cannot extract the final, large quantum of electronic excitation energy. Moreover, the molecule cannot radiate its energy because return to the ground state is forbidden: a triplet state cannot convert into a singlet state because the spin of one electron cannot reverse in direction relative to the other electron during a transition.[4] The radiative transition, however, is not totally forbidden because the spin–orbit coupling responsible for the intersystem crossing also breaks this rule. The molecules are therefore able to emit weakly and the emission may continue long after the original excited state was formed.

The mechanism of phosphorescence summarized in Fig 18.24 accounts for the observation that the excitation energy seems to become trapped in a slowly leaking reservoir. It also suggests (as is confirmed experimentally) that phosphorescence should be most intense from solid samples: energy transfer is then less efficient and the intersystem crossing has time to occur as the singlet excited state loses vibrational energy. The mechanism also suggests that the phosphorescence efficiency should depend on the presence of a moderately heavy atom (with its ability to flip electron spins), which is in fact the case.

---

[3] In the language introduced in Section 13.16, a triplet state has $S = 1$ and M_S has one of the three values $+1$, 0, and -1; a singlet state has $S = 0$ and M_S has the single value 0.

[4] A selection rule valid for light atoms is $\Delta S = 0$.

18.8 Lasers

Lasers have transformed experimental chemistry as much as they have the everyday world. The word *laser* is an acronym formed from *light amplification by stimulated emission of radiation*. As this name suggests, it is a process that depends on *stimulated* emission as distinct from the spontaneous emission processes characteristic of fluorescence and phosphorescence. In **stimulated emission,** an excited state is stimulated to emit a photon by the presence of radiation of the same frequency and, the more photons there are present, the greater the probability of the emission. To picture the process, we can think of the oscillations of the electromagnetic field as periodically (at the frequency at which the field oscillates) distorting the excited molecule at the frequency of the transition and hence encouraging the molecule to generate a photon of the same frequency. The essential feature of laser action is the strong **gain,** or growth of intensity, that results: the more photons present of the appropriate frequency, the more photons of that frequency the excited molecules will be stimulated to form, and so the laser medium fills with photons.

One requirement for laser action is the existence of an excited state that has a long enough lifetime for it to participate in stimulated emission. Another requirement is the existence of a greater population in the upper state than in the lower state where the transition terminates. Because at thermal equilibrium the population is greater in the lower energy state, it is necessary to achieve a **population inversion** in which there are more molecules in the upper state than in the lower.

One way to achieve population inversion is illustrated in Fig 18.25. The inversion is achieved indirectly through an intermediate state I. Thus, the molecule is excited to I, which then gives up some of its energy nonradiatively (by passing energy on to vibrations of the surroundings) and changes into a lower state B; the laser transition is the return of B to a lower state A. Because four levels are involved

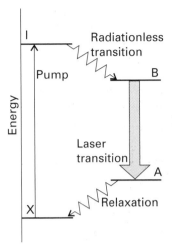

Fig 18.25 The transitions involved in a four-level laser. Because the laser transition terminates in an excited state (A), the population inversion between A and B is much easier to achieve than when the lower state of the laser transition is the ground state.

overall, this arrangement leads to a **four-level laser.** The transition from X to I is caused by an intense flash of light in the process called **pumping.** In some cases the pumping flash is achieved with an electric discharge through xenon or with the radiation from another laser.

Laser radiation has a number of advantages for applications in chemistry. One advantage is its highly monochromatic character, which enables very precise spectroscopic observations to be made. Another advantage is the ability of laser radiation to be produced in very short pulses (currently, as brief as about 1 fs): as a result, very fast chemical events, such as the individual transfers of atoms during a chemical reaction, can be followed (see Box 10.1). Laser radiation is also very intense, which reduces the time needed for spectroscopic observations: this characteristic is particularly useful in Raman spectroscopy, where the scattered radiation is of very low intensity.

Photoelectron spectroscopy

T HE exposure of a molecule to high-frequency radiation can result in the ejection of an electron. This **photoejection** is the basis of another type of spectroscopy in which we monitor the energies of the ejected photoelectrons. If the incident radiation has frequency v, the photon that causes photoejection has energy hv. If the ionization energy of the molecule is I, the difference in energy, $hv - I$, is carried away as kinetic energy. Because the kinetic energy of an electron of speed v is $\frac{1}{2}m_e v^2$, we can write $hv - I = \frac{1}{2}m_e v^2$, so

$$hv = I + \tfrac{1}{2}m_e v^2 \qquad (18.13)$$

Therefore, by monitoring the velocity of the photoelectron, and knowing the frequency of the incident radiation, we can determine the ionization energy of the molecule and hence the strength with which the electron was bound (Fig 18.26). In this context the 'ionization energy' of the molecule has different values depending on the orbital that the photoelectron occupied and, the slower the ejected electron, the lower in energy the orbital from which it was ejected. The apparatus is a modification of a mass spectrometer (Fig 18.27), in which the velocity of the photoelectrons is measured by determining the strength of the electric field required to bend their paths on to the detector.

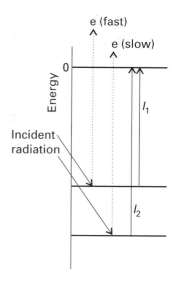

Fig 18.26 The basic principle of photoelectron spectroscopy. An incoming photon of known energy collides with an electron in one of the orbitals and expels it with a kinetic energy that is equal to the difference between the energy supplied by the photon and the ionization energy from the occupied orbital. An electron from an orbital with a low ionization energy will emerge with a high kinetic energy (and high speed) whereas an electron from an orbital with a high ionization energy will be ejected with a low kinetic energy (and low speed).

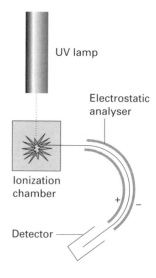

Fig 18.27 A photoelectron spectrometer consists of a source of ionizing radiation (such as a helium discharge lamp for ultraviolet photoelectron spectroscopy, UPS, and an X-ray source for X-ray photoelectron spectroscopy, XPS), an electrostatic analyser, and an electron detector. The deflection of the electron paths caused by the analyser depends on the electrons' speed.

Self-test 18.4

What is the velocity of photoelectrons that are ejected from a molecule with radiation of energy 21 eV (from a helium discharge lamp) and are known to come from an orbital of ionization energy 12 eV?

[*Answer*: 1.8×10^3 km s^{-1}]

Figure 18.28 shows a typical photoelectron spectrum (of HBr). If we disregard the fine structure, we see that the HBr lines fall into two main groups. The least tightly bound electrons (with the lowest ionization energies and hence highest kinetic energies when ejected) are those in the lone pairs of the Br atom. The next ionization energy lies at 15.2 eV, and corresponds to the removal of an electron from the HBr σ bond.

The HBr spectrum shows that ejection of a σ electron is accompanied by a considerable amount of vibrational excitation. The Franck–Condon principle would account for this observation if ejection were accompanied by an appreciable change of equilibrium bond length between HBr and HBr⁺. When that is so, the ion is formed in a bond-compressed state, which is consistent with the important bonding effect of the σ electrons. The lack of much vibrational structure in the other band is consistent with the nonbonding role of the $Br4p_x$ and $Br4p_y$ lone-pair electrons, for the equilibrium bond length is little changed when one is removed.

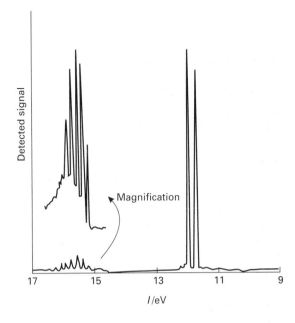

Fig 18.28 The photoelectron spectrum of HBr. The lowest ionization energy band corresponds to the ionization of a Br lone-pair electron. The higher ionization energy band corresponds to the ionization of a bonding electron. The structure on the latter is due to the vibrational excitation of HBr⁺ that results from the ionization.

Example 18.3 *Interpreting a UV photoelectron spectrum*

The highest kinetic energy electrons in the spectrum of H_2O using 21.22 eV He radiation are at about 9 eV and show a large vibrational spacing of 0.41 eV. The symmetric stretching mode of the neutral H_2O molecule lies at 3652 cm⁻¹. What conclusions can be drawn from the nature of the orbital from which the electron is ejected?

Strategy To convert from electronvolts to reciprocal centimetres, use 1 eV = 8065.5 cm⁻¹. If the vibrational separation in the ion is similar to that in the molecule, then the ejected electron had little influence on bonding in the molecule. A lot of vibrational structure would suggest that the electron had been heavily involved in bonding.

Solution Because 0.41 eV corresponds to 3.3×10^3 cm⁻¹, which is similar to the 3652 cm⁻¹ of the nonionized molecule, we can suspect that the electron is ejected from an orbital that has little influence on the bonding in the molecule. That is, photoejection is from a largely nonbonding orbital.

Self-test 18.5

In the same spectrum of H_2O, the band near 7.0 eV shows a long vibrational series with spacing 0.125 eV. The bending mode of H_2O lies at 1596 cm⁻¹. What conclusions can you draw about the characteristics of the orbital occupied by the photoelectron?

[*Answer*: the electron contributes to long distance HH bonding across the molecule]

Exercises

18.1 The molar absorption coefficient of a substance dissolved in hexane is known to be $743 \, mol^{-1} \, L \, cm^{-1}$ at $285 \, nm$. Calculate the percentage reduction in intensity when light of that wavelength passes through $2.5 \, mm$ of a solution of concentration $3.25 \times 10^{-3} \, mol \, L^{-1}$.

18.2 When light of wavelength $410 \, nm$ passes through $2.5 \, mm$ of a solution of the dye responsible for the yellow of daffodils at a concentration $4.33 \times 10^{-4} \, mol \, L^{-1}$, the transmission is 71.5 per cent. Calculate the molar absorption coefficient of the colouring matter at this wavelength and express the answer in centimetre-squared per mole $(cm^2 \, mol^{-1})$.

18.3 The molar absorption coefficient of cytochrome P450, one of the compounds involved in photosynthesis, at $522 \, nm$ is $291 \, L \, mol^{-1} \, cm^{-1}$. When light of that wavelength passes through a cell of length $6.5 \, mm$ containing a solution of the solute, 39.8 per cent of the light was absorbed. What is the molar concentration of the solution?

18.4 An aqueous solution of a triphosphate derivative of molar mass $602 \, g \, mol^{-1}$ was prepared by dissolving $30.2 \, mg$ in $500 \, mL$ of water and a sample was transferred to a cell of length $1.00 \, cm$. The absorbance (optical density) was measured as 1.011. (a) Calculate the molar absorption coefficient. (b) Calculate the transmittance, expressed as a percentage, for a solution of twice the concentration.

18.5 A *Dubosq colorimeter* consists of a cell of fixed path length and a cell of variable path length. By adjusting the length of the latter until the transmission through the two cells is the same, the concentration of the second solution can be inferred from that of the former. Suppose that a plant dye of concentration $25 \, \mu g \, L^{-1}$ is added to the fixed cell, the length of which is $1.55 \, cm$. Then a solution of the same dye, but of unknown concentration, is added to the second cell. It is found that the same transmittance is obtained when the length of the second cell is adjusted to $1.18 \, cm$. What is the concentration of the second solution?

18.6 The Beer–Lambert law is derived on the basis that the concentration of absorbing species is uniform (see *Derivation 18.1*). Suppose, instead, that the concentration falls exponentially as $[J] = [J]_0 e^{-x/\lambda}$. Derive an expression

for the variation of I with sample length: suppose that $l \gg \lambda$. (*Hint.* Work through *Derivation 18.1*, but use this expression for the concentration.)

18.7 The following data were obtained for the absorption by Br_2 in carbon tetrachloride using a cell of length $2.0 \, mm$. Calculate the molar absorption coefficient (ε) of bromine at the wavelength employed:

$[Br_2]/(mol \, L^{-1})$	0.0010	0.0050	0.0100	0.0500
$T/(per \, cent)$	81.4	35.6	12.7	3.0×10^{-3}

18.8 A cell of length $2.0 \, mm$ was filled with a solution of benzene in a nonabsorbing solvent. The concentration of the benzene was $0.010 \, L \, mol^{-1}$ and the wavelength of the radiation was $256 \, nm$ (where there is a maximum in the absorption). Calculate the molar absorption coefficient of benzene at this wavelength given that the transmission was 48 per cent. What will the transmittance be in a cell of length $4.0 \, mm$ at the same wavelength?

18.9 A swimmer enters a gloomier world (in one sense) on diving to greater depths. Given that the mean molar absorption coefficient of sea water in the visible region is $6.2 \times 10^{-5} \, L \, mol^{-1} \, cm^{-1}$, calculate the depth at which a diver will experience (a) half the surface intensity of light, (b) one-tenth that intensity.

18.10 The molar absorption coefficients of tryptophan and tyrosine at $240 \, nm$ are $2.00 \times 10^3 \, L \, mol^{-1} \, cm^{-1}$ and $1.12 \times 10^4 \, L \, mol^{-1} \, cm^{-1}$, respectively, and at $280 \, nm$ they are $5.40 \times 10^3 \, L \, mol^{-1} \, cm^{-1}$ and $1.50 \times 10^3 \, L \, mol^{-1} \, cm^{-1}$. The absorbance of a sample obtained by hydrolysis of a protein was measured in a cell of thickness $1.00 \, cm$, and was found to be 0.660 at $240 \, nm$ and 0.221 at $280 \, nm$. What are the concentrations of the two amino acids?

18.11 Consider a solution of two unrelated substances A and B. Let their molar absorption coefficients be equal at a certain wavelength, and write their total absorbance A. Show that we can infer the concentration of A and B from the total absorbance at some other wavelength provided we know the molar absorption coefficients at that different wavelength (see Fig 18.29).

18.12 A solution was prepared by dissolving tryptophan and tyrosine in $0.15 \, M \, NaOH(aq)$ and a sample was transferred to a cell of length $1.00 \, cm$. The two amino acids share the same molar absorption coefficient at $294 \, nm$ (2.38×10^3

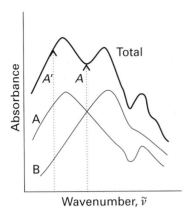

Fig 18.29

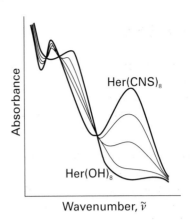

Fig 18.30

L mol^{-1} cm^{-1}) and the absorbance of the solution at that wavelength is 0.468. At 280 nm the molar absorption coefficients are 5.23×10^3 and 1.58×10^3 L mol^{-1} cm^{-1}, respectively and the total absorbance of the solution is 0.676. What are the concentrations of the two amino acids? (*Hint.* It would be sensible to use the result derived in Exercise 18.11, but this specific example could be worked through without using that general case.)

18.13 Figure 18.30 shows the UV–visible absorption spectrum of a derivative of haemerythrin in the presence of different concentrations of CNS$^-$ ions. What may be inferred from the spectrum?

18.14 The compound $CH_3CH=CHCHO$ has a strong absorption in the ultraviolet at 46 950 cm^{-1} and a weak absorption at 30 000 cm^{-1}. Justify these features in terms of the structure of the compound.

18.15 Figure 18.31 shows the UV–visible absorption spectra of a selection of amino acids. Suggest reasons for their different appearances in terms of the structures of the molecules.

18.16 The fluorescence of a solution of a plant pigment illuminated by 330 nm radiation was studied in the presence of a quenching agent, with the following results:

[Q]/(mmol L^{-1})	1.0	2.0	3.0	4.0	5.0
I_f/I_{abs}	0.31	0.18	0.13	0.10	0.081

In a second series of experiments, the incident radiation was extinguished and the lifetime of the decay of the fluorescence was observed:

[Q]/(mmol L^{-1})	1.0	2.0	3.0	4.0	5.0
τ/ns	76	45	32	25	20

Determine the quenching rate constant and the half-life of the fluorescence.

18.17 What is the kinetic energy of an electron that has been accelerated through a potential difference of 5.0 V?

18.18 What is the energy of an electron that has been ejected from an orbital of ionization energy 11.0 eV by a photon of radiation of wavelength 100 nm?

18.19 In a particular photoelectron spectrum using 21.21 eV photons, electrons were ejected with kinetic energies of 11.01 eV, 8.23 eV, and 5.22 eV. Sketch the molecular orbital energy level diagram for the species, showing the ionization energies of the three identifiable orbitals.

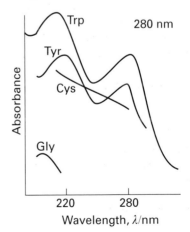

Fig 18.31

Magnetic resonance

Contents

ONE of the most widely used and helpful forms of spectroscopy, and a technique that has transformed the practice of chemistry and its dependent disciplines, makes use of an effect that is familiar from classical physics. When two pendulums are joined by the same slightly flexible support and one is set in motion, the other is forced into oscillation by the motion of the common axle, and energy flows between the two. The energy transfer occurs most efficiently when the frequencies of the two oscillators are identical. The condition of strong effective coupling when the frequencies are identical is called **resonance**, and the excitation energy is said to 'resonate' between the coupled oscillators.

Resonance is the basis of a number of everyday phenomena, including the response of radios to the weak oscillations of the electromagnetic field generated by a distant transmitter. In this chapter we explore a spectroscopic application that when originally developed (and in some cases still) depended on matching a set of energy levels to a source of monochromatic radiation and observing the strong absorption that occurs at resonance.

Principles of magnetic resonance

THE application of resonance that we describe here depends on the fact that many nuclei possess spin angular momentum (just as an electron does, but nuclei possess a wider range of values). A nucleus with **spin quantum number** I (the analogue of s for electrons, and which may be an integer or a half-integer) may take $2I + 1$ different orientations relative to an arbitrary axis. These orientations are distinguished by the quantum number m_I, which may take on the values $m_I = I, I - 1, \ldots, -I$. A proton has $I = \frac{1}{2}$ (the same spin as an electron) and may adopt either of two orientations ($m_I = +\frac{1}{2}$ and $-\frac{1}{2}$). A ^{14}N nucleus has $I = 1$ and may adopt any of three orientations ($m_I = +1, 0, -1$). In this section we shall consider only **spin-$\frac{1}{2}$ nuclei**, those with $I = \frac{1}{2}$. Spin-$\frac{1}{2}$ nuclei include protons (^{1}H), ^{13}C, ^{19}F, and ^{31}P nuclei (Table 19.1). As for electrons, the state with $m_I = +\frac{1}{2}$ (↑) is denoted α, and that with $m_I = -\frac{1}{2}$ (↓) is denoted β.

19.1 Nuclei in magnetic fields

A nucleus with nonzero spin behaves like a tiny magnet. The orientation of this magnet is determined by the value of m_I, and in a magnetic field B the $2I + 1$ orientations of the nucleus have different energies. These energies are given by

$$E_{m_I} = -\gamma_N \hbar B m_I \tag{19.1}$$

where γ_N is the **magnetogyric ratio** of the nucleus. This quantity is sometimes written in terms of the **nuclear magneton**, μ_N, and an empirical constant called the **nuclear g-factor**, g_I:

$$\gamma_N \hbar = g_I \mu_N \qquad n_N = \frac{e \hbar}{2 m_p} \tag{19.2}$$

The numerical value of μ_N is 5.051×10^{-27} J T^{-1}.[1] Nuclear g-factors are experimentally determined

---

[1] The symbol T denotes the unit tesla, which is used to measure the intensity of a magnetic field (1 T = 1 kg s^{-2} A^{-1}).

Table 19.1 *Nuclear spin properties*

Nucleus	Natural abundance/ per cent	Spin, I	$\gamma_N/(10^7$ T^{-1} s$^{-1})$
^{1}H	99.98	$\frac{1}{2}$	26.752
^{1}H (D)	0.0156	1	4.1067
^{12}C	98.99	0	—
^{13}C	1.11	$\frac{1}{2}$	6.7272
^{14}N	99.64	1	1.9328
^{16}O	99.96	0	—
^{17}O	0.037	$\frac{5}{2}$	−3.627
^{19}F	100	$\frac{1}{2}$	25.177
^{31}P	100	$\frac{1}{2}$	10.840
^{35}Cl	75.4	$\frac{3}{2}$	2.624
^{37}Cl	24.6	$\frac{3}{2}$	2.184

dimensionless quantities of the order of 1: for protons, $g_I = 5.5857$. Positive values of γ_N (and g_I) indicate that the nuclear magnet lies in the same direction as the nuclear spin (this is the case for protons). Negative values indicate that the magnet points in the opposite direction. A nuclear magnet is about 2000 times weaker than the magnet associated with electron spin. Two very common nuclei, ^{12}C and ^{16}O, have zero spin and hence are invisible in magnetic resonance.

The energy separation of the two states of a spin-$\frac{1}{2}$ nucleus (Fig 19.1) is

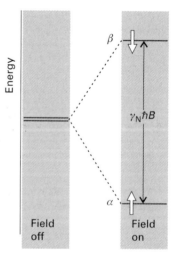

Fig 19.1 The energy levels of a spin-$\frac{1}{2}$ nucleus (e.g. ^{1}H or ^{13}C) in a magnetic field. Resonance occurs when the energy separation of the levels matches the energy of the photons in the electromagnetic field.

$$\Delta E = E_\beta - E_\alpha = \tfrac{1}{2}\gamma_N\hbar B - (-\tfrac{1}{2}\gamma_N\hbar B) = \gamma_N\hbar B \qquad (19.3)$$

For most nuclei, γ_N is positive, so the β state lies above the α state and, according to the Boltzmann distribution, there are slightly more α spins than β spins. If the sample is exposed to radiation of frequency ν, the energy separations come into resonance with the radiation when the frequency satisfies the **resonance condition**:

$$h\nu = \gamma_N\hbar B \quad \text{or} \quad \nu = \frac{c_N B}{2\pi} \qquad (19.4)$$

At resonance there is strong coupling between the nuclear spins and the radiation, and strong absorption occurs as the spins flip from $\uparrow$ (low energy) to $\downarrow$ (high energy).

> **Self-test 19.1**
>
> Calculate the frequency at which radiation comes into resonance with proton spins in a 12 T magnetic field.
>
> [*Answer*: 510 MHz]

19.2 **The technique**

In its simplest form, **nuclear magnetic resonance** (NMR) is the observation of the frequency at which magnetic nuclei in molecules come into resonance with an electromagnetic field when the molecule is exposed to a strong magnetic field. When applied to proton spins, the technique is occasionally called **proton magnetic resonance** (^{1}H-NMR). In the early days of the technique the only nuclei that could be studied were protons (which behave like relatively strong magnets because γ_N is large), but now a wide variety of nuclei (especially ^{13}C and ^{31}P) are investigated routinely.

An NMR spectrometer consists of a magnet that can produce a uniform, intense field and the appropriate sources of radiofrequency radiation. In simple instruments the magnetic field is provided by an electromagnet; for serious work, a superconducting magnet capable of producing fields of the order of 10 T and more is used.[2] The use of high magnetic fields has two advantages. One is that the field increases the energy separation and therefore the population difference between the two spin states, and hence a stronger net absorption is obtained. Secondly, a high field simplifies the appearance of certain spectra. Proton resonance occurs at about 400 MHz in fields of 9.4 T, so NMR is a radiofrequency technique (400 MHz corresponds to a wavelength of 75 cm).

All modern work is done using a modification of this basic technique. In **Fourier transform NMR** (FT-NMR), the sample is held in a strong magnetic field generated by a superconducting magnet and exposed to one or more carefully controlled brief bursts of radiofrequency radiation. This radiation changes the orientations of the nuclear spins in a controlled way, and the radiofrequency radiation they emit as they return to equilibrium is monitored and analysed mathematically (the latter is the 'Fourier transform' part of the technique). The detected radiation contains all the information in the spectrum obtained by the earlier technique, but it is a much more efficient way of obtaining the spectrum and hence is much more sensitive. Moreover, by choosing different sequences of exciting pulses, the data can be analysed much more closely.

Electrons also have a magnetic moment, and adopt either of two orientations in a magnetic field. The resonance technique is then called **electron paramagnetic resonance** (EPR) or **electron spin resonance** (ESR). Because electron magnetic moments are much bigger than nuclear magnetic moments, even quite modest fields can require high frequencies to achieve resonance. Much work is done using fields of about 0.3 T, when resonance occurs at about 9 GHz, corresponding to 3 cm microwave radiation. Electron paramagnetic resonance is much more limited than NMR because it is applicable only to species with unpaired electrons, which include radicals (perhaps prepared by radiation damage or photolysis) d-metal complexes, and triplet states. It gives valuable information about electron distributions and can be used to monitor, for instance, the uptake of oxygen by haemoglobin and biological electron transfer processes.

---

[2] A magnetic field of 10 T is very strong: a small magnet, for example, gives a magnetic field of only a few millitesla.

The information in NMR spectra

NUCLEAR spins interact with the *local* magnetic field. The local field may differ from the applied field either on account of the local electronic structure of the molecule or because there is another magnetic nucleus nearby.

19.3 The chemical shift

The applied magnetic field can induce a circulating motion of the electrons in the molecule, and that motion gives rise to a small additional magnetic field, δB. This additional field is proportional to the applied field, and it is conventional to express it as

$$\delta B = -\sigma B \qquad (19.5)$$

where σ is the **shielding constant**. The constant σ may be positive or negative according to whether the induced field lies in the opposite or same direction as the applied field. The ability of the applied field to induce the circulation of electrons through the nuclear framework of the molecule depends on the details of the electronic structure near the magnetic nucleus of interest, so nuclei in different chemical groups have different shielding constants.

Because the total local field is

$$B_{loc} = B + \delta B = (1 - \sigma)B$$

the resonance condition is

$$\nu = \frac{\gamma_N B_{loc}}{2\pi} = \frac{\gamma_N}{2\pi}(1-\sigma)B \qquad (19.6)$$

Because σ varies with the environment, different nuclei (even of the same element so long as they are in different parts of a molecule) come into resonance at different frequencies.

The **chemical shift** of a nucleus is the difference between its resonance frequency and that of a reference standard. The standard for protons is the proton resonance in tetramethylsilane, $Si(CH_3)_4$, commonly referred to as TMS, which bristles with protons and dissolves without reaction in many solutions. Other references are used for other nuclei. For ^{13}C, the reference frequency is the ^{13}C resonance in TMS, and for ^{31}P it is the ^{31}P resonance in 85 per cent H_3PO_4(aq). The separation of the resonance of a particular group of nuclei from the standard increases with the strength of the applied magnetic field because the induced field is proportional to the applied field, and the stronger the latter, the greater the shift.

Chemical shifts are reported on the δ scale, which is defined as[3]

$$\delta = \frac{\nu - \nu^o}{\nu^o} \times 10^6 \qquad (19.7)$$

where ν^o is the resonance frequency of the standard. The advantage of the δ scale is that shifts reported on it are independent of the applied field (because both numerator and denominator are proportional to the applied field). The resonance frequencies themselves, however, do depend on the applied field through

$$\nu = \nu^o + (\nu^o/10^6)\delta \qquad (19.8)$$

If $\delta > 0$, we say that the nucleus is **deshielded**; if $\delta < 0$, then it is **shielded**. A positive δ indicates that the resonance frequency of the group of nuclei in question is higher than that of the standard. Hence $\delta > 0$ indicates that the local magnetic field is stronger than that experienced by the nuclei in the standard under the same conditions. Some typical chemical shifts are given in Fig 19.2.

Self-test 19.2

What is the shift of the resonance from TMS of a group of nuclei with $\delta = 3.50$ and an operating frequency of 350 MHz?

[*Answer*: 1.23 kHz]

---

[3] In much of the literature that uses NMR, chemical shifts are reported in parts per million, ppm, in recognition of the factor of 10^6 in the definition. This practice is unnecessary.

Illustration 19.1

The existence of a chemical shift explains the general features of the spectrum of ethanol shown in Fig 19.3. The CH₃ protons form one group of nuclei with $\delta = 1$. The two CH₂ are in a different part of the molecule, experience a different local magnetic field, and hence resonate at $\delta = 3$. Finally, the OH proton is in another environment, and has a chemical shift of $\delta = 4$.

We can use the relative intensities of the signal (the areas under the absorption lines) to help distinguish which group of lines corresponds to which chemical group, and spectrometers can integrate the absorption—that is, determine the areas under the absorption signal—automatically (as is shown in Fig 19.3). In ethanol the group intensities are in the ratio 3:2:1 because there are three CH₃ protons,

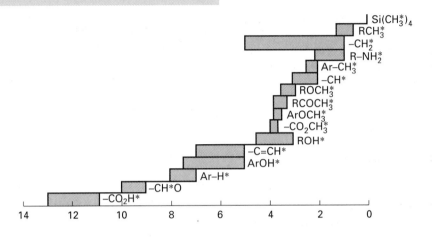

Fig 19.2 The range of typical chemical shifts for ¹H resonances.

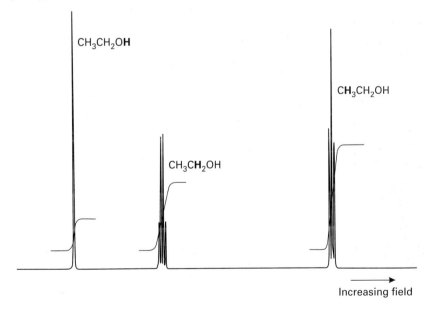

Fig 19.3 The NMR spectrum of ethanol. The bold letters denote the protons giving rise to the resonance peak, and the step-like curves are the integrated signals for each group of lines.

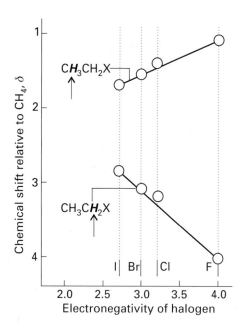

Fig 19.4 The variation of chemical shift with the electronegativity of the halogen in the haloalkanes. Note that, although the chemical shift of the immediately adjacent protons becomes more positive (the protons are deshielded) as the electronegativity increases, that of the next nearest protons decreases.

two CH_2 protons, and one OH proton in each molecule. Counting the number of magnetic nuclei as well as noting their chemical shifts is valuable analytically because it helps us identify the compound present in a sample and to identify substances in different environments (Box 19.1).

Broadly speaking, the value of δ depends on the electron density around the nucleus in question. It is therefore responsive to the effects of electronegativity in the molecule. The proton chemical shifts of the haloalkanes, for instance, increase with the electronegativity of the halogen. The electronegative atom removes electron density from the nearby protons and deshields them, implying that their resonance shifts to higher δ (Fig 19.4).

19.4 The fine structure

The splitting of the groups of resonances into individual lines in Fig 19.3 is called the **fine structure** of the spectrum. It arises because each magnetic nu-

cleus contributes to the local field experienced by the other nuclei and modifies their resonance frequencies. The strength of the interaction is expressed in terms of the **spin–spin coupling constant**, J, and reported in hertz (Hz). Spin coupling constants are an intrinsic property of the molecule and independent of the strength of the applied field.

Consider first a molecule that contains two spin-$\frac{1}{2}$ nuclei A and X. Suppose the spin of X is α; then A will resonate at a certain frequency as a result of the combined effect of the external field, the shielding constant, and the spin–spin interaction of nucleus A with nucleus X. As we show in the *Derivation* below, instead of a single line from A, we get a doublet of lines separated by a frequency J (Fig 19.5). The same splitting occurs in the X resonance: instead of a single line it is a doublet with splitting J (the same value as for the splitting of A).

Derivation 19.1 *The structure of an AX spectrum*

First, neglect spin–spin coupling. The total energy of two protons in a magnetic field B is the sum of two terms like eqn 19.1, but with B modified to $(1-\sigma)B$:

$$E = -\gamma_N \hbar (1-\sigma_A) Bm_A - \gamma_N \hbar (1-\sigma_X) Bm_X$$

Here σ_A and σ_X are the shielding constants of A and X. The four energy levels predicted by this formula are shown on the left of Fig 19.6. The spin–spin coupling energy is normally written

$$E_{\text{spin–spin}} = hJm_A m_X$$

There are four possibilities, depending on the values of the quantum numbers m_A and m_X:

	$\alpha_A\alpha_X$	$\alpha_A\beta_X$	$\beta_A\alpha_X$	$\beta_A\beta_X$
$E_{\text{spin–spin}}$	$+\frac{1}{4}hJ$	$-\frac{1}{4}hJ$	$-\frac{1}{4}hJ$	$+\frac{1}{4}hJ$

The resulting energy levels are shown on the right in Fig 19.6. Now consider the transitions. When an A nucleus changes its spin from α to β, the X nucleus remains in its same spin state, which may be either α or β. The two transitions are shown in the illustration, and we see that they differ in frequency by J. Alternatively, the X nucleus can undergo a transition from α to β; now the A nucleus remains in its same spin state, which may be either α or β and we again get two transitions that differ in frequency by J.

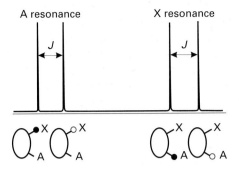

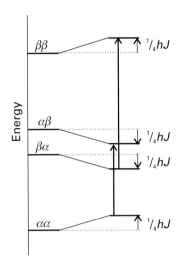

Fig 19.5 The effect of spin–spin coupling on a NMR spectrum of two spin-$\frac{1}{2}$ nuclei with widely different chemical shifts. Each resonance is split into two lines separated by J. Full circles indicate α spins, open circles indicate β spins.

If there is another X nucleus in the molecule with the same chemical shift as the first X (giving an AX_2 species), the resonance of A is split into a doublet by one X, and each line of the doublet is split again by the same amount (Fig 19.7) by the second X. This splitting results in three lines in the intensity ratio 1:2:1 (because the central frequency can be obtained in two ways). As in the AX case discussed above, the X resonance of the AX_2 species is split into a doublet by A.

Three equivalent X nuclei (an AX_3 species) split the resonance of A into four lines of intensity ratio 1:3:3:1 (Fig 19.8). The X resonance remains a doublet as a result of the splitting caused by A. In general, N equivalent spin-$\frac{1}{2}$ nuclei split the resonance of a nearby spin or group of equivalent spins into $N + 1$ lines with an intensity distribution given by Pascal's triangle (**1**). Subsequent rows of this triangle are formed by adding together the two adjacent numbers in the line above.

Fig 19.6 The energy levels of a two-proton system in the presence of a magnetic field. The levels on the left apply in the absence of spin–spin coupling. Those on the right are the result of allowing for spin–spin coupling. The only allowed transitions differ in frequency by J.

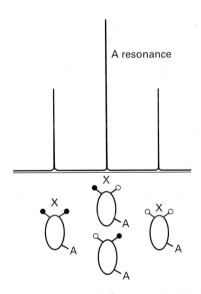

A resonance

Fig 19.7 The origin of the 1:2:1 triplet in the A resonance of an AX_2 species. The two X nuclei may have the $2^2 = 4$ spin arrangements ($\uparrow\uparrow$); ($\uparrow\downarrow$); ($\downarrow\uparrow$); ($\downarrow\downarrow$). The middle two arrangements are responsible for the coincident resonances of A.

```
              1
          1       1
      1       2       1
  1       3       3       1
1       4       6       4       1
```

1 Pascal's triangle

Box 19.1 *Magnetic resonance imaging*

One of the most striking applications of nuclear magnetic resonance is in medicine. The technique of *magnetic resonance imaging* (MRI) is a simple portrayal of the concentrations of protons in a solid object. If an object containing hydrogen nuclei (a flask of water or a human body) is placed in an NMR spectrometer and exposed to a *uniform* magnetic field, then a single resonance signal will be detected. However, if the magnetic field varies linearly across the object, then the protons in different regions will be resonant at different frequencies, and the intensity of the signal will be proportional to the numbers of protons at each magnetic field. The intensity of the signal will therefore be a map of the distribution of protons in the sample. More precisely, it will be a projection of the numbers of protons on a line parallel to the field gradient, as shown in the first illustration. If the body is rotated into different orientations, another projection can be determined. Then, the data are analysed on a computer to reconstruct the three-dimensional distribution of protons (for example, as water) in the living tissue, to detect abnormalities, and to observe metabolic processes. With *functional MRI*, blood flow in different regions of the brain can be studied and related to the mental activities of the subject.

The special advantage of MRI is that it can image *soft* tissues (see the second illustration), whereas X-rays are largely used for imaging hard, bony structures and abnormally dense regions, such as tumours. In fact, the invisibility of hard structures in MRI is an advantage, as it allows the imaging of structures encased by bone, such as the brain and the spinal cord. X-rays are known to be dangerous on account of the ionization they cause; the high magnetic fields used in MRI may also be dangerous but, apart from anecdotes about the extraction of loose fillings from teeth, there is no convincing evidence of their harmfulness, and the technique is considered safe.

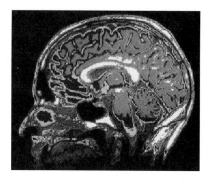

The great advantage of MRI is that it can display soft tissue, such as in this cross-section through a patient's head. (Courtesy of the University of Manitoba.)

In a magnetic field that varies linearly over a sample, all the protons within a given slice (that is, at a given field value) come into resonance and give a signal of the corresponding intensity. The resulting intensity pattern is a map of the numbers in all the slices, and portrays the shape of the sample. Changing the orientation of the field shows the shape along the corresponding direction, and computer manipulation can be used to build up the three-dimensional shape of the sample.

Exercise 1 You are designing an MRI spectrometer. What field gradient (in microtesla per metre, $\mu T\ m^{-1}$) is required to produce a separation of 100 Hz between two protons separated by the long diameter of a human kidney (taken as 8 cm) given that they are in environments with $\delta = 3.4$? The radiofrequency field of the spectrometer is at 400 MHz and the applied field is 9.4 T.

Exercise 2 Suppose a uniform disk-shaped organ is in a linear field gradient, and that the MRI signal is proportional to the number of protons in a slice of width δx at each horizontal distance x from the centre of the disk. Sketch the shape of the absorption intensity for the MRI image of the disk before any computer manipulation has been carried out.

Self-test 19.3

Complete the next line of the triangle, the pattern arising from five equivalent protons.

[*Answer*: 1:5:10:10:5:1]

Example 19.1 *Accounting for the fine structure in a spectrum*

Account for the fine structure in the ^{1}H-NMR spectrum of the C–H protons of ethanol.

Strategy Refer to Pascal's triangle to determine the effect of a group of N equivalent protons on a proton, or (equivalently) a group of protons, of interest.

Solution The three protons of the CH_3 group split the single resonance of the CH_2 protons into a 1:3:3:1 quartet with a splitting J. Likewise, the two protons of the CH_2 group split the single resonance of the CH_3 protons into a 1:2:1 triplet. Each of these lines is split into a doublet to a small extent by the OH proton.

Self-test 19.4

What fine-structure can be expected for the protons in $^{14}NH_4^+$?

[*Answer*: a 1:1:1 triplet from ^{14}N]

The spin–spin coupling constant of two nuclei joined by N bonds is normally denoted ^NJ, with subscripts for the types of nuclei involved. Thus, $^1J_{CH}$ is the coupling constant for a proton joined directly to a ^{13}C atom, and $^2J_{CH}$ is the coupling constant when the same two nuclei are separated by two bonds (as in ^{13}C–C–H). A typical value of $^1J_{CH}$ is between 10^2 to 10^3 Hz; the value of $^2J_{CH}$ is about 10 times less, between about 10 and 10^2 Hz. Both ^3J and ^4J give detectable effects in a spectrum, but couplings over larger numbers of bonds can generally be ignored.

Example 19.2 *Interpreting an NMR spectrum*

Suggest an interpretation of the ^{1}H-NMR spectrum in Fig 19.9.

Strategy We need to look for groups with characteristic chemical shifts (Fig 19.2) and account for the fine structure, as we did for ethanol.

Solution The resonance at $\delta = 3.4$ corresponds to CH_2 in an ether; that at $\delta = 1.2$ corresponds to CH_3 in CH_3CH_2. The fine structure of the CH_2 group (a 1:3:3:1 quartet) is characteristic of splitting caused by CH_3; the fine structure of the CH_3 resonance is characteristic of splitting caused by CH_2. The fine structure constant is $J = 60$ Hz (the same for each group). The compound is probably $(CH_3CH_2)_2O$.

Self-test 19.5

What changes in the spectrum would be observed on recording it in a spectrometer that operates at a magnetic field that is five times stronger?

[*Answer*: groups of lines 5 times further apart in frequency (but the same δ values); no change in spin–spin splitting]

The magnitude of $^3J_{HH}$ depends on the angle, ϕ, between the two C–H bonds (**2**). The variation is expressed quite well by the **Karplus equation**:

$$^3J_{HH} = A + B \cos \phi + C \cos 2\phi \qquad (19.9)$$

2

Typical values of A, B, and C are +7 Hz, –1 Hz, and +5 Hz, respectively. Figure 19.10 shows the angular variation the equation predicts.

Illustration 19.2

The investigation of H–N–C–H couplings in polypeptides can help reveal their conformation. For $^3J_{HH}$ coupling in such a group, $A = +5.1$ Hz, $B = -1.4$ Hz, and $C = +3.2$ Hz. For an α helix, ϕ is close to 120°, which would give $^3J_{HH} \approx 4$ Hz. For a β sheet, ϕ is close to 180°, which would give $^3J_{HH} \approx 10$ Hz. Consequently, small coupling constants indicate an α helix whereas large couplings indicate a β sheet.

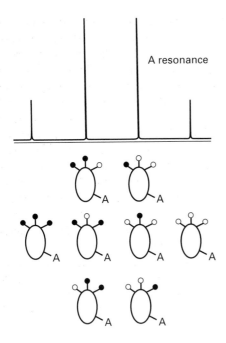

Fig 19.8 The origin of the 1:3:3:1 quartet in the A resonance of an AX$_3$ species where A and X are spin-$\frac{1}{2}$ nuclei with widely different chemical shifts. There are $2^3 = 8$ arrangements of the spins of the three X nuclei, and their effects on the A nucleus give rise to four groups of resonances.

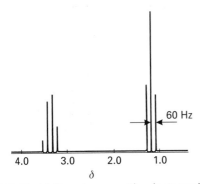

Fig 19.9 The NMR spectrum considered in Example 19.2.

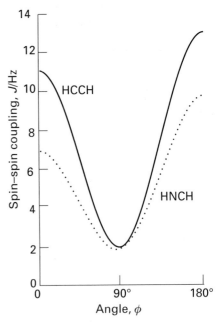

Fig 19.10 The variation of $^3J_{HH}$ with angle, according to the Karplus equation. The solid line is for H–C–C–H and the dotted line for H–N–C–H.

19.5 **Spin relaxation**

As resonant absorption continues, the population of the upper state rises to match that of the lower state. The net rate of absorption is the difference between the rate at which absorption occurs and the rate at which the incident radiation stimulates the spins in the upper state to return to the ground state and emit radiation. Therefore, we can expect the net rate of absorption, and hence the strength of the absorption signal, to decrease with time. This decrease due to the progressive equalization of populations is called **saturation**.

The fact that saturation is often not observed must mean that there are nonradiative processes by which β spins can become α spins again, and hence help to maintain the population difference between the two sites. The nonradiative return to a Boltzmann distribution of populations in a system is an aspect of **relaxation**. If we were to imagine forming a system of spins in which all the nuclei were in their β state, then the system returns exponentially to the equilibrium distribution (a small excess of α spins over β spins) with a time constant called the **spin–lattice relaxation time**, T_1 (Fig 19.11).

However, there is another, more subtle aspect of relaxation. In a classical picture of magnetic nuclei, we imagine each one as a tiny bar magnet. A bar magnet in an externally applied magnetic field undergoes the motion called **precession** as it twists round the direction of the field (Fig 19.12). The rate

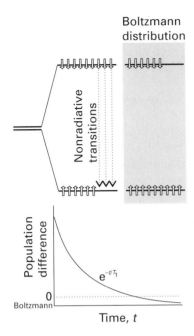

Boltzmann distribution

Fig 19.11 The spin–lattice relaxation time is the time constant for the exponential return of the population of the spin states to their equilibrium (Boltzmann) distribution.

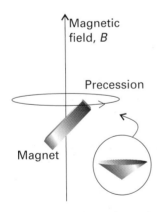

Fig 19.12 A bar magnet in a magnetic field undergoes the motion called *precession*. A nuclear spin (and an electron spin) has an associated magnetic moment, and behaves in the same way. The frequency of precession is called the Larmor precession frequency, and is proportional to the applied field and the magnitude of the magnetic moment.

of precession is proportional to the strength of the applied field, and is in fact equal to $(\gamma_N/2\pi)B$, which in this context is called the **Larmor precession**

frequency.[4] Now imagine that somehow we arrange all the spins in a sample to have exactly the same angle around the field direction at an instant. If each spin has a slightly different Larmor frequency (because they experience slightly different local magnetic fields), they will gradually fan out. At thermal equilibrium, all the bar magnets lie at *random* angles round the direction of the applied field, and the time constant for the exponential return of the system into this random arrangement is called the **spin–spin relaxation time**, T_2 (Fig 19.13). For spins to be truly at thermal equilibrium, not only must the populations have a Boltzmann distribution but the spin orientations must be random around the field direction.

What causes each type of relaxation? In each case the spins are responding to local magnetic fields that act to twist them into different orientations. However, there is a crucial difference between the two processes.

Random phases

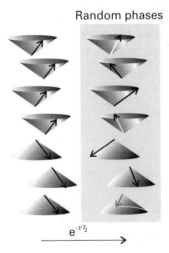

Fig 19.13 The spin–spin relaxation time is the time constant for the exponential return of the spins to a random distribution around the direction of the magnetic field. No change in populations of the two spin states is involved in this type of relaxation, so no energy is transferred from the spins to the surroundings.

[4] Note that resonance absorption occurs when the Larmor precession frequency is the same as the frequency of the applied electromagnetic field.

The best kind of local magnetic field for inducing a transition from β to α (as in spin–lattice relaxation) is one that fluctuates at a frequency close to the resonance frequency. Such a field can arise from the tumbling motion of the molecule in the fluid sample. If the tumbling motion of the molecule is slow compared to the resonance frequency, it will give rise to a fluctuating magnetic field that oscillates too slowly to induce transitions, so T_1 will be long. If the molecule tumbles much faster than the resonance frequency, then it will give rise to a fluctuating magnetic field that oscillates too rapidly to induce transitions, so T_1 will again be long. Only if the molecule tumbles at about the resonance frequency will the fluctuating magnetic field be able to induce transitions effectively, and only then will T_1 be short. The rate of molecular tumbling increases with temperature and with reducing viscosity of the solvent, so we can expect a dependence like that shown in Fig 19.14.

The best kind of local magnetic field for causing spin–spin relaxation is one that does not change very rapidly. Then each molecule in the sample lingers in its particular local magnetic environment for a long time, and the orientations of the spins have time to become randomized around the applied field direction. If the molecules move rapidly from one magnetic environment to another, the effects of different magnetic fields average out, and the randomization does not take place as quickly. In other words, slow molecular motion corresponds to short T_2 and fast motion corresponds to long T_2 (as shown in Fig 19.14). Detailed calculation shows that, when the motion is fast, the two relaxation times are equal, as has been drawn in the illustration.

Spin relaxation studies—using advanced techniques that utilize complicated sequences of pulses of radiofrequency energy to stimulate the spins into special orientations, and then monitoring their return to equilibrium—have two main applications. First, they reveal information about the mobility of molecules or parts of molecules. For example, by studying spin relaxation times of protons in the hydrocarbon chains of micelles and bilayers (Section 16.14) we can build up a detailed picture of the motion of these chains, and hence come to an understanding of the dynamics of cell membranes. Second, relaxation times depend on the separation of the nucleus from the source of the magnetic field that is causing its relaxation: that source may be another magnetic nucleus in the same molecule. By studying the relaxation times, we can determine the internuclear distances within the molecule and use them to build up a model of its shape.

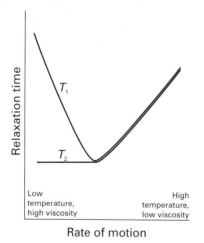

Fig 19.14 The variation of the two relaxation times with the rate at which the molecules move (either by tumbling or migrating through the solution). The horizontal axis can be interpreted as representing temperature or viscosity. Note that, at rapid rates of motion, the two relaxation times coincide.

19.6 The nuclear Overhauser effect

An effect that makes use of spin relaxation is of considerable usefulness for the determination of the conformations of proteins and other biological macromolecules in their natural aqueous environment. To introduce the effect, we consider a very simple AX system in which the two spins interact by a dipole–dipole interaction. We expect two lines in the spectrum, one from A and the other from X. However, when we irradiate the system with radiofrequency radiation at the resonance frequency of X using such a high intensity that we saturate the transition, we find that the A resonance is modified. It may be enhanced, diminished, or even converted into an emission rather than an absorption. That modification of one resonance when we saturate

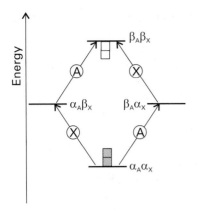

Fig 19.15 The energy levels of an AX system and an indication of their relative populations. Each grey square above the line represents an excess population and each white square below the line represents a population deficit. The transitions of A and X are marked.

another is called the **nuclear Overhauser effect** (NOE).

To understand the effect, we need to think about the populations of the four levels of an AX system (Fig 19.15). At thermal equilibrium, the population of the $\alpha_A\alpha_X$ level is the greatest, and that of the $\beta_A\beta_X$ level is the least; the other two levels have the same energy and an intermediate population. The thermal equilibrium absorption intensities reflect these populations as Fig 19.15 shows. Now consider the combined effect of saturating the X transition and spin relaxation. When we saturate the X transition, the populations of the X levels are equalized, but at this stage there is no change in the populations of the A levels. If that were all there were to happen, all we would see would be the loss of the X resonance and no effect on the A resonance.

Now consider the effect of spin relaxation. Relaxation can occur in a variety of ways if there is a dipolar interaction between the A and X spins. One possibility is for the magnetic field acting between the two spins to cause them both to flop from β to α, so the $\alpha_A\alpha_X$ and $\beta_A\beta_X$ states regain their thermal equilibrium populations. However, the populations of the $\alpha_A\beta_X$ and $\beta_A\alpha_X$ levels remain unchanged at the values characteristic of saturation. As we see from Fig 19.16, the population difference between the states joined by transitions of A is now greater than at equilibrium, so the resonance absorption is enhanced. Another possibility is for the dipolar interaction between the two spins to cause α to flip to β and β to flop to α. This transition equilibrates the populations of $\alpha_A\beta_X$ and $\beta_A\alpha_X$ but leaves the $\alpha_A\alpha_X$ and $\beta_A\beta_X$ populations unchanged (Fig 19.17). Now we see from the illustration that the population differences in the states involved in the A transitions are decreased, so the resonance absorption is diminished.

Which effect wins? Does NOE enhance the A absorption or does it diminish it? As in the discussion of relaxation times in Section 19.5, the efficiency of

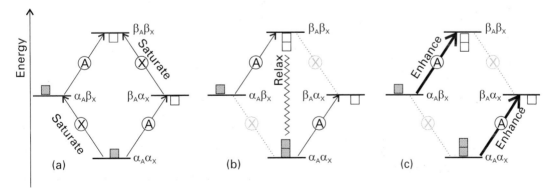

Fig 19.16 (a) When the X transition is saturated, the populations of its two states are equalized and the population excess and deficit become as shown (using the same symbols as in Fig 19.15). (b) Dipole–dipole relaxation relaxes the populations of the highest and lowest states, and they regain their original populations. (c) The A transitions reflect the difference in populations resulting from the preceding changes, and are enhanced compared with those shown in Fig 19.15.

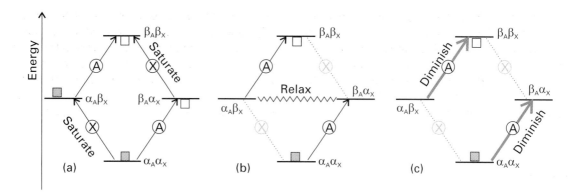

Fig 19.17 (a) When the X transition is saturated, just as in Fig 19.16 the populations of its two states are equalized and the population excess and deficit become as shown. (b) Dipole–dipole relaxation relaxes the populations of the two intermediate states, and they regain their original populations. (c) The A transitions reflect the difference in populations resulting from the preceding changes, and are diminished compared with those shown in Fig 19.15.

the intensity-enhancing $\beta_A\beta_X \leftrightarrow \alpha_A\alpha_X$ relaxation is high if the dipole field is modulated at the transition frequency, which in this case is close to 2ω; likewise, the efficiency of the intensity-diminishing $\alpha_A\beta_X \leftrightarrow \beta_A\alpha_X$ relaxation is high if the dipole field is stationary (as there is no frequency difference between the initial and final states). A large molecule rotates so slowly that there is very little motion at 2ω, so we expect intensity decrease (Fig 19.18). A small molecule rotating rapidly can be expected to

have substantial motion at 2ω, and a consequent enhancement of the signal. In practice, the enhancement lies somewhere between the two extremes and is reported in terms of the parameter η (eta), where

$$\eta = \frac{I - I_0}{I_0} \tag{19.10}$$

Here I_0 is the normal intensity and I is the NOE intensity of a particular transition; theoretically, η lies between -1 (diminution) and $+\frac{1}{2}$ (enhancement).

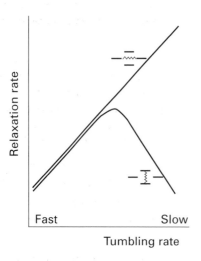

Fig 19.18 The relaxation rates of the two types of relaxation (as indicated by the small diagrams) as a function of the tumbling rate of the molecule.

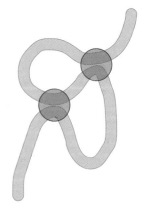

Fig 19.19 If a NOE experiment shows that the protons within each of the two circles are coupled by a dipolar interaction, we can be confident that those protons are close together, and therefore infer the conformation of the polypeptide chain.

The value of η depends strongly on the separation of the two spins involved in the NOE, for the strength of the dipolar interaction between two spins separated by a distance r is proportional to $1/r^3$ and its effect depends on the square of that strength, and therefore on $1/r^6$. This sharp dependence on separation is used to build up a picture of the conformation of a protein by using NOE to identify which nuclei can be regarded as neighbours (Fig 19.19). The enormous importance of this procedure is that we can determine the conformation of polypeptides in an aqueous environment and do not need to try to make the single crystals that are essential for an X-ray diffraction investigation.

Exercises

19.1 The nucleus ^{32}S has a spin of $\frac{3}{2}$ and a nuclear g-factor of 0.4289. Calculate the energies of the nuclear spin states in a magnetic field of 7.500 T.

19.2 Equation 19.2 defines the g-value and the magnetogyric ratio of a nucleus. Given that g is a dimensionless number, what are the units of γ_N expressed in (a) tesla and hertz, (b) SI base units?

19.3 The magnetogyric ratio of ^{31}P is 1.0840×10^8 $T^{-1} s^{-1}$. What is the g-value of the nucleus?

19.4 The magnetogyric ratio of ^{19}F is 2.5177×10^8 $T^{-1} s^{-1}$. Calculate the frequency of the nuclear transition in a field of 8.200 T.

19.5 Calculate the resonance frequency of a ^{14}N nucleus ($I = 1$, $g = 0.4036$) in a 15.00 T magnetic field.

19.6 Calculate the magnetic field needed to satisfy the resonance condition for unshielded protons in a 550.0 MHz radiofrequency field.

19.7 Calculate the resonance frequency and the corresponding wavelength for an electron in a magnetic field of 0.330 T, the magnetic field commonly used in EPR.

19.8 What is the shift of the resonance from TMS of a group of protons with $\delta = 6.33$ in a polypeptide in a spectrometer operating at 420 MHz?

19.9 The chemical shift of the CH_3 protons in acetaldehyde (ethanal) is $\delta = 2.20$ and that of the CHO proton is 9.80. What is the difference in local magnetic field between the two regions of the molecule when the applied field is (a) 1.5 T, (b) 6.0 T?

19.10 Using the information in Fig 19.2, state the splitting (in hertz, Hz) between the methyl and aldehydic proton resonances in a spectrometer operating at (a) 300 MHz, (b) 550 MHz.

19.11 What would be the nuclear magnetic resonance spectrum for a proton resonance line that was split by interaction with seven identical protons?

19.12 What would be the nuclear magnetic resonance spectrum for a proton resonance line that was split by interaction with (a) two, (b) three equivalent nitrogen nuclei (the spin of a nitrogen nucleus is 1)?

19.13 Repeat *Derivation 19.1* for an AX_2 spin-$\frac{1}{2}$ system and deduce the pattern of lines expected in the spectrum.

19.14 Sketch the appearance of the ^1H-NMR spectrum of acetaldehyde using $J = 2.90$ Hz and the data in Fig 19.2 in a spectrometer operating at (a) 300 MHz, (b) 550 MHz.

19.15 Sketch the form of the ^{19}F-NMR spectra of a natural sample of $^{10}BF_4^-$ and $^{11}BF_4^-$.

19.16 Sketch the form of an $A_3M_2X_4$ spectrum, where A, M, and X are protons with distinctly different chemical shifts and $J_{AM} > J_{AX} > J_{MX}$.

19.17 Show that the coupling constant as expressed by the Karplus equation passes through a minimum when $\cos \phi = B/4C$. (*Hint*. Use calculus: evaluate the first derivative with respect to ϕ and set the result equal to 0. To confirm that the extremum is a minimum, go on to evaluate the second derivative and show that it is positive.)

19.18 Suggest a reason why the relaxation times of ^{13}C nuclei are typically much longer than those of ^1H nuclei.

19.19 Why does the removal of dissolved oxygen from a solution commonly increase the spin relaxation time of the solute?

19.20 Suggest a reason why the spin–lattice relaxation time of benzene (a small molecule) in a mobile, deuterated hydrocarbon solvent increases with temperature whereas that of an oligopeptide (a large molecule) decreases.

Statistical thermodynamics

Contents

THERE are two great rivers in physical chemistry. One is the river of thermodynamics, which deals with the relations between bulk properties of matter, particularly properties related to the transfer of energy. The other is the river of molecular structure, including spectroscopy, which deals with the structures and properties of individual atoms and molecules. These two great rivers flow together in the part of physical chemistry called **statistical thermodynamics**, which shows how thermodynamic properties emerge from the properties of atoms and molecules. The first nine chapters of this book dealt with thermodynamic properties; the nine chapters preceding this chapter have dealt with atomic and molecular structure and its investigation. This is the chapter where the two great rivers merge.

A great problem with statistical thermodynamics is that it is highly mathematical. Many of the derivations—even the most fundamental—are beyond the scope of this text.[1] All we can hope to see is some of the key concepts and the key results. Where possible the treatment will be qualitative.

---

[1] See my *Physical chemistry*, Oxford University Press and W.H. Freeman & Co, New York (1998), for details.

The partition function

THE key concept of quantum mechanics is the existence of a wavefunction that contains in principle all the dynamical information about a system, such as its energy, the electron density, the dipole moment, and so on. Once we know the wavefunction of an atom or molecule, we can extract from it all the dynamical information possible about the system—provided we know how to manipulate it. There is a similar concept in statistical thermodynamics. The *partition function*, q, contains all the *thermodynamic* information about the system, such as its internal energy, entropy, heat capacity, and so on. Our task here is to see how to calculate the partition function and how to extract the information it contains.

20.1 The Boltzmann distribution

The single most important result in the whole of statistical thermodynamics is the **Boltzmann distribution**, the formula that tells us how to calculate the numbers of molecules in each state of a system at any temperature:[2]

$$N_i = \frac{N e^{-E_i/kT}}{q} \tag{20.1}$$

Here N_i is the number of molecules in a state with energy E_i, N is the total number of molecules, k is the Boltzmann constant, a fundamental constant with the value $1.380\,66 \times 10^{-23}$ J K^{-1}, and T is the absolute temperature. The term in the denominator, q, is the **partition function**:

$$q = \text{sum over states}(e^{-\text{Energy of state}/kT})$$

More succinctly,

$$q = \sum_i e^{-E_i/kT} \tag{20.2}$$

---

[2] Throughout this chapter 'molecules' will include single atoms or any other entities.

We shall have much more to say about q later, and shall see how it can be calculated and given physical meaning.

The conceptual basis of the derivation of the Boltzmann distribution is very simple. We imagine a stack of energy levels arranged like bookshelves, one above the other. Then we imagine being blindfolded and throwing balls (the molecules) at the shelves and letting them land on the available shelves entirely at random, apart from one condition. That condition is that the total energy, E, of the final arrangement must have the actual energy of the sample of matter we are seeking to describe. So, provided the temperature is above absolute zero, not all the balls are allowed to land on the bottom shelf, for that would give a total energy of zero. Some of the balls may land on the bottom shelf, but there must be others ending up on higher shelves to ensure that the total energy is E. If we imagine throwing 100 balls at a set of shelves, then we will end up with one particular valid distribution. If we repeated the experiment with the same number of balls, we would end up with a different but still valid distribution. If we went on repeating the experiment, we would get many different distributions, but some of them would occur more often than others (Fig 20.1).

When this game is analysed mathematically, it turns out that the *most probable* distribution—the arrangement that turns up most often—is that given by the Boltzmann distribution. In other words, *the Boltzmann distribution is the outcome of blind chance occupation of energy levels, subject to the requirement that the total energy has a particular value.* When we deal with about 10^{23} molecules and repeat the experiment millions of times, the Boltzmann distribution turns out to be very accurate, and we can use it with confidence for all typical samples of matter.

The simplest application of the Boltzmann distribution is to calculate the relative numbers of molecules in two states separated in energy by ΔE. Suppose the energies of the two states are E_1 and E_2, then from eqn 20.1 we can write

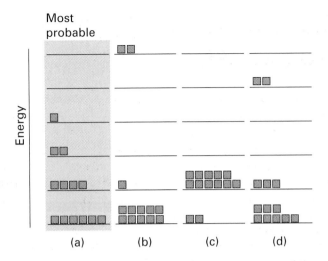

Fig 20.1 The derivation of the Boltzmann distribution involves imagining that the molecules of a system (the squares) are distributed at random over the available energy levels subject to the requirements that the number of molecules and the total energy is constant, and then looking for the most probable arrangement. Of the four shown here, the numbers of ways of achieving each arrangement are (a) 181 180, (b) 858, (c) 78, (d) 12 870.[3] The number of ways of achieving (a) is by far the greatest, so this distribution is the most probable; it corresponds to the Boltzmann distribution.

$$\frac{N_2}{N_1} = \frac{N e^{-E_2/kT}/q}{N e^{-E_1/kT}/q} = \frac{e^{-E_2/kT}}{e^{-E_1/kT}}$$

$$= e^{-(E_2 - E_1)/kT} = e^{-\Delta E/kT}$$

(20.3)

where $\Delta E = E_2 - E_1$. This important result tells us that *the relative population of the upper state decreases exponentially with its energy above the lower state.* In this expression, the energy difference is in joules. If the energy difference is given in kilojoules per mole, we simply use the gas constant in eqn 20.3 in place of the Boltzmann constant (because $R = kN_A$).

1

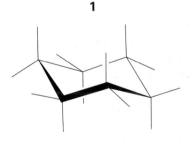

2

Illustration 20.1

The boat conformation of cyclohexane (**1**) lies 22 kJ mol^{-1} higher in energy than the chair conformation (**2**). To find the relative populations of the two conformations in a sample of cyclohexane at 20°C, we set ΔE = 22 kJ mol^{-1} and T = 293 K and use eqn 20.3 with R in the exponent. We obtain:

$$\frac{N_{boat}}{N_{chair}} = e^{-\frac{22 \times 10^3 \text{ J mol}^{-1}}{(8.31451 \text{ J K}^{-1}\text{mol}^{-1}) \times (293 \text{ K})}} = 1.2 \times 10^{-4}$$

[3] To calculate the number of ways, W, of arranging N molecules with N_1 in state 1, N_2 in state 2, etc., use $W = N!/N_1!N_2!\cdots$, with $n! = n(n-1)(n-2)\cdots1$, and $0! = 1$.

One very important feature about the Boltzmann distribution is that it applies to the populations of *states*. We have seen that in some cases (the hydrogen atom is a good example) several different states have the same energy. That is, some energy levels are *degenerate* (Section 12.10). The Boltzmann distribution tells us, for instance, the number of atoms with an electron in a $2p_x$ orbital. Because a $2p_y$ orbital has exactly the same energy, the number of atoms with an electron in a $2p_y$ orbital is the same as the number with an electron in a $2p_x$ orbital. The same is true of atoms with an electron in a $2p_z$ orbital. Therefore, if we want the *total* number of atoms with electrons in $2p$ orbitals, we have to multiply the number in *one* of them by a factor of 3. In general, if the degeneracy of an energy level (that is, the number of states of that energy) is g, we use a factor of g to get the population of the level (as distinct from an individual state). It is obviously very important to decide whether you want to express the population of an individual state or the population of an entire degenerate energy level.

Illustration 20.2

We saw in Section 17.3 that the rotational energy of a linear rotor is $hBJ(J + 1)$. Given that the degeneracy of the level with $J = 2$ (and energy $6hB$) is 5 and that of the level with $J = 1$ (and energy $2hB$) is 3,[4] the relative numbers of molecules with $J = 2$ and 1 are

$$\frac{N_2}{N_1} = \frac{g_2\,Ne^{-E_2/kT}/q}{g_1\,Ne^{-E_1/kT}/q} = \frac{g_2}{g_1}e^{-(E_2-E_1)/kT} = \frac{5}{3}e^{-4hB/kT}$$

For HCl, $B = 318.0$ GHz, so at 25°C,

$$\frac{N_2}{N_1} = \frac{5}{3} \times e^{-\frac{4 \times (6.62608 \times 10^{-34}\text{ J s}) \times (3.180 \times 10^{11}\text{ Hz})}{(1.38066 \times 10^{-23}\text{ J K}^{-1}) \times (298\text{ K})}} = 1.36$$

We see that there are *more* molecules in the level with $J = 2$ than in the level with $J = 1$, even though $J = 2$ corresponds to a higher energy. Each individual *state* with $J = 2$ has a lower population than each state with $J = 1$, but there are more states in the level with $J = 2$.

One important convention that we adopt (largely for convenience) is that *all energies are measured relative to the ground state of the molecule*. So, we set the ground-state energy equal to zero, even if there is a zero-point energy. Likewise, the energies of the states of the hydrogen atom are measured from zero for the $1s$ orbital. This convention greatly simplifies our interpretation of the significance of q.

20.2 The interpretation of the partition function

When we are interested only in the relative populations of levels and states, we do not need to know the partition function because it cancels in eqn 20.3. However, if we want to know the actual population of a state, then we use eqn 20.1, which requires us to know q. We also need to know q when we derive thermodynamic functions, as we shall see.

The definition of q is the sum over states (not levels), as given in eqn 20.2. We can write out the first few terms as follows:

$$q = 1 + e^{-E_1/kT} + e^{-E_2/kT} + e^{-E_3/kT} + \cdots$$

The first term is 1 because the energy of the ground state (E_0) is 0, according to our convention, and $e^0 = 1$. In principle, we just substitute the values of the energies, evaluate each term for the temperature of interest, and add them together to get q. However, that procedure does not give much insight.

To see the physical significance of q, let's suppose first that $T = 0$. Then, because $e^{-\infty} = 0$, all terms other than the first are equal to 0, and

$$q = 1$$

At $T = 0$ only the ground state is occupied and $q = 1$. We can begin to suspect that the partition function is telling us the number of states that are occupied at a given temperature. Now consider the other extreme: a temperature so high that all the $E_i/kT = 0$. Then, because $e^0 = 1$, the partition function is[5]

---

[4] In general, the degeneracy of a linear rotor is $2J + 1$.

[5] We are supposing that the molecule has an infinite number of states. If there are only N states, then there are N contributions of 1, and $q = N$.

$$q \approx 1 + 1 + 1 + 1 + \cdots = \infty$$

At very high temperatures, all the states of the system are thermally accessible, and the partition function rises to infinity.

Now consider an intermediate temperature, at which only some of the states are occupied significantly. Let's suppose that the temperature is such that kT is large compared to E_1 and E_2 but small compared to E_3 and all subsequent terms (Fig 20.2). Because E_1/kT and E_2/kT are both small compared to 1, and $e^{-x} \approx 1$ when x is very small, the first three terms are all close to 1. However, because E_3/kT is large compared to 1, and $e^{-x} \approx 0$ when x is large, all the remaining terms are close to 0. Therefore,

$$q \approx 1 + 1 + 1 + 0 + \cdots = 3$$

Once again, we see that the partition function is telling us the number of significantly occupied states at the temperature of interest. That is the principal meaning of the partition function: *q tells us the number of thermally accessible states at the temperature of interest.*

Once we grasp the significance of q, statistical thermodynamics becomes much easier to understand. We can anticipate, even before we do any calculations, that q increases with temperature,

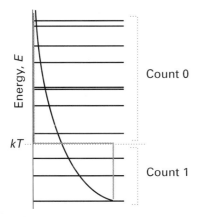

Fig 20.2 The physical interpretation of the partition function is that it is a measure of the number of thermally accessible states. Thus, for all states with $E < kT$ the exponential term is reasonably close to 1, whereas for all states with $E > kT$ the exponential term is close to 0. The states with $E < kT$ are significantly thermally accessible.

because more states become accessible as the temperature is raised. At low temperatures q is small, and falls to 1 as the temperature approaches absolute zero (when only one state, the ground state, is accessible). Molecules with numerous, closely spaced energy levels (like the rotational states of bulky molecules) can be expected to have very large partition functions. Molecules with widely spaced energy levels can be expected to have small partition functions, because only the few lowest states will be occupied at low temperatures.

Example 20.1 *Calculating a partition function*

Calculate the partition function for the cyclohexane molecule, confining attention to the chair and boat conformations mentioned in Illustration 20.1. Show how the partition function varies with temperature.

Strategy Whenever calculating a partition function, start at the definition in eqn 20.2 and write out the individual terms. Remember to set the ground-state energy equal to 0. When the energies of states are given in joules per mole, replace the k in the definition of q by R.

Solution There are only two states, so the partition function has only two terms. The energy of the chair form is set at 0 and that of the boat form is $E = 22 \text{ kJ mol}^{-1}$. Therefore:

$$q = 1 + e^{-E/RT}$$

This function is plotted in Fig 20.3. We see that it rises from $q = 1$ (only the chair form accessible at $T = 0$) to $q = 2$ (both states are thermally accessible at high temperatures). At 20°C, $q = 1.0001$. As we saw in Illustration 20.1, the boat form is only slightly populated and so q differs very little from 1.

Self-test 20.1

The ground configuration of a fluorine atom gives rise to a ^2P term with two levels,[6] the $J = \frac{3}{2}$ level (of degeneracy 4) and the $J = \frac{1}{2}$ level (of degeneracy 2) at an energy corresponding to 404.0 cm^{-1} above the ground state. Write down an expression for the partition function and plot it as a function of temperature. (*Hint.* Take $E = hc\tilde{v}$ for the energy of the upper level.)
[*Answer:* $q = 4 + 2e^{-hc\tilde{v}/kT}$; Fig 20.4]

----

[6] The notation used here was introduced in Section 13.16.

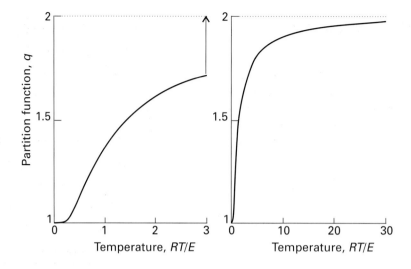

Fig 20.3 The partition function for a two-level system with states at the energies 0 and E. At 20°C (293 K) and for $E = 22$ kJ mol^{-1}, RT/E = 0.11, where $q = 1.0001$. Note how the partition function rises from 1 and approaches 2 at high temperatures.

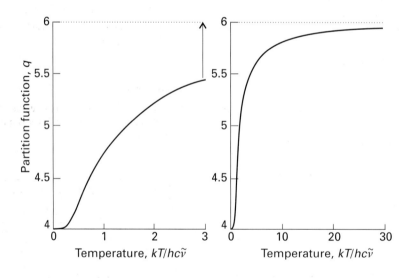

Fig 20.4 The partition function for the six-level system treated in Self-test 20.1. Note how q rises from 4 (when only the four states of the $^3P_{3/2}$ level are occupied) and approaches 6 (when the two states of the $^3P_{1/2}$ level are also accessible). At 20°C, $kT/hc\tilde{v}$ = 0.504, corresponding to $q = 4.28$.

20.3 Examples of partition functions

In a number of cases we can derive simple expressions for partition functions. For example, the energy levels of a harmonic oscillator form a simple ladder-like array (Fig 20.5). If we set the energy of the lowest state equal to zero, the energies of the states are

$$E_0 = 0, \ E_1 = h\nu, \ E_2 = 2h\nu, \ E_3 = 3h\nu, \text{ etc.}$$

Therefore, the vibrational partition function is

$$q = 1 + e^{-h\nu/kT} + e^{-2h\nu/kT} + e^{-3h\nu/kT} + \cdots$$

The sum of this infinite series is[7]

$$q = \frac{1}{1 - e^{-h\nu/kT}} \tag{20.4}$$

This is the partition function for a harmonic oscillator, or any vibrating diatomic molecule. Figure 20.6 shows how q varies with temperature. Note that $q = 1$ at $T = 0$, when only the lowest state is occupied, and that, as T becomes high, so q becomes infinite because all the states of the infinite ladder are thermally accessible. At room temperature, and for typical molecular vibrational frequencies, q is very close to 1 because only the vibrational ground state is occupied.

We can do similar calculations for certain other types of motion. For example, suppose a molecule of mass m is confined in a flask of volume V at a temperature T, then to a good approximation,[8] the **translational partition function** is

$$q = \frac{(2\pi m kT)^{3/2} V}{h^3} \tag{20.5}$$

We see that the partition function increases with temperature, as we have come to expect. However, notice that q also increases with the volume of the flask. That we should expect too: the en-

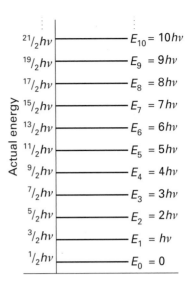

Fig 20.5 The energy levels of a harmonic oscillator. When calculating a partition function, set the zero of energy at the lowest level, as shown on the right.

ergy levels of a particle in a box become closer together as the size of the box increases (Section 12.9) so, at a given temperature, more states are thermally accessible.

Illustration 20.3

Suppose we have an oxygen molecule (of mass 32 u) in a 100 mL flask at 25°C. The translational partition function is

$$q = \frac{\left(2\pi \times \overbrace{32 \times (1.660\,54 \times 10^{-27}\,\text{kg})}^{m} \times \overbrace{(1.380\,66 \times 10^{-23}\,\text{J K}^{-1}) \times (298\,\text{K})}^{kT} \right)^{3/2} \times \overbrace{(1.00 \times 10^{-4}\,\text{m}^3)}^{V}}{(6.626\,08 \times 10^{-34}\,\text{J s})^3}$$

$$= 1.75 \times 10^{28}$$

Note that a huge number of translational states are accessible at room temperature. Note also that all the units cancel: all partition functions are dimensionless numbers.

---

[7] If we set $x = e^{-h\nu/kT}$, the series is $1 + x + x^2 + x^3 + \cdots$, which sums to $1/(1-x)$.

[8] The approximation is valid for large containers and $T > 0$.

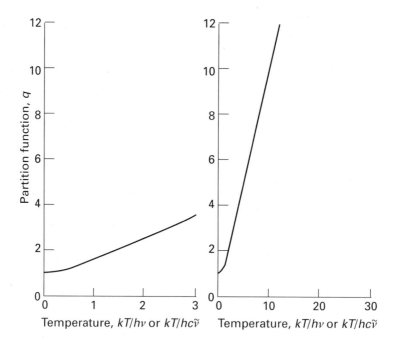

Fig 20.6 The partition function for a harmonic oscillator. For an oscillator with $\tilde{v} = 1000$ cm^{-1}, at 20°C, $kT/hc\tilde{v} = 0.204$, corresponding to $q = 1.01$.

The **rotational partition function** can also be approximated when the temperature is high enough for many rotational states to be occupied. For a linear rotor it turns out that[9]

$$q = \frac{kT}{\sigma h B} \qquad (20.6)$$

In this expression, B is the rotational constant (Section 17.3) and σ is the **symmetry number**: $\sigma = 1$ for an unsymmetrical linear rotor (such as HCl or HCN) and $\sigma = 2$ for a symmetrical linear rotor (such as H_2 or CO_2).[10] The rotational partition of HCl at 25°C works out to 19.6, so about 20 rotational states[11] are significantly occupied at that temperature.

No closed form can be given for the **electronic partition function**, the partition function for the distribution of electrons over their available states. However, for closed-shell molecules the excited states are so high in energy that only the ground state is occupied, and for them $q = 1$. Special care has to be taken for atoms and molecules that do not have closed shells (as we saw in Self-test 20.1 for the fluorine atom).

Thermodynamic properties

THE principal reason for calculating the partition function is to use it to calculate thermodynamic properties of systems. There are two fundamental relations we need. We can deal with First-Law quantities (such as heat capacity and enthalpy) once we

---

[9] This is an approximation valid for heavy molecules and $T > 0$.

[10] The symmetry number reflects the fact that an unsymmetrical molecule is distinguishable after rotation by 180° but a symmetrical molecule is not. When evaluating q we have to count only distinguishable states.

[11] *Not* levels: remember the $(2J + 1)$-fold degeneracy of each rotational level.

know how to calculate the internal energy. We can deal with Second-Law quantities (such as the Gibbs energy and equilibrium constants) once we know how to calculate the entropy.

20.4 The internal energy and the heat capacity

To calculate the total energy, E, of the system, we note the energy of each state (E_i), multiply that energy by the number of molecules in the state (N_i), and then add together all these products:

$$E = N_0 E_0 + N_1 E_1 + N_2 E_2 + \cdots = \sum_i N_i E_i$$

However, the Boltzmann distribution tells us the number of molecules in each state of a system, so we can replace the N_i in this expression by eqn 20.1:

$$E = \sum_i \frac{Ne^{-E_i/kT}}{q} \times E_i = \frac{N}{q} \sum_i E_i e^{-E_i/kT} \quad (20.7)$$

If we know the individual energies of the states (from spectroscopy, for instance), then we just substitute their values into this expression. However, there is a much simpler method available when we have an expression for the partition function, such as those given in Section 20.3.

Derivation 20.1 *The internal energy from the partition function*

The expression on the right of eqn 20.7 resembles the definition of the partition function, but differs from it by having the E_i factor multiplying each term. However, we can recognize the following mathematical relation:

$$\frac{d}{dT} e^{-E_i/kT} = \frac{d}{dT}\left(-\frac{E_i}{kT}\right) \times e^{-E_i/kT} = \frac{E_i}{kT^2} e^{-E_i/kT}$$

In other words,

$$E_i e^{-E_i/kT} = kT^2 \frac{d}{dT} e^{-E_i/kT}$$

With this substitution, the expression for the total energy becomes

$$E = \frac{N}{q} \sum_i kT^2 \frac{d}{dT} e^{-E_i/kT} = \frac{NkT^2}{q} \frac{d}{dT} \sum_i e^{-E_i/kT}$$

because the sum of derivatives is the derivative of the sum. Magically (or more precisely, mathematically), the expression for the partition function has appeared, so we can write

$$E = \frac{NkT^2}{q} \frac{dq}{dT} \quad (20.8)$$

The remarkable feature of eqn 20.8 is that it is an expression for the total energy in terms of the partition function alone. The partition function is starting to fulfil its promise to deliver all thermodynamic information about the system.

There is one more detail to take into account before we use eqn 20.8. Recall that we have set the zero of energy at the energy of the lowest state of the molecule. However, the internal energy of the system might be nonzero on account of zero-point energy, and the E in eqn 20.8 is the energy *above* the zero-point energy. That is, the internal energy at a temperature T is

$$U = U(0) + E \quad (20.9)$$

with E given by eqn 20.8.

Example 20.2 *Calculating the internal energy*

Calculate the molar internal energy of a monatomic gas.

Strategy The only mode of motion of a monatomic gas is translation (we ignore electronic excitation). Therefore, substitute the translational partition function in eqn 20.5 into eqn 20.8 and then insert the result into eqn 20.9. The partition function has the form $q = aT^{3/2}$, where a is a collection of constants.

Solution When we substitute eqn 20.5 into eqn 20.8 we get

$$E = \frac{NkT^2}{q} \frac{d}{dT}(aT^{3/2}) = \frac{aNkT^2}{q} \frac{d}{dT} \overset{\frac{3}{2}T^{1/2}}{T^{3/2}}$$

$$= \frac{aNkT^2}{aT^{3/2}} \times \frac{3}{2} T^{1/2} = \frac{3}{2} NkT$$

The molar internal energy is obtained by replacing N by Avogadro's constant and using eqn 20.9:

$$U_m = U_m(0) + \tfrac{3}{2}N_A kT = U_m(0) + \tfrac{3}{2}RT$$

The term $U_m(0)$ contains all the contributions from the binding energy of the electrons and of the nucleons in the nucleus. The term $\tfrac{3}{2}RT$ is the contribution to the internal energy from the translational motion of the atoms in their container.

Self-test 20.2

Calculate the molar internal energy of a gas of diatomic molecules. (*Hint*. Add the rotational contribution to the translational contribution calculated in the example.)

[*Answer*: $U_m = U_m(0) + \tfrac{5}{2}RT$]

Once we have calculated the internal energy of a sample of molecules, it is a simple matter to calculate the heat capacity. It should be recalled that the heat capacity at constant volume, C_V, is defined as the slope of the plot of internal energy against temperature:

$$C_V = \frac{dU}{dT} \quad \text{at constant volume}$$

Therefore, all we need do is to differentiate the expression for the internal energy.

Illustration 20.4

The constant-volume molar heat capacity of a monatomic gas is

$$C_{V,m} = \frac{d}{dT}\left(U_m(0) + \tfrac{3}{2}RT\right) = \tfrac{3}{2}R$$

To calculate $C_{p,m}$, we use eqn 2.17 ($C_{p,m} - C_{V,m} = R$) and obtain $\tfrac{5}{2}R$.

Self-test 20.3

Calculate the contribution to the molar heat capacity of a two-state system, like the chair–boat interconversion of cyclohexane (Illustration 20.1) and show how the heat capacity varies with temperature.

[*Answer*: $C_{V,m} = \dfrac{R(E/RT)^2\, e^{E/RT}}{(1 + e^{E/RT})^2}$, Fig 20.7]

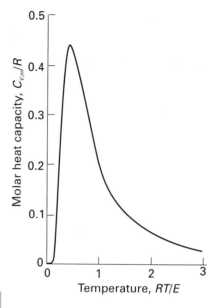

Fig 20.7 The variation of the heat capacity of a two-level system with states at energies 0 and E. Note how the heat capacity is zero at $T = 0$, passes through a maximum at $T = 0.417E/R$, and approaches 0 at high temperatures.

20.5 The entropy and the Gibbs energy

The entry point into the calculation of properties arising from the Second Law of thermodynamics is the proposal made by Boltzmann that the entropy of a system can be calculated from the expression

$$S = k \ln W \tag{20.10}$$

Here W is the number of different ways in which the molecules of a system can be arranged yet result in the same total energy. This expression is the **Boltzmann formula** for the entropy. The entropy is zero if there is only one way of achieving a given total energy (because ln 1 = 0). The entropy is high if there are many ways of achieving the same energy.

In certain cases, W may differ from 1 at $T = 0$. This is the case if disorder survives down to absolute zero because there is no energy advantage in adopting a particular orientation. For instance, there may be no energy difference between the arrangements ... AB AB AB ... and ... BA AB BA ..., so $W > 1$ even at $T = 0$. If $S > 0$ at $T = 0$ we say that the substance has a **residual entropy**. Ice has a residual entropy of 3.4 J K^{-1} mol^{-1}. It stems from the disorder in the hydrogen bonds between neighbouring water molecules: a given O atom has two short O–H bonds and two long O$\cdots$H bonds to its neighbours, but there is a degree of randomness in which two bonds are short and which two are long.

The analogous term for indistinguishable molecules (identical molecules free to exchange places, as in a gas) is

$$S = \frac{U - U(0)}{T} + Nk \ln q - Nk(\ln N - 1) \quad (20.11b)$$

Because we can calculate the first term on the right from q, we now have a method for calculating the entropy of any system of noninteracting molecules once we know its partition function.

Illustration 20.5

Consider a sample of solid carbon monoxide containing N CO molecules. We saw in Chapter 16 that CO has a very small dipole moment. In fact, the dipolar interactions between CO molecules are so weak in a solid that even at $T = 0$ they lie either head-to-tail or head-to-head with approximately equal energies to give randomly orientated arrangements such as $\cdots$CO CO OC CO OC OC$\cdots$ with the same energy. Because each CO molecule can lie in either of two orientations (CO or OC) with equal energy, there are $2 \times 2 \times 2 \times \cdots = 2^N$ ways of achieving the same energy. The residual entropy of the sample, its entropy at $T = 0$, where there is this orientation disorder but no motional disorder, is therefore

$$S = k \ln 2^N = Nk \ln 2$$

(We have used $\ln x^a = a \ln x$.) The molar entropy is therefore

$$S_m = N_A k \ln 2 = R \ln 2$$

or 5.8 J K^{-1} mol^{-1}, close to the experimental value of 5 J K^{-1} mol^{-1}.

Boltzmann went on to show that there is a close relation between the entropy and the partition function: both are measures of the number of arrangements available to the molecules. The precise connection for distinguishable molecules (those locked in place in a solid) and for internal contributions is

$$S = \frac{U - U(0)}{T} + Nk \ln q \quad (20.11a)$$

Example 20.3 *Calculating the entropy*

Calculate the contribution that rotational motion makes to the molar entropy of a gas of HCl molecules at 25°C.

Strategy We have already calculated the contribution to the internal energy (Self-test 20.2), and we have the rotational partition function in eqn 20.6 (with $\sigma = 1$). We need to combine the two parts. We use eqn 20.11a because we are concentrating on the internal motion (the rotation) of the molecules, not their translational motion.

Solution We substitute $U - U(0) = RT$ and $q = kT/hB$ into eqn 20.11a, and obtain[12]

$$S_m = \frac{RT}{T} + R \ln \frac{kT}{hB} = R\left(1 + \ln \frac{kT}{hB}\right)$$

Notice that the entropy increases with temperature (Fig 20.8). At a given temperature, the entropy is larger the smaller the value of B. That is, bulky molecules (which have large moments of inertia and therefore small rotational constants) have a higher rotational entropy than small molecules. Substitution of the numerical values gives $S_m = 3.98R$, or 33.1 J K^{-1} mol^{-1}.

Self-test 20.4

The rotational partition function of an ethene molecule is 661 at 25°C. What is the contribution to its rotational entropy?

[*Answer*: 7.49R]

The Gibbs energy, G, was central to most of the thermodynamic discussions in the early chapters of this book, so to show that statistical thermodynamics is really useful we have to see how to calculate G

[12] This result is valid only for $T > 0$.

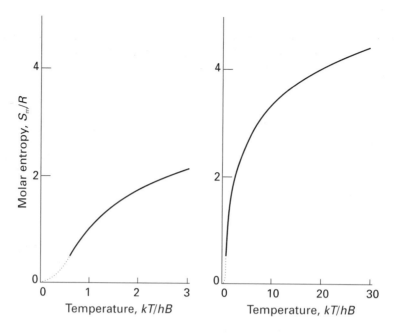

Fig 20.8 The variation of the rotational contribution to the molar entropy with temperature. Note that eqn 20.6 is valid only for high temperatures, so the formula derived in Example 20.3 cannot be used at low temperatures (so we have terminated the curves before they become invalid). The dotted lines show the correct behaviour.

from the partition function, q. We shall confine our attention to a perfect gas, because it is difficult to take molecular interactions into account.

Derivation 20.2 *Calculating the Gibbs energy from the partition function*

To set up the calculation, we go back to first principles. The Gibbs energy is defined as $G = H - TS$, and the enthalpy, H, is defined as $H = U + pV$. Therefore

$$G = U - TS + pV$$

For a perfect gas we can replace pV by nRT and note that, at $T = 0$, $G(0) = U(0)$ (because the terms TS and nRT vanish at $T = 0$). Therefore,

$$G - G(0) = U - U(0) - TS + nRT$$

Now we substitute eqn 20.11b for S, and obtain

$$G - G(0) = -NkT \ln q + kT(N \ln N - N) + nRT$$
$$= -NkT(\ln q - \ln N)$$

That is,

$$G - G(0) = -NkT \ln \frac{q}{N} \qquad (20.12$$

We can convert eqn 20.12 into an expression for the molar Gibbs energy. First, we write $N = nN_A$. Then we introduce the **molar partition function**, $q_m = q/n$, with units 1/mole (mol^{-1}). On dividing both sides of eqn 20.12 by n, we get

$$G_m - G_m(0) = -RT \ln \frac{q_m}{N_A} \qquad (20.13)$$

Example 20.4 *Calculating the Gibbs energy*

Calculate the molar Gibbs energy of a monatomic perfect gas and express it in terms of the pressure of the gas.

Strategy The calculation is based on eqn 20.13 with $q_m = q/n$. All we need to know is the translational partition function, which is given in eqn 20.5. Convert from V to p by using the perfect gas law.

Solution When we substitute $q = (2\pi mkT)^{3/2} V/h^3$ into eqn 20.13 we get

$$G_m - G_m(0) = -RT \ln \frac{(2\pi mkT)^{3/2} V}{nh^3 N_A}$$

Next, we replace V by nRT/p (notice that the ns cancel), and obtain (after a little tidying up, including writing $R = kN_A$)

$$G_m - G_m(0) = -RT \ln \frac{(2\pi m)^{3/2}(kT)^{5/2}}{ph^3}$$

$$= RT \ln ap \qquad a = \frac{h^3}{(2\pi m)^{3/2}(kT)^{5/2}}$$

(20.14)

The Gibbs energy increases logarithmically (as $\ln p$) as p increases, just as we saw in Section 5.2 (eqn 5.4).

Self-test 20.5

The molar partition function of a diatomic molecule is $q_m^{trans}q^{rot}$, where q_m^{trans} is the same as q_m in the Example and q^{rot} is the rotational contribution (we are ignoring vibration). What is the molar Gibbs energy of such a gas?

[*Answer*: as in eqn 20.14, but with a replaced by $a\sigma hB/kT$]

The only further piece of information we require is the expression for the *standard* molar Gibbs energy, for that played such an important role in the discussion of equilibrium properties. All we need do is to use the partition function calculated at $p^{\ominus}$. For instance, for a monatomic gas, we use $p = 1$ bar in eqn 14 and obtain the standard value of the molar Gibbs energy. In general, we write

$$G_m^{\ominus} - G_m^{\ominus}(0) = -RT \ln \frac{q_m^{\ominus}}{N_A}$$

(20.15)

where the standard state sign on q simply reminds us to calculate its value at $p^{\ominus}$; to do so, we use $V^{\ominus} = nRT/p^{\ominus}$ wherever it appears in q.

20.6 The statistical basis of equilibrium

Once we have the standard molar Gibbs energy of a substance in terms of its partition function, we are very close to having a statistical thermodynamic expression for that most chemical of quantities, the equilibrium constant.

Derivation 20.3 *The equilibrium constant from the partition function*

We know from thermodynamics (Section 7.6) that the equilibrium constant for a reaction is related to the standard reaction Gibbs energy by

$$\Delta_r G^{\ominus} = -RT \ln K$$

Let's consider a reaction of the form $A + B \rightleftharpoons C$. From eqn 20.15 we know that the standard reaction Gibbs energy is

$$\Delta_r G^{\ominus} = G_m^{\ominus}(C) - \left\{ G_m^{\ominus}(A) + G_m^{\ominus}(B) \right\}$$

$$= \left\{ G_m^{\ominus}(C,0) - RT \ln \frac{q_m^{\ominus}(C)}{N_A} \right\}$$

$$- \left[\left\{ G_m^{\ominus}(A,0) - RT \ln \frac{q_m^{\ominus}(A)}{N_A} \right\} \right.$$

$$\left. + \left\{ G_m^{\ominus}(B,0) - RT \ln \frac{q_m^{\ominus}(B)}{N_A} \right\} \right]$$

$$= \left\{ G_m^{\ominus}(C,0) - \left(G_m^{\ominus}(A,0) + G_m^{\ominus}(B,0) \right) \right\}$$

$$- RT \left[\ln \frac{q_m^{\ominus}(C)}{N_A} - \left\{ \ln \frac{q_m^{\ominus}(A)}{N_A} + \ln \frac{q_m^{\ominus}(B)}{N_A} \right\} \right]$$

We can simplify this somewhat alarming expression. First, note that

$$G_m^{\ominus}(C,0) - \{ G_m^{\ominus}(A,0) + G_m^{\ominus}(B,0) \}$$
$$= U_m^{\ominus}(C,0) - \{ U_m^{\ominus}(A,0) + U_m^{\ominus}(B,0) \}$$
$$= \Delta E$$

where ΔE is the difference in energy between the ground state of the product and that of the reactants. Next, we combine the logarithms as follows:[13]

$$\ln \frac{q_m^{\ominus}(C)}{N_A} - \left\{ \ln \frac{q_m^{\ominus}(A)}{N_A} + \ln \frac{q_m^{\ominus}(B)}{N_A} \right\} = \ln \frac{q_m^{\ominus}(C)N_A}{q_m^{\ominus}(A)q_m^{\ominus}(B)}$$

At this stage we have reached

$$\Delta_r G^{\ominus} = \Delta E - RT \ln \frac{q_m^{\ominus}(C)N}{q_m^{\ominus}(A)q_m^{\ominus}(B)}$$

---

[13] Here we use $\ln x - \ln y - \ln z = \ln x/yz$.

This expression is the same as[14]

$$\Delta_r G^{\ominus} = -RT \ln \left\{ \frac{q_m^{\ominus}(C) N_A}{q_m^{\ominus}(A) q_m^{\ominus}(B)} e^{-\Delta E/RT} \right\}$$

All we have to do now is to compare this expression with the thermodynamic expression for the equilibrium constant, and obtain

$$K = \frac{q_m^{\ominus}(C) N_A}{q_m^{\ominus}(A) q_m^{\ominus}(B)} e^{-\Delta E/RT} \qquad (20.16)$$

This expression is easy to remember: it has the same form as the equilibrium constant written in terms of the activities (Section 7.6), but with $q_m^{\ominus}/N_A$ replacing each activity (and an additional exponential factor).

Equation 20.16 is quite extraordinary, as it provides a key link between partition functions, which can be derived from spectroscopy, and the equilibrium constant, which is central to the analysis of chemical reactions at equilibrium. It represents the merging of the two rivers that have flowed through this text.

Example 20.5 *Calculating an equilibrium constant*

Calculate the equilibrium constant for the gas-phase ionization $Cs(g) \rightleftharpoons Cs^+(g) + e^-(g)$ at 1000 K.

Strategy This is a reaction of the form $A \rightleftharpoons B + C$ rather than $A + B \rightleftharpoons C$, so we need to modify eqn 20.16 slightly, but the form to use should be clear. Analyse each species individually, and write its partition function as the product of partition functions for each mode of motion. Evaluate these partition functions at the standard pressure (1 bar), and combine them as specified in eqn 20.16. For the difference in energy ΔE, use the ionization energy of $Cs(g)$.

Solution The equilibrium constant is

$$K = \frac{q_m^{\ominus}(B) q_m^{\ominus}(C)}{q_m^{\ominus}(A) N_A} e^{-\Delta E/RT}$$

The electron has translational motion, so we need its translational partition function. It also has two spin states, which are equally occupied in the absence of any magnetic field, so they contribute a factor of 2 to the partition function. Therefore

$$q_m^{\ominus}(e^-) = 2 \times \frac{(2\pi m_e kT)^{3/2} V^{\ominus}}{n h^3} = \frac{2(2\pi m_e kT)^{3/2} RT}{p^{\ominus} h^3}$$

The Cs^+ ion, a closed-shell species, has only translational freedom:

$$q_m^{\ominus}(Cs^+) = \frac{(2\pi m_{Cs} kT)^{3/2} RT}{p^{\ominus} h^3}$$

The Cs atom has translational freedom and its electron can exist in either of two spin states (up or down), so we need a factor of 2:

$$q_m^{\ominus}(Cs) = \frac{2(2\pi m_{Cs} kT)^{3/2} RT}{p^{\ominus} h^3}$$

(We are not distinguishing the masses of the Cs atom and the Cs^+ ion.) Then, with $\Delta E = I$, the ionization energy of the atom, we find

$$K = \frac{\overbrace{(2\pi m_{Cs} kT)^{3/2} RT/p^{\ominus} h^3}^{Cs^+} \times \overbrace{2(2\pi m_e kT)^{3/2} RT/p^{\ominus} h^3}^{e^-}}{\underbrace{2(2\pi m_{Cs} kT)^{3/2} RT/p^{\ominus} h^3 \times N_A}_{Cs}} \times e^{-I/RT}$$

$$= \frac{(2\pi m_e kT)^{3/2} kT}{p^{\ominus} h^3} e^{-I/RT} = \frac{(2\pi m_e)^{3/2} (kT)^{5/2}}{p^{\ominus} h^3} e^{-I/RT}$$

When we substitute the data (only the ionization energy is specific to the element), we find:

$$K = \frac{\left(2\pi \times \overbrace{(9.10939 \times 10^{-31} \text{kg})}^{m_e}\right)^{3/2} \times \left(\overbrace{(1.38066 \times 10^{-23} \text{JK}^{-1})}^{k} \times \overbrace{(1000\,\text{K})}^{T}\right)^{5/2}}{\underbrace{(10^5\,\text{Pa})}_{p^{\ominus}} \times \underbrace{(6.62608 \times 10^{-34}\,\text{Js})^3}_{h}}$$

$$\times e^{-\frac{376 \times 10^3\,\text{J mol}^{-1}}{(8.31451\,\text{J K}^{-1}\text{mol}^{-1}) \times (1000\,\text{K})}}$$

$$= 2.42 \times 10^{-19}$$

Self-test 20.6

Calculate the equilibrium constant for the dissociation $Na_2(g) \rightleftharpoons 2\,Na(g)$ at 1000 K. You will need the following information: $B = 46.38$ MHz, $\tilde{\nu} = 159.2$ cm^{-1}, $D_0 = 70.4$ kJ mol^{-1}; the ground state of a sodium atom is ^2S.

[*Answer*: 2.42]

---

[14] In this step we use $\ln a + x = \ln a + \ln(e^x) = \ln(ae^x)$.

Box 20.1 *The helix–coil transition in polypeptides*

The helical form of a polypeptide may collapse cata-strophically into a random coil when certain conditions are changed. For instance, poly-γ-benzyl-L-glutamate in a non-hydrogen-bonding solvent adopts a helical secondary structure, but in a hydrogen-bonding solvent it forms a random coil. In a mixed solvent it undergoes a helix–coil transition suddenly in a narrow temperature range, the coil being the stable form at low temperatures and the helix the stable form at high temperatures. The sudden reverse transition may also occur when the temperature is lowered again. What is the origin of this transition and how can it be expressed in terms of statistical thermodynamics?

We have seen that the secondary structure of a polypeptide is controlled by the hydrogen bonds between the –NH and –CO groups of the peptide link (Section 16.8), in which –NH of group 1 is linked to –CO of group 4, –NH of group 2 is linked to –CO of group 5, and so on. When the solvent can intrude into this bonding pattern, a lower Gibbs energy may be achieved by forming hydrogen bonds to the solvent molecules, and as a result the secondary structure collapses. The origin of the suddenness of the conversion of the coil to the helix can be traced to the difficulty of forming one hydrogen bond but the ease of forming the neighbouring hydrogen bond once the first has been formed. Thus, it may be difficult to form the 1–4 link. However, once that link is formed, groups 2 and 5 are brought into a position in which it is relatively easy for them to form the 2–5 link. Similarly, groups 3 and 6 are then brought into position, and that link is easy to form. Therefore, provided *one* link can be formed, the cascade of events that result in the formation of the helix is facilitated. In other words, the formation of the helix from the random coil is a *cooperative process* in which each bonding event renders the next more likely.[15] The reverse is also true: once one bond has been broken, it is relatively easy to break the neighbouring bond, and the collapse is cooperative.

The reaction we are considering has the form

Coil(solvent)$_x$ ⇌ helix + x solvent

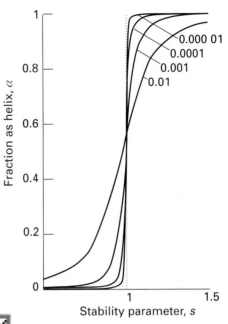

The fraction of molecules present as helices as a function of the stability parameter for different values of the initiation parameter σ.

The forward reaction is endothermic (the polypeptide–solvent bonds are strong) and the entropy change is positive as solvent molecules are released from their binding sites. Thermodynamically, therefore, we can see that the helix form can be expected to be stable at high temperatures.

To calculate the fraction of polypeptide molecules present as helix or coil we need to set up the partition function for the various states of the molecule. Broadly speaking, a molecule is characterized by the number of groups that are linked by hydrogen bonds and the number that are not. Care has to be taken to allow for the presence of a nonbonded group following a bonded group, for the latter is more likely to form a bond than one that does not follow a bonded group (because the process is cooperative). When this calculation is followed through, we find that the fraction of hydrogen bonds formed, α, is given by

----

[15] A zip fastener works on much the same principle.

$$\alpha = \frac{1}{2}\left\{1 + \frac{(s-1) + 2\sigma}{[(s-1)^2 + 4s\sigma]^{1/2}}\right\}$$

where s is a measure of the ease of forming an *individual* hydrogen bond and the *initiation parameter σ* is a measure of the difficulty of forming an isolated bond—that is, the difficulty of starting off a strip of helix in a random coil: the smaller the value of σ, the greater the difficulty. The illustration shows the variation of the fraction of helix predicted by this expression, and we see that, as expected, there is a sudden surge of transition as s passes through 1, and the smaller the initiation parameter the greater the sharpness. That is, the harder it is to get the helix formation started, the sharper the transition from coil to helix.

Exercise 1 Helix–coil transitions cannot be studied experimentally for many polypeptide homopolymers because they are insoluble in aqueous solutions. However, helix-forming tendency can be studied by investigating the effect of the residue on a water-soluble homopolymer that has a well characterized helix–coil transition. Such studies have been used to determine σ and s for polyamino acids. Estimate the fraction of hydrogen bonds formed (helix-forming tendency) in glycine ($\sigma = 1.0 \times 10^{-5}$; $s = 0.62$) and L-leucine ($\sigma = 33 \times 10^{-4}$; $s = 1.14$).

Exercise 2 A thermodynamic treatment allows predictions to be made of the temperature for a helix–coil transition, T_m, based on chain length, n:

$$T_m = \left(\frac{n-4}{n-2}\right)\frac{\Delta_r H}{\Delta_r S}$$

provided $n \geq 5$. Plot $T_m/(\Delta_r H/\Delta_r S)$ for $5 \leq n \leq 20$. At what value of n does T_m change by less than 1 per cent when n increases by 1?

Exercises

20.1 Suppose polyethylene molecules in solution can exist either as a random coil or fully stretched out, with the latter conformation 2.4 kJ mol^{-1} higher in energy. What is the ratio of the two conformations at 20°C?

20.2 What is the ratio of populations of proton spin orientations in a magnetic field of (a) 1.5 T, (b) 15 T in a sample at 20°C? (*Hint*. For the energy difference, refer to Chapter 19.)

20.3 What is the ratio of populations of electron spin orientations in a magnetic field of 0.33 T in a sample at 20°C? (*Hint*. For the energy difference, refer to Chapter 19.)

20.4 Calculate the ratio of populations of CO_2 molecules with $J = 5$ and $J = 1$ at 20°C. The rotational constant of CO_2 is 11.70 GHz. (*Hint*. Molecular rotations are discussed in Section 17.3.)

20.5 Calculate the ratio of populations of CH_4 molecules with $J = 5$ and $J = 1$ at 20°C. The rotational constant of CH_4 is 157 GHz. (*Hint*. The degeneracy of a spherical rotor in a state with quantum number J is $(2J + 1)^2$.)

20.6 (a) Write down the expression for the partition function of a molecule that has three energy levels at 0, ε, and 3ε with degeneracies 1, 5, and 3, respectively. What are the values of q at (b) $T = 0$, (c) $T = \infty$?

20.7 The ground configuration of carbon gives rise to a triplet with the three levels $^3P_0$, $^3P_1$, and $^3P_2$ at wavenumbers 0, 16.4, and 43.5 cm^{-1}, respectively. Evaluate the partition function of carbon at (a) 10 K, (b) 298 K. (*Hint*. Remember that a level with quantum number J has $2J + 1$ states.)

20.8 The ground configuration of oxygen gives rise to the three levels $^3P_2$, $^3P_1$, and $^3P_0$ at wavenumbers 0, 158.5, and 226.5 cm^{-1}, respectively. (a) Before doing any calculation, state the value of the partition function at $T = 0$. (b) Evaluate the partition function at 298 K and confirm that its value at $T = 0$ is what you anticipated in (a).

20.9 Evaluate the vibrational partition function for $^{35}Cl_2$ at 298 K. For data, see Table 17.1.

20.10 Evaluate the translational partition function at 298 K of (a) a methane molecule trapped in the pore of a zeolite catalyst: take the pore to be spherical with a radius that allows the molecule to move through 1 nm in any direction (that is, the *effective* diameter is 1 nm), (b) a methane molecule in a flask of volume 100 mL.

20.11 Evaluate the rotational partition function at 298 K of (a) $^1H^{35}Cl$, for which the rotational constant is 318 GHz, (b) $^{12}C^{16}O_2$, for which the rotational constant is 11.70 GHz.

20.12 N_2O and CO_2 have similar rotational constants (12.6 and 11.7 GHz, respectively) but strikingly different rotational partition functions. Why?

20.13 Derive an expression for the internal energy of a collection of harmonic oscillators. (*Hint*. Substitute eqn 20.4 for the partition function into eqn 20.8 for the energy.)

20.14 Derive an expression for the energy of a molecule that has three energy levels at 0, 2ε, and 4ε with degeneracies 1, 6, and 8, respectively.

20.15 The states arising from the ground configuration of a carbon atom are described in Exercise 20.7. (a) Derive an expression for the electronic contribution to the molar internal energy and plot it as a function of temperature. (b) Evaluate the expression at 25°C.

20.16 (a) Derive an expression for the electronic contribution to the molar heat capacity of an oxygen atom and plot it as a function of temperature. (b) Evaluate the expression at 25°C. The structure of the atom is described in Exercise 20.8.

20.17 Suppose that the $FClO_3$ molecule can take up any of four orientations in the solid at $T = 0$. What is its residual molar entropy?

20.18 Calculate the molar entropy of nitrogen (N_2) at 298 K. (*Hint*. Ignore the vibration of the molecule. Write the overall partition function as the product of the translational and rotational partition functions.) For data, see Table 17.1.

20.19 Estimate the change in molar entropy when a micelle consisting of 100 molecules disperses. (*Hint*. Treat the transition as the expansion of a gas-like substance that initially occupies a volume V_{micelle} and spreads into a volume V_{solution}.)

20.20 The average end-to-end distance of a flexible polymer (such as a fully denatured polypeptide or a strand of DNA) is $N^{1/2}l$, where N is the number of groups (residues or bases) and l is the length of each group. Initially, therefore, one end of the polymer can be found anywhere within a sphere of radius $N^{1/2}l$ centred on the other end. When the ends join to form a circle, they are confined to a volume of radius l. What is the change in molar entropy? Plot the function you derive as a function of N.

20.21 Calculate the standard molar Gibbs energy of nitrogen (N_2) at 298 K relative to its value at $T = 0$.

20.22 Calculate the equilibrium constant for the ionization equilibrium of sodium atoms at 1000 K.

20.23 Calculate the equilibrium constant for the dissociation of $I_2(g)$ at 500 K. Use $D_e = 151$ kJ mol^{-1}.

FURTHER INFORMATION 1 *Mathematical techniques*

THE art of doing mathematics correctly is to do nothing at each step of a calculation. That is, it is permissible to develop an equation by ensuring that the left-hand side of an expression remains equal to the right-hand side. There are several ways of modifying the *appearance* of an expression without upsetting its balance.

1.1 Algebraic equations and graphs

The simplest types of equation we have to deal with have the form

$$y = ax + b$$

This expression may be modified by subtracting b from both sides, to give

$$y - b = ax$$

It may be modified further by dividing both sides by a, to give

$$\frac{y-b}{a} = x$$

This series of manipulations is called **rearranging** the expression for y in terms of x to give an expression for x in terms of y. A short cut, as can be seen by inspecting these two steps, is that an added term can be moved through the equals sign provided that as it passes $=$ it changes sign (that happened to b in the example). Similarly, a multiplying factor becomes a divisor (and vice versa) when it passes through the $=$ sign (as happened to a).

There are several more complicated manipulations that are required in certain cases. The only one that we make use of in this text is to find the values of x that satisfy an equation of the form

$$ax^2 + bx + c = 0$$

or any equation that can be rearranged into this form by the steps we have already illustrated. An equation in which x occurs as its square is called a **quadratic equation**. Its solutions are found by inserting the values of the constants a, b, and c into the expression

$$x = \frac{-b \pm (b^2 - 4ac)^{1/2}}{2a}$$

where the two values of x given by this expression (one by using the $+$ sign and the other by using the $-$ sign) are called the two **roots** of the original quadratic equation.

A **function**, f, tells us how something changes as a variable is changed. For example, we might write

$$f(x) = ax + b$$

to show how a property f changes as x is changed. The variation of f with x is best shown by drawing a graph in which f is plotted on the vertical axis and x is plotted horizontally. The graph of the function we have just written is shown in Fig 1. The important point about this graph is that it is **linear** (that is, it is a straight line); its **intercept** with the vertical axis (the value of f when $x = 0$) is b, and its **slope** is a. That is, a straight line has the form

$$f = \text{slope} \times x + \text{intercept}$$

A positive value of a indicates an upward slope from left to right (increasing x); a change of sign of a results in a negative slope, down from left to right.

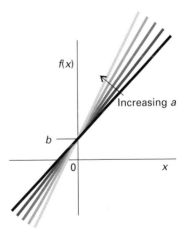

Fig 1 A straight line is described by the equation $f(x) = ax + b$, where a is the slope and b is the intercept.

The solutions of the equation $f(x) = 0$ can be visualized graphically: they are the values of x for which f cuts through the horizontal axis (the axis corresponding to $f = 0$). For example, the solution of the quadratic equation given earlier is depicted in Fig 2. In general, a quadratic equation has a graph that cuts through the horizontal axis at two points (the equation has two roots), a cubic equation (an equation in which x^3 is the highest power of x) cuts through it three times (the equation has three roots), and so on.

1.2 Logarithms, exponentials, and powers

Some equations are most readily solved by using logarithms and related functions. The **natural logarithm** of a number x is denoted $\ln x$, and is defined as the power to which a certain number designated e must be raised for the result to be equal to x. The number e, which is equal to 2.718. . . may seem to be decidedly unnatural; however, it falls out naturally from various manipulations in mathematics and its use greatly simplifies calculations. On a calculator, $\ln x$ is obtained simply by entering the number x and pressing the 'ln' key or its equivalent. It follows from the definition of logarithms that

$$\ln x + \ln y = \ln xy$$

$$\ln x - \ln y = \ln \frac{x}{y} \qquad a \ln x = \ln x^a$$

Thus, $\ln 5 + \ln 3$ is the same as $\ln 15$ and $\ln 6 - \ln 2$ is the same as $\ln 3$, as may readily be checked with a calculator. The last of these three relations is very useful for finding an awkward root of a number. For example, suppose we wanted the fifth root of 28. We write the required root as x, with $x^5 = 28$. We take logarithms of both sides, which gives $\ln x^5 = \ln 28$, and then rewrite the left-hand side of this equation as $5 \ln x$. At this stage we see that we have to solve

$$5 \ln x = \ln 28$$

To do so, we divide both sides by 5, which gives

$$\ln x = \frac{\ln 28}{5} = 0.6664...$$

All we need do at this stage is find the **antilogarithm** of the number on the right, the value of x for which the natural logarithm is the number quoted. The natural antilogarithm of a number is obtained by pressing the 'exp' key on a calculator (where 'exp' is an abbreviation for exponential), and in this case the answer is 1.947 . . . .

There are a number of useful points to remember about logarithms, and they are summarized in Fig 3. We see how logarithms increase only very slowly as x increases. For instance, when x increases from 1 to 1000, $\ln x$ increases from 0 to only 6.9. Another point is that the logarithm of 1 is 0: $\ln 1 = 0$. The logarithms of numbers less than 1 are negative, and in elementary mathematics the logarithms of negative numbers are not defined.[1]

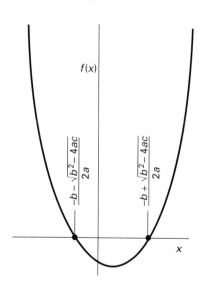

Fig 2 The roots of a quadratic equation are given by the values of x where the parabola intersects the x-axis.

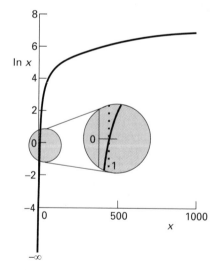

Fig 3 The graph of $\ln x$. Note that $\ln x$ approaches $-\infty$ as x approaches 0.

---

[1] The logarithm of a negative number is complex (that is, involves i, the square-root of -1): $\ln(-x) = i\pi + \ln x$.

We also encounter the **common logarithm** of a number, the logarithm compiled with 10 in place of e; they are denoted log x. For example, log 5 is the power to which 10 must be raised to obtain 5, and is 0.69897 . . . . Common logarithms follow the same rules of addition and subtraction as natural logarithms. They are largely of historical interest now that calculators are so readily available, but they survive in the context of acid–base chemistry and pH. Common and natural logarithms (log and ln, respectively) are related by

$$\ln x = \ln 10 \times \log x = (2.303 \ldots) \times \log x$$

The **exponential function**, e^x, plays a very special role in the mathematics of chemistry. It is evaluated by entering x and pressing the 'exp' or e^x key on a calculator. The following properties are important:

$$e^x \times e^y = e^{x+y} \qquad \frac{e^x}{e^y} = e^{x-y} \qquad (e^x)^a = e^{ax}$$

(These relations are the analogues of the relations for logarithms.) A graph of e^x is shown in Fig 4. As we see, it is positive for all values of x. It is less than 1 for all negative values of x, is equal to 1 when x = 0, and rises ever more rapidly towards infinity as x increases. This sharply rising character of e^x is the origin of the colloquial expression 'exponentially increasing' widely but loosely used in the media. (Strictly, a function increases exponentially if its rate of change is proportional to its current value.)

A useful expansion of the exponential function is

$$e^x = 1 + x + \frac{x^2}{2!} + \frac{x^3}{3!} + \cdots = \sum_n \frac{x^n}{n!}$$

where n!, or **n factorial**, is $n(n-1)(n-2)\cdots 1$ and 0! = 1 by definition. The corresponding expansion for logarithms is

$$\ln(1+x) = x - \tfrac{1}{2}x^2 + \tfrac{1}{3}x^3 - \cdots = \sum_{n>0}(-1)^{n+1}\left(\frac{x^n}{n}\right)$$

which is valid provided $x \le 1$. These expressions are useful for making approximations when x is very small ($x \ll 1$). Thus, we may write $e^x \approx 1 + x$ and $\ln(1+x) \approx x$.

1.3 Differentiation and integration

Rates of change of functions—slopes—are best discussed in terms of the infinitesimal calculus. The slope of a function, like the slope of a hill, is obtained by dividing the rise of the hill by the horizontal distance (Fig 5). However, because the slope may vary from point to point, we should make the horizontal distance between the points as small as possible. In fact, we let it become infinitesimally small—hence the name *infinitesimal* calculus. The values of a function f at two locations x and x + δx are f(x) and f(x + δx), respectively. Therefore, the slope of the function f at x is the vertical distance, which we write δf, divided by the horizontal distance, which we write δx:

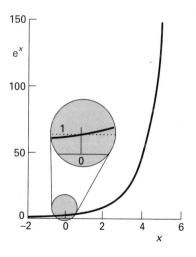

Fig 4 The graph of e^x. Note that e^x approaches 0 as x approaches $-\infty$.

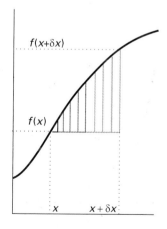

Fig 5 The slope of f(x) at x, df/dx, is obtained by making a series of approximations to the value of f(x + δx) − f(x) divided by change in x, denoted δx, and allowing δx to approach 0 (as denoted by the vertical lines getting closer to x).

$$\text{Slope} = \frac{\text{rise in value}}{\text{horizontal distance}} = \frac{\delta f}{\delta x} = \frac{f(x+\delta x) - f(x)}{\delta x}$$

The slope exactly at x itself is obtained by letting the horizontal distance become zero, which we write $\lim \delta x \to 0$. In this limit, the δ is replaced by a d, and we write

$$\text{Slope at } x = \frac{\mathrm{d} f}{\mathrm{d} x} = \lim_{\delta x \to 0} \frac{f(x+\delta x) - f(x)}{\delta x}$$

To work out the slope of any function, we work out the expression on the right: this process is called **differentiation**. It leads to the following important expressions:

$$\frac{\mathrm{d} x^n}{\mathrm{d} x} = n x^{n-1} \quad \frac{\mathrm{d} e^{ax}}{\mathrm{d} x} = a e^{ax} \quad \frac{\mathrm{d} \ln ax}{\mathrm{d} x} = \frac{1}{x}$$

$$\frac{\mathrm{d} \sin ax}{\mathrm{d} x} = a \cos ax \quad \frac{\mathrm{d} \cos ax}{\mathrm{d} x} = -a \sin ax$$

Most of the functions encountered in chemistry can be differentiated by using these relations in conjunction with the following rules:

$$\frac{\mathrm{d}(f+g)}{\mathrm{d} x} = \frac{\mathrm{d} f}{\mathrm{d} x} + \frac{\mathrm{d} g}{\mathrm{d} x} \quad \frac{\mathrm{d} f g}{\mathrm{d} x} = f \frac{\mathrm{d} g}{\mathrm{d} x} + g \frac{\mathrm{d} f}{\mathrm{d} x}$$

$$\frac{\mathrm{d} f(g)}{\mathrm{d} x} = \frac{\mathrm{d} g}{\mathrm{d} x} \cdot \frac{\mathrm{d} f}{\mathrm{d} g} \quad \frac{\mathrm{d}(f/g)}{\mathrm{d} x} = \frac{1}{g}\frac{\mathrm{d} f}{\mathrm{d} x} - \frac{f}{g^2}\frac{\mathrm{d} g}{\mathrm{d} x}$$

In the second row, $f(g)$ is a 'function of a function', as in $\ln(1 + x^2)$ or $\ln(\sin x)$.

The area under a graph of any function f is found by the techniques of **integration**. For instance, the area under the graph of the function f drawn in Fig 6 can be written as the value of f evaluated at a point multiplied by the width of the region, δx, and then all those products $f(x)\delta x$ summed over all the regions:

$$\text{Area between } a \text{ and } b = \sum f(x)\delta x$$

When we allow δx to become infinitesimally small, written dx, and sum an infinite number of strips, we write

$$\text{Area between } a \text{ and } b = \int_a^b f(x)\,\mathrm{d} x$$

The elongated S symbol on the right is called the **integral** of the function f. When written as $\int$ alone, it is the **indefinite integral** of the function. When written with limits (as in the expression above), it is the **definite integral** of the function. The definite integral is the indefinite integral evaluated at the upper limit (b) minus the indefinite integral evaluated at the lower limit (a).

Some important integrals are[2]

$$\int x^n \, \mathrm{d} x = \frac{x^{n+1}}{n+1} \quad \int e^{ax} \, \mathrm{d} x = \frac{e^{ax}}{a}$$

$$\int \ln \, ax \, \mathrm{d} x = x \, \ln \, ax - x$$

$$\int \sin \, ax \, \mathrm{d} x = -\frac{\cos \, ax}{a} \quad \int \cos \, ax \, \mathrm{d} x = \frac{\sin \, ax}{a}$$

It may be verified from these examples—and this is a very deep result of infinitesimal calculus—that *integration is the inverse of differentiation*. That is, if we integrate a function and then differentiate the result, we get back the original function.

A (first-order) **differential equation** is an equation that tells us how the slope of a function varies from place to place. For example, if the slope increased in proportion to x, we would write

$$\frac{\mathrm{d} f}{\mathrm{d} x} = ax$$

where a is a constant. To solve a differential equation, we have to look for the function f that satisfies it: the process is called **integrating** the equation. In this case we would multiply each side by dx, to obtain

$$\mathrm{d} f = ax\mathrm{d} x$$

and then integrate both sides:

$$\int \mathrm{d} f = \int ax \, \mathrm{d} x$$

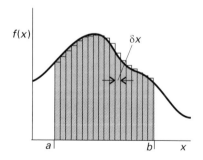

Fig 6 The shaded area is equal to the definite integral of $f(x)$ between the limits a and b.

---

[2] Strictly, an indefinite integral should be written with an arbitrary constant on the right, so $\int x \, \mathrm{d} x = \frac{1}{2}x^2 + \text{constant}$. However, tables of integrals commonly omit the constant. It cancels when the definite integral is evaluated.

The integral on the left is f (because integration is the inverse of differentiation) and that on the right is $\frac{1}{2}ax^2$ (plus a constant in each case). Therefore:

$$f(x) = \tfrac{1}{2}ax^2 + \text{constant}$$

This is the **general solution** of the equation (Fig 7). To fix the value of the constant and to find the **particular solution**, we take note of the **boundary conditions** that the function must satisfy, the value that we know the function has at a particular point. Thus, if we know that $f(0) = 1$, then we can write

$$1 = \text{constant}$$

The particular solution that satisfies the boundary condition is therefore

$$f(x) = \tfrac{1}{2}ax^2 + 1$$

In chemical kinetics, for instance, we may know that the reaction rate is proportional to the concentration of a reactant, and look for a general solution of the rate equation (a differential equation) which tells us how the concentration varies with time as the reaction proceeds. The particular solution is then obtained by making sure that the concentration has the correct value initially.[3]

A differential equation that is expressed in terms of first derivatives is a **first-order differential equation**. Rate laws are first-order differential equations.[4] A differential equation that is expressed in terms of second derivatives is a **second-order differential equation**. The Schrödinger equation is a second-order differential equation. The solution of differential equations is a very powerful technique in the physical sciences, but is often very difficult.

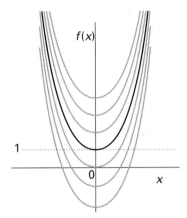

Fig 7 The *general* soloution of the differential equation $df/dx = ax$ is any one of the parabolas shown here (and others like them); the *particular* solution, which is identified by the boundary condition that f must satisfy, is shown by the dark line.

---

[3] A boundary condition is called an *initial condition* if the variable is time, as in a rate law.

[4] Do not confuse this use of the term 'order' with the order of the rate law: even a second-order rate law is a first-order differential equation!

FURTHER INFORMATION 2 *Quantities and units*

THE result of a measurement is a **physical quantity** (such as mass or density) that is reported as a numerical multiple of an agreed **unit**:

Physical quantity = numerical value × unit

For example, the mass of an object may be reported as $m = 2.5$ g and its density as $d = 1.01$ g cm^{-3} where the units are, respectively, 1 gram (1 g) and 1 gram per centimetre cubed (1 g cm^{-3}). Units are treated like algebraic quantities, and may be multiplied, divided, and cancelled. Thus, the expression (physical quantity)/unit is simply the numerical value of the measurement in the specified units, and hence is a dimensionless quantity. For instance, the mass reported above could be denoted $m/g = 2.5$ and the density as $d/(g\ cm^{-3}) = 1.01$.

Physical quantities are denoted by italic or Greek letters (as in m for mass and Π for osmotic pressure). Units are denoted by Roman letters (as in m for metre). In the **International System** of units (SI, from the French *Système International d'Unités*), the units are formed from seven **base units** listed in Table 1. All other physical quantities may be expressed as combinations of these physical quantities and reported in terms of **derived units**. Thus, volume is (length)^3 and may be reported as a multiple of 1 metre cubed (1 m^3), and density, which is mass/volume, may be reported as a multiple of 1 kilogram per metre cubed (1 kg m^{-3}).

A number of derived units have special names and symbols. The names of units derived from names of people are lower case (as in torr, joule, pascal, and kelvin), but their symbols are upper case (as in Torr, J, Pa, and K). The most important for our purposes are listed in Table 2. In all cases (both for base and derived quantities), the units may be modified by a prefix that denotes a factor of a power of 10. In a perfect world, Greek prefixes of units are upright (as in μm) and sloping for physical properties (as in μ for chemical potential), but available typefaces are not always so obliging. Among the most common prefixes are those listed in Table 3. Examples of the use of these prefixes are

$$1\ nm = 10^{-9}\ m \quad 1\ ps = 10^{-12}\ s \quad 1\ \mu mol = 10^{-6}\ mol$$

The kilogram (kg) is anomalous: although it is a base unit, it is interpreted as 10^3 g, and prefixes are attached to the gram (as in 1 mg = 10^{-3} g). Powers of units apply to the prefix as well as the unit they modify:

$$1\ cm^3 = 1\ (cm)^3 = 1\ (10^{-2}\ m)^3 = 10^{-6}\ m^3$$

Note that 1 cm^3 does not mean 1 c(m^3). When carrying out numerical calculations, it is usually safest to write out the numerical value of an observable as powers of 10.

There are a number of units that are in wide use but are not a part of the International System. Some are exactly

Table 2 *A selection of derived units*

Physical quantity	Derived unit	Name of derived unit
Force	1 kg m s^{-2}	newton, N
Pressure	1 kg m^{-1} s^{-2}	pascal, P
Energy	1 kg m^2 s^{-2}	joule, J
Power	1 kg m^2 s^{-3}	watt, W

Table 1 *The SI base units*

Physical quantity	Symbol for quantity	Base unit
Length	l	metre, m
Mass	m	kilogram, kg
Time	t	second, s
Electric current	I	ampere, A
Thermodynamic temperature	T	kelvin, K
Amount of substance	n	mole, mol
Luminous intensity	I	candela, cd

Table 3 *Common SI prefixes*

Prefix	f	p	n	μ	m	c	d
Name	femto	pico	nano	micro	milli	centi	deci
Factor	10^{-15}	10^{-12}	10^{-9}	10^{-6}	10^{-3}	10^{-2}	10^{-1}

Prefix	k	M	G	T
Name	kilo	mega	giga	tera
Factor	10^3	10^6	10^9	10^{12}

equal to multiples of SI units. These include the *litre* (L), which is exactly 10^3 cm^3 (or 1 dm^3) and the *atmosphere* (atm), which is exactly 101.325 kPa. Others rely on the values of fundamental constants, and hence are liable to change when the values of the fundamental constants are modified by more accurate or more precise measurements. Thus, the size of the energy unit *electronvolt* (eV), the energy acquired by an electron that is accelerated through a potential difference of exactly 1 V, depends on the value of the charge of the electron, and the present (2000) conversion factor is 1 eV = 1.602 177 33 $\times$ 10^{-19} J. Table 4 gives the conversion factors for a number of these convenient units.

Table 4 *Some common units*

Physical quantity	Name of unit	Symbol for unit	Value
Time	minute	min	60 s
	hour	h	3600 s
	day	d	86 400 s
Length	ångström	Å	10^{-10} m
Volume	litre	L	1 dm^3
Mass	tonne	t	10^3 kg
Pressure	bar	bar	10^5 Pa
	atmosphere	atm	101.325 kPa
Energy	electronvolt	eV	1.602 177 33 $\times 10^{-19}$ J
			96.485 31 kJ mol^{-1}

All values in the final column are exact, except for the definition of 1 eV.

FURTHER INFORMATION 3: *Energy and force*

KINETIC energy, E_K, is the energy that a body (a block of matter, an atom, or an electron) possesses by virtue of its motion. The formula for calculating the kinetic energy of a body of mass m that is travelling at a speed v is

$$E_K = \tfrac{1}{2}mv^2$$

This expression shows that a body may have a high kinetic energy if it is heavy (m large) and is travelling rapidly (v large). A stationary body ($v = 0$) has zero kinetic energy, whatever its mass. The energy of a sample of perfect gas is entirely due to the kinetic energy of its molecules: they travel more rapidly (on average) at high temperatures than at low, so raising the temperature of a gas increases the kinetic energy of its molecules.

Potential energy, E_P or V, is the energy that a body has by virtue of its position. A body on the surface of the Earth has a potential energy on account of the gravitational force it experiences: if the body is raised, then its potential energy is increased. There is no general formula for calculating the potential energy of a body because there are several kinds of force. For a body of mass m at a height h above (but close to) the surface of the Earth, the gravitational potential energy is

$$E_P = mgh$$

where g is the acceleration of free fall (g = 9.81 m s^{-2}). A heavy object at a certain height has a greater potential energy than a light object at the same height. One very important contribution to the potential energy is encountered when a charged particle is brought up to another charge. In this case the potential energy is inversely proportional to the distance between the charges (see *Further information 6*):

$$E_P \propto \frac{1}{r} \quad \text{specifically,} \quad E_P = \frac{q_1 q_2}{4\pi\varepsilon_0 r}$$

This **Coulomb potential energy** decreases with distance, and two infinitely widely separated charged particles have zero potential energy of interaction. The Coulomb potential energy plays a central role in the structures of atoms, molecules, and solids.

The **total energy**, E, of a body is the sum of its kinetic and potential energies. It is a central feature of physics that *the total energy of a body that is free from external influences is constant*. Thus, a stationary ball at a height h above the surface of the Earth has a potential energy of magnitude mgh; if it is released and begins to fall to the ground, it loses potential energy (as it loses height), but gains the same amount of kinetic energy (and therefore accelerates). Just before it hits the surface, it has lost all its potential energy, and all its energy is kinetic.

The state of motion of a body is changed by a **force**, F. According to Newton's second law of motion, a force changes the momentum of a body such that the acceleration, a, of the body (its rate of change of velocity) is proportional to the strength of the force:

$$\text{Force} = \text{mass} \times \text{acceleration, or } F = ma$$

Thus, to accelerate a heavy particle by a given amount requires a stronger force than to accelerate a light particle by the same amount. A force can be used to change the kinetic energy of a body, by accelerating the body to a higher speed. It may also be used to change the potential energy of a body by moving it to another position (for example, by raising it near the surface of the Earth).

The **work**, w, done on an object is the product of the distance, s, moved and the force opposing the motion:

$$w = -Fs$$

It requires a lot of work to move a long distance against a strong opposing force (think of cycling into a strong wind). If the opposing force changes at different points on the path, then we consider the force as a function of position, $F(s)$, and write

$$w = -\int F(s)\, ds$$

The integral is evaluated along the path traversed by the particle.

The units of energy and force are given in *Further information 2*.

FURTHER INFORMATION 4: *The kinetic theory of gases*

CONSIDER the arrangement in Fig 1. When a particle of mass m collides with the wall on the right, its component of linear momentum (the product of its mass and its velocity) parallel to the x-axis changes from mv_x (when it is travelling to the right) to $-mv_x$ (when it is travelling to the left). Its momentum therefore changes by $2mv_x$ on each collision. The number of collisions in an interval Δt is equal to the number of particles able to reach the wall in that interval. Because a particle with speed v_x can travel a distance $v_x\Delta t$ in an interval Δt, all the particles within a distance $v_x\Delta t$ of the wall will strike it if they are travelling towards it. Therefore, if the wall has area A, then all the particles in a volume $Av_x\Delta t$ will reach the wall (if they are travelling towards it). If the number density, the number of particles divided by the total volume, is $\mathcal{N}$, the number in the volume $Av_x\Delta t$ is $\mathcal{N}Av_x\Delta t$.

On average, half the particles are moving to the right, and half are moving to the left. Therefore, the average number of collisions with the wall during the interval Δt is $\frac{1}{2}\mathcal{N}Av_x\Delta t$. The total momentum change in that interval is the product of this number and the change $2mv_x$ that an individual molecule experiences:

$$\text{Momentum change} = \tfrac{1}{2}\mathcal{N}Av_x\Delta t \times 2mv_x$$
$$= m\mathcal{N}Av_x^2\Delta t$$

The rate of change of momentum is this change of momentum divided by the interval Δt during which it occurs:

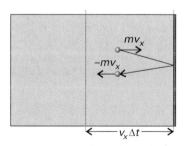

Fig 1 When a molecule of mass m travelling to the right with velocity v_x parallel to the x-axis strikes the right-hand wall and reverses its direction, it imparts a force of magnitude $2mv_x$ on the wall.

$$\text{Rate of change of momentum} = m\mathcal{N}Av_x^2$$

The rate of change of momentum is equal to the force (by Newton's second law of motion), so the force exerted by the gas on the wall is also $m\mathcal{N}Av_x^2$. It follows that the pressure, the force divided by the area, A, on which the force acts, is

$$\text{Pressure} = m\mathcal{N}v_x^2$$

The detected pressure, p, is the average (denoted $\langle\ldots\rangle$) of the quantity just calculated:

$$p = m\mathcal{N}\langle v_x^2\rangle$$

The root mean square speed, c, of the particles is

$$c = \langle v^2\rangle^{1/2} = (\langle v_x^2\rangle + \langle v_y^2\rangle + \langle v_z^2\rangle)^{1/2}$$

However, because the particles are moving randomly, the average of v_x^2 is the same as the average of the analogous quantities in the y- and z-directions. Because $\langle v_x^2\rangle$, $\langle v_y^2\rangle$, and $\langle v_z^2\rangle$ are all equal,

$$c = (3\langle v_x^2\rangle)^{1/2}$$

which implies that

$$\langle v_x^2\rangle = \tfrac{1}{3}c^2$$

Therefore,

$$p = \tfrac{1}{3}\mathcal{N}mc^2$$

The number density, $\mathcal{N}$, is the product of the amount (the number of moles, n) and the Avogadro constant, N_A, divided by the volume, V, so the last equation becomes

$$pV = \tfrac{1}{3}nN_Amc^2 = \tfrac{1}{3}nMc^2$$

where $M = mN_A$ is the molar mass of the molecules. This expression is the equation used in the text.

FURTHER INFORMATION 5: *The variation of Gibbs energy with the conditions*

OUR aim here is to find an expression for the change in Gibbs energy when the temperature and pressure change. Although the calculation might look a bit lengthy, it teaches us a lot about the manipulation of thermodynamic expressions and leads to a conclusion of enormous importance.

Lesson 1 When in doubt about where to start, go back to a definition. So, we start with the definition of G:

$$G = H - TS$$

When the state of the system changes by an infinitesimal amount, its enthalpy, temperature, and entropy all change: H changes to $H + dH$, T changes to $T + dT$, and S changes to $S + dS$. As a result, G changes to $G + dG$, where

$$G + dG = H + dH - (T + dT)(S + dS)$$
$$= H - TS + dH - TdS - SdT - dTdS$$

We recognize $H - TS$ as G on the right, which cancels G on the left, and, as always in calculus, we ignore the doubly infinitesimal term $dTdS$. We are left with

$$dG = dH - TdS - SdT$$

Lesson 2 We could have arrived at this result in a single step by thinking of a change in $G = H - TS$ as arising from changes in H, S, and T individually, and then adding the effects together.

Now we deal with the dH in this expression. *Lesson 1* advises us to go back to the definition $H = U + pV$. *Lesson 2* then lets us write a change in dH as

$$dH = dU + pdV + Vdp$$

(We could use the same argument as for G to confirm this conclusion.) When this expression is inserted into dG we get

$$dG = dU + pdV + Vdp - TdS - SdT$$

Finally, we launch an attack on dU. First (*Lesson 1*), we use $dU = dw + dq$ to write

$$dG = dw + dq + pdV + Vdp - TdS - SdT$$

Lesson 3 is that expressions are often much simpler for reversible changes. For a reversible change, $dq = TdS$ and $dw = -pdV$, so we are left with

$$\text{For a reversible change: } dG = Vdp - SdT$$

Lesson 4 is to use the property of being a state function to generalize a result. Because G is a state function, a change in G, including an infinitesimally small change, has the same value regardless of how the change is brought about. Therefore, for *any* type of infinitesimal change

$$dG = Vdp - SdT$$

FURTHER INFORMATION 6: *Concepts of electrostatics*

THE fundamental expression in **electrostatics**, the interactions of stationary electric charges, is the Coulomb potential energy of one charge of magnitude q at a distance r from another charge q':

$$V = \frac{qq'}{4\pi\varepsilon_0 r}$$

That is, the potential energy is inversely proportional to the separation of the charges. The fundamental constant ε_0 is the **vacuum permittivity**; its value is

$$\varepsilon_0 = 8.854\ 187\ 816 \times 10^{-12}\ \mathrm{J}^{-1}\ \mathrm{C}^2\ \mathrm{m}^{-1}$$

With r in metres and the charges in coulombs, the potential energy is in joules. The potential energy is equal to the work that must be done to bring up a charge q from infinity to a distance r from a charge q'. The implication is then that the *force* exerted by a charge q on a charge q' is inversely proportional to the *square* of their separation:

$$F = \frac{qq'}{4\pi\varepsilon_0 r^2}$$

This expression is **Coulomb's inverse-square law of force**.

We can express the potential energy of a charge q in the presence of another charge q' in terms of the **Coulomb potential**,[1] ϕ, due to q':

$$V = q\phi, \qquad \phi = \frac{q'}{4\pi\varepsilon_0 r}$$

The units of potential are joules per coulomb ($\mathrm{J\,C}^{-1}$) so, when ϕ is multiplied by a charge in coulombs, the result is in joules. The combination joules per coulomb occurs widely in electrostatics, and is called a *volt*, V:

$$1\ \mathrm{V} = 1\ \mathrm{J\,C}^{-1}$$

(which implies that $1\ \mathrm{V\,C} = 1\ \mathrm{J}$). If there are several charges $q_1, q_2, \ldots$ present in the system, then the total potential experienced by the charge q is the sum of the potential generated by each charge:

$$\phi = \phi_1 + \phi_2 + \cdots$$

For example, the potential generated by a dipole is the sum of the potentials of the two equal and opposite charges: these potentials do not in general cancel because the point of interest is at different distances from the two charges (Fig 1).

The motion of charge gives rise to an **electric current**, I. Electric current is measured in amperes, A, where

$$1\ \mathrm{A} = 1\ \mathrm{C\,s}^{-1}$$

If the electric charge is that of electrons (as it is through metals and semiconductors), then a current of 1 A represents the flow of 6×10^{18} electrons per second. If the current flows from a region of potential ϕ_i to ϕ_f, through a potential difference $\Delta\phi = \phi_f - \phi_i$, then the rate of doing work is the current (the rate of transfer of charge) multiplied by the potential difference, $I \times \Delta\phi$. The rate of doing work is called **power**, P, so

$$P = I\Delta\phi$$

With current in amperes and the potential difference in volts, the power works out in joules per second, or watts, W:

$$1\ \mathrm{W} = 1\ \mathrm{J\,s}^{-1}$$

The total energy supplied in a time t is the power (the energy per second) multiplied by the time:

$$E = Pt = It\Delta\phi$$

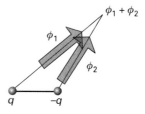

Fig 1 The electric potential at a point is equal to the sum of the potentials due to each charge.

---

[1] Distinguish potential from potential energy.

The energy is obtained in joules with the current in amperes, the potential difference in volts, and the time in seconds.

The current flowing through a conductor is proportional to the potential difference between the ends of the conductor and inversely proportional to the **resistance**, R, of the conductor:

$$I = \frac{\Delta \phi}{R}$$

This empirical relation is called **Ohm's law**. With the current in amperes and the potential difference in volts, the resistance is measured in *ohms*, Ω, with $1\ \Omega = 1\ \text{V A}^{-1}$.

FURTHER INFORMATION 7: *Electromagnetic radiation and photons*

ELECTROMAGNETIC radiation, which includes γ radiation, ultraviolet radiation, visible light, infrared radiation, microwave radiation, and radio waves, is a wave-like, oscillating electric and magnetic field that propagates through space with a constant speed c, the **speed of light**. The radiation is depicted in Fig 1: the electric field and magnetic field are perpendicular to each other and vary sinusoidally with a **wavelength**, λ (lambda), and **frequency**, ν (nu), that are related by

$$\lambda \nu = c$$

Therefore, the shorter the wavelength, the higher the frequency of the radiation. The rate of change of the fields is also commonly reported as the **wavenumber**, $\tilde{\nu}$ (nu tilde), which is defined as

$$\tilde{\nu} = \frac{\nu}{c}$$

The wavenumber is the number of complete wavelengths in a given region divided by the length of the region (typically, 1 cm). According to classical physics, the **intensity** of a ray is proportional to the square of the amplitude of the wave, so an intense ray would be a wave of electromagnetic field that oscillated with a large amplitude. The classification of the electromagnetic spectrum into different regions according to the frequency and wavelength of the radiation is summarized in Table 1.

The wave shown in the illustration is **plane polarized**: it is so called because the electric field oscillates in a single plane. The plane may be orientated in any direction around the direction of propagation (with the electric field perpendicular to that direction). An alternative mode of polarization is **circular polarization**, in which the electric field rotates around the direction of propagation in either a clockwise or a counterclockwise sense.

According to quantum theory, a ray of frequency ν consists of a stream of **photons**, each one of which has energy

$$E = h\nu$$

where h is Planck's constant (Section 12.1). Thus, a photon of high-frequency radiation has more energy than a photon of low-frequency radiation. The greater the intensity of the ray, the greater the number of photons in it. In a vacuum, each photon travels with the speed of light. The frequency of the radiation determines the colour of visible light because different visual receptors in the eye respond to photons of different energy. The relation between colour and frequency is shown in Table 2, which also gives the energy carried by each type of photon.

Photons may also be polarized. A plane polarized ray of light consists of plane polarized photons and a circularly polarized ray consists of circularly polarized

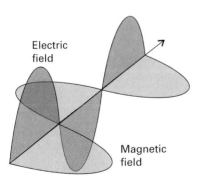

Fig 1 Electromagnetic radiation consists of a wave of electric and magnetic fields perpendicular to the direction of propagation, and mutually perpendicular to each other. This illustration shows a plane-polarized wave.

Electric field

Magnetic field

Table 1 *The regions of the electromagnetic spectrum**

Regions	Wavelength	Frequency/Hz
Radiofrequency	> 30 cm	< 10^9
Microwave	3 mm to 30 cm	10^9 to 10^{11}
Infrared	1000 nm to 3 mm	10^{11} to 3×10^{14}
Visible	400 nm to 800 nm	4×10^{14} to 8×10^{14}
Ultraviolet	3 nm to 300 nm	10^{15} to 10^{17}
X-Rays, γ-rays	< 3 nm	> 10^{17}

* The boundaries of the regions are only approximate.

photons. The latter can be regarded as spinning either clockwise (for left-circularly polarized radiation) or coun-terclockwise (for right-circularly polarized radiation) about their direction of propagation.

Table 2 *Colour, frequency, and wavelength of light*[†]

	Frequency/ $(10^{14}$ Hz)	Wavelength/ nm	Energy of photon/$(10^{-19}$ J)
X-rays and γ-rays	10^3 and above	3 and below	660 and above
Ultraviolet	10	300	6.6
Visible light			
Violet	7.1	420	4.7
Blue	6.4	470	4.2
Green	5.7	530	3.7
Yellow	5.2	580	3.4
Orange	4.8	620	3.2
Red	4.3	700	2.8
Infrared	3.0	1000	1.9
Microwaves and radiowaves	3×10^{-11} Hz and below	3×10^6 and above	2.0×10^{-22} J and below

[†] The values given are approximate but typical.

FURTHER INFORMATION 8: *Oxidation numbers*

A simple way of judging whether a monatomic species has undergone oxidation or reduction is to note if the charge number of the species has changed. For example, an increase in the charge number of a monatomic ion (which corresponds to electron loss), as in the conversion of Fe^{2+} to Fe^{3+}, is an oxidation. A decrease in charge number (to a less positive or more negative value, as a result of electron gain), as in the conversion of Br to Br^-, is a reduction.

It is possible to assign to an atom in a polyatomic species an effective charge number, called the **oxidation number**, ω. (There is no standard symbol for this quantity.) The oxidation number is defined so that an increase in its value ($\Delta\omega > 0$) corresponds to oxidation, and a decrease ($\Delta\omega < 0$) corresponds to reduction.

An oxidation number is assigned to an element in a compound by supposing that it is present as an ion with a characteristic charge; for instance, oxygen is present as O^{2-} in most of its compounds, and fluorine is present as F^- (Fig 1). The more electronegative element is supposed to be present as the anion. This procedure implies that:

1 The oxidation number of an elemental substance is zero: $\omega(\text{element}) = 0$.

2 The oxidation number of a monatomic ion is equal to the charge number of that ion: $\omega(E^{z\pm}) = \pm z$.

3 The sum of the oxidation numbers of all the atoms in a species is equal to the overall charge number of the species.

Thus, hydrogen, oxygen, iron, and all the elements have $\omega = 0$ in their elemental forms; $\omega(Fe^{3+}) = +3$ and $\omega(Br^-) = -1$. It follows that the conversion of Fe to Fe^{3+} is an oxidation (because $\Delta\omega > 0$) and the conversion of Br to Br^- is a reduction (because $\Delta\omega < 0$). The definition of oxidation number and its relation to oxidation and reduction are consistent with the definitions in terms of electron loss and gain.

As an illustration, consider the oxidation numbers of the elements in SO_2 and SO_4^{2-}. The sum of oxidation numbers of the atoms in SO_2 must be 0, so we can write

$$\omega(S) + 2\omega(O) = 0$$

Each O atom has $\omega = -2$. Hence,

$$\omega(S) + 2 \times (-2) = 0$$

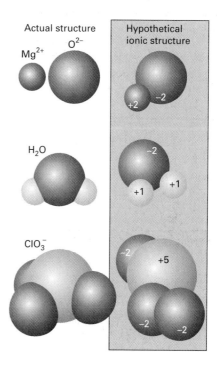

Fig 1 To calculate the oxidation number of an element in an oxide or oxoacid, we suppose that each O atom is present as an O^{2-} ion, and then identify the charge of the element required to give the actual overall charge of the species. The more electronegative element plays a similar role in other compounds.

which solves to $\omega(S) = +4$. Now consider SO_4^{2-}. The sum of oxidation numbers of the atoms in the ion is -2, so we can write

$$\omega(S) + 4\omega(O) = -2$$

Because $\omega(O) = -2$,

$$\omega(S) + 4 \times (-2) = -2$$

which solves to $\omega(S) = +6$. The sulfur is more highly oxidized in the sulfate ion than in sulfur dioxide. To calculate the oxidation number of compounds containing hydrogen we use $\omega(H) = +1$ unless the compound is a metal hydride, in which case $\omega(H) = -1$.

Self-test 1

Calculate the oxidation numbers of the elements in (a) H_2S, (b) PO_4^{3-}, (c) NO_3^-.

[*Answer:* (a) $\omega(H) = +1$, $\omega(S) = -2$; (b) $\omega(P) = +5$, $\omega(O) = -2$; (c) $\omega(N) = +5$, $\omega(O) = -2$]

FURTHER INFORMATION 9: *The Lewis theory of covalent bonding*

IN his original formulation of a theory of the covalent bond, G.N. Lewis proposed that each bond consisted of one electron pair. Each atom in a molecule shared electrons until it had acquired an octet characteristic of a noble gas atom near it in the periodic table. (Hydrogen is an exception: it acquires a duplet of electrons.) Thus, to write down a Lewis structure:

1 Arrange the atoms as they are found in the molecule.

2 Add one electron pair (represented by dots, :) between each bonded atom.

3 Use the remaining electron pairs to complete the octets of all the atoms present either by forming lone pairs or by forming multiple bonds.

4 Replace bonding electron pairs by bond lines (—) but leave lone pairs as dots (:).

A Lewis structure does not (except in very simple cases), portray the actual geometrical structure of the molecule; it is a topological map of the arrangement of bonds.

As an example, consider the Lewis structure of methanol, CH_3OH, in which there are $4 \times 1 + 4 + 6 = 14$ valence electrons (and hence seven electron pairs) to accommodate. The first step is to write the atoms in the arrangement (1); the pale rectangles have been included to indicate which atoms are linked. The next step is to add electron pairs to denote bonds (2). The C atom now has a complete octet and all four H atoms have complete duplets. There are two unused electron pairs, which are used as lone pairs to complete the octet of the O atom (3). Finally, replace the bonding pairs by lines to indicate bonds (4). An example of a species with a multiple bond is acetic acid (5).

In some cases, more than one structure can be written in which the only difference is the location of multiple bonds or lone pairs. In such cases, the molecule's structure is interpreted as a **resonance hybrid,** a quantum mechanical blend, of the individual structures. Resonance is depicted by a double-headed arrow. For example, the ozone molecule, O_3, is a resonance hybrid of two structures (6). Resonance distributes multiple-bond character over the participating atoms.

Many molecules cannot be written in a way that conforms to the octet rule. Those classified as **hypervalent molecules** require an expansion of the octet. Although it is often stated that octet expansion requires the involvement of *d* orbitals, and is therefore confined to Period 3 and subsequent elements, there is good evidence to suggest that octet expansion is a consequence of an atom's size, not its intrinsic orbital structure. Whatever the reason, octet expansion is needed to account for the structures of PCl_5 with expansion to ten electrons (7), SF_6, expansion to 12 electrons (8), and XeO_4, expansion to 16 electrons (9). Octet expansion is also encountered in species that do not necessarily require it, but which, if it is permitted, may acquire a lower energy. Thus, of the structures (10a) and (10b) of the SO_4^{2-} ion, the second has a lower energy than the first. The actual structure of the ion is a resonance hybrid of both structures (together with analogous structures with double bonds in different locations), but the latter structure makes the dominant contribution.

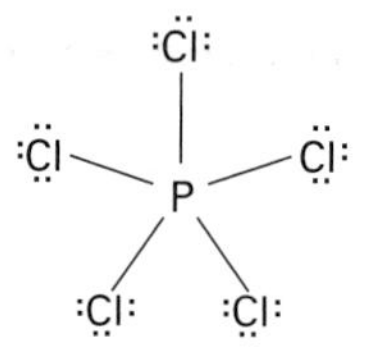

7

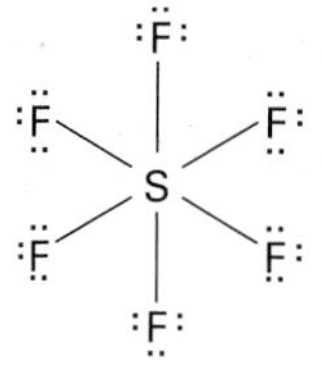

8

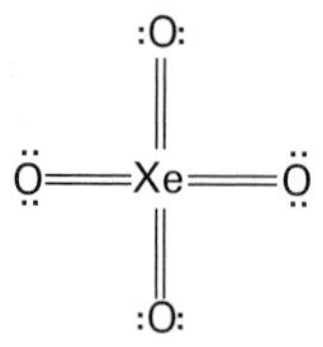

9

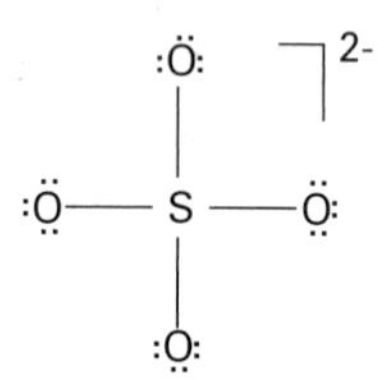

10a 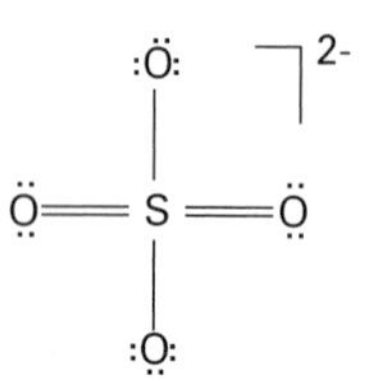   **10b**

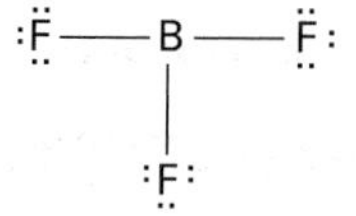

11a 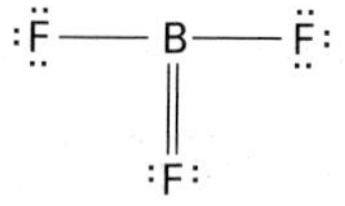   **11b**

Octet completion is not always energetically appropriate. Such is the case with boron trifluoride, BF_3. Two of the possible Lewis structures for this molecule are (**11a**) and (**11b**). In the former, the B atom has an **incomplete octet**. Nevertheless, it has a lower energy than the other structure, for to form the latter structure one F atom has had partially to relinquish an electron pair, which is energetically demanding for such an electronegative element. The actual molecule is a resonance hybrid of the two structures (and of others with the double bond in different locations), but the overwhelming contribution is from the former structure. Consequently, we regard BF_3 as a molecule with an incomplete octet. This feature is responsible for its ability to act as a Lewis acid (an electron pair acceptor).

The Lewis approach fails for a class of **electron-deficient compounds**, which are molecules that have too few electrons for a Lewis structure to be written. The most famous example is diborane, B_2H_6, which requires at least seven pairs of electrons to bind the eight atoms together but has only twelve valence electrons in all. The structures of such molecules can be explained in terms of molecular orbital theory and the concept of delocalized electron pairs, in which the influence of an electron pair is distributed over several atoms.

FURTHER INFORMATION 10: *The VSEPR model*

IN the **valence shell electron pair repulsion model** (VSEPR) we focus on a single, central atom and consider the local arrangement of atoms that are linked to it. For example, in considering the H_2O molecule, we concentrate on the electron pairs in the valence shell of the central O atom. This procedure can be extended to molecules in which there is no obvious central atom, such as in benzene, C_6H_6, or hydrogen peroxide, H_2O_2, by focusing attention on a group of atoms, such as a C–CH–C fragment of benzene or an H–O–O fragment of hydrogen peroxide, and considering the arrangement of electron pairs around the central atom of the fragment.

The basic assumption of the VSEPR model is that *the valence-shell electron pairs of the central atom adopt positions that maximize their separations*. Thus, if the atom has four electron pairs in its valence shell, then the pairs adopt a tetrahedral arrangement around the atom; if the atom has five pairs, then the arrangement is trigonal bipyramidal. The arrangements adopted by electron pairs are summarized in Table 1.

Once the basic shape of the arrangement of electron pairs has been identified, the pairs are identified as bonding or nonbonding. For instance, in the H_2O molecule, two of the tetrahedrally arranged pairs are bonding pairs and two are nonbonding pairs. Then the shape of the molecule is classified by noting the arrangement of the atoms around the central atom. The H_2O molecule, for instance, has an underlying tetrahedral arrangement of lone pairs but, as only two of the pairs are bonding pairs, the molecule is classified as angular (Fig 1) It is important to keep in mind the distinction between the arrangement of electron pairs and the shape of the resulting molecule: the latter is identified by noting the relative locations of the atoms, not the lone pairs (Fig 2).

For example, to predict the shape of an ethane molecule we concentrate on one of the C atoms initially. That atom has four electron pairs in its valence shell (in the molecule), and they adopt a tetrahedral arrangement. All four electron pairs are bonding: three bond H atoms and the fourth bonds the second C atom. Therefore, the arrangement of atoms is tetrahedral around the C atom. The second C atom has the same environment, so we conclude that the ethane molecule consists of two tetrahedral CH_3 groups (**1**).

The next stage in the application of the VSEPR model is to accommodate the greater repelling effect of lone pairs

---

Table 1 *Electron pair arrangements*

Number of electron pairs	Arrangement
2	Linear
3	Trigonal planar
4	Tetrahedral
5	Trigonal bipyramidal
6	Octahedral
7	Pentagonal bipyramidal

---

compared with that of bonding pairs. That is, *bonding pairs tend to move away from lone pairs even though that might reduce their separation from other bonding pairs*. The NH_3 molecule provides a simple example. The N atom has four electron

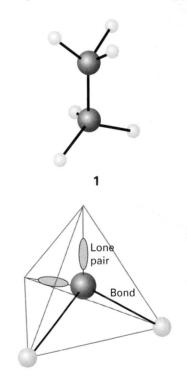

1

Fig 1 The shape of a molecule is identified by noting the arrangement of its atoms, not its lone pairs. This molecule is angular even though the electron-pair distribution is tetrahedral.

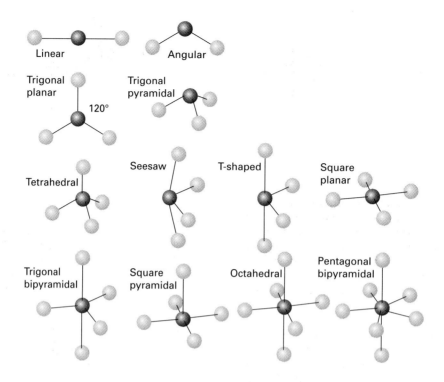

Fig 2 The classification of molecular shapes according to the relative locations of atoms.

pairs in its valence shell and they adopt a tetrahedral arrangement. Three of the pairs are bonding pairs, and the fourth is a lone pair. The basic shape of the molecule is therefore trigonal pyramidal. However, a lower energy is achieved if the three bonding pairs move away from the lone pair, even though they are brought slightly closer together (**2**). We therefore predict an HNH bond angle of slightly less than the tetrahedral angle of 109.5°, which is consistent with the observed angle of 107°.

As an example, consider the shape of an SF_4 molecule. The first step is to write a Lewis (electron dot) structure for the molecule to identify the number of lone pairs in the valence shell of the S atom (**3**). This structure shows that

there are five electron pairs on the S atom. Reference to Table 1 shows that the five pairs are arranged as a trigonal bipyramid. Four of the pairs are bonding pairs and one is a lone pair. The repulsions stemming from the lone pair are minimized if the lone pair is placed in an equatorial position: then it is close to the axial pairs (**4**), whereas if it had adopted an axial position it would have been close to three equatorial pairs (**5**). Finally, the four bonding pairs are allowed to relax away from the single lone pair, to give a distorted seesaw arrangement (**6**).

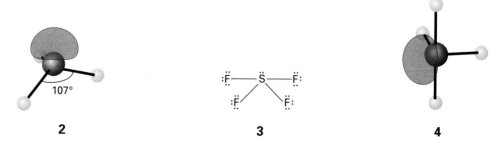

2

3

4

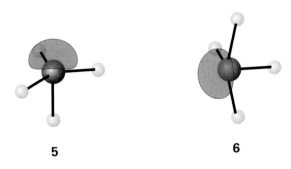

5

6

7

To take into account multiple bonds, each set of two or three electron pairs is treated as a single region of high electron density, a kind of 'superpair'. For example, each C atom in an ethene molecule, $CH_2=CH_2$, is regarded as having three pairs (one of them the superpair of two electron pairs of the double bond); these regions of high electron density adopt a trigonal planar arrangement around each atom, so the shape of the molecule is trigonal planar at each C atom (**7**). Another example is the SO_3^{2-} ion: if we adopt the Lewis structure in (**8**), then we see that there are four regions of high electron density around the S atom, indicating a tetrahedral arrangement. One region is a lone pair, so overall the ion is trigonal pyramidal (**9**). We would reach the same conclusion if we adopted the alternative Lewis structure (**10**) in which there are four electron pairs (none of them a 'superpair').

8

9

10

APPENDIX 1

Table A1.1 Thermodynamic data for organic compounds (all values relate to 298.15 K)

	$M/$ (g mol^{-1})	$\Delta_f H^{\ominus}/$ (kJ mol^{-1})	$\Delta_f G^{\ominus}/$ (kJ mol^{-1})	$S_m^{\ominus}/$ (J K^{-1}mol^{-1})	$C_{p,m}/$ (J K^{-1}mol^{-1})	$\Delta_c H^{\ominus}/$ (kJ mol^{-1})
C(s) (graphite)	12.011	0	0	5.740	8.527	−393.51
C(s) (diamond)	12.011	+1.895	+2.900	2.377	6.113	−395.40
CO$_2$(g)	44.010	−393.51	−394.36	213.74	37.11	
Hydrocarbons						
CH$_4$(g), methane	16.04	−74.81	−50.72	186.26	35.31	−890
CH$_3$(g), methyl	15.04	+145.69	+147.92	194.2	38.70	
C$_2$H$_2$(g), ethyne	26.04	+226.73	+209.20	200.94	43.93	−1300
C$_2$H$_4$(g), ethene	28.05	+52.26	+68.15	219.56	43.56	−1411
C$_2$H$_6$(g), ethane	30.07	−84.68	−32.82	229.60	52.63	−1560
C$_3$H$_6$(g), propene	42.08	+20.42	+62.78	267.05	63.89	−2058
C$_3$H$_6$(g), cyclopropane	42.08	+53.30	+104.45	237.55	55.94	−2091
C$_3$H$_8$(g), propane	42.10	−103.85	−23.49	269.91	73.5	−2220
C$_4$H$_8$(g), 1-butene	56.11	−0.13	+71.39	305.71	85.65	−2717
C$_4$H$_8$(g), cis-2-butene	56.11	−6.99	+65.95	300.94	78.91	−2710
C$_4$H$_8$(g), trans-2-butene	56.11	−11.17	+63.06	296.59	87.82	−2707
C$_4$H$_{10}$(g), butane	58.13	−126.15	−17.03	310.23	97.45	−2878
C$_5$H$_{12}$(g), pentane	72.15	−146.44	−8.20	348.40	120.2	−3537
C$_5$H$_{12}$(l)	72.15	−173.1				
C$_6$H$_6$(l), benzene	78.12	+49.0	+124.3	173.3	136.1	−3268
C$_6$H$_6$(g)	78.12	+82.93	+129.72	269.31	81.67	−3320
C$_6$H$_{12}$(l), cyclohexane	84.16	−156	+26.8		156.5	−3902
C$_6$H$_{14}$(l), hexane	86.18	−198.7		204.3		−4163
C$_6$H$_5$CH$_3$(g), methylbenzene (toluene)	92.14	+50.0	+122.0	320.7	103.6	−3953
C$_7$H$_{16}$(l), heptane	100.21	−224.4	+1.0	328.6	224.3	
C$_8$H$_{18}$(l), octane	114.23	−249.9	+6.4	361.1		−5471
C$_8$H$_{18}$(l), iso-octane	114.23	−255.1				−5461
C$_{10}$H$_8$(s), naphthalene	128.18	+78.53				−5157
Alcohols and phenols						
CH$_3$OH(l), methanol	32.04	−238.66	−166.27	126.8	81.6	−726
CH$_3$OH(g)	32.04	−200.66	−161.96	239.81	43.89	−764
C$_2$H$_5$OH(l), ethanol	46.07	−277.69	−174.78	160.7	111.46	−1368
C$_2$H$_5$OH(g)	46.07	−235.10	−168.49	282.70	65.44	−1409
C$_6$H$_5$OH(s), phenol	94.12	−165.0	−50.9	146.0		−3054
Carboxylic acids, hydroxy acids, and esters						
HCOOH(l), formic	46.03	−424.72	−361.35	128.95	99.04	−255
CH$_3$COOH(l), acetic	60.05	−484.5	−389.9	159.8	124.3	−875
CH$_3$COOH(aq)	60.05	−485.76	−396.46	178.7		

	$M|$ (g mol^{-1})	$\Delta_f H^{\ominus}|$ (kJ mol^{-1})	$\Delta_f G^{\ominus}|$ (kJ mol^{-1})	$S_m^{\ominus}|$ J K^{-1}mol^{-1}	$C_{p,m}|$ J K^{-1}mol^{-1}	$\Delta_c H^{\ominus}|$ (kJ mol^{-1})
CH$_3$CO$_2^-$(aq)	59.05	−486.01	−369.31	86.6	−6.3	
(COOH)$_2$(s), oxalic	90.04	−827.2			117	−254
C$_6$H$_5$COOH(s), benzoic	122.13	−385.1	−245.3	167.6	146.8	−3227
CH$_3$CH(OH)COOH(s), lactic	90.08	−694.0				−1344
CH$_3$COOC$_2$H$_5$(l), ethyl acetate	88.11	−479.0	−332.7	259.4	170.1	−2231
Alkanals and alkanones						
HCHO(g), methanal	30.03	−108.57	−102.53	218.77	35.40	−571
CH$_3$CHO(l), ethanal	44.05	−192.30	−128.12	160.2		−1166
CH$_3$CHO(g)	44.05	−166.19	−128.86	250.3	57.3	−1192
CH$_3$COCH$_3$(l), propanone	58.08	−248.1	−155.4	200.4	124.7	−1790
Sugars						
C$_6$H$_{12}$O$_6$(s), α-D-glucose	180.16	−1274				−2808
C$_6$H$_{12}$O$_6$(s), β-D-glucose	180.16	−1268	−910	212		
C$_6$H$_{12}$O$_6$(s), β-D-fructose	180.16	−1266				−2810
C$_{12}$H$_{22}$O$_{11}$(s), sucrose	342.30	−2222	−1543	360.2		−5645
Nitrogen compounds						
CO(NH$_2$)$_2$(s), urea	60.06	−333.51	−197.33	104.60	93.14	−632
CH$_3$NH$_2$(g), methylamine	31.06	−22.97	+32.16	243.41	53.1	−1085
C$_6$H$_5$NH$_2$(l), aniline	93.13	+31.1				−3393
CH$_2$(NH$_2$)COOH(s), glycine	75.07	−532.9	−373.4	103.5	99.2	−969

Table A1.2　Thermodynamic data (all values relate to 298.15 K)*

	$M|$ (g mol^{-1})	$\Delta_f H^{\ominus}|$ (kJ mol^{-1})	$\Delta_f G^{\ominus}|$ (kJ mol^{-1})	$S_m^{\ominus}|$ (J K^{-1}mol^{-1})	$C_{p,m}|$ (J K^{-1}mol^{-1})
Aluminium (aluminum)					
Al(s)	26.98	0	0	28.33	24.35
Al(l)	26.98	+10.56	+7.20	39.55	24.21
Al(g)	26.98	+326.4	+285.7	164.54	21.38
Al^{3+}(g)	26.98	+5483.17			
Al^{3+}(aq)	26.98	−531	−485	−321.7	
Al$_2$O$_3$(s, α)	101.96	−1675.7	−1582.3	50.92	79.04
AlCl$_3$(s)	133.24	−704.2	−628.8	110.67	91.84
Argon					
Ar(g)	39.95	0	0	154.84	20.786
Antimony					
Sb(s)	121.75	0	0	45.69	25.23
SbH$_3$(g)	153.24	+145.11	+147.75	232.78	41.05
Arsenic					
As(s, α)	74.92	0	0	35.1	24.64
As(g)	74.92	+302.5	+261.0	174.21	20.79
As$_4$(g)	299.69	+143.9	+92.4	314	
AsH$_3$(g)	77.95	+66.44	+68.93	222.78	38.07

	$M/$ (g mol^{-1})	$\Delta_f H^\ominus/$ (kJ mol^{-1})	$\Delta_f G^\ominus/$ (kJ mol^{-1})	$S_m^\ominus/$ (J K^{-1} mol^{-1})	$C_{p,m}/$ (J K^{-1} mol^{-1})
Barium					
Ba(s)	137.34	0	0	62.8	28.07
Ba(g)	137.34	+180	+146	170.24	20.79
Ba^{2+}(aq)	137.34	−537.64	−560.77	+9.6	
BaO(s)	153.34	−553.5	−525.1	70.43	47.78
BaCl$_2$(s)	208.25	−858.6	−810.4	123.68	75.14
Beryllium					
Be(s)	9.01	0	0	9.50	16.44
Be(g)	9.01	+324.3	+286.6	136.27	20.79
Bismuth					
Bi(s)	208.98	0	0	56.74	25.52
Bi(g)	208.98	+207.1	+168.2	187.00	20.79
Bromine					
Br$_2$(l)	159.82	0	0	152.23	75.689
Br$_2$(g)	159.82	+30.907	+3.110	245.46	36.02
Br(g)	79.91	+111.88	+82.396	175.02	20.786
Br$^-$(g)	79.91	−219.07			
Br$^-$(aq)	79.91	−121.55	−103.96	+82.4	−141.8
HBr(g)	90.92	−36.40	−53.45	198.70	29.142
Cadium					
Cd(s, γ)	112.40	0	0	51.76	25.98
Cd(g)	112.40	+112.01	+77.41	167.75	20.79
Cd^{2+}(aq)	112.40	−75.90	−77.612	−73.2	
CdO(s)	128.40	−258.2	−228.4	54.8	43.43
CdCO$_3$(s)	172.41	−750.6	−669.4	92.5	
Caesium (cesium)					
Cs(s)	132.91	0	0	85.23	32.17
Cs(g)	132.91	+76.06	+49.12	175.60	20.79
Cs$^+$(aq)	132.91	−258.28	−292.02	+133.05	−10.5
Calcium					
Ca(s)	40.08	0	0	41.42	25.31
Ca(g)	40.08	+178.2	+144.3	154.88	20.786
Ca^{2+}(aq)	40.08	−542.83	−553.58	−53.1	
CaO(s)	56.08	−635.09	−604.03	39.75	42.80
CaCO$_3$(s) (calcite)	100.09	−1206.9	−1128.8	92.9	81.88
CaCO$_3$(s) (aragonite)	100.09	−1207.1	−1127.8	88.7	81.25
CaF$_2$(s)	78.08	1219.6	−1167.3	68.87	67.03
CaCl$_2$(s)	110.99	−795.8	−748.1	104.6	72.59
CaBr$_2$(s)	199.90	−682.8	−663.6	130	
Carbon (for 'organic' compounds of carbon, see Table A1.1)					
C(s) (graphite)	12.011	0	0	5.740	8.527
C(s) (diamond)	12.011	+1.895	+2.900	2.377	6.133
C(g)	12.011	+716.68	+671.26	158.10	20.838
C$_2$(g)	24.022	+831.90	+775.89	199.42	43.21

	$M/$ (g mol^{-1})	$\Delta_f H^{\ominus}/$ (kJ mol^{-1})	$\Delta_f G^{\ominus}/$ (kJ mol^{-1})	$S_m^{\ominus}/$ (J K^{-1}mol^{-1})	$C_{p,m}/$ (J K^{-1}mol^{-1})
CO(g)	28.011	−110.53	−137.17	197.67	29.14
CO$_2$(g)	44.010	−393.51	−394.36	213.74	37.11
CO$_2$(aq)	44.010	−413.80	−385.98	117.6	
H$_2$CO$_3$(aq)	62.03	−699.65	−623.08	187.4	
HCO$_3^-$(aq)	61.02	−691.99	−586.77	+91.2	
CO$_3^{2-}$(aq)	60.01	−677.14	−527.81	−56.9	
CCl$_4$(l)	153.82	−135.44	−65.21	216.40	131.75
CS$_2$(l)	76.14	+89.70	+65.27	151.34	75.7
HCN(g)	27.03	+135.1	+124.7	201.78	35.86
HCN(l)	27.03	+108.87	+124.97	112.84	70.63
CN$^-$(aq)	26.02	+150.6	+172.4	+94.1	
Chlorine					
Cl$_2$(g)	70.91	0	0	223.07	33.91
Cl(g)	35.45	+121.68	+105.68	165.20	21.840
Cl$^-$(g)	35.45	−233.13			
Cl$^-$(aq)	35.45	−167.16	−131.23	+56.5	−136.4
HCl(g)	36.46	−92.31	−95.30	186.91	29.12
HCl(aq)	36.46	−167.16	−131.23	56.5	−136.4
Chromium					
Cr(s)	52.00	0	0	23.77	23.35
Cr(g)	52.00	+396.6	+351.8	174.50	20.79
CrO$_4^{2-}$(aq)	115.99	−881.15	−727.75	+50.21	
Cr$_2$O$_7^{2-}$(aq)	215.99	−1490.3	−1301.1	+261.9	
Copper					
Cu(s)	63.54	0	0	33.150	24.44
Cu(g)	63.54	+338.32	+298.58	166.38	20.79
Cu$^+$(aq)	63.54	+71.67	+49.98	+40.6	
Cu^{2+}(aq)	63.54	+64.77	+65.49	−99.6	
Cu$_2$O(s)	143.08	−168.6	−146.0	93.14	63.64
CuO(s)	79.54	−157.3	−129.7	42.63	42.30
CuSO$_4$(s)	159.60	−771.36	−661.8	109	100.0
CuSO$_4$·H$_2$O(s)	177.62	−1085.8	−918.11	146.0	134
CuSO$_4$·5H$_2$O(s)	249.68	−2279.7	−1879.7	300.4	280
Deuterium					
D$_2$(g)	4.028	0	0	144.96	29.20
HD(g)	3.022	+0.318	−1.464	143.80	29.196
D$_2$O(g)	20.028	−249.20	−234.54	198.34	34.27
D$_2$O(l)	20.028	−294.60	−243.44	75.94	84.35
HDO(g)	19.022	−245.30	−233.11	199.51	33.81
HDO(l)	19.022	−289.89	−241.86	79.29	
Fluorine					
F$_2$(g)	38.00	0	0	202.78	31.30
F(g)	19.00	+78.99	+61.91	158.75	22.74
F$^-$(aq)	19.00	−332.63	−278.79	−13.8	−106.7
HF(g)	20.01	−271.1	−273.2	173.78	29.13

	$M/$ (g mol^{-1})	$\Delta_f H^{\ominus}/$ (kJ mol^{-1})	$\Delta_f G^{\ominus}/$ (kJ mol^{-1})	$S_m^{\ominus}/$ (J K^{-1}mol^{-1})	$C_{p,m}/$ (J K^{-1}mol^{-1})
Gold					
Au(s)	196.97	0	0	47.40	25.42
Au(g)	196.97	+366.1	+326.3	180.50	20.79
Helium					
He(g)	4.003	0	0	126.15	20.786
Hydrogen (see also deuterium)					
H_2(g)	2.016	0	0	130.684	28.824
H(g)	1.008	+217.97	+203.25	114.71	20.784
H^+(aq)	1.008	0	0	0	0
H_2O(l)	18.015	−285.83	−237.13	69.91	75.291
H_2O(g)	18.015	−241.82	−228.57	188.83	33.58
H_2O_2(l)	34.015	−187.78	−120.35	109.6	89.1
Iodine					
I_2(s)	253.81	0	0	116.135	54.44
I_2(g)	253.81	+62.44	+19.33	260.69	36.90
I(g)	126.90	+106.84	+70.25	180.79	20.786
I^-(aq)	126.90	−55.19	−51.57	+111.3	−142.3
HI(g)	127.91	+26.48	+1.70	206.59	29.158
Iron					
Fe(s)	55.85	0	0	27.28	25.10
Fe(g)	55.85	+416.3	+370.7	180.49	25.68
Fe^{2+}(aq)	55.85	−89.1	−78.90	−137.7	
Fe^{3+}(aq)	55.85	−48.5	−4.7	−315.9	
Fe_3O_4(s) (magnetite)	231.54	−1184.4	−1015.4	146.4	143.43
Fe_2O_3(s) (haematite)	159.69	−824.2	−742.2	87.40	103.85
FeS(s, α)	87.91	−100.0	−100.4	60.29	50.54
FeS_2(s)	119.98	−178.2	−166.9	52.93	62.17
Krypton					
Kr(g)	83.80	0	0	164.08	20.786
Lead					
Pb(s)	207.19	0	0	64.81	26.44
Pb(g)	207.19	+195.0	+161.9	175.37	20.79
Pb^{2+}(aq)	207.19	−1.7	−24.43	+10.5	
PbO(s, yellow)	223.19	−217.32	−187.89	68.70	45.77
PbO(s, red)	223.19	−218.99	−188.93	66.5	45.81
PbO_2(s)	239.19	−277.4	−217.33	68.6	64.64
Lithium					
Li(s)	6.94	0	0	29.12	24.77
Li(g)	6.94	+159.37	+126.66	138.77	20.79
Li^+(aq)	6.94	−278.49	−293.31	+13.4	+68.6
Magnesium					
Mg(s)	24.31	0	0	32.68	24.89
Mg(g)	24.31	+147.70	+113.10	148.65	20.786

	$M/$ (g mol^{-1})	$\Delta_f H^{\ominus}/$ (kJ mol^{-1})	$\Delta_f G^{\ominus}/$ (kJ mol^{-1})	$S_m^{\ominus}/$ (J K^{-1}mol^{-1})	$C_{p,m}/$ (J K^{-1}mol^{-1})
$Mg^{2+}(aq)$	24.31	−466.85	−454.8	−138.1	
$MgO(s)$	40.31	−601.70	−569.43	26.94	37.15
$MgCO_3(s)$	84.32	−1095.8	−1012.1	65.7	75.52
$MgCl_2(s)$	95.22	−641.32	−591.79	89.62	71.38
$MgBr_2(s)$	184.13	−524.3	−503.8	117.2	
Mercury					
$Hg(l)$	200.59	0	0	76.02	27.983
$Hg(g)$	200.59	+61.32	+31.82	174.96	20.786
$Hg^{2+}(aq)$	200.59	+171.1	+164.40	−32.2	
$Hg_2^{2+}(aq)$	401.18	+172.4	+153.52	+84.5	
$HgO(s)$	216.59	−90.83	−58.54	70.29	44.06
$Hg_2Cl_2(s)$	472.09	−265.22	−210.75	192.5	102
$HgCl_2(s)$	271.50	−224.3	−178.6	146.0	
$HgS(s, black)$	232.65	−53.6	−47.7	88.3	
Neon					
$Ne(g)$	20.18	0	0	146.33	20.786
Nitrogen					
$N_2(g)$	28.013	0	0	191.61	29.125
$N(g)$	14.007	+472.70	+455.56	153.30	20.786
$NO(g)$	30.01	+90.25	+86.55	210.76	29.844
$N_2O(g)$	44.01	+82.05	+104.20	219.85	38.45
$NO_2(g)$	46.01	+33.18	+51.31	240.06	37.20
$N_2O_4(g)$	92.01	+9.16	+97.89	304.29	77.28
$N_2O_5(s)$	108.01	−43.1	+113.9	178.2	143.1
$N_2O_5(g)$	108.01	+11.3	+115.1	355.7	84.5
$HNO_3(l)$	63.01	−174.10	−80.71	155.60	109.87
$HNO_3(aq)$	63.01	−207.36	−111.25	146.4	−86.6
$NO_3^-(aq)$	62.01	−205.0	−108.74	+146.4	−86.6
$NH_3(g)$	17.03	−46.11	−16.45	192.45	35.06
$NH_3(aq)$	17.03	−80.29	−26.50	113.3	
$NH_4^+(aq)$	18.04	−132.51	−79.31	+113.4	+79.9
$NH_2OH(s)$	33.03	−114.2			
$HN_3(l)$	43.03	+264.0	+327.3	140.6	43.68
$HN_3(g)$	43.03	+294.1	+328.1	238.97	98.87
$N_2H_4(l)$	32.05	+50.63	+149.43	121.21	139.3
$NH_4NO_3(s)$	80.04	−365.56	−183.87	151.08	84.1
$NH_4Cl(s)$	53.49	−314.43	−202.87	94.6	
Oxygen					
$O_2(g)$	31.999	0	0	205.138	29.355
$O(g)$	15.999	+249.17	+231.73	161.06	21.912
$O_3(g)$	47.998	+142.7	+163.2	238.93	39.20
$OH^-(aq)$	17.007	−229.99	−157.24	−10.75	−148.5
Phosphorus					
$P(s, wh)$	30.97	0	0	41.09	23.840
$P(g)$	30.97	+314.64	+278.25	163.19	20.786
$P_2(g)$	61.95	+144.3	+103.7	218.13	32.05

	$M/$ (g mol^{-1})	$\Delta_f H^{\ominus}/$ (kJ mol^{-1})	$\Delta_f G^{\ominus}/$ (kJ mol^{-1})	$S_m^{\ominus}/$ (J K^{-1} mol^{-1})	$C_{p,m}/$ (J K^{-1} mol^{-1})
Phosphorus (continued)					
$P_4(g)$	123.90	+58.91	+24.44	279.98	67.15
$PH_3(g)$	34.00	+5.4	+13.4	210.23	37.11
$PCl_3(g)$	137.33	−287.0	−267.8	311.78	71.84
$PCl_3(l)$	137.33	−319.7	−272.3	217.1	
$PCl_5(g)$	208.24	−374.9	−305.0	364.6	112.8
$PCl_5(s)$	208.24	−443.5			
$H_3PO_3(s)$	82.00	−964.4			
$H_3PO_3(aq)$	82.00	−964.8			
$H_3PO_4(s)$	94.97	−1279.0	−1119.1	110.50	106.06
$H_3PO_4(l)$	94.97	−1266.9			
$H_3PO_4(aq)$	94.97	−1277.4	−1018.7	−222	
$PO_4^{3-}(aq)$	94.97	−1277.4	−1018.7	−222	
$P_4O_{10}(s)$	283.89	−2984.0	−2697.0	228.86	211.71
$P_4O_6(s)$	219.89	−1640.1			
Potassium					
$K(s)$	39.10	0	0	64.18	29.58
$K(g)$	39.10	+89.24	+60.59	160.336	20.786
$K^+(g)$	39.10	+514.26			
$K^+(aq)$	39.10	−252.38	−283.27	+102.5	+21.8
$KOH(s)$	56.11	−424.76	−379.08	78.9	64.9
$KF(s)$	58.10	−576.27	−537.75	66.57	49.04
$KCl(s)$	74.56	−436.75	−409.14	82.59	51.30
$KBr(s)$	119.01	−393.80	−380.66	95.90	52.30
$KI(s)$	166.01	−327.90	−324.89	106.32	52.93
Silicon					
$Si(s)$	28.09	0	0	18.83	20.00
$Si(g)$	28.09	+455.6	+411.3	167.97	22.25
$SiO_2(s, \alpha)$	60.09	−910.93	−856.64	41.84	44.43
Silver					
$Ag(s)$	107.87	0	0	42.55	25.351
$Ag(g)$	107.87	+284.55	+245.65	173.00	20.79
$Ag^+(aq)$	107.87	+105.58	+77.11	+72.68	+21.8
$AgBr(s)$	187.78	−100.37	−96.90	107.1	52.38
$AgCl(s)$	143.32	−127.07	−109.79	96.2	50.79
$Ag_2O(s)$	231.74	−31.05	−11.20	121.3	65.86
$AgNO_3(s)$	169.88	−124.39	−33.41	140.92	93.05
Sodium					
$Na(s)$	22.99	0	0	51.21	28.24
$Na(g)$	22.99	+107.32	+76.76	153.71	20.79
$Na^+(aq)$	22.99	−240.12	−261.91	+59.0	+46.4
$NaOH(s)$	40.00	−425.61	−379.49	64.46	59.54
$NaCl(s)$	58.44	−411.15	−384.14	72.13	50.50
$NaBr(s)$	102.90	−361.06	−348.98	86.82	51.38
$NaI(s)$	149.89	−287.78	−286.06	98.53	52.09

	$M/$ (g mol^{-1})	$\Delta_f H^{\ominus}/$ (kJ mol^{-1})	$\Delta_f G^{\ominus}/$ (kJ mol^{-1})	$S_m^{\ominus}/$ (J K^{-1}mol^{-1})	$C_{p,m}/$ (J K^{-1}mol^{-1})
Sulfur					
S(s, α) (rhombic)	32.06	0	0	31.80	22.64
S(s, β) (monoclinic)	32.06	+0.33	+0.1	32.6	23.6
S(g)	32.06	+278.81	+238.25	167.82	23.673
S$_2$(g)	64.13	+128.37	+79.30	228.18	32.47
S^{2-}(aq)	32.06	+33.1	+85.8	−14.6	
SO$_2$(g)	64.06	−296.83	−300.19	248.22	39.87
SO$_3$(g)	80.06	−395.72	−371.06	256.76	50.67
H$_2$SO$_4$(l)	98.08	−813.99	−690.00	156.90	138.9
H$_2$SO$_4$(aq)	98.08	−909.27	−744.53	20.1	−293
SO$_4^{2-}$(aq)	96.06	−909.27	−744.53	+20.1	−293
HSO$_4^-$(aq)	97.07	−887.34	−755.91	+131.8	−84
H$_2$S(g)	34.08	−20.63	−33.56	205.79	34.23
H$_2$S(aq)	34.08	−39.7	−27.83	121	
HS$^-$(aq)	33.072	−17.6	+12.08	+62.08	
SF$_6$(g)	146.05	−1209	−1105.3	291.82	97.28
Tin					
Sn(s, β)	118.69	0	0	51.55	26.99
Sn(g)	118.69	+302.1	+267.3	168.49	20.26
Sn^{2+}(aq)	118.69	−8.8	−27.2	−17	
SnO(s)	134.69	−285.8	−256.8	56.5	44.31
SnO$_2$(s)	150.69	−580.7	+519.6	52.3	52.59
Xenon					
Xe(g)	131.30	0	0	169.68	20.786
Zinc					
Zn(s)	65.37	0	0	41.63	25.40
Zn(g)	65.37	+130.73	+95.14	160.98	20.79
Zn^{2+}(aq)	65.37	−153.89	−147.06	−112.1	+46
ZnO(s)	81.37	−348.28	−318.30	43.64	40.25

* Entropies and heat capacities of ions are relative to H$^+$(aq) and are given with a sign.

APPENDIX 2

Table A2.1a   Standard potentials at 298.15 K in electrochemical order

Reduction half-reaction	$E^{\ominus}/V$	Reduction half-reaction	$E^{\ominus}/V$
Strongly oxidizing		$BrO^- + H_2O + 2e^- \rightarrow Br^- + 2OH^-$	+0.76
$H_4XeO_6 + 2H^+ + 2e^- \rightarrow XeO_3 + 3H_2O$	+3.0	$Hg_2SO_4 + 2e^- \rightarrow 2Hg + SO_4^{2-}$	+0.62
$F_2 + 2e^- \rightarrow 2F^-$	+2.87	$MnO_4^{2-} + 2H_2O + 2e^- \rightarrow MnO_2 + 4OH^-$	+0.60
$O_3 + 2H^+ + 2e^- \rightarrow O_2 + H_2O$	+2.07	$MnO_4^- + e^- \rightarrow MnO_4^{2-}$	+0.56
$S_2O_8^{2-} + 2e^- \rightarrow 2SO_4^{2-}$	+2.05	$I_2 + 2e^- \rightarrow 2I^-$	+0.54
$Ag^{2+} + e^- \rightarrow Ag^+$	+1.98	$Cu^+ + e^- \rightarrow Cu$	+0.52
$Co^{3+} + e^- \rightarrow Co^{2+}$	+1.81	$I_3^- + 2e^- \rightarrow 3I^-$	+0.53
$HO_2 + 2H^+ + 2e^- \rightarrow 2H_2O$	+1.78	$NiOOH + H_2O + e^- \rightarrow Ni(OH)_2 + OH^-$	+0.49
$Au^+ + e^- \rightarrow Au$	+1.69	$Ag_2CrO_4 + 2e^- \rightarrow 2Ag + CrO_4^{2-}$	+0.45
$Pb^{4+} + 2e^- \rightarrow Pb^{2+}$	+1.67	$O_2 + 2H_2O + 4e^- \rightarrow 4OH^-$	+0.40
$2HClO + 2H^+ + 2e^- \rightarrow Cl_2 + 2H_2O$	+1.63	$ClO_4^- + H_2O + 2e^- \rightarrow ClO_3^- + 2OH^-$	+0.36
$Ce^{4+} + e^- \rightarrow Ce^{3+}$	+1.61	$[Fe(CN)_6]^{3-} + e^- \rightarrow [Fe(CN)_6]^{4-}$	+0.36
$2HBrO + 2H^+ + 2e^- \rightarrow Br_2 + 2H$	+1.60	$Cu^{2+} + 2e^- \rightarrow Cu$	+0.34
$MnO_4^- + 8H^+ + 5e^- \rightarrow Mn^{2+} + 4H_2O$	+1.51	$Hg_2Cl_2 + 2e^- \rightarrow 2Hg + 2Cl^-$	+0.27
$Mn^{3+} + e^- \rightarrow Mn^{2+}$	+1.51	$AgCl + e^- \rightarrow Ag + Cl^-$	+0.22
$Au^{3+} + 3e^- \rightarrow Au$	+1.40	$Bi^{3+} + 3e^- \rightarrow Bi$	+0.20
$Cl_2 + 2e^- \rightarrow 2Cl^-$	+1.36	$Cu^{2+} + e^- \rightarrow Cu^+$	+0.16
$Cr_2O_7^{2-} + 14H^+ + 6e^- \rightarrow 2Cr^{3+} + 7H_2O$	+1.33	$Sn^{4+} + 2e^- \rightarrow Sn^{2+}$	+0.15
$O_3 + H_2O + 2e^- \rightarrow O_2 + 2OH^-$	+1.24	$AgBr + e^- \rightarrow Ag + Br^-$	+0.07
$O_2 + 4H^+ + 4e^- \rightarrow 2H_2O$	+1.23	$Ti^{4+} + e^- \rightarrow Ti^{3+}$	0.00
$ClO_4^- + 2H^+ + 2e^- \rightarrow ClO_3^- + H_2O$	+1.23	$2H^+ + 2e^- \rightarrow H$	0, by definition
$MnO_2 + 4H^+ + 2e^- \rightarrow Mn^{2+} + 2H_2O$	+1.23	$Fe^{3+} + 3e^- \rightarrow Fe$	−0.04
$Br_2 + 2e^- \rightarrow 2Br^-$	+1.09	$O_2 + H_2O + 2e^- \rightarrow HO_2^- + OH^-$	−0.08
$Pu^{4+} + e^- \rightarrow Pu^{3+}$	+0.97	$Pb^{2+} + 2e^- \rightarrow Pb$	−0.13
$NO_3^- + 4H^+ + 3e^- \rightarrow NO + 2H_2O$	+0.96	$In^+ + e^- \rightarrow In$	−0.14
$2Hg^{2+} + 2e^- \rightarrow Hg_2^{2+}$	+0.92	$Sn^{2+} + 2e^- \rightarrow Sn$	−0.14
$ClO^- + H_2O + 2e^- \rightarrow Cl^- + 2OH^-$	+0.89	$AgI + e^- \rightarrow Ag + I^-$	−0.15
$Hg^{2+} + 2e^- \rightarrow Hg$	+0.86	$Ni^{2+} + 2e^- \rightarrow Ni$	−0.23
$NO_3^- + 2H^+ + e^- \rightarrow NO_2 + H_2O$	+0.80	$Co^{2+} + 2e^- \rightarrow Co$	−0.28
$Ag^+ + e^- \rightarrow Ag$	+0.80	$In^{3+} + 3e^- \rightarrow In$	−0.34
$Hg_2^{2+} + 2e^- \rightarrow 2Hg$	+0.79	$Tl^+ + e^- \rightarrow Tl$	−0.34
$Fe^{3+} + e^- \rightarrow Fe^{2+}$	+0.77	$PbSO_4 + 2e^- \rightarrow Pb + SO_4^{2-}$	−0.36

Table A2.1a (continued)

Reduction half-reaction	$E^{\ominus}/V$	Reduction half-reaction	$E^{\ominus}/V$
$Ti^{3+} + e^- \rightarrow Ti^{2+}$	−0.37	$Ti^{2+} + 2e^- \rightarrow Ti$	−1.63
$Cd^{2+} + 2e^- \rightarrow Cd$	−0.40	$Al^{3+} + 3e^- \rightarrow Al$	−1.66
$In^{2+} + e^- \rightarrow In^+$	−0.40	$U^{3+} + 3e^- \rightarrow U$	−1.79
$Cr^{3+} + e^- \rightarrow Cr^{2+}$	−0.41	$Mg^{2+} + 2e^- \rightarrow Mg$	−2.36
$Fe^{2+} + 2e^- \rightarrow Fe$	−0.44	$Ce^{3+} + 3e^- \rightarrow Ce$	−2.48
$In^{3+} + 2e^- \rightarrow In^+$	−0.44	$La^{3+} + 3e^- \rightarrow La$	−2.52
$S + 2e^- \rightarrow S^{2-}$	−0.48	$Na^+ + e^- \rightarrow Na$	−2.71
$In^{3+} + e^- \rightarrow In^{2+}$	−0.49	$Ca^{2+} + 2e^- \rightarrow Ca$	−2.87
$U^{4+} + e^- \rightarrow U^{3+}$	−0.61	$Sr^{2+} + 2e^- \rightarrow Sr$	−2.89
$Cr^{3+} + 3e^- \rightarrow Cr$	−0.74	$Ba^{2+} + 2e^- \rightarrow Ba$	−2.91
$Zn^{2+} + 2e^- \rightarrow Zn$	−0.76	$Ra^{2+} + 2e^- \rightarrow Ra$	−2.92
$Cd(OH)_2 + 2e^- \rightarrow Cd + 2OH^-$	−0.81	$Cs^+ + e^- \rightarrow Cs$	−2.92
$2H_2O + 2e^- \rightarrow H_2 + 2OH^-$	−0.83	$Rb^+ + e^- \rightarrow Rb$	−2.93
$Cr^{2+} + 2e^- \rightarrow Cr$	−0.91	$K^+ + e^- \rightarrow K$	−2.93
$Mn^{2+} + 2e^- \rightarrow Mn$	−1.18	$Li^+ + e^- \rightarrow Li$	−3.05
$V^{2+} + 2e^- \rightarrow V$	−1.19	**Strongly reducing**	

Table A2.1b Standard potentials at 298.15 K in alphabetical order

Reduction half-reaction	$E^{\ominus}/V$	Reduction half-reaction	$E^{\ominus}/V$
$Ag^+ + e^- \rightarrow Ag$	+0.80	$Co^{2+} + 2e^- \rightarrow Co$	−0.28
$Ag^{2+} + e^- \rightarrow Ag^+$	+1.98	$Co^{3+} + e^- \rightarrow Co^{2+}$	+1.81
$AgBr + e^- \rightarrow Ag + Br^-$	+0.0713	$Cr^{2+} + 2e^- \rightarrow Cr$	−0.91
$AgCl + e^- \rightarrow Ag + Cl^-$	+0.22	$Cr_2O_7^{2-} + 14H^+ + 6e^- \rightarrow 2Cr^{3+} + 7H_2O$	+1.33
$Ag_2CrO_4 + 2e^- \rightarrow 2Ag + CrO_4^{2-}$	+0.45	$Cr^{3+} + 3e^- \rightarrow Cr$	−0.74
$AgF + e^- \rightarrow Ag + F^-$	+0.78	$Cr^{3+} + e^- \rightarrow Cr^{2+}$	−0.41
$AgI + e^- \rightarrow Ag + I^-$	−0.15	$Cs^+ + e^- \rightarrow Cs$	−2.92
$Al^{3+} + 3e^- \rightarrow Al$	−1.66	$Cu^+ + e^- \rightarrow Cu$	+0.52
$Au^+ + e^- \rightarrow Au$	+1.69	$Cu^{2+} + 2e^- \rightarrow Cu$	+0.34
$Au^{3+} + 3e^- \rightarrow Au$	+1.40	$Cu^{2+} + e^- \rightarrow Cu^+$	+0.16
$Ba^{2+} + 2e^- \rightarrow Ba$	−2.91	$F_2 + 2e^- \rightarrow 2F^-$	+2.87
$Be^{2+} + 2e^- \rightarrow Be$	−1.85	$Fe^{2+} + 2e^- \rightarrow Fe$	−0.44
$Bi^{3+} + 3e^- \rightarrow Bi$	+0.20	$Fe^{3+} + 3e^- \rightarrow Fe$	−0.04
$Br_2 + 2e^- \rightarrow 2Br^-$	+1.09	$Fe^{3+} + e^- \rightarrow Fe^{2+}$	+0.77
$BrO^- + H_2O + 2e^- \rightarrow Br^- + 2OH^-$	+0.76	$[Fe(CN)_6]^{3-} + e^- \rightarrow [Fe(CN)_6]^{4-}$	+0.36
$Ca^{2+} + 2e^- \rightarrow Ca$	−2.87	$2H^+ + 2e^- \rightarrow H_2$	0, by definition
$Cd(OH)_2 + 2e^- \rightarrow Cd + 2OH^-$	−0.81	$2H_2O + 2e^- \rightarrow H_2 + 2OH^-$	−0.83
$Cd^{2+} + 2e^- \rightarrow Cd$	−0.40	$2HBrO + 2H^+ + 2e^- \rightarrow Br_2 + 2H_2O$	+1.60
$Ce^{3+} + 3e^- \rightarrow Ce$	−2.48	$2HClO + 2H^+ + 2e^- \rightarrow Cl_2 + 2H_2O$	+1.63
$Ce^{4+} + e^- \rightarrow Ce^{3+}$	+1.61	$H_2O_2 + 2H^+ + 2e^- \rightarrow 2H_2O$	+1.78
$Cl_2 + 2e^- \rightarrow 2Cl^-$	+1.36	$H_4XeO_6 + 2H^+ + 2e^- \rightarrow XeO_3 + 3H_2O$	+3.0
$ClO^- + H_2O + 2e^- \rightarrow Cl^- + 2OH^-$	+0.89	$Hg_2^{2+} + 2e^- \rightarrow 2Hg$	+0.79
$ClO_4^- + 2H^+ + 2e^- \rightarrow ClO_3^- + H_2O$	+1.23	$Hg_2Cl_2 + 2e^- \rightarrow 2Hg + 2Cl^-$	+0.27
$ClO_4^- + H_2O + 2e^- \rightarrow ClO_3^- + 2OH^-$	+0.36	$Hg^{2+} + 2e^- \rightarrow Hg$	+0.86

Table A2.1b   (continued)

Reduction half-reaction	$E^{\ominus}/V$	Reduction half-reaction	$E^{\ominus}/V$
$2Hg^{2+} + 2e^- \rightarrow Hg_2^{2+}$	+0.92	$O_2 + 4H^+ + 4e^- \rightarrow 2H_2O$	+1.23
$Hg_2SO_4 + 2e^- \rightarrow 2Hg + SO_4^{2-}$	+0.62	$O_2 + e^- \rightarrow O_2^-$	−0.56
$I_2 + 2e^- \rightarrow 2I^-$	+0.54	$O_2 + H_2O + 2e^- \rightarrow HO_2^- + OH^-$	−0.08
$I_3^- + 2e^- \rightarrow 3I^-$	+0.53	$O_3 + 2H^+ + 2e^- \rightarrow O_2 + H_2O$	+2.07
$In^+ + e^- \rightarrow In$	−0.14	$O_3 + H_2O + 2e^- \rightarrow O_2 + 2OH^-$	+1.24
$In^{2+} + e^- \rightarrow In^+$	−0.40	$Pb^{2+} + 2e^- \rightarrow Pb$	−0.13
$In^{3+} + 2e^- \rightarrow In^+$	−0.44	$Pb^{4+} + 2e^- \rightarrow Pb^{2+}$	+1.67
$In^{3+} + 3e^- \rightarrow In$	−0.34	$PbSO_4 + 2e^- \rightarrow Pb + SO_4^{2-}$	−0.36
$In^{3+} + e^- \rightarrow In^{2+}$	−0.49	$Pt^{2+} + 2e^- \rightarrow Pt$	+1.20
$K^+ + e^- \rightarrow K$	−2.93	$Pu^{4+} + e^- \rightarrow Pu^{3+}$	+0.97
$La^{3+} + 3e^- \rightarrow La$	−2.52	$Ra^{2+} + 2e^- \rightarrow Ra$	−2.92
$Li^+ + e^- \rightarrow Li$	−3.05	$Rb^+ + e^- \rightarrow Rb$	−2.93
$Mg^{2+} + 2e^- \rightarrow Mg$	−2.36	$S + 2e^- \rightarrow S^{2-}$	−0.48
$Mn^{2+} + 2e^- \rightarrow Mn$	−1.18	$S_2O_8^{2-} + 2e^- \rightarrow 2SO_4^{2-}$	+2.05
$Mn^{3+} + e^- \rightarrow Mn^{2+}$	+1.51	$Sn^{2+} + 2e^- \rightarrow Sn$	−0.14
$MnO_2 + 4H^+ + 2e^- \rightarrow Mn^{2+} + 2H_2O$	+1.23	$Sn^{4+} + 2e^- \rightarrow Sn^{2+}$	+0.15
$MnO_4^- + 8H^+ + 5e^- \rightarrow Mn^{2+} + 4H_2O$	+1.51	$Sr^{2+} + 2e^- \rightarrow Sr$	−2.89
$MnO_4^- + e^- \rightarrow MnO_4^{2-}$	+0.56	$Ti^{2+} + 2e^- \rightarrow Ti$	−1.63
$MnO_4^{2-} + 2H_2O + 2e^- \rightarrow MnO_2 + 4OH^-$	+0.60	$Ti^{3+} + e^- \rightarrow Ti^{2+}$	−0.37
$Na^+ + e^- \rightarrow Na$	−2.71	$Ti^{4+} + e^- \rightarrow Ti^{3+}$	0.00
$Ni^{2+} + 2e^- \rightarrow Ni$	−0.23	$Tl^+ + e^- \rightarrow Tl$	−0.34
$NiOOH + H_2O + e^- \rightarrow Ni(OH)_2 + OH^-$	+0.49	$U^{3+} + 3e^- \rightarrow U$	−1.79
$NO_3^- + 2H^+ + e^- \rightarrow NO_2 + H_2O$	+0.80	$U^{4+} + e^- \rightarrow U^{3+}$	−0.61
$NO_3^- + 3H^+ + 3e^- \rightarrow NO + 2H_2O$	+0.96	$V^{2+} + 2e^- \rightarrow V$	−1.19
$NO_3^- + H_2O + 2e^- \rightarrow NO_2^- + 2OH^-$	+0.10	$V^{3+} + e^- \rightarrow V^{2+}$	−0.26
$O_2 + 2H_2O + 4e^- \rightarrow 4OH^-$	+0.40	$Zn^{2+} + 2e^- \rightarrow Zn$	−0.76

APPENDIX 3: THE AMINO ACIDS

Amino acid, R–CH(NH$_2$)COOH	Structure, R =	Abbreviations Three-letter	One-letter
Alanine	CH$_3$–	Ala	A
Arginine	(guanidino structure)	Arg	R
Asparagine	H$_2$NCOCH$_2$–	Asn	N
Aspartic acid	HOOCCH$_2$–	Asp	D
Asparagine or aspartic acid		Asx	B
Cysteine	HSCH$_2$–	Cys	C
Glutamine	H$_2$NCO(CH$_2$)$_2$–	Gln	Q
Glutamic acid	HOOC(CH$_2$)$_2$–	Glu	E
Glutamine or glutamic acid		Glx	Z
Glycine	H–	Gly	G
Histidine	(imidazole structure, H$_2$C–)	His	H
Isoleucine	CH$_3$CH$_2$CH(CH$_3$)–	Ile	I
Leucine	(CH$_3$)$_2$CHCH$_2$–	Leu	L
Lysine	NH$_2$(CH$_2$)$_4$–	Lys	K
Methionine	CH$_3$SCH$_2$CH$_2$–	Met	M
Phenylalanine	C$_6$H$_5$CH$_2$–	Phe	F
Proline	(pyrrolidine structure, HN–, COOH) [complete acid]	Pro	P
Serine	HOCH$_2$–	Ser	S
Threonine	HOCH(CH$_2$)–	Thr	T
Tryptophan	(indole structure, CH$_2$)	Trp	W
Tyrosine	(hydroxyphenyl structure, CH$_2$, OH)	Tyr	Y
Valine	(CH$_3$)$_2$CH–	Val	V

ANSWERS TO EXERCISES

0.1 (a) 824 Torr; (b) 0.984 atm; (c) 0.212 atm; (d) 9.64×10^4 Pa

0.2 1.24×10^8 Pa ($= 1.22 \times 10^3$ atm)

0.3 1.6×10^{-2}

0.4 1.5×10^3 Pa ($= 1.5 \times 10^{-2}$ atm)

0.5 755 Torr

0.6 $-459.67\ ^\circ$F

0.7 $671.67\ ^\circ$R

0.8 3.74×10^{19} molecules

0.9 0.97 or 97%

1.1 89.2 kPa

1.2 4.22×10^{-2} atm

1.3 2.52×10^{-3} mol

1.4 6.64×10^3 kPa

1.5 10.0 atm

1.6 418 kPa

1.7 173 kPa

1.8 29.5 K

1.9 388 K

1.10 (a) 3.6 m^3; (b) 178 m^3

1.11 0.50 m^3

1.12 3.2×10^{-2} atm

1.13 (a) 1.32 L; (b) 61.2 kPa

1.14 713 Torr

1.15 132 g mol^{-1}

1.16 16.4 g mol^{-1}

1.17 $p(H_2) = 2.0$ atm, $p(N_2) = 1.0$ atm; (b) 3.0 atm

1.18 (a) (i) 693 m s^{-1}, (ii) 1363 m s^{-1}, (iii) 2497 m s^{-1}; (b) (i) 346 m s^{-1}, (ii) 681 m s^{-1}, (iii) 1247 m s^{-1}

1.19 0.065 Pa

1.20 2.4×10^6 Pa

1.21 0.97 μm

1.22 (a) 5.3×10^{10}; (b) 5.3×10^9 collisions; (c) 5.3×10^4

1.23 (a) 6.4×10^{33} collisions; (b) 6.4×10^{31} collisions; (c) 6.4×10^{21} collisions

1.24 4.5×10^8 collisions/s

1.25 (a) 6.8 nm; (b) 68 nm; (c) 7 mm

1.26 0.91%

1.27 independent of temperature

1.28 (a) 8.3×10^2 atm; (b) (i) 0.99 atm, (ii) 1.7×10^3 atm

1.29 4.37 MPa

1.30
$$p = \frac{RT}{V_m}\left(1 + \frac{B}{V_m} + \frac{C}{V_m^2} + \ldots\right)$$
$$B = b - \frac{a}{RT} \text{ and } C = b^2$$

1.31 1.26 L^2 atm mol^{-2}, 34.6 cm^3 mol^{-1}

1.32 1.02×10^3 K

2.1 (a) 98 J; (b) 16 J

2.2 39 J

2.3 2.6 kJ

2.4 3.03 J

2.5 -14 J

2.6 -1.0×10^2 J

2.7 (a) 895 J; (b) 899 J

2.8 (a) -88 J; (b) -167 J

2.9 $+123$ J

2.10 $+2.99$ kJ

2.11 -1.25 kJ

2.12 4.4×10^4 J

2.13 23.7 J K^{-1}

2.14 42 kJ

2.15 30 J K^{-1} mol^{-1}, 38 J K^{-1} mol^{-1}

2.16 9.2×10^2 k J, 6.1×10^2 s

2.17 1.86×10^3 J

2.18 773 J

2.19 -8.0 J

2.20 29.3 J

2.21 (a) 2.479 kJ mol^{-1}; (b) decreased

2.22 -1.2 kJ, 80 J K^{-1}

2.23 $q = +2.2$ kJ; $\Delta H = +2.2$ kJ; $\Delta U = +1.6$ kJ

2.24 20.83 J K^{-1} mol^{-1}

2.25 641 J mol^{-1}, 458 J mol^{-1}

2.26 818 J mol^{-1}

3.1 2.53×10^4 kJ

3.2 (a) 2.44×10^3 kJ; (b) 2.26×10^3 kJ

3.3 39.8 kJ mol^{-1}

3.4 $q = +39.0$ kJ; $w = -3.12$ kJ; $\Delta H = +39.0$ kJ (constant pressure); $\Delta U = 35.9$ kJ

3.5 301 kJ

3.6 478 kJ

3.7 (a) 388 kJ mol^{-1}; (b) smaller

3.8 (a) 16 kJ mol^{-1}; (b) -3028 kJ mol^{-1}

3.9 (a) -92.22 kJ, (b) -46.11 kJ

3.10 (a) -1560 kJ mol^{-1}; (b) -2340 kJ

3.11 -4564.7 kJ mol^{-1}

3.12 -85 kJ mol^{-1}

3.13 -432 kJ mol^{-1}

3.14 $+79$ kJ mol^{-1}

3.15 4.22 kJK^{-1}, 0.769 K

3.16 (a) -2.80 MJ mol^{-1}; (b) -2.80 MJ mol^{-1}; (c) -1.28 MJ mol^{-1}

3.17 (a) -1333 kJ mol^{-1}; (b) -1331 kJ mol^{-1}; (c) -815 kJ mol^{-1}

3.18 $+84.40$ kJ mol^{-1}

3.19 -383 kJ mol^{-1}

3.20 $+1.9$ kJ mol^{-1}

3.21 $+30.6$ kJ mol^{-1}

3.22 -25 kJ, 7.5 m

3.23 (a) -2205 kJ mol^{-1}; (b) -2200 kJ mol^{-1}

3.24 (a) exothermic, $\Delta_r H^\ominus$ = negative; (b) endothermic, $\Delta H^\ominus$ = positive; (c) endothermic, $\Delta_{vap} H^\ominus$ = positive; (d) endothermic, $\Delta_{fus} H^\ominus$ = positive; (e) endothermic, $\Delta_{sub} H^\ominus$ = positive

3.25 (a) -57.20 kJ mol^{-1}; -28.6 kJ mol^{-1};
(c) -138.2 kJ mol^{-1}; (d) -32.88 kJ mol^{-1};
(e) -55.84 kJ mol^{-1}

3.26 $+11.3$ kJ mol^{-1}

3.27 -56.98 kJ mol^{-1}

3.28 (a) decrease; (b) decrease; (c) increase

3.29 (a) increase; (c) increase

3.30 higher

4.1 0.410 J K^{-1}

4.2 -0.12 kJ K^{-1}

4.3 -161 J K^{-1}

4.4 $q = -45.1$ kJ; $\Delta S = -161$ J K^{-1}

4.5 $+9.1$ J K^{-1} mol^{-1}

4.6 2.91 L

4.7 56 J K^{-1}

4.8 23.6 J K^{-1}

4.9 -7.9 J K^{-1}

4.10 0.630 T_i

4.12 5.11 J K^{-1}

4.13 0.95 J K^{-1} mol^{-1}

4.14 (a) $+87.8$ J K^{-1} mol^{-1}; (b) -87.8 J K^{-1} mol^{-1}

4.15 (a) $+85$ J K^{-1} mol^{-1}; (b) $+34$ kJ K^{-1} mol^{-1}

4.16 4.0×10^{-4} J K^{-1} mol^{-1}

4.17 (a) positive; (b) negative; (c) positive

4.18 (a) -386.1 J K^{-1} mol^{-1}; (b) $+92.6$ J K^{-1} mol^{-1}; (c) -153.1 J K^{-1} mol^{-1}; (d) -21.0 J K^{-1} mol^{-1}; (e) $+512.0$ J K^{-1} mol^{-1}

4.19 (a) -0.75 J K^{-1}; (b) $+0.15$ J K^{-1}

4.20 -5.03 kJ K^{-1}

4.21 (a) -86 kJ mol^{-1}; (b) Yes, ΔG is negative;
(c) $+0.28$ kJ K^{-1} mol^{-1}

4.22 0.41 g

4.23 (a) Yes; (b) 0.46 mol ATP

4.24 8.1×10^{23} molecules of ATP

5.1 rhombic sulfur

5.2 No

5.3 (a) $+2.03$ kJ mol^{-1}; (b) $+1.50$ J mol^{-1}

5.4 $+19$ kJ mol^{-1}

5.5 (a) $+1.7$ kJ mol^{-1}; (b) -20 kJ mol^{-1}

5.6 $+4.3$ kJ mol^{-1}

5.7 (a) lowered; (b) raise

5.8 (a) $\Delta G_m = RT\ln\dfrac{p_f}{p_i} + b(p_f - p_i)$; (b) greater;
(c) 0.68%

5.9 (a) yes; (b) 7×10^2 K

5.10 -5.2 kJ mol^{-1}

5.12 (a) 2.37 kg; (b) 41.9 kg; (c) 1.87 g

5.13 (a) -134.6 bar K^{-1}; (b) 134.6 bar

5.14 31.69 kJ mol^{-1}

5.15 1.53 Pa

5.16 36.7 kJ mol^{-1}

5.17 356 K

5.18 Yes; 3 Torr or more

6.1 5.04 g

6.2 5.04 g

6.3 0.460 g

6.4 2.41×10^{-3}

6.5 269 g sucrose

6.6 $x_1 = x_2 = 0.500$

6.7 0.451; 0.549

6.8 886.8 cm^3

6.9 96 cm^3

6.10 $V_{ethanol} = 53.1$ mL mol^{-1}; $x_{ethanol} = 0.072$

6.11 (a) $V_W/(\text{mL mol}^{-1}) = 18.067 + 6.556 \times 10^{-3}\, b^2 - 1.018 \times 10^{-3}\, b^3$

6.12 (a) $\Delta G_m = -1.31$ kJ mol^{-1}; spontaneous;
(b) $\Delta S_m = +4.38$ J K^{-1} mol^{-1}-

6.13 $\Delta G_m = -1.40$ kJ mol^{-1}; spontaneous; $\Delta S_m = +4.71$ J K^{-1} mol^{-1}

6.14 4.99 kPa

6.15 2.30 kPa

6.16 6.4×10^3 kPa

6.17 4.8×10^{-3}

6.18 (a) 1.3 mmol kg^{-1}; (b) 33 mmol kg^{-1}

6.19 5.1×10^{-4} mol kg^{-1} in N_2, 27 mmol kg^{-1} in O_2

6.20 0.101 mol L^{-1}

6.21 $p = x_A\, p_A^{\bullet} + (1 - x_A)\, p_B^{\bullet}$
If A = toluene and B = *o*-xylene
$x_A = 0.920$, $x_B = 0.080$
The composition of the vapor is given by
$y_A = 0.968$, $y_B = 0.032$

6.22 53.8 g mol^{-1}

6.23 $-0.27°C$

6.24 207 g mol^{-1}

6.25 $K = \dfrac{1 - \dfrac{r(P* - P)}{cP}}{c\left(1 - \dfrac{2r(P* - P)}{cP}\right)^2}$

6.26 $-0.09°C$

6.27 88.3 kg mol^{-1}

6.28 13.94 kg mol^{-1}

6.29 (a) 0.36; (b) 0.81

6.36 (a) no

7.1 (a) $Q = \dfrac{[G][P_i]}{[G6P]}$

(b) $Q = \dfrac{[Gly - Ala]}{[Gly][Ala]}$

(c) $Q = \dfrac{[MgATP^{2-}]}{[Mg^{2+}][ATP^{4-}]}$

(d) $Q = \dfrac{P_{CO_2}^{\,6}}{P_{CO_2}^{\,5}[CH_3COCOOH]^2}$

7.2 (a) $Q = 1.3 \times 10^{-2}$; $\Delta G = -29.5$ kJ mol^{-1} (b) yes

7.3 -14.38 kJ mol^{-1}

7.4 (a) $\dfrac{p_{COCl}\, p_{Cl}}{p_{CO}\, p_{Cl_2}}$

(b) $\dfrac{p_{SO_3}^2}{p_{O_2}\, p_{SO_3}^2}$

(c) $\dfrac{p_{HBr}^2}{p_{H_2}\, p_{Br_2}}$

(d) $\dfrac{p_{O_3}^2}{p_{O_2}^3}$

7.5 4.46

7.6 (a) 1.2×10^9; (b) 1.8×10^2

7.7 -2.42 kJ mol^{-1}

7.8 3.01

7.9 $K_1^{1/2}$

7.10 -294 kJ mol^{-1}

7.11 $K = 1$

7.12 $K(G1P) = 3.5 \times 10^3$, $K(G6P) = 2.3 \times 10^2$, $K(G3P) = 36$

7.13 (a) -48.3 kJ mol^{-1}; (b) -66.1 kJ mol^{-1}

7.14 380

7.15 6.8 kJ mol^{-1}

7.16 (a) 1110 K, (b) 398 K (125 °C)

7.17 1.5×10^3 K

7.18 9.5×10^{-3}

7.19 (a) exergonic; (b) endergonic; (c) endergonic; (d) exergonic

7.20 (a) -91.14 kJ mol^{-1}; (b) $+594.6$ kJ mol^{-1}; (c) -66.82 kJ mol^{-1}; (d) $+99.82$ kJ mol^{-1}; (e) -415.80 kJ mol^{-1}

7.21 (a) -522.1 kJ mol^{-1}, $K > 1$; (b) $+25.78$ kJ mol^{-1}, $K < 1$; (c) -178.6 kJ mol^{-1}, $K > 1$; (d) -212.55 kJ mol^{-1}, $K > 1$; (e) -5798 kJ mol^{-1}, $K > 1$

7.22 (a) 5.5×10^4 kJ; (b) -5.1×10^4 kJ

7.23 -118 kJ

7.24 (a) -1.4×10^4 kJ; (b) -1.57×10^4 kJ or 1.57×10^4 kJ of non-expansion work

7.25 (a) total work done $= 1.68 \times 10^4$ kJ; (b) total work done $= 1.69 \times 10^4$ kJ

7.26 50 kg

7.27 -49.8 kJ mol^{-1}

7.28 817.90 kJ mol^{-1}

7.29 -53 kJ mol^{-1}

7.30 smaller

7.31 smaller

7.32 -25.0 kJ mol^{-1}

7.33 $+26$ kJ mol^{-1}

7.34 -16.8 J K^{-1} mol^{-1}

7.35 $+12.3$ kJ mol^{-1}

7.36 $x_I = 0.096$, $x_B = 0.904$

7.37 $x_{NH_3} = 0.632$, $x_{N_2} = 0.010$, $x_{H_2} = 0.358$

7.38 0.0456

7.39 2.6×10^{-4} bar

7.40 $[PCl_3] = [Cl_2] = 1.58 \times 10^{-2}$ mol L^{-1}, $[PCl_5] = 2.3 \times 10^{-2}$ mol L^{-1}; (b) 42%

7.41 $p_{N_2} = 0.020$ bar, $p_{H_2} = 0.020$ bar

7.43 -41.0 kJ mol^{-1}

7.44 (b) and (d)

7.45 (a) $+53$ kJ mol^{-1}; (b) -53 kJ mol^{-1}

7.46 (a) 9.24; (b) -12.9 kJ mol^{-1}; (c) $+161$ kJ mol^{-1}; (d) $+248$ J K^{-1} mol^{-1}

7.47 10^{746}

7.48 $+82.8$ kJ mol^{-1}

8.2 (a) $CH_3CH(OH)COOH + H_2O \rightleftharpoons$
$\quad CH_3CH(OH)COO^- + H_3O^+$

(b) $HOOCH_2C(NH_2)COOH + H_2O \rightleftharpoons$
$\quad HOOCH_2C(NH_2)COO^- + H_3O^+$
$\quad HOOCH_2C(NH_2)COO^- + H_2O \rightleftharpoons$
$\quad {}^-OOCH_2C(NH_2)COO^- + H_3O^+$

(c) $NH_2CH_2COOH + H_2O \rightleftharpoons$
$\quad {}^+NH_3CH_2COO^- + H_2O$

(d) $HOOCCOOH + H_2O \rightleftharpoons HOOCCOO^- + H_3O^+$
$\quad HOOCCOO^- + H_2O \rightleftharpoons {}^-OOCCOO^- + H_3O^+$

8.3 (a) 1.6×10^{-7} m, 6.8; (b) 6.8

8.4 (a) $D_2O + D_2O \rightleftharpoons D_3O^+ + OD^-$; (b) 14.87; (c) 3.67×10^{-8} mol L^{-1}; (d) pD $= 7.45 =$ pOD; (e) pD + pOD $= pK_w(D_2O) = 14.87$

8.5 (a) pH $= 4.82$, pOH $= 9.18$; (b) pH $= 2.82$, pOH $= 11.18$; (c) pH $=$ pOH $= 7.0$; (d) pH $= 4.30$, pOH $= 9.70$

8.6 (a) 9.60×10^{-3} mol L^{-1}, pH $= 2.02$; (b) 4.0×10^{-3} mol L^{-1}, 12.40; (c) 5.35×10^{-2} M, pH $= 1.27$

8.7 (a) acidic; (b) basic; (c) basic; (d) neutral; (e) acidic; (f) acidic

8.8 (a) 9.15; (b) 4.84; (c) none

8.9 (a) $pK_a = 3.08$, $K_a = 8.3 \times 10^{-4}$; (b) 2.8

8.10 (a) 13.48 (b) 34 mL

8.11 (a) 1.6%; (b) 0.33%; (c) 2.4%

8.12 (a) pH $= 2.00$, pOH $= 12.00$, fraction $= 0.083$; (b) pH $= 3.9$, pOH $= 10.09$, fraction $= 0.8$; (c) pH $= 1.1$, pOH $= 12.9$, fraction $= 0.73$

8.15 2.71

8.16 (a) 6.54, (b) 2.12 (c) 1.49

8.17 pH $= pK_a$

8.18 (a) 1.58×10^{-5}; (b) 1; (c) 5.0

8.19 (a) 0.069 mol L^{-1}; (b) $H_2S = 0.065$ mol L^{-1}, $HS^- = 9.2 \times 10^{-5}$ mol L$^{-1} = H_3O^+$, $S^{2-} = 7.1 \times 10^{-15}$ mol L^{-1}, $OH^- = 1.1 \times 10^{-10}$ mol L^{-1}

8.20 (a) 2.89; (b) 4.57; (c) 12.5 mL; (d) 4.74; (e) 25.0 mL; (f) 8.72

8.21 (a) 4.74 (b) 5.03 (c) 4.14

8.22 (a) 2–4; (b) 3–5; (c) 11.5–13.5; (d) 6–8; (e) 5–7

8.23 $pK_a = 4.66$, $K_a = 2.19 \times 10^{-5}$, pH $= 3.24$

8.24 (a) 5.04; (b) 8.96; (c) 2.78

8.25 7.94

8.27 (a) H_3PO_4 and NaH_2PO_4; (b) NaH_2PO_4 and Na_2HPO_4

8.28 (a) $K_s = [Ag^+] [I^-]$; (b) $K_s = [Hg_2^{+2}] [S^{2-}]$; (c) $K_s = [Fe^{3+}] [OH^-]^3$; (d) $K_s = [Ag^+]^2 [CrO_4^{2-}]$

8.29 (a) 1.0×10^{-8}; (b) 6.2×10^{-12}; (c) 2.0×10^{-39}; (d) 7.5×10^{-18}

8.30 (a) 5.5×10^{-10} mol L^{-1}; (b) 3.2×10^{-3} mol L^{-1}; (c) 1.6×10^{-7} mol L^{-1}; (d) 1.5×10^{-7} mol L^{-1}

8.31 1.25×10^{-5} mol L^{-1}

9.1 Yes

9.2 $CH_3COCO_2^- + 2H^+ + 2e^- \rightarrow CH_3CH(OH)CO_2^-$, $NAD^+ + H^+ + 2e^- \rightarrow NADH$

9.6 $+1.14$V, -440 kJ mol^{-1}

9.7 41 mV

9.8 21 mV

9.9 Standard potential $= -1.18$V

9.10 (a) R: $Ag^+(aq, b_R) + e^- \rightarrow Ag(s)$
$\quad$ L: $Ag^+(aq, b_L) + e^- \rightarrow Ag(s)$

R–L: $Ag^+(aq, b_R) \rightarrow Ag^+ (aq, b_L)$

(b) R: $2H^+(aq) + 2e^- \rightarrow H_2(g, P_R)$
   L: $2H^+(aq) + 2e^- \rightarrow H_2(g, P_L)$
   R–L: $H_2(G, P_L) \rightarrow H_2(g, P_R)$

(c) R: $MnO_2(s) + 4H^+(aq) + 2e^- \rightarrow Mn^{2+}(aq) + 2H_2O$
   L: $[Fe(CN)_6]^{3-} = e^- \rightarrow Fe(CN)_6]^{4-}$
   R–L: $MnO_2(s) + 4H^+(aq) + 2[Fe(CN)_6]^{4-} (aq) \rightarrow Mn^{2+}(aq) + 2[Fe(CN)_6]^{3-} (aq) + 2H_2O(l)$

(d) R: $Br_2(l) + 2e^- \rightarrow 2Br- (aq)$
   L: $Cl_2(g) + 2e^- \rightarrow 2Cl^-(aq)$
   R–L: $Br_2(l) + 2Cl^- (aq) \rightarrow Cl_2(g) + 2Br^-(aq)$

(e) R: $Sn^{4+}(aq) + 2e^- \rightarrow Sn^{2+}(aq)$
   L: $2Fe^{3+}(aq) + 2e^- \rightarrow 2Fe^{2+}(aq)$
   R–L: $Sn^{4+} (aq) + 2Fe^{2+}(aq) \rightarrow Sn^{2+}(aq) + 2Fe^{3+}(aq)$

(f) R: $MnO_2(s) + 4H^+(aq) + 2e^- \rightarrow Mn^{2+}(aq) + 2H_2O(l)$
   L: $Fe^{2+} (aq) + 2e^- \rightarrow Fe(s)$
   R–L: $Fe(s) + MnO_2(s) + 4H^+(aq) \rightarrow Fe^{2+} (aq) + Mn^{2+} (aq) + 2H_2O(l)$

9.11 (a) $E = E^\ominus - (RT/F) \ln b_L/b_R$

(b) $E = E^\ominus - (RT/2F) \ln p_R/p_L$

(c) $E = E^\ominus - \dfrac{RT}{2F} \ln \dfrac{[Mn^{2+}] [Fe(CN)_6^{3-}]^2}{[H^+]^4 [Fe(CN)_6^{4-}]^2}$

(d) $E = E^\ominus - \dfrac{RT}{2F} \ln \dfrac{[P_{Cl_Z}[Br^-]^2}{[Cl^-]^2}$

(e) $E = E^\ominus - \dfrac{RT}{2F} \ln \dfrac{[Sn^{2+}][Fe^{3+}]^2}{[Sn^{4+}][Fe^{2+}]^2}$

(f) $E = E^\ominus - \dfrac{RT}{2F} \ln \dfrac{[Fe^{2+}][Mn^{2+}]}{[H^+]^4}$

9.12 (a) $Fe(s)|FeSO_4(aq)|PbSO_4(s)|Pb(s)$ $v=2$
(b) $Pt|H_2(g)|H^+(aq)|Hg_2Cl_2(s)|Hg(l)$ $v=2$
(c) $Pt|H_2(g)|H^+(aq), H_2O_2|O_2(g)|Pt$ $v=4$
(d) $Pt|H_2(g)|H^+(aq), H_2O_2|O_2(aq)|O_2(g)|Pt$ $v=2$
(e) $Pt|H_2(g)|H_+(aq), I^-(aq)|I_2(s)|Pt$ $v=2$
(f) $CuCl_2(aq) | CuCl(aq)|Cu(s)$ $v=2$

9.13 (a) $Pt|CH_3CH_2OH(aq), CH_3CHO(aq), H^+(aq)||NAD^+(aq), NADH(aq)|Pt$ $v=2$
(b) $Mg(s)|ATP^{4-}(aq), MgATP^{2-}||Mg^{2+}(aq)|Mg(s)$ $v=2$
(c) $Pt|Cy-c (red, aq), Cyt-c(ox,aq)||CH_3CH(OH)CO_2^-(aq), CH_3COCO_2^-(aq)|Pt$

9.14 (a) 0; (b) 0; (c) + 0.87 V; (d) – 0.27 V;
(e) –0.62 V; (f) +1.67 V

9.15 (a) +0.08 V; (b) +0.27 V; (c) +1,23 V; (d) +0.695 V;
(e) +0.54 V; (f) +0.37 V

9.16 (a) $E^\ominus(NAD^+, NADH) - E^\ominus(CH_3CHO, CH_3CH_2OH)$;
(b) $E^\ominus(Mg^{2+}, Mg) - E^\ominus(Mg, MgATP^{4-})$;
(c) $E^\ominus CH_3COCO_2^-, CH_3CH(OH)CO_2^- - E^\ominus(Cyt-c(ox), Cyt-c(red))$

9.17 (a) E decreases, |E| increases; (b) E increases, |E| increases;
(c) E increases, |E| increases; (d) E increases, |E| decreases;
(e) E increases, |E| decreases; (f) E decreases, |E| decreases

9.18 (a) decreases E; (b) decreases E; (c) increases E;
(d) decreases E; (e) decreases E; (f) has no effect on E

9.19 (a) cell potential decreases; (b) cell potential increases;
(c) cell potential decreases

9.20 (a) –1.20 V (b) –1.18 V

9.21 (a) –394 kJ mol^{-1}; (b) –787 kJ mol^{-1};
(c) +75 kJ mol^{-1}; (d) –291 kJ mol^{-1};
(e) as (d); (f) +498 kJ mol^{-1}

9.22 (a) – 440 kJ mol^{-1}; (b) 29.7 kJ mol^{-1}; (c) +316 kJ mol^{-1}

9.23 (a) + 0.324 V; (b) + 0.45 V

9.24 0.37 V

9.25 –0.46 V, +89.7 kJ mol^{-1}, 146 kJ mol^{-1}, +87.9 kJ mol^{-1}

9.26 –606 kJ mol^{-1}

9.27 +0.80 V, –0.77 V, +1.57 V

9.28 3.6×10^{-8}

9.29 +0.24 V

9.30 (a) + 0.3108 V; (b) +1.14 V; (c) $E = E^\ominus - \dfrac{RT}{2F} \ln \dfrac{a_{HCO_3^-} a_{OH^-}}{a_{CO_3^-}}$
(d) 207 m V; (e) 10.3

9.31 (a) 1.6×10^{-8} [mol L^{-1}];
(b) + 0.12 V

9.32 (a) 6.5×10^9; (b) 1.2×10^7;
(c) 4×10^{69}; (d) 1.0×10^{25};
(e) 8.2×10^{-7}

9.33 1.8×10^{-10}, 9.04×10^{-7}, $E^\ominus(AgCl) = +0.22$ V, $E^\ominus(BaSO4) = -1.63$ V

9.34 $E = E^\ominus - \dfrac{RT}{6F} \ln \dfrac{a(Cr^{3+})^2}{a(Cr_2O_7^{2-}) a(H^+)^{14}}$

9.35 0.78

9.36 $E = 0$

9.37 8.5×10^{-17}, 9.19×10^{-9} mol L^{-1}

9.38 +1.03 V

9.39 (a) + 1.23 V (b) +1.108 V

9.40 (a) cathode will be left side; (b) 0.088 V

9.41 (a) + 0.94 V, (b) –0.77 V, +1.57 V

10.1 $A = 1.5$ mol L^{-1} s^{-1}
$B = 0.73$ mol L^{-1} s^{-1}
$D = 1.5$ mol L^{-1} s^{-1}

10.2 $\dfrac{\text{mol L}^{-1} \text{ s}^{-1}}{(\text{mol L}^{-1})^3} = \text{mol}^{-2} \text{ L}^2 \text{ s}^{-1}$

10.3 (a) (i) second-order, m^{3}s^{-1}
(ii) third-order, m^{6}s^{-1}
(b) (i) second-order, kPa^{-1}s^{-1}
(ii) third-order, kPa^{-2} s^{-1}

10.4 (a) 1; (b) 4.89×10^3 s^{-1}

10.5 (a) 1; (b) 2.33×10^9 L^{2}mol^{-2} s^{-1}

10.6 2.05×10^4 s; (a) 500 Torr; (b) 515 Torr

10.7 1.12×10^{-4} s^{-1}

10.8 4.2×10^{-5} s^{-1}

10.9 1.09×10^{-3} L mol^{-1} s^{-1}

10.10 1.12×10^{-4} s^{-1}

10.11 3.19×10^{-6} Pa^{-1} s^{-1} (for pseudosecond-order process)

10.12 (a) 0.014 kPa s^{-1}; (b) 1.5×10^3s

10.13 1330 s

10.14 3.1×10^3 y

10.15 (a) 0.63 µg; (b) 0.16 µg

10.16 (a) 0.138 mol L^{-1}; (b) 0.095 mol L^{-1}

10.17 $2.8 \times 10^4 s$

10.18 first-order, $5.6 \times 10^{-4} s^{-1}$

10.19 $E_a = 85.4$ kJ mol^{-1}; $A = 3.66 \times 10^{11}$ mol $L^{-1} s^{-1}$

10.20 299 K

10.22 21.6 kJ mol^{-1}

10.23 120 kJ mol^{-1}

10.24 -21.6 kJ mol^{-1}

10.25 48 kJ mol^{-1}

10.26 $5.4 \times 10^4 s$

10.27 1.09×10^6 L $mol^{-1} s^{-1}$

10.28 2.1 mm^2

10.29 126 kJ mol^{-1}

10.30 2.7 J $K^{-1} mol^{-1}$

11.1 3.1×10^5 L $mol^{-1} s^{-1}$

11.7 39.1d

11.8 first-order in H_2O_2 and in Br^-, and second-order overall

11.9 rate $= k_{eff}[A_2]^{1/2}[B]$

11.10 rate equation $= k$ [A] [B]

rate constant, $k = \dfrac{k_1 \, k_2}{k_1'}$

11.11 rate $= \dfrac{k_1 \, k_2 [O_3]^2}{k_1' [O_2] + k_2 [O_3]}$

11.12 $[A^-] = \dfrac{k_1 [AH][B]}{k_2 [BH^+] + k_3 [AH]}$

rate of formation of product $= \dfrac{k_1 \, k_3 [AH]^2 [B]}{k_2 [BH^+] + k_3 [AH]}$

11.13 rate law $= \dfrac{k_1 \, k_2}{k_1'}$ [HA] [H^+][B]

11.15 (a) 998 Torr; (b) 111 cm^3

11.16 -24.5 kJ mol^{-1}

11.17 rate $= \dfrac{k \, k_A \, k_B \, p_A \, p_B}{(1 + k_A \, p_A + k_B \, p_B)^2}$

11.18 rate $= \dfrac{k_1 \, k_3 [A]^2}{k_{-1}[A] + k_2 S + k_3}$

11.19 1.62×10^{-3} mol $L^{-1} s^{-1}$

11.20 $[S] = K_M$

11.22 1.5×10^{15}

11.23 $K_m = 3.4$ mmol L^{-1}, 0.38 s^{-1}

11.24 $K_m = 1.08$ µmol L^{-1}, $k_b = 1.1 \times 10^2 \, s^{-1}$

11.25 non−competitive

11.27 $-k_1[R_2] - k_2\left(\dfrac{k_1}{k_4}\right)^{1/2} [R_2]^{3/2}$

11.28 (a) does not occur; (b) between 0.16 kPa and 4.0 kPa; (c) lower limit $= 1.3 \times 10^2$ Pa

11.29 3.1×10^{18}

11.30 0.412

12.1 1.0×10^{-6} m

12.2 8.4×10^{11}

12.3 4.5×10^3 K, this radiation is not thermal radiation

12.4 4.6×10^3 W

12.6 (a) 1.6×10^{-33} J m^{-3}; (b) 2.5×10^{-4} J m^{-3}

12.7 $\lambda_{max} T = hc/5k = 2.88$ mm K

12.8 6.29×10^{-34} Js

12.9 (a) 4.0×10^2 kJ mol^{-1}; (b) 20 kJ mol^{-1}; (c) 8.0×10^{-13} kJ mol^{-1}

12.10 sodium

12.11 (a) $1.8 \times 10^{18} s^{-1}$; (b) $1.8 \times 10^{20} s^{-1}$

12.12 $6.90 \times 10^{29} s^{-1}$

12.13 (a) no ejection occurs; (b) 4.52×10^{-19} J, 996 km s^{-1}

12.14 1.32×10^6 m s^{-1}

12.15 (a) 6.6×10^{-31} m; (b) 6.6×10^{-39} m; (c) 99.7 pm

12.16 (a) 1.23 nm; (b) 39 pm; (c) 3.88pm

12.17 (a) 9.14×10^{-28} kg m s^{-1}; (b) 8.8×10^{-24} kg m s^{-1}; (c) 3.3×10^{-35} kg m s^{-1}

12.19 2.2×10^{-24} m s^{-1}

12.20 (a) 3.34×10^{-6} N; (b) 3.34×10^{-12} Pa; (c) 83 h

12.21 50.6 nm

12.22 1.11×10^{-15} J

12.23 (a) 1.84×10^{-4}; (b) 2.36×10^{-4}

12.24 0.90 nm

12.25 2.1×10^{-29} m s^{-1}

12.26 1×10^{-26} m

12.27 5.8×10^{-5} m s^{-1}

12.28 9.85×10^{-23} J

12.29 $L/4$ or $3L/4$

12.30 (a) 2.17×10^{-20} J; (b) 9.16×10^{-6} m

12.31 $\Psi = (1 \, / \, L)^{1/2}$

12.32 1.24×10^{-6} m

12.33 (a) 4.34×10^{-47} kg m^2; (b) 1.55 mm

12.34 0.04 N m^{-1}

12.35 (a) $6.89 \times 10^{13} s^{-1}$; (b) 4.35 µm

13.1 434 nm

13.2 $n = 6$

13.3 (a) 6842 cm^{-1}; (b) 1.36×10^{-19} J

13.4 $2.24 \times 10^3 \, cm^{-1}$

13.5 $4p \rightarrow 2s$

13.6 122.31 e V

13.7 16

13.8 $n_2 \rightarrow 6$

13.9 3092 nm

13.10 (a) 397.13nm; (b) 3.40 eV

13.11 $R_{Li}^{2+} = 987663 \, cm^{-1}$

137175 cm^{-1} 185187 cm^{-1}

122.5eV

13.12 $r = 0.693$ a_0 (36.7 pm)

13.13 (a) 40 pm or 13 pm; (b) 29 or 24 pm

13.14 1/6

13.15 1.1×10^{-5}

13.16 (a) 2.7×10^{-7} pm^{-3}; (b) 2.5×10^{-8} pm^3; (c) 0

13.17 $5.2 a_0$

13.18 (a) 101 pm and 376; (b) 99 pm, 349 ppm, and 821 pm

13.19 $0°$, $180°$; $90°$, $270°$

13.21 (a) $g = 1$; (b) $g = 9$; (c) $g = 49$

13.22 (a) forbidden; (b) allowed; (c) allowed; (d) forbidden; (e) allowed

13.23 all d and g orbitals

13.24 (a) 2; (b) 14; (c) 22

13.26 14.0 eV

13.28 Fe^{2+}

13.29 $^1F_3$; $^3F_4, ^3F_3, ^3F_2$; $^1D_2$; $^3D_3, ^3D_2, ^3D_1$; $^1P_1, ^3P_1, ^3P_0$

13.30 2,1, and 0

13.31 (a) 1; (b) 3; (c) 1; (d) 3

13.32 (a) ^3F; (b) 3

14.3 3.11×10^{-18} J, 1187×10^6 J mol^{-1}

14.5 (a) linear; (b) angular; (c)linear; (d) angular; (e) angular; (f) angular; (g) angular

14.6 (a) angular; (b) octahedral; (c) square planar; (d) see-saw

14.10 $1 - 1 = 0$ [orthogonality]

14.11 $N = \left(\dfrac{1}{1 + 2\lambda S + \lambda^2} \right)^{1/2}$

14.12 978 and 22, respectively

14.14 (c) 45°

14.16 (a) b = 1; (b) b = 0; (c) b = 2

14.17 (a) b = 1/2; (b) b = 3; (c) b = 2

14.18 (a) b = 3; (b) b = 5/2; (c) b = 3

14.19 C_2

14.20 C_2 and CN stabilized by anion formation; NO, O_2 and F_2 stabilized by cation formation

14.21 no

14.22 (a) g; (b) inapplicable; (c) g; (d) u

14.23 g, u, g, u

14.24 g, u, g, u; (b) for v even, g; for v odd, u

14.26 N_2

14.27 $F_2^+ < F_2 < F_2^-$

14.29 (a) $7\alpha + 7\beta$; (b) $5\alpha = 7\beta$

14.30 2.7 eV

15.1 (a) n–type; (b) p–type

15.2 metallic conductor

15.4 3888 kJ mol^{-1}

15.5 2258 kJ mol^{-1}

15.6 $V = \dfrac{qNze\pi}{48\,\varepsilon_0 d}$

15.7 1.06

15.10 (326), (11), (122), (3$\bar{2}\bar{2}$)

15.13 (111) = 307 pm, (211) = 217 pm, (100) = 532 pm

15.14 (123) = 0.129 nm, (236) = 0.0671 nm

15.15 66.1 pm

15.16 0.215 cm

15.17 bcc

15.18 8.97 g cm^{-3}

15.20 (a) 0.9069; (b) 0.5236

15.21 0.740 g cm^{-3}

15.22 (a) eight nearest neighbours, 433 nm; (b) six next-nearest neighbours, 500 nm

15.23 (a) 12 nearest neighbours, 354 nm; (b) six next-nearest neighbours, 500 nm

15.24 (a) less dense; (b) 92%

15.25 $V = 3.96 \times 10^{-28}$ m^3; D = 2.40×10^6 g m^{-3}

15.26 N = 4, 4.01 g cm^{-3}

15.27 (a) (321) = 220 pm; (b) (642) = 110 pm

15.28 rock salt structure

16.1 dipole moment = 0.9 D, 3.0×10^{-30} C m

16.2 SF_4 is polar

16.3 (a) ortho-xylene = 0.7 D; (b) meta-xylene = 0.4 D;

(c) para-xylene = 0; the para-xylene value is exact by symmetry

16.4 (a) 1,2,3-trimethylbenzene = 0.8 D; (b) 1,2,4-trimethylbenzene = 0.2 D; (C) 1,3,5-trimethylbenzene = 0; (c) should cancel exactly by symmetry

16.5 1.4 D

16.6 (a) 1.7 D; (b) 2.6 D; (c) 1.3 D; (d) 2.1 D

16.7 3.10 D

16.9 (a) 1070 kJ mol^{-1}; (b) 119 kJ mol^{-1}

16.11 (a) $E = 3/2\ RT = 3.7$ kJ mol^{-1} at 298K; (b) 0.14 J; the kinetic theory of gases is justifiable

16.12 196 pm

16.13 -2.4×10^{-4} J mol^{-1}

16.14 4.2×10^{-3} J mol^{-1}

16.15 -8.2 kJ mol^{-1}

16.16 $r = 2^{1/6}\sigma$

16.18 (a) $V_0 = 11.6$ kJ mol^{-1}; (c) 4×10^{12} s^{-1}

16.22 (a) 5.22×10^{-8} mol m^{-2} s^{-1}; (b) 1.6×10^{-11} mol

16.23 (a) 1×10^3 s; (b) 1×10^5 s; (c) $1m = 1 \times 10^9$ s

16.24 (a) 6.2×10^{-9} m^2 s^{-1}; (b) 1.6×10^{-9} m^2 s^{-1}

16.25 1.1×10^{-9} s

16.26 10^6

16.28 (a) $d = \lambda \sqrt{t/r}$; (b) 4.2×10^5 s

17.1 (a) 4.47×10^{14} s^{-1}; (b) 1.49×10^4 cm^{-1}

17.2 (a) 3.07×10^{-3} cm^{-1}; (b) 3.26 m

17.3 0.999 999 918 $\times$ 660 mm, green at 6.36×10^7 m s^{-1}

17.4 8.4×10^5 K

17.5 (a) 53 ps; (b) 5 ps; (c) 1.6×10^{-2} ps

17.6 0.27cm^{-1}

17.7 4×10^{33}

17.8 (a) 4.601×10^{-48} kg m^2; (b) 9.196×10^{-48} kg m^2; (c) 6.67×10^{-46} kg m^2; (d) 6.67×10^{-46} kg m^2

17.9 (a) 1.825×10^{12} Hz; (b) 9.128×10^{11} Hz; (c) 1.26×10^{10} Hz; (d) 1.26×10^{10} Hz

17.10 (a) $I = 4m_BR^2$; (b) 1.583×10^9 Hz

17.11 (a) $I_{\parallel} = 4mR^2$, $I_{\perp} = 2mR^2$

17.12 (a) 5.152×10^9 Hz; (b) microwave spectroscopy could not be used

17.13 (a), (b), (c), and (d)

17.14 all

17.15 one state with J = 10

17.16 21 states and 11 energy levels

17.17 (a) 636 GHz, 1272 GHz, 1908 GHz . . . (b) 21.21 cm^{-1}, 42.42 cm^{-1}, 63.63 cm^{-1} . . .

17.18 lower

17.19 1.162×10^{-10} m = 116.2 pm

17.20 separation of lines = 194 GHz

17.21 (a) 4.49×10^{13}Hz; (b) 4.39×10^{13} Hz

17.22 329 Nm^{-1}

17.23 HF = 967.1, HCl = 515.6, HBr = 411.8. HI = 314.2

17.24 ^{2}HF = 3002, ^{2}HCl = 2144, ^{2}HBr = 1886, ^{2}HI = 1640

17.25 (b) HCl, (c) CO_2, (d) H_2O, (e) CH_3CH_3, (f) CH_4 and (g) CH_3Cl

17.26 (a) 3; (b) 4; (c) 48; (d) 54

17.27 (a) Raman active

17.28 (a) aromatic C–H stretch; (b) C–H aldehyde; (c) C=O; (d) aromatic C=C; (e) aromatic C=C; (f) aromatic C–H out of plane. This compound could be benzaldehyde

18.1 75.1%

18.2 $1.35 \times 10^3\ \mathrm{L\ mol^{-1}\ cm^{-1}}$

18.3 $1.16 \times 10^{-3}\ \mathrm{mol\ L^{-1}}$

18.4 (a) $1.01 \times 10^4\ \mathrm{L\ mol^{-1}\ cm^{-1}}$; (b) 0.965%

18.5 $33\ \mathrm{\mu g\ L^{-1}}$

18.6 $\log \dfrac{I}{I_0} = -\varepsilon \lambda [J]_0$

18.7 $450\ \mathrm{L\ mol^{-1}\ cm^{-1}}$

18.8 $159\ \mathrm{L\ mol^{-1}\ cm^{-1}}$, 23%

18.9 (a) 0.9 m; (b) 3 m

18.10 $c_1 = 2.58 \times 10^{-5}\ \mathrm{mol\ L^{-1}}$, $c_2 = 5.43 \times 10^{-5}\ \mathrm{mol\ L^{-1}}$

18.11 $c_A = \dfrac{\varepsilon^\circ A' - \varepsilon_B A}{(\varepsilon_A - \varepsilon_B)\varepsilon^\circ l}$, $c_B = \dfrac{\varepsilon_A A - \varepsilon^\circ A'}{(\varepsilon_A - \varepsilon_B)\varepsilon^\circ l}$

18.12 $c_A = 1.00 \times 10^{-4}\ \mathrm{mol\ L^{-1}}$, tyrosine $= c_B = 9.66 \times 10^{-5}\ \mathrm{mol\ L^{-1}}$

18.13 only two solutes in equilibrium with each other are present

18.16 quenching rate constant $= 9.2 \times 10^9\ \mathrm{L\ mol^{-1}\ s^{-1}}$, half-life 1.9×10^{-7} s

18.17 8.0×10^{-19} J

18.18 2.3×10^{-19} J

19.1 $^-1.625 \times 10^{-26}\ \mathrm{J} \times m_r$

19.2 (a) $\mathrm{T^{-1}}$ Hz; (b) $\mathrm{A\ s\ kg^{-1}}$

19.3 2.263

19.4 328.6 MHz

19.5 46.15 MHz

19.6 12.92 T

19.7 9.248 GHz

19.8 2.66 kHz

19.9 $46\ \mathrm{\mu T}$

19.10 (a) 2.4 kHz; (b) 4.4 kHz

19.11 1: 7: 21: 35: 35: 21: 7: 1

19.12 (a) quintet 1: 2: 3: 2: 1; (b) septet 1: 3: 6: 7: 6: 3: 1

20.1 0.37

20.2 (a) 0.9999895; (b) 0.9998955

20.3 0.99849

20.4 3.475

20.5 6.55

20.6 (a) $1 + 5e^{-\epsilon/kT} + 3e^{-3\epsilon/kT}$; (b) $5e^{-\epsilon/kT}$, $q = 1$; (c) $q = 9$

20.7 (a) 1.29; (b) 7.82

20.8 (a) 9; (b) 6.731

20.9 1.072

20.10 (a) 3.2×10^4; (b) 6.2×10^{27}

20.11 (a) 19.5; (b) 265

20.13 $\dfrac{Nh\nu}{e^{h\nu/kT} - 1}$

20.14 $\dfrac{5\epsilon\ e^{-\epsilon/kT} + 9\epsilon\ e^{-3\epsilon/kT}}{q}$

20.15 (a) $E = \dfrac{R(70.77\ \mathrm{K})\,e^{-2359\mathrm{K}/T} + R(3129\ \mathrm{K})\,e^{-6258\mathrm{K}/T}}{1 + 3e^{-2359\mathrm{K}/T} + 5e^{-6258\mathrm{K}/T}}$

(b) $339\ \mathrm{J\ mol^{-1}}$

20.16 (a) $E = \dfrac{R(6840\ \mathrm{K})\,e^{-2280\mathrm{K}/T} + 3259\ \mathrm{K}\,e^{-3259\mathrm{K}/T}}{5 + 3e^{-2280\mathrm{K}/T} + e^{-3259\mathrm{K}/T}}$

(b) $0.993\ \mathrm{kJ\ mol^{-1}}$

20.17 $11.5\ \mathrm{J\ K^{-1}\ mol^{-1}}$

20.18 $191.4\ \mathrm{J\ K^{-1}\ mol^{-1}}$

20.19 $40\ \mathrm{kJ\ K^{-1}\ mol^{-1}}$

20.20 1.37×10^{-25}

20.21 $-48.4 \times \mathrm{kJ\ mol^{-1}}$

20.22 1.37×10^{-25}

20.23 1.93×10^{-11}

Answers to Box Exercises

Box 1.1 (1) 0.069, 135 kg; (2) 3.5×10^8 L

Box 1.2 (1) 2.1×10^7; (2) 3.0×10^6 Pa

Box 3.1 (1) (a) -2.3×10^3 kJ; (b) 9.2 K; (2) 6.7%

Box 6.1 (1) 4.5 mL; (2) $0.056\ \mathrm{mgN_2}$ compared to 0.014 mg N_2, 0.17 mg N_2

Box 6.2 (1) $K = 1.167\ \mathrm{L\ \mu mol^{-1}}$, $N = 5.24$ model is applicable

Box 7.1 (1) yes; (2) 7.6 kg

Box 9.1 (1) 1.29 mol; (2) ≈ 2 mol

Box 9.2 (1) $\Delta G(\mathrm{Na^+}) = +13.6\ \mathrm{kJ\ mol^{-1}}$; $\Delta G(\mathrm{K^+}) = +1.0\ \mathrm{kJ\ mol^{-1}}$; (2) yes

Box 9.3 (1) (a) $1.085 \times 10^5\ \mathrm{J\ mol^{-1}}$, (b) 9.60×10^{-20}; (2) $+473\ \mathrm{kJ\ mol^{-1}}$

Box 10.1 (2) $\tau = \dfrac{1}{k' + \dfrac{2k}{K}[(1 + 4K[A]_{total})^{1/2} - 1]}$

Box 11.1 (1) 1.64 min; (2) yes, but concentrations are low $8 \times 10^{-6}\ \mathrm{mol\ L^{-1}}$ for any given tretrapeptide, $2 \times 10^{-4}\ \mathrm{mol\ L^{-1}}$ total for all tetrapeptides

Box 11.2 (1) $110\ \mathrm{kcal\ mol^{-1}}$; (2) 1.0%, 40 mg L^{-1}

Box 12.1 (1) (a) 5.36 pm, (b) 2.5%; (2) (a) $1.1 \times 10^{10}\ \mathrm{s^{-1}}$, (b) $2.7 \times 10^9\ \mathrm{s^{-1}}$

Box 13.1 (1) (a) $1s^2 2s^2 2p^6 3s^2 3p^6 3d^{10}$, (b) no; (2) 1.5 L, 4.5×10^{14} L

Box 14.1 (1) -4.36×10^{-14} J, (a) -4.2×10^{-20} J, (b) -1.0×10^{-21} J

Box 16.1 (1) $1.3 \times 10^{-22}\ \mathrm{J\ deg^2}$; (2) $20.4\ \mathrm{kJ\ mol^{-1}}$

Box 16.2 (1) 2.5×10^{-5} s (plasma), 2.5×10^{-6} s (lipid); (2) $t_{ribo}/t_{cat} = 4.27$

Box 18.1 (1) 3×10^{-3}; (2) 587 nm

Box 19.1 $29\ \mathrm{\mu T\ m^{-1}}$

Box 20.1 (1) 6.9×10^{-5} (glycine), 0.89 (L-leucine); (2) $n = 16.1$

INDEX